advanced calculus

WILLIAM F. TRENCH
Drexel University

HARPER & ROW, PUBLISHERS
NEW YORK, HAGERSTOWN, SAN FRANCISCO, LONDON

To Lucille and Bud, Randy, John, and Gina

Sponsoring Editor: Charlie Dresser
Project Editor: Penelope Schmukler
Designer: Katrine Stevens
Cover Design: Helen Iranyi
Production Supervisor: Will C. Jomarron
Compositor: Syntax International Pte. Ltd.
Printer and Binder: The Maple Press Company
Art Studio: Danmark & Michaels Inc.

ADVANCED CALCULUS

Library of Congress Cataloging in Publication Data

Trench, William F., Date—
    Advanced calculus.

    Includes index.
    1. Calculus. I. Title.
QA303.T755   515   77-26264
ISBN 0-06-046665-0

contents

preface

This is a text for a two-term course in advanced calculus for junior or senior mathematics majors and students of science and engineering. It is designed to fill the gaps left in the development of calculus as it is usually presented in the elementary course, and to provide the background required for insight into more advanced courses in pure and applied mathematics.

The standard elementary calculus sequence is the only specific prerequisite for Chapters 1–5, which deal with real-valued functions. Chapters 6–8 require a working knowledge of determinants, matrices, and linear transformations (reviewed in Section 6.1), which are integral parts of the study of vector-valued functions as presented there.

Chapter 1 is concerned with the real number system. It begins with a brief discussion of the axioms for a complete ordered field, but no attempt is made to develop the reals from them; rather, it is assumed that the reader is familiar with the consequences of these axioms, except for one: the completeness axiom. Since the difference between a rigorous and a nonrigorous treatment of calculus can be described largely in terms of the attitude taken toward this axiom, considerable effort is expended in developing its consequences. Section 1.2 is devoted to mathematical induction. Although this may seem incongruous in an era when most high school algebra courses include induction, I have found that the average beginning student of advanced calculus simply cannot do an induction proof without reviewing the method at the start of the course. Section 1.3 is devoted to elementary set theory and the topology of the real line, ending with the theorems of Heine-Borel and Bolzano-Weierstrass.

Chapter 2 covers the differential calculus of functions of one variable: limits, continuity, differentiability, l'Hospital's rule, and Taylor's theorem. The emphasis is on rigorous presentation of principles; no attempt is made to develop the properties of specific elementary functions. Although a case can be made that this is properly within the province of advanced calculus, since it is usually done nonrigorously in the elementary course, I believe that the student's time is better spent on principles than on reestablishing familiar formulas and relationships.

Chapter 3 is devoted to the Riemann integral of functions of one variable. The integral is defined in Section 3.1, in the standard way in terms of Riemann sums. Upper and lower integrals are also defined there, and used in Section 3.2 to study the existence of the integral. Section 3.3 is devoted to its properties. Improper integrals are studied in Section 3.4. I believe that my treatment of improper integrals is more detailed than is to be found in most comparable textbooks. A more advanced look at the existence of the proper integral is given in Section 3.5, which concludes with Lebesgue's existence criterion. This section can be omitted without compromising the reader's preparedness for subsequent sections.

Sequences are discussed in Section 4.1. I have chosen to make the concepts of limit inferior and limit superior an integral part of this development, mainly because this permits greater flexibility and generality with little extra effort in the study of infinite series. Section 4.2 provides a brief introduction to the way in which continuity and differentiability can be studied by means of sequences. Sections 4.3, 4.4, and 4.5 treat infinite series of constants, infinite series of functions, and power series—again in more detail than is to be found in most comparable textbooks. The instructor who chooses not to cover these sections completely can omit the less standard topics without loss in subsequent sections.

Chapter 5 is devoted to real-valued functions of several variables. It begins with a discussion of the topology of $\mathcal{R}^n$ in Section 5.1. Continuity and differentiability of functions of several variables are discussed in Sections 5.2 and 5.3. The chain rule and Taylor's theorem are discussed in Section 5.4.

Chapter 6 covers the differential calculus of vector-valued functions of several variables. Since they form an integral part of the treatment of this subject as presented here, matrices, determinants, and linear transformations are reviewed in Section 6.1. In Section 6.2 the differential of a vector-valued function is defined as a certain linear transformation, and the chain rule is discussed in terms of the composition of such transformations. The inverse function theorem is taken up in Section 6.3, where the notion of branches of an inverse is introduced. In Section 6.4 the implicit function theorem is motivated by first considering linear transformations, and then stated and proved in general. The method of Lagrange multipliers is discussed in Section 6.5. This section is not a prerequisite for anything that follows.

Multiple integrals are defined in Section 7.1, first over rectangles and then over more general sets. Although Lebesgue's existence criterion is stated, the main part of the discussion deals with the multiple integral of a function whose discontinuities form a set of Jordan content zero, over a set whose boundary has Jordan content zero. Section 7.2 deals with the evaluation of multiple integrals by means of iterated integrals. Section 7.3 begins with the definition of Jordan measurability, followed by a derivation of the rule for change of content under a linear transformation, an intuitive formulation of the rule for change of variables in multiple integrals, and finally a careful statement and proof of the rule. This is a complicated proof, and some of its more tedious details are left to the exercises. Improper multiple integrals and ordinary integrals involving a parameter are discussed in Sections 7.4 and 7.5.

Chapter 8 deals with line and surface integrals. The level of rigor in this chapter is lower than in others, simply because I believe that the price for complete rigor in the presentation of line and surface integrals—and in particular the divergence theorem and the theorems of Green and Stokes—exceeds the value of the return that it yields to students at this level. However, the reader is informed of the corner cutting that takes place in some of the definitions and proofs, and why it is done.

In writing this text I have emphasized careful statement of definitions and theorems and have tried to be complete and detailed in proofs, except for omissions explicitly left to the exercises. In making the transition from one to several independent variables, and from real-valued to vector-valued functions, I have left to the reader those proofs that are essentially repetitions of proofs of earlier theorems. Some proofs, especially those of the inverse and implicit function theorems and the rule for change of variable in a multiple integral, are difficult. Since this is in the nature of the subject, there is no apology to be made here; however, virtually every definition and theorem is followed by at least one example of which there are 377 in all.

Great care has gone into the preparation of the 996 numbered exercises, many of which have several parts. Answers to exercises that have answers are given in the back of the book, but solutions to exercises calling for proofs are not.

For their helpful comments and criticism of parts of the manuscript I thank Professors Norman Bloch of SUNY at Brockport; Frank Kocher of Pennsylvania State University; Roger B. Nelson of Lewis and Clark College; and Elliot Wolk of the University of Connecticut at Storrs.

WILLIAM F. TRENCH

to the reader

Equations are numbered 1, 2, 3, . . . within each section. Other numbered items, such as definitions, theorems, and examples, are also identified by the sections in which they appear; thus, theorems in the second section of any chapter are numbered 2.1, 2.2, 2.3, and so on. Since this means that there are eight (one in each chapter) theorems numbered 2.1, the following convention applies: unless stated to the contrary, references are to the chapters in which they appear. For example, a reference in Section 3.4 to Theorem 2.1 means Theorem 2.1 of Section 3.2; if it meant Theorem 2.1 of Section 1.2, it would be so stated explicitly.

Exercises preceded by asterisks are referenced in later sections.

W. F. T.

advanced calculus

the real numbers

1.1 THE REAL NUMBER SYSTEM

The student of advanced calculus has had considerable experience with the real number system but may not realize that all its properties follow from a few basic ones. Although we will not carry out the development of the real number system from these basic properties, it is nevertheless useful to state them, as a starting point for the study of advanced calculus and also to focus attention on one—completeness—with which the reader may be less familiar.

Field properties

The real number system (which we usually call simply the *reals*) is first of all a nonempty set $\{a, b, c, \ldots\}$ on which the operations of addition and multiplication are defined so that every pair of real numbers has a unique sum and product, both real numbers, such that:

A. $a + b = b + a$ and $ab = ba$ (commutative laws).
B. $(a + b) + c = a + (b + c)$ and $(ab)c = a(bc)$ (associative laws).

C. $a(b + c) = ab + ac$ (distributive law).

D. There are distinct elements 0 and 1 such that $a + 0 = a$ and $a(1) = a$ for all a.

E. For each a there is a real number, $-a$, such that $a + (-a) = 0$, and, if $a \neq 0$, there is a real number, $1/a$, such that $a(1/a) = 1$.

The manipulative properties of the real numbers, such as the relations

$$(a + b)^2 = a^2 + 2ab + b^2,$$
$$(3a + 2b)(4c + 2d) = 12ac + 6ad + 8bc + 4bd,$$
$$(-a) = (-1)a, \qquad a(-b) = (-a)b = -ab,$$

and

$$\frac{a}{b} + \frac{c}{d} = \frac{ad + bc}{bd} \qquad (b, d \neq 0),$$

all follow from A–E. We assume that the reader is familiar with these properties.

A set of elements on which two operations are defined so as to have properties A–E is called a *field*. The real number system is by no means the only field. The *rational numbers* (which are the real numbers that can be written as $r = p/q$, where p and q are integers, $q \neq 0$) also form a field under addition and multiplication. The simplest possible field consists of two elements, which we denote by 0 and 1, with addition defined by

$$0 + 0 = 1 + 1 = 0, \qquad 1 + 0 = 0 + 1 = 1, \tag{1}$$

and multiplication by

$$0 \cdot 0 = 0 \cdot 1 = 1 \cdot 0 = 0, \qquad 1 \cdot 1 = 1 \tag{2}$$

(*Exercise 1*).

The order relation

The real number system is ordered by the relation $<$, as follows.

F. For each pair of real numbers a and b, exactly one of the following is true:

$a = b, \quad a < b, \quad$ or $\quad b < a.$

G. If $a < b$ and $b < c$, then $a < c$. (The relation $<$ is *transitive*.)

H. If $a < b$, then $a + c < b + c$ for any c, and, if $0 < c$, then $ac < bc$.

A field with an order relation satisfying F–H is an *ordered* field. Thus, the real numbers form an ordered field. The rational numbers also form an ordered field, but it is impossible to define an order on the field with two elements defined by (1) and (2) so as to make it into an ordered field (*Exercise 1*).

We assume that the reader understands the elementary properties of the order relation, and other definitions connected with it: for example, that $a > b$ means $b < a$ and that $a \geq b$ means that either $a = b$ or $a > b$; that $|a| = a$ if $a \geq 0$ and $|a| = -a$ if $a < 0$; and that

$$|a + b| \leq |a| + |b|$$

and

$$|a - b| \geq ||a| - |b||$$

(*Exercise 2*). The first of these inequalities is the *triangle inequality*, which appears in various forms in many contexts; it is one of the most important inequalities in mathematics, and we will use it often.

Least upper bound of a set

A set S of real numbers is *bounded above* if there is a real number b such that $x \leq b$ whenever x is in S. In this case, b is an *upper bound* for S. If b is an upper bound for S, then so is any larger number, because of property G. If β is an upper bound for S, but no number less than β is, then β is a *least upper bound of S*, and we write

$$\beta = \text{l.u.b. } S.$$

(Some authors call β the *supremum* of S, and write $\beta = \sup S$.)

With the real numbers associated in the usual way with the points on a line, these definitions can be interpreted geometrically as follows: b is an upper bound of S if no point of S is to the right of b; $\beta = \text{l.u.b. } S$ if no point of S is to the right of β, but there is at least one point of S to the right of any number less than β (Figure 1.1).

Example 1.1 If S is the set of negative numbers, then any nonnegative number is an upper bound of S, and

$$\text{l.u.b. } S = 0.$$

If S_1 is the set of negative integers, then any number a such that $a \geq -1$ is an upper bound for S_1, and

$$\text{l.u.b. } S_1 = -1.$$

β b

(S = dark line segments)

FIGURE 1.1

This example shows that a least upper bound of a set may or may not be in the set, since S_1 contains its least upper bound but S does not.

Some definitions concerning sets

This is a good place to introduce some notation. We will often define a set S by writing

$$S = \{x | \cdots\},$$

which means that S consists of all x that satisfy the conditions to the right of the vertical bar; thus, in Example 1.1,

$$S = \{x | x < 0\} \tag{3}$$

and

$$S_1 = \{x | x = \text{negative integer}\}.$$

We will sometimes abbreviate "x is an element of S" by

$$x \in S,$$

and "x is not an element of S" by

$$x \notin S.$$

For example, if S is defined by (3), then

$$-1 \in S \quad \text{but} \quad 0 \notin S.$$

A *nonempty* set is a set that contains at least one element. The *empty set*, denoted by $\varnothing$, is the set that contains no elements; although it may seem foolish to speak of such a set, we will see that it is nevertheless a useful idea.

The completeness axiom

It is one thing to define an object, and another to show that there is an object which satisfies the definition. (For example, does it make sense to define the smallest positive real number?) This observation is particularly appropriate in connection with the definition of the least upper bound of a set. For example, the empty set is bounded above by every real number, and therefore it has no least upper bound. (Think about this.) More importantly, we will see below that properties A–H do not guarantee that every nonempty set which is bounded above has a least upper bound. Nevertheless, this property is indispensable to the rigorous development of calculus, and we therefore take it as an axiom for the real numbers.

I. Every nonempty set of real numbers that is bounded above has a least upper bound.

This property is called *completeness*, and we say that the real number system is a *complete ordered field*. It can be shown that the real number system is essentially the only complete ordered field; that is, if a being from another planet were to construct a mathematical system with properties A–I, then its system would differ from the real number system only in that it might use different symbols for the real numbers and $+, \cdot$, and $<$.

Theorem 1.1 *If S is a nonempty set of reals which is bounded above, then* l.u.b. S *is the unique real number β such that*

(a) $x \leq \beta$ *for all x in S;*

and

(b) *if $\varepsilon > 0$, there is an x_0 in S such that $x_0 > \beta - \varepsilon$.*

Proof. Since any real number a which is less than β can be written as $\beta - \varepsilon$, with $\varepsilon = \beta - a > 0$, (b) is just another way of saying that no number less than β is an upper bound for S. Hence, by definition, $\beta =$ l.u.b. S if and only if β satisfies (a) and (b), and we need only show that there cannot be more than one real number with these properties. If $\beta_1 < \beta_2$ and β_2 has property (b), then by taking $\beta = \beta_2$ and $\varepsilon = \beta_2 - \beta_1$, we see that there is an x_0 in S such that

$$x_0 > \beta_2 - (\beta_2 - \beta_1) = \beta_1,$$

so that β_1 cannot have property (a). Therefore, only one real number satisfies both (a) and (b). This completes the proof.

The Archimedean property

The property of the real numbers described in the next theorem is called the *Archimedean property*. Intuitively, it states that it is possible to exceed any positive number—no matter how large—by adding an arbitrary positive number—no matter how small—to itself sufficiently often.

Theorem 1.2 *If ρ and ε are positive, there is an integer n such that $n\varepsilon > \rho$.*

Proof. If the statement were false, ρ would be an upper bound for the set

$$S = \{x \mid x = n\varepsilon, n \text{ an integer}\},$$

and therefore S would have a least upper bound, by property I. Let $\beta =$ l.u.b. S. Then

$$n\varepsilon \leq \beta \qquad \text{for all integers } n. \tag{4}$$

Since $n + 1$ is an integer whenever n is, (4) implies that

$$(n + 1)\varepsilon \le \beta \qquad \text{for all integers } n,$$

which implies that

$$n\varepsilon \le \beta - \varepsilon \qquad \text{for all integers } n,$$

so that $\beta - \varepsilon$ is an upper bound for S. Since $\beta - \varepsilon < \beta$, this contradicts the definition of β, and proves the theorem.

Density of the rationals

The rational numbers are said to be *dense* in the real numbers because they have the property stated in the following theorem.

Theorem 1.3 *Between any two real numbers there is a rational number; that is, if $a < b$, there is a rational number p/q such that $a < p/q < b$.*

Proof.   From Theorem 1.2, with $\rho = 1$ and $\varepsilon = b - a$, there is a positive integer q such that $q(b - a) > 1$. There is also an integer j such that $j > qa$; this is obvious if $a \le 0$, and it follows from applying Theorem 1.2 with $\varepsilon = 1$ and $\rho = qa$ if $a > 0$. Let p be the smallest integer such that $p > qa$. Then $p - 1 \le qa$ and

$$qa < p \le qa + 1 < qa + q(b - a) = qb,$$

so that $qa < p < qb$ and therefore

$$a < \frac{p}{q} < b.$$

This completes the proof.

We can use this theorem to show that the rational number system is not complete; that is, a set of rational numbers may be bounded above (by rationals), but not have a rational upper bound which is less than any other rational upper bound. For example, let

$$S = \{x \,|\, x^2 < 2\}.$$

Then l.u.b. $S = \sqrt{2}$; however, $\sqrt{2}$ is irrational (*Exercise 3*), so that if r is any rational upper bound of S, then $\sqrt{2} < r$. By Theorem 1.3, there is a rational number r_1 such that $\sqrt{2} < r_1 < r$. Since r_1 is also an upper bound of S, we have shown that there is no least rational upper bound of S.

Since the rational numbers have properties A–H, but not I, it can be seen from this example that I does not follow from A–H.

Greatest lower bound of a set

A set S is *bounded below* if there is a real number a such that each x in S satisfies the inequality $x \geq a$, in which case a is a *lower bound* of S. If α is a lower bound of S such that no number greater than α is a lower bound of S, we say that α is the *greatest lower bound* of S, and write

$$\alpha = \text{g.l.b. } S.$$

(Some authors call α the *infimum* of S, and write $\alpha = \inf S$.) Geometrically, this means that there are no points of S to the left of α, but there is at least one point of S to the left of any number greater than α.

Theorem 1.4 *Every nonempty set S of real numbers which is bounded below has exactly one greatest lower bound α, which is the unique real number with the following properties:*

(a) *If $x \in S$, then $x \geq \alpha$.*
(b) *If $\varepsilon > 0$, there is an x_0 in S such that $x_0 < \alpha + \varepsilon$.*

Proof. *Exercise 6.*

A set S is *bounded* if there are numbers a and b such that $a \leq x \leq b$ for all x in S. A bounded set has a unique least upper bound and a unique greatest lower bound, and

$$\text{g.l.b. } S \leq \text{l.u.b. } S \tag{5}$$

(*Exercise 7*).

Points at infinity

It is convenient to adjoin to the real number system two fictitious points, $+\infty$ (or simply ∞) and $-\infty$, and define the order relationships between them and any real number x by

$$-\infty < x < \infty. \tag{6}$$

In terms of these symbols we will often indicate that a set is not bounded above by writing

$$\text{l.u.b. } S = \infty, \tag{7}$$

or is not bounded below by writing

$$\text{g.l.b. } S = -\infty. \tag{8}$$

Example 1.2   If

$$S = \{x \,|\, x < 2\},$$

then l.u.b. $S = 2$ and g.l.b. $S = -\infty$. If $S = \{x \,|\, x \geq -2\}$, then l.u.b. $S = \infty$ and g.l.b. $S = -2$. If S is the set of all integers, then l.u.b. $S = \infty$ and g.l.b. $S = -\infty$.

The real number system with ∞ and $-\infty$ adjoined is called the *extended real number system,* or simply the *extended reals.* An element of the extended reals which is not $-\infty$ or ∞ is *finite;* that is, an ordinary real number is finite. However, the word "finite" in "finite real number" is redundant, and used only for emphasis, since we would never refer to ∞ or $-\infty$ as real numbers. Also, let us agree once and for all that whenever we speak of a set S of real numbers, we will mean a set of *finite* real numbers.

The arithmetic relationships among ∞, $-\infty$, and the real numbers are defined as follows:

1. If a is any real number, then

$$a + \infty = \infty + a = \infty,$$
$$-\infty = -\infty + a = -\infty,$$

$$\frac{a}{\infty} = \frac{a}{-\infty} = 0.$$

2. If $a > 0$, then

$$a\infty = \infty a = \infty,$$
$$a(-\infty) = (-\infty)a = -\infty.$$

3. If $a < 0$, then

$$a\infty = \infty a = -\infty,$$
$$a(-\infty) = (-\infty)a = \infty.$$

We also define

$$\infty + \infty = \infty\infty = (-\infty)(-\infty) = \infty$$

and

$$-\infty - \infty = \infty(-\infty) = (-\infty)\infty = -\infty.$$

Finally, we define

$$|\infty| = |-\infty| = \infty.$$

The introduction of ∞ and $-\infty$, along with the arithmetic and order relationships defined above, leads to simplifications in the statements of theorems. For example, the inequality (5), first stated only for bounded sets, holds for any nonempty set S if it is interpreted properly in accordance with (6) and the definition of (7) and (8). *Exercises 11(b) and 12(b) illustrate the convenience* afforded by some of the arithmetic relationships with extended reals, and other examples will illustrate this further in future sections.

We will see in Section 2.4 that it is not useful to define $\infty - \infty$, $0 \cdot \infty$, ∞/∞, and $0/0$. They are called *indeterminate forms*, and left undefined. Other indeterminate forms also occur in Section 2.4.

1.1 EXERCISES

1. Verify that the set consisting of two elements, 0 and 1, with operations defined by Eqs. (1) and (2), is a field. Then show that it is impossible to define an order $<$ on this field which has properties F, G, and H.

2. (a) Prove: If a and b are any real numbers, then

   $$|a + b| \leq |a| + |b|.$$

   (You may use only properties A–H and the definitions of $|\ |$, $\leq$, and $\geq$.)

   (b) Use the result of part (a) and appropriate manipulative properties of the reals to prove that

   $$|a - b| \geq \big||a| - |b|\big|.$$

3. Show that $\sqrt{2}$ is irrational. (*Hint:* Show that if $\sqrt{2} = m/n$, where m and n are integers, then both m and n must be even. Obtain a contradiction from this.)

4. Show that $\sqrt{p}$ is irrational if p is prime.

5. Find the least upper bound and greatest lower bound of each of the following sets. State whether they are in the set.
   (a) $S = \{x | x = -(1/n) + [1 + (-1)^n]n^2, n = 1, 2, 3, \ldots\}$
   (b) $S = \{x | x^2 < 9\}$
   (c) $S = \{x | x^2 \leq 7\}$
   (d) $S = \{x | |2x + 1| < 5\}$
   (e) $S = \{x | (x^2 + 1)^{-1} > \frac{1}{2}\}$
   (f) $S = \{x | x = \text{rational and } x^2 \leq 7\}$

6. Prove Theorem 1.4. (*Hint:* The set $T = \{x | -x \in S\}$ is bounded above if S is bounded below. Apply property I and Theorem 1.1 to T.)

7. (a) Show that

   $$\text{g.l.b. } S \leq \text{l.u.b. } S \tag{A}$$

   for any nonempty set S of reals, and give necessary and sufficient conditions for equality.

   (b) Show that if (A) is interpreted according to Eq. (6) and the definitions of Eqs. (7) and (8), then it holds even if S is unbounded.

8. Let S and T be nonempty sets such that every real number is in S or T and if $s \in S$ and $t \in T$, then $s < t$. Prove that there is a unique real number β such that every number less than β is in S and every number greater than β is in T. (A decomposition of the reals into two sets with these properties is a *Dedekind cut*. This result is known as Dedekind's theorem.)

9. Using properties A–H of the real numbers and taking Dedekind's theorem (Exercise 8) as given, show that every nonempty set U of reals which is bounded above has a least upper bound. (*Hint:* Let T be the set of upper bounds of U and S be the set of reals that are not in T.)

*10. (a) Suppose $a < b$ and $c > 0$. Show that there are rational numbers r_1 and r_2 such that $a < r_1 + r_2 c < b$.
 (b) Use part (a) to prove that there is an irrational between any two real numbers.

11. Let S and T be nonempty sets of reals and define
$$S + T = \{x \,|\, x = s + t, s \in S, t \in T\}.$$
 (a) Show that
$$\text{l.u.b.}(S + T) = \text{l.u.b. } S + \text{l.u.b. } T \tag{A}$$
  if S and T are bounded above, and
$$\text{g.l.b.}(S + T) = \text{g.l.b. } S + \text{g.l.b. } T \tag{B}$$
  if S and T are bounded below.
 (b) Show that (A) and (B) hold for arbitrary sets S and T, provided they are properly interpreted in the extended reals.

12. Let S and T be nonempty sets of reals and define
$$S - T = \{x \,|\, x = s - t, s \in S, t \in T\}.$$
 (a) Show that if S and T are bounded, then
$$\text{l.u.b.}(S - T) = \text{l.u.b. } S - \text{g.l.b. } T \tag{A}$$
  and
$$\text{g.l.b.}(S - T) = \text{g.l.b. } S - \text{l.u.b. } T. \tag{B}$$
 (b) Show that (A) and (B) hold for arbitrary sets S and T, provided they are properly interpreted in the extended reals.

1.2   MATHEMATICAL INDUCTION

If a flight of stairs is designed so that falling off any step inevitably leads to falling off the next, then falling off the first step is a sure way to end up at the bottom. Crudely expressed, this is the essence of the *principle of mathematical induction:* If the truth of a statement for a given integer n implies the truth of the corresponding statement for $n + 1$, then the statement is true for all positive integers n if it is true for $n = 1$. The reader has probably studied this principle before, since it is usually included in an elementary algebra course; nevertheless, the principle is so important that it merits reviewing here.

Peano's postulates and induction

The rigorous construction of the real number system starts with a set $\mathcal{N}$ of undefined elements called *natural numbers*, with the following properties:

I. $\mathcal{N}$ is nonempty.

II. Associated with each natural number n there is a unique natural number n', called the *successor of n*.

III. There is a natural number $\bar{n}$ which is not the successor of any natural number.

IV. Distinct natural numbers have distinct successors; that is, if $n \neq m$, then $n' \neq m'$.

V. The only subset of $\mathcal{N}$ which contains $\bar{n}$ and the successors of all its elements is $\mathcal{N}$ itself.

These axioms are known as *Peano's postulates*. The real numbers can be constructed from the natural numbers by definitions and arguments based on them. Although this is a task for a course in the theory of functions of a real variable rather than advanced calculus, we mention it here to show how little one needs to start with to construct the reals and, more important, to draw attention to postulate V, which is the basis for the principle of mathematical induction.

It can be shown that the positive integers form a subset of the reals which satisfies Peano's postulates (with $\bar{n} = 1$ and $n' = n + 1$), and it is customary to regard the positive integers and the natural numbers as identical. From this point of view, the principle of mathematical induction is basically a restatement of postulate V.

Theorem 2.1 (Principle of Mathematical Induction)   *Let $P_1, P_2, \ldots, P_n, \ldots$ be propositions, one corresponding to each positive integer, such that:*

(a) P_1 is true.

(b) For each positive integer n, the assumption that P_n is true leads to the conclusion that P_{n+1} is true.

Then P_n is true for each positive integer n.

Proof.   Let

$$\mathcal{M} = \{n \,|\, n \in \mathcal{N} \text{ and } P_n \text{ is true}\}.$$

From (a), $1 \in \mathcal{M}$, and, from (b), $n + 1 \in \mathcal{M}$ whenever $n \in \mathcal{M}$; therefore, $\mathcal{M} = \mathcal{N}$, by postulate V.

Example 2.1   Let P_n be the proposition that

$$1 + 2 + \cdots + n = \frac{n(n + 1)}{2}. \tag{1}$$

Then P_1 is the proposition that $1 = 1$, which is certainly true. If P_n is true, then adding $n + 1$ to both sides of (1) yields

$$(1 + 2 + \cdots + n) + (n + 1) = \frac{n(n + 1)}{2} + (n + 1)$$

$$= (n + 1)\left(\frac{n}{2} + 1\right)$$

$$= \frac{(n + 1)(n + 2)}{2},$$

or

$$1 + 2 + \cdots + (n + 1) = \frac{(n + 1)(n + 2)}{2},$$

which is P_{n+1} [since it has the form of (1), with n replaced by $n + 1$]; hence, P_n implies P_{n+1}, and therefore (1) holds for all n, by Theorem 2.1.

A proof based on Theorem 2.1 is an *induction proof*, or *proof by induction*. The assumption that P_n is true is the *induction assumption*. Theorem 2.3, below, permits a kind of induction proof in which the induction assumption takes a different form.

Induction, by definition, can be used only to verify results conjectured by other means. Thus, in Example 2.1 we did not use induction to *find* the sum

$$s_n = 1 + 2 + \cdots + n; \tag{2}$$

rather, we *verified* that

$$s_n = \frac{n(n + 1)}{2}. \tag{3}$$

How one guesses what to prove by induction depends upon the problem and one's approach to it. For example, (3) might be conjectured after observing that

$$s_1 = 1 = \frac{1 \cdot 2}{2}, \qquad s_2 = 3 = \frac{2 \cdot 3}{2}, \qquad s_3 = 6 = \frac{4 \cdot 3}{2};$$

however, this requires sufficient insight to recognize that these results are of the form (3) for $n = 1$, 2, and 3. In fact, although it is easy to prove (3) by induction once it has been conjectured, induction is not the most efficient way to find s_n, which can be obtained quickly by rewriting (2) as

$$s_n = n + (n - 1) + \cdots + 1,$$

and adding this to (2) to obtain

$$2s_n = [n + 1] + [(n - 1) + 2] + \cdots + [1 + n].$$

There are n bracketed expressions on the right, and the terms in each add up to $n + 1$; hence,

$$2s_n = n(n + 1),$$

which yields (3).

The next two examples concern problems for which induction is a natural and efficient method of solution.

Example 2.2 Let $a_1 = 1$ and

$$a_{n+1} = \frac{1}{n + 1} a_n, \qquad n \geq 1 \tag{4}$$

(we say that a_n is defined *inductively*), and suppose we wish to find an explicit formula for a_n. By considering $n = 1$, 2, and 3, we find that

$$a_1 = \frac{1}{1}, \quad a_2 = \frac{1}{1 \cdot 2}, \quad \text{and} \quad a_3 = \frac{1}{1 \cdot 2 \cdot 3},$$

and therefore conjecture that

$$a_n = \frac{1}{n!}. \tag{5}$$

This is given for $n = 1$. If we assume it is true for some n, then substituting it into (4) yields

$$a_{n+1} = \frac{1}{n + 1} \frac{1}{n!} = \frac{1}{(n + 1)!},$$

which is proposition (5) with n replaced by $n + 1$. Therefore, (5) is valid for every positive integer n, by Theorem 2.1.

Example 2.3 For each nonnegative integer n, let x_n be a real number, and suppose

$$|x_{n+1} - x_n| \leq r|x_n - x_{n-1}|, \qquad n \geq 1, \tag{6}$$

where r is a fixed positive number. By considering (6) for $n = 1$, 2, and 3 we find that

$$|x_2 - x_1| \leq r|x_1 - x_0|,$$
$$|x_3 - x_2| \leq r|x_2 - x_1| \leq r^2|x_1 - x_0|,$$

and

$$|x_4 - x_3| \leq r|x_3 - x_2| \leq r^3|x_1 - x_0|;$$

therefore, we conjecture that

$$|x_n - x_{n-1}| \leq r^{n-1}|x_1 - x_0| \tag{7}$$

if $n \geq 1$. This is trivial for $n = 1$. If it is true for some n, then (6) and (7) imply that

$$|x_{n+1} - x_n| \leq r(r^{n-1}|x_1 - x_0|)$$

or

$$|x_{n+1} - x_n| \leq r^n|x_1 - x_0|,$$

which is proposition (7) with n replaced by $n + 1$. Hence, (7) is valid for every positive integer n, by Theorem 2.1.

The major effort in an induction proof (after $P_1, P_2, \ldots, P_n, \ldots$ have been formulated) is usually directed toward showing that P_n implies P_{n+1}. However, it is equally important to verify P_1, since P_n may imply P_{n+1} even if some or all of the propositions $P_1, P_2, \ldots, P_n, \ldots$ are false.

Example 2.4   Let P_n be the proposition that $2n - 1$ is divisible by 2. If P_n is true, then P_{n+1} is also, since

$$2n + 1 = (2n - 1) + 2.$$

However, we cannot conclude that P_n is true for $n \geq 1$; in fact, P_n is false for every n.

The following formulation of the induction principle permits us to start induction proofs with an arbitrary integer, rather than 1, as required in Theorem 2.1.

Theorem 2.2   *Let n_0 be any integer (positive, negative, or 0), and suppose P_{n_0}, $P_{n_0+1}, \ldots, P_n, \ldots$ is a collection of propositions, one for each integer $n \geq n_0$, such that:*
   (a) P_{n_0} is true.
   (b) For $n \geq n_0$, the assumption that P_n is true implies that P_{n+1} is true. Then P_n is true for $n \geq n_0$.

We leave it to the reader to prove this theorem, using Theorem 2.1 (*Exercise 10*).

Example 2.5 Consider the proposition P_n that

$$3n + 16 > 0.$$

If P_n is true, then so is P_{n+1}, since

$$3(n + 1) + 16 = 3n + 16 + 3$$
$$> 0 + 3 \quad \text{(by the induction assumption)}$$
$$> 0.$$

The smallest n_0 for which P_{n_0} is true is $n_0 = -5$; hence, P_n holds for $n \geq -5$, by Theorem 2.2.

Example 2.6 Let P_n be the proposition that

$$n! - 3^n > 0.$$

If P_n is true, then

$$(n + 1)! - 3^{n+1} = n!(n + 1) - 3^{n+1}$$
$$> 3^n(n + 1) - 3^{n+1} \quad \text{(by the induction assumption)}$$
$$= 3^n(n - 2).$$

Therefore, P_n implies P_{n+1} if $n > 2$. By trial and error, $n_0 = 7$ is the smallest integer such that P_{n_0} is true; hence, P_n holds for $n \geq 7$, by Theorem 2.2.

The next theorem presents another useful formulation of the induction principle. It follows from Theorem 2.1 (*Exercise 12*).

Theorem 2.3 *Let n_0 be any integer and suppose $P_{n_0}, P_{n_0+1}, \ldots, P_n, \ldots$ is a collection of propositions, one for each integer $n \geq n_0$, such that:*
  (a) P_{n_0} is true.
  (b) For $n \geq n_0$, the assumption that $P_{n_0}, P_{n_0+1}, \ldots, P_n$ are all true implies that P_{n+1} is true.
  Then P_n is true for $n \geq n_0$.

Example 2.7 An integer $p > 1$ is *prime* if it cannot be factored as $p = rs$, where r and s are integers and $1 < r, s < p$. Thus, 2, 3, 5, 7, and 11 are primes, and, although 4, 6, 8, 9, and 10 are not, they are products of primes:

$$4 = 2 \cdot 2, \quad 6 = 2 \cdot 3, \quad 8 = 2 \cdot 2 \cdot 2, \quad 9 = 3 \cdot 3, \quad 10 = 2 \cdot 5.$$

These observations suggest for every integer n the proposition P_n that n is a prime or a product of primes. Then P_2 is true, but Theorem 2.2 does not apply, since P_n does not imply P_{n+1} in any obvious way. (For example, it is not evident from $24 = 2 \cdot 2 \cdot 2 \cdot 3$ that 25 is a product of primes.) However, Theorem 2.3 yields the desired result, as follows. Suppose $n \geq 2$ and $P_2, \ldots, P_n$ are true. Either $n + 1$ is a prime or

$$n + 1 = rs, \tag{8}$$

where r and s are integers and $1 < r, s < n$, so that P_r and P_s are true, by assumption. Hence, r and s are primes or products of primes, and (8) implies that $n + 1$ is a product of primes. We have now proved P_{n+1} ($n + 1$ is a prime or a product of primes), and therefore P_n holds for all $n \geq 2$, by Theorem 2.3.

1.2   EXERCISES

1. Prove by induction: The sum of the first n odd integers is n^2.

2. Prove by induction:

$$1^2 + 2^2 + \cdots + n^2 = \frac{n(n + 1)(2n + 1)}{6}.$$

3. Prove by induction: If $a_i \geq 0$ for $i \geq 1$, then

$$(1 + a_1)(1 + a_2) \cdots (1 + a_n) \geq 1 + a_1 + a_2 + \cdots + a_n.$$

4. Prove by induction: If $0 \leq a_i \leq 1$ for $i \geq 1$, then

$$(1 - a_1)(1 - a_2) \cdots (1 - a_n) \geq 1 - a_1 - a_2 \cdots - a_n.$$

5. Suppose $s_0 > 0$ and $s_n = 1 - e^{-s_{n-1}}$, $n \geq 1$. Show that $0 < s_n < 1$ for $n = 1, 2, \ldots$.

6. Suppose $R > 0$, $x_0 > 0$, and

$$x_{n+1} = \frac{1}{2}\left(\frac{R}{x_n} + x_n\right), \qquad n \geq 0.$$

Prove: For $n \geq 1$, $x_n > x_{n+1} > \sqrt{R}$ and

$$x_n - \sqrt{R} \leq \frac{1}{2^n} \frac{(x_0 - \sqrt{R})^2}{x_0}.$$

7. Find and justify by induction an explicit formula for a_n if $a_1 = 1$ and, for $n \geq 1$,

(a) $a_{n+1} = \dfrac{a_n}{(n + 1)(2n + 1)}$ (b) $a_{n+1} = \dfrac{3a_n}{(2n + 3)(2n + 2)}$

(c) $a_{n+1} = \dfrac{2n + 1}{n + 1} a_n$ (d) $a_{n+1} = \left(1 + \dfrac{1}{n}\right)^n a_n$

8. Let $a_1 = 0$ and $a_{n+1} = (n+1)a_n$ for $n \geq 1$, and let P_n be the proposition that $a_n = n!$.
   (a) Show that P_n implies P_{n+1}.
   (b) Is there an integer n for which P_n is true?

9. Let P_n be the proposition that

$$1 + 2 + \cdots + n = \frac{(n+2)(n-1)}{2}.$$

   (a) Show that P_n implies P_{n+1}.
   (b) Is there an integer n for which P_n is true?

10. Use Theorem 2.1 to prove Theorem 2.2.

11. For what integers n is

$$\frac{1}{n!} > \frac{8^n}{(2n)!}?$$

   Prove your answer by induction.

12. Use Theorem 2.1 to prove Theorem 2.3. (*Hint:* What is the induction assumption in Theorem 2.3?)

13. (a) Let a and b be integers, $a > 0$, and $b \geq 0$. Show by induction that there are integers q (quotient) and r (remainder) such that

$$b = aq + r \quad \text{and} \quad 0 \leq r < a.$$

   (b) Show that the result of part (a) holds if b is an arbitrary integer (not necessarily nonnegative).

14. Take the following statement as given: If p is a prime and a and b are integers such that p divides the product ab, then p divides a or b.
   (a) Prove: If $p, p_1, \ldots, p_n$ are positive primes and p divides the product $p_1 \cdots p_n$, then $p = p_i$ for some i.
   (b) Let n be an integer > 1. Show that the prime factorization of n found in Example 2.7 is unique in the following sense: If

$$n = p_1 \cdots p_r \quad \text{and} \quad n = q_1 q_2 \cdots q_s,$$

   where $p_1, \ldots, p_r, q_1, \ldots, q_s$ are positive primes, then $r = s$ and $\{q_1, \ldots, q_r\}$ is a permutation of $\{p_1, \ldots, p_r\}$.

15. Let $a_1 = a_2 = 5$ and

$$a_{n+1} = a_n + 6a_{n-1}, \qquad n \geq 2.$$

   Show that $a_n = 3^n - (-2)^n$ if $n \geq 1$.

16. Let $a_1 = 2$, $a_2 = 0$, $a_3 = -14$, and

$$a_{n+1} = 9a_n - 23a_{n-1} + 15a_{n-2}, \qquad n \geq 3.$$

   Show that $a_n = 3^{n-1} - 5^{n-1} + 2$ if $n \geq 1$.

*17. Suppose n and k are integers and $n \geq 0$. The *binomial coefficient*

$$\binom{n}{k}$$

   is the coefficient of x^k in the expansion of $(1 + x)^n$; that is,

$$(1 + x)^n = \sum_k \binom{n}{k} x^k.$$

   (a) Show that

$$\binom{n+1}{k} = \binom{n}{k} + \binom{n}{k-1}$$

   and use this to show that

$$\binom{n}{k} = \begin{cases} \dfrac{n!}{k!(n-k)!}, & k = 0, 1, \ldots, n, \\[2mm] 0, & \begin{aligned} &k = n+1, n+2, \ldots, \\ &k = -1, -2, \ldots. \end{aligned} \end{cases}$$

   (b) Show that

$$\sum_{k=0}^{n} (-1)^k \binom{n}{k} = 0.$$

   (Do not use induction here.)

1.3 THE REAL LINE

One objective of an advanced calculus course is to develop rigorously the concepts of limit, continuity, differentiability, and integrability, which the student has encountered in previous calculus courses. To do this properly requires a better understanding of the real numbers than is usually provided by those courses. It is the purpose of this section to develop this understanding. Since the utility of the concepts introduced here will not become apparent until we are well into the study of limits and continuity, the reader is advised to reserve judgment on their value until they are applied. As this occurs, it would be good to reread the applicable parts of this section. This suggestion applies especially to the concept of an open covering, and to the Heine–Borel and Bolzano–Weierstrass theorems, which will undoubtedly seem mysterious at first.

We assume that the reader is familiar with the geometric interpretation of the real numbers as points on a line. We will not prove that this interpretation is legitimate, for two reasons: first, the proof requires an excursion into the axiomatic foundations of Euclidean geometry, which is not the purpose of this book; and second, although we will use geometric terminology and intuition in discussing the reals, we will base all proofs on properties A–I and their consequences, not on geometric arguments.

Henceforth, we will use the terms *real number system* and *real line* synonymously, and denote both by the symbol $\mathscr{R}$; also, we will refer to a real number as a *point* (on the real line) when it is convenient.

Some set theory

Although we are interested in this section in sets of points on the real line, we will also consider other kinds of sets in later sections. The following definition applies to arbitrary sets, with the understanding that the elements that comprise the sets come from a specific collection of objects, identified in advance.

Definition 3.1 *Let S and T be sets. Then:*

(a) S contains T, and we write $S \supset T$ or $T \subset S$, if every element in T is also in S. In this case T is said to be a subset of S, and $S - T$ denotes the set of elements that are in S but not in T.

(b) S equals T, and we write $S = T$, if S contains T and T contains S; thus, $S = T$ if and only if S and T consist of the same elements.

(c) S strictly contains T, and we write $S \supsetneq T$ or $T \subsetneq S$, if S contains T but T does not contain S, that is, if every element in T is also in S, but at least one element in S is not in T [Figure 3.1(a)].

(d) The complement of S, denoted by S^c, is the set of elements not in S.

(e) The union of S and T, denoted by $S \cup T$, is the set of elements in at least one of S and T [Figure 3.1(b)].

(f) The intersection of S and T, denoted by $S \cap T$, is the set of elements in both S and T [Figure 3.1(c)]. If $S \cap T = \varnothing$ (the empty set), then S and T are said to be disjoint [Figure 3.1(d)].

(g) A set containing only one element, x_0, is called a singleton set, and denoted by $\{x_0\}$.

Example 3.1 Let

$$S = \{x \,|\, 0 < x < 1\}, \qquad T = \{x \,|\, 0 < x < 1 \text{ and } x \text{ is rational}\},$$

and

$$U = \{x \,|\, 0 < x < 1 \text{ and } x \text{ is irrational}\}.$$

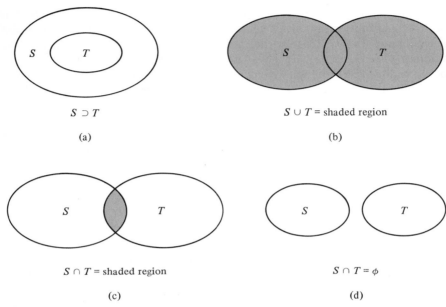

$S \supset T$

(a)

$S \cup T$ = shaded region

(b)

$S \cap T$ = shaded region

(c)

$S \cap T = \phi$

(d)

FIGURE 3.1

Then $S \supset T$ and $S \supset U$, and the inclusion is strict in both cases, so we can also write $S \supsetneq T$ and $S \supsetneq U$. The unions of pairs of these sets are

$$S \cup T = S, \quad S \cup U = S, \quad \text{and} \quad T \cup U = S,$$

and their intersections are

$$S \cap T = T, \quad S \cap U = U, \quad \text{and} \quad T \cap U = \varnothing.$$

Also,

$$S \cup T \cup U = S \quad \text{and} \quad S \cap T \cap U = \varnothing,$$
$$S - U = T \quad \text{and} \quad S - T = U.$$

Every set S contains the empty set, for to say that $\varnothing$ is not contained in S is to say that some element of $\varnothing$ is not in S, which is absurd, since $\varnothing$ has no elements. Notice that if S is any set, then

$$(S^c)^c = S \quad \text{and} \quad S \cap S^c = \varnothing.$$

If S is a set of real numbers, then $S \cup S^c = \mathscr{R}$.

The definitions of union and intersection have obvious generalizations: If $\mathscr{F}$ is an arbitrary collection of sets, then $\cup \{S \,|\, S \in \mathscr{F}\}$ is the set of all elements that belong to at least one of the sets in $\mathscr{F}$ and $\cap \{S \,|\, S \in \mathscr{F}\}$ is the set of all elements that belong to every set in $\mathscr{F}$. The union and intersection of finitely many sets $S_1, \ldots, S_n$ are also written as $\bigcup_{k=1}^{n} S_k$ and $\bigcap_{k=1}^{n} S_k$. The union and intersection of an infinite sequence of sets are written as $\bigcup_{k=1}^{\infty} S_k$ and $\bigcap_{k=1}^{\infty} S_k$.

Example 3.2 If $\mathscr{F}$ is the collection of sets

$$S = \{x \mid \rho < x \le 1 + \rho\}, \qquad 0 < \rho \le \tfrac{1}{2},$$

then

$$\cup \{S \mid S \in \mathscr{F}\} = \{x \mid 0 < x \le \tfrac{3}{2}\}$$

and

$$\cap \{S \mid S \in \mathscr{F}\} = \{x \mid \tfrac{1}{2} < x \le 1\}.$$

If for each positive integer k, S_k is the set of real numbers that can be written in the form $x = m/k$ for some integer m, then $\bigcup_{k=1}^{\infty} S_k$ is the set of rational numbers and $\bigcap_{k=1}^{\infty} S_k$ is the set of integers.

Open and closed sets

If a and b are elements of the extended reals and $a < b$, then the *open interval* (a, b) is defined by

$$(a, b) = \{x \mid a < x < b\}.$$

The open intervals (c, ∞) and $(-\infty, d)$ are *semiinfinite* if c and d are finite, and, of course, $(-\infty, \infty)$ is the entire real line.

Definition 3.2 *If x_0 is a real number and $\varepsilon > 0$, the open interval $(x_0 - \varepsilon, x_0 + \varepsilon)$ is an ε-neighborhood of x_0. If some ε-neighborhood of x_0 is contained in S, then S is a neighborhood of x_0 and x_0 is an interior point of S (Figure 3.2). The set of interior points of S is the interior of S, denoted by S^0. If every point of S is an interior point (that is, $S^0 = S$), then S is open. The complement of an open set is said to be closed.*

The idea of neighborhood is fundamental and occurs in many other contexts, some of which we will see later in this book. Whatever the context, the idea is the same: some definition of "closeness" is given (for example, two real

S = four line segments
x_0 = interior point of S

FIGURE 3.2

numbers are "close" if their difference is "small") and a neighborhood of a point x_0 is a set that contains all points sufficiently close to x_0.

Example 3.3   An open interval (a, b) is an open set, because if $x_0 \in (a, b)$ and $\varepsilon \leq \min\{x_0 - a, b - x_0\}$, then

$$(x_0 - \varepsilon, x_0 + \varepsilon) \subset (a, b).$$

The entire line $\mathscr{R} = (-\infty, \infty)$ is open, and therefore $\varnothing \, (= \mathscr{R}^c)$ is closed. However, $\varnothing$ is also open, for to deny this is to say that it contains a point which is not an interior point, which is absurd because $\varnothing$ contains no points. Since $\varnothing$ is open, $\mathscr{R} \, (= \varnothing^c)$ is closed. Thus, $\mathscr{R}$ and $\varnothing$ are both open and closed. They are the only subsets of $\mathscr{R}$ with this property (*Exercise 16*).

A *deleted neighborhood* of a point x_0 is a set that contains every point of some neighborhood of x_0 except for x_0 itself. For example,

$$S = \{x \,|\, 0 < |x - x_0| < \varepsilon\}$$

is a deleted neighborhood of x_0; we also say that it is a *deleted ε-neighborhood*.

Theorem 3.1   (*a*) *The union of open sets is open.* (*b*) *The intersection of closed sets is closed.* (*These statements apply to arbitrary collections—finite or infinite— of open and closed sets.*)

Proof.   (a) Let $\mathscr{G}$ be a collection of open sets and

$$S = \cup \, \{G \,|\, G \in \mathscr{G}\}.$$

If $x_0 \in S$, then $x_0 \in G_0$ for some G_0 in $\mathscr{G}$, and, since G_0 is open, it contains some ε-neighborhood of x_0. Since $G_0 \subset S$, this ε-neighborhood is in S, which is consequently a neighborhood of x_0. Thus, S is a neighborhood of each of its points, and therefore open, by definition.

(b) Let $\mathscr{F}$ be a collection of closed sets and

$$T = \cap \, \{F \,|\, F \in \mathscr{F}\}.$$

Then

$$T^c = \cup \, \{F^c \,|\, F \in \mathscr{F}\}$$

(*Exercise 5*), and, since each F^c is open (because each F is closed), T^c is open, from (a). Therefore, T is closed, by definition.

Example 3.4   If $-\infty < a < b < \infty$, the set

$$[a, b] = \{x \,|\, a \leq x < b\}$$

is closed, since its complement is the union of the open sets $(-\infty, a)$ and (b, ∞). We say that $[a, b]$ is a *closed interval*. The set

$$[a, b) = \{x \mid a \le x < b\}$$

is a *half-closed* or *half-open interval* if $-\infty < a < b < \infty$, as is

$$(a, b] = \{x \mid a < x \le b\};$$

however, neither of these sets is open or closed. (Why not?) *Semiinfinite closed intervals* are sets of the form

$$[a, \infty) = \{x \mid a \le x\} \quad \text{and} \quad (-\infty, a] = \{x \mid x \le a\},$$

where a is finite; they are closed sets, since their complements are the open intervals $(-\infty, a)$ and (a, ∞), respectively.

Example 3.3 shows that a set may be both open and closed, and Example 3.4 shows that a set may be neither. Thus, open and closed are not opposites in this context, as they are in everyday speech.

Example 3.5 From Theorem 3.1 and Example 3.3, the union of any collection of open intervals is an open set. (In fact, it can be shown that every nonempty open subset of $\mathscr{R}$ is the union of open intervals.) From Theorem 3.1 and Example 3.4, the intersection of any collection of closed intervals is closed.

It can be shown that the intersection of finitely many open sets is open, and the union of finitely many closed sets is closed; however, the intersection of infinitely many open sets need not be open, and the union of infinitely many closed sets need not be closed (*Exercises 6 and 7*).

Definition 3.3 *Let S be a subset of $\mathscr{R}$. Then:*

(a) x_0 is a limit point of S if every neighborhood of x_0 contains points of S distinct from x_0.

(b) x_0 is a boundary point of S if every neighborhood of x_0 contains at least one point in S and one not in S. The set of boundary points of S is the boundary of S, denoted by ∂S. The closure of S, denoted by $\bar{S}$, is defined by $\bar{S} = S \cup \partial S$.

(c) x_0 is an isolated point of S if $x_0 \in S$ and there is a neighborhood of x_0 which contains no other point of S.

(d) x_0 is exterior to S if x_0 is in the interior of S^c. The collection of such points is called the exterior of S.

Example 3.6 Let $S = (-\infty, -1] \cup (1, 2) \cup \{3\}$. Then (a) the set of limit points of S is $(-\infty, -1] \cup [1, 2]$; (b) $\partial S = \{-1, 1, 2, 3\}$ and $\bar{S} = (-\infty, -1] \cup [1, 2] \cup \{3\}$; (c) $\{3\}$ is the only isolated point of S; (d) the exterior of S is $(-1, 1) \cup (2, 3) \cup (3, \infty)$. Also, $S^0 = (-\infty, -1) \cup (1, 2)$.

Example 3.7   Let I_n be the closed interval

$$\left[\frac{1}{2n+1}, \frac{1}{2n} \right] \quad \text{and} \quad S = \bigcup_{n=1}^{\infty} I_n.$$

Then

$$S^0 = \bigcup_{n=1}^{\infty} \left(\frac{1}{2n+1}, \frac{1}{2n} \right)$$

and (a) the set of limit points of S is $S \cup \{0\}$; (b) $\partial S = \{x \,|\, x = 0 \text{ or } x = 1/n$ $(n = 2, 3, \ldots)\}$ and $\bar{S} = S \cup \{0\}$; (c) S has no isolated points; (d) the exterior of S is

$$(-\infty, 0) \cup \left[\bigcup_{n=1}^{\infty} \left(\frac{1}{2n+2}, \frac{1}{2n+1} \right) \right] \cup \left(\frac{1}{2}, \infty \right).$$

Example 3.8   Let S be the set of rational numbers. Since every interval contains a rational number (Theorem 1.3), every real number is a limit point of S; thus, $\bar{S} = \mathscr{R}$. Since every interval also contains an irrational number (*Exercise 10, Section 1.1*), every real number is a boundary point of S; thus $\partial S = \mathscr{R}$. The set S has no isolated points, and its interior and exterior are both empty. Clearly, S is neither open nor closed.

The next theorem says that S is closed if and only if $S = \bar{S}$ (*Exercise 12*).

Theorem 3.2   *A set S is closed if and only if no point of S^c is a limit point of S.*

Proof.   Suppose S is closed and $x_0 \in S^c$. Since S^c is open, there is a neighborhood of x_0 which is contained in S^c and therefore contains no points of S; hence, x_0 cannot be a limit point of S. For the converse, if no point of S^c is a limit point of S, then every point in S^c must have a neighborhood contained in S^c. Therefore, S^c is open, and S is closed.

Theorem 3.2 is usually stated as follows.

Corollary 3.1   *A set is closed if and only if it contains all its limit points.*

Theorem 3.2 and Corollary 3.1 are equivalent; however, we stated the theorem as we did because students sometimes incorrectly conclude from the corollary that a closed set must have limit points. The corollary does not say this. If S has no limit points, then the set of limit points is empty, and therefore contained in S; hence, a set with no limit points is closed according to the corollary, in agreement with Theorem 3.2. For example, any finite set is closed, and so is an infinite set comprised entirely of isolated points, such as the set of integers.

Open coverings

A collection $\mathcal{H}$ of open sets is an *open covering* for a set S if every point in S is contained in a set H belonging to $\mathcal{H}$, that is, if $S \subset \cup \{H | H \in \mathcal{H}\}$.

Example 3.9   Let

$$S_1 = [0, 1], \qquad S_2 = \{1, 2, \ldots, n, \ldots\},$$

$$S_3 = \left\{1, \frac{1}{2}, \ldots, \frac{1}{n}, \ldots\right\}, \quad \text{and} \quad S_4 = (0, 1).$$

These four sets are covered by the families of open intervals

$$\mathcal{H}_1 = \left\{\left(x - \frac{1}{N}, x + \frac{1}{N}\right) \bigg| 0 < x < 1\right\} \quad \text{(where } N \text{ is a positive integer)},$$

$$\mathcal{H}_2 = \left\{\left(n - \frac{1}{4}, n + \frac{1}{4}\right) \bigg| n = 1, 2, \ldots\right\},$$

$$\mathcal{H}_3 = \left\{\left(\frac{1}{n + \frac{1}{2}}, \frac{1}{n - \frac{1}{2}}\right) \bigg| n = 1, 2, \ldots, \right\}$$

and

$$\mathcal{H}_4 = \{(0, \rho) | 0 < \rho < 1\},$$

respectively.

Theorem 3.3 (Heine–Borel Theorem for the Real Line)   *If $\mathcal{H}$ is an open covering for a closed and bounded subset S of the real line, then S has an open covering $\hat{\mathcal{H}}$ consisting of finitely many open sets belonging to $\mathcal{H}$.*

Proof.   Since S is bounded, it has a greatest lower bound α and a least upper bound β, and, since S is closed, α and β belong to S (*Exercise 15*). If $\alpha \le t$, define

$$S_t = S \cap [\alpha, t]$$

and let

$$F = \{t | \alpha \le t \le \beta \text{ and finitely many sets from } \mathcal{H} \text{ cover } S_t\}.$$

Since $S_\beta = S$, the theorem will be proved if we can show that $\beta \in F$. To do this, we use the completeness of the reals.

Since $\alpha \in S$, S_α is the singleton set $\{\alpha\}$, which is contained in some open set H_α from $\mathscr{H}$, because $\mathscr{H}$ covers S; therefore, $\alpha \in F$. Since F is nonempty and bounded above by β, it has a least upper bound γ. First we wish to show that $\gamma = \beta$, but since $\gamma \leq \beta$ by definition of F, it suffices to rule out the possibility that $\gamma < \beta$. We consider two cases.

Case 1. Suppose $\gamma < \beta$ and $\gamma \notin S$. Then, since S is closed, γ is not a limit point of S (Theorem 3.2); consequently, there is a positive ε such that

$$[\gamma - \varepsilon, \gamma + \varepsilon] \cap S = \varnothing,$$

and therefore $S_{\gamma - \varepsilon} = S_{\gamma + \varepsilon}$. However, the definition of γ implies that $S_{\gamma - \varepsilon}$ has a finite subcovering from $\mathscr{H}$ while $S_{\gamma + \varepsilon}$ does not. This is a contradiction.

Case 2. Suppose $\gamma < \beta$ and $\gamma \in S$. Then there is an open set H_γ in $\mathscr{H}$ which contains γ, and, along with γ, an interval $[\gamma - \varepsilon, \gamma + \varepsilon]$, for some positive ε. Since $S_{\gamma - \varepsilon}$ has a finite covering $\{H_1, \ldots, H_n\}$ of sets from $\mathscr{H}$, it follows that $S_{\gamma + \varepsilon}$ has the finite covering $\{H_1, \ldots, H_n, H_\gamma\}$. However, this contradicts the definition of γ.

Now we know that $\gamma = \beta$, which is in S, and therefore there is an open set H_β in $\mathscr{H}$ which contains β, and, along with β, an interval of the form $[\beta - \varepsilon, \beta + \varepsilon]$, for some positive ε. Since $S_{\beta - \varepsilon}$ is covered by a finite collection of sets $\{H_1, \ldots, H_k\}$, S_β is covered by the finite collection $\{H_1, \ldots, H_k, H_\beta\}$. But $S_\beta = S$, so we are finished.

Henceforth we will say that a closed and bounded set is *compact*. The Heine–Borel theorem says that any open covering of a compact set S contains a finite collection which also covers S. This theorem and its converse (*Exercise 19*) show that we could just as well define a set S of reals to be compact if it has the Heine–Borel property, that is, if every open covering of S contains a finite subcovering. The same is true of $\mathscr{R}^n$, which we study in Section 5.1. This definition generalizes to more abstract spaces (called *topological spaces*) for which the concept of boundedness need not be defined.

Example 3.10 In Example 3.9 S_1 is compact, and therefore Theorem 3.3 implies that S_1 can be covered by a finite number of intervals from $\mathscr{H}_1$. This is easily verified, since, for example, the $2N + 1$ intervals from $\mathscr{H}_1$ centered at the points $x_k = k/2N$ $(0 \leq k \leq 2N)$ cover S_1.

The Heine–Borel theorem does not apply to the other sets in Example 3.9, which are not compact: S_2 is unbounded, and S_3 and S_4 are not closed, since they do not contain all their limit points (Corollary 3.1). The conclusion of the Heine–Borel theorem does not hold for these sets: each point in S_2 is contained in exactly one set from $\mathscr{H}_2$, so removing even one of these sets leaves a point of S_2 uncovered; if $\hat{\mathscr{H}}_3$ is any finite collection of sets from $\mathscr{H}_3$, then

$$\frac{1}{n} \notin \cup \{H \mid H \in \hat{\mathscr{H}}_3\}$$

for n sufficiently large; and any finite collection $\{(0, \rho_1), \ldots, (0, \rho_n)\}$ from $\mathscr{H}_4$ covers only the interval $(0, \rho_{max})$, where

$$\rho_{max} = \max\{\rho_1, \ldots, \rho_n\} < 1.$$

The Bolzano–Weierstrass theorem

As an application of Theorem 3.3, we now prove the following theorem of Bolzano and Weierstrass.

Theorem 3.4 (Bolzano–Weierstrass Theorem for the Real Line)   *Every bounded infinite set of real numbers has at least one limit point.*

Proof.   We will show that a bounded nonempty set without a limit point can contain only a finite number of points. If S has no limit points, then S is closed (Theorem 3.2) and every point x of S has an open neighborhood N_x which contains no point of S other than x. The family

$$\mathscr{H} = \{N_x \mid x \in S\}$$

is an open covering for S. Since S is also bounded, Theorem 3.3 implies that S can be covered by a finite collection of sets from $\mathscr{H}$, say $N_{x_1}, \ldots, N_{x_n}$. Since these sets contain only $x_1, \ldots, x_n$ from S, it follows that $S = \{x_1, \ldots, x_n\}$. This completes the proof.

1.3   EXERCISES

1. Find $S \cap T, (S \cap T)^c, S^c \cap T^c, S \cup T, (S \cup T)^c, S^c \cup T^c$:
   (a) $S = (0, 1), T = \left[\frac{1}{2}, \frac{3}{2}\right]$
   (b) $S = \{x \mid x^2 > 4\}, T = \{x \mid x^2 < 9\}$
   (c) $S = (-\infty, \infty), T = \varnothing$
   (d) $S = (-\infty, -1), T = (1, \infty)$

2. Let $S_k = (1 - 1/k, 2 + 1/k], k = 1, 2, \ldots$. Find:

   (a) $\displaystyle\bigcup_{k=1}^{\infty} S_k$ (b) $\displaystyle\bigcap_{k=1}^{\infty} S_k$ (c) $\displaystyle\bigcup_{k=1}^{\infty} S_k^c$ (d) $\displaystyle\bigcap_{k=1}^{\infty} S_k^c$

3. Find the largest ε such that S contains an ε-neighborhood of x_0:
   (a) $x_0 = \frac{3}{4}$, $S = [\frac{1}{2}, 1)$ (b) $x_0 = \frac{2}{3}$, $S = [\frac{1}{2}, \frac{3}{2}]$
   (c) $x_0 = 5$, $S = (-1, \infty)$ (d) $x_0 = 1$, $S = (0, 2)$

4. Describe the following sets as open, closed, or neither, and find S^0, $(S^c)^0$, and $(S^0)^c$.
   (a) $S = (-1, 2) \cup [3, \infty)$ (b) $S = (-\infty, 1) \cup (2, \infty)$
   (c) $S = [-3, -2] \cup [7, 8]$ (d) $S = \{x \mid x = \text{integer}\}$

5. Let $\mathscr{F}$ be a family of sets and define

   $$I = \cap \{F \mid F \in \mathscr{F}\} \text{ and } U = \cup \{F \mid F \in \mathscr{F}\}.$$

   Prove that (a) $I^c = \cup \{F^c \mid F \in \mathscr{F}\}$; (b) $U^c = \{\cap F^c \mid F \in \mathscr{F}\}$.

6. (a) Show that the intersection of finitely many open sets is open.
   (b) Give an example showing that the intersection of infinitely many open sets may fail to be open.

7. Use Exercises 5 and 6 to show that:
   (a) The union of finitely many closed sets is closed.
   (b) The union of infinitely many closed sets may fail to be closed.

8. Prove: (a) If U is a neighborhood of x_0 and $U \subset V$, then V is a neighborhood of x_0. (b) If $U_1, \ldots, U_n$ are neighborhoods of x_0, so is $\bigcap_{i=1}^n U_i$.

9. Find: the set of limit points of S; ∂S; $\bar{S}$; the set of isolated points of S; the exterior of S.
   (a) $S = (-\infty, -2) \cup (2, 3) \cup \{4\}, \cup (7, \infty)$
   (b) $S = \{\text{all integers}\}$
   (c) $S = \cup \{(n, n + 1) \mid n = \text{integer}\}$
   (d) $S = \{x \mid x = 1/n, n = 1, 2, 3, \ldots\}$

10. Prove: A limit point of a set S is either an interior point or a boundary point of S.

11. Prove: An isolated point of S is a boundary point of S^c.

12. Prove: (a) A boundary point of a set S is either a limit point or an isolated point of S. (b) A set S is closed if and only if $S = \bar{S}$.

13. Prove or disprove: A set has no limit points if and only if each of its points is isolated.

14. (a) Prove: If S is bounded above and $\beta = \text{l.u.b. } S$, then $\beta \in \partial S$.
    (b) State the analogous result for a set bounded below.

15. Prove: If S is closed and bounded, then g.l.b. S and l.u.b. S are both in S.

16. If a nonempty subset S of $\mathscr{R}$ is both open and closed, then $S = \mathscr{R}$.

17. Let S be an arbitrary set. Prove: (a) ∂S is closed. (b) S^0 is open. (c) The exterior of S is open. (d) The limit points of S form a closed set. (e) $(\bar{\bar{S}}) = \bar{S}$.

*18. Give counter examples to the following false statements.
  (a) The isolated points of a set form a closed set.
  (b) Every open set contains at least two points.
  (c) If S_1 and S_2 are arbitrary sets, then $\partial(S_1 \cup S_2) = \partial S_1 \cup \partial S_2$.
  (d) If S_1 and S_2 are arbitrary sets, then $\partial(S_1 \cap S_2) = \partial S_1 \cap \partial S_2$.
  (e) The least upper bound of a bounded nonempty set is the greatest of its limit points.
  (f) If S is any set, then $\partial(\partial S) = \partial S$.
  (g) If S is any set, then $\partial \bar{S} = \partial S$.
  (h) If S_1 and S_2 are arbitrary sets, then $(S_1 \cup S_2)^0 = S_1^0 \cup S_2^0$.

19. Let S be a nonempty subset of $\mathscr{R}$ such that if $\mathscr{H}$ is any open covering of S, then S has an open covering $\mathscr{H}$ comprised of finitely many open sets from $\mathscr{H}$. Show that S is closed and bounded.

*20. A set S is *dense* in a set T if $S \subset T \subset \bar{S}$.
  (a) Prove: If S and T are sets of real numbers and $S \subset T$, then S is dense in T if and only if every neighborhood of each point in T contains a point from S.
  (b) State how the result of part (a) indicates that the definition given here is consistent with the restricted definition of a dense subset of the reals, given in Theorem 1.3.

21. Show that:
  (a) $(S_1 \cap S_2)^0 = S_1^0 \cap S_2^0$ (b) $S_1^0 \cup S_2^0 \subset (S_1 \cup S_2)^0$

22. Show that:
  (a) $\partial(S_1 \cup S_2) \subset \partial S_1 \cup \partial S_2$ (b) $\partial(S_1 \cap S_2) \subset \partial S_1 \cup \partial S_2$
  (c) $\partial \bar{S} \subset \partial S$ (d) $\partial S = \partial S^c$

*23. Prove: If $T \subset S$, then $\partial(S - T) \subset \partial S \cup \partial T$.

differential calculus of functions of one variable

2.1 FUNCTIONS AND LIMITS

In this section we study limits of real-valued functions of a real variable. The concept of limit is an important part of an elementary calculus course, and so not new to the student of advanced calculus; however, the emphasis in the elementary course is usually on the manipulative properties of limits, often accepted without proof. Here we will look more carefully at the definition of limit and prove theorems usually left unproved in the elementary course.

A rule f that assigns to each element of a nonempty set D a unique element in a set Y is said to be a *function from D to Y*. We write the relationship between an element x in D and the element y in Y that f assigns to x as

$$y = f(x).$$

The set D is the *domain* of f. We will often denote it by D_f. The elements in Y are the possible *values* of f (even if they are not numerical quantities); if $y_0 \in Y$ and there is an x_0 in D such that $f(x_0) = y_0$, we say that f *attains* or *assumes* the value y_0. The set of values attained by f is the *range* of f. A *real-valued function of a real variable* is a function whose domain and range are both subsets of the reals. Although we are concerned only with real-valued functions of a real variable in this section, notice that our definitions are not restricted to this situation. In later sections we will consider situations where the range or domain, or both, of the function are subsets of vector spaces.

Example 1.1   The functions f, g, and h defined on $(-\infty, \infty)$ by

$$f(x) = x^2, \quad g(x) = \sin x, \quad \text{and} \quad h(x) = e^x$$

have ranges $[0, \infty)$, $[-1, 1]$, and $(0, \infty)$, respectively.

Example 1.2   The equation

$$[f(x)]^2 = x \tag{1}$$

does not define a function except on the singleton set $\{0\}$; if $x < 0$, no real number satisfies (1), while if $x > 0$, two real numbers satisfy (1). However, the conditions

$$[f(x)]^2 = x \quad \text{and} \quad f(x) \geq 0$$

define a function f on $D_f = [0, \infty)$, with values usually written as

$$f(x) = \sqrt{x}.$$

Similarly, the conditions

$$[g(x)]^2 = x \quad \text{and} \quad g(x) \leq 0$$

define a function g on $D_g = [0, \infty)$, with values

$$g(x) = -\sqrt{x}.$$

The ranges of f and g are $[0, \infty)$ and $(-\infty, 0]$, respectively.

It is important to understand that the definition of a function includes the specification of its domain, and that there is a difference between f, the *name* of the function, and $f(x)$, its *value* at x. However, strict observance of these points leads to annoying verbosity, such as "the function f with domain $(-\infty, \infty)$ and values $f(x) = x$." We will avoid this in two ways: first, by agreeing that if a function f is introduced without explicitly defining D_f, its domain will be understood to consist of all points x for which the rule defining $f(x)$ makes sense, and, second, by bearing in mind the distinction between f and $f(x)$, but not emphasizing it when it would be a nuisance to do so. For example, we will write "consider the function $f(x) = \sqrt{1 - x^2}$," rather than "consider the function f defined on $[-1, 1]$ by $f(x) = \sqrt{1 - x^2}$," or "consider the function $g(x) = 1/\sin x$," rather than "consider the function g defined for $x \neq k\pi$ (k = integer), by $g(x) = 1/\sin x$." We will also write $f = c$ (constant) to denote the function f defined by $f(x) = c$ for all x.

Our definition of function is somewhat intuitive, but adequate for our purposes; moreover, it is the "working" form of the definition even if the idea is introduced more rigorously to begin with. For a more precise definition, we first define the *Cartesian product* $X \times Y$ of two nonempty sets X and Y to be the set of all ordered pairs (x, y) such that $x \in X$ and $y \in Y$; thus,

$$X \times Y = \{(x, y) \mid x \in X, y \in Y\}.$$

A nonempty subset f of $X \times Y$ is said to be a *function* if no x in X occurs more than once as a first member among the elements of f; put another way, this means that if (x, y) and (x, y_1) are in f, then $y = y_1$. The set of x's that occur as first members of f is called the *domain of f*. If x is in the domain of f, then the unique y in Y such that $(x, y) \in f$ is called the *value of f at x*, and we write $y = f(x)$. The set of all such values, a subset of Y, is the *range of f*.

Arithmetic operations on functions

Definition 1.1   *Suppose $D_f \cap D_g \neq \varnothing$. Then $f + g$, $f - g$, and fg are defined on $D_f \cap D_g$ by*

$$(f + g)(x) = f(x) + g(x),$$
$$(f - g)(x) = f(x) - g(x),$$

and

$$(fg)(x) = f(x)g(x).$$

The quotient f/g is defined by

$$\left(\frac{f}{g}\right)(x) = \frac{f(x)}{g(x)}$$

for x in $D_f \cap D_g$ such that $g(x) \neq 0$.

Example 1.3   If $f(x) = \sqrt{4 - x^2}$ and $g(x) = \sqrt{x - 1}$, then $D_f = [-2, 2]$ and $D_g = [1, \infty)$, so that $f + g$, $f - g$, and fg are defined on $D_f \cap D_g = [1, 2]$ by

$$(f + g)(x) = \sqrt{4 - x^2} + \sqrt{x - 1},$$
$$(f - g)(x) = \sqrt{4 - x^2} - \sqrt{x - 1},$$

and

$$(fg)(x) = (\sqrt{4 - x^2})(\sqrt{x - 1}) = \sqrt{(4 - x^2)(x - 1)} \tag{2}$$

The quotient f/g is defined on $(1, 2]$ by

$$\left(\frac{f}{g}\right)(x) = \sqrt{\frac{4 - x^2}{x - 1}}.$$

Notice that although the last expression in (2) is also defined for $-\infty < x < -2$, it does not represent fg for such x, since f and g are not defined on $(-\infty, -2]$.

Example 1.4 If c is a real number, the function cf defined by $(cf)(x) = cf(x)$ can be regarded as the product of f and a constant function; its domain is D_f. The sum and product of $n(\geq 2)$ functions $f_1, \ldots, f_n$ are defined by

$$(f_1 + f_2 + \cdots + f_n)(x) = f_1(x) + f_2(x) + \cdots + f_n(x)$$

and

$$(f_1 f_2 \cdots f_n)(x) = f_1(x) f_2(x) \cdots f_n(x) \tag{3}$$

on $D = \bigcap_{i=1}^{n} D_{f_i}$, provided D is nonempty. If $f_1 = f_2 = \cdots = f_n$, then (3) defines the nth *power of* f:

$$(f^n)(x) = [f(x)]^n.$$

From the definition given in this example, it is possible to build up the set of all *polynomials*,

$$p(x) = a_0 + a_1 x + \cdots + a_n x^n,$$

starting from the constant functions and $f(x) = x$. Taking quotients of polynomials yields the *rational functions*,

$$r(x) = \frac{a_0 + a_1 x + \cdots + a_n x^n}{b_0 + b_1 x + \cdots + b_m x^m} \qquad (b_m \neq 0).$$

The domain of r is the set of points where its denominator is nonzero.

Limits

The essence of the concept of limit for real-valued functions of a real variable is this: $\lim_{x \to x_0} f(x) = L$ means that the value $f(x)$ can be made as close to L as we wish by taking x sufficiently close to x_0. This is made precise in the following definition.

Definition 1.2 *We say that $f(x)$ approaches the limit L as x approaches x_0, and write*

$$\lim_{x \to x_0} f(x) = L,$$

if f is defined on some deleted neighborhood of x_0, and if for every $\varepsilon > 0$ there is a $\delta > 0$ such that

$$|f(x) - L| < \varepsilon \tag{4}$$

if

$$0 < |x - x_0| < \delta. \tag{5}$$

Figure 1.1 depicts the graph of a function for which $\lim_{x \to x_0} f(x)$ exists.

Example 1.5 If c and x_0 are arbitrary real numbers and $f(x) = cx$, then

$$\lim_{x \to x_0} f(x) = cx_0.$$

To prove this, we write

$$|f(x) - cx_0| = |cx - cx_0| = |c| \, |x - x_0|.$$

If $c \neq 0$, this yields

$$|f(x) - cx_0| < \varepsilon \tag{6}$$

if

$$|x - x_0| < \delta,$$

where δ is any number such that $0 < \delta \le \varepsilon/|c|$; if $c = 0$, then $f(x) - cx_0 = 0$ for all x, so (6) holds for all x.

We emphasize that Definition 1.2 does not involve $f(x_0)$, or even require that it be defined, since (5) excludes the case where $x = x_0$.

Example 1.6 If

$$f(x) = x \sin \frac{1}{x}, \qquad x \neq 0,$$

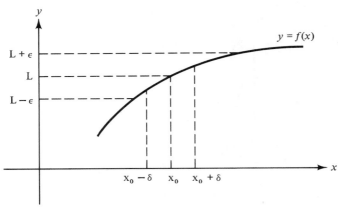

FIGURE 1.1

then

$$\lim_{x \to 0} f(x) = 0$$

even though f is not defined at $x_0 = 0$, because if

$$0 < |x| < \varepsilon,$$

then

$$|f(x) - 0| = \left| x \sin \frac{1}{x} \right| \le |x| < \varepsilon,$$

so that we can take $\delta = \varepsilon$ and satisfy the requirements of Definition 1.2. On the other hand, the function

$$g(x) = \sin \frac{1}{x}, \qquad x \neq 0,$$

has no limit as x approaches 0, since it assumes all values between -1 and 1 in every neighborhood of the origin (*Exercise 25*).

The next theorem says that a function cannot have more than one limit at a point.

Theorem 1.1   *If $\lim_{x \to x_0} f(x)$ exists, then it is unique; that is, if*

$$\lim_{x \to x_0} f(x) = L_1 \quad and \quad \lim_{x \to x_0} f(x) = L_2,$$

then $L_1 = L_2$.

Proof.   Definition 1.2 implies that if $\varepsilon > 0$, there are positive numbers δ_1 and δ_2 such that

$$|f(x) - L_i| < \varepsilon \quad \text{if} \quad 0 < |x - x_0| < \delta_i, \qquad i = 1, 2.$$

If $\delta = \min(\delta_1, \delta_2)$, then

$$|L_1 - L_2| = |L_1 - f(x) + f(x) - L_2|$$
$$\le |L_1 - f(x)| + |f(x) - L_2| < 2\varepsilon \quad \text{if} \quad 0 < |x - x_0| < \delta.$$

We have established an inequality that does not depend on x at all; namely, that

$$|L_1 - L_2| < 2\varepsilon.$$

Since this has been shown to hold for any positive ε, it follows that $L_1 = L_2$, which completes the proof.

Definition 1.2 is not changed by replacing (4) with

$$|f(x) - L| < K\varepsilon, \tag{7}$$

where K is a positive constant, because if either (4) or (7) can be made to hold for any $\varepsilon > 0$ by making $|x - x_0|$ sufficiently small and positive, then so can the other (*Exercise 5*). This may seem to be a minor point, but it is often convenient to work with (7) rather than (4), as we will see in the proof of the following theorem.

A useful theorem about limits

Theorem 1.2 *If*

$$\lim_{x \to x_0} f(x) = L_1 \quad and \quad \lim_{x \to x_0} g(x) = L_2, \tag{8}$$

then

$$\lim_{x \to x_0} (f + g)(x) = L_1 + L_2; \tag{9}$$

$$\lim_{x \to x_0} (f - g)(x) = L_1 - L_2; \tag{10}$$

$$\lim_{x \to x_0} (fg)(x) = L_1 L_2; \tag{11}$$

and, if $L_2 \neq 0$,

$$\lim_{x \to x_0} \left(\frac{f}{g}\right)(x) = \frac{L_1}{L_2}. \tag{12}$$

Proof.   Definition 1.2 and (8) imply that there is a $\delta_1 > 0$ such that

$$|f(x) - L_1| < \varepsilon \tag{13}$$

if $0 < |x - x_0| < \delta_1$, and a $\delta_2 > 0$ such that

$$|g(x) - L_2| < \varepsilon \tag{14}$$

if $0 < |x - x_0| < \delta_2$. Suppose

$$0 < |x - x_0| < \delta = \min(\delta_1, \delta_2), \tag{15}$$

so that (13) and (14) both hold. Then

$$|(f \pm g)(x) - (L_1 \pm L_2)| = |(f(x) - L_1) \pm (g(x) - L_2)|$$
$$\leq |f(x) - L_1| + |g(x) - L_2| < 2\varepsilon,$$

which proves (9) and (10).

To prove (11), we assume (15) and write

$$
\begin{aligned}
\left|(fg)(x) - L_1 L_2\right| &= |f(x)g(x) - L_1 L_2| \\
&= |f(x)(g(x) - L_2) + L_2(f(x) - L_1)| \\
&\leq |f(x)|\,|g(x) - L_2| + |L_2|\,|f(x) - L_1| \\
&\leq (|f(x)| + |L_2|)\varepsilon \\
&\leq (|f(x) - L_1| + |L_1| + |L_2|)\varepsilon \\
&\leq (\varepsilon + |L_1| + |L_2|)\varepsilon.
\end{aligned}
$$

It is no restriction to assume that $\varepsilon < 1$ (why not?); hence, the last inequality implies that

$$
\left|(fg)(x) - L_1 L_2\right| < (1 + |L_1| + |L_2|)\varepsilon
$$

if x satisfies (15). This proves (11).

To prove (12), we first observe that if $L_2 \neq 0$, there is a $\delta_3 > 0$ such that

$$
|g(x) - L_2| < \frac{|L_2|}{2},
$$

and, consequently,

$$
|g(x)| > \frac{|L_2|}{2}, \tag{16}
$$

if $\quad 0 < |x - x_0| < \delta_3$.

[This can be seen by taking $L = L_2$ and $\varepsilon = |L_2|/2$ in (4).] Now suppose $0 < |x - x_0| < \min(\delta_1, \delta_2, \delta_3)$, so that (13), (14), and (16) all hold; then

$$
\begin{aligned}
\left|\left(\frac{f}{g}\right)(x) - \frac{L_1}{L_2}\right| &= \left|\frac{f(x)}{g(x)} - \frac{L_1}{L_2}\right| \\
&= \frac{|L_2 f(x) - L_1 g(x)|}{|g(x)L_2|} \\
&\leq \frac{2}{|L_2|^2}|L_2 f(x) - L_1 g(x)| \\
&= \frac{2}{|L_2|^2}|L_2[f(x) - L_1] + L_1[L_2 - g(x)]| \\
&\leq \frac{2}{|L_2|^2}[|L_2|\,|f(x) - L_1| + |L_1|\,|L_2 - g(x)|] \\
&\leq \frac{2}{|L_2|^2}(|L_2| + |L_1|)\varepsilon,
\end{aligned}
$$

which proves (12).

Successive applications of the various parts of Theorem 1.2 permit us to find limits without the "epsilon–delta" arguments required by Definition 1.2.

Example 1.7 If c is a constant, then $\lim_{x \to x_0} c = c$, and, from Example 1.5, $\lim_{x \to x_0} x = x_0$. Therefore, from Theorem 1.2,

$$\lim_{x \to 2} (9 - x^2) = \lim_{x \to 2} 9 - \lim_{x \to 2} x^2$$

$$= \lim_{x \to 2} 9 - \left(\lim_{x \to 2} x \right)^2$$

$$= 9 - 2^2 = 5,$$

and

$$\lim_{x \to 2} (x + 1) = \lim_{x \to 2} x + \lim_{x \to 2} 1 = 2 + 1 = 3.$$

Therefore,

$$\lim_{x \to 2} \frac{9 - x^2}{x + 1} = \frac{\lim_{x \to 2} (9 - x^2)}{\lim_{x \to 2} (x + 1)} = \frac{5}{3}$$

and

$$\lim_{x \to 2} (9 - x^2)(x + 1) = \lim_{x \to 2} (9 - x^2) \lim_{x \to 2} (x + 1) = (5)(3) = 15.$$

One-sided limits

The function

$$f(x) = 2x \sin \sqrt{x}$$

satisfies the inequality

$$|f(x)| < \varepsilon$$

if $0 < x < \delta = \varepsilon/2$. However, this does not mean that $\lim_{x \to 0} f(x) = 0$, since f is not defined for negative x, as it must be to satisfy the conditions of Definition 1.2 with $x_0 = 0$ and $L = 0$. The function

$$g(x) = x + \frac{|x|}{x}, \qquad x \neq 0,$$

can be rewritten as

$$g(x) = \begin{cases} x + 1, & x > 0, \\ x - 1, & x < 0; \end{cases}$$

hence, every open interval containing $x_0 = 0$ also contains points x_1 and x_2 such that $|g(x_1) - g(x_2)|$ is as close to 2 as we please, and therefore $\lim_{x \to x_0} g(x)$ does not exist (*Exercise 25*).

Although $f(x)$ and $g(x)$ here do not approach limits as x approaches zero, they each exhibit a definite sort of limiting behavior for small positive values of x, as does $g(x)$ for small negative values of x. The kind of behavior we have in mind here is defined precisely as follows.

Definition 1.3   (a) *We say that $f(x)$ approaches the left-hand limit λ as x approaches x_0 from the left, and write*

$$\lim_{x \to x_0-} f(x) = \lambda,$$

if f is defined on some interval (a, x_0) and, for each $\varepsilon > 0$, there is a $\delta > 0$ such that

$$|f(x) - \lambda| < \varepsilon \quad if \quad x_0 - \delta < x < x_0.$$

   (b) *We say that $f(x)$ approaches the right-hand limit μ as x approaches x_0 from the right, and write*

$$\lim_{x \to x_0+} f(x) = \mu,$$

if f is defined on some interval (x_0, b), and, for every $\varepsilon > 0$, there is a $\delta > 0$ such that

$$|f(x) - \mu| < \varepsilon \quad if \quad x_0 < x < x_0 + \delta.$$

Figure 1.2 shows the graph of a function which has distinct left- and right-hand limits at a point x_0.

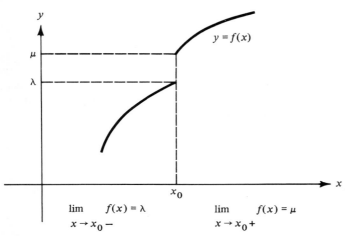

FIGURE 1.2

Example 1.8 Let

$$f(x) = \frac{x}{|x|}, \qquad x \neq 0.$$

If $x < 0$, then $f(x) = -x/x = -1$, so that

$$\lim_{x \to 0-} f(x) = -1;$$

if $x > 0$, then $f(x) = x/x = 1$, so that

$$\lim_{x \to 0+} f(x) = 1.$$

Example 1.9 Let

$$g(x) = \frac{x + |x|(1 + x)}{x} \sin \frac{1}{x}, \qquad x \neq 0.$$

If $x < 0$, then

$$g(x) = -x \sin \frac{1}{x},$$

and therefore

$$\lim_{x \to 0-} g(x) = 0,$$

since

$$|g(x) - 0| = \left| x \sin \frac{1}{x} \right| \leq |x| < \varepsilon$$

if $-\varepsilon < x < 0$; that is, Definition 1.3(a) is satisfied, with $\delta = \varepsilon$. If $x > 0$, then

$$g(x) = (2 + x) \sin \frac{1}{x},$$

which takes on every value between -2 and 2 in every interval $(0, \delta)$; hence, $g(x)$ does not approach a right-hand limit at x approaches 0 from the right. This shows that a function may have a limit from one side at a point but fail to have a limit from the other side.

Example 1.10 We leave it to the reader to verify that

$$\lim_{x \to 0+} \left(\frac{|x|}{x} + x \right) = 1,$$

$$\lim_{x \to 0-} \left(\frac{|x|}{x} + x \right) = -1,$$

$$\lim_{x \to 0+} x \sin \sqrt{x} = 0,$$

and $\lim_{x \to 0-} x \sin \sqrt{x}$ does not exist.

Left- and right-hand limits are also called *one-sided limits*. We will often simplify the notation by writing

$$\lim_{x \to x_0 -} f(x) = f(x_0 -) \quad \text{and} \quad \lim_{x \to x_0 +} f(x) = f(x_0 +).$$

The following theorem states the connection between limits and one-sided limits. We leave its proof to the reader (*Exercises 12 and 26*).

Theorem 1.3   *A function f has a limit at x_0 if and only if it has right- and left-hand limits at x_0 and they are equal. More specifically,*

$$\lim_{x \to x_0} f(x) = L$$

if and only if

$$f(x_0 +) = f(x_0 -) = L.$$

With only minor modifications of their proofs (replacing the inequality $0 < |x - x_0| < \delta$ by $x_0 - \delta < x < x_0$ or $x_0 < x < x_0 + \delta$), it can be shown that the assertions of Theorems 1.1 and 1.2 remain valid if "$\lim_{x \to x_0}$" is replaced by "$\lim_{x \to x_0 -}$" or "$\lim_{x \to x_0 +}$" throughout (*Exercise 13*).

Limits at $\pm \infty$

Limits and one-sided limits have to do with the behavior of a function f near a limit point of D_f. It is equally reasonable to study f for large positive values of x if D_f is unbounded above, or for large negative values of x if D_f is unbounded below.

Definition 1.4   *We say that $f(x)$ approaches the limit L as x approaches ∞, and write*

$$\lim_{x \to \infty} f(x) = L,$$

if f is defined on an interval (a, ∞) and, for each $\varepsilon > 0$, there is a number β such that

$$|f(x) - L| < \varepsilon \quad \text{if } x > \beta.$$

Figure 1.3 depicts the situation described in Definition 1.4.

We leave it to the reader to define the statement "$\lim_{x \to -\infty} f(x) = L$" (*Exercise 14*) and to show that Theorems 1.1 and 1.2 remain valid if x_0 is replaced throughout by ∞ or $-\infty$ (*Exercise 16*).

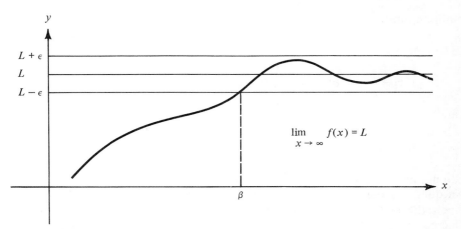

FIGURE 1.3

Example 1.11   Let

$$f(x) = 1 - \frac{1}{x^2}, \quad g(x) = \frac{2|x|}{1 + x}, \quad \text{and} \quad h(x) = \sin x.$$

Then

$$\lim_{x \to \infty} f(x) = 1,$$

since

$$|f(x) - 1| = \frac{1}{x^2} < \varepsilon \qquad \text{if } x > \frac{1}{\sqrt{\varepsilon}},$$

and

$$\lim_{x \to \infty} g(x) = 2,$$

since

$$|g(x) - 2| = \left| \frac{2x}{1 + x} - 2 \right| = \frac{2}{1 + x} < \frac{2}{x} < \varepsilon \quad \text{if} \quad x > \frac{2}{\varepsilon};$$

however, $\lim_{x \to \infty} h(x)$ does not exist, since h assumes all values between -1 and 1 in any semiinfinite interval (ρ, ∞).

We leave it to the reader to show that

$$\lim_{x \to -\infty} f(x) = 1, \qquad \lim_{x \to -\infty} g(x) = -2,$$

and $\lim_{x \to -\infty} h(x)$ does not exist (*Exercise 17*).

It will often be convenient to write $\lim_{x \to \infty} f(x)$ and $\lim_{x \to -\infty} f(x)$ as $f(\infty)$ and $f(-\infty)$, respectively.

Infinite limits

The functions

$$f(x) = \frac{1}{x}, \qquad g(x) = \frac{1}{x^2}, \qquad p(x) = \sin\frac{1}{x},$$

and

$$q(x) = \frac{1}{x^2}\sin\frac{1}{x}$$

do not have limits, or even one-sided limits, at $x_0 = 0$. They fail to have limits in quite different ways: $f(x)$ increases beyond bound as x approaches 0 from the right, and decreases beyond bound as x approaches 0 from the left; $g(x)$ increases beyond bound; $p(x)$ oscillates with ever-increasing frequency; and $q(x)$ oscillates with ever-increasing amplitude and frequency as x approaches 0. The kind of behavior exhibited by f and g near $x_0 = 0$ is sufficiently common and simple to lead us to define *infinite limits*.

Definition 1.5   *We say that $f(x)$ approaches ∞ as x approaches x_0 from the left, and write*

$$\lim_{x\to x_0-} f(x) = \infty \quad or \quad f(x_0-) = \infty,$$

if f is defined on an interval (a, x_0) and, for every real number M, there is a $\delta > 0$ such that

$$f(x) > M \qquad if \; x_0 - \delta < x < x_0.$$

We leave it to the reader to define the other kinds of infinite limits (*Exercise 19*) and verify the results stated in the next example.

Example 1.12   We leave it to the reader to show that

$$\lim_{x\to 0-}\frac{1}{x} = -\infty, \qquad \lim_{x\to 0+}\frac{1}{x} = \infty;$$

$$\lim_{x\to 0-}\frac{1}{x^2} = \lim_{x\to 0+}\frac{1}{x^2} = \lim_{x\to 0}\frac{1}{x^2} = \infty;$$

$$\lim_{x\to\infty} x^2 = \lim_{x\to-\infty} x^2 = \infty;$$

and

$$\lim_{x\to\infty} x^3 = \infty, \qquad \lim_{x\to-\infty} x^3 = -\infty.$$

Throughout this book, the statement "$\lim_{x \to x_0} f(x)$ exists" will mean that

$$\lim_{x \to x_0} f(x) = L,$$

where L is *finite*. To leave open the possibility that $L = \pm\infty$, we will say "$\lim_{x \to x_0} f(x)$ exists in the extended reals." This convention also applies to one-sided limits and limits as x approaches $\pm\infty$.

We mentioned earlier that Theorems 1.1 and 1.2 remain valid if "$\lim_{x \to x_0}$" is replaced by "$\lim_{x \to x_0-}$" or "$\lim_{x \to x_0+}$." They are also valid with x_0 replaced by $\pm\infty$. Moreover, the counterparts of (9), (10), and (11) in all of these versions of Theorem 1.2 remain valid if either or both of L_1 and L_2 is infinite, provided their right sides are not indeterminate (*Exercises 27 and 28*). Equation (12) and its counterparts remain valid if L_1/L_2 is not indeterminate *and* $L_2 \neq 0$ (*Exercise 29*).

Example 1.13 Results such as Theorem 1.2 yield

$$\lim_{x \to \infty} \sinh x = \lim_{x \to \infty} \frac{e^x - e^{-x}}{2} = \frac{1}{2}\left(\lim_{x \to \infty} e^x - \lim_{x \to \infty} e^{-x} \right)$$

$$= \frac{1}{2}(\infty - 0) = \infty;$$

$$\lim_{x \to -\infty} \sinh x = \lim_{x \to -\infty} \frac{e^x - e^{-x}}{2} = \frac{1}{2}\left(\lim_{x \to -\infty} e^x - \lim_{x \to -\infty} e^{-x} \right)$$

$$= \frac{1}{2}(0 - \infty) = -\infty;$$

and

$$\lim_{x \to \infty} \frac{e^{-x}}{x} = \frac{\lim_{x \to \infty} e^{-x}}{\lim_{x \to \infty} x} = \frac{0}{\infty} = 0;$$

however, $\lim_{x \to -\infty} e^{-x}/x$ cannot be obtained this way, since

$$\frac{\lim_{x \to -\infty} e^{-x}}{\lim_{x \to -\infty} x} = \frac{\infty}{-\infty},$$

which is an indeterminate form. (We will see how to find this limit in Section 2.4.)

Example 1.14 If

$$f(x) = e^{2x} - e^x,$$

we cannot obtain $\lim_{x \to \infty} f(x)$ by writing

$$\lim_{x \to \infty} f(x) = \lim_{x \to \infty} e^{2x} - \lim_{x \to \infty} e^x,$$

because this produces the indeterminate form $\infty - \infty$; however, by writing

$$f(x) = e^{2x}(1 - e^{-x}),$$

we find that

$$\lim_{x \to \infty} f(x) = \left(\lim_{x \to \infty} e^{2x} \right) \left(\lim_{x \to \infty} 1 - \lim_{x \to \infty} e^{-x} \right)$$
$$= \infty(1 - 0) = \infty.$$

Example 1.15 Let

$$g(x) = \frac{2x^2 - x + 1}{3x^2 + 2x - 1}.$$

Trying to find $\lim_{x \to \infty} g(x)$ by applying a version of Theorem 1.2 to this fraction as it is written leads to an indeterminate form (try it!); however, by rewriting it as

$$g(x) = \frac{2 - 1/x + 1/x^2}{3 + 2/x - 1/x^2}, \qquad x \neq 0,$$

we find that

$$\lim_{x \to \infty} g(x) = \frac{\lim_{x \to \infty} 2 - \lim_{x \to \infty} 1/x + \lim_{x \to \infty} 1/x^2}{\lim_{x \to \infty} 3 + \lim_{x \to \infty} 2/x - \lim_{x \to \infty} 1/x^2}$$
$$= \frac{2 - 0 + 0}{3 + 0 - 0} = \frac{2}{3}.$$

Monotonic functions

A function f is *nondecreasing* on an interval I if

$$f(x_1) \leq f(x_2) \qquad \text{whenever } x_1 \text{ and } x_2 \text{ are in } I \text{ and } x_1 < x_2, \tag{17}$$

or *nonincreasing* on I if

$$f(x_1) \geq f(x_2) \qquad \text{whenever } x_1 \text{ and } x_2 \text{ are in } I \text{ and } x_1 < x_2. \tag{18}$$

In either case we say that f is *monotonic* on I. If $\leq$ can be replaced by $<$ in (17), we say that f is *increasing* on I; if $\geq$ can be replaced by $>$ in (18), f is *decreasing* on I. In either of these two cases f is *strictly monotonic* on I.

Example 1.16 The function

$$f(x) = \begin{cases} x, & 0 \leq x \leq 1, \\ 2, & 1 < x \leq 2, \end{cases}$$

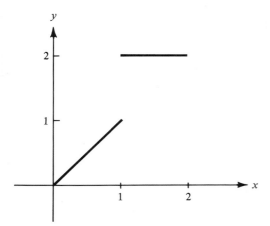

FIGURE 1.4

is nondecreasing on $I = [0, 2]$ (Figure 1.4), and $-f$ is nonincreasing on the same interval. The function

$$g(x) = x^2$$

is increasing on $[0, \infty)$ (Figure 1.5), and the function

$$h(x) = -x^3$$

is decreasing on $(-\infty, \infty)$ (Figure 1.6).

We leave the proof of the following theorem to the reader (*Exercise 33*).

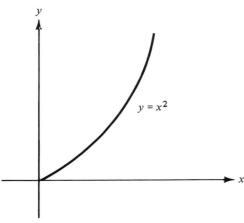

FIGURE 1.5

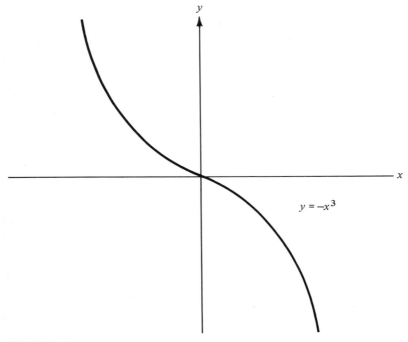

$$y = -x^3$$

FIGURE 1.6

Theorem 1.4   *Suppose f is monotonic on $I = (a, b)$, where $-\infty < a < b < \infty$, and let $R_f = \{f(x)\,|\,x \in I\}$ be the range of f. Then:*

(a) *If f is nondecreasing on I, then*

$$f(a+) = \text{g.l.b. } R_f \quad and \quad f(b-) = \text{l.u.b. } R_f.$$

(b) *If f is nonincreasing on I, then*

$$f(a+) = \text{l.u.b. } R_f \quad and \quad f(b-) = \text{g.l.b. } R_f.$$

(If R_f is unbounded, these relations hold in the extended reals.)

(c) *If $a < x_0 < b$, then $f(x_0+)$ and $f(x_0-)$ exist and are finite; $f(x_0-) \le f(x_0+)$ if f is nondecreasing, and $f(x_0-) \ge f(x_0+)$ if f is nonincreasing.*

2.1   EXERCISES

1. Each of the following conditions fails to define a function on any domain.
   State why.
   (a) $\sin f(x) = x$ (b) $e^{f(x)} = -|x|$
   (c) $1 + x^2 + [f(x)]^2 = 0$ (d) $f(x)[f(x) - 1] = x^2$

2. Find D_f:

(a) $f(x) = \tan x$

(b) $f(x) = \dfrac{1}{\sqrt{1 - |\sin x|}}$

(c) $f(x) = \dfrac{1}{x(x - 1)}$

(d) $f(x) = \dfrac{\sin x}{x}$

(e) $e^{[f(x)]^2} = x, \; f(x) \geq 0$

3. If

$$f(x) = \sqrt{\frac{(x - 3)(x + 2)}{x - 1}} \quad \text{and} \quad g(x) = \frac{x^2 - 16}{x - 7}\sqrt{x^2 - 9},$$

find D_f, D_g, $D_{f \pm g}$, D_{fg}, and $D_{f/g}$.

4. Find $\lim_{x \to x_0} f(x)$, and justify your answers with an "epsilon–delta" proof:

(a) $x^2 + 2x + 1, \; x_0 = 1$

(b) $\dfrac{x^3 - 8}{x - 2}, \; x_0 = 2$

(c) $\dfrac{1}{x^2 - 1}, \; x_0 = 0$

(d) $\sqrt{x}, \; x_0 = 4$

5. Prove: Definition 1.2 is unchanged if Eq. (4) is replaced by

$$|f(x) - L| < K\varepsilon,$$

where K is any positive constant. [That is, $\lim_{x \to x_0} f(x) = L$ according to Definition 1.2 if and only if $\lim_{x \to x_0} f(x) = L$ according to the revised definition.]

6. Use Theorem 1.2 and the known limits $\lim_{x \to x_0} x = x_0$, $\lim_{x \to x_0} c = c$ to find

(a) $\lim\limits_{x \to 2} \dfrac{x^2 + 2x + 3}{2x^3 + 1}$

(b) $\lim\limits_{x \to 2} \left(\dfrac{1}{x + 1} - \dfrac{1}{x - 1} \right)$

(c) $\lim\limits_{x \to 1} \dfrac{x - 1}{x^3 + x^2 - 2x}$

(d) $\lim\limits_{x \to 1} \dfrac{x^8 - 1}{x^4 - 1}$

7. Find $\lim_{x \to x_0-} f(x)$ and $\lim_{x \to x_0+} f(x)$, if they exist, and use epsilon–delta proofs, where applicable, to justify your answers in parts (a)–(d):

(a) $\dfrac{x + |x|}{x}, \; x_0 = 0$

(b) $x \cos \dfrac{1}{x} + \sin \dfrac{1}{x} + \sin \dfrac{1}{|x|}, \; x_0 = 0$

(c) $\dfrac{|x - 1|}{x^2 + x - 2}, \; x_0 = 1$

(d) $\dfrac{x^2 + x - 2}{\sqrt{x + 2}}, \; x_0 = -2$

(e) $\log(1 - x) - \log(1 - x^2), \; x_0 = 1$

(f) $\sin\left(\dfrac{\pi |x|}{x^2 + 2x} \right), \; x_0 = 0$

8. Prove: If $h(x) \geq 0$ for $a < x < x_0$ and $\lim_{x \to x_0-} h(x)$ exists, then $\lim_{x \to x_0-} h(x) \geq 0$.

9. (a) Prove: If $\lim_{x \to x_0} f(x)$ exists, then there is a constant M and a $\rho > 0$ such that $|f(x)| \leq M$ if $0 < |x - x_0| < \rho$. (We say then that f is *bounded* on $\{x \mid 0 < |x - x_0| < \rho\}$.)
   (b) State similar results with "$\lim_{x \to x_0}$" replaced by "$\lim_{x \to x_0-}$" and "$\lim_{x \to x_0+}$."

10. Suppose $\lim_{x \to x_0} f(x) = L$ and n is a positive integer. Prove that $\lim_{x \to x_0} [f(x)]^n = L^n$ (a) by using Theorem 1.2 and induction; (b) directly from Definition 1.2. [*Hint for (b):* You will find Exercise 9 useful.]

11. Prove: If $\lim_{x \to x_0} f(x) = L > 0$, then $\lim_{x \to x_0} \sqrt{f(x)} = \sqrt{L}$.

12. Prove Theorem 1.3.

13. (a) Using the hint stated after Theorem 1.3, prove that Theorem 1.1 remains valid with "$\lim_{x \to x_0}$" replaced by "$\lim_{x \to x_0-}$."
    (b) Repeat part (a) for Theorem 1.2.

14. Define the statement "$\lim_{x \to -\infty} f(x) = L$."

15. Find $\lim_{x \to \infty} f(x)$, if it exists, and justify your answers directly from Definition 1.4:

   (a) $\dfrac{1}{x^2 + 1}$ (b) $\dfrac{\sin x}{|x|^\alpha}$ $(\alpha > 0)$ (c) $\dfrac{\sin x}{|x|^\alpha}$ $(\alpha \leq 0)$

   (d) $e^{-x} \sin x$ (e) $\tan x$ (f) $e^{-x^2} e^{2x}$

16. Theorems 1.1 and 1.2 remain valid with "$\lim_{x \to x_0}$" replaced throughout by "$\lim_{x \to \infty}$" ("$\lim_{x \to -\infty}$"). How would their proofs have to be changed?

17. Using the definition you gave in Exercise 14, show that

   (a) $\lim_{x \to -\infty} \left(1 - \dfrac{1}{x^2}\right) = 1$ (b) $\lim_{x \to -\infty} \dfrac{2|x|}{1 + x} = -2$

   (c) $\lim_{x \to -\infty} \sin x$ does not exist

18. Find $\lim_{x \to -\infty} f(x)$, if it exists, for each of the functions in Exercise 15, and justify your answers directly from the definition you gave in Exercise 14.

19. Define:

   (a) $\lim_{x \to x_0-} f(x) = -\infty$ (b) $\lim_{x \to x_0+} f(x) = \infty$ (c) $\lim_{x \to x_0+} f(x) = -\infty$

20. Find:

   (a) $\lim_{x \to 0+} \dfrac{1}{x^3}$ (b) $\lim_{x \to 0-} \dfrac{1}{x^3}$

(c) $\lim\limits_{x\to 0+} \dfrac{1}{x^6}$

(d) $\lim\limits_{x\to 0-} \dfrac{1}{x^6}$

(e) $\lim\limits_{x\to x_0+} \dfrac{1}{(x-x_0)^{2k}}$

(f) $\lim\limits_{x\to x_0-} \dfrac{1}{(x-x_0)^{2k+1}}$

(k = positive integer)

21. Define:

(a) $\lim\limits_{x\to x_0} f(x) = \infty$

(b) $\lim\limits_{x\to x_0} f(x) = -\infty$

22. Find:

(a) $\lim\limits_{x\to 0} \dfrac{1}{x^3}$

(b) $\lim\limits_{x\to 0} \dfrac{1}{x^6}$

(c) $\lim\limits_{x\to x_0} \dfrac{1}{(x-x_0)^{2k}}$

(d) $\lim\limits_{x\to x_0} \dfrac{1}{(x-x_0)^{2k+1}}$

(k = positive integer)

23. Define:

(a) $\lim\limits_{x\to \infty} f(x) = \infty$

(b) $\lim\limits_{x\to -\infty} f(x) = -\infty$

24. Find:

(a) $\lim\limits_{x\to \infty} x^{2k}$

(b) $\lim\limits_{x\to -\infty} x^{2k}$

(c) $\lim\limits_{x\to \infty} x^{2k+1}$

(d) $\lim\limits_{x\to -\infty} x^{2k+1}$

(k = positive integer)

(e) $\lim\limits_{x\to \infty} \sqrt{x}\,\sin x$

(f) $\lim\limits_{x\to \infty} e^x$

25. (a) Prove: $\lim_{x\to x_0} f(x)$ does not exist (finite) if there is an $\varepsilon_0 > 0$ such that every deleted neighborhood of x_0 contains points x_1 and x_2 such that

$$|f(x_1) - f(x_2)| \geq \varepsilon_0.$$

(b) Give analogous statements about the nonexistence of $\lim_{x\to x_0+} f(x)$, $\lim_{x\to x_0-} f(x)$, $\lim_{x\to \infty} f(x)$, and $\lim_{x\to -\infty} f(x)$.

26. Prove: If $-\infty < x_0 < \infty$, then $\lim_{x\to x_0} f(x)$ exists in the extended reals if and only if $\lim_{x\to x_0-} f(x)$ and $\lim_{x\to x_0+} f(x)$ both exist in the extended reals and are equal, in which case all three are equal.

In Exercises 27–29, consider only the case where at least one of L_1 and L_2 is infinite.

27. Prove: If $\lim_{x \to x_0} f(x) = L_1$, $\lim_{x \to x_0} g(x) = L_2$, and $L_1 + L_2$ is not indeterminate, then

$$\lim_{x \to x_0} (f + g)(x) = L_1 + L_2.$$

28. Prove: If $\lim_{x \to \infty} f(x) = L_1$, $\lim_{x \to \infty} g(x) = L_2$, and $L_1 L_2$ is not indeterminate, then

$$\lim_{x \to \infty} (fg)(x) = L_1 L_2.$$

29. (a) Prove: If $\lim_{x \to x_0} f(x) = L_1$, $\lim_{x \to x_0} g(x) = L_2 \neq 0$, and L_1/L_2 is not indeterminate, then

$$\lim_{x \to x_0} \left(\frac{f}{g} \right)(x) = \frac{L_1}{L_2}.$$

   (b) Show that it is necessary to assume that $L_2 \neq 0$ in (a), by considering $f(x) = \sin x$, $g(x) = \cos x$, and $x_0 = \pi/2$.

30. Find:

   (a) $\displaystyle \lim_{x \to 0+} \frac{x^3 + 2x + 3}{2x^4 + 3x^2 + 2}$

   (b) $\displaystyle \lim_{x \to 0-} \frac{x^3 + 2x + 3}{2x^4 + 3x^2 + 2}$

   (c) $\displaystyle \lim_{x \to \infty} \frac{2x^4 + 3x^2 + 2}{x^3 + 2x + 3}$

   (d) $\displaystyle \lim_{x \to -\infty} \frac{2x^4 + 3x^2 + 2}{x^3 + 2x + 3}$

   (e) $\displaystyle \lim_{x \to \infty} (e^{x^2} - e^x)$

   (f) $\displaystyle \lim_{x \to \infty} \frac{x + \sqrt{x} \sin x}{2x + e^{-x}}$

31. Find $\lim_{x \to \infty} r(x)$ and $\lim_{x \to -\infty} r(x)$ for the rational function

$$r(x) = \frac{a_0 + a_1 x + \cdots + a_n x^n}{b_0 + b_1 x + \cdots + b_m x^m},$$

   where $a_n \neq 0$ and $b_m \neq 0$.

32. Suppose $\lim_{x \to x_0} f(x)$ exists for every x_0 in (a, b) and $g(x) = f(x)$ except on a set S with no limit points in (a, b). What can be said about $\lim_{x \to x_0} g(x)$ for x_0 in (a, b)? Justify your answer.

33. Prove Theorem 1.4.

34. We say that f is *bounded above* on a set S if $f(x) \leq M$ for some constant M and all x in S, that f is *bounded below* on S if $f(x) \geq m$ for some constant m and all x in S, and that f is *bounded* on S if it is bounded above and below on S. If f is defined on (a, x_0), we define $\overline{\lim}_{x \to x_0-} f(x)$ and $\underline{\lim}_{x \to x_0-} f(x)$ (the *left limit superior and left limit inferior of f at x_0*) as follows:

   (i) If f is bounded above on some interval (α_0, x_0), then

$$\overline{\lim_{x \to x_0-}} f(x) = \lim_{x \to x_0-} L_f(x; x_0),$$

where

$$L_f(x:x_0) = \text{l.u.b.}\{f(t)\,|\,x \le t < x_0\}, \alpha_0 < x < x_0.$$

If f is unbounded above on every interval (α, x_0), then $\overline{\lim}_{x \to x_0-} f(x) = \infty$.

(ii) If f is bounded below on some interval (α_0, x_0), then

$$\underline{\lim}_{x \to x_0-} f(x) = \lim_{x \to x_0-} G_f(x; x_0),$$

where

$$G_f(x; x_0) = \text{g.l.b.}\{f(t)\,|\,x \le t < x_0\}, \alpha_0 < x < x_0.$$

If f is unbounded below on every interval (α, x_0), then $\underline{\lim}_{x \to x_0-} f(x) = -\infty$.

Show that:

(a) $\overline{\lim}_{x \to x_0-} f(x)$ and $\underline{\lim}_{x \to x_0-} f(x)$ exist in the extended reals. (*Hint:* Use Theorem 1.4.)

(b) $\underline{\lim}_{x \to x_0-} f(x) \le \overline{\lim}_{x \to x_0-} f(x).$

(c) $\overline{\lim}_{x \to x_0-} f(x) = -\underline{\lim}_{x \to x_0-} (-f)(x)$ and $\underline{\lim}_{x \to x_0-} f(x) = -\overline{\lim}_{x \to x_0-} (-f)(x).$

(d) $\overline{\lim}_{x \to x_0-} f(x) = -\infty$ if and only if $\lim_{x \to x_0-} f(x) = -\infty.$

(e) $\underline{\lim}_{x \to x_0-} f(x) = \infty$ if and only if $\lim_{x \to x_0-} f(x) = \infty.$

(f) $\underline{\lim}_{x \to x_0-} f(x) = \overline{\lim}_{x \to x_0-} f(x)$ if and only if $\lim_{x \to x_0-} f(x)$ exists in the extended

reals, in which case

$$\lim_{x \to x_0-} f(x) = \underline{\lim}_{x \to x_0-} f(x) = \overline{\lim}_{x \to x_0-} f(x).$$

35. Show that

(a) $\overline{\lim}_{x \to x_0-} (f + g)(x) \le \overline{\lim}_{x \to x_0-} f(x) + \overline{\lim}_{x \to x_0-} g(x)$

and

(b) $\underline{\lim}_{x \to x_0-} (f + g)(x) \ge \underline{\lim}_{x \to x_0-} f(x) + \underline{\lim}_{x \to x_0-} g(x)$

provided the right sides are not indeterminate.

(c) State inequalities analogous to those in parts (a) and (b) for

$$\underline{\lim}_{x \to x_0-} (f - g)(x) \text{ and } \overline{\lim}_{x \to x_0-} (f - g)(x).$$

*36. Prove: $\lim_{x \to x_0-} f(x)$ exists (finite) if and only if for each $\varepsilon > 0$ there is a $\delta > 0$ such that $|f(x_1) - f(x_2)| < \varepsilon$ if $x_0 - \delta < x_1$, $x_2 < x_0$. (*Hint:* For sufficiency, show that f is bounded on some interval (α, x_0) and $\overline{\lim}_{x \to x_0-} f(x) = \underline{\lim}_{x \to x_0-} f(x)$. Then use Exercise 34(f).)

37. Using Exercise 34 as a guide, define $\overline{\lim}_{x \to x_0+} f(x)$ and $\underline{\lim}_{x \to x_0+} f(x)$. Then state and prove results analogous to those of Exercise 34.

38. Prove: $\lim_{x \to x_0} f(x)$ exists in the extended reals if and only if

$$\overline{\lim_{x \to x_0-}} f(x) = \overline{\lim_{x \to x_0+}} f(x) = \lim_{x \to x_0-} f(x) = \lim_{x \to x_0+} f(x).$$

*39. (a) Using Exercise 34 as a guide, define $\underline{\lim}_{x \to \infty} f(x)$ and $\overline{\lim}_{x \to \infty} f(x)$.
    (b) Show that $\lim_{x \to \infty} f(x)$ exists in the extended reals if and only if

$$\overline{\lim_{x \to \infty}} f(x) = \underline{\lim_{x \to \infty}} f(x).$$

   (c) Show that $\lim_{x \to \infty} f(x)$ exists (finite) if and only if, for each $\varepsilon > 0$, there is a real number α such that $|f(x_1) - f(x_2)| < \varepsilon$ if $x_1, x_2 > \alpha$.

2.2 CONTINUITY

In this section we study continuous functions of a real variable. We will prove some important theorems about continuous functions which, although intuitively plausible, are beyond the scope of the elementary calculus course. They are accessible now because of our better understanding of the real number system—especially of those properties that stem from the completeness axiom.

The definitions of $f(x_0-)$, $f(x_0+)$, and $\lim_{x \to x_0} f(x)$ do not involve $f(x_0)$, or even require that it be defined. However, the case where $f(x_0)$ is defined and equal to one or more of these quantities is important.

Definition 2.1 *Suppose $x_0 \in D_f$. Then f is said to be (a) continuous at x_0 if $\lim_{x \to x_0} f(x) = f(x_0)$; (b) continuous from the left at x_0 if $f(x_0-) = f(x_0)$; (c) continuous from the right at x_0 if $f(x_0+) = f(x_0)$.*

The following theorem provides a method for determining whether these definitions are satisfied. Its proof, which we leave to the reader (*Exercise 1*), rests on Definitions 1.2, 1.3, and 2.1.

Theorem 2.1 *A function f is*
    (a) continuous at x_0 if and only if f is defined on an open interval containing x_0 and, for each $\varepsilon > 0$, there is a $\delta > 0$ such that

$$|f(x) - f(x_0)| < \varepsilon \tag{1}$$

whenever $|x - x_0| < \delta$;

 (b) *continuous from the right at* x_0 *if and only if* f *is defined on an interval of the form* $[x_0, \beta)$ *and, for each* $\varepsilon > 0$, *there is a* $\delta > 0$ *such that* (1) *holds whenever* $x_0 \leq x < x_0 + \delta$;

 (c) *continuous from the left at* x_0 *if and only if* f *is defined on some interval of the form* $(\alpha, x_0]$ *and, for each* $\varepsilon > 0$, *there is a* $\delta > 0$ *such that* (1) *holds whenever* $x_0 - \delta < x \leq x_0$.

From Definition 2.1 and Theorem 2.1, f is continuous at x_0 if and only if

$$f(x_0 -) = f(x_0 +) = f(x_0),$$

or, equivalently, if and only if it is continuous from the right and left at x_0 (*Exercise* 2).

Example 2.1   Let f be defined on $[0, 2]$ by

$$f(x) = \begin{cases} x^2, & 0 \leq x < 1, \\ x + 1, & 1 \leq x \leq 2 \end{cases}$$

(Figure 2.1); then

$$f(0+) = 0 = f(0),$$
$$f(1-) = 1 \neq 2 = f(1),$$
$$f(1+) = 2 = f(1),$$
$$f(2-) = 3 = f(2).$$

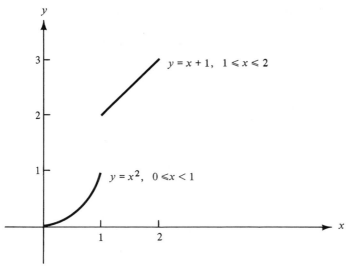

FIGURE 2.1

Therefore, f is continuous from the right at 0 and 1, and continuous from the left at 2, but not at 1. If $0 < x$, $x_0 < 1$, then

$$|f(x) - f(x_0)| = |x^2 - x_0^2| = |x - x_0| \, |x + x_0|$$
$$\le 2|x - x_0| < \varepsilon \qquad \text{if } |x - x_0| < \frac{\varepsilon}{2};$$

hence f is continuous at each x_0 in $(0, 1)$. If $1 < x$, $x_0 < 2$, then

$$|f(x) - f(x_0)| = |(x + 1) - (x_0 + 1)| = |x - x_0|$$
$$< \varepsilon \qquad \text{if } |x - x_0| < \varepsilon;$$

hence, f is continuous at each x_0 in $(1, 2)$.

Definition 2.2   *A function f is continuous on an open interval (a, b) if it is continuous at every point in (a, b). If, in addition,*

$$f(b-) = f(b) \tag{2}$$

or

$$f(a+) = f(a), \tag{3}$$

then f is continuous on $(a, b]$ or $[a, b)$, respectively; if f is continuous on (a, b) and (2) and (3) both hold, then f is continuous on $[a, b]$. More generally, if S is a subset of D_f consisting of finitely or infinitely many disjoint intervals, then f is continuous on S if f is continuous on every interval in S. (It should be understood henceforth—in connection with functions of one variable—that whenever we say "f is continuous on S," we mean that S is a set of this kind.)

Example 2.2   Let $f(x) = \sqrt{x}$, $0 \le x < \infty$. Then

$$|f(x) - f(0)| = \sqrt{x} < \varepsilon \qquad \text{if } 0 \le x < \varepsilon^2, \tag{4}$$

and, if $x_0 > 0$ and $x \ge 0$,

$$|f(x) - f(x_0)| = |\sqrt{x} - \sqrt{x_0}| = \frac{|x - x_0|}{\sqrt{x} + \sqrt{x_0}}$$
$$\le \frac{|x - x_0|}{\sqrt{x_0}} < \varepsilon \quad \text{if } |x - x_0| < \varepsilon \sqrt{x_0}. \tag{5}$$

Since (4) implies that $f(0+) = f(0)$ and (5) implies that $\lim_{x \to x_0} f(x) = f(x_0)$ if $x_0 > 0$, f is continuous on $[0, \infty)$.

Example 2.3   The function

$$g(x) = \frac{1}{\sin \pi x}$$

is continuous on the set $S = \bigcup_{n=-\infty}^{\infty} (n, n+1)$; however, g is not continuous at any point $x_0 = n$ (integer), since it is not defined at such points.

The function f defined in Example 2.1 (see also Figure 2.1) is continuous on $[0, 1)$ and $[1, 2]$, but not on any open interval containing 1. The discontinuity of f there is of the simplest kind, described in the following definition.

Definition 2.3   *A function f is said to be piecewise continuous on $[a, b]$ if*
  (a) $f(x_0 +)$ exists for all x_0 in $[a, b)$;
  (b) $f(x_0 -)$ exists for all x_0 in $(a, b]$;
and
  (c) $f(x_0 +) = f(x_0 -) = f(x_0)$ for all but finitely many points x_0 in (a, b).
If (c) fails to hold at some x_0 in (a, b), f is said to have a jump discontinuity at x_0. Also, f is said to have a jump discontinuity at a if $f(a+) \neq f(a)$, or at b if $f(b-) \neq f(b)$.

Example 2.4   The function

$$f(x) = \begin{cases} 1, & x = 0, \\ x, & 0 < x < 1, \\ 2, & x = 1, \\ x, & 1 < x \leq 2, \\ -1, & 2 < x < 3, \\ 0, & x = 3, \end{cases}$$

(Figure 2.2) is piecewise continuous on $[0, 3]$, with jump discontinuities at $x_0 = 0, 1, 2$, and 3.

The geometrical justification for the adjective "jump" can be seen in Figures 2.1 and 2.2, where the graphs exhibit a definite jump at each point of discontinuity. Not all discontinuities are of this kind, as we see in the next example.

Example 2.5   The function

$$f(x) = \begin{cases} \sin \dfrac{1}{x}, & x \neq 0 \\ 0, & x = 0, \end{cases}$$

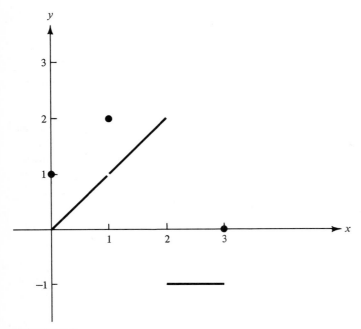

FIGURE 2.2

is continuous at all x_0 except $x_0 = 0$. As x approaches 0 from either side, $f(x)$ oscillates between -1 and 1 with ever-increasing frequency, and so neither $f(0+)$ nor $f(0-)$ exists. Therefore, the discontinuity of f at 0 is not a jump discontinuity, and f is not piecewise continuous on any interval of the form $[-\rho, 0], [-\rho, \rho],$ or $[0, \rho]$ $(\rho > 0)$.

The next theorem follows immediately from Theorems 1.2 and 2.1 (*Exercise 18*).

Theorem 2.2   *If f and g are continuous on a set S, then so are $f + g$, $f - g$, and fg; in addition, f/g is continuous at each x_0 in S such that $g(x_0) \neq 0$.*

Example 2.6   Since the constant functions and the function $f(x) = x$ are continuous for all x, successive applications of the various parts of Theorem 2.2 imply that the function

$$r(x) = \frac{9 - x^2}{x + 1}$$

is continuous for all x except $x = -1$ (see Example 1.7). More generally, by starting from Theorem 2.2 and using induction, it can be shown that if $f_1, f_2, \ldots, f_n$ are continuous on a set S, then so are $f_1 + f_2 + \cdots + f_n$ and $f_1 f_2 \cdots f_n$; therefore, any rational function (Example 1.4) is continuous for all values of x except those for which its denominator vanishes.

Removable discontinuities

Let f be defined in a deleted neighborhood of x_0 and discontinuous (perhaps even undefined) at x_0. We say that f has a *removable* discontinuity at x_0 if $\lim_{x \to x_0} f(x)$ exists. In this case the function

$$g(x) = \begin{cases} f(x) & \text{if } x \in D_f \text{ and } x \neq x_0, \\ \lim_{x \to x_0} f(x) & \text{if } x = x_0, \end{cases}$$

is continuous at x_0.

Example 2.7 The function

$$f(x) = x \sin \frac{1}{x}$$

is not defined at $x_0 = 0$—and therefore certainly is not continuous there—but $\lim_{x \to 0} f(x) = 0$ (Example 1.6). Therefore, f has a removable discontinuity at 0. The function

$$f_1(x) = \sin \frac{1}{x}$$

is also undefined at 0, but its discontinuity there is not removable, since $\lim_{x \to 0} f_1(x)$ does not exist (Example 2.5).

Composite functions

We have seen that the investigation of limits and continuity can be simplified by regarding a given function as the result of addition, subtraction, multiplication, and division of simpler functions. Another operation useful in this connection is *composition* of functions, which is, loosely speaking, the substitution of one function into another.

Definition 2.4 *If D_g has a nonempty subset T such that the number $g(x)$ is in D_f whenever $x \in T$, then the composite function $f \circ g$ is defined on T by*

$$f \circ g(x) = f(g(x)).$$

Example 2.8   Let

$$g(x) = \frac{1}{1 - x^2} \quad \text{and} \quad f(x) = \log x;$$

then

$$D_g = \{x \mid x \neq \pm 1\} \quad \text{and} \quad D_f = (0, \infty),$$

and, since $g(x) > 0$ if $x \in T = (-1, 1)$, the function

$$f \circ g(x) = \log \frac{1}{1 - x^2}$$

is defined on $(-1, 1)$. Similarly, the composite function $g \circ f$ is defined on $(0, 1/e) \cup (1/e, e) \cup (e, \infty)$ by

$$g \circ f(x) = \frac{1}{1 - (\log x)^2}.$$

The next theorem says that the composition of a continuous function with a continuous function is continuous.

Theorem 2.3   *Suppose g is continuous at x_0, $g(x_0)$ is an interior point of D_f, and f is continuous at $g(x_0)$. Then the composite function $f \circ g$ is continuous at x_0.*

Proof.   Suppose $\varepsilon > 0$. Since $g(x_0)$ is an interior point of D_f at which f is continuous, there is an $\varepsilon_1 > 0$ such that $f(t)$ is defined and

$$|f(t) - f(g(x_0))| < \varepsilon \quad \text{if} \quad |t - g(x_0)| < \varepsilon_1. \tag{6}$$

Since g is continuous at x_0, there is a $\delta > 0$ such that $g(x)$ is defined and

$$|g(x) - g(x_0)| < \varepsilon_1 \tag{7}$$

if

$$|x - x_0| < \delta \quad \text{and} \quad x \in D_g. \tag{8}$$

Now (6) and (7) imply that

$$|f(g(x)) - f(g(x_0))| < \varepsilon$$

if x satisfies (8). Therefore, $f \circ g$ is continuous at x_0.

For a related result concerning limits, see *Exercise 22*.

Example 2.9   In Examples 2.2 and 2.6 we saw that the function

$$f(x) = \sqrt{x}$$

is continuous for $x > 0$ and the function

$$g(x) = \frac{9 - x^2}{x + 1}$$

is continuous for $x \neq -1$. Since $g(x) > 0$ if $x < -3$ or $-1 < x < 3$, Theorem 2.3 implies that the function

$$f \circ g(x) = \sqrt{\frac{9 - x^2}{x + 1}}$$

is continuous on $(-\infty, -3) \cup (-1, 3)$. It is also continuous from the left at -3 and 3.

Bounded functions

A function f is *bounded below* on a set S if there is a real number m such that

$$f(x) \geq m \text{ for all } x \text{ in } S.$$

In this case, the set

$$V = \{f(x) \mid x \in S\}$$

has a greatest lower bound α, and we write

$$\alpha = \underset{x \in S}{\text{g.l.b. }} f(x).$$

If there is a point $\bar{x}$ in S such that $f(\bar{x}) = \alpha$, then we also say that α is the *minimum of f on S*, and write

$$\alpha = \underset{x \in S}{\min} f(x).$$

Similarly, f is *bounded above on S* if there is a real number M such that $f(x) \leq M$ for all x in S. In this case V has a least upper bound β, and we write

$$\beta = \underset{x \in S}{\text{l.u.b. }} f(x),$$

or, if there is a point $\bar{\bar{x}}$ in S such that $f(\bar{\bar{x}}) = \beta$,

$$\beta = \underset{x \in S}{\max} f(x).$$

If a function is bounded above and below on a set S, we say that it is *bounded on S*.

Figure 2.3 illustrates the geometrical meaning of these definitions for a function bounded on an interval $S = [a, b]$. The graph of the function lies in the strip bounded by the lines $y = M$ and $y = m$, where M and m are, respectively,

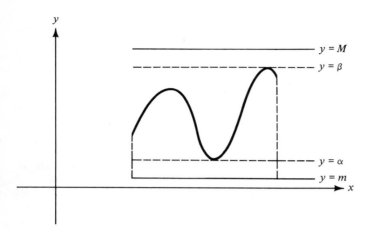

FIGURE 2.3

any upper and lower bounds for f on $[a, b]$. The narrowest strip containing the graph is the one bounded above by $y = \beta = \text{l.u.b.}_{x \in [a,b]} f(x)$ and below by $y = \alpha = \text{g.l.b.}_{x \in [a,b]} f(x)$

Example 2.10   The function

$$g(x) = \begin{cases} \frac{1}{2}, & x = 0 \text{ or } x = 1, \\ 1 - x, & 0 < x < 1, \end{cases}$$

[Figure 2.4(a)] is bounded on $[0, 1]$, and

$$\text{l.u.b.}_{x \in [0,1]} g(x) = 1, \qquad \text{g.l.b.}_{x \in [0,1]} g(x) = 0.$$

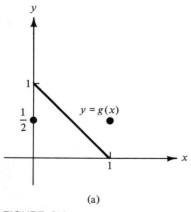

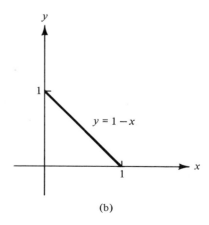

(a)

(b)

FIGURE 2.4

Therefore, g has no maximum or minimum on $[0, 1]$, since it does not assume either of the values 0 and 1.

The function

$$h(x) = 1 - x, \qquad 0 \le x \le 1,$$

which differs from g only at 0 and 1 [Figure 2.4(b)], has the same least upper bound and greatest lower bound as g, but it attains these values at $x = 0$ and $x = 1$, respectively; therefore,

$$\max_{x \in [0,1]} h(x) = 1 \quad \text{and} \quad \min_{x \in [0,1]} h(x) = 0.$$

Example 2.11 The function

$$f(x) = e^{x(x-1)} \sin \frac{1}{x(x-1)}, \qquad 0 < x < 1,$$

oscillates between $\pm e^{x(x-1)}$ infinitely often in every interval of the form $(0, \rho)$ or $(1 - \rho, 1), 0 < \rho < 1$, and

$$\underset{x \in (0,1)}{\text{l.u.b.}} \ f(x) = 1, \qquad \underset{x \in (0,1)}{\text{g.l.b.}} \ f(x) = -1;$$

however, f does not assume these values, and therefore has no maximum or minimum on $(0, 1)$.

Theorem 2.4 *If f is continuous on a closed interval $[a, b]$, then f is bounded on $[a, b]$.*

Proof. Suppose $x_0 \in [a, b]$. Since f is continuous at x_0, there is an open interval I_{x_0}, containing x_0, such that

$$|f(x) - f(x_0)| < 1 \qquad \text{if } x \in I_{x_0} \cap [a, b]. \tag{9}$$

[To see this, set $\varepsilon = 1$ in (1), Theorem 2.1.] The collection

$$\mathscr{H} = \{I_x | a \le x \le b\}$$

is an open covering for $[a, b]$, and the Heine–Borel theorem implies that finitely many intervals $I_{x_1}, \ldots, I_{x_n}$ from $\mathscr{H}$ also cover $[a, b]$. According to (9),

$$|f(x) - f(x_i)| < 1,$$

and therefore

$$|f(x)| \le |f(x_i)| + 1 \qquad \text{if } x \in I_{x_i} \cap [a, b]. \tag{10}$$

Since $[a, b] \subset \bigcup_{i=1}^{n} (I_{x_i} \cap [a, b])$, (10) implies that

$$|f(x)| \le M \quad \text{if} \quad x \in [a, b],$$

where

$$M = 1 + \max_{1 \le i \le n} \{|f(x_i)|\}.$$

This completes the proof.

This proof illustrates the value of the Heine–Borel theorem, which guarantees that M is well defined, being the largest of a *finite* set of numbers.

The next theorem shows that a function continuous on a closed interval assumes its minimum and maximum values on the interval.

Theorem 2.5   *Suppose f is continuous on a closed interval $[a, b]$ and*

$$\alpha = \underset{x \in [a,b]}{\text{g.l.b.}} f(x), \qquad \beta = \underset{x \in [a,b]}{\text{l.u.b.}} f(x).$$

Then

$$f(\bar{x}) = \alpha \quad \text{and} \quad f(\bar{\bar{x}}) = \beta.$$

for some $\bar{x}$ and $\bar{\bar{x}}$ in $[a, b]$.

Proof.   We show that $\bar{x}$ exists and leave the rest of the proof to the reader (*Exercise 24*). The proof is by contradiction. If $f(x) > \alpha$ for every x in $[a, b]$, then, because of the continuity of f, there is an open interval I_x about each point x in $[a, b]$ such that

$$f(t) > \frac{f(x) + \alpha}{2} > \alpha \quad \text{if} \quad t \in I_x \cap [a, b] \quad (\textit{Exercise 15}). \tag{11}$$

By the Heine–Borel theorem, finitely many of these intervals, say $I_{x_1}, \ldots, I_{x_n}$, cover $[a, b]$. Define

$$\alpha_1 = \min_{1 \le i \le n} \frac{f(x_i) + \alpha}{2};$$

then, since $[a, b] \subset \bigcup_{i=1}^{n} (I_{x_i} \cap D_f)$, (11) implies that

$$f(t) > \alpha_1, \qquad a \le t \le b.$$

But $\alpha_1 > \alpha$, and therefore this contradicts the definition of α. Since the assumption that $f(x) > \alpha$ for all x in $[a, b]$ led to the contradiction, we conclude that $f(\bar{x}) = \alpha$ for some $\bar{x}$ in $[a, b]$.

Example 2.12   We used the compactness of the closed interval $[a, b]$ in the proof of Theorem 2.5 when we invoked the Heine–Borel theorem. To see that this is essential to the proof, consider the function

$$g(x) = 1 - (1 - x) \sin \frac{1}{x},$$

which is continuous and has least upper bound equal to 2 on the noncompact interval $(0, 1]$, but does not assume its least upper bound on the interval, since

$$g(x) \le 1 + (1 - x)\left|\sin\frac{1}{x}\right|$$

$$\le 1 + (1 - x) < 2 \qquad \text{if } 0 < x \le 1.$$

As another example, consider the function

$$f(x) = e^{-x},$$

which is continuous and has greatest lower bound 0—which it does not attain—on the noncompact interval $(0, \infty)$.

The next theorem shows that if f is continuous on a closed interval $[a, b]$, then f assumes every value between $f(a)$ and $f(b)$ as x varies from a to b (Figure 2.5).

Theorem 2.6 (Intermediate Value Theorem)   *Suppose f is continuous on $[a, b]$, $f(a) \ne f(b)$, and μ is between $f(a)$ and $f(b)$. Then $f(\bar{x}) = \mu$ for some $\bar{x}$ in (a, b).*

Proof.   Suppose $f(a) < u < f(b)$. The set

$$S = \{x \mid a \le x \le b \text{ and } f(x) \le \mu\}$$

is bounded and nonempty. Let $\bar{x} = $ l.u.b. S. If $f(\bar{x}) > \mu$, then $\bar{x} > a$ and, since f is continuous at $\bar{x}$, there is an $\varepsilon > 0$ such that $f(x) > \mu$ for $\bar{x} - \varepsilon < x \le \bar{x}$. Therefore, $\bar{x} - \varepsilon$ is an upper bound for S, which contradicts the definition of $\bar{x}$ as its least upper bound. If $f(\bar{x}) < \mu$, then $\bar{x} < b$ and there is an $\varepsilon > 0$ such

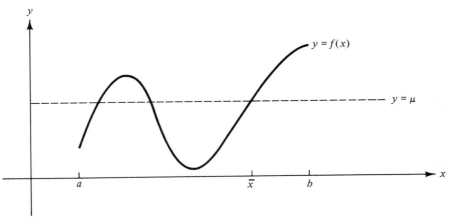

FIGURE 2.5

that $f(x) < \mu$ for $\bar{x} \leq x < \bar{x} + \varepsilon$, so that $\bar{x}$ is not an upper bound for S. This is also a contradiction. Therefore, $f(\bar{x}) = \mu$, which completes the proof in this case. The proof for the case where $f(b) < \mu < f(a)$ can be obtained by applying this result to $-f$.

Uniform continuity

Theorem 2.1 and Definition 2.2 imply that a function f is continuous on a subset S of its domain if, for each $\varepsilon > 0$ and each x_0 in S, there is a $\delta > 0$ *which may depend upon x_0 as well as ε* such that

$$|f(x) - f(x_0)| < \varepsilon \quad \text{if} \quad |x - x_0| < \delta \quad \text{and} \quad x \in D_f.$$

The next definition introduces another kind of continuity on a set S.

Definition 2.5   *A function f is said to be uniformly continuous on a subset S of its domain if, for every $\varepsilon > 0$, there is a $\delta > 0$ such that*

$$|f(x) - f(x')| < \varepsilon \quad \text{whenever} \quad |x - x'| < \delta \quad \text{and} \quad x, x' \in S.$$

We emphasize that in this definition δ depends only on ε and S, and not on the particular choice of x and x'.

Example 2.13   The function

$$f(x) = 2x$$

is uniformly continuous on $(-\infty, \infty)$, since

$$|f(x) - f(x')| = 2|x - x'| < \varepsilon \quad \text{if} \quad |x - x'| < \frac{\varepsilon}{2}.$$

Example 2.14   If $0 < r < \infty$, the function

$$g(x) = x^2$$

is uniformly continuous on $[-r, r]$, since

$$|g(x) - g(x')| = |x^2 - (x')^2| = |x - x'|\,|x + x'| \leq 2r|x - x'|;$$

hence

$$|g(x) - g(x')| < \varepsilon \quad \text{if} \quad |x - x'| < \delta = \frac{\varepsilon}{2r} \quad \text{and} \quad -r \leq x, x' \leq r.$$

Often a concept is clarified by considering its negation: a function f is *not* uniformly continuous on S if, for some $\varepsilon_0 > 0$, no matter what $\delta > 0$ is chosen, there are points x and x' in S such that

$$|x - x'| < \delta \quad \text{but} \quad |f(x) - f(x')| \geq \varepsilon_0.$$

Example 2.15 The function $g(x) = x^2$, although uniformly continuous on $[-r, r]$ for any finite r (Example 2.14), is not uniformly continuous on $(-\infty, \infty)$, since if $\varepsilon_0 = 1$ (for example) and $\rho > 0$, then

$$|g(x) - g(x')| = |x^2 - (x')^2| = |x - x'|\,|x + x'| > \frac{\rho}{2}\left(\frac{1}{\rho} + \frac{1}{\rho}\right) = 1$$

if $|x - x'| = \rho/2$ and $x, x' > 1/\rho$.

Example 2.16 The function

$$f(x) = \cos \frac{1}{x}$$

is continuous on $(0.1]$ [*Exercise 23(i)*], but not uniformly, because

$$\left| f\left(\frac{1}{n\pi}\right) - f\left(\frac{1}{(n+1)\pi}\right) \right| = 2, \qquad n = 1, 2, \ldots .$$

From Examples 2.15 and 2.16 it can be seen that a function may be continuous on an interval I and yet fail to be uniformly continuous on I. The next theorem shows that this cannot happen if I is closed and bounded, and therefore compact.

Theorem 2.7 *If f is continuous on a closed and bounded interval $I = [a, b]$, then it is uniformly continuous on I.*

Proof. Suppose $\varepsilon > 0$. Since f is continuous on I, there is for each x in I a positive number δ_x such that

$$|f(t) - f(x)| < \frac{\varepsilon}{2} \tag{12}$$

if

$$|t - x| < 2\delta_x \quad \text{and} \quad t \in I. \tag{13}$$

If $I_x = (x - \delta_x, x + \delta_x)$, then the family

$$\mathcal{H} = \{I_x \,|\, x \in I\}$$

is an open covering of I. Since I is compact, the Heine–Borel theorem implies that there are finitely many points $x_1, \ldots, x_n$ in I such that $I_{x_1}, \ldots, I_{x_n}$ cover I.

Now define

$$\delta = \min_{1 \leq i \leq n} \delta_{x_i},$$ (14)

and suppose that

$$|x - x'| < \delta \quad \text{and} \quad x, x' \in I.$$ (15)

We will show that

$$|f(x) - f(x')| < \varepsilon.$$ (16)

Since $I_{x_1}, \ldots, I_{x_n}$ cover I, x must be in one of these intervals. If $x \in I_{x_r}$, then

$$|x - x_r| < \delta_{x_r}$$ (17)

and, because of this and (15),

$$|x' - x_r| \leq |x' - x| + |x - x_r| < \delta + \delta_{x_r} \leq 2\delta_{x_r}.$$ (18)

Now (17), (18), and the definition of δ_{x_r} [see (12) and (13)] imply that

$$|f(x) - f(x_r)| < \frac{\varepsilon}{2} \quad \text{and} \quad |f(x') - f(x_r)| < \frac{\varepsilon}{2};$$

Therefore,

$$|f(x) - f(x')| \leq |f(x) - f(x_r)| + |f(x_r) - f(x')| < \frac{\varepsilon}{2} + \frac{\varepsilon}{2} = \varepsilon.$$

Thus, (15) implies (16), and the proof is complete.

This proof again shows the value of the Heine–Borel theorem, which allowed us to define δ in (14) as the smallest of a *finite* set of positive numbers, so that δ is sure to be positive. [An infinite set of positive numbers may fail to have a smallest positive member; for example, the open interval $(0, 1)$.]

Corollary 2.1   *If f is continuous on a set T, then it is uniformly continuous on any finite closed interval contained in T.*

Applied to Example 2.16, Corollary 2.1 implies that the function $g(x) = \cos 1/x$ is uniformly continuous on $[\delta, 1]$ if $0 < \delta < 1$.

More about monotonic functions

Theorem 1.4 implies that if f is monotonic on an interval I, then f is either continuous or has a jump discontinuity at each x_0 in I. This is the key to the proof of the following theorem.

Theorem 2.8 *If f is monotonic on $I = [a, b]$, then f is continuous on I if and only if its range $R_f = \{f(x) | x \in I\}$ is the closed interval with end points $f(a)$ and $f(b)$.*

Proof. *Exercise 34.*

This leads to the following theorem.

Theorem 2.9 *Suppose f is increasing and continuous on $[a, b]$, and let $f(a) = c$ and $f(b) = d$. Then there is a unique function g defined on $[c, d]$ such that*

$$g(f(x)) = x, \qquad a \le x \le b, \tag{19}$$

and

$$f(g(y)) = y, \qquad c \le y \le d. \tag{20}$$

Moreover, g is continuous and increasing on $[c, d]$.

Proof. Since f is continuous, Theorem 2.8 implies that for each y_0 in $[c, d]$ there is an x_0 in $[a, b]$ such that

$$f(x_0) = y_0, \tag{21}$$

and, since f is increasing, there is only one such x_0. Define

$$g(y_0) = x_0. \tag{22}$$

The definition of x_0 is illustrated in Figure 2.6: with $[c, d]$ drawn on the y axis, find the intersection of the line $y = y_0$ with the curve $y = f(x)$ and drop a vertical from the intersection to the x axis to find x_0.

Substituting (22) into (21) yields

$$f(g(y_0)) = y_0,$$

and substituting (21) into (22) yields

$$g(f(x_0)) = x_0.$$

Dropping the subscripts in these two equations yields (19) and (20). Notice that the uniqueness of the function g follows from the fact that f is increasing, and therefore only one value of x_0 can satisfy (21) for each y_0.

If $y_1 < y_2$, $f(x_1) = y_1$, and $f(x_2) = y_2$, then $x_1 < x_2$, again because f is increasing; hence,

$$g(y_1) = x_1 < x_2 = g(y_2),$$

so g is increasing. Since every x in $[a, b]$ is of the form $x = g(y)$ for some y in $[c, d]$, Theorem 2.8 implies that g is continuous on $[c, d]$.

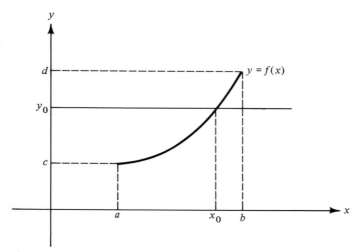

FIGURE 2.6

The function g of Theorem 2.9 is called the *inverse* of f, and denoted by f^{-1}. Since (19) and (20) are symmetric in f and g, we can also regard f as the inverse of g, and denote it by g^{-1}.

Example 2.17   If

$$f(x) = x^2, \qquad 0 \le x \le R,$$

then

$$f^{-1}(y) = g(y) = \sqrt{y}, \qquad 0 \le y \le R^2.$$

Example 2.18   If

$$f(x) = 2x + 4, \qquad 0 \le x \le 2,$$

then

$$f^{-1}(y) = g(y) = \frac{y - 4}{2}, \qquad 4 \le y \le 8.$$

2.2 EXERCISES

1. Prove Theorem 2.1.

2. Prove: A function f is continuous at x_0 if and only if

$$\lim_{x \to x_0 -} f(x) = \lim_{x \to x_0 +} f(x) = f(x_0).$$

3. Determine whether f is continuous or continuous from the right or left at x_0:

(a) $f(x) = \sqrt{x}$ $(x_0 = 0)$ (b) $f(x) = \sqrt{x}$ $(x_0 > 0)$

(c) $f(x) = \dfrac{1}{x}$ $(x_0 = 0)$ (d) $f(x) = x^2$ $(x_0$ arbitrary$)$

(e) $f(x) = \begin{cases} x \sin 1/x, & x \neq 0 \\ 1, & x = 0 \end{cases}$ $(x_0 = 0)$

(f) $f(x) = \begin{cases} x \sin 1/x, & x \neq 0 \\ 0, & x = 0 \end{cases}$ $(x_0 = 0)$

(g) $f(x) = \begin{cases} \dfrac{x + |x|(1 + x)}{x} \sin \dfrac{1}{x}, & x \neq 0 \\ 0, & x = 0 \end{cases}$ $(x_0 = 0)$

4. Let f be defined on $[0, 2]$ by

$$f(x) = \begin{cases} x^2, & 0 \leq x < 1, \\ x + 1, & 1 \leq x \leq 2. \end{cases}$$

On which of the following intervals is f continuous, according to Definition 2.2: $[0, 1)$, $(0, 1)$, $(0, 1]$, $[0, 1]$, $[1, 2)$, $(1, 2)$, $(1, 2]$, $[1, 2]$?

5. Let

$$g(x) = \frac{\sqrt{x}}{x - 1}.$$

On which of the following intervals is g continuous, according to Definition 2.2: $[0, 1)$, $(0, 1)$, $(0, 1]$, $[1, \infty)$, $(1, \infty)$?

6. Let

$$f(x) = \begin{cases} 0 & \text{if } x \text{ is irrational,} \\ 1 & \text{if } x \text{ is rational.} \end{cases}$$

Show that f is not continuous anywhere.

* 7. Let $f(x) = 0$ if x is irrational and $f(p/q) = 1/q$ if p and q are positive integers with no common factors. Show that f is discontinuous at every rational and continuous at every irrational on $(0, \infty)$.

8. Prove: If f assumes only finitely many values, then f is continuous at a point x_0 in D_f^0 if and only if f is constant on some interval $(x_0 - \delta, x_0 + \delta)$.

* 9. The *characteristic function* ψ_T of a set T is defined by

$$\psi_T(x) = \begin{cases} 1, & x \in T, \\ 0, & x \notin T. \end{cases}$$

Show that ψ_T is continuous at a point x_0 if and only if $x_0 \in T^0 \cup (T^c)^0$.

10. Prove: If f and g are continuous on (a, b) and $f(x) = g(x)$ for every x in a dense subset (Exercise 20, Section 1.3) of (a, b), then $f(x) = g(x)$ for all x in (a, b).

11. Prove that the function $g(x) = \log x$ is continuous on $(0, \infty)$. Take the following properties as given:

    (i) $\lim_{x \to 1} g(x) = 0.$

    (ii) $g(x_1) + g(x_2) = g(x_1 x_2)$, $x_1, x_2 > 0.$

12. Prove that the function $f(x) = e^{ax}$ is continuous on $(-\infty, \infty)$. Take the following properties as given:

    (i) $\lim_{x \to 0} f(x) = 1.$

    (ii) $f(x_1 + x_2) = f(x_1)f(x_2)$, $-\infty < x_1, x_2 < \infty.$

13. (a) Prove that the functions $\sinh x$ and $\cosh x$ are continuous for all x.
    (b) For what values of x are $\tanh x$ and $\coth x$ continuous?

14. Prove that the functions $s(x) = \sin x$ and $c(x) = \cos x$ are continuous on $(-\infty, \infty)$. Take the following properties as given:

    (i) $\lim_{x \to 0} c(x) = 1.$

    (ii) $c(x_1 - x_2) = c(x_1)c(x_2) + s(x_1)s(x_2)$, $-\infty < x_1, x_2 < \infty.$
    (iii) $s^2(x) + c^2(x) = 1$, $-\infty < x < \infty.$

15. (a) Prove: If f is continuous at x_0 and $f(x_0) > \mu$, there is a neighborhood N of x_0 such that $f(x) > \mu$ if $x \in N$.
    (b) State the corresponding result for the case where $f(x_0) < \mu$.
    (c) Prove: If $f(x) \le \mu$ for all x in S and x_0 is a limit point of S at which f is continuous, then $f(x_0) \le \mu$.
    (d) State results analogous to parts (a), (b), and (c) for continuity from the right and the left.

16. Let $|f|$ be the function whose value at each x in D_f is $|f(x)|$.
    (a) Prove: If f is continuous at x_0, then so is $|f|$.
    (b) Is the converse true?

17. Prove: If f is monotonic on $[a, b]$, it is piecewise continuous if and only if it has only finitely many discontinuities.

18. Prove Theorem 2.2.

19. (a) Show that if $f_1, f_2, \ldots, f_n$ are continuous on a set S, then so are $f_1 + f_2 + \cdots + f_n$ and $f_1 f_2 \cdots f_n$.
    (b) Use part (a) to show that a rational function is continuous for all values of x except the zeros of its denominator.

20. (a) Let f_1 and f_2 be continuous at x_0 and define

$$F(x) = \max(f_1(x), f_2(x)).$$

Show that F is continuous at x_0.
(b) Let $f_1, f_2, \ldots, f_n$ be continuous at x_0 and define

$$F(x) = \max(f_1(x), f_2(x), \ldots, f_n(x)).$$

Show that F is continuous at x_0.

21. Find the domains of $f \circ g$ and $g \circ f$:
    (a) $f(x) = \sqrt{x}$, $g(x) = 1 - x^2$
    (b) $f(x) = \log x$, $g(x) = \sin x$

    (c) $f(x) = \dfrac{1}{1 - x^2}$, $g(x) = \cos x$

    (d) $f(x) = \sqrt{x}$, $g(x) = \sin 2x$

*22. (a) Suppose $\lim_{x \to x_0+} g(x)$ exists and is an interior point of D_f, and f is continuous at the point $y_0 = \lim_{x \to x_0+} g(x)$. Show that

$$\lim_{x \to x_0+} f \circ g(x) = f(y_0).$$

(b) State an analogous result for limits from the left.
(c) State an analogous result for limits.

23. Use Theorem 2.3 to find all points x_0 at which the following functions are continuous:

(a) $\sqrt{1 - x^2}$ (b) $\sin e^{-x^2}$ (c) $\log(1 + \sin x)$

(d) $e^{-1/(1 - 2x)}$ (e) $\sin \dfrac{1}{(x - 1)^2}$ (f) $\sin\left(\dfrac{1}{\cos x}\right)$

(g) $(1 - \sin^2/x)^{-1/2}$ (h) $\cot(1 - e^{-x^2})$ (i) $\cos \dfrac{1}{x}$

24. Complete the proof of Theorem 2.5.

*25. Prove: If f is nonconstant and continuous on an interval I, the set $S = \{y \mid y = f(x), x \in I\}$ is an interval. Moreover, if I is a finite closed interval, so is S.

26. Suppose f and p are defined on $(-\infty, \infty)$, f is increasing, and $h = f \circ p$ is continuous on $(-\infty, \infty)$. Show that p is continuous on $(-\infty, \infty)$.

27. Let f be continuous on $[a, b)$ and define

$$F(x) = \max_{a \le t \le x} f(t), \qquad a \le x < b.$$

(How do we know that F is well defined?) Show that F is continuous on $[a, b)$.

28. Let f and g be uniformly continuous on an interval I.
    (a) Show that $f + g$ and $f - g$ are uniformly continuous on I.
    (b) Show that fg is uniformly continuous on I if I is compact.
    (c) Show that f/g is uniformly continuous on I if I is compact and g has no zeros in S.
    (d) Give examples showing that the conclusions of (b) and (c) may fail to hold if I is not compact.
    (e) State additional conditions on f and g which guarantee that fg is uniformly continuous on I even if I is not compact. Do the same for f/g.

29. Suppose f is uniformly continuous on a set S, g is uniformly continuous on a set T, and $g(x) \in S$ for every x in T. Show that $f \circ g$ is uniformly continuous on T.

30. (a) Prove: If f is uniformly continuous on disjoint closed intervals $I_1, \ldots, I_n$, then it is uniformly continuous on $\bigcup_{j=1}^{n} I_j$.
    (b) Is part (a) valid without the word "closed?"

31. (a) Prove: If f is uniformly continuous on a bounded open interval (a, b), then $f(a+)$ and $f(b-)$ exist and are finite. (*Hint:* See Exercise 36, Section 2.1.)
    (b) Show that the conclusion in part (a) does not follow if (a, b) is unbounded.

32. Prove: If f is continuous on $[a, \infty)$ and $f(\infty)$ exists (finite), then f is uniformly continuous on $[a, \infty)$.

33. Let f be a function with properties (i) and (ii) of Exercise 12. Prove:
    (a) $f(r) = [f(1)]^r$ if r is rational.
    (b) If $f(1) = 1$, then f is constant.
    (c) If $f(1) = \rho > 1$, then f is increasing, $\lim_{x \to \infty} f(x) = \infty$, and $\lim_{x \to -\infty} f(x) = 0$. [Thus, the function $f(x) = e^{ax}$ has these properties if $a > 0$.]

34. Prove Theorem 2.8.

2.3 DIFFERENTIABLE FUNCTIONS OF ONE VARIABLE

Much of the elementary calculus course is devoted to differentiation, with emphasis on developing formal rules for calculating derivatives. Here we consider the theoretical properties of differentiable functions. In doing this we assume that the reader knows how to differentiate elementary functions such as x^n, e^x, and $\sin x$, and we will not hesitate to use such functions in examples.

Definition of the derivative

Definition 3.1   *A function f is differentiable at an interior point x_0 of its domain if the difference quotient*

$$\frac{f(x) - f(x_0)}{x - x_0}, \qquad x \neq x_0, \tag{1}$$

approaches a limit as x approaches x_0, in which case the limit is called the derivative of f at x_0, and is denoted by $f'(x_0)$; thus,

$$f'(x_0) = \lim_{x \to x_0} \frac{f(x) - f(x_0)}{x - x_0}. \tag{2}$$

It is sometimes convenient to write (2) *as*

$$f'(x_0) = \lim_{h \to 0} \frac{f(x_0 + h) - f(x_0)}{h}.$$

A function f is *differentiable on a set S* if it is differentiable at every point of S. This definition assumes that $S \subset D_f^0$, since f' is defined only at interior points of D_f; for example, if we say that f is differentiable on the closed interval $[a, b]$, it is to be understood that f is defined on an open interval containing $[a, b]$.

If f is differentiable on S, then f' is a function on S. We say that f is *continuously differentiable* on S if f' is continuous on S. If f is differentiable on a neighborhood of x_0, it is reasonable to ask if f' is differentiable at x_0. If so, we denote its derivative there by $f''(x_0)$. This is the *second derivative of f at x_0*, and it is also denoted by $f^{(2)}(x_0)$. Continuing inductively, if $f^{(n-1)}$ is defined on a neighborhood of x_0, then the *nth derivative of f at x_0*, denoted by $f^{(n)}(x_0)$, is the derivative of $f^{(n-1)}$ at x_0. For convenience we define the *zeroth derivative* of f to be f itself; thus,

$$f^{(0)} = f.$$

We assume that the reader is familiar with the other standard notations for derivatives; for example,

$$f^{(2)} = f'', \qquad f^{(3)} = f''',$$

and so on, and

$$\frac{d^n f}{dx^n} = f^{(n)}.$$

Example 3.1   If n is a positive integer and

$$f(x) = x^n,$$

then

$$\frac{f(x) - f(x_0)}{x - x_0} = \frac{x^n - x_0^n}{x - x_0}$$

$$= \frac{x - x_0}{x - x_0} \sum_{k=0}^{n-1} x^{n-k-1} x_0^k,$$

and so

$$f'(x_0) = \lim_{x \to x_0} \sum_{k=0}^{n-1} x^{n-k-1} x_0^k = n x_0^{n-1}.$$

Since this holds for every x_0, we drop the subscript and write

$$f'(x) = n x^{n-1} \quad \text{or} \quad \frac{d}{dx}(x^n) = n x^{n-1}.$$

To derive differentiation formulas for elementary functions such as $\sin x$, $\cos x$, and e^x directly from Definition 3.1 requires estimates based on the properties of these functions. Since this is done in the elementary calculus course, we will not repeat it here.

Interpretations of the derivative

If $f(x)$ is the position of a particle at time x, the difference quotient (1) is its *average velocity between times* x_0 *and* x. As x approaches x_0 this average applies to shorter and shorter intervals, and therefore it makes sense to regard the limit (2)—if it exists—as the particle's *instantaneous velocity at time* x_0. This interpretation may be useful even if x is not time, so we often regard $f'(x_0)$ as the *instantaneous rate of change of* $f(x)$ *at* x_0, regardless of the specific nature of the variable x.

The derivative also has a geometric interpretation. The equation of the line through two points $(x_0, f(x_0))$ and $(x_1, f(x_1))$ on the curve $y = f(x)$ (Figure 3.1) is

$$y = f(x_0) + \frac{f(x_1) - f(x_0)}{x_1 - x_0}(x - x_0).$$

Varying x_1 generates lines through $(x_0, f(x_0))$ which rotate into the line

$$y = f(x_0) + f'(x_0)(x - x_0) \tag{3}$$

as x_1 approaches x_0. This is the *tangent* to the curve $y = f(x)$ at the point $(x_0, f(x_0))$. Figure 3.2 depicts the situation for various values of x_1.

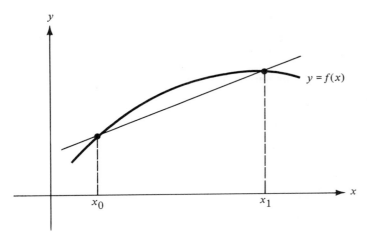

FIGURE 3.1

A less intuitive definition of the tangent line is as follows: If the function

$$T(x) = f(x_0) + m(x - x_0)$$

approximates f so well near x_0 that

$$\lim_{x \to x_0} \frac{f(x) - T(x)}{x - x_0} = 0,$$

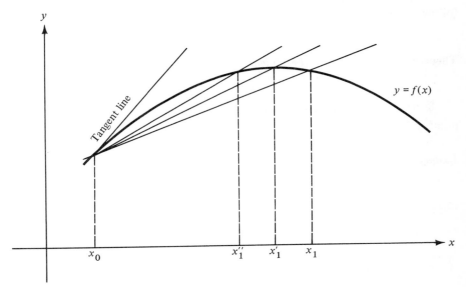

FIGURE 3.2

then we say that the line $y = T(x)$ is tangent to the curve $y = f(x)$ at $(x_0, f(x_0))$. This tangent line exists if and only if $f'(x_0)$ exists, in which case m is uniquely determined by $m = f'(x_0)$ (*Exercise 1*); thus, (3) is the equation of the tangent line as defined here.

We will use the following lemma to study differentiable functions.

Lemma 3.1 *If f is differentiable at x_0, then*

$$f(x) = f(x_0) + [f'(x_0) + E(x)](x - x_0), \qquad (4)$$

where E is defined on a neighborhood of x_0 and

$$\lim_{x \to x_0} E(x) = E(x_0) = 0.$$

Proof.   Define

$$E(x) = \begin{cases} \dfrac{f(x) - f(x_0)}{x - x_0} - f'(x_0), & x \in D_f \text{ and } x \neq x_0 \\[2mm] 0, & x = x_0. \end{cases} \qquad (5)$$

Solving (5) for $f(x)$ yields (4) if $x \neq x_0$, and (4) is obvious if $x = x_0$. Definition 3.1 implies that $\lim_{x \to x_0} E(x) = 0$. We define $E(x_0) = 0$ to make E continuous at x_0.

Since the right side of (4) is continuous at x_0, so is the left. This yields the following theorem.

Theorem 3.1 *If f is differentiable at x_0, then it is continuous at x_0.*

The converse of this theorem is false; a function may be continuous at a point without being differentiable.

Example 3.2   The function

$$f(x) = |x|$$

can be written as

$$f(x) = x, \qquad x > 0, \qquad (6)$$

or as

$$f(x) = -x, \qquad x < 0. \qquad (7)$$

From (6),

$$f'(x) = 1, \qquad x > 0,$$

and from (7),

$$f'(x) = -1, \qquad x < 0.$$

Neither (6) nor (7) holds throughout any neighborhood of 0, so neither can be used alone to calculate $f'(0)$. In fact, since the one-sided limits

$$\lim_{x \to 0+} \frac{f(x) - f(0)}{x - 0} = \lim_{x \to 0+} \frac{x}{x} = 1 \tag{8}$$

and

$$\lim_{x \to 0-} \frac{f(x) - f(0)}{x - 0} = \lim_{x \to 0-} \frac{-x}{x} = -1 \tag{9}$$

are different,

$$\lim_{x \to 0} \frac{f(x) - f(0)}{x - 0}$$

does not exist (Theorem 1.3); thus, f is not differentiable at 0 even though it is continuous there.

Interchanging differentiation and arithmetic operations

The following theorem should be familiar from the elementary calculus course.

Theorem 3.2 *If f and g are differentiable at x_0, then so are $f + g$, $f - g$, and fg, with*

(a) $(f + g)'(x_0) = f'(x_0) + g'(x_0)$;
(b) $(f - g)'(x_0) = f'(x_0) - g'(x_0)$;
(c) $(fg)'(x_0) = f'(x_0)g(x_0) + f(x_0)g'(x_0)$.

The quotient f/g is differentiable at x_0 if $g(x_0) \neq 0$, with

(d) $\left(\dfrac{f}{g}\right)'(x_0) = \dfrac{f'(x_0)g(x_0) - f(x_0)g'(x_0)}{[g(x_0)]^2}.$

Proof. The proof is obtained by forming the appropriate difference quotients and applying Definition 3.1 and Theorem 1.2. We will prove (c) and leave the rest to the reader (*Exercises 9, 10, and 11*).

The trick is to add and subtract the right quantity in the numerator of the difference quotient for $(fg)'(x_0)$; thus,

$$\frac{f(x)g(x) - f(x_0)g(x_0)}{x - x_0} = \frac{f(x)g(x) - f(x_0)g(x) + f(x_0)g(x) - f(x_0)g(x_0)}{x - x_0}$$

$$= \frac{f(x) - f(x_0)}{x - x_0} g(x) + f(x_0) \frac{g(x) - g(x_0)}{x - x_0}$$

The difference quotients on the right approach $f'(x_0)$ and $g'(x_0)$ as x approaches x_0, and $\lim_{x \to x_0} g(x) = g(x_0)$ (Theorem 3.1). This proves (c).

The chain rule

The rule for differentiating a composite function is as follows.

Theorem 3.3 (Chain Rule)   *Suppose g is differentiable at x_0 and f is differentiable at $g(x_0)$. Then the composite function $h = f \circ g$, defined by*

$$h(x) = f(g(x)),$$

is differentiable at x_0, with

$$h'(x_0) = f'(g(x_0))g'(x_0).$$

Proof.   Since f is differentiable at $g(x_0)$, Lemma 3.1 implies that

$$f(t) - f(g(x_0)) = [f'(g(x_0)) + E(t)][t - g(x_0)],$$

where

$$\lim_{t \to g(x_0)} E(t) = E(g(x_0)) = 0. \tag{10}$$

Letting $t = g(x)$ yields

$$f(g(x)) - f(g(x_0)) = [f'(g(x_0)) + E(g(x))][g(x) - g(x_0)].$$

Since $h(x) = f(g(x))$ (by definition), this implies that

$$\frac{h(x) - h(x_0)}{x - x_0} = [f'(g(x_0)) + E(g(x))] \frac{g(x) - g(x_0)}{x - x_0}. \tag{11}$$

Since g is continuous at x_0 (Theorem 3.1), (10) and Theorem 2.3 imply that

$$\lim_{x \to x_0} E(g(x)) = 0;$$

hence, we conclude from (11) that

$$h'(x_0) = \lim_{x \to x_0} \frac{h(x) - h(x_0)}{x - x_0} = f'(g(x_0))g'(x_0).$$

This completes the proof.

Example 3.3 If

$$f(x) = \sin x \quad \text{and} \quad g(x) = \frac{1}{x}, \qquad x \neq 0,$$

then

$$h(x) = f(g(x)) = \sin \frac{1}{x}$$

and

$$h'(x) = f'(g(x))g'(x) = \left(\cos \frac{1}{x}\right)\left(-\frac{1}{x^2}\right), \qquad x \neq 0.$$

It may seem reasonable to justify the chain rule by writing

$$\frac{h(x) - h(x_0)}{x - x_0} = \frac{f(g(x)) - f(g(x_0))}{x - x_0}$$

$$= \frac{f(g(x)) - f(g(x_0))}{g(x) - g(x_0)} \frac{g(x) - g(x_0)}{x - x_0}$$

and arguing that

$$\lim_{x \to x_0} \frac{f(g(x)) - f(g(x_0))}{g(x) - g(x_0)} = f'(g(x_0))$$

[because $\lim_{x \to x_0} g(x) = g(x_0)$] and

$$\lim_{x \to x_0} \frac{g(x) - g(x_0)}{x - x_0} = g'(x_0).$$

However, this is not a valid proof (*Exercise 13*).

One-sided derivatives

One-sided limits of difference quotients such as (8) and (9) in Example 3.2 are called *one-sided*, or right- and left-hand derivatives. That is, if f is defined on some interval $[x_0, b)$, its *right-hand derivative at* x_0 is defined to be

$$f'_+(x_0) = \lim_{x \to x_0 +} \frac{f(x) - f(x_0)}{x - x_0}$$

if the limit exists, while if f is defined on an interval $(a, x_0]$, its *left-hand derivative* at x_0 is defined to be

$$f'_-(x_0) = \lim_{x \to x_0-} \frac{f(x) - f(x_0)}{x - x_0}$$

if the limit exists. Theorem 1.3 implies that f is differentiable at x_0 if and only if $f'_+(x_0)$ and $f'_-(x_0)$ exist and are equal, in which case

$$f'(x_0) = f'_+(x_0) = f'_-(x_0).$$

In Example 3.2, $f'_+(0) = 1$ and $f'_-(0) = -1$.

Example 3.4   If

$$f(x) = \begin{cases} x^3, & x \le 0, \\ x^2 \sin \dfrac{1}{x}, & x > 0, \end{cases} \tag{12}$$

then

$$f'(x) = \begin{cases} 3x^2, & x < 0, \\ 2x \sin \dfrac{1}{x} - \cos \dfrac{1}{x}, & x > 0. \end{cases} \tag{13}$$

Since neither formula in (12) holds for all x in any neighborhood of 0, we cannot simply differentiate either to obtain $f'(0)$; instead, we calculate

$$f'_+(0) = \lim_{x \to 0+} \frac{x^2 \sin \dfrac{1}{x} - 0}{x - 0} = \lim_{x \to 0+} x \sin \frac{1}{x} = 0,$$

$$f'_-(0) = \lim_{x \to 0-} \frac{x^3 - 0}{x - 0} = \lim_{x \to 0-} x^2 = 0;$$

hence,

$$f'(0) = f'_+(0) = f'_-(0) = 0.$$

This example shows that there is a difference between a one-sided derivative and a one-sided limit of a derivative, since $f'_+(0) = 0$, but, from (13), $f'(0+) = \lim_{x \to 0+} f'(x)$ does not exist. It also shows that a derivative may exist in a neighborhood of a point x_0 ($x_0 = 0$ in this case), but be discontinuous at x_0.

Exercise 5 justifies the method used in Example 3.4 to compute $f'(x)$ for $x \ne 0$.

Extreme values

We say that $f(x_0)$ is a *local extreme value* of f if there is a $\delta > 0$ such that $f(x) - f(x_0)$ does not change sign on

$$(x_0 - \delta, x_0 + \delta) \cap D_f. \tag{14}$$

More specifically, $f(x_0)$ is a *local maximum value* of f if

$$f(x) \leq f(x_0) \tag{15}$$

or a *local minimum value* of f if

$$f(x) \geq f(x_0) \tag{16}$$

for all x in the set (14). The point x_0 is called a *local extreme point* of f, or, more specifically, a *local maximum* or *local minimum point* of f.

Example 3.5   If

$$f(x) = \begin{cases} 1, & -1 < x \leq -\tfrac{1}{2}, \\ |x|, & -\tfrac{1}{2} < x \leq \tfrac{1}{2}, \\ \dfrac{1}{\sqrt{2}} \sin \dfrac{\pi x}{2}, & \tfrac{1}{2} < x \leq 4, \end{cases}$$

(Figure 3.3), then 0, 3, and every x in $(-1, -\tfrac{1}{2})$ are local minimum points of f, while 1, 4, and every x in $(-1, -\tfrac{1}{2}]$ are local maximum points.

It is geometrically plausible that if the curve $y = f(x)$ has a tangent at a local extreme point of f, then the tangent must be horizontal, that is, have zero slope. [For example, see $x = 1$, $x = 3$, and every x in $(-1, -\tfrac{1}{2})$, Figure 3.3.] The following theorem shows that this must be so.

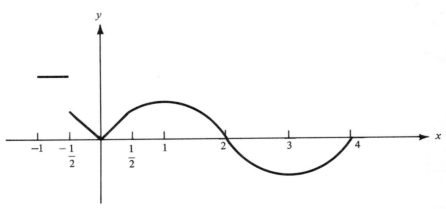

FIGURE 3.3

Theorem 3.4   *If f is differentiable at a local extreme point x_0, then $f'(x_0) = 0$.*

Proof.   From Lemma 3.1,

$$\frac{f(x) - f(x_0)}{x - x_0} = f'(x_0) + E(x), \tag{17}$$

where $\lim_{x \to x_0} E(x) = 0$. Therefore, if $f'(x_0) \neq 0$, there is a $\delta > 0$ such that

$$|E(x)| < |f'(x_0)| \quad \text{if} \quad |x - x_0| < \delta,$$

and the right side of (17) must have the same sign as $f'(x_0)$ for $|x - x_0| < \delta$. Since the same is true of the left side, $f(x) - f(x_0)$ must change sign in every neighborhood of x_0 (since $x - x_0$ does), and therefore neither (15) nor (16) can hold for all x in any interval about x_0. Hence, x_0 is not a local extreme point of f if $f'(x_0) \neq 0$. This completes the proof.

If $f'(x_0) = 0$, we say that x_0 is a *critical point* of f. Theorem 3.4 says that every local extreme point of f at which f is differentiable is a critical point of f. The converse is false; for example, 0 is a critical point of $f(x) = x^3$, but not a local extreme point.

Rolle's theorem

The use of Theorem 3.4 for finding local extreme point is covered in most elementary calculus courses, so we will not pursue it here. However, we will use it to prove the following fundamental theorem, which says that if a curve $y = f(x)$ intersects the x axis at $x = a$ and $x = b$ and has a tangent for every x in (a, b), then the tangent must be horizontal at some point c in (a, b) (Figure 3.4).

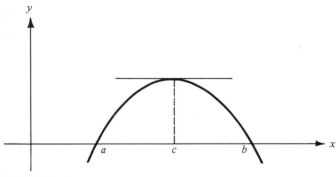

FIGURE 3.4

Theorem 3.5 (Rolle's Theorem) *Suppose f is continuous on the closed interval $[a, b]$ and differentiable on the open interval (a, b), and $f(a) = f(b) = 0$. Then $f'(c) = 0$ for some c in the open interval (a, b).*

Proof. Since f is continuous on $[a, b]$, it attains a maximum and a minimum value on $[a, b]$ (Theorem 2.5). If they are both zero, then f is identically zero on (a, b), and so is f'. If one of the extreme values is nonzero, it must be attained at some point c in the open interval (a, b), and $f'(c) = 0$, by Theorem 3.4. This completes the proof.

Intermediate values of derivatives

A derivative may exist on an interval $[a, b]$ without being continuous there; nevertheless, it must satisfy an intermediate value property similar to Theorem 2.6.

Theorem 3.6 *Suppose f is differentiable on $[a, b]$ and $f'(a) \neq f'(b)$. Let μ be a number between $f'(a)$ and $f'(b)$. Then $f'(c) = \mu$ for some c in (a, b).*

Proof. Suppose

$$f'(a) < \mu < f'(b) \tag{18}$$

and define

$$g(x) = f(x) - \mu x.$$

Then

$$g'(x) = f'(x) - \mu, \qquad a \leq x \leq b, \tag{19}$$

and (18) implies that

$$g'(a) < 0 \quad \text{and} \quad g'(b) > 0. \tag{20}$$

Since g is continuous on $[a, b]$, it attains a minimum at some point c in $[a, b]$. Lemma 3.1 and (20) imply that there is a $\delta > 0$ such that

$$g(x) < g(a), \qquad a < x < a + \delta,$$

and

$$g(x) < g(b), \qquad b - \delta < x < b$$

(*Exercise 3*), and therefore $c \neq a$ and $c \neq b$; hence, $a < c < b$ and consequently $g'(c) = 0$, by Theorem 3.4. From (19), $f'(c) = \mu$, which completes the proof if

(18) holds. The proof for the case where

$$f'(b) < \mu < f'(a)$$

can be obtained by applying this to the function $g = -f$.

Mean value theorems

By applying Rolle's theorem to the function

$$h(x) = [g(b) - g(a)][f(x) - f(a)] - [f(b) - f(a)][g(x) - g(a)],$$

we obtain the following theorem. We leave the details to the reader (*Exercise 27*).

Theorem 3.7 (Generalized Mean Value Theorem)   *If f and g are continuous on the closed interval $[a, b]$ and differentiable on the open interval (a, b), then*

$$[g(b) - g(a)]f'(c) = [f(b) - f(a)]g'(c)$$

for some c in the open interval (a, b).

The special case where $g(x) = x$ is important enough to be stated separately. We leave the proof to the reader (*Exercise 28*).

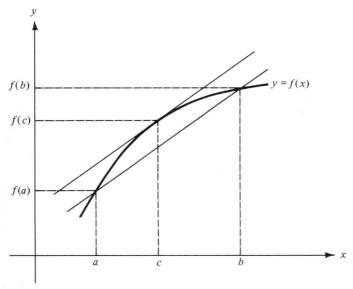

FIGURE 3.5

Theorem 3.8 (Mean Value Theorem) *If f is continuous on the closed interval $[a, b]$ and differentiable on the open interval (a, b), then*

$$f'(c) = \frac{f(b) - f(a)}{b - a}$$

for some c in the open interval (a, b).

Theorem 3.8 implies that the tangent to the curve $y = f(x)$ at $(c, f(c))$ is parallel to the line connecting the points $(a, f(a))$ and $(b, f(b))$ on the curve (Figure 3.5).

Consequences of the mean value theorem

If f is differentiable on (a, b) and $x_1, x_2 \in (a, b)$, then f is continuous on the closed interval with end points x_1 and x_2, and differentiable on its interior; hence, the mean value theorem implies that

$$f(x_2) - f(x_1) = f'(x_3)(x_2 - x_1)$$

for some x_3 between x_1 and x_2. (This is true whether $x_1 < x_2$ or $x_2 < x_1$.) The next three theorems follow easily from this (*Exercises 29, 30, and 31*).

Theorem 3.9 *If $f'(x) = 0$ for all x in (a, b), then f is constant on (a, b).*

Theorem 3.10 *If f' exists and does not change sign on (a, b), then f is monotonic on (a, b): increasing, nondecreasing, decreasing, or nonincreasing as*

$$f'(x) > 0, \quad f'(x) \geq 0, \quad f'(x) < 0, \quad or \quad f'(x) \leq 0,$$

respectively, for all x in (a, b).

Theorem 3.11 *If*

$$|f'(x)| \leq M, \qquad a < x < b,$$

then

$$|f(x) - f(x')| \leq M|x - x'|, \qquad x, x' \in (a, b).$$

A function which satisfies an inequality like this for all x and x' in an interval I is said to satisfy a *Lipschitz condition* on I.

2.3   EXERCISES

* 1. Prove: A function f is differentiable at x_0 if and only if

$$\lim_{x \to x_0} \frac{f(x) - f(x_0) - m(x - x_0)}{x - x_0} = 0$$

for some constant m. In this case, $f'(x_0) = m$.

2. Prove: If f is defined on a neighborhood of x_0, then f is differentiable at x_0 if and only if the discontinuity of

$$h(x) = \frac{f(x) - f(x_0)}{x - x_0}$$

at x_0 is removable.

3. Prove: If $f'(x_0) > 0$, there is a $\delta > 0$ such that

$$f(x) < f(x_0) \quad \text{if} \quad x_0 - \delta < x < x_0,$$

and

$$f(x) > f(x_0) \quad \text{if} \quad x_0 < x < x_0 + \delta.$$

4. Find all derivatives of $f(x) = x^{n-1}|x|$, where n is a positive integer.

5. Let

$$f(x) = \begin{cases} p(x), & a < x \le c, \\ q(x), & c < x < b. \end{cases}$$

where p and q are differentiable.
(a) Show that

$$f'(x) = \begin{cases} p'(x), & a < x < c, \\ q'(x), & c < x < b. \end{cases}$$

(b) State necessary and sufficient conditions for existence of $f'(c)$.

6. Suppose $f'(0)$ exists and $f(x + y) = f(x)f(y)$ for all x and y. Prove that f' exists for all x.

7. Suppose $c'(0) = a$ and $s'(0) = b$ $(a^2 + b^2 \ne 0)$, and

$$c(x + y) = c(x)c(y) - s(x)s(y)$$
$$s(x + y) = s(x)c(y) + c(x)s(y)$$

for all x and y.
(a) Show that c and s are differentiable on $(-\infty, \infty)$, and find c' and s' in terms of c, s, a, and b.

(b) (For readers who have studied differential equations.) Find c and s explicitly.

8. (a) Suppose f and g are differentiable at x_0, $f(x_0) = g(x_0) = 0$, and $g'(x_0) \neq 0$. Show that

$$\lim_{x \to x_0} \frac{f(x)}{g(x)} = \frac{f'(x_0)}{g'(x_0)}.$$

(b) State the corresponding results for one-sided limits.

9. Prove Theorem 3.2(a).

10. Prove Theorem 3.2(b).

11. Prove Theorem 3.2(d).

*12. Prove: If $f^{(n)}(x_0)$ and $g^{(n)}(x_0)$ exist, then so does $(fg)^{(n)}(x_0)$, and

$$(fg)^{(n)}(x_0) = \sum_{k=0}^{n} \binom{n}{k} f^{(k)}(x_0) g^{(n-k)}(x_0).$$

(*Hint:* See Exercise 17(a), Section 1.2.) This is *Leibniz's rule* for differentiating a product.

13. What is wrong with the "proof" of the chain rule indicated after Example 3.3? Correct it.

*14. Suppose f is increasing and continuous on $[a, b]$, $a < x_0 < b$, and f is differentiable at x_0, with $f'(x_0) \neq 0$. If g is the inverse of f (Theorem 2.9), show that $g'(y_0) = 1/f'(x_0))$, where $y_0 = f(x_0)$.

15. (a) Show that $f'_+(a) = f'(a+)$ if both quantities exist.
    (b) Example 3.4 shows that $f'_+(a)$ may exist even if $f'(a+)$ does not. Give an example where $f'(a+)$ exists, but $f'_+(a)$ does not.
    (c) Complete the following statement so that it becomes a theorem, and prove the theorem: "If $f'(a+)$ exists and f is ———————— at a, then $f'_+(a) = f'(a+)$."

16. Show that $f(a+)$ and $f(b-)$ exist (finite) if f' is bounded on (a, b). (*Hint:* See Exercise 36, Section 2.1.)

17. Suppose f is continuous on $[a, b]$, $f'_+(a)$ exists, and μ is between it and $[f(b) - f(a)]/(b - a)$. Show that $f(c) - f(a) = \mu(c - a)$ for some c in (a, b).

18. Suppose f is continuous on $[a, b]$, $f'_+(a) < \mu < f'_-(b)$, and $\mu \neq \alpha = [f(b) - f(a)]/(b - a)$. Show that either $f(c) - f(a) = \mu(c - a)$ or $f(c) - f(b) = \mu(c - b)$ for some c in (a, b).

19. Let n be a positive integer. Give an example of a function f such that $f^{(n)}$ exists for all x, but is discontinuous at 0.

20. Let n be a positive integer and

$$f(x) = \frac{\sin nx}{n \sin x}, \qquad x \neq k\pi \quad (k = \text{integer}).$$

(a) Define $f(k\pi)$ so that f is continuous at $k\pi$. (*Hint:* Use Exercise 8.)
(b) Show that if $\bar{x}$ is a local extreme point of f and $x \neq k\pi$ where k is an integer, then

$$\left| f(\bar{x}) \right| = \left[1 + (n^2 - 1) \sin^2 \bar{x} \right]^{-1/2}.$$

[*Hint:* Express $\sin nx$ and $\cos nx$ in terms of $f(x)$ and $f'(x)$, and add their squares to obtain a useful identity.]
(c) Show that $\left| f(x) \right| \leq 1$ for all x. For what values of x is equality attained?

*21. Let

$$f(x) = \frac{\sin x}{x}, \qquad x \neq 0.$$

(a) Define $f(0)$ so that f is continuous at $x = 0$.
(b) Calculate $f'(0)$.
(c) Show that if $\bar{x}$ is a local extreme point of f, then

$$\left| f(\bar{x}) \right| = (1 + \bar{x}^2)^{-1/2}.$$

(*Hint:* Express $\sin x$ and $\cos x$ in terms of $f(x)$ and $f'(x)$, and add their squares to obtain a useful identity.]
(d) Show that $\left| f(x) \right| \leq 1$ for all x. For what values of x is equality attained?

22. We say that f has at least n zeros, *counting multiplicities*, on an interval I if there are distinct points $x_1, \ldots, x_p$ such that

$$f^{(j)}(x_i) = 0, \qquad 0 \leq j \leq n_i - 1,$$

and $n_1 + \cdots + n_p = n$. Prove: If f is differentiable and has at least n zeros, counting multiplicities, on an interval I, then f' has at least $n - 1$ zeros, counting multiplicities, on I.

23. Give an example of a function f such that f' exists on an interval (a, b) and has a jump discontinuity at a point x_0 of (a, b), or show that there is no such function.

24. Let $x_1, \ldots, x_n$ and $y_1, \ldots, y_n$ be in (a, b) and $y_i < x_i$ ($i = 1, \ldots, n$). Show that if f is differentiable on (a, b), then

$$\sum_{i=1}^{n} \left[f(x_i) - f(y_i) \right] = f'(c) \sum_{i=1}^{n} (x_i - y_i)$$

for some c in (a, b).

25. Prove or give a counterexample: If f is differentiable on a neighborhood of x_0, then f satisfies a Lipschitz condition on some neighborhood of x_0.

26. Let

$$f''(x) + p(x)f(x) = 0 \quad \text{and} \quad g''(x) + p(x)g(x) = 0, \qquad a < x < b.$$

   (a) Show that the function $W = f'g - fg'$ is constant on (a, b).
   (b) Prove: If $W \neq 0$ and $f(x_1) = f(x_2) = 0$, where $a < x_1 < x_2 < b$, then $g(c) = 0$ for some c in (x_1, x_2). (*Hint*: Consider f/g.)

27. Prove Theorem 3.7.

28. Prove Theorem 3.8.

29. Prove Theorem 3.9.

30. Prove Theorem 3.10.

31. Prove Theorem 3.11.

32. Suppose we extend the definition of differentiability by saying that f is differentiable at x_0 if

$$f'(x_0) = \lim_{x \to x_0} \frac{f(x) - f(x_0)}{x - x_0}$$

exists in the extended reals. Show that if

$$f(x) = \begin{cases} \sqrt{x}, & x \geq 0, \\ -\sqrt{-x}, & x < 0 \end{cases}$$

then $f'(0) = \infty$.

33. Prove or give a counterexample: If f is differentiable at x_0 in the extended sense of Exercise 32, then f is continuous at x_0.

2.4 L'HOSPITAL'S RULE

The method of Theorem 1.2 for finding limits of the sum, difference, product, and quotient of functions breaks down if it leads to indeterminate forms. The generalized mean value theorem provides a method for solving some problems of this kind.

Theorem 4.1 (l'Hospital's Rule) *Suppose f and g are differentiable and g' has no zeros on (a, b). Let*

$$\lim_{x \to b-} f(x) = \lim_{x \to b-} g(x) = 0 \tag{1}$$

or

$$\lim_{x\to b-} f(x) = \lim_{x\to b-} g(x) = \infty, \tag{2}$$

and suppose

$$\lim_{x\to b-} \frac{f'(x)}{g'(x)} = L \qquad (\textit{finite or } \pm\infty). \tag{3}$$

Then

$$\lim_{x\to b-} \frac{f(x)}{g(x)} = L. \tag{4}$$

Proof.   We prove the theorem for finite L and leave the case where $L = \pm\infty$ to the reader (*Exercise 1*). Suppose $\varepsilon > 0$. From (3), there is an x_0 in (a, b) such that

$$\left|\frac{f'(c)}{g'(c)} - L\right| < \varepsilon \quad \text{if} \quad x_0 < c < b. \tag{5}$$

Theorem 3.7 implies that if x and x' are in $[x_0, b)$, there is a c between them—and therefore in (x_0, b)—such that

$$g'(c)[f(x) - f(x')] = f'(c)[g(x) - g(x')]. \tag{6}$$

Since g' has no zeros in (a, b), Theorem 3.8 implies that

$$g(x) - g(x') \neq 0 \quad \text{if} \quad x, x' \in (a, b).$$

This means that g cannot have more than one zero in (a, b), and so we can also choose x_0 so that, in addition to (5), g has no zeros in $[x_0, b)$. It also means that (6) can be rewritten as

$$\frac{f(x) - f(x')}{g(x) - g(x')} = \frac{f'(c)}{g'(c)},$$

and now (5) implies that

$$\left|\frac{f(x) - f(x')}{g(x) - g(x')} - L\right| < \varepsilon \quad \text{if} \quad x, x' \in [x_0, b). \tag{7}$$

If (1) holds, think of x as fixed in $[x_0, b)$ and consider the function $G(x')$ defined by

$$G(x') = \frac{f(x) - f(x')}{g(x) - g(x')} - L.$$

From (1),

$$\lim_{x'\to b-} G(x') = \frac{f(x)}{g(x)} - L. \tag{8}$$

Since

$$|G(x')| < \varepsilon \quad \text{if} \quad x_0 < x' < b,$$

because of (7), (8) implies that

$$\left| \frac{f(x)}{g(x)} - L \right| \le \varepsilon.$$

This holds for all x in (x_0, b), which implies (4).

The proof under assumption (2) is more complicated. We now choose x_0 to satisfy the requirements stated above, and, in addition, so that $f(x) \ne 0$ if $x_0 < x < b$. Letting $x' = x_0$ in (7), we see that

$$\left| \frac{f(x)}{g(x)u(x)} - L \right| < \varepsilon \quad \text{if} \quad x_1 < x < b, \tag{9}$$

where $x_1 > x_0$ and $f(x) \ne f(x_0)$ if $x_1 < x < b$, and

$$u(x) = \frac{1 - g(x_0)/g(x)}{1 - f(x_0)/f(x)}.$$

Note that

$$\lim_{x \to b-} u(x) = 1 \tag{10}$$

because of (2). Now rewrite (9) as

$$\left| \frac{f(x)}{g(x)} - Lu(x) \right| < \varepsilon |u(x)| \quad \text{if} \quad x_1 < x < b.$$

Then

$$\left| \frac{f(x)}{g(x)} - L \right| \le \left| \frac{f(x)}{g(x)} - Lu(x) \right| + |Lu(x) - L| \tag{11}$$

$$\le \varepsilon |u(x)| + |L| \, |u(x) - 1|.$$

Because of (10), there is a point x_2 in (x_1, b) such that

$$|u(x) - 1| < \varepsilon \quad \text{if} \quad x_2 < x < b,$$

This and (11) imply that

$$\left| \frac{f(x)}{g(x)} - L \right| < \varepsilon(1 + \varepsilon) + |L|\varepsilon \quad \text{if} \quad x_2 < x < b,$$

which proves (4) under assumption (2).

Theorem 4.1 and the proof given here remain valid if $b = \infty$ and "$x \to b-$" is replaced by "$x \to \infty$" throughout. Only minor changes in the proof are needed to show that similar theorems are valid for limits from the right, limits at $-\infty$, and ordinary (two-sided) limits. We will take these as given.

The indeterminate forms 0/0 and ∞/∞

We say that f/g has the indeterminate form $0/0$ as $x \to b-$ if

$$\lim_{x \to b-} f(x) = \lim_{x \to b-} g(x) = 0,$$

or the indeterminate form ∞/∞ if

$$\lim_{x \to b-} f(x) = \pm\infty \quad \text{and} \quad \lim_{x \to b-} g(x) = \pm\infty.$$

The corresponding definitions for $x \to b+$ and $x \to \pm\infty$ are obvious. If f/g is of one of these forms as $x \to b-$ and as $x \to b+$, then we say that it is of that form as $x \to b$.

Example 4.1   The ratio $(\sin x)/x$ is of the form $0/0$ as $x \to 0$, and l'Hospital's rule yields

$$\lim_{x \to 0} \frac{\sin x}{x} = \lim_{x \to 0} \frac{\cos x}{1} = 1.$$

Example 4.2   The ratio e^{-x}/x is of the form ∞/∞ as $x \to -\infty$, and l'Hospital's rule yields

$$\lim_{x \to -\infty} \frac{e^{-x}}{x} = \lim_{x \to -\infty} \frac{-e^{-x}}{1} = -\infty.$$

Example 4.3   Using l'Hospital's rule may lead to another indeterminate form; thus,

$$\lim_{x \to \infty} \frac{e^x}{x^2} = \lim_{x \to \infty} \frac{e^x}{2x}$$

if the latter exists in the extended reals. Applying l'Hospital's rule again yields

$$\lim_{x \to \infty} \frac{e^x}{2x} = \lim_{x \to \infty} \frac{e^x}{2} = \infty.$$

Therefore, we conclude that

$$\lim_{x \to \infty} \frac{e^x}{x^2} = \infty.$$

More generally,

$$\lim_{x \to \infty} \frac{e^x}{x^\alpha} = \infty$$

for any real number α (*Exercise 33*).

Example 4.4 Sometimes it pays to combine l'Hospital's rule with other manipulations. For example,

$$\lim_{x\to 0} \frac{4 - 4\cos x - 2\sin^2 x}{x^4} = \lim_{x\to 0} \frac{4\sin x - 4\sin x \cos x}{4x^3}$$

$$= \lim_{x\to 0} \frac{\sin x}{x} \lim_{x\to 0} \frac{1 - \cos x}{x^2}$$

$$= \lim_{x\to 0} \frac{\sin x}{x} \lim_{x\to 0} \frac{\sin x}{2x}$$

$$= \frac{1}{2}\left(\lim_{x\to 0} \frac{\sin x}{x}\right)^2$$

$$= \frac{1}{2}(1)^2 \qquad \text{(Example 4.1)}$$

$$= \frac{1}{2}.$$

As another example, l'Hospital's rule yields

$$\lim_{x\to 0} \frac{e^{-x^2}\log(1 + x)}{x} = \lim_{x\to 0} \frac{-2xe^{-x^2}\log(1 + x) + e^{-x^2}(1 + x)^{-1}}{1}$$

$$= 1.$$

However, it is better to remove the "determinate" part of the ratio before using l'Hospital's rule:

$$\lim_{x\to 0} \frac{e^{-x^2}\log(1 + x)}{x} = \lim_{x\to 0} e^{-x^2} \lim_{x\to 0} \frac{\log(1 + x)}{x}$$

$$= (1)\lim_{x\to 0} \frac{\log(1 + x)}{x}$$

$$= \lim_{x\to 0} \frac{1/(1 + x)}{1}$$

$$= 1.$$

In using l'Hospital's rule, we usually write, for example,

$$\lim_{x\to b} \frac{f(x)}{g(x)} = \lim_{x\to b} \frac{f'(x)}{g'(x)} \tag{12}$$

and then try to find the limit on the right. This is convenient but technically incorrect; (12) is true only if the limit on the right exists in the extended reals. It may happen that the limit on the left exists but the one on the right does not. In this case (12) is obviously incorrect.

Example 4.5   If

$$f(x) = x - x^2 \sin \frac{1}{x} \quad \text{and} \quad g(x) = \sin x,$$

then

$$f'(x) = 1 - 2x \sin \frac{1}{x} + \cos \frac{1}{x} \quad \text{and} \quad g'(x) = \cos x.$$

Therefore, $\lim_{x \to 0} f'(x)/g'(x)$ does not exist; however,

$$\lim_{x \to 0} \frac{f(x)}{g(x)} = \lim_{x \to 0} \frac{1 - x \sin(1/x)}{\sin x/x} = \frac{1}{1} = 1.$$

The indeterminate form $0 \cdot \infty$

We say that a product fg is *indeterminate of the form* $0 \cdot \infty$ *as* $x \to b-$ if one of the factors approaches 0 and the other approaches $\pm \infty$ as $x \to b-$. In this case it may be useful to apply l'Hospital's rule after writing

$$f(x)g(x) = \frac{f(x)}{1/g(x)} \quad \text{or} \quad f(x)g(x) = \frac{g(x)}{1/f(x)},$$

since one of these ratios is of the form $0/0$ and the other is of the form ∞/∞ as $x \to b-$.

Similar statements apply to limits as $x \to b+$, $x \to b$, and $x \to \pm \infty$.

Example 4.6

$$\lim_{x \to 0+} x \log x = \lim_{x \to 0+} \frac{\log x}{1/x}$$

$$= \lim_{x \to 0+} \frac{1/x}{-1/x^2}$$

$$= -\lim_{x \to 0+} x = 0.$$

Here we converted a $0 \cdot \infty$ form into an ∞/∞ form. Converting it to a $0/0$ form leads to a more complicated problem:

$$\lim_{x \to 0+} x \log x = \lim_{x \to 0+} \frac{x}{1/\log x}$$

$$= \lim_{x \to 0+} \frac{1}{-1/x(\log x)^2}$$

$$= -\lim_{x \to 0+} x(\log x)^2 = ?$$

Example 4.7

$$\lim_{x \to \infty} x \log(1 + 1/x) = \lim_{x \to \infty} \frac{\log(1 + 1/x)}{1/x}$$

$$= \lim_{x \to \infty} \frac{[1/(1 + 1/x)](-1/x^2)}{-1/x^2}$$

$$= \lim_{x \to \infty} \frac{1}{1 + 1/x} = 1.$$

Here we converted from $0 \cdot \infty$ to $0/0$. In this case, converting to ∞/∞ complicates the problem:

$$\lim_{x \to \infty} x \log(1 + 1/x) = \lim_{x \to \infty} \frac{x}{1/\log(1 + 1/x)}$$

$$= \lim_{x \to \infty} \frac{1}{\dfrac{-1}{[\log(1 + 1/x)]^2} \dfrac{-1/x^2}{1 + 1/x}}$$

$$= \lim_{x \to \infty} x(x + 1)[\log(1 + 1/x)]^2 = \,?$$

The indeterminate form $\infty - \infty$

A difference $f - g$ is *indeterminate of the form* $\infty - \infty$ *as* $x \to b-$ if

$$\lim_{x \to b-} f(x) = \lim_{x \to b-} g(x) = \pm \infty.$$

In this case it may be possible to manipulate $f - g$ into an expression which is no longer indeterminate, or of the form $0/0$ or ∞/∞ as $x \to b-$. Similar remarks apply to limits as $x \to b+$, $x \to b$, or $x \to \pm \infty$. (We say that $f - g$ is of the form $\infty - \infty$ as $x \to b$ if it is of that form as $x \to b-$ and as $x \to b+$.)

Example 4.8 The difference

$$\frac{\sin x}{x^2} - \frac{1}{x}$$

is of the form $\infty - \infty$ as $x \to 0$, but it can be rewritten as the $0/0$ form

$$\frac{\sin x - x}{x^2}.$$

Hence,

$$\lim_{x \to 0} \left(\frac{\sin x}{x^2} - \frac{1}{x} \right) = \lim_{x \to 0} \frac{\sin x - x}{x^2} = \lim_{x \to 0} \frac{\cos x - 1}{2x}$$

$$= \lim_{x \to 0} \frac{-\sin x}{2} = 0.$$

***Example* 4.9**   The difference

$$x^2 - x$$

is of the form $\infty - \infty$ as $x \to \infty$. Rewriting it as

$$x^2\left(1 - \frac{1}{x}\right),$$

which is no longer indeterminate as $x \to \infty$, we find that

$$\lim_{x \to \infty} (x^2 - x) = \lim_{x \to \infty} x^2\left(1 - \frac{1}{x}\right)$$

$$= \lim_{x \to \infty} x^2 \lim_{x \to \infty}\left(1 - \frac{1}{x}\right)$$

$$= (\infty)(1) = \infty.$$

The indeterminate forms 0^0, 1^∞, and ∞^0

The function f^g is defined by

$$f(x)^{g(x)} = e^{g(x)\,\log f(x)} = \exp[g(x)\log f(x)]$$

for all x such that $f(x) > 0$. Therefore, if f and g are defined and $f(x) > 0$ on an interval (a, b), Exercise 22 of Section 2.2 implies that

$$\lim_{x \to b-} [f(x)]^{g(x)} = \exp\left[\lim_{x \to b-} g(x)\log f(x)\right] \tag{13}$$

if $\lim_{x \to b-} g(x)\log f(x)$ exists in the extended reals. [If this limit is $\pm\infty$, (13) is valid if we define $e^{-\infty} = 0$ and $e^\infty = \infty$.] The product $g \log f$ can be indeterminate of the form $0 \cdot \infty$ in three ways as $x \to b-$:

1. If $\lim_{x \to b-} g(x) = 0$ and $\lim_{x \to b-} f(x) = 0$.
2. If $\lim_{x \to b-} g(x) = \pm\infty$ and $\lim_{x \to b-} f(x) = 1$.
3. If $\lim_{x \to b-} g(x) = 0$ and $\lim_{x \to b-} f(x) = \infty$.

In these three cases, we say that f^g is indeterminate of the form 0^0, 1^∞, and ∞^0, respectively, as $x \to b-$. Similar definitions apply to limits as $x \to b+$ and $x \to \pm\infty$. If f^g is of one of these forms as $x \to b-$ and as $x \to b+$, then we say that it is of that form as $x \to b$.

***Example* 4.10**   The function x^x is of the form 0^0 as $x \to 0+$. Since

$$x^x = e^{x\,\log x}$$

and $\lim_{x\to 0+} x \log x = 0$ (Example 4.6),

$$\lim_{x\to 0+} x^x = e^0 = 1.$$

Example 4.11   The function

$$x^{1/(x-1)}$$

is of the form 1^∞ as $x \to 1$. Since

$$x^{1/(x-1)} = \exp\left(\frac{\log x}{x-1}\right)$$

and

$$\lim_{x\to 1} \frac{\log x}{x-1} = \lim_{x\to 1} \frac{1/x}{1} = 1,$$

it follows that

$$\lim_{x\to 1} x^{1/(x-1)} = e^1 = e.$$

Example 4.12   The function $x^{1/x}$ is of the form ∞^0 as $x \to \infty$. Since

$$x^{1/x} = \exp\left(\frac{\log x}{x}\right)$$

and

$$\lim_{x\to\infty} \frac{\log x}{x} = \lim_{x\to\infty} \frac{1/x}{1} = 0,$$

it follows that

$$\lim_{x\to\infty} x^{1/x} = e^0 = 1.$$

2.4   EXERCISES

1. Prove Theorem 4.1 for the case where $\lim_{x\to b-} f'(x)/g'(x) = \pm\infty$.

In Exercises 2–53, find the indicated limits.

2. $\lim_{x\to 0} \dfrac{\tan^{-1} x}{\sin^{-1} x}$

3. $\lim_{x\to 0} \dfrac{1 - \cos x}{\log(1 + x^2)}$

4. $\lim_{x\to 0+} \dfrac{1 + \cos x}{e^x - 1}$

5. $\lim_{x\to\pi} \dfrac{\sin nx}{\sin x}$

6. $\lim_{x\to 0} \dfrac{\log(1 + x)}{x}$

7. $\lim_{x\to\infty} e^x \sin e^{-x^2}$

8. $\lim\limits_{x \to \infty} x \sin 1/x$

9. $\lim\limits_{x \to \infty} \sqrt{x}(e^{-1/x} - 1)$

10. $\lim\limits_{x \to 0+} \tan x \log x$

11. $\lim\limits_{x \to \pi} \sin x \log(|\tan x|)$

12. $\lim\limits_{x \to 0+} \left[\dfrac{1}{x} + \log(\tan x) \right]$

13. $\lim\limits_{x \to \infty} (\sqrt{x + 1} - \sqrt{x})$

14. $\lim\limits_{x \to 0} \left(\dfrac{1}{e^x - 1} - \dfrac{1}{x} \right)$

15. $\lim\limits_{x \to 0} (\cot x - \csc x)$

16. $\lim\limits_{x \to 0} \left(\dfrac{1}{\sin x} - \dfrac{1}{x} \right)$

17. $\lim\limits_{x \to \pi} |\sin x|^{\tan x}$

18. $\lim\limits_{x \to \pi/2} |\tan x|^{\cos x}$

19. $\lim\limits_{x \to 0} |\sin x|^x$

20. $\lim\limits_{x \to 0} (1 + x)^{1/x}$

21. $\lim\limits_{x \to \infty} x^{\sin 1/x}$

22. $\lim\limits_{x \to 0} \left(\dfrac{x}{1 - \cos x} - \dfrac{2}{x} \right)$

23. $\lim\limits_{x \to 0+} x^\alpha \log x$

24. $\lim\limits_{x \to e} \dfrac{\log(\log x)}{\sin(x - e)}$

25. $\lim\limits_{x \to \infty} \left(\dfrac{x + 1}{x - 1} \right)^{\sqrt{x^2 - 1}}$

26. $\lim\limits_{x \to 1+} \left(\dfrac{x + 1}{x - 1} \right)^{\sqrt{x^2 - 1}}$

27. $\lim\limits_{x \to \infty} \dfrac{(\log x)^\beta}{x}$

28. $\lim\limits_{x \to \infty} (\cosh x - \sinh x)$

29. $\lim\limits_{x \to \infty} (x^\alpha - \log x)$

30. $\lim\limits_{x \to -\infty} e^{x^2} \sin e^x$

31. $\lim\limits_{x \to \infty} x(x + 1)[\log(1 + 1/x)]^2$

32. $\lim\limits_{x \to 0} \dfrac{\sin x - x + x^3/6}{x^5}$

33. $\lim\limits_{x \to \infty} \dfrac{e^x}{x^\alpha}$

34. $\lim\limits_{x \to 3\pi/2 -} e^{\tan x} \cos x$

35. $\lim\limits_{x \to 1+} (\log x)^\alpha \log(\log x)$

36. $\lim\limits_{x \to \infty} \left[xe - (x + 1)^x x^{-x+1} \right]$

37. $\lim\limits_{x \to \infty} \dfrac{x^x}{x \log x}$

38. $\lim\limits_{x \to \pi/2} (\sin x)^{\tan x}$

39. $\lim\limits_{x \to e+} (x - e)^{1/\log[\log(\log x)]}$

40. $\lim\limits_{x \to \pi} (n \csc nx - m \csc mx)$ $(m, n = \text{positive integers})$

41. $\lim\limits_{x \to \infty} \left[\dfrac{x}{\log x} - \log(\log x) \right]$

42. $\lim\limits_{x \to \infty} \dfrac{\log[\log(1 + 1/x)]}{1 + x}$

43. $\lim\limits_{x\to\infty} [(\log x)^x - x^{\log x}]$

44. $\lim\limits_{x\to\infty} (1 + 1/x)^{x^{3/2}}$

45. $\lim\limits_{x\to\infty} \dfrac{e^x}{\cosh x}$

46. $\lim\limits_{x\to 1} (\cos 2\pi x)^{1/(x^2-1)}$

47. $\lim\limits_{x\to\infty} (e^x)^{e^{-x}}$

48. $\lim\limits_{x\to 1-} (\sqrt{1-x^2})^{1-x}$

49. $\lim\limits_{x\to\pi/2} (\tan x - \sec x)$

50. $\lim\limits_{x\to\pi/4} (\tan x)^{\sec 2x}$

51. $\lim\limits_{x\to 0+} [\log(x+1)]^{\log x}$

*52. $\lim\limits_{x\to 0} \dfrac{e^x - \sum\limits_{r=0}^{n} \dfrac{x^r}{r!}}{x^k}$ $(k = \text{integer} \geq 0)$

53. $\lim\limits_{x\to 0} \dfrac{\sin x - \sum\limits_{r=0}^{n} (-1)^r \dfrac{x^{2r+1}}{(2r+1)!}}{x^{2k+1}}$ $(k = \text{integer} \geq 0)$

*54. Show that

$$\lim\limits_{x\to 0} \dfrac{e^{-1/x^2}}{x^k} = 0 \qquad (k = \text{integer}).$$

55. (a) Prove: If f is continuous at x_0 and $\lim_{x\to x_0} f'(x)$ exists, then $f'(x_0)$ exists and f' is continuous at x_0.

(b) Give an example showing that it is necessary to assume in part (a) that f is continuous at x_0.

*56. The *iterated logarithms* are defined by $L_0(x) = x$ and

$$L_n(x) = \log[L_{n-1}(x)], \qquad x > a_n,$$

where $a_1 = 0$ and $a_n = e^{a_{n-1}}, n \geq 1$. Show that:

(a) $\lim\limits_{x\to a_n+} [L_{n-1}(x)]^x L_n(x) = 0$ if $\alpha > 0$ and $n \geq 1$

(b) $\lim\limits_{x\to\infty} [L_n(x)]^\alpha/L_{n-1}(x) = 0$ if α is arbitrary and $n \geq 1$

57. Let f be positive on $(0, \infty)$ and

$$\lim\limits_{x\to\infty} \dfrac{f'(x)}{f(x)} = L \qquad (0 < L \leq \infty).$$

Define $f_0(x) = x$ and

$$f_n(x) = f(f_{n-1}(x)), \qquad n = 1, 2, \ldots.$$

Show that

$$\lim_{x \to \infty} \frac{[f_n(x)]^\alpha}{f_{n-1}(x)} = \infty, \qquad \alpha > 0, n \geq 1.$$

(Do not use integration.)

58. Let f' be defined and have no zeros in some deleted neighborhood of x_0. Find
    (a) $\lim_{x \to x_0} |f(x)|^{f(x)}$ if f has no zeros in some neighborhood of x_0 and $\lim_{x \to x_0} f(x) = 0$

    (b) $\lim_{x \to x_0} [f(x)]^{1/[f(x)-1]}$ if $\lim_{x \to x_0} f(x) = 1$

    (c) $\lim_{x \to x_0} [f(x)]^{1/f(x)}$ if $\lim_{x \to x_0} f(x) = \infty$

59. Fill in the blank and prove the resulting statement: If f' and g' exist and g' has no zeros on (a, b),

$$\lim_{x \to b-} f(x) = \lim_{x \to b-} g(x) = 0, \quad \text{and} \quad \lim_{x \to b-} f'(x)/g'(x) = L,$$

then

$$\lim_{x \to b-} [1 + f(x)]^{1/g(x)} = \underline{\hspace{2cm}}.$$

60. We distinguish between $\infty \cdot \infty \ (= \infty)$ and $(-\infty)\infty \ (= -\infty)$, and between $\infty + \infty \ (= \infty)$ and $-\infty - \infty \ (= -\infty)$. Why don't we distinguish between $0 \cdot \infty$ and $0 \cdot (-\infty)$, $\infty - \infty$ and $-\infty + \infty$, ∞/∞ and $-\infty/\infty$, 1^∞ and $1^{-\infty}$?

2.5 TAYLOR'S THEOREM

A *polynomial* is a function of the form

$$p(x) = a_0 + a_1(x - x_0) + \cdots + a_n(x - x_0)^n, \tag{1}$$

where $a_0, \ldots, a_n$ and x_0 are constants. Since it is easy to calculate the values of a polynomial, considerable effort has been devoted to using them to approximate more complicated functions. Taylor's theorem is one of the oldest and most important results on this question.

Before turning to the main task of this section, which is to discuss Taylor's theorem and some of its applications, we review some terminology concerning polynomials. The polynomial (1) is said to be written *in powers* of $x - x_0$, and is *of degree* n if $a_n \neq 0$. If we wish to leave open the possibility that $a_n = 0$, we say that p is of degree $\leq n$. In particular, a constant polynomial

$p(x) = a_0$ is of degree zero if $a_0 \neq 0$. If $a_0 = 0$, so that p vanishes identically, then p has no degree according to the definition given above, and we say for convenience that p has degree $-\infty$. (Any negative number would do as well as $-\infty$; the point is that with this convention, the statement that p is a polynomial of degree $\leq n$ includes the possibility that p is identically zero.)

Taylor polynomials

We saw in Lemma 3.1 that if f is differentiable at x_0, then

$$f(x) = f(x_0) + f'(x_0)(x - x_0) + E(x)(x - x_0),$$

where

$$\lim_{x \to x_0} E(x) = 0.$$

To generalize this result, we first restate it: The polynomial

$$T_1(x) = f(x_0) + f'(x_0)(x - x_0),$$

which is of degree ≤ 1 and satisfies

$$T_1(x_0) = f(x_0), \qquad T_1'(x_0) = f'(x_0),$$

approximates f so well near x_0 that

$$\lim_{x \to x_0} \frac{f(x) - T_1(x)}{x - x_0} = 0.$$

Now suppose f has n derivatives at x_0 and T_n is the polynomial of degree $\leq n$ such that

$$T_n^{(r)}(x_0) = f^{(r)}(x_0), \qquad 0 \leq r \leq n. \tag{2}$$

How well does T_n approximate f near x_0?

Before answering this question, let us find T_n. Since T_n is a polynomial of degree $\leq n$, it can be written as

$$T_n(x) = a_0 + a_1(x - x_0) + \cdots + a_n(x - x_0)^n, \tag{3}$$

where $a_0, \ldots, a_n$ are constants. Differentiating (3) yields

$$T_n^{(r)}(x_0) = r! \, a_r, \qquad 0 \leq r \leq n,$$

so (2) determines a_r uniquely as

$$a_r = \frac{f^{(r)}(x_0)}{r!}, \qquad 0 \leq r \leq n,$$

and therefore

$$T_n(x) = f(x_0) + \frac{f'(x_0)}{1!}(x - x_0) + \cdots + \frac{f^{(n)}(x_0)}{n!}(x - x_0)^n$$

$$= \sum_{r=0}^{n} \frac{f^{(r)}(x_0)}{r!}(x - x_0)^r.$$

We call T_n the nth *Taylor polynomial of* f *about* x_0.

The following theorem describes how T_n approximates f near x_0.

Theorem 5.1   *If* $f^{(n)}(x_0)$ *exists for some integer* $n \geq 1$ *and* T_n *is the nth Taylor polynomial of* f *about* x_0, *then*

$$\lim_{x \to x_0} \frac{f(x) - T_n(x)}{(x - x_0)^n} = 0. \tag{4}$$

Proof.   The proof is by induction. From Lemma 3.1, Section 2.3, (4) is true if $n = 1$. Now suppose $n > 1$ and (4) is true with n replaced with $n - 1$. Since the ratio in (4) is indeterminate of the form $0/0$ as $x \to x_0$, l'Hospital's rule implies that

$$\lim_{x \to x_0} \frac{f(x) - T_n(x)}{(x - x_0)^n} = \frac{1}{n} \lim_{x \to x_0} \frac{f'(x) - T_n'(x)}{(x - x_0)^{n-1}} \tag{5}$$

if the latter exists. But f' has an $(n - 1)$st derivative at x_0, and T_n' is the $(n - 1)$st Taylor polynomial of f' about x_0, so that f' satisfies the hypotheses of Theorem 5.1 with n replaced by $n - 1$. Hence, our induction assumption implies that the second limit in (5)—and consequently also the first—is zero. This proves (4) and completes the proof of the theorem.

Example 5.1   If $f(x) = e^x$, then $f^{(n)}(x) = e^x$, and the nth Taylor polynomial of f about $x_0 = 0$ is

$$T_n(x) = \sum_{r=0}^{n} \frac{x^r}{r!} = 1 + \frac{x}{1!} + \frac{x^2}{2!} + \cdots + \frac{x^n}{n!}.$$

Theorem 5.1 implies that

$$\lim_{x \to 0} \frac{e^x - \sum_{r=0}^{n} \frac{x^r}{r!}}{x^n} = 0.$$

(See also *Exercise 52, Section 2.4*.)

Example 5.2   If $f(x) = \log x$, then $f(1) = 0$ and

$$f^{(r)}(x) = (-1)^{r-1} \frac{(r-1)!}{x^r}, \qquad r = 1, 2, \ldots,$$

and the nth Taylor polynomial of f about $x_0 = 1$ is

$$T_n(x) = \sum_{r=1}^{n} \frac{(-1)^{r-1}}{r}(x-1)^r$$

if $n \geq 1$. ($T_0 = 0$.) Theorem 5.1 implies that

$$\lim_{x \to 1} \frac{\log x - \sum_{r=0}^{n} \frac{(-1)^{r-1}}{r}(x-1)^r}{(x-1)^n} = 0, \qquad n = 1, 2, \ldots.$$

Example 5.3   If $f(x) = (1 + x)^q$, then

$$f'(x) = q(1 + x)^{q-1}$$
$$f''(x) = q(q-1)(1 + x)^{q-2}$$
$$\vdots \qquad \qquad \vdots$$
$$f^{(n)}(x) = q(q-1) \cdots (q - n + 1)(1 + x)^{q-n}.$$

If we define

$$\binom{q}{0} = 1 \quad \text{and} \quad \binom{q}{n} = \frac{q(q-1) \cdots (q-n+1)}{n!}, \qquad n \geq 1,$$

then

$$\frac{f^{(n)}(0)}{n!} = \binom{q}{n},$$

and the nth Taylor polynomial of f about 0 can be written as

$$T_n(x) = \sum_{r=0}^{n} \binom{q}{r} x^r. \tag{6}$$

Theorem 5.1 implies that

$$\lim_{x \to 0} \frac{(1 + x)^q - \sum_{r=0}^{n} \binom{q}{r} x^r}{x^n} = 0.$$

If q is a nonnegative integer, then $\binom{q}{n}$ is the binomial coefficient defined in *Exercise 17, Section 1.2*. In this case we see from (6) that

$$T_n(x) = (1 + x)^q = f(x), \qquad n \geq q.$$

Applications to finding local extrema

Theorem 5.1 can be restated as a generalization of Lemma 3.1.

Lemma 5.1   *If $f^{(n)}(x_0)$ exists, then*

$$f(x) = \sum_{r=0}^{n} \frac{f^{(r)}(x_0)}{r!}(x - x_0)^r + E(x)(x - x_0)^n, \tag{7}$$

where

$$\lim_{x \to x_0} E(x) = E(x_0) = 0.$$

This lemma yields the following theorem.

Theorem 5.2   *Suppose f has n derivatives at x_0 and n is the smallest positive integer such that $f^{(n)}(x_0) \neq 0$. Then:*
   (a) If n is odd, x_0 is not a local extreme point of f.
   (b) If n is even, x_0 is a local maximum of f if $f^{(n)}(x_0) < 0$, or a local minimum if $f^{(n)}(x_0) > 0$.

Proof.   From (7) and the definition of n,

$$f(x) - f(x_0) = \left[\frac{f^{(n)}(x_0)}{n!} + E(x) \right](x - x_0)^n \tag{8}$$

in some interval containing x_0. Since $\lim_{x \to x_0} E(x) = 0$ and $f^{(n)}(x_0) \neq 0$, there is a $\delta > 0$ such that

$$|E(x)| < \left| \frac{f^{(n)}(x_0)}{n!} \right| \quad \text{if} \quad |x - x_0| < \delta.$$

This and (8) imply that

$$\frac{f(x) - f(x_0)}{(x - x_0)^n} \tag{9}$$

has the same sign as $f^{(n)}(x_0)$ if $0 < |x - x_0| < \delta$. If n is odd, the denominator of (9) changes sign in every neighborhood of x_0, and therefore so must the numerator (since the ratio has constant sign for $0 < |x - x_0| < \delta$); consequently, $f(x_0)$ cannot be a local extreme value of f. This proves (a). If n is even, the denominator of (9) is positive for $x \neq x_0$, and therefore $f(x) - f(x_0)$ must have the same sign as $f^{(n)}(x_0)$ for $0 < |x - x_0| < \delta$. This proves (b).

For $n = 2$, (b) is called the *second derivative test* for local extreme points.

Example 5.4 If $f(x) = e^{x^3}$, then $f'(x) = 3x^2 e^{x^3}$, and 0 is the only critical point of f. Since

$$f''(x) = (6x + 9x^4)e^{x^3}$$

and

$$f'''(x) = (6 + 54x^3 + 27x^6)e^{x^3},$$

it follows that $f''(0) = 0$ and $f'''(0) \neq 0$, so Theorem 5.2 implies that 0 is not a local extreme point of f. Since f is differentiable everywhere, it has no local maxima or minima.

Example 5.5 If $f(x) = \sin x^2$, then $f'(x) = 2x \cos x^2$, so the critical points of f are 0 and $\pm\sqrt{(k + \frac{1}{2})\pi}$ $(k = 0, 1, 2, \ldots)$. Since

$$f''(x) = 2 \cos x^2 - 4x^2 \sin x^2,$$

we have

$$f''(0) = 2 \quad \text{and} \quad f''(\pm\sqrt{(k + \tfrac{1}{2})\pi}) = (-1)^{k+1}(4k + 2)\pi.$$

Therefore, Theorem 5.2 implies that f attains local minima at 0 and $\pm\sqrt{(k + \frac{1}{2})\pi}$ for odd integers k, and local maxima at $\pm\sqrt{(k + \frac{1}{2})\pi}$ for even integers k.

Taylor's theorem

Theorem 5.1 implies that the error in approximating $f(x)$ by $T_n(x)$ approaches zero faster than $(x - x_0)^n$ as x approaches x_0; however, it gives no estimate of the error in approximating $f(x)$ by $T_n(x)$ for a *fixed* x. For instance, it provides no estimate of the error in the approximation

$$e^{0.1} \approx T_2(0.1) = 1 + \frac{0.1}{1!} + \frac{(0.1)^2}{2!} = 1.1050, \tag{10}$$

obtained by setting $n = 2$ and $x = 0.1$ in the result of Example 5.1. The following theorem provides a way of estimating errors of this kind under the additional assumption that $f^{(n+1)}$ exists in a neighborhood of x_0.

Theorem 5.3 (Taylor's Theorem) *Suppose $f^{(n+1)}$ exists in an open interval I about x_0 and let x be in I. Then the remainder*

$$R_n(x) = f(x) - T_n(x)$$

can be written as

$$R_n(x) = \frac{f^{(n+1)}(c)}{(n + 1)!}(x - x_0)^{n+1},$$

where c depends upon x and is between x and x_0.

This theorem follows from an extension of the mean value theorem which we will prove below. For now let us assume that Theorem 5.3 is correct, and apply it.

Example 5.6 If $f(x) = e^x$, then $f'''(x) = e^x$, and Theorem 5.3 implies that

$$e^x = 1 + x + \frac{x^2}{2!} + \frac{e^c x^3}{3!},$$

where c is between 0 and x; hence, from (10),

$$e^{0.1} = 1.1050 + \frac{e^c (0.1)^3}{6},$$

where $0 < c < 0.1$. Since $0 < e^c < e^{0.1}$, we know from this that

$$1.1050 < e^{0.1} < 1.1050 + \frac{e^{0.1}(0.1)^3}{6}.$$

The second inequality implies that

$$e^{0.1}\left[1 - \frac{(0.1)^3}{6}\right] < 1.1050,$$

or

$$e^{0.1} < 1.1052.$$

Therefore,

$$1.1050 < e^{0.1} < 1.1052$$

and the error in (10) is less than 0.0002.

Example 5.7 In many problems in numerical analysis *forward differences* are used to approximate derivatives. If $h > 0$, the first and second forward differences (with spacing h) are defined by

$$\Delta f(x) = f(x + h) - f(x)$$

and

$$\Delta^2 f(x) = \Delta[\Delta f(x)] = \Delta f(x + h) - \Delta f(x)$$
$$= f(x + 2h) - 2f(x + h) + f(x).$$

Higher forward differences are defined inductively (*Exercise 17*). The approximations in question are

$$f'(x) \approx \frac{\Delta f(x)}{h} \tag{11}$$

and

$$f''(x) \approx \frac{\Delta^2 f(x)}{h^2}. \tag{12}$$

If f'' exists on an open interval containing x and $x + h$, we can use Theorem 5.3 to estimate the error in (1·1) by writing

$$f(x + h) = f(x) + f'(x)h + \frac{f''(c)h^2}{2},$$

where $x < c < x + h$. Routine manipulations yield

$$\frac{\Delta f(x)}{h} - f'(x) = \frac{f''(c)h}{2},$$

so that

$$\left| \frac{\Delta f(x)}{h} - f'(x) \right| \leq \frac{M_2 h}{2}$$

if M_2 is an upper bound for $|f''|$ on $(x, x + h)$.

If f''' exists on an open interval containing x and $x + 2h$, we can use Theorem 5.3 to estimate the error in (12) by writing

$$f(x + h) = f(x) + hf'(x) + \frac{h^2}{2} f''(x) + \frac{h^3}{6} f'''(c_0)$$

and

$$f(x + 2h) = f(x) + 2hf'(x) + 2h^2 f''(x) + \frac{4h^3}{3} f'''(c_1),$$

where $x < c_0 < x + h$ and $x < c_1 < x + 2h$. Routine manipulations yield

$$\frac{\Delta^2 f(x)}{h^2} - f''(x) = [\tfrac{4}{3} f'''(c_1) - \tfrac{1}{3} f'''(c_0)]h,$$

so that

$$\left| \frac{\Delta^2 f(x)}{h^2} - f''(x) \right| \leq \tfrac{5}{3} M_3 h$$

if M_3 is an upper bound for $|f'''|$ on $(x, x + 2h)$.

The extended mean value theorem

We now consider the extended mean value theorem, which implies Theorem 5.3 (*Exercise 23*). In the following theorem, a and b are the end points of an interval, but we do not assume that $a < b$.

Theorem 5.4 (Extended Mean Value Theorem)   *Suppose f is continuous on a finite closed interval I with end points a and b, $f^{(n+1)}$ exists on the open interval I^0, and if $n > 0, f', \ldots, f^{(n)}$ exist and are continuous at a. Then*

$$f(b) - \sum_{r=0}^{n} \frac{f^{(r)}(a)}{r!}(b-a)^r = \frac{f^{(n+1)}(c)}{(n+1)!}(b-a)^{n+1} \tag{13}$$

for some c in I^0.

Proof.   The proof is by induction. For $n = 0$ the conclusion is the same as Theorem 3.8, and therefore true. Now suppose $n \geq 1$ and assume the conclusion is true with n replaced by $n - 1$. The left side of (13) can be written as

$$f(b) - \sum_{r=0}^{n} \frac{f^{(r)}(a)}{r!}(b-a)^r = K \frac{(b-a)^{n+1}}{(n+1)!} \tag{14}$$

for some number K; what we must prove is that $K = f^{(n+1)}(c)$ for some c in I^0. To this end, consider the auxiliary function

$$h(x) = f(x) - \sum_{r=0}^{n} \frac{f^{(r)}(a)}{r!}(x-a)^r - K \frac{(x-a)^{n+1}}{(n+1)!},$$

which satisfies

$$h(a) = 0, \qquad h(b) = 0,$$

[the latter because of (14)] and is continuous on the closed interval I and differentiable on I^0, with

$$h'(x) = f'(x) - \sum_{r=0}^{n-1} \frac{f^{(r+1)}(a)}{r!}(x-a)^r - K \frac{(x-a)^n}{n!}. \tag{15}$$

Therefore, Rolle's theorem implies that $h'(b_1) = 0$ for some b_1 in I^0; thus, from (15),

$$f'(b_1) - \sum_{r=0}^{n-1} \frac{f^{(r+1)}(a)}{r!}(b_1-a)^r - K \frac{(b_1-a)^n}{n!} = 0.$$

If we temporarily write $f' = g$, this becomes

$$g(b_1) - \sum_{r=0}^{n-1} \frac{g^{(r)}(a)}{r!}(b_1-a)^r - K \frac{(b_1-a)^n}{n!} = 0. \tag{16}$$

Since $b_1 \in I^0$, the hypotheses on f imply that g is continuous on the closed J with end points a and b_1, $g^{(n)}$ exists on J^0, and, if $n \geq 1, g', \ldots, g^{(n-1)}$ exist and are continuous at a (also at b_1, but this is not important). The induction hypothesis applied to g on the interval J implies that

$$g(b_1) - \sum_{r=0}^{n-1} \frac{g^{(r)}(a)}{r!}(b_1-a)^r = \frac{g^{(n)}(c)}{n!}(b_1-a)^n$$

for some c in J^0. Comparing this with (16) and recalling that $g = f'$ yields

$$K = g^{(n)}(c) = f^{(n+1)}(c).$$

Since c is in I^0, this completes the induction.

2.5 EXERCISES

* 1. Let

$$f(x) = \begin{cases} e^{-1/x^2}, & x \neq 0, \\ 0, & x = 0. \end{cases}$$

Show that f has derivatives of all orders on $(-\infty, \infty)$ and that every Taylor polynomial of f about 0 is identically zero. (*Hint:* See Exercise 54, Section 2.4.)

2. (a) Prove: If $f''(x_0)$ exists, then

$$\lim_{h \to 0} \frac{f(x_0 + h) - 2f(x_0) + f(x_0 - h)}{h^2} = f''(x_0).$$

   (b) Prove or give a counterexample: If the limit in part (a) exists, then so does $f''(x_0)$, and they are equal.

3. Let k be a positive integer. A function f is said to have a zero *of multiplicity* k at x_0 if $f^{(j)}(x_0) = 0 \ (0 \leq j \leq k - 1)$ and $f^{(k)}(x_0) \neq 0$.
   (a) Prove that f has a zero of multiplicity k at x_0 if and only if

$$f(x) = g(x)(x - x_0)^k,$$

   where g is continuous and nonzero at x_0 and has at least $k - 1$ derivatives on a deleted neighborhood of x_0 which satisfy

$$\lim_{x \to x_0} (x - x_0)^j g^{(j)}(x) = 0, \qquad 1 \leq j \leq k - 1.$$

   (b) Give an example showing that g in part (a) need not be differentiable at x_0.

4. Suppose

$$p(x) = a_0 + a_1(x - x_0) + \cdots + a_k(x - x_0)^k$$

and, for a nonnegative integer $n \leq k$, $f^{(n)}(x_0)$ exists and

$$\lim_{x \to x_0} \frac{f(x) - p(x)}{(x - x_0)^n} = 0.$$

Show that

$$a_r = \frac{f^{(r)}(x_0)}{r!} \qquad \text{if} \quad 0 \leq r \leq n.$$

5. Show that if $f^{(n)}(x_0)$ and $g^{(n)}(x_0)$ exist and

$$\lim_{x \to x_0} \frac{f(x) - g(x)}{(x - x_0)^n} = 0,$$

   then $f^{(r)}(x_0) = g^{(r)}(x_0)$, $0 \leq r \leq n$.

6. (a) Let F_n, G_n, and H_n be the nth Taylor polynomials about x_0 of f, g, and $h = fg$ (product). Show that H_n can be obtained by multiplying F_n by G_n, retaining powers of $x - x_0$ through the nth, and discarding higher powers. (*Hint:* See Exercise 4.)
   (b) Use the method suggested by part (a) to compute $h^{(r)}(x_0)$, $r = 1, 2, 3, 4$:
   (i) $h(x) = e^x \sin x$, $x_0 = 0$
   (ii) $h(x) = (\cos \pi x/2)(\log x)$, $x_0 = 1$
   (iii) $h(x) = x^2 \cos x$, $x_0 = \pi/2$
   (iv) $h(x) = (1 + x)^{-1}e^{-x}$, $x_0 = 0$

7. (a) Suppose $g^{(n)}(x_0)$ and $f^{(n)}(g(x_0))$ exist and let $h(x) = f(g(x))$. Prove that $h^{(n)}(x_0)$ exists and can be written as

$$h^{(n)}(x_0) = \sum_{j=1}^{n} f^{(j)}(g(x_0))Q_{nj}(x_0),$$

   where Q_{nj} is a sum of products of powers of $g', \ldots, g^{(n-j+1)}$ and constants which do not depend on f, g, or x_0.
   (b) Compute the first four derivatives of $h(x) = \cos(\sin x)$ at $x_0 = 0$, using the method suggested by part (a).
   (c) Under the assumptions of part (a), let F_n be the nth Taylor polynomial of f about $y_0 = g(x_0)$, and let G_n and H_n be the nth Taylor polynomials of g and h about x_0. Show that H_n can be obtained by substituting G_n into F_n and retaining only powers of $x - x_0$ through the nth. (*Hint:* See Exercise 4.)
   (d) Compute the first four derivatives of $h(x) = \cos(\sin x)$ at $x_0 = 0$, using the method suggested by part (c).

8. (a) If $g(x_0) \neq 0$ and $g^{(n)}(x_0)$ exists, then the reciprocal $h = 1/g$ is also n times differentiable at x_0. Use Exercise 7(c) to prove this and to show that if $g(x_0) = 1$, then the nth Taylor polynomial of h about x_0, $H_n(x)$, can be obtained by expanding the polynomial

$$\sum_{r=1}^{n} (-1)^r [G_n(x) - 1]^r$$

   in powers of $x - x_0$ and retaining only powers through the nth.
   (b) Use the method of part (a) to compute the first four derivatives of the following functions at x_0:
   (i) $h(x) = \csc x$, $x_0 = \pi/2$
   (ii) $h(x) = (1 + x + x^2)^{-1}$, $x_0 = 0$

(iii) $h(x) = \sec x, x_0 = \pi/4$

(iv) $h(x) = [1 + \log(1 + x)]^{-1}, x_0 = 0$

(c) Use Exercise 6 to justify the following alternate procedure for obtaining H_n, again assuming that $g(x_0) = 1$: If

$$G_n(x) = 1 + a_1(x - x_0) + \cdots + a_n(x - x_0)^n$$

(where, of course, $a_r = g^{(r)}(x_0)/r!$) and

$$H_n(x) = b_0 + b_1(x - x_0) + \cdots + b_n(x - x_0)^n,$$

then

$$b_0 = 1, b_n = -\sum_{r=1}^{n} a_r b_{n-r}, n \geq 1.$$

(d) Use the method of part (c) to calculate the derivatives required in part (b).

9. Suppose f is increasing and continuous on $[a, b]$, $f(a) = c$, and $f(b) = d$. Let $g = f^{-1}$; thus, $g(f(x)) = x$, $a \leq x \leq b$, and $f(g(y)) = y$, $c \leq y \leq d$ (Theorem 2.9). Suppose further that $a < x_0 < b$, $f(x_0) = y_0$, $f'(x_0) = a_1 \neq 0$, and $f^{(n)}(x_0)$ exists for some $n \geq 1$.

(a) Show that $g^{(n)}(y_0)$ exists. (*Hint:* See Exercise 14, Section 2.3.)

(b) Let

$$F_n(x) = y_0 + a_1(x - x_0) + \cdots + a_n(x - x_0)^n$$

be the nth Taylor polynomial of f about x_0. Show that the nth Taylor polynomial of g about y_0 is

$$G_n(y) = x_0 + b_1(y - y_0) + \cdots + b_n(y - y_0)^n,$$

with

$$b_1 = \frac{1}{a_1}, \qquad b_2 = \frac{-a_2}{a_1^3},$$

and, if $n \geq 3$,

$$b_j = -\frac{a_j}{a_1^{j+1}} - \frac{c_j}{a_1}, \qquad 3 \leq j \leq n,$$

where c_j is the coefficient of $(y - y_0)^j$ in $F_{j-1}(G_{j-1}(y))$. (*Hint:* See Exercise 7.)

10. Each of the following functions has an inverse g such that $g(f(x)) = x$ for x near x_0 and $f(g(y)) = y$ for y near $y_0 = f(x_0)$. Use a method based on Exercise 9 to find $g^{(j)}(y_0), j = 1, 2, 3, 4$:

(a) $f(x) = \sin x, x_0 = 0$

(b) $f(x) = \cos x, x_0 = \pi/4$

(c) $f(x) = \sinh x + \sin x^2$, $x_0 = 0$
(d) $f(x) = xe^x$, $x_0 = 0$
(e) $f(x) = e^{2(x-1)}$, $x_0 = 1$
(f) $f(x) = \sin x + x^2 \cos x$, $x_0 = 0$

11. Determine whether $x_0 = 0$ is a local maximum, local minimum, or neither:

(a) $f(x) = x^2 e^{x^3}$ (b) $f(x) = x^3 e^{x^2}$

(c) $f(x) = \dfrac{1 + x^2}{1 + x^3}$ (d) $f(x) = \dfrac{1 + x^3}{1 + x^2}$

(e) $f(x) = x^2 \sin^3 x + x^2 \cos x$ (f) $f(x) = e^{x^2} \sin x$
(g) $f(x) = e^x \sin x^2$ (h) $f(x) = e^{x^2} \cos x$

12. Give an example of a function which has zero derivatives of all orders at a local minimum.

13. Find the critical points of

$$f(x) = \frac{x^3}{3} + \frac{bx^2}{2} + cx + d$$

and identify them as local maxima, local minima, or neither.

14. Let m and n be integers ≥ 2, $f^{(i)}(y_0) = 0$ $(1 \leq i \leq m - 1)$, $g^{(j)}(x_0) = 0$ $(1 \leq j \leq n - 1)$, $f^{(m)}(y_0) \neq 0$, and $g^{(n)}(x_0) \neq 0$. Suppose also that $y_0 = g(x_0)$ and define $h(x) = f(g(x))$. For each of the nine possible ways to construct a sentence by choosing first a phrase from a, b, and c and then one from A, B, and C, pick the phrase from α, β, and γ that completes the sentence correctly.
    If x_0 is
(a) not a critical point of g
(b) a critical point, but not a local extreme point of g
(c) a local extreme point of g
and y_0 is
(A) not a critical point of f,
(B) a critical point, but not a local extreme point of f,
(C) a local extreme point of f,
then x_0 is
(α) not a critical point of h.
(β) a critical point, but not a local extreme point of h.
(γ) a local extreme point of h.

15. Find an upper bound for the magnitude of the error in the approximation:

(a) $\sin x \approx x$, $|x| < \dfrac{\pi}{20}$

(b) $\sqrt{1 + x} \approx 1 + \dfrac{x}{2}, |x| < \dfrac{1}{8}$

(c) $\cos x \approx \dfrac{1}{\sqrt{2}}\left[1 - \left(x - \dfrac{\pi}{4}\right)\right], \dfrac{\pi}{4} < x < \dfrac{5\pi}{16}$

(d) $\log x \approx (x - 1) - \dfrac{(x - 1)^2}{2} + \dfrac{(x - 1)^3}{3}, |x - 1| < \dfrac{1}{64}$

16. Prove: If

$$T_n(x) = \sum_{r=0}^{n} \frac{x^r}{r!}$$

then

$$T_n(x) < T_{n+1}(x) < e^x < \left[1 - \frac{x^{n+1}}{(n+1)!}\right]^{-1} T_n(x), \quad 0 < x < [(n+1)!]^{1/(n+1)}.$$

17. The forward difference operators with spacing $h > 0$ are defined by

$$\Delta^0 f(x) = f(x), \qquad \Delta f(x) = f(x + h) - f(x),$$
$$\Delta^{n+1} f(x) = \Delta[\Delta^n f(x)], \qquad n \geq 1.$$

(a) Prove: If $c_1, \ldots, c_k$ are constants, then

$$\Delta^n(c_1 f_1 + \cdots + c_k f_k) = c_1 \Delta^n f_1 + \cdots + c_k \Delta^n f_k.$$

(b) Prove by induction:

$$\Delta^n f(x) = \sum_{k=0}^{n} (-1)^{n-k} \binom{n}{k} f(x + kh).$$

(*Hint:* See Exercise 17, Section 1.2.)

18. Prove: If

$$p(x) = a_0 + a_1(x - x_0) + \cdots + a_n(x - x_0)^n,$$

then

$$\Delta^n p(x) = n! \, h^n a_n.$$

(*Hint:* Use induction.)

19. Estimate the error in the approximation

$$f''(x) \approx \frac{\Delta^2 f(x - h)}{h^2}$$

(a) assuming that f''' is bounded on $(x - h, x + h)$;
(b) assuming that $f^{(4)}$ is bounded on $(x - h, x + h)$.

20. Let f''' be bounded on an open interval containing x and $x + 2h$. Choose the constant k so that the magnitude of the error in the approximation

$$f'(x) \approx \frac{\Delta f(x)}{h} + k \frac{\Delta^2 f(x)}{h^2}$$

is not greater than Mh^2, where $M = \text{l.u.b.}\{|f'''(c)| \, | \, x < c < x + 2h\}$.

21. Prove: If $f^{(n+1)}$ is bounded on an open interval containing x and $x + nh$, then

$$\left| \frac{\Delta^n f(x)}{h^n} - f^{(n)}(x) \right| \leq A_n M_{n+1} h,$$

where A_n is a constant independent of f and

$$M_{n+1} = \text{l.u.b.}\{|f^{(n+1)}(c)| \, | \, x < c < x + nh\}.$$

(*Hint:* See Exercises 17 and 18.)

22. Suppose $f^{(n+1)}$ exists on (a, b), $x_0, \ldots, x_n$ are in (a, b), and p is the polynomial of degree $\leq n$ such that $p(x_i) = f(x_i)$, $0 \leq i \leq n$. Prove: If $x \in (a, b)$, then

$$f(x) = p(x) + \frac{f^{(n+1)}(c)}{(n+1)!}(x - x_0)(x - x_1) \cdots (x - x_n),$$

where c, which depends on x, is a point in (a, b). [*Hint:* Let x be fixed, distinct from $x_0, x_1, \ldots, x_n$, and consider the function

$$g(y) = f(y) - p(y) - \frac{K}{(n+1)!}(y - x_0)(y - x_1) \cdots (y - x_n),$$

where K is chosen so that $g(x) = 0$. Use Rolle's theorem to show that $K = f^{(n+1)}(c)$ for some c in (a, b).]

23. Deduce Theorem 5.3 from Theorem 5.4.

integral calculus of functions of one variable

3.1 DEFINITION OF THE INTEGRAL

The integral studied in the basic calculus course is called the *Riemann integral*, in honor of the German mathematician Bernhard Riemann, who provided a rigorous formulation to replace the intuitive notion of integral due to Newton and Leibniz. Since Riemann's time other kinds of integrals have been defined and studied; however, they are all generalizations of the Riemann integral, and it is hardly possible to understand them or appreciate the reasons for developing them without a thorough understanding of the Riemann integral.

In this section we deal with functions defined on a finite interval $[a, b]$. A *partition of* $[a, b]$ is a set of subintervals

$$[x_0, x_1], \quad [x_1, x_2], \quad \ldots, \quad [x_{n-1}, x_n], \tag{1}$$

where

$$a = x_0 < x_1 \cdots < x_n = b. \tag{2}$$

Thus, any $n + 1$ points satisfying (2) define a partition P of $[a, b]$, which we denote by

$$P = \{x_0, x_1, \ldots, x_n\}.$$

The points $x_0, x_1, \ldots, x_n$ are the *partition points* of P. The largest of the lengths of the subintervals (1) is the *norm* of P, written as $\|P\|$; thus,

$$\|P\| = \max_{1 \le i \le n} \{x_i - x_{i-1}\}.$$

If P and P' are partitions of $[a, b]$, then P' is a *refinement* of P if every partition point of P is also a partition point of P'; that is, if P' is obtained by inserting additional points between those of P.

If f is defined on $[a, b]$, then a sum

$$\sigma = \sum_{j=1}^{n} f(c_j)(x_j - x_{j-1}),$$

where

$$x_{j-1} \le c_j \le x_j, \qquad j = 1, 2, \ldots, n, \tag{3}$$

is *called a Riemann sum of f over the partition* $P = \{x_0, x_1, \ldots, x_n\}$. Since c_j can be chosen arbitrarily in $[x_j, x_{j-1}]$, there are infinitely many Riemann sums for a given function f over a given partition P.

Definition 1.1   *A function f is Riemann integrable over $[a, b]$ if there is a number L with the following property: For every $\varepsilon > 0$ there is a $\delta > 0$ such that every Riemann sum of f over any partition $P = \{x_0, x_1, \ldots, x_n\}$ of $[a, b]$ with $\|P\| < \delta$ satisfies the inequality*

$$\left| \sum_{j=1}^{n} f(c_j)(x_j - x_{j-1}) - L \right| < \varepsilon$$

for every choice of $c_1, c_2, \ldots, c_n$ satisfying (3). The number L is called the Riemann integral of f over $[a, b]$ and is denoted by $\int_a^b f(x)\, dx$; thus,

$$L = \int_a^b f(x)\, dx.$$

We leave it to the reader (*Exercise 1*) to show that $\int_a^b f(x)\, dx$ is unique, if it exists; that is, there cannot be more than one number L that satisfies Definition 1.1.

For brevity we will simply say "integrable" and "integral" when we mean "Riemann integrable" and "Riemann integral."

Example 1.1   If

$$f(x) = 1, \qquad a \le x \le b,$$

then

$$\sum_{j=1}^{n} f(c_j)(x_j - x_{j-1}) = \sum_{j=1}^{n} (x_j - x_{j-1}).$$

Most of the terms in the sum on the right cancel in pairs; that is,

$$\sum_{j=1}^{n} (x_j - x_{j-1}) = (x_1 - x_0) + (x_2 - x_1) + \cdots + (x_n - x_{n-1})$$
$$= -x_0 + (x_1 - x_1) + (x_2 - x_2) + \cdots$$
$$+ (x_{n-1} - x_{n-1}) + x_n$$
$$= x_n - x_0$$
$$= b - a.$$

Thus, every Riemann sum of f over any partition of $[a, b]$ equals $b - a$, and therefore surely

$$\int_a^b dx = b - a.$$

Example 1.2   Riemann sums for the function

$$f(x) = x, \qquad a \le x \le b,$$

are of the form

$$\sigma = \sum_{j=1}^{n} c_j(x_j - x_{j-1}). \tag{4}$$

Since $x_{j-1} \le c_j \le x_j$ and $(x_j + x_{j-1})/2$ is the midpoint of $[x_{j-1}, x_j]$, we can write

$$c_j = \frac{x_j + x_{j-1}}{2} + d_j, \tag{5}$$

where

$$|d_j| \le \frac{x_j - x_{j-1}}{2} \le \frac{\|P\|}{2} \tag{6}$$

Substituting (5) into (4) yields

$$\sigma = \sum_{j=1}^{n} \frac{x_j + x_{j-1}}{2} (x_j - x_{j-1}) + \sum_{j=1}^{n} d_j(x_j - x_{j-1})$$
$$= \frac{1}{2} \sum_{j=1}^{n} (x_j^2 - x_{j-1}^2) + \sum_{j=1}^{n} d_j(x_j - x_{j-1}). \tag{7}$$

Because of cancellations similar to those in Example 1.1,

$$\sum_{j=1}^{n} (x_j^2 - x_{j-1}^2) = b^2 - a^2,$$

so (7) can be rewritten as

$$\sigma = \frac{b^2 - a^2}{2} + \sum_{j=1}^{n} d_j(x_j - x_{j-1}).$$

Hence,

$$\left| \sigma - \frac{b^2 - a^2}{2} \right| \le \sum_{j=1}^{n} |d_j|(x_j - x_{j-1})$$

$$\le \frac{\|P\|}{2} \sum_{j=1}^{n} (x_j - x_{j-1}) \qquad [\text{see (6)}]$$

$$= \frac{\|P\|}{2}(b - a).$$

Therefore, every Riemann sum of f over a partition P of $[a, b]$ satisfies

$$\left| \sigma - \frac{b^2 - a^2}{2} \right| < \varepsilon \quad \text{if} \quad \|P\| < \frac{2\varepsilon}{b - a} = \delta.$$

Hence,

$$\int_a^b x \, dx = \frac{b^2 - a^2}{2}.$$

The integral as the area under a curve

An important application of the integral—indeed, the one invariably used to motivate its definition—is to the computation of the area bounded by a curve $y = f(x)$, the x axis, and the lines $x = a$ and $x = b$ ("the area under the curve"), as in Figure 1.1.

For simplicity, suppose $f(x) > 0$; then $f(c_j)(x_j - x_{j-1})$ is the area of a rectangle with base $x_j - x_{j-1}$ and height $f(c_j)$, and so the Riemann sum

$$\sum_{j=1}^{n} f(c_j)(x_j - x_{j-1})$$

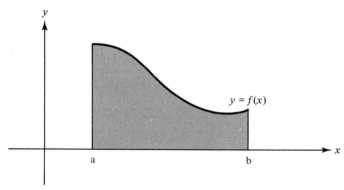

FIGURE 1.1

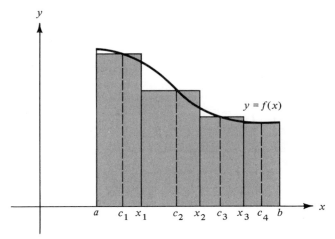

FIGURE 1.2

can be interpreted as the sum of the areas of rectangles related to the curve $y = f(x)$, as shown in Figure 1.2. An apparently plausible argument, that the Riemann sums approximate the area under the curve more and more closely as the number of rectangles increases and the largest of their widths is made smaller, seems to support the assertion that $\int_a^b f(x)\,dx$ equals the area under the curve. This argument is unquestionably useful as a motivation for Definition 1.1, which without it would seem mysterious indeed to the beginning student; nevertheless, the logic is incorrect, since it is based on the implicit assumption that the area under the curve has been previously defined in some other way. Although this is true for certain curves such as, for example, those consisting of line segments or circular arcs, it is not true in general. In fact, the area under a more complicated curve is *defined* to be equal to the integral, if the integral exists. That this new definition is consistent with the old one, where the latter applies, is evidence that the integral provides a useful generalization of the definition of area.

Example 1.3   Let

$$f(x) = x, \qquad 1 \le x \le 2$$

(Figure 1.3). The region under the curve consists of a square of unit area, surmounted by a triangle of area $\frac{1}{2}$; thus, the total area is $\frac{3}{2}$. From Example 1.2,

$$\int_1^2 x\,dx = \frac{1}{2}(2^2 - 1^2) = \frac{3}{2},$$

and so the integral equals the area under the curve.

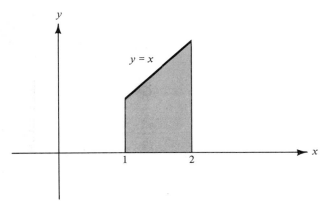

FIGURE 1.3

Example 1.4 Let

$$f(x) = x^2, \qquad 1 \le x \le 2$$

(Figure 1.4); then

$$\int_1^2 f(x)\,dx = \frac{1}{3}(2^3 - 1^3) = \frac{7}{3}$$

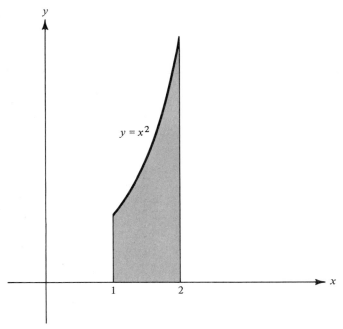

FIGURE 1.4

(*Exercise 4*), and so we say that the area under the curve is $\frac{7}{3}$; however, this is the *definition* of the area rather than a confirmation of a previously known fact, as in Example 1.3.

Theorem 1.1 *If f is unbounded on $[a, b]$, then f is not integrable on $[a, b]$ in the sense of Definition 1.1.*

Proof. We will show that if f is unbounded on $[a, b]$, P is any partition of $[a, b]$, and $M > 0$, then there are Riemann sums σ and σ' of f over P such that

$$|\sigma - \sigma'| \geq M.$$

We leave it to the reader to complete the proof by showing from this that f cannot satisfy Definition 1.1 (*Exercise 2*).

Let

$$\sigma = \sum_{j=1}^{n} f(c_j)(x_j - x_{j-1})$$

be a Riemann sum of f over a partition P of $[a, b]$. There must be an integer i in $\{1, 2, \ldots, n\}$ such that

$$|f(c) - f(c_i)|(x_i - x_{i-1}) \geq M \tag{8}$$

for some c in $[x_{i-1}, x_i]$, because if this were not so, we would have

$$|f(x) - f(c_j)| < \frac{M}{x_j - x_{j-1}}, \qquad x_{j-1} \leq x \leq x_j, \quad j = 1, 2, \ldots, n,$$

which implies that f is bounded on $[a, b]$, a contradiction. Now consider the Riemann sum

$$\sigma' = \sum_{j=1}^{n} f(c_j')(x_j - x_{j-1})$$

over the same partition P, where

$$c_j' = \begin{cases} c_j, & j \neq i, \\ c, & j = i. \end{cases}$$

Since

$$|\sigma - \sigma'| = |f(c) - f(c_i)|(x_i - x_{i-1}),$$

(8) implies that $|\sigma - \sigma'| \geq M$.

Upper and lower integrals

Because of Theorem 1.1 we will consider only bounded functions throughout the rest of this section.

To prove directly from Definition 1.1 that $\int_a^b f(x)\,dx$ exists, it is necessary to discover its value L in one way or another, and to show that L has the properties required by the definition. For a specific function it may happen that this can be done by straightforward calculation, as in Examples 1.1 and 1.2; however, this is not so if the objective is to find general conditions which imply that $\int_a^b f(x)\,dx$ exists. The following approach avoids this difficulty of having to discover L in advance (without knowing whether it exists in the first place), and requires only that we compare two numbers which must exist if f is bounded on $[a, b]$; as we will see, $\int_a^b f(x)\,dx$ exists if and only if the numbers are equal.

Definition 1.2   *If f is bounded on $[a, b]$ and $P = \{x_0, x_1, \ldots, x_n\}$ is a partition of $[a, b]$, let*

$$M_j = \underset{x_{j-1} \le x \le x_j}{\text{l.u.b.}}\ f(x) \quad \text{and} \quad m_j = \underset{x_{j-1} \le x \le x_j}{\text{g.l.b.}}\ f(x).$$

The upper sum of f over P is defined to be

$$S(P) = \sum_{j=1}^{n} M_j(x_j - x_{j-1})$$

and the upper integral of f over $[a, b]$, denoted by

$$\overline{\int_a^b} f(x)\,dx,$$

is defined to be the greatest lower bound of all upper sums. The lower sum of f over P is defined to be

$$s(P) = \sum_{j=1}^{n} m_j(x_j - x_{j-1})$$

and the lower integral, denoted by

$$\underline{\int_a^b} f(x)\,dx,$$

is defined to be the least upper bound of all lower sums.

If

$$m \le f(x) \le M, \qquad a \le x \le b,$$

then

$$m(b - a) \le s(P) \le S(P) \le M(b - a)$$

for every partition P; thus, the set of upper sums of f over all partitions P of $[a, b]$ is bounded, as is the set of lower sums. Theorems 1.1 and 1.4 of Section 1.1 imply that $\overline{\int_a^b} f(x)\, dx$ and $\underline{\int_a^b} f(x)\, dx$ exist, are unique, and satisfy the inequalities

$$m(b-a) \le \overline{\int_a^b} f(x)\, dx \le M(b-a)$$

and

$$m(b-a) \le \underline{\int_a^b} f(x)\, dx \le M(b-a).$$

Example 1.5 Let

$$f(x) = \begin{cases} 0 & \text{if } x \text{ is irrational,} \\ 1 & \text{if } x \text{ is rational,} \end{cases}$$

and $P = \{x_0, x_1, \ldots, x_n\}$ be a partition of $[a, b]$. Since every interval contains both rational and irrational numbers,

$$m_j = 0 \quad \text{and} \quad M_j = 1, \qquad j = 1, 2, \ldots, n.$$

Hence,

$$S(P) = \sum_{j=1}^{n} 1(x_j - x_{j-1}) = b - a$$

and

$$s(P) = \sum_{j=1}^{n} 0(x_j - x_{j-1}) = 0.$$

Since all upper sums equal $b - a$ and all lower sums vanish, Definition 1.2 implies that

$$\overline{\int_a^b} f(x)\, dx = b - a \quad \text{and} \quad \underline{\int_a^b} f(x)\, dx = 0.$$

Example 1.6 Let f be defined on $[1, 2]$ by $f(x) = 0$ if x is irrational and $f(p/q) = 1/q$ if p and q are positive integers with no common factors. (*See Exercise 7, Section 2.2.*) If $P = \{x_0, x_1, \ldots, x_n\}$ is any partition of $[1, 2]$, then $m_j = 0$, $j = 1, 2, \ldots, n$, and so $s(P) = 0$; hence,

$$\underline{\int_1^2} f(x)\, dx = 0.$$

We now show that

$$\overline{\int_1^2} f(x)\, dx = 0 \tag{9}$$

also. Since $S(P) > 0$ for every P, Definition 1.2 implies that

$$\overline{\int_1^2} f(x)\, dx \ge 0,$$

and so we need only show that

$$\overline{\int_1^2} f(x)\, dx \leq 0,$$

which will follow if we show that no positive number is less than every upper sum. To this end we observe that if $0 < \varepsilon < 2$, then $f(x) \geq \varepsilon/2$ for only finitely many values of x in $[1, 2]$. Let P_0 be a partition of $[1, 2]$ such that

$$\|P_0\| < \frac{\varepsilon}{2k}, \tag{10}$$

where k is the number of such points, and consider the upper sum

$$S(P_0) = \sum_{j=1}^{n} M_j(x_j - x_{j-1}).$$

There are at most k values of j in this sum for which $M_j \geq \varepsilon/2$, and $M_j \leq 1$ even for these. The contribution of these terms to the sum is less than $k(\varepsilon/2k) = \varepsilon/2$ [because of (10)]. Since $M_j < \varepsilon/2$ for all other values of j, the sum of the other terms is surely less than

$$\frac{\varepsilon}{2} \sum_{j=1}^{n} (x_j - x_{j-1}) = \frac{\varepsilon}{2}(x_n - x_0) = \frac{\varepsilon}{2}(2 - 1) = \frac{\varepsilon}{2}.$$

Therefore, $S(P_0) < \varepsilon$ and, since ε can be chosen as small as we wish, no positive number is less than all upper sums. This proves (9).

The motivation for Definition 1.2 can be seen by again considering the idea of area under a curve. Figure 1.5 shows the graph of a positive function

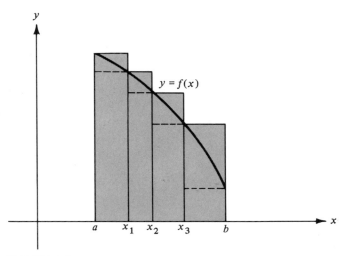

FIGURE 1.5

$y = f(x)$, $a \le x \le b$, with $[a, b]$ partitioned into four subintervals. The upper and lower sums of f over this partition can be interpreted as the sums of the areas of the rectangles surmounted by the solid and dashed lines, respectively. This indicates that a sensible definition of area A under the curve must admit the inequalities

$$s(P) \le A \le S(P)$$

for every partition P of $[a, b]$. Thus, A must be an upper bound for all lower sums and a lower bound for all upper sums of f over partitions of $[a, b]$. If

$$\overline{\int_a^b} f(x)\,dx = \underline{\int_a^b} f(x)\,dx, \tag{11}$$

there is only one number—the common value of the upper and lower integrals— with this property, and we define A to be that number; if (11) does not hold, then A is not defined. We will see below that this definition of area is consistent with the definition stated earlier in terms of Riemann sums.

Example 1.7 Returning to Example 1.3, consider the function

$$f(x) = x, \qquad 1 \le x \le 2.$$

If $P = \{x_0, x_1, \ldots, x_n\}$ is a partition of $[a, b]$, then, since f is increasing,

$$M_j = f(x_j) = x_j \quad \text{and} \quad m_j = f(x_{j-1}) = x_{j-1}.$$

Hence,

$$S(P) = \sum_{j=1}^n x_j(x_j - x_{j-1}) \tag{12}$$

and

$$s(P) = \sum_{j=1}^n x_{j-1}(x_j - x_{j-1}). \tag{13}$$

By writing

$$x_j = \frac{x_j + x_{j-1}}{2} + \frac{x_j - x_{j-1}}{2},$$

we see from (12) that

$$S(P) = \frac{1}{2} \sum_{j=1}^n (x_j^2 - x_{j-1}^2) + \frac{1}{2} \sum_{j=1}^n (x_j - x_{j-1})^2$$

$$= \frac{1}{2}(2^2 - 1^2) + \frac{1}{2} \sum_{j=1}^n (x_j - x_{j-1})^2. \tag{14}$$

Since

$$0 < \sum_{j=1}^n (x_j - x_{j-1})^2 \le \|P\| \sum_{j=1}^n (x_j - x_{j-1}) = \|P\|(2 - 1),$$

(14) implies that

$$\frac{3}{2} < S(P) \leq \frac{3}{2} + \frac{\|P\|}{2}.$$

Since $\|P\|$ can be made as small as we please, Definition 1.2 implies that

$$\overline{\int_a^b} f(x) \, dx = \frac{3}{2}.$$

A similar argument starting from (13) shows that

$$\frac{3}{2} - \frac{\|P\|}{2} \leq s(P) < \frac{3}{2},$$

so that

$$\underline{\int_a^b} f(x) \, dx = \frac{3}{2}.$$

Since the upper and lower integrals both equal $\frac{3}{2}$, the area under the curve is $\frac{3}{2}$, according to our new definition. This is consistent with the result of Example 1.3.

The Riemann–Stieltjes integral

The *Riemann–Stieltjes integral* is an important generalization of the Riemann integral. We define it here but confine our study of it to the exercises in this and other sections of this chapter.

Definition 1.3   *Let f and g be defined on $[a, b]$. We say that f is Riemann–Stieltjes-integrable with respect to g over $[a, b]$ if there is a number L with the following property: For every $\varepsilon > 0$ there is a $\delta > 0$ such that*

$$\left| \sum_{j=1}^{n} f(c_j)[g(x_j) - g(x_{j-1})] - L \right| < \varepsilon \tag{15}$$

provided only that $P = \{x_0, x_1, \ldots, x_n\}$ is a partition of $[a, b]$ such that $\|P\| < \delta$, and

$$x_{j-1} \leq c_j \leq x_j, \qquad j = 1, 2, \ldots, n.$$

In this case we say that L is the Riemann–Stieltjes integral of f with respect to g over $[a, b]$, and write

$$L = \int_a^b f(x) \, dg(x).$$

The sum in (15) is called a Riemann–Stieltjes sum of f with respect to g over the partition P.

3.1   EXERCISES

1. Show that there cannot be more than one number L that satisfies Definition 1.1.

2. (a) Prove: If $\int_a^b f(x)\,dx$ exists, then for every $\varepsilon > 0$ there is a $\delta > 0$ such that $|\sigma_1 - \sigma_2| < \varepsilon$ if σ_1 and σ_2 are Riemann sums of f over partitions P_1 and P_2 of $[a, b]$ with norms less than δ.
   (b) Use part (a) to complete the proof of Theorem 1.1.

* 3. Suppose $\int_a^b f(x)\,dx$ exists and there is a number A such that for every $\varepsilon > 0$ and $\delta > 0$ there is a partition P of $[a, b]$ with $\|P\| < \delta$ and a Riemann sum σ of f over P which satisfies the inequality $|\sigma - A| < \varepsilon$. Show that $\int_a^b f(x)\,dx = A$.

4. Prove directly from Definition 1.1 that

$$\int_a^b x^2\,dx = \frac{b^3 - a^3}{3}.$$

Do not assume in advance that the integral exists; the proof of this is part of the problem. [*Hint*: Use the mean value theorem to show that

$$\frac{b^3 - a^3}{3} = \sum_{j=1}^n d_j^2(x_j - x_{j-1}), \quad \text{where} \quad x_{j-1} < d_j < x_j.$$

Then relate this sum to arbitrary Riemann sums for $f(x) = x^2$ over the partition $P = \{x_0, x_1, \ldots, x_n\}$.]

5. Generalize the proof of Exercise 4 to show directly from Definition 1.1 that

$$\int_a^b x^m\,dx = \frac{b^{m+1} - a^{m+1}}{m + 1}$$

if m is an integer ≥ 0.

* 6. (a) Prove: The upper sum $S(P)$ is the least upper bound of the set of all Riemann sums of f over P.
   (b) State the corresponding characterization of $s(P)$.

7. (a) Prove: If f is continuous on $[a, b]$, then $s(P)$ and $S(P)$ are Riemann sums of f over P.
   (b) Name another class of functions for which the conclusion of part (a) is valid.
   (c) Give an example where $s(P)$ and $S(P)$ are not Riemann sums.

8. Find $\underline{\int_0^1} f(x)\,dx$ and $\overline{\int_0^1} f(x)\,dx$ if

   (a) $f(x) = \begin{cases} x & \text{if } x \text{ is rational,} \\ -x & \text{if } x \text{ is irrational.} \end{cases}$

(b) $f(x) = \begin{cases} 1 & \text{if } x \text{ is rational,} \\ x & \text{if } x \text{ is irrational.} \end{cases}$

9. Given that $\int_a^b e^x \, dx$ exists, evaluate it by using the formula

$$1 + r + r^2 + \cdots + r^n = \frac{1 - r^{n+1}}{1 - r} \qquad (r \neq 1)$$

to calculate certain Riemann sums. (*Hint:* See Exercise 3.)

10. Given that $\int_0^b \sin x \, dx$ exists, evaluate it by using the identity

$$\cos(j - 1)\theta - \cos(j + 1)\theta = 2 \sin \theta \sin j\theta$$

to calculate certain Riemann sums. (*Hint:* See Exercise 3.)

11. Given that $\int_0^b \cos x \, dx$ exists, evaluate it by using the identity

$$\sin(j + 1)\theta - \sin(j - 1)\theta = 2 \sin \theta \cos j\theta$$

to calculate certain Riemann sums. (*Hint:* See Exercise 3.)

12. Show that

$$\int_a^b f(x) \, dg(x) = \int_a^b f(x) \, dx$$

if $g(x) = x + c$ ($c = $ constant).

13. Suppose

$$g(x) = \begin{cases} g_1, & a < x < c, \\ g_2, & c < x < b, \end{cases} \qquad (g_1, g_2 = \text{constants})$$

and let $g(a)$, $g(b)$, and $g(c)$ be arbitrary. Show that if f is defined on $[a, b]$, continuous from the right at a and from the left at b, and continuous at c, then

$$\int_a^b f(x) \, dg(x) = f(a)[g_1 - g(a)] + f(c)(g_2 - g_1) + f(b)[g(b) - g_2].$$

14. Suppose $a = a_0 < a_1 < \cdots < a_n = b$, let $g(x) = g_j$ (constant) on (x_{j-1}, x_j) ($j = 1, 2, \ldots, n$), and let $g(a_0), g(a_1), \ldots, g(a_n)$ be arbitrary. Suppose f is defined on $[a, b]$ and continuous at $a_0, a_1, \ldots, a_n$. Evaluate $\int_a^b f(x) \, dg(x)$. (*Hint:* See Exercise 13.)

15. (a) Give an example where $\int_a^b f(x) \, dg(x)$ exists even though f is unbounded on $[a, b]$. (Thus, the analog of Theorem 1.1 does not hold for the Riemann–Stieltjes integral.)
    (b) State and prove an analog of Theorem 1.1 for the case where g is increasing.

16. For the case where g is nondecreasing and f is bounded on $[a, b]$, define upper and lower Riemann–Stieltjes integrals in a way analogous to Definition 1.2.

3.2   EXISTENCE OF THE INTEGRAL

The following lemma is the starting point for our study of existence of $\int_a^b f(x)\, dx$.

Lemma 2.1   *Suppose*

$$|f(x)| \le K, \qquad a \le x \le b,$$

and let P' be a partition of $[a, b]$ obtained by adding m points to a partition P. Then

$$S(P) - 2Km\|P\| \le S(P') \le S(P) \tag{1}$$

and

$$s(P) \le s(P') \le s(P) + 2Km\|P\|. \tag{2}$$

Proof.   We prove (1) by induction. Let $m = 1$, let c be the additional partition point, and suppose

$$x_{i-1} < c < x_i,$$

where $P = \{x_0, x_1, \ldots, x_n\}$. If $j \ne i$, the product $M_j(x_j - x_{j-1})$ appears in both $S(P)$ and $S(P')$, and cancels out of the difference $S(P) - S(P')$. Therefore, if

$$M_{i1} = \underset{x_{i-1} \le x \le c}{\text{l.u.b.}} f(x) \quad \text{and} \quad M_{i2} = \underset{c \le x \le x_i}{\text{l.u.b.}} f(x),$$

then

$$\begin{aligned}
S(P) - S(P') &= M_i(x_i - x_{i-1}) - M_{i1}(c - x_{i-1}) - M_{i2}(x_i - c) \\
&= (M_i - M_{i1})(c - x_{i-1}) + (M_i - M_{i2})(x_i - c).
\end{aligned} \tag{3}$$

Since

$$0 \le M_i - M_{ir} \le 2K, \qquad r = 1, 2,$$

(3) implies that

$$0 \le S(P) - S(P') \le 2K(x_i - x_{i-1}) \le 2K\|P\|.$$

This proves (1) for $m = 1$.

Now suppose $m \ge 2$ and let P_1 be a refinement of P obtained by deleting one point not in P from P'; thus, P_1 results from adding $m - 1$ partition points to P, and P' results from adding one point to P_1. If the proposition concerning upper sums is assumed to be true for $m - 1$, then

$$S(P) - 2K(m - 1)\|P\| \le S(P_1) \le S(P). \tag{4}$$

Since we have shown it to be true for $m = 1$,

$$S(P_1) - 2K\|P_1\| \le S(P') \le S(P_1). \tag{5}$$

Since $\|P_1\| \le \|P\|$, (4) and (5) imply (1). This completes the induction proof of (1).

The proof of (2) is similar; it can also be accomplished by applying (1) to the function $g = -f$ (*Exercise 1*).

Theorem 2.1   *If f is bounded on $[a, b]$, then*

$$\underline{\int_a^b} f(x)\, dx \le \overline{\int_a^b} f(x)\, dx. \tag{6}$$

Proof.   Suppose P_1 and P_2 are partitions of $[a, b]$ and P' is a refinement of both. Letting $P = P_1$ in (2) and $P = P_2$ in (1) yields

$$s(P_1) \le s(P') \quad \text{and} \quad S(P') \le S(P_2).$$

Since $s(P') \le S(P')$, this implies that

$$s(P_1) \le S(P_2).$$

Thus, every lower sum is a lower bound for the set of all upper sums; since $\overline{\int_a^b} f(x)\, dx$ is the greatest lower bound of this set, it follows that

$$s(P_1) \le \overline{\int_a^b} f(x)\, dx$$

for every partition P_1. This means that $\overline{\int_a^b} f(x)\, dx$ is an upper bound for the set of all lower sums; since $\underline{\int_a^b} f(x)\, dx$ is the least upper bound of this set, this implies (6).

Theorem 2.2   *If f is integrable on $[a, b]$, then*

$$\underline{\int_a^b} f(x)\, dx = \overline{\int_a^b} f(x)\, dx = \int_a^b f(x)\, dx.$$

Proof.   We prove the second equality and leave the first to the reader (*Exercise 2*). Suppose P is a partition of $[a, b]$ and σ is a Riemann sum of f over P; by the triangle inequality,

$$\left| \overline{\int_a^b} f(x)\, dx - \int_a^b f(x)\, dx \right| \le \left| \overline{\int_a^b} f(x)\, dx - S(P) \right| + |S(P) - \sigma|$$

$$+ \left| \sigma - \int_a^b f(x)\, dx \right|. \tag{7}$$

Now suppose $\varepsilon > 0$. From Definition 1.2 there is a partition P_0 of $[a, b]$ such that

$$\left| S(P_0) - \overline{\int_a^b} f(x)\, dx \right| < \frac{\varepsilon}{3}. \tag{8}$$

From Definition 1.1 there is a $\delta > 0$ such that

$$\left| \sigma - \int_a^b f(x)\, dx \right| < \frac{\varepsilon}{3} \tag{9}$$

if $\|P\| < \delta$. Now suppose $\|P\| < \delta$ and P is a refinement of P_0; then (8) and Lemma 2.1 imply that

$$\left| S(P) - \overline{\int_a^b} f(x)\, dx \right| < \frac{\varepsilon}{3} \tag{10}$$

in addition to (9). Now (7), (9), and (10) imply that

$$\left| \overline{\int_a^b} f(x)\, dx - \int_a^b f(x)\, dx \right| < \frac{2\varepsilon}{3} + |S(P) - \sigma| \tag{11}$$

for every Riemann sum σ of f over P. Since $S(P)$ is the least upper bound of these sums (*Exercise 6, Section 3.1*), we may choose σ so that

$$|S(P) - \sigma| < \frac{\varepsilon}{3},$$

and now (11) implies that

$$\left| \overline{\int_a^b} f(x)\, dx - \int_a^b f(x)\, dx \right| < \varepsilon.$$

Since ε is an arbitrary positive number, it follows that

$$\overline{\int_a^b} f(x)\, dx = \int_a^b f(x)\, dx.$$

Lemma 2.2 *If f is bounded on $[a, b]$, there is for each $\varepsilon > 0$ a $\delta > 0$ such that*

$$S(P) < \overline{\int_a^b} f(x)\, dx + \varepsilon \tag{12}$$

and

$$\underline{\int_a^b} f(x)\, dx - \varepsilon < s(P)$$

if $\|P\| < \delta$.

Proof. We show that (12) holds if $\|P\|$ is sufficiently small, and leave the rest of the proof to the reader (*Exercise 3*).

Suppose $\varepsilon > 0$. From Definition 1.2 there is a partition $P_0 = \{x_0, x_1, \ldots, x_{m+1}\}$ of $[a, b]$ such that

$$S(P_0) \leq \overline{\int_a^b} f(x)\, dx + \frac{\varepsilon}{2}. \tag{13}$$

If P is any partition of $[a, b]$, let P' be constructed from the partition points of P_0 and P. Lemma 2.1 implies that

$$S(P') \le S(P_0) \tag{14}$$

because P' is a refinement of P_0 and, if

$$|f(x)| \le K, \qquad a \le x \le b,$$

that

$$S(P) - 2Km\|P\| \le S(P'), \tag{15}$$

because P' is obtained by inserting at most m additional partition points into P. Now (13), (14), and (15) imply that

$$S(P) \le \int_a^b f(x)\, dx + \frac{\varepsilon}{2} + 2Km\|P\|$$

for every partition P; therefore, (12) holds if

$$\|P\| < \delta = \frac{\varepsilon}{4Km}.$$

Theorem 2.3   *If f is bounded on $[a, b]$ and*

$$\underline{\int_a^b} f(x)\, dx = \overline{\int_a^b} f(x)\, dx = L, \tag{16}$$

then f is integrable on $[a, b]$ and

$$\int_a^b f(x)\, dx = L. \tag{17}$$

Proof.   If $\varepsilon > 0$, there is a $\delta > 0$ such that

$$\underline{\int_a^b} f(x)\, dx - \varepsilon < s(P) \le S(P) < \overline{\int_a^b} f(x)\, dx + \varepsilon \tag{18}$$

if $\|P\| < \delta$ (Lemma 2.2). Since any Riemann sum σ of f over P satisfies

$$s(P) \le \sigma \le S(P),$$

(16) and (18) imply that

$$L - \varepsilon < \sigma < L + \varepsilon$$

if $\|P\| < \delta$. Now (17) follows from Definition 1.1.

Theorems 2.2 and 2.3 together imply that a bounded function f is integrable over $[a, b]$ if and only if

$$\underline{\int_a^b} f(x)\, dx = \overline{\int_a^b} f(x)\, dx.$$

The next theorem translates this result into a test that can be conveniently applied.

Theorem 2.4 *If f is bounded on* $[a, b]$*, then* $\int_a^b f(x)\, dx$ *exists if and only if for each* $\varepsilon > 0$ *there is a partition P of* $[a, b]$ *for which*

$$S(P) - s(P) < \varepsilon. \tag{19}$$

Proof. We leave it to the reader (*Exercise 4*) to show that (19) holds for $\|P\|$ sufficiently small if $\int_a^b f(x)\, dx$ exists. This implies that the given condition is necessary for integrability. For sufficiency we observe that since

$$s(P) \leq \underline{\int_a^b} f(x)\, dx \leq \overline{\int_a^b} f(x)\, dx \leq S(P)$$

for all P, the fact that (19) holds for some P implies that

$$0 \leq \overline{\int_a^b} f(x)\, dx - \underline{\int_a^b} f(x)\, dx < \varepsilon.$$

Since ε can be any positive number this means that

$$\overline{\int_a^b} f(x)\, dx = \underline{\int_a^b} f(x)\, dx,$$

and so $\int_a^b f(x)\, dx$ exists, by Theorem 2.3.

The next two theorems are important applications of Theorem 2.4.

Theorem 2.5 *If f is continuous on* $[a, b]$*, then f is integrable on* $[a, b]$*.*

Proof. Let $P = \{x_0, x_1, \ldots, x_n\}$ be a partition of $[a, b]$. Since f is continuous there are points c_j and c'_j in $[x_{j-1}, x_j]$ such that

$$f(c_j) = \underset{x_{j-1} \leq x \leq x_j}{\text{l.u.b.}} f(x) \quad \text{and} \quad f(c'_j) = \underset{x_{j-1} \leq x \leq x_j}{\text{g.l.b.}} f(x)$$

(Theorem 2.5, Section 2.2); therefore,

$$S(P) - s(P) = \sum_{j=1}^n [f(c_j) - f(c'_j)](x_j - x_{j-1}). \tag{20}$$

Since f is uniformly continuous on $[a, b]$ (Theorem 2.7, Section 2.2), there is for each $\varepsilon > 0$ a $\delta > 0$ such that

$$|f(x') - f(x)| < \varepsilon$$

if x and x' are in $[a, b]$ and $|x - x'| < \delta$. If $\|P\| < \delta$, then $|c_j - c'_j| < \delta$ and, from (20),

$$S(P) - s(P) < \varepsilon \sum_{j=1}^{n} (x_j - x_{j-1}) = \varepsilon(b - a).$$

Hence f is integrable on $[a, b]$, by Theorem 2.4.

Theorem 2.6   *If f is monotonic on $[a, b]$, then f is integrable on $[a, b]$.*

Proof.   Let f be nondecreasing. If $P = \{x_0, x_1, \ldots, x_n\}$ is a partition of $[a, b]$, then

$$\underset{x_{j-1} \leq x \leq x_j}{\text{l.u.b.}} \ f(x) = f(x_j) \quad \text{and} \quad \underset{x_{j-1} \leq x \leq x_j}{\text{g.l.b.}} \ f(x) = f(x_{j-1}).$$

Hence,

$$S(P) - s(P) = \sum_{j=1}^{n} [f(x_j) - f(x_{j-1})](x_j - x_{j-1}).$$

Since $0 < x_j - x_{j-1} \leq \|P\|$ and $f(x_j) - f(x_{j-1}) \geq 0$,

$$S(P) - s(P) \leq \|P\| \sum_{j=1}^{n} [f(x_j) - f(x_{j-1})]$$

$$= \|P\|[f(b) - f(a)].$$

Therefore,

$$S(P) - s(P) < \varepsilon$$

if

$$\|P\|[f(b) - f(a)] < \varepsilon,$$

and so f is integrable over $[a, b]$, by Theorem 2.4.

We will also use Theorem 2.4 in the next section to establish properties of the integral. In Section 3.5 we will study more general conditions for integrability.

3.2   EXERCISES

1. Complete the proof of Lemma 2.1.

2. Show that if f is integrable on $[a, b]$, then

$$\underline{\int_a^b} f(x)\,dx = \int_a^b f(x)\,dx.$$

3. Complete the proof of Lemma 2.2.

*4. Prove: If f is integrable over $[a, b]$ and $\varepsilon > 0$, then $S(P) - s(P) < \varepsilon$ if $\|P\|$ is sufficiently small. (*Hint:* Use Exercise 6, Section 3.1.)

5. Suppose f is integrable and g bounded on $[a, b]$, and g differs from f only at points in a set H with the following property: For each $\varepsilon > 0$, H can be covered by a finite number of closed subintervals of $[a, b]$, the sum of whose lengths is less than ε. Show that g is integrable on $[a, b]$ and

$$\int_a^b g(x)\, dx = \int_a^b f(x)\, dx.$$

(*Hint:* Use Exercise 3, Section 3.1.)

*6. A function f is *of bounded variation* on $[a, b]$ if there is a number K such that

$$\sum_{j=1}^n |f(a_j) - f(a_{j-1})| \leq K$$

whenever $a = a_0 < a_1 < \cdots < a_n = b$. (The smallest number with this property is the *total variation of f on* $[a, b]$.) Prove that if f is of bounded variation on $[a, b]$, then f is integrable on $[a, b]$. (*Hint:* Use Exercise 6, Section 3.1, and Theorem 2.4.)

7. Let $P = \{x_0, x_1, \ldots, x_n\}$ be a partition of $[a, b]$, $c_0 = x_0 = a$, $c_{n+1} = x_n = b$, and $x_{j-1} \leq c_j \leq x_j$ $(j = 1, 2, \ldots, n)$. Verify that

$$\sum_{j=1}^n g(c_j)[f(x_j) - f(x_{j-1})] = g(b)f(b) - g(a)f(a)$$

$$- \sum_{j=0}^n f(x_j)[g(c_{j+1}) - g(c_j)],$$

and use this to prove that if $\int_a^b f(x)\, dg(x)$ exists, then so does $\int_a^b g(x)\, df(x)$, and

$$\int_a^b g(x)\, df(x) = f(b)g(b) - f(a)g(a) - \int_a^b f(x)\, dg(x).$$

(This is the *integration-by-parts formula* for Riemann–Stieltjes integrals.)

8. Let f be continuous and g of bounded variation on $[a, b]$.
   (a) Show that if $\varepsilon > 0$, there is a $\delta > 0$ such that $|\sigma - \sigma'| < \varepsilon/2$ if σ and σ' are Riemann–Stieltjes sums of f with respect to g over partitions P and P' of $[a, b]$, where P' is a refinement of P and $\|P\| < \delta$. Conclude from this that $|\sigma - \sigma'| < \varepsilon$ if P and P' are any two partitions of $[a, b]$ with norm less than δ. [*Hint:* Use Theorem 2.7, Section 2.2.]
   (b) If $\rho > 0$, let $L(\rho)$ be the least upper bound of all Riemann–Stieltjes sums of f with respect to g over partitions of $[a, b]$ with norms less than ρ. Show that $L(\rho)$ is finite. Then show that $L = \lim_{\rho \to 0+} L(\rho)$ exists. [*Hint:* Use Theorem 1.4, Section 2.1.]
   (c) Show that $\int_a^b f(x)\, dg(x) = L$.

9. Show that $\int_a^b f(x)\, dg(x)$ exists if f is of bounded variation and g is continuous on $[a, b]$. (*Hint:* See Exercises 7 and 8.)

3.3   PROPERTIES OF THE INTEGRAL

We now use the results of Sections 3.1 and 3.2 to establish the properties of the integral. The reader is no doubt familiar with most of these properties, but perhaps not with their proofs.

Theorem 3.1   *If f and g are integrable on $[a, b]$, then so is $f + g$, and*

$$\int_a^b (f + g)(x)\, dx = \int_a^b f(x)\, dx + \int_a^b g(x)\, dx.$$

Proof.   Any Riemann sum of $f + g$ over a partition $P = \{x_0, x_1, \ldots, x_n\}$ of $[a, b]$ can be written as

$$\sigma_{f+g} = \sum_{j=1}^n [f(c_j) + g(c_j)](x_j - x_{j-1})$$

$$= \sum_{j=1}^n f(c_j)(x_j - x_{j-1}) + \sum_{j=1}^n g(c_j)(x_j - x_{j-1}) \tag{1}$$

$$= \sigma_f + \sigma_g,$$

where σ_f and σ_g are Riemann sums for f and g. Definition 1.1 implies that if $\varepsilon > 0$ there are positive numbers δ_1 and δ_2 such that

$$\left| \sigma_f - \int_a^b f(x)\, dx \right| < \frac{\varepsilon}{2} \quad \text{if} \quad ||P|| < \delta_1,$$

and

$$\left| \sigma_g - \int_a^b g(x)\, dx \right| < \frac{\varepsilon}{2} \quad \text{if} \quad ||P|| < \delta_2.$$

Therefore, (1) and the triangle inequality yield

$$\left| \sigma_{f+g} - \int_a^b f(x)\, dx - \int_a^b g(x)\, dx \right| < \varepsilon \quad \text{if} \quad ||P|| < \delta = \min(\delta_1, \delta_2),$$

and the conclusion follows from Definition 1.1.

The next theorem also follows directly from Definition 1.1 (*Exercise 1*).

Theorem 3.2   *If f is integrable on $[a, b]$ and c is a constant, then cf is integrable on $[a, b]$ and*

$$\int_a^b cf(x)\, dx = c \int_a^b f(x)\, dx.$$

Theorems 3.1 and 3.2 and induction yield the following result (*Exercise 2*).

Theorem 3.3   *If* $f_1, f_2, \ldots, f_n$ *are integrable on* $[a, b]$ *and* $c_1, c_2, \ldots, c_n$ *are constants, then* $c_1 f_1 + c_2 f_2 + \cdots + c_n f_n$ *is integrable on* $[a, b]$ *and*

$$\int_a^b (c_1 f_1 + c_2 f_2 + \cdots + c_n f_n)(x)\, dx = c_1 \int_a^b f_1(x)\, dx + c_2 \int_a^b f_2(x)\, dx$$

$$+ \cdots + c_n \int_a^b f_n(x)\, dx.$$

Lemma 3.1   *If* f *and* g *are integrable on* $[a, b]$ *and* $f(x) \le g(x)$ *for* $a \le x \le b$, *then*

$$\int_a^b f(x)\, dx \le \int_a^b g(x)\, dx. \tag{2}$$

Proof.   Since $g(x) - f(x) \ge 0$, every lower sum $g - f$ over any partition of $[a, b]$ is nonnegative, and therefore

$$\int_a^b [g(x) - f(x)]\, dx \ge 0.$$

Hence,

$$\int_a^b g(x)\, dx - \int_a^b f(x)\, dx = \int_a^b [g(x) - f(x)]\, dx$$

$$= \int_a^b [g(x) - f(x)]\, dx \ge 0, \tag{3}$$

which yields (2). [The first equality in (3) follows from Theorems 3.1 and 3.2; the second, from Theorem 2.2.]

Theorem 3.4   *If* f *is integrable on* $[a, b]$, *then so is* $|f|$, *and*

$$\left| \int_a^b f(x)\, dx \right| \le \int_a^b |f(x)|\, dx.$$

Proof.   Let P be a partition of $[a, b]$ and

$$M_j = \text{l.u.b.}\{f(x)|x_{j-1} \le x \le x_j\}, \quad m_j = \text{g.l.b.}\{f(x)|x_{j-1} \le x \le x_j\},$$
$$\bar{M}_j = \text{l.u.b.}\{|f(x)||x_{j-1} \le x \le x_j\}, \quad \bar{m}_j = \text{g.l.b.}\{|f(x)||x_{j-1} \le x \le x_{j-1}\}.$$

Then

$$\bar{M}_j - \bar{m}_j = \text{l.u.b.}\{|f(x)| - |f(x')||x, x' \in [x_{j-1}, x_j]\}$$
$$\le \text{l.u.b.}\{|f(x) - f(x')||x, x' \in [x_{j-1}, x_j]\}$$
$$= M_j - m_j.$$

Therefore

$$\overline{S}(P) - \overline{s}(P) \le S(P) - s(P),$$

where the upper and lower sums on the left are associated with $|f|$ and those on the right are associated with f. Since $\int_a^b f(x)\,dx$ exists, Theorem 2.4 implies that the right side of this inequality is less than any positive ε for suitable P, and consequently so is the left; therefore $\int_a^b |f(x)|\,dx$ exists, again by Theorem 2.4. The proof can be completed by using Lemma 3.1 (*Exercise 5*).

Theorem 3.5   *If f and g are integrable over $[a, b]$, then so is the product fg.*

Proof.   We consider the case where f and g are nonnegative, and leave the rest of the proof to the reader (*Exercise 3*). The subscripts f, g, and fg in the following identify the functions with which the various quantities are associated. We assume that neither f nor g vanishes identically, since the result is obviously true if one of them does.

If $P = \{x_0, x_1, \ldots, x_n\}$ is a partition of $[a, b]$, then

$$S_{fg}(P) - s_{fg}(P) = \sum_{j=1}^{n} (M_{fg,j} - m_{fg,j})(x_j - x_{j-1}). \tag{4}$$

Since f and g are nonnegative, $M_{fg,j} \le M_{f,j}M_{g,j}$ and $m_{fg,j} \ge m_{f,j}m_{g,j}$; hence,

$$\begin{aligned}
M_{fg,j} - m_{fg,j} &\le M_{f,j}M_{g,j} - m_{f,j}m_{g,j} \\
&= (M_{f,j} - m_{f,j})M_{g,j} + m_{f,j}(M_{g,j} - m_{g,j}) \\
&\le M_g(M_{f,j} - m_{f,j}) + M_f(M_{g,j} - m_{g,j}),
\end{aligned}$$

where M_f and M_g are upper bounds for f and g on $[a, b]$. From (4) and the last inequality,

$$S_{fg}(P) - s_{fg}(P) \le M_g[S_f(P) - s_f(P)] + M_f[S_g(P) - s_g(P)]. \tag{5}$$

Now suppose $\varepsilon > 0$. Theorem 2.4 implies that there are partitions P_1 and P_2 such that

$$S_f(P_1) - s_f(P_1) < \frac{\varepsilon}{2M_g} \quad \text{and} \quad S_g(P_2) - s_g(P_2) < \frac{\varepsilon}{2M_f}, \tag{6}$$

because f and g are integrable on $[a, b]$. From Lemma 2.1 the inequalities in (6) also hold for any partition P that is a refinement of P_1 and P_2; hence, (5) yields

$$S_{fg}(P) - s_{fg}(P) < \frac{\varepsilon}{2} + \frac{\varepsilon}{2} = \varepsilon$$

for any such partition, and so fg is integrable over $[a, b]$, by Theorem 2.4.

Theorem 3.6 (First Mean Value Theorem for Integrals) *Suppose u is contin-uous and v integrable and nonnegative on* $[a, b]$. *Then*

$$\int_a^b u(x)v(x)\,dx = u(c) \int_a^b v(x)\,dx \tag{7}$$

for some c in $[a, b]$.

Proof. Theorems 2.5 and 3.5 imply that the integral on the left exists. If $m = \min_{a \le x \le b} u(x)$ and $M = \max_{a \le x \le b} u(x)$, then

$$m \le u(x) \le M$$

and, since $v(x) \ge 0$,

$$mv(x) \le u(x)v(x) \le Mv(x).$$

Therefore Theorem 3.2 and Lemma 3.1 imply that

$$m \int_a^b v(x)\,dx \le \int_a^b u(x)v(x)\,dx \le M \int_a^b v(x)\,dx. \tag{8}$$

From this it is clear that (7) holds for any c in $[a, b]$ if $\int_a^b v(x)\,dx = 0$. If $\int_a^b v(x)\,dx \ne 0$, let

$$\bar{u} = \frac{\int_a^b u(x)v(x)\,dx}{\int_a^b v(x)\,dx}. \tag{9}$$

Since $\int_a^b v(x)\,dx > 0$ in this case (why?), (8) implies that $m \le \bar{u} \le M$, and the intermediate value theorem (Theorem 2.6, Section 2.2) implies that $f(c) = \bar{u}$ for some c in $[a, b]$. This implies (7) and completes the proof.

If $v(x) \equiv 1$, then (9) reduces to

$$\bar{u} = \frac{1}{b-a} \int_a^b u(x)\,dx,$$

so $\bar{u}$ is the average of $u(x)$ over $[a, b]$. More generally, if v is any nonnegative integrable function such that $\int_a^b v(x)\,dx \ne 0$, $\bar{u}$ in (9) is the *weighted average of* $u(x)$ *over* $[a, b]$ *with respect to* v. Theorem 3.6 says that a continuous function assumes any such weighted average at some point in $[a, b]$.

Theorem 3.7 *If f is integrable over* $[a, b]$ *and* $a \le a_1 < b_1 \le b$, *then f is integrable over* $[a_1, b_1]$.

Proof.   Suppose $\varepsilon > 0$. From Theorem 2.4 there is a partition $P = \{x_0, x_1, \ldots, x_n\}$ of $[a, b]$ such that

$$S(P) - s(P) = \sum_{j=1}^{n} (M_j - m_j)(x_j - x_{j-1}) < \varepsilon. \tag{10}$$

We may assume that a_1 and b_1 are partition points of P, because if not they could simply be inserted to obtain a refinement P' of P such that $S(P') - s(P') \le S(P) - s(P)$ (Lemma 2.1). Let $a_1 = x_r$ and $b_1 = x_s$. Since every term in (10) is nonnegative,

$$\sum_{j=r+1}^{s} (M_j - m_j)(x_j - x_{j-1}) < \varepsilon.$$

Thus, $\bar{P} = \{x_r, x_{r+1}, \ldots, x_s\}$ is a partition of $[a_1, b_1]$ over which the upper and lower sums of f satisfy

$$S(\bar{P}) - s(\bar{P}) < \varepsilon.$$

Therefore, f is integrable over $[a_1, b_1]$, by Theorem 2.4.

We leave the proof of the next theorem to the reader (*Exercise 10*).

Theorem 3.8   *If f is integrable over $[a, b]$ and $[b, c]$, then it is integrable over $[a, c]$, and*

$$\int_a^c f(x)\, dx = \int_a^b f(x)\, dx + \int_b^c f(x)\, dx. \tag{11}$$

So far we have defined $\int_\alpha^\beta f(x)\, dx$ only for the case where $\alpha < \beta$. Now we define

$$\int_\beta^\alpha f(x)\, dx = -\int_\alpha^\beta f(x)\, dx$$

if $\alpha < \beta$, and

$$\int_\alpha^\alpha f(x)\, dx = 0.$$

With these conventions, (11) holds no matter what the relative order of a, b, and c, provided f is integrable on some closed interval containing them (*Exercise 11*).

Theorem 3.7 and these definitions enable us to define a function F on $[a, b]$ by

$$F(x) = \int_c^x f(t)\, dt, \tag{12}$$

where c is an arbitrary—but fixed—point in $[a, b]$.

Theorem 3.9 *If f is integrable on $[a, b]$ and $a \le c \le b$, then F in (12) satisfies a Lipschitz condition, and is therefore continuous, on $[a, b]$.*

Proof. If x and x' are in $[a, b]$, then

$$F(x) - F(x') = \int_a^x f(t)\, dt - \int_a^{x'} f(t)\, dt = \int_{x'}^x f(t)\, dt,$$

by Theorem 3.8 and the conventions just adopted. Since $|f(t)| \le K (a \le t \le b)$ for some constant K,

$$\left| \int_{x'}^x f(t)\, dt \right| \le K |x - x'|, \qquad a \le x, x' \le b$$

(Theorem 3.4), and so

$$|F(x) - F(x')| \le K |x - x'|, \qquad a \le x, x' \le b,$$

which completes the proof.

Theorem 3.10 *If f is integrable on $[a, b]$ and $a \le c \le b$, then the function $F(x) = \int_c^x f(t)\, dt$ is differentiable at any point in $[a, b]$ where f is continuous, with $F'(x_0) = f(x_0)$. If f is continuous from the right at a, then $F'_+(a) = f(a)$, and $F'_-(b) = f(b)$ if f is continuous from the left at b.*

Proof. We consider the case where $a < x_0 < b$ and leave the rest to the reader (*Exercise 13*). Noting that

$$\frac{1}{x - x_0} \int_{x_0}^x f(x_0)\, dt = f(x_0),$$

we write

$$\frac{F(x) - F(x_0)}{x - x_0} - f(x_0) = \frac{1}{x - x_0} \int_{x_0}^x [f(t) - f(x_0)]\, dt$$

and, from Theorem 3.4,

$$\left| \frac{F(x) - F(x_0)}{x - x_0} - f(x_0) \right| \le \frac{1}{|x - x_0|} \left| \int_{x_0}^x |f(t) - f(x_0)|\, dt \right|. \tag{13}$$

(Why do we need the absolute value bars outside the integral?) Since f is continuous at x_0, there is for each $\varepsilon > 0$ a $\delta > 0$ such that

$$|f(t) - f(x_0)| < \varepsilon \quad \text{if} \quad |x - x_0| < \delta.$$

and so, from (13),

$$\left| \frac{F(x) - F(x_0)}{x - x_0} - f(x_0) \right| < \varepsilon \frac{|x - x_0|}{|x - x_0|} = \varepsilon \quad \text{if} \quad 0 < |x - x_0| < \delta.$$

This proves that $F'(x_0) = f(x_0)$.

Example 3.1   If

$$f(x) = \begin{cases} x, & 0 \le x \le 1, \\ x + 1, & 1 < x \le 2, \end{cases}$$

then the function

$$F(x) = \int_0^x f(t) \, dt = \begin{cases} \dfrac{x^2}{2}, & 0 < x \le 1, \\ \dfrac{x^2}{2} + x - 1, & 1 < x \le 2, \end{cases}$$

is continuous on $[0, 2]$. As implied by Theorem 3.10,

$$F'(x) = \begin{cases} x = f(x), & 0 < x < 1, \\ x + 1 = f(x), & 1 < x < 2, \end{cases}$$

$$F'_+(0) = \lim_{x \to 0+} \frac{F(x) - F(0)}{x} = \lim_{x \to 0+} \frac{(x^2/2) - 0}{x} = 0 = f(0),$$

$$F'_-(2) = \lim_{x \to 2-} \frac{F(x) - F(2)}{x - 2} = \lim_{x \to 2-} \frac{(x^2/2) + x - 1 - 3}{x - 2}$$

$$= \lim_{x \to 2-} \frac{x + 4}{2} = 3 = f(2).$$

The function F does not have a derivative at 1, since

$$F'_-(1) = 1 \quad \text{and} \quad F'_+(1) = 2.$$

The next theorem relates integration and differentiation in another way.

Theorem 3.11   *Suppose that F is continuous on $[a, b]$ and differentiable on (a, b), and f is integrable on $[a, b]$. Suppose also that*

$$F'(x) = f(x), \qquad a < x < b.$$

Then

$$\int_a^b f(x) \, dx = F(b) - F(a). \tag{14}$$

Proof. If $P = \{x_0, x_1, \ldots, x_n\}$ is a partition of $[a, b]$, then

$$F(b) - F(a) = \sum_{j=1}^{n} [F(x_j) - F(x_{j-1})]. \tag{15}$$

From the mean value theorem there is in each interval (x_{j-1}, x_j) a point c_j such that

$$F(x_j) - F(x_{j-1}) = f(c_j)(x_j - x_{j-1});$$

hence, (15) can be written as

$$F(b) - F(a) = \sum_{j=1}^{n} f(c_j)(x_j - x_{j-1}) = \sigma,$$

where σ is a Riemann sum for f over P. Since f is integrable over $[a, b]$, there is for each $\varepsilon > 0$ a $\delta > 0$ such that

$$\left| \sigma - \int_a^b f(x)\, dx \right| < \varepsilon \text{ if } \ \|P\| < \delta.$$

Therefore,

$$\left| F(b) - F(a) - \int_a^b f(x)\, dx \right| < \varepsilon$$

for every $\varepsilon > 0$, and so (14) follows.

A function F is an *antiderivative* (also, an *indefinite integral* or *primitive*) of f on $[a, b]$ if F is continuous on $[a, b]$ and differentiable on (a, b), with

$$F'(x) = f(x), \qquad a < x < b.$$

If F is an antiderivative of f on $[a, b]$ then so is $F + c$ for any constant c; conversely, if F_1 and F_2 are antiderivatives of f on $[a, b]$, then $F_1 - F_2$ is constant on $[a, b]$ (Theorem 3.9, Section 2.3). When we wish to emphasize the difference between an indefinite integral of f and the integral $\int_a^b f(x)\, dx$, we refer to the latter as a *definite* integral. (An indefinite integral is a function, while a definite integral is a number.) Theorem 3.11 shows that indefinite integrals can be used to evaluate definite integrals.

We leave the proof of the following theorem to the reader (*Exercise 15*).

Theorem 3.12 **(Fundamental Theorem of the Calculus)** *If f is continuous on $[a, b]$, then it has an antiderivative F on $[a, b]$, and*

$$\int_a^b f(x)\, dx = F(b) - F(a).$$

In applying this theorem we will use the familiar notation

$$F(b) - F(a) = F(x)\Big|_a^b.$$

The following theorem on change of variable is especially useful for evaluating definite integrals.

Theorem 3.13   *Suppose the transformation* $x = \phi(t)$ *maps the interval* $c \le t \le d$ *into the interval* $a \le x \le b$, *with* $\phi(c) = \alpha$ *and* $\phi(d) = \beta$, *and let* f *be continuous on* $[a, b]$. *Let* ϕ' *be integrable over* $[c, d]$. *Then*

$$\int_\alpha^\beta f(x)\, dx = \int_c^d f(\phi(t))\phi'(t)\, dt. \tag{16}$$

Proof.   Both integrals in (16) exist: the one on the left by Theorem 2.5; the one on the right by Theorems 2.5 and 3.5, and the continuity of $f(\phi(t))$. The function

$$F(x) = \int_a^x f(y)\, dy$$

is an antiderivative of f on $[a, b]$ (Theorem 3.10) and, therefore, also on the closed interval with end points α and β. Hence, by Theorem 3.11,

$$\int_\alpha^\beta f(x)\, dx = F(\beta) - F(\alpha). \tag{17}$$

The function

$$G(t) = F(\phi(t))$$

is an antiderivative of $f(\phi(t))\phi'(t)$ on $[c, d]$, and Theorem 3.11 implies that

$$\int_c^d f(\phi(t))\phi'(t)\, dt = G(d) - G(c) = F(\phi(d)) - F(\phi(c))$$
$$= F(\beta) - F(\alpha).$$

Comparing this with (17) yields (16).

Example 3.2   To evaluate the integral

$$I = \int_{-1/\sqrt{2}}^{1/\sqrt{2}} (1 - 2x^2)(1 - x^2)^{-1/2}\, dx,$$

we let

$$f(x) = (1 - 2x^2)(1 - x^2)^{-1/2}, \qquad -1/\sqrt{2} \le x \le 1/\sqrt{2},$$

and

$$\phi(t) = \sin t, \qquad -\pi/4 \le t \le \pi/4.$$

Then $\phi'(t) = \cos t$ and

$$I = \int_{-1/\sqrt{2}}^{1/\sqrt{2}} f(x)\, dx = \int_{-\pi/4}^{\pi/4} f(\sin t)\cos t\, dt$$
$$= \int_{-\pi/4}^{\pi/4} (1 - 2\sin^2 t)(1 - \sin^2 t)^{-1/2} \cos t\, dt.$$

Since

$$(1 - \sin^2 t)^{1/2} = \cos t, \qquad -\pi/4 \le t \le \pi/4,$$

and

$$1 - 2 \sin^2 t = \cos 2t,$$

this yields

$$I = \int_{-\pi/4}^{\pi/4} \cos 2t \, dt = \frac{\sin 2t}{2} \Big|_{-\pi/4}^{\pi/4} = 1.$$

Example 3.3   To evaluate the integral

$$I = \int_0^{5\pi} \frac{\sin t}{2 + \cos t} \, dt,$$

we take $\phi(t) = \cos t$; then $\phi'(t) = -\sin t$ and

$$I = -\int_0^{5\pi} \frac{\phi'(t)}{2 + \phi(t)} \, dt = -\int_0^{5\pi} f(\phi(t))\phi'(t) \, dt,$$

where

$$f(x) = \frac{1}{2 + x};$$

therefore, since $\phi(0) = 1$ and $\phi(5\pi) = -1$,

$$I = -\int_1^{-1} \frac{dx}{2 + x} = -\log(2 + x) \Big|_1^{-1} = \log 3.$$

These examples illustrate two ways to use Theorem 3.13. In Example 3.2 we evaluated the left side of (16) by transforming it to the right side with a suitable substitution $x = \phi(t)$, while in Example 3.3 we evaluated the right side of (16) by recognizing that it could be obtained from the left side by a suitable substitution.

The following theorem shows that the rule for change of variable remains valid under weaker assumptions on f if ϕ is monotonic.

Theorem 3.14   *Suppose ϕ' is integrable and ϕ monotonic on $[c, d]$, and the transformation $x = \phi(t)$ maps $[c, d]$ onto $[a, b]$. Let f be bounded on $[a, b]$. Then the function*

$$g(t) = f(\phi(t))\phi'(t)$$

is integrable over $[c, d]$ if and only if f is integrable over $[a, b]$, and in this case

$$\int_a^b f(x) \, dx = \int_c^d f(\phi(t))|\phi'(t)| \, dt.$$

Proof.   We sketch the proof to show how the monotonicity of ϕ is used, and leave some of the details to the reader (*Exercises 19–22*).

First assume that ϕ is increasing and let $\bar{P} = \{t_0, t_1, \ldots, t_n\}$ be a partition of $[c, d]$ and $P = \{x_0, x_1, \ldots, x_n\}$, with $x_j = \phi(t_j)$, be the corresponding partition of $[a, b]$. Define

$$U_j = \operatorname*{l.u.b.}_{t_{j-1} \leq t \leq t_j} \phi'(t), \qquad u_j = \operatorname*{g.l.b.}_{t_{j-1} \leq t \leq t_j} \phi'(t),$$

$$\bar{M}_j = \operatorname*{l.u.b.}_{t_{j-1} \leq t \leq t_j} f(\phi(t))\phi'(t), \qquad M_j = \operatorname*{l.u.b.}_{x_{j-1} \leq x \leq x_j} f(x).$$

We show first that

$$\int_a^{\bar{b}} f(x) \, dx = \int_c^{\bar{d}} f(\phi(t))\phi'(t) \, dt. \tag{18}$$

To this end, consider the upper sums

$$\bar{S}(\bar{P}) = \sum_{j=1}^n \bar{M}_j(t_j - t_{j-1}) \quad \text{and} \quad S(P) = \sum_{j=1}^n M_j(x_j - x_{j-1}). \tag{19}$$

Since ϕ is increasing,

$$0 \leq u_j \leq \phi'(t) \leq U_j, \qquad t_{j-1} \leq t \leq t_j,$$

and this implies that

$$\bar{M}_j = M_j \rho_j, \tag{20}$$

where

$$u_j \leq \rho_j \leq U_j \tag{21}$$

(*Exercise 19*).

From the mean value theorem,

$$x_j - x_{j-1} = \phi(t_j) - \phi(t_{j-1}) = \phi'(\tau_j)(t_j - t_{j-1}), \tag{22}$$

where $t_{j-1} < \tau_j < t_j$, and so

$$u_j \leq \phi'(\tau_j) \leq U_j. \tag{23}$$

From (19), (20), and (22),

$$\bar{S}(\bar{P}) - S(P) = \sum_{j=1}^n M_j[\rho_j - \phi'(\tau_j)](t_j - t_{j-1}). \tag{24}$$

Now suppose $|f(x)| \leq M(a \leq x \leq b)$; then (21), (23), and (24) imply that

$$|\bar{S}(\bar{P}) - S(P)| \leq M \sum_{j=1}^n (U_j - u_j)(t_j - t_{j-1}).$$

The sum on the right is the difference between the upper and lower sums of ϕ' over $\bar{P}$; since ϕ' is integrable over $[c, d]$, this can be made as small as we please by choosing $\|\bar{P}\|$ sufficiently small (*Exercise 4, Section 3.2*). This implies (18) (*Exercise 20*).

If ϕ is nondecreasing (rather than increasing), then it may happen that $x_{j-1} = x_j$ for some values of j; however, this is no real complication, since it simply means that some terms in $S(P)$ vanish.

By applying (18) to $-f$ we infer that

$$\int_a^b f(x)\,dx = \int_c^d f(\phi(t))\phi'(t)\,dt \qquad (25)$$

(*Exercise 21*), and the conclusion of the theorem follows from this and (18) (*Exercise 22*).

The proof of Theorem 3.14 for the case where ϕ is nonincreasing is sketched in *Exercise 23*.

Integration by parts

Theorem 3.15 (Integration by Parts)   *If u' and v' are integrable on $[a, b]$, then*

$$\int_a^b u(x)v'(x)\,dx = u(x)v(x)\Big|_a^b - \int_a^b v(x)u'(x)\,dx. \qquad (26)$$

Proof.   Since u and v are continuous on $[a, b]$, they are integrable over $[a, b]$. Therefore, Theorems 3.1 and 3.5 imply that the function

$$(uv)' = u'v + uv'$$

is integrable over $[a, b]$, and Theorem 3.11 implies that

$$\int_a^b [u(x)v'(x) + u'(x)v(x)]\,dx = u(x)v(x)\Big|_a^b,$$

which yields (26).

The student of advanced calculus must know that integration by parts is useful for the practical task of evaluating integrals; we will use it now and in the next section to obtain theoretical results.

Theorem 3.16 (Second Mean Value Theorem for Integrals)   *Suppose f' is nonnegative and integrable and g is continuous on $[a, b]$. Then*

$$\int_a^b f(x)g(x)\,dx = f(a)\int_a^c g(x)\,dx + f(b)\int_c^b g(x)\,dx \qquad (27)$$

for some c in $[a, b]$.

Proof.   From Theorems 2.5 and 3.5, the integrals in (27) exist. If

$$G(x) = \int_a^x g(t)\, dx, \tag{28}$$

then $G'(x) = g(x)$, $a < x < b$ (Theorem 3.10); therefore, Theorem 3.15, with $u = f$ and $v = G$, yields

$$\int_a^b f(x)g(x)\, dx = f(x)G(x)\Big|_a^b - \int_a^b f'(x)G(x)\, dx. \tag{29}$$

Since f' is nonnegative and G is continuous, the first mean value theorem for integrals (Theorem 3.6) implies that

$$\int_a^b f'(x)G(x)\, dx = G(c) \int_a^b f'(x)\, dx \tag{30}$$

for some c in $[a, b]$. From Theorem 3.11,

$$\int_a^b f'(x)\, dx = f(b) - f(a).$$

From this and (28), (30) can be rewritten as

$$\int_a^b f'(x)G(x)\, dx = [f(b) - f(a)] \int_a^c g(x)\, dx.$$

Substituting this in (29) and noting that $G(a) = 0$ yields

$$\int_a^b f(x)g(x)\, dx = f(b) \int_a^b g(x)\, dx - [f(b) - f(a)] \int_a^c g(x)\, dx,$$

and now Theorem 3.8 implies (27).

3.3   EXERCISES

1. Prove Theorem 3.2.

2. Prove Theorem 3.3.

3. Complete the proof of Theorem 3.5.

4. Prove: If f is integrable over $[a, b]$ and $|f(x)| \geq \rho > 0$ for $a \leq x \leq b$, then $1/f$ is integrable over $[a, b]$.

5. (a) Show that if f is integrable on $[a, b]$, then

$$\left| \int_a^b f(x)\, dx \right| \leq \int_a^b |f(x)|\, dx.$$

(b) Can $|f|$ be integrable on $[a, b]$ if f is not?

6. Suppose f is integrable on $[a, b]$ and define

$$f^+(x) = \begin{cases} f(x) & \text{if } f(x) \geq 0 \\ 0 & \text{if } f(x) < 0 \end{cases} \quad \text{and} \quad f^-(x) = \begin{cases} 0 & \text{if } f(x) \geq 0 \\ f(x) & \text{if } f(x) < 0. \end{cases}$$

Show that f^+ and f^- are integrable over $[a, b]$, and

$$\int_a^b f(x)\, dx = \int_a^b f^+(x)\, dx + \int_a^b f^-(x)\, dx.$$

7. Find the weighted average $\bar{u}$ of $u(x)$ over $[a, b]$ with respect to v, and find a point c in $[a, b]$ such that $u(c) = \bar{u}$.
   (a) $u(x) = x$, $v(x) = x$, $[a, b] = [0, 1]$
   (b) $u(x) = \sin x$, $v(x) = x^2$, $[a, b] = [-1, 1]$
   (c) $u(x) = x^2$, $v(x) = e^x$, $[a, b] = [0, 1]$

8. Suppose f is continuous on $[a, b]$ and $P = \{x_0, x_1, \ldots, x_n\}$ is a partition of $[a, b]$. Show that there is a Riemann sum of f over P that equals $\int_a^b f(x)\, dx$.

9. Suppose f' exists and $|f'(x)| \le M$ on $[a, b]$. Show that any Riemann sum σ of f over any partition P of $[a, b]$ satisfies

$$\left| \sigma - \int_a^b f(x)\, dx \right| \le M(b - a)\|P\|.$$

(*Hint:* See Exercise 8).

10. Prove Theorem 3.8.

11. Show that

$$\int_a^c f(x)\, dx = \int_a^b f(x)\, dx + \int_b^c f(x)\, dx$$

for all possible relative orderings of a, b, and c, provided f is integrable on a closed interval containing them.

12. (a) Prove: If f is integrable and $f(x) > 0$ on $[a, b]$, then $\int_a^b f(x)\, dx > 0$.
    (b) Prove: If f is integrable and $f(x) \ge 0$ on $[a, b]$, then $\int_a^b f(x)\, dx \ge 0$, with strict inequality if f is continuous and positive at some point in $[a, b]$.

13. Complete the proof of Theorem 3.10.

14. State theorems analogous to Theorems 3.9 and 3.10 for the function

$$G(x) = \int_x^b f(t)\, dt$$

and show how your theorems can be obtained from them.

15. Prove Theorem 3.12.

16. Show that if f is differentiable and f' integrable on $[a, b]$, then

$$\int_a^b f'(x)\, dx = f(b) - f(a).$$

17. The symbol $\int f(x)\, dx$ denotes an indefinite integral (antiderivative) of f. A plausible analog of Theorem 3.1 would state that if f and g have indefinite

integrals on $[a, b]$, then so does $f + g$—which is true—and

$$\int (f + g)(x)\, dx = \int f(x)\, dx + \int g(x)\, dx. \tag{A}$$

However, this is not true in the usual sense.
(a) Why not?
(b) State a correct interpretation of (A).

18. (See Exercise 17.) Formulate a valid interpretation of the relation

$$\int (cf)(x)\, dx = c \int f(x)\, dx \qquad (c \neq 0).$$

Is your interpretation valid if $c = 0$?

Exercises 19–23 refer to the proof of Theorem 3.14.

19. Verify Eqs. (20) and (21). (*Hint:* Consider separately the cases where $M_j > 0$ and $M_j \leq 0$.)

20. Show that Eq. (18) follows from the fact that $|\bar{S}(\bar{P}) - S(P)|$ can be made as small as we please by choosing $\|\bar{P}\|$ sufficiently small. (*Hint:* Use Lemma 2.2.)

21. Show that Eq. (25) follows from Eq. (18).

22. Show that the conclusions of Theorem 3.14 follow from Eqs. (18) and (25).

23. (a) Show directly from Definition 1.1 that $h(x)$ is integrable on $[a, b]$ if and only if $h(-x)$ is integrable on $[-b, -a]$, and that in this case

$$\int_a^b h(x)\, dx = \int_{-b}^{-a} h(-x)\, dx.$$

(b) Use part (a) and the known validity of Theorem 3.14 for nondecreasing ϕ to show that Theorem 3.14 holds for nonincreasing ϕ.

24. (a) Let $f^{(n+1)}$ be integrable on $[a, b]$. Show that

$$f(b) = \sum_{r=0}^{n} \frac{f^{(r)}(a)}{r!}(b - a)^r + \frac{1}{n!} \int_a^b f^{(n+1)}(t)(b - t)^n\, dt.$$

(*Hint:* Integrate by parts and use induction.)
(b) Do you see a connection between part (a) and Theorem 5.4, Section 2.5?

25. In addition to the assumptions of Theorem 3.16, suppose $f(a) = 0$, $f \not\equiv 0$, and $g(x) > 0$ $(a < x < b)$. Show that there is only one point c in $[a, b]$ with the property stated in Theorem 3.16. (*Hint:* Use Exercise 12(b).)

26. Suppose g' is integrable and f continuous on $[a, b]$. Show that $\int_a^b f(x)\, dg(x)$ exists and equals $\int_a^b f(x)g'(x)\, dx$.

27. Suppose f and g'' are bounded and fg' is integrable on $[a, b]$. Show that $\int_a^b f(x)\,dg(x)$ exists and equals $\int_a^b f(x)g'(x)\,dx$. (*Hint:* Use Theorem 5.3, Section 2.5.)

3.4   IMPROPER INTEGRALS

So far we have confined our study of the integral to bounded functions on finite closed intervals. This was for good reason: Theorem 1.1 showed that $\int_a^b f(x)\,dx$ cannot exist in the sense of Definition 1.1 if f is unbounded on $[a, b]$, and attempting to extend Definition 1.1 directly to an infinite or semiinfinite interval would introduce questions concerning convergence of the resulting Riemann sums, which would then be infinite series. In this section we extend the definition of integral to include these cases.

We say that f is *locally integrable* on an interval I if f is integrable on every finite closed subinterval of I. (See *Exercise 34* for an alternative definition.) For example, the function

$$f(x) = \sin x$$

is locally integrable on $(-\infty, \infty)$; the function

$$g(x) = \frac{1}{x(x-1)}$$

is locally integrable on $(-\infty, 0)$, $(0, 1)$, and $(1, \infty)$; and the function

$$h(x) = \sqrt{x}$$

is locally integrable on $[0, \infty)$.

Definition 4.1   *If f is locally integrable on $[a, b)$, we define*

$$\int_a^b f(x)\,dx = \lim_{c \to b-} \int_a^c f(x)\,dx \tag{1}$$

if the limit exists (finite).

The limit in (1) always exists if $[a, b)$ is finite and f is locally integrable and *bounded* on $[a, b)$, and, no matter how $f(b)$ is defined, Definitions 1.1 and 4.1 assign the same value to $\int_a^b f(x)\,dx$ in this case (*Exercise 1*). However, the limit may also exist in cases where $b = \infty$ or $b < \infty$ and f is unbounded on $[a, b)$. In these cases Definition 4.1 assigns a value to an integral which does not exist in the sense of Definition 1.1, and $\int_a^b f(x)\,dx$ is said to be an *improper integral* which *converges* to the limit in (1); we also say in this case that f is *integrable on* $[a, b]$. If the limit in (1) does not exist (finite) we say that the improper integral $\int_a^b f(x)\,dx$ *diverges*, and f is *nonintegrable on* $[a, b]$. When we wish to distinguish between improper integrals and integrals in the sense of Definition 1.1, we will call the latter *proper integrals*.

Similar comments apply to the next two definitions.

Definition 4.2   *If f is locally integrable on $(a, b]$, we define*

$$\int_a^b f(x)\,dx = \lim_{c \to a+} \int_c^b f(x)\,dx$$

provided the limit exists (finite).

Definition 4.3   *If f is locally integrable on (a, b), we define*

$$\int_a^b f(x)\,dx = \int_a^\alpha f(x)\,dx + \int_\alpha^b f(x)\,dx$$

where $a < \alpha < b$, provided the improper integrals on the right exist (finite).

The existence and value of $\int_a^b f(x)\,dx$ according to this definition do not depend on the particular choice of α in (a, b) (*Exercise 2*).

In stating and proving theorems on improper integrals, we will consider integrals of the kind introduced in Definition 4.1. Similar results apply to the integrals of Definitions 4.2 and 4.3. We leave it to the reader to formulate and use them in the examples and exercises as the need arises.

Example 4.1   The function

$$f(x) = 2x \sin \frac{1}{x} - \cos \frac{1}{x}$$

is locally integrable and the derivative of

$$F(x) = x^2 \sin \frac{1}{x}$$

on $[-2/\pi, 0)$. Hence,

$$\int_{-2/\pi}^c f(x)\,dx = x^2 \sin \frac{1}{x}\Big|_{-2/\pi}^c = c^2 \sin \frac{1}{c} + \frac{4}{\pi^2}$$

and

$$\int_{-2/\pi}^0 f(x)\,dx = \lim_{c \to 0-} \left(c^2 \sin \frac{1}{c} + \frac{4}{\pi^2} \right) = \frac{4}{\pi^2}$$

according to Definition 4.1; however, this is not an improper integral, even though $f(0)$ is not defined and cannot be defined so as to make f continuous at 0. If we define $f(0)$ arbitrarily [say $f(0) = 10$], then f is bounded on the closed interval $[-2/\pi, 0]$ and continuous except at 0. Therefore, $\int_{-2/\pi}^{0} f(x)\,dx$ exists and equals $4/\pi^2$ as a proper integral (*Exercise 1*) in the sense of Definition 1.1.

Example 4.2 The function

$$f(x) = (1 - x)^{-p}$$

is locally integrable on $[0, 1)$ and, if $p \neq 1$ and $0 < c < 1$,

$$\int_0^c (1 - x)^{-p}\,dx = \frac{(1-x)^{-p+1}}{p-1}\bigg|_0^c = \frac{(1-c)^{-p+1} - 1}{p-1}.$$

Hence,

$$\lim_{c \to 1-} \int_0^c (1-x)^{-p}\,dx = \begin{cases} \infty, & p > 1, \\ (1-p)^{-1}, & p < 1. \end{cases}$$

For $p = 1$,

$$\lim_{c \to 1-} \int_0^c (1-x)^{-1}\,dx = -\lim_{c \to 1-} \log(1-c) = \infty.$$

Therefore, $\int_0^1 (1-x)^{-p}\,dx$ converges to $(1-p)^{-1}$ if $p < 1$, and we write

$$\int_0^1 (1-x)^{-p}\,dx = (1-p)^{-1}, \qquad p < 1.$$

If $p \geq 1$, $\int_0^1 (1-x)^{-p}\,dx$ diverges to ∞.

Example 4.3 The function

$$f(x) = x^{-p}$$

is locally integrable on $[1, \infty)$ and, if $p \neq 1$ and $c > 1$,

$$\int_1^c x^{-p}\,dx = \frac{x^{-p+1}}{-p+1}\bigg|_1^c = \frac{c^{-p+1} - 1}{-p+1}.$$

Hence,

$$\lim_{c \to \infty} \int_1^c x^{-p}\,dx = \begin{cases} (p-1)^{-1}, & p > 1, \\ \infty, & p < 1. \end{cases}$$

For $p = 1$,

$$\lim_{c \to \infty} \int_1^c x^{-1}\,dx = \lim_{c \to \infty} \log c = \infty.$$

Therefore, $\int_1^\infty x^{-p}\,dx$ converges to $(p-1)^{-1}$ if $p > 1$, and we write

$$\int_1^\infty x^{-p}\,dx = (p-1)^{-1}$$

in this case. If $p \le 1$, $\int_1^\infty x^{-p}\,dx$ diverges to ∞.

Example 4.4   If $1 < c < \infty$, then

$$\int_1^c \frac{1}{x} \log \frac{1}{x}\,dx = -\int_1^c \frac{1}{x} \log x\,dx = -\frac{1}{2}(\log x)^2 \bigg|_1^c = -\frac{1}{2}(\log c)^2.$$

Hence,

$$\lim_{c\to\infty} \int_1^c \frac{1}{x} \log \frac{1}{x}\,dx = -\infty.$$

Thus, $\int_1^\infty 1/x \log(1/x)\,dx$ diverges to $-\infty$.

Example 4.5   The function $f(x) = \cos x$ is locally integrable on $[0, \infty)$ and

$$\lim_{c\to\infty} \int_0^c \cos x\,dx = \lim_{c\to\infty} \sin c$$

does not exist; thus, $\int_0^\infty \cos x\,dx$ diverges.

Example 4.6   The function $f(x) = \log x$ is locally integrable on $(0, 1]$, but unbounded as $x \to 0+$. Since

$$\lim_{c\to 0+} \int_c^1 \log x\,dx = \lim_{c\to 0+} (x \log x - x) \bigg|_c^1 = -1 - \lim_{c\to 0+} (c \log c - c) = -1,$$

Definition 4.2 yields

$$\int_0^1 \log x\,dx = -1.$$

Example 4.7   In connection with Definition 4.3 it is important to recognize that the improper integrals $\int_a^z f(x)\,dx$ and $\int_a^b f(x)\,dx$ must converge *separately* for $\int_a^b f(x)\,dx$ to converge. For example, the existence of the symmetric limit

$$\lim_{R\to\infty} \int_{-R}^R f(x)\,dx$$

[which is called the *principal value* of $\int_{-\infty}^\infty f(x)\,dx$] does not imply that $\int_{-\infty}^\infty f(x)\,dx$ converges; thus,

$$\lim_{R\to\infty} \int_{-R}^R x\,dx = \lim_{R\to\infty} 0 = 0,$$

but $\int_0^\infty x\,dx$ and $\int_{-\infty}^0 x\,dx$ diverge and therefore so does $\int_{-\infty}^\infty x\,dx$.

If $\int_a^b f(x)\,dx$ diverges to ∞ or $-\infty$, we usually write

$$\int_a^b f(x)\,dx = \infty \quad \text{or} \quad \int_a^b f(x)\,dx = -\infty.$$

Thus, from Examples 4.2, 4.3, and 4.4,

$$\int_0^1 (1-x)^{-p}\, dx = \infty, \qquad p \geq 1,$$

$$\int_1^\infty x^{-p}\, dx = \infty, \qquad p \leq 1$$

and

$$\int_1^\infty \frac{1}{x} \log \frac{1}{x}\, dx = -\infty.$$

There is no such convenient notation for $\int_0^\infty \cos x\, dx$, which diverges, but not to $\pm\infty$ (Example 4.5).

Improper integrals of nonnegative functions

The theory of improper integrals with nonnegative integrands is particularly simple.

Theorem 4.1 *If f is nonnegative and locally integrable on $[a, b)$, then $\int_a^b f(x)\, dx$ converges if the function*

$$F(x) = \int_a^x f(t)\, dt$$

is bounded on $[a, b)$ and $\int_a^b f(x)\, dx = \infty$ if it is not. These are the only possibilities, and in either case

$$\int_a^b f(t)\, dt = \underset{a \leq x < b}{\text{l.u.b. }} F(x).$$

Proof. *Exercise 9.*

We often write

$$\int_a^b f(x)\, dx < \infty$$

to indicate that an improper integral of a nonnegative function converges. Theorem 4.1 justifies this convention, since it asserts that a divergent integral of this kind can only diverge to ∞. Similarly, if f is nonpositive and $\int_a^b f(x)\, dx$ converges, we write

$$\int_a^b f(x)\, dx > -\infty,$$

because a divergent integral of this kind can only diverge to $-\infty$. (To see this, apply Theorem 4.1 to the function $g = -f$.) These conventions do not apply to improper integrals of functions which assume both positive and negative values in (a, b), since they may diverge without diverging to $\pm\infty$.

Theorem 4.2 (Comparison Test)   *If f and g are locally integrable on $[a, b)$ and*

$$0 \le f(x) \le g(x), \qquad a \le x < b, \tag{2}$$

then:

(a) $\displaystyle \int_a^b f(x)\, dx < \infty \quad \text{if} \quad \int_a^b g(x)\, dx < \infty.$

(b) $\displaystyle \int_a^b g(x)\, dx = \infty \quad \text{if} \quad \int_a^b f(x)\, dx = \infty.$

Proof.   Assumption (2) implies that

$$\int_a^x f(t)\, dt \le \int_a^x g(t)\, dt, \qquad a \le x < b$$

(Lemma 3.1), and therefore

$$\operatorname*{l.u.b.}_{a \le x < b} \int_a^x f(t)\, dt \le \operatorname*{l.u.b.}_{a \le x < b} \int_a^x g(t)\, dt.$$

If $\int_a^b g(x)\, dx < \infty$, the right side of this inequality is finite, by Theorem 4.1, and so the left side is also. This implies that $\int_a^b f(x)\, dx < \infty$, again by Theorem 4.1. This proves (a), and (b) follows from (a) (*Exercise 10*).

The comparison test is particularly useful if the integrand of the improper integral is complicated but can be compared with a function that is easy to integrate.

Example 4.8   The improper integral

$$I = \int_0^1 \frac{2 + \sin \pi x}{(1 - x)^p}\, dx$$

converges if $p < 1$, since

$$0 < \frac{2 + \sin \pi x}{(1 - x)^p} \le \frac{3}{(1 - x)^p}, \qquad 0 \le x < 1,$$

and, from Example 4.2,

$$\int_0^1 \frac{3\, dx}{(1 - x)^p} < \infty, \qquad p < 1.$$

However, I diverges if $p \ge 1$, since

$$0 < \frac{1}{(1 - x)^p} \le \frac{2 + \sin \pi x}{(1 - x)^p}, \qquad 0 \le x < 1,$$

and

$$\int_0^1 \frac{dx}{(1-x)^p} = \infty, \qquad p \geq 1.$$

If f is any function (not necessarily nonnegative) locally integrable on $[a, b)$, then

$$\int_a^c f(x)\, dx = \int_a^{a_1} f(x)\, dx + \int_{a_1}^c f(x)\, dx$$

if a_1 and c are in $[a, b)$. Letting $c \to b-$, we conclude that if either of the improper integrals $\int_a^b f(x)\, dx$ and $\int_{a_1}^b f(x)\, dx$ converges, then so does the other, and in this case

$$\int_a^b f(x)\, dx = \int_a^{a_1} f(x)\, dx + \int_{a_1}^b f(x)\, dx,$$

where, of course, $\int_a^{a_1} f(x)\, dx$ is a proper integral. This means that any theorem implying convergence or divergence of an improper integral $\int_a^b f(x)\, dx$ in the sense of Definition 4.1 remains valid if its hypotheses are satisfied on a sub-interval $[a_1, b)$ of $[a, b)$ rather than on all of $[a, b)$. For example, Theorem 4.1 remains valid if (2) is replaced by

$$0 \leq f(x) \leq g(x), \qquad a_1 \leq x < b,$$

where a_1 is any point in $[a, b)$.

Example 4.9 If $p \geq 0$, then

$$\frac{x^{-p}}{2} \leq \frac{(x-1)^p(2 + \sin x)}{(x - \frac{1}{3})^{2p}} \leq 4x^{-p}$$

for x sufficiently large. Therefore, Theorem 4.2 and Example 4.3 imply that

$$\int_1^\infty \frac{(x-1)^p(2 + \sin x)}{(x - \frac{1}{3})^{2p}}\, dx$$

converges if $p > 1$ and diverges if $p \leq 1$.

Theorem 4.3 *Suppose f and g are locally integrable on $[a, b)$, $g(x) > 0$ and $f(x) \geq 0$ on some subinterval $[a_1, b)$ of $[a, b)$, and*

$$\lim_{x \to b-} \frac{f(x)}{g(x)} = M. \qquad (3)$$

Then:

　(a) *If $0 < M < \infty$, then $\int_a^b f(x)\, dx$ and $\int_a^b g(x)\, dx$ converge or diverge together.*

　(b) *If $M = \infty$ and $\int_a^b g(x)\, dx = \infty$, then $\int_a^b f(x)\, dx = \infty$.*

　(c) *If $M = 0$ and $\int_a^b g(x)\, dx < \infty$, then $\int_a^b f(x)\, dx < \infty$.*

Proof.   We prove (a) and leave the rest to the reader (*Exercise 12*). From (3) there is a point a_2 in $[a_1, b)$ such that

$$0 < \frac{M}{2} < \frac{f(x)}{g(x)} < \frac{3M}{2}, \qquad a_2 \leq x < b,$$

and therefore

$$\frac{M}{2} g(x) < f(x) < \frac{3M}{2} g(x), \qquad a_2 \leq x < b.$$

Theorem 4.2 and the first inequality imply that $\int_{a_2}^{b} g(x)\, dx < \infty$ if $\int_{a_2}^{b} f(x)\, dx < \infty$; Theorem 4.2 and the second inequality imply that $\int_{a_2}^{b} f(x)\, dx < \infty$ if $\int_{a_2}^{b} g(x)\, dx < \infty$. Therefore, $\int_{a_2}^{b} f(x)\, dx$ and $\int_{a_2}^{b} g(x)\, dx$ converge or diverge together, and in the latter case they must diverge to ∞, since their integrands are nonnegative (Theorem 4.1). Since the same is true of $\int_{a}^{b} g(x)\, dx$ and $\int_{a}^{b} f(x)\, dx$, the proof is complete.

Example 4.10   Let

$$f(x) = (1 + x)^{-p} \quad \text{and} \quad g(x) = x^{-p}.$$

Since

$$\lim_{x \to \infty} \frac{f(x)}{g(x)} = 1$$

and we know from Example 4.3 that $\int_{1}^{\infty} x^{-p}\, dx$ converges if $p > 1$ and diverges if $p \leq 1$, Theorem 4.3 implies that the same is true of

$$\int_{1}^{\infty} (1 + x)^{-p}\, dx$$

and, because of the remark preceding Theorem 4.3, of

$$\int_{a}^{\infty} (1 + x)^{-p}\, dx, \qquad a > -1.$$

Example 4.11   The function

$$f(x) = x^{-p}(1 + x)^{-q}$$

is locally integrable on $(0, \infty)$. To see whether

$$I = \int_{0}^{\infty} x^{-p}(1 + x)^{-q}\, dx$$

converges according to Definition 4.3, we consider separately the improper integrals

$$I_1 = \int_{0}^{1} x^{-p}(1 + x)^{-q}\, dx \quad \text{and} \quad I_2 = \int_{1}^{\infty} x^{-p}(1 + x)^{-q}\, dx.$$

(The choice of 1 as the upper limit of I_1 and the lower limit of I_2 is completely arbitrary; any other positive number would do just as well.) Since

$$\lim_{x \to 0+} \frac{f(x)}{x^{-p}} = \lim_{x \to 0+} (1 + x)^{-q} = 1$$

and

$$\int_0^1 x^{-p} \, dx = \begin{cases} (1 - p)^{-1}, & p < 1, \\ \infty, & p \geq 1, \end{cases}$$

Theorem 4.3 implies that I_1 converges if and only if $p < 1$. Since

$$\lim_{x \to \infty} \frac{f(x)}{x^{-p-q}} = \lim_{x \to \infty} (1 + x)^{-q} x^q = 1$$

and

$$\int_1^\infty x^{-p-q} \, dx = \begin{cases} (p + q - 1)^{-1}, & p + q > 1, \\ \infty, & p + q \leq 1, \end{cases}$$

Theorem 4.3 implies that I_2 converges if and only if $p + q > 1$. Combining these results, we conclude that I converges according to Definition 4.3 if and only if $p < 1$ and $p + q > 1$.

Absolute integrability

Definition 4.4 *We say that f is absolutely integrable on $[a, b)$ if f is locally integrable on $[a, b)$ and $\int_a^b |f(x)| \, dx < \infty$. In this case we also say that $\int_a^b f(x) \, dx$ converges absolutely, or is absolutely convergent.*

Example 4.12 If f is nonnegative and integrable on $[a, b)$, then it is absolutely integrable on $[a, b)$, since $|f| = f$.

Example 4.13 Since

$$\left| \frac{\sin x}{x^p} \right| \leq \frac{1}{x^p}$$

and $\int_1^\infty x^{-p} \, dx < \infty$ if $p > 1$ (Example 4.3), Theorem 4.2 implies that

$$\int_1^\infty \frac{|\sin x|}{x^p} \, dx < \infty, \qquad p > 1;$$

that is, the function

$$f(x) = \frac{\sin x}{x^p}$$

is absolutely integrable on $[1, \infty)$ if $p > 1$. It is not absolutely integrable on $[1, \infty)$ if $p \leq 1$; to see this, we let $p = 1$ and write

$$\int_1^{k\pi} \frac{|\sin x|}{x} \, dx > \int_\pi^{k\pi} \frac{|\sin x|}{x} \, dx$$

$$= \sum_{j=1}^{k-1} \int_{j\pi}^{(j+1)\pi} \frac{|\sin x|}{x} \, dx \tag{4}$$

$$> \sum_{j=1}^{k-1} \frac{1}{(j+1)\pi} \int_{j\pi}^{(j+1)\pi} |\sin x| \, dx$$

if $k \geq 2$. But

$$\int_{j\pi}^{(j+1)\pi} |\sin x| \, dx = \int_0^\pi \sin x \, dx = 2,$$

so (4) yields

$$\int_1^{k\pi} \frac{|\sin x|}{x} \, dx > \frac{2}{\pi} \sum_{j=1}^{k-1} \frac{1}{j+1}. \tag{5}$$

However,

$$\frac{1}{j+1} \geq \int_{j+1}^{j+2} \frac{dx}{x}, \qquad j = 1, 2, \ldots,$$

so (5) implies that

$$\int_1^{k\pi} \frac{|\sin x|}{x} > \frac{2}{\pi} \sum_{j=1}^{k-1} \int_{j+1}^{j+2} \frac{dx}{x}$$

$$= \frac{2}{\pi} \int_2^{k+1} \frac{dx}{x} = \frac{2}{\pi} \log \frac{(k+1)}{2}.$$

Since $\lim_{k \to \infty} \log(k+1)/2 = \infty$, Theorem 4.1 implies that

$$\int_1^\infty \frac{|\sin x|}{x} \, dx = \infty,$$

and now (b) of Theorem 4.2 implies that

$$\int_1^\infty \frac{|\sin x|}{x^p} \, dx = \infty, \qquad p \leq 1. \tag{6}$$

Theorem 4.4 *If f is locally integrable on $[a, b]$ and $\int_a^b |f(x)| \, dx < \infty$, then $\int_a^b f(x) \, dx$ converges; that is, an absolutely convergent integral is convergent.*

Proof. If

$$g(x) = |f(x)| - f(x),$$

then

$$0 \le g(x) \le 2|f(x)|$$

and $\int_a^b g(x) \, dx < \infty$, because of Theorem 4.2 and the absolute integrability of f. Since

$$f = |f| - g,$$

it follows that $\int_a^b f(x) \, dx$ converges (*Exercise 6*).

Conditional convergence

We say that f is *nonoscillatory* at $b- (= \infty$ if $b = \infty)$ if f is defined on $[a, b)$ and does not change sign on some subinterval $[a_1, b)$ of $[a, b)$. If f changes sign on every such subinterval, it is *oscillatory* at $b-$. For a function that is locally integrable on $[a, b)$ and nonoscillatory at $b-$, convergence and absolute convergence of $\int_a^b f(x) \, dx$ amount to the same thing (*Exercise 20*), so absolute convergence is not an interesting concept in connection with such functions. However, an oscillatory function may be integrable, but not absolutely, on $[a, b)$, as the next example shows. We then say that f is *conditionally* integrable on $[a, b)$, and that $\int_a^b f(x) \, dx$ converges *conditionally*.

Example 4.14 We saw in Example 4.13 that the integral

$$I(p) = \int_1^\infty \frac{\sin x}{x^p} \, dx$$

is not absolutely convergent if $0 < p \le 1$. We will show that it converges for these values of p.

Integration by parts yields

$$\int_1^c \frac{\sin x}{x^p} \, dx = \frac{-\cos c}{c^p} + \cos 1 - p \int_1^c \frac{\cos x}{x^{p+1}} \, dx. \tag{7}$$

Since

$$\left| \frac{\cos x}{x^{p+1}} \right| \le \frac{1}{x^{p+1}}$$

and $\int_1^\infty x^{-p-1} \, dx < \infty$ if $p > 0$, Theorem 4.2 implies that $x^{-p-1} \cos x$ is absolutely integrable—and therefore integrable—on $[1, \infty)$ if $p > 0$. Letting $c \to \infty$ in (7), we find that $I(p)$ converges, and

$$I(p) = \cos 1 - p \int_1^\infty \frac{\cos x}{x^{p+1}} \, dx \quad \text{if} \quad p > 0.$$

This and (6) imply that $I(p)$ converges conditionally if $0 < p \le 1$.

The method used in Example 4.14 is a special case of the following test for convergence of improper integrals.

Theorem 4.5 (Dirichlet's Test)   *Suppose f is continuous and the antiderivative $F(x) = \int_a^x f(t)\, dt$ is bounded on $[a, b)$. Let g' be absolutely integrable on $[a, b)$ and*

$$\lim_{x \to b-} g(x) = 0. \tag{8}$$

Then $\int_a^b f(x)g(x)\, dx$ converges.

Proof.   Our assumptions imply that fg is locally integrable on $[a, b)$. Integration by parts yields

$$\int_a^c f(x)g(x)\, dx = F(c)g(c) - \int_a^c F(x)g'(x)\, dx, \qquad a \le c < b. \tag{9}$$

Theorem 4.2 implies that the integral on the right converges absolutely as $c \to b-$, since $\int_a^b |g'(x)|\, dx < \infty$ by assumption and

$$|F(x)g'(x)| \le M|g'(x)|,$$

where M is an upper bound for F on $[a, b)$; moreover, (8) and the boundedness of F imply that $\lim_{c \to b-} F(c)g(c) = 0$. Letting $c \to b-$ in (9) yields

$$\int_a^b f(x)g(x)\, dx = -\int_a^b F(x)g'(x)\, dx$$

(where the integral on the right converges absolutely). This completes the proof.

Dirichlet's test is of interest only if f is oscillatory at $b-$, since it can be shown that if f is nonoscillatory at $b-$ and F is bounded on $[a, b)$, then $\int_a^b |f(x)g(x)|\, dx < \infty$ if only g is locally integrable and bounded on $[a, b)$ (*Exercise 18*).

Dirichlet's test can also be used to show that certain integrals diverge. For example,

$$\int_1^\infty x^q \sin x\, dx$$

diverges if $q > 0$, but none of the other tests that we have studied so far implies this. (It is not enough to argue that the integrand does not approach zero as $x \to \infty$—a common mistake—since this does not imply divergence; see *Exercise 35, Section 4.4*) To see that the integral diverges, we observe that if it converged for some $q > 0$, then we could take

$$f(x) = x^q \sin x \quad \text{and} \quad g(x) = x^{-q}$$

in Theorem 4.5, and conclude that

$$\int_1^\infty \sin x \, dx$$

also converges. This is a contradiction.

If Dirichlet's test shows that $\int_a^b f(x)g(x) \, dx$ converges, there remains the question of whether it converges absolutely or conditionally. The next theorem sometimes supplies the answer to this question. Its proof can be modeled after the method of Example 4.13 (*Exercise 21*). (The idea of an infinite sequence, which we will discuss in Section 4.1, enters into the statement of this theorem; we assume that the student of advanced calculus recalls the concept sufficiently well to understand the meaning of the theorem.)

Theorem 4.6 *Suppose g is monotonic on $[a, b)$ and $\int_a^b g(x) \, dx = \infty$. Let f be locally integrable on $[a, b)$ and*

$$\int_{x_j}^{x_{j+1}} |f(x)| \, dx \ge \rho, \qquad j = 0, 1, 2, \dots,$$

for some positive ρ, where $\{x_j\}$ is an increasing infinite sequence of points in $[a, b)$ such that $\lim_{j \to \infty} x_j = b$ and $x_{j+1} - x_j \le M$ $(j = 0, 1, \dots)$ for some M. Then

$$\int_a^b |f(x)g(x)| \, dx = \infty.$$

Change of variable in an improper integral

The following theorem enables us to investigate an improper integral by transforming it into another whose convergence or divergence is known. It follows from Theorem 3.14 and Definitions 4.1, 4.2, and 4.3. We leave the proof to the reader (*Exercise 29*).

Theorem 4.7 *Suppose ϕ' is locally integrable and ϕ monotonic on either of the half-open intervals $I = [c, d)$ or $(c, d]$, and let $x = \phi(t)$ map I onto either of the half-open intervals $J = [a, b)$ or $(a, b]$. Let f be locally integrable over J. Then the improper integrals*

$$\int_a^b f(x) \, dx \quad and \quad \int_c^d f(\phi(t))|\phi'(t)| \, dt$$

diverge or converge together, in the latter case to the same value. The same conclusion holds if ϕ has the stated properties only on the open interval (a, b), the transformation $x = \phi(t)$ maps (c, d) onto (a, b), and f is locally integrable on (a, b).

Example 4.15   To apply Theorem 4.7 to the integral

$$\int_0^\infty \sin x^2 \, dx,$$

we use the change of variable $x = \phi(t) = \sqrt{t}$, which takes $[c, d) = [0, \infty)$ into $[a, b) = [0, \infty)$, with $\phi'(t) = 1/2\sqrt{t}$. Theorem 4.7 implies that

$$\int_0^\infty \sin x^2 \, dx = \frac{1}{2} \int_0^\infty \frac{\sin t}{\sqrt{t}} \, dt.$$

Since the integral on the right converges (Example 4.14), so does the one on the left.

Example 4.16   The integral

$$\int_1^\infty x^{-p} \, dx$$

converges if and only if $p > 1$ (Example 4.3). Defining $\phi(t) = 1/t$ and applying Theorem 4.7 yields

$$\int_1^\infty x^{-p} \, dx = \int_0^1 t^p \left| -t^{-2} \right| dt = \int_0^1 t^{p-2} \, dt,$$

which implies that $\int_0^1 t^q \, dt$ converges if and only if $q > -1$.

3.4   EXERCISES

1. (a) Let f be locally integrable and bounded on $[a, b)$, and define $f(b)$ arbitrarily, if it is not defined already. Show that f is then properly integrable on $[a, b]$, that $\int_a^b f(x) \, dx$ does not depend on the value assigned to $f(b)$, and that

$$\int_a^b f(x) \, dx = \lim_{c \to b-} \int_a^c f(x) \, dx.$$

   (b) State a result analogous to part (a) that ends with the conclusion that

$$\int_a^b f(x) \, dx = \lim_{c \to a+} \int_c^b f(x) \, dx.$$

2. Show that neither the existence nor the value of the improper integral of Definition 4.3 depends on the choice of the intermediate point α.

3. Prove: If f is integrable on $[a, b]$ according to Definition 4.1 or 4.2, then it is also integrable according to Definition 4.3.

4. Find all values of p for which the following integrals exist (i) as proper integrals (perhaps after defining f at the end points of the interval), or (ii) as improper integrals.

(a) $\int_0^{1/\pi} \left(px^{p-1} \sin \frac{1}{x} - x^{p-2} \cos \frac{1}{x} \right) dx$

(b) $\int_0^{2/\pi} \left(px^{p-1} \cos \frac{1}{x} + x^{p-2} \sin \frac{1}{x} \right) dx$

(c) $\int_0^\infty e^{-px} dx$ (d) $\int_0^1 x^{-p} dx$ (e) $\int_0^\infty x^{-p} dx$

5. Evaluate:

(a) $\int_0^\infty e^{-x} x^n \, dx$ $(n = 0, 1, \ldots)$ (b) $\int_0^\infty e^{-x} \sin x \, dx$

(c) $\int_{-\infty}^\infty \frac{x \, dx}{x^2 + 1}$ (d) $\int_0^1 \frac{x \, dx}{\sqrt{1 - x^2}}$

(e) $\int_0^\pi \left(\frac{\cos x}{x} - \frac{\sin x}{x^2} \right) dx$ (f) $\int_{\pi/2}^\infty \left(\frac{\sin x}{x} + \frac{\cos x}{x^2} \right) dx$

6. Show that the relations

$$\int_a^b (f + g)(x) \, dx = \int_a^b f(x) \, dx + \int_a^b g(x) \, dx$$

and

$$\int_a^b (kf)(x) \, dx = k \int_a^b f(x) \, dx \qquad (k \text{ constant})$$

remain valid for improper integrals if the integrals on the right exist (finite). What if one or both is infinite?

* 7. Prove: If $\int_a^b f(x) \, dx$ exists—as a proper or improper integral—then

$$\lim_{x \to b-} \int_x^b f(t) \, dt = 0.$$

8. Prove: If f is locally integrable on $[a, b)$, then $\int_a^b f(x) \, dx$ exists if and only if for each $\varepsilon > 0$ there is a point r in (a, b) such that

$$\left| \int_{x_1}^{x_2} f(t) \, dt \right| < \varepsilon$$

whenever $r \le x_1, x_2 < b$. (*Hint*: See Exercises 36 and 39(c), Section 2.1.)

9. Prove Theorem 4.1.

10. Prove that (b) of Theorem 4.2 follows from (a).

11. Use Theorem 4.2 to determine whether the following integrals converge.

(a) $\int_1^\infty \frac{\log x + \sin x}{\sqrt{x}} dx$ (b) $\int_{-\infty}^\infty \frac{(x^2 + 3)^{3/2}}{(x^4 + 1)^{3/2}} \sin^2 x \, dx$

(c) $\int_0^\infty \dfrac{1 + \cos^2 x}{\sqrt{1 + x^2}}\, dx$ (d) $\int_0^\infty \dfrac{4 + \cos x}{(1 + x)\sqrt{x}}\, dx$

(e) $\int_0^\infty (x^{27} + \sin x)e^{-x}\, dx$ (f) $\int_0^\infty x^{-p}(2 + \sin x)\, dx$

12. Prove (b) and (c) of Theorem 4.3. Show by examples that the stronger conclusion of (a) does not hold in these cases.

13. Find all values of p for which the following integrals converge.

(a) $\int_0^{\pi/2} \dfrac{\sin x}{x^p}\, dx$ (b) $\int_0^{\pi/2} \dfrac{\cos x}{x^p}\, dx$ (c) $\int_0^\infty x^p e^{-x}\, dx$

(d) $\int_0^{\pi/2} \dfrac{\sin x}{(\tan x)^p}\, dx$ (e) $\int_1^\infty \dfrac{dx}{x(\log x)^p}$ (f) $\int_0^1 \dfrac{dx}{x(|\log x|)^p}$

(g) $\int_0^\pi \dfrac{x\, dx}{(\sin x)^p}$

*14. Let $L_n(x)$ be the iterated logarithm defined in Exercise 56, Section 2.4. Show that

$$\int_a^\infty \frac{dx}{L_0(x)L_1(x)\cdots L_k(x)[L_{k+1}(x)]^p}$$

converges if and only if $p > 1$. Here a is any number such that $L_{k+1}(x) > 0$ for $x \geq a$.

15. Find conditions on p and q such that the integral converges:

(a) $\int_{-1}^1 \dfrac{(\cos \pi x/2)^q}{(1 - x^2)^p}\, dx$ (b) $\int_{-1}^1 \dfrac{dx}{(1 - x)^p(1 + x)^q}$

(c) $\int_0^\infty \dfrac{x^p\, dx}{(1 + x^2)^q}$ (d) $\int_1^\infty \dfrac{[\log(1 + x)]^p(\log x)^q}{x^{p+q}}\, dx$

(e) $\int_1^\infty \dfrac{(\log(1 + x) - \log x)^q}{x^p}\, dx$ (f) $\int_0^\infty \dfrac{(x - \sin x)^q}{x^p}\, dx$

16. Let f and g be polynomials and suppose g has no real roots. Find necessary and sufficient conditions for convergence of

$$\int_{-\infty}^\infty \frac{f(x)}{g(x)}\, dx.$$

17. Prove: If f and g are locally integrable on $[a, b)$ and the improper integrals $\int_a^b f^2(x)\, dx$ and $\int_a^b g^2(x)\, dx$ converge, then $\int_a^b f(x)g(x)\, dx$ converges absolutely. [Hint: $(f \pm g)^2 \geq 0$.]

18. Suppose f is locally integrable and $F(x) = \int_a^x f(t)\, dt$ is bounded on $[a, b)$, and let f be nonoscillatory at $b-$. Let g be locally integrable and bounded

on $[a, b)$. Show that

$$\int_a^b |f(x)g(x)| \, dx < \infty.$$

19. Suppose g is positive and nonincreasing on $[a, b)$ and $\int_a^b f(x) \, dx$ exists as a proper or absolutely convergent improper integral. Show that $\int_a^b f(x)g(x) \, dx$ exists and

$$\lim_{x \to b-} \frac{1}{g(x)} \int_x^b f(t)g(t) \, dt = 0.$$

(*Hint*: Use Exercise 7.)

20. Show that if f is locally integrable on $[a, b)$ and nonoscillatory at $b-$, then $\int_a^b f(x) \, dx$ exists if and only if $\int_a^b |f(x)| \, dx < \infty$.

21. (a) Prove Theorem 4.6. (*Hint*: See Example 4.13.)
     (b) Show that g satisfies the assumptions of Theorem 4.5 if g' is locally integrable and g is monotonic on $[a, b)$, and $\lim_{x \to b-} g(x) = 0$.

22. Find all values of p for which the integral converges (i) absolutely; (ii) conditionally:

(a) $\int_1^\infty \frac{\cos x}{x^p} \, dx$

(b) $\int_2^\infty \frac{\sin x}{x(\log x)^p} \, dx$

(c) $\int_2^\infty \frac{\sin x}{x^p \log x} \, dx$

(d) $\int_1^\infty \frac{\sin 1/x}{x^p} \, dx$

(e) $\int_0^\infty \frac{\sin^2 x \sin 2x}{x^p} \, dx$

(f) $\int_{-\infty}^\infty \frac{\sin x}{(1 + x^2)^p} \, dx$

(g) $\int_1^\infty \frac{\cos x}{(\log x)^p} \, dx$

23. Suppose g'' is absolutely integrable on $[0, \infty)$, $\lim_{x \to \infty} g'(x) = 0$, and $\lim_{x \to \infty} g(x) = L$ (finite or infinite). Show that $\int_0^\infty g(x) \sin x \, dx$ converges if and only if $L = 0$. (*Hint*: Integrate by parts.)

24. Let h be continuous on $[0, \infty)$. Prove:
     (a) If $\int_0^\infty e^{-s_0 x} h(x) \, dx$ converges absolutely, then $\int_0^\infty e^{-sx} h(x) \, dx$ converges absolutely if $s > s_0$.
     (b) If $\int_0^\infty e^{-s_0 x} h(x) \, dx$ converges, then $\int_0^\infty e^{-sx} h(x) \, dx$ converges if $s > s_0$.

25. Suppose f is locally integrable on $[0, \infty)$, $\lim_{x \to \infty} f(x) = A$, and $\alpha > -1$. Find $\lim_{x \to \infty} x^{-\alpha - 1} \int_0^x f(t)t^\alpha \, dt$, and prove your answer.

26. Suppose f is continuous and $F(x) = \int_a^x f(t) \, dt$ is bounded on $[a, b)$. Suppose $g > 0$, g' is nonnegative and locally integrable on $[a, b)$, and $\lim_{x \to b-} g(x) = \infty$. Show that

$$\lim_{x \to b-} \frac{1}{[g(x)]^p} \int_a^x f(t)g(t) \, dt = 0, \qquad p > 1.$$

(*Hint*: Integrate by parts.)

27. In addition to the assumptions of Exercise 26, assume that $\int_a^b f(t)\,dt$ converges. Show that

$$\lim_{x \to b-} \frac{1}{g(x)} \int_a^x f(t)g(t)\,dt = 0.$$

[*Hint:* Let $F_1(x) = -\int_x^b f(t)\,dt$, integrate by parts, and use Exercise 7.]

28. Suppose f is continuous, $g'(x) \leq 0$, and $g(x) > 0$ on $[a, b)$. Show that if g' is integrable on $[a, b)$ and $\int_a^b f(x)\,dx$ exists, then $\int_a^b f(x)g(x)\,dx$ exists and

$$\lim_{x \to b-} \frac{1}{g(x)} \int_x^b f(t)g(t)\,dt = 0.$$

[*Hint:* Let $F(x) = \int_x^b f(t)\,dt$, integrate by parts, and use Exercise 7.]

29. Prove Theorem 4.7. (*Hint:* Use Theorem 3.14.)

30. Find all values of p for which the integral converges (i) absolutely; (ii) conditionally.

(a) $\int_0^1 x^p \sin 1/x\,dx$ (b) $\int_0^1 |\log x|^p\,dx$ (c) $\int_1^\infty x^p \cos(\log x)\,dx$

(d) $\int_1^\infty (\log x)^p\,dx$ (e) $\int_0^\infty \sin x^p\,dx$

31. Let u_1 be positive and satisfy the differential equation

$$u'' + p(x)u = 0, \qquad 0 \leq x < \infty. \tag{A}$$

(a) Prove: If

$$\int_0^\infty \frac{dx}{u_1^2(x)} < \infty,$$

then the function

$$u_2(x) = u_1(x) \int_x^\infty \frac{dt}{u_1^2(t)}$$

also satisfies (A), while if

$$\int_0^\infty \frac{dx}{u_1^2(x)} = \infty,$$

then the function

$$u_2(x) = u_1(x) \int_0^x \frac{dt}{u_1^2(t)}$$

also satisfies (A).

(b) Prove: If (A) has a solution which is positive on $[0, \infty)$, then it has solutions y_1 and y_2 which are positive on $(0, \infty)$ and have the following properties:

$$y_1(x)y_2'(x) - y_1'(x)y_2(x) = 1, \quad x > 0,$$

$$\left[\frac{y_1(x)}{y_2(x)}\right]' < 0, \quad x > 0,$$

and

$$\lim_{x \to \infty} \frac{y_1(x)}{y_2(x)} = 0.$$

32. (a) Prove: If h is continuous on $[0, \infty)$, the function

$$u(x) = c_1 e^{-x} + c_2 e^x + \int_0^x h(t) \sinh(x - t) \, dt$$

satisfies the differential equation

$$u'' - u = h(x), \quad x > 0.$$

(It is not necessary to know how to differentiate under the integral sign to do this. Why not?)

(b) Rewrite u in the form

$$u(x) = a(x)e^{-x} + b(x)e^x$$

and show that

$$u'(x) = -a(x)e^{-x} + b(x)e^x.$$

(c) Show that if $\lim_{x \to \infty} a(x) = A$ (finite), then

$$\lim_{x \to \infty} e^{2x}[b(x) - B] = 0$$

for some constant B. (*Hint:* Use Exercise 28.) Show also that

$$\lim_{x \to \infty} e^x[u(x) - Ae^{-x} - Be^x] = 0.$$

(d) Prove: If $\lim_{x \to \infty} b(x) = B$ (finite), then

$$\lim_{x \to \infty} u(x)e^{-x} = \lim_{x \to \infty} u'(x)e^{-x} = B.$$

(*Hint:* Use Exercise 27.)

33. Suppose the differential equation

$$u'' + p(x)u = 0 \qquad \text{(A)}$$

has a solution which is positive on $[0, \infty)$, and therefore has two solutions y_1 and y_2 with the properties given in Exercise 31(b).

(a) Prove: If h is continuous on $[0, \infty)$ and c_1 and c_2 are constants, the function

$$u(x) = c_1 y_1(x) + c_2 y_2(x) + \int_0^x h(t)[y_1(t)y_2(x) - y_1(x)y_2(t)]\, dt \tag{B}$$

satisfies the differential equation

$$u'' + p(x)u = h(x).$$

For convenience in parts (b) and (c), rewrite (B) as

$$u(x) = a(x)y_1(x) + b(x)y_2(x).$$

(b) Prove: If $\int_0^\infty h(t)y_2(t)\, dt$ converges, then $\int_0^\infty h(t)y_1(t)\, dt$ converges, and

$$\lim_{x \to \infty} \frac{u(x) - Ay_1(x) - By_2(x)}{y_1(x)} = 0$$

for some constants A and B. (*Hint:* Use Exercise 28 with $f = hy_2$ and $g = y_1/y_2$.)

(c) Prove: If $\int_0^\infty h(t)y_1(t)\, dt$ converges, then

$$\lim_{x \to \infty} \frac{u(x)}{y_2(x)} = B$$

for some constant B. (*Hint:* Use Exercise 27 with $f = hy_1$ and $g = y_2/y_1$.)

34. Show that the following definition of local integrability is equivalent to the one given at the beginning of the section: f is locally integrable on an interval I if for every x in I there is a closed interval I_x such that $x \in I_x^0$ and f is integrable on $I_x \cap I$. (*Hint:* If $[\alpha, \beta] \subset I$, let

$$S = \{t | \alpha \le t \le \beta \quad \text{and } f \text{ is integrable on } [\alpha, t]\}.$$

Show that $\beta \in S$.)

35. Suppose f, f_1, and g are continuous, $f > 0$, and $(f_1/f)'$ is absolutely integrable on $[a, b]$. Show that $\int_a^b f_1(x)g(x)\, dx$ converges if $\int_a^b f(x)g(x)\, dx$ does.

36. Let g be locally integrable and f continuous, with $f(x) \ge \rho > 0$ on $[a, b]$. Suppose that for some positive M and for every r in $[a, b)$ there are points x_1 and x_2 such that (a) $r < x_1 < x_2 < b$; (b) g does not change sign in $[x_1, x_2]$; and (c) $\int_{x_1}^{x_2}|g(x)|\, dx \ge M$. Show that $\int_a^b f(x)g(x)\, dx$ diverges. (*Hint:* Use Exercise 8 and Theorem 3.6.)

3.5 A MORE ADVANCED LOOK AT THE EXISTENCE OF THE INTEGRAL

This section is more advanced than the other sections in this chapter; since the rest of the book does not depend on it, it may be skipped.

In Section 3.2 we found necessary and sufficient conditions for existence of the proper Riemann integral, and in Section 3.3 we used them to study its properties. However, it is awkward to apply these conditions to a specific function and determine whether it is integrable, since they require computations—of upper and lower sums and upper and lower integrals—which are hard to do. The main result of this section is an integrability criterion due to Lebesgue, which does not require computation but has to do with how badly discontinuous a function may be and still be integrable.

We emphasize that we are again considering proper integrals of bounded functions on finite intervals.

Definition 5.1 *If f is defined on $[a, b]$, the oscillation of f on $[a, b]$ is defined by*

$$W_f[a, b] = \underset{a \leq x \leq b}{\text{l.u.b.}} \ f(x) - \underset{a \leq x \leq b}{\text{g.l.b.}} \ f(x).$$

If $a < x < b$, the oscillation of f at x is defined by

$$w_f(x) = \lim_{h \to 0+} W_f[x - h, x + h].$$

The corresponding definitions for $x = a$ and $x = b$ are

$$w_f(a) = \lim_{h \to 0+} W_f[a, a + h] \quad and \quad w_f(b) = \lim_{h \to 0+} W_f[b - h, b].$$

From this definition, $W_f[a, b]$ and $w_f(x)$ are nonnegative; moreover,

$$W_f[a, b] = \text{l.u.b.}\{|f(x) - f(x')| \, \big| \, x, x' \in [a, b]\}$$

(*Exercise 1*).

Lemma 5.1 *Let f be defined on $[a, b]$. Then f is discontinuous at x_0 in $[a, b]$ if and only if $w_f(x_0) > 0$. (Here discontinuity at a or b means discontinuity from the left or right, respectively.)*

Proof. Consider x_0 in (a, b); a similar argument applies if $x_0 = a$ or $x_0 = b$. Since $w_f(x_0) \geq 0$ in any case (Exercise 1(b)), it suffices to show that f is continuous at x_0 if and only if $w_f(x_0) = 0$. First suppose $w_f(x_0) = 0$ and $\varepsilon > 0$. Then

$$W_f[x_0 - h, x_0 + h] < \varepsilon$$

for some $h > 0$, and consequently

$$|f(x) - f(x')| < \varepsilon \qquad \text{if} \quad x_0 - h \leq x, x' \leq x_0 + h$$

[*Exercise 1(a)*]. Letting $x' = x_0$, we conclude that

$$|f(x) - f(x_0)| < \varepsilon \qquad \text{if} \quad |x - x_0| < h.$$

Therefore, f is continuous at x_0.

Conversely, if f is continuous at x_0 and $\varepsilon > 0$, there is a $\delta > 0$ such that

$$|f(x) - f(x_0)| < \frac{\varepsilon}{2} \quad \text{and} \quad |f(x') - f(x_0)| < \frac{\varepsilon}{2}$$

if $x_0 - \delta \leq x, x' \leq x_0 + \delta$. The triangle inequality implies that

$$|f(x) - f(x')| < \varepsilon$$

for such x and x', and this implies that

$$W_f[x_0 - h, x_0 + h] \leq \varepsilon \quad \text{if} \quad h < \delta$$

[*Exercise 1(a)*], so that $w_f(x_0) = 0$. This completes the proof.

Lemma 5.2   *If $w_f(x) < \varepsilon$ for $a \leq x \leq b$, there is a $\delta > 0$ such that $W_f[a_1, b_1] \leq \varepsilon$ provided $[a_1, b_1] \subset [a, b]$ and $b_1 - a_1 < \delta$.*

Proof.   We use the Heine–Borel theorem. The assumption implies that for each x in $[a, b]$, there is a $\delta_x > 0$ such that

$$|f(x') - f(x'')| < \varepsilon \tag{1}$$

if

$$x - 2\delta_x < x', \quad x'' < x + 2\delta_x \quad \text{and} \quad x', x'' \in [a, b]. \tag{2}$$

If $I_x = (x - \delta_x, x + \delta_x)$, the family

$$\mathcal{H} = \{I_x \mid a \leq x \leq b\}$$

is an open covering of $[a, b]$, and the Heine–Borel theorem implies that there are finitely many points $x_1, x_2, \ldots, x_n$ in $[a, b]$ such that $I_{x_1}, I_{x_2}, \ldots, I_{x_n}$ cover $[a, b]$. Let

$$\delta = \min_{1 \leq i \leq n} \{\delta_{x_i}\}$$

and suppose $[a_1, b_1] \subset [a, b]$ and $b_1 - a_1 < \delta$. If x' and x'' are in $[a_1, b_1]$, then $x' \in I_{x_r}$ for some r $(1 \leq r \leq n)$, so

$$|x' - x_r| < \delta_{x_r},$$

and therefore

$$\begin{aligned}
|x'' - x_r| &\leq |x'' - x'| + |x' - x_r| \\
&< b_1 - a_1 + \delta_{x_r} \\
&< \delta + \delta_{x_r} \\
&\leq 2\delta_{x_r}.
\end{aligned}$$

Thus, any two points x' and x'' in $[a_1, b_1]$ satisfy (2) with $x = x_r$, and so they also satisfy (1). Therefore, ε is an upper bound for the set

$$\{|f(x') - f(x'')| \mid x', x'' \in [a_1, b_1]\},$$

which has least upper bound $W_f[a_1, b_1]$ [*Exercise 1 (a)*]; hence $W_f[a_1, b_1] \leq \varepsilon$ and the proof is complete.

In the following $L(I)$ is the length of the interval I.

Lemma 5.3   *Let f be bounded on $[a, b]$ and define*

$$E_\rho = \{x \in [a, b] \mid w_f(x) \geq \rho\}.$$

Then f is integrable on $[a, b]$ if and only if for every pair of positive numbers ρ and δ, E_ρ can be covered by finitely many open intervals $I_1, I_2, \ldots, I_p$ such that

$$\sum_{j=1}^{p} L(I_j) < \delta. \tag{3}$$

Proof.   Suppose the conditions is not satisfied; that is, there is a $\rho > 0$ and a $\delta > 0$ such that

$$\sum_{j=1}^{p} L(I_j) \geq \delta$$

for every finite set of open intervals covering E_ρ. If $P = \{x_0, x_1, \ldots, x_n\}$ is a partition of $[a, b]$, then

$$S(P) - s(P) = \sum_{j \in A} (M_j - m_j)(x_j - x_{j-1}) + \sum_{j \in B} (M_j - m_j)(x_j - x_{j-1}), \tag{4}$$

where

$$A = \{j \mid [x_{j-1}, x_j] \cap E_\rho \neq \varnothing\} \quad \text{and} \quad B = \{j \mid [x_{j-1}, x_j] \cap E_\rho = \varnothing\}.$$

Since $\bigcup_{j \in A} (x_{j-1}, x_j)$ contains all points of E_ρ except any of $x_0, x_1,$ $\ldots, x_n$ that may be in E_ρ, and each of these finitely many possible exceptions can be covered by an open interval of length as small as we please, our assumption on E_ρ implies that

$$\sum_{j \in A} (x_j - x_{j-1}) \geq \delta.$$

Moreover, if $j \in A$, then

$$M_j - m_j \geq \rho,$$

and so (4) implies that

$$S(P) - s(P) \geq \rho \sum_{j \in A} (x_j - x_{j-1}) \geq \rho\delta.$$

Since this holds for every partition of $[a, b]$, f is not integrable on $[a, b]$, by Theorem 2.4. This proves that the stated condition is necessary for integrability.

For sufficiency, let ρ and δ be positive numbers and let $I_1, I_2, \ldots, I_p$ be open intervals which cover E_ρ and satisfy (3). Let

$$\tilde{I}_j = [a, b] \cap \bar{I}_j.$$

$(\bar{I}_j = $ closure of I.) After combining any of $\tilde{I}_1, \tilde{I}_2, \ldots, \tilde{I}_p$ that overlap, we obtain a set of pairwise disjoint closed subintervals

$$C_j = [\alpha_j, \beta_j] \qquad j = 1, 2, \ldots, q \, (\leq p),$$

of $[a, b]$ such that

$$a \leq \alpha_1 < \beta_1 < \alpha_2 < \cdots < \beta_{q-1} < \alpha_q < \beta_q \leq b, \tag{5}$$

$$\sum_{i=1}^{q} (\beta_i - \alpha_i) < \delta \tag{6}$$

and

$$w_f(x) < \rho, \qquad \beta_j \leq x \leq \alpha_{j+1}.$$

[Also, $w_f(x) < \rho$ for $a \leq x \leq \alpha_1$ if $a < \alpha_1$ and for $\beta_q \leq x \leq b$ if $\beta_q < b$.]

Let P_0 be the partition of $[a, b]$ with the partition points indicated in (5), and refine P_0 by partitioning each subinterval $[\beta_j, \alpha_{j+1}]$ (as well as $[a, \alpha_1]$ if $a < \alpha_1$ and $[\beta_q, b]$ if $\beta_q < b$) into subintervals on which the oscillation of f is not greater than ρ. This is possible by Lemma 5.2. In this way, after renaming the entire collection of partition points, we obtain a partition $P = \{x_0, x_1, \ldots, x_n\}$ of $[a, b]$ for which $S(P) - s(P)$ can be written as in (4), with

$$\sum_{j \in A} (x_j - x_{j-1}) = \sum_{i=1}^{q} (\beta_i - \alpha_i) < \delta$$

[see (6)] and

$$M_j - m_j \leq \rho, \qquad j \in B.$$

For this partition,

$$\sum_{j \in A} (M_j - m_j)(x_j - x_{j-1}) < 2K\delta,$$

where K is an upper bound for $|f|$ on $[a, b]$, and

$$\sum_{j \in B} (M_j - m_j)(x_j - x_{j-1}) \leq \rho(b - a).$$

We have now shown that if ρ and δ are arbitrary positive numbers, there is a partition P of $[a, b]$ such that

$$S(P) - s(P) < 2K\delta + \rho(b - a). \tag{7}$$

If $\varepsilon > 0$, let

$$\delta = \frac{\varepsilon}{4K} \quad \text{and} \quad \rho = \frac{\varepsilon}{2(b-a)}.$$

Then (7) yields

$$S(P) - s(P) < \varepsilon,$$

and Theorem 2.4 implies that f is integrable on $[a, b]$. This completes the proof.

We need the next definition to state Lebesgue's integrability condition.

Definition 5.2 *A subset S of the real line is said to be of Lebesgue measure zero if for every $\varepsilon > 0$ there is a finite or infinite sequence of open intervals $I_1, I_2, \ldots$ such that*

$$S \subset \bigcup_j I_j \tag{8}$$

and

$$\sum_{j=1}^{n} L(I_j) < \varepsilon, \qquad n = 1, 2, \ldots. \tag{9}$$

Example 5.1 The empty set is of Lebesgue measure zero, since it is contained in any open interval.

Example 5.2 Any finite set $S = \{x_1, x_2, \ldots, x_n\}$ is of Lebesgue measure zero, since we can choose open intervals $I_1, I_2, \ldots, I_n$ such that $x_j \in I_j$ and $L(I_j) < \varepsilon/n$ $(j = 1, 2, \ldots, n)$.

Example 5.3 An infinite set is said to be *denumerable* if its members can be listed in a sequence (that is, in a one-to-one correspondence with the positive integers); thus,

$$S = \{x_1, x_2, \ldots, x_n, \ldots\}. \tag{10}$$

An infinite set which does not have this property is *nondenumerable*. Any denumerable set (10) is of Lebesgue measure zero, since if $\varepsilon > 0$ it is possible to choose open intervals $I_1, I_2, \ldots$ so that

$$x_j \in I_j \quad \text{and} \quad L(I_j) < \frac{\varepsilon}{2^j}, \qquad j = 1, 2, \ldots.$$

Then (8) holds because

$$\frac{1}{2} + \frac{1}{2^2} + \frac{1}{2^3} + \cdots + \frac{1}{2^n} = 1 - \frac{1}{2^n} < 1. \tag{11}$$

There are also nondenumerable sets of Lebesgue measure zero, but it would take us beyond the scope of this book to discuss examples.

The next theorem is the main result of this section.

Theorem 5.1   *A bounded function f is integrable on a finite interval $[a, b]$ if and only if the set S of discontinuities of f in $[a, b]$ is of Lebesgue measure zero.*

Proof.   From Lemma 5.1,

$$S = \{x \in [a, b] \,|\, w_f(x) > 0\}.$$

Since $w_f(x) > 0$ if and only if $w_f(x) \geq 1/i$ for some positive integer i, we can write

$$S = \bigcup_{i=1}^{\infty} S_i, \tag{12}$$

where

$$S_i = \{x \in [a, b] \,|\, w_f(x) \geq 1/i\}.$$

From Lemma 5.3, if f is integrable on $[a, b]$ and $\varepsilon > 0$, then each S_i can be covered by a finite number of open intervals $I_{i1}, I_{i2}, \ldots, I_{in_i}$ of total length less than $\varepsilon/2^i$. We simply renumber these intervals consecutively; thus

$$\{I_1, I_2, \ldots\} = \{I_{11}, \ldots, I_{1n_1}, I_{21}, \ldots, I_{2n_2}, \ldots, I_{i1}, \ldots, I_{in_i}, \ldots\}.$$

Now (8) and (9) hold because of (11) and (12), and we have shown that the stated condition is necessary for integrability.

For sufficiency, suppose the stated condition holds and $\varepsilon > 0$. Then S can be covered by open intervals $\{I_1, I_2, \ldots\}$ which satisfy (9). If $\rho > 0$, then the set E_ρ of Lemma 5.3 is contained in S (Lemma 5.1), and therefore E_ρ is covered by $\{I_1, I_2, \ldots\}$. Since E_ρ is closed (*Exercise 2*) and bounded, the Heine–Borel theorem implies that E_ρ is covered by a finite number of intervals from $\{I_1, I_2, \ldots\}$. The sum of the lengths of the latter is less than ε, so Lemma 5.3 implies that f is integrable on $[a, b]$. This completes the proof.

3.5   EXERCISES

1. In connection with Definition 5.1 show that
   (a) $W_f[a, b] = \text{l.u.b.}\{|f(x) - f(x')| \,|\, x, x' \in [a, b]\}$.
   (b) $w_f(x)$ exists and is nonnegative for $a \leq x \leq b$. (*Hint:* Use Theorem 1.4, Section 2.1.)

2. Show that the set E_ρ introduced in Lemma 5.3 is closed if $\rho > 0$.

3. Use Theorem 5.1 to show that if f is integrable on $[a, b]$, then so is $|f|$ and, if $f(x) \geq \rho > 0$ $(a \leq x \leq b)$, so is $1/f$.

4. Prove: The union of two sets of Lebesgue measure zero is of Lebesgue measure zero.

5. Use Theorem 5.1 and Exercise 4 to show that if f and g are integrable on $[a, b]$, then so are $f + g$ and fg.

6. Prove: If S is a set of Lebesgue measure zero, then every interval contains a point not in S.

7. Let $h(x) = 0$ for all x in $[a, b]$ except for x in a set of Lebesgue measure zero. Show that if $\int_a^b h(x)\, dx$ exists, it equals zero. (*Hint:* Use Exercise 6.)

8. Suppose f and g are integrable on $[a, b]$ and $f(x) = g(x)$ except for x in a set of Lebesgue measure zero. Show that

$$\int_a^b f(x)\, dx = \int_a^b g(x)\, dx.$$

infinite sequences and series

4.1 SEQUENCES OF REAL NUMBERS

An *infinite sequence* (more briefly, a *sequence*) of real numbers is a real-valued function defined on a set of integers $\{n \,|\, n \geq k\}$. We denote a sequence by listing its values in order; thus,

$$\{s_n\}_k^\infty = \{s_k, s_{k+1}, \ldots\}.$$

For example,

$$\left\{\frac{1}{n^2 + 1}\right\}_0^\infty = \left\{1, \frac{1}{2}, \frac{1}{5}, \ldots, \frac{1}{n^2 + 1}, \ldots\right\},$$

$$\{(-1)^n\}_0^\infty = \{1, -1, 1, \ldots, (-1)^n, \ldots\},$$

and

$$\left\{\frac{1}{n - 2}\right\}_3^\infty = \left\{1, \frac{1}{2}, \frac{1}{3}, \ldots, \frac{1}{n - 2}, \ldots\right\}.$$

The real number s_n is the nth *element* of the sequence. Usually we are interested only in the elements of a sequence and the order in which they appear, but not in the particular value of k. Therefore, we regard the sequences

$$\left\{\frac{1}{n-2}\right\}_3^\infty \quad \text{and} \quad \left\{\frac{1}{n}\right\}_1^\infty$$

as identical. In the absence of any statement to the contrary, we will take $k = 0$ unless s_n is given by a rule that is invalid for some nonnegative integer, in which case k will be the smallest positive integer such that s_n is defined for all $n \ge k$. (For example, if

$$s_n = \frac{1}{(n-1)(n-5)},$$

then $k = 6$.) We will usually write $\{s_n\}$ rather than $\{s_n\}_k^\infty$.

The interesting questions about a sequence $\{s_n\}$ concern the behavior of s_n for large n.

Limit of a sequence

Definition 1.1   *A sequence $\{s_n\}$ is said to converge to a limit s if for every positive number ε there is an integer N such that*

$$|s_n - s| < \varepsilon \text{ if } n \ge N. \tag{1}$$

In this case we say that $\{s_n\}$ is convergent and write

$$\lim_{n \to \infty} s_n = s.$$

A sequence that does not converge is said to diverge, or be divergent.

Example 1.1   If $s_n = c$ for $n \ge k$, then $|s_n - c| = 0$ for $n \ge k$, and $\lim_{n \to \infty} s_n = c$.

Example 1.2   The sequence

$$\left\{\frac{2n+1}{n+1}\right\}$$

converges to 2, since

$$\left|\frac{2n+1}{n+1} - 2\right| = \left|\frac{2n+1}{n+1} - \frac{2n+2}{n+1}\right| = \frac{1}{n+1};$$

hence, if $\varepsilon > 0$, then (1) is satisfied, with $s = 2$, if $N \ge 1/\varepsilon$.

Definition 1.1 does not require that there be an integer N such that (1) holds for all ε; rather, it requires that for each positive ε there be an integer N which satisfies (1) for that particular ε. Usually, N depends on ε, and must be increased if ε is decreased. The constant sequences (Example 1.1) are essentially the only ones for which N does not depend on ε (*Exercise 3*).

We say that the elements of a sequence $\{s_n\}_1^\infty$ satisfy a given condition *for all n* if s_n satisfies the condition for all $n \geq k$, or *for large n* if there is an integer $N \geq k$ such that s_n satisfies the condition whenever $n \geq N$. For example, the terms of $\{1/n\}_1^\infty$ are positive for all n, while those of $\{1 - 7/n\}_1^\infty$ are positive for large n (take $N = 8$).

Sequences diverging to $\pm \infty$

We will say that

$$\lim_{n \to \infty} s_n = \infty$$

if for any real number b, $s_n > b$ for large n. Similarly,

$$\lim_{n \to \infty} s_n = -\infty$$

if for any real number a, $s_n < a$ for large n. However, we will not regard $\{s_n\}$ as convergent unless $\lim_{n \to \infty} s_n$ is finite, as required by Definition 1.1. To emphasize this distinction we will say that $\{s_n\}$ *diverges to* $\infty (-\infty)$ if $\lim_{n \to \infty} s_n = \infty (-\infty)$.

Example 1.3 The sequence $\{n/2 + 1/n\}$ diverges to ∞, since if b is any real number,

$$\frac{n}{2} + \frac{1}{n} > b \qquad \text{if} \quad n \geq 2b.$$

The sequence $\{n - n^2\}$ diverges to $-\infty$, since, if a is any real number,

$$-n^2 + n = -n(n - 1) < a \qquad \text{if} \quad n > 1 + \sqrt{|a|}.$$

Therefore, we write

$$\lim_{n \to \infty} \left(\frac{n}{2} + \frac{1}{n}\right) = \infty$$

and

$$\lim_{n \to \infty} (-n^2 + n) = -\infty.$$

The sequence $\{(-1)^n n^3\}$ diverges, but not to $-\infty$ or ∞.

Uniqueness of the limit

Theorem 1.1   *The limit of a convergent sequence is unique.*

Proof.   Suppose $\{s_n\}$ has distinct limits s and s', and let

$$\varepsilon = \frac{|s - s'|}{2}.$$

Then $\varepsilon > 0$ and, from Definition 1.1, there are integers N_1 and N_2 such that

$$|s_n - s| < \varepsilon \qquad \text{if} \quad n \geq N_1$$

(because $\lim_{n \to \infty} s_n = s$), and

$$|s_n - s'| < \varepsilon \qquad \text{if} \quad n \geq N_2$$

(because $\lim_{n \to \infty} s_n = s'$). These inequalities both hold if $n \geq N = \max(N_1, N_2)$, which implies that

$$2\varepsilon = |s - s'| = |(s - s_N) + (s_N - s')|$$
$$\leq |s - s_N| + |s_N - s'| < \varepsilon + \varepsilon = 2\varepsilon.$$

Therefore, the assumption that $\{s_n\}$ has two distinct limits leads to the absurd conclusion that $2\varepsilon < 2\varepsilon$, which means the assumption is false. This proves the theorem.

Bounded sequences

Definition 1.2   *A sequence $\{s_n\}$ is said to be bounded above if there is a real number b such that*

$$s_n \leq b \qquad \text{for all } n,$$

bounded below if there is a real number a such that

$$s_n \geq a \qquad \text{for all } n,$$

and bounded if there is a positive number r such that

$$|s_n| \leq r \qquad \text{for all } n.$$

Example 1.4   If $s_n = [1 + (-1)^n]n$, then $\{s_n\}$ is bounded below ($s_n \geq 0$) but unbounded above, and $\{-s_n\}$ is bounded above ($-s_n \leq 0$) but unbounded below. If $s_n = (-1)^n$, then $\{s_n\}$ is bounded. If $s_n = (-1)^n n$, then $\{s_n\}$ is not bounded above or below.

Theorem 1.2   *A convergent sequence is bounded.*

Proof.   If $\lim_{n\to\infty} s_n = s$, there is an integer N such that

$$|s_n - s| < 1 \qquad \text{if} \quad n \geq N$$

[take $\varepsilon = 1$ in (1)]; therefore,

$$|s_n| = |(s_n - s) + s| \leq |s_n - s| + |s| < 1 + |s| \quad \text{if} \quad n \geq N,$$

and

$$|s_n| \leq \max\{|s_0|, |s_1|, \ldots, |s_{N-1}|, 1 + |s|\}$$

for all n, so $\{s_n\}$ is bounded.

Monotonic sequences

Definition 1.3   *A sequence $\{s_n\}$ is said to be nondecreasing if $s_n \geq s_{n-1}$ for all n, or nonincreasing if $s_n \leq s_{n-1}$ for all n. A monotonic sequence is one which is either nonincreasing or nondecreasing. If $s_n > s_{n-1}$ for all n, then $\{s_n\}$ is said to be increasing, while if $s_n < s_{n-1}$ for all n, it is said to be decreasing.*

Theorem 1.3   *(a) If $\{s_n\}$ is nondecreasing, then $\lim_{n\to\infty} s_n = \text{l.u.b.}\{s_n\}$. (b) If $\{s_n\}$ is nonincreasing, then $\lim_{n\to\infty} s_n = \text{g.l.b.}\{s_n\}$.*

Proof of (a).   We consider the case where $\beta = \text{l.u.b.}\{s_n\} < \infty$, and leave the rest to the reader (*Exercise 6*). Theorem 1.1 of Section 1.1 implies that if $\varepsilon > 0$, then

$$\beta - \varepsilon < s_N \leq \beta$$

for some integer N. Since $s_N \leq s_n \leq \beta$ if $n \geq N$, it follows that

$$\beta - \varepsilon < s_n \leq \beta \qquad \text{if} \quad n \geq N.$$

Hence, $\lim_{n\to\infty} s_n = \beta$, by Definition 1.1.

Example 1.5   If $s_0 = 1$ and $s_n = 1 - e^{-s_{n-1}}$, then $0 < s_n \leq 1$ for all n, by induction. Since

$$s_{n+1} - s_n = -(e^{-s_n} - e^{-s_{n-1}}) \qquad \text{if} \quad n \geq 1,$$

the mean value theorem implies that

$$s_{n+1} - s_n = e^{-t_n}(s_n - s_{n-1}) \qquad \text{if} \quad n \geq 1, \tag{2}$$

where t_n is between s_{n-1} and s_n. Since $s_1 - s_0 = -1/e < 0$, it follows by induction from (2) that $s_{n+1} - s_n < 0$ for all n. Hence, $\{s_n\}$ is bounded and decreasing, and therefore convergent.

The next theorem enables us to apply to sequences some of the theory of limits developed in Section 2.1. We leave the proof to the reader (*Exercise 10*).

Sequences of functional values

Theorem 1.4   *Let* $\lim_{x \to \infty} f(x) = L$, *where* L *is any element in the extended reals, and suppose* $s_n = f(n)$ *for large* n. *Then*

$$\lim_{n \to \infty} s_n = L.$$

Example 1.6   Let

$$s_n = \frac{\log n}{n} \quad \text{and} \quad f(x) = \frac{\log x}{x}.$$

By l'Hospital's rule,

$$\lim_{x \to \infty} \frac{\log x}{x} = \lim_{x \to \infty} \frac{1/x}{1} = 0;$$

hence, $\lim_{n \to \infty} \log n/n = 0$.

Example 1.7   Let $s_n = (1 + 1/n)^n$ and

$$f(x) = \left(1 + \frac{1}{x}\right)^x = e^{x \, \log(1 + 1/x)}.$$

By l'Hospital's rule,

$$\lim_{x \to \infty} x \log\left(1 + \frac{1}{x}\right) = \lim_{x \to \infty} \frac{\log(1 + 1/x)}{1/x}$$

$$= \lim_{x \to \infty} \frac{-\dfrac{1}{x^2}\dfrac{1}{1 + 1/x}}{-1/x^2} = 1;$$

hence,

$$\lim_{x \to \infty} \left(1 + \frac{1}{x}\right)^x = e^1 = e \quad \text{and} \quad \lim_{n \to \infty} \left(1 + \frac{1}{n}\right)^n = e.$$

The last equation is sometimes used to define e.

Example 1.8 Suppose $\rho > 0$ and $s_n = \rho^n$, and let $f(x) = \rho^x = e^{x \log \rho}$. Since

$$\lim_{x \to \infty} e^{x \log \rho} = \begin{cases} 0 & \text{if } \log \rho < 0 \ (0 < \rho < 1), \\ 1 & \text{if } \log \rho = 0 \ (\rho = 1), \\ \infty & \text{if } \log \rho > 1 \ (1 < \rho), \end{cases}$$

it follows that

$$\lim_{n \to \infty} \rho^n = \begin{cases} 0, & 0 < \rho < 1, \\ 1, & \rho = 1, \\ \infty, & 1 < \rho. \end{cases}$$

Therefore,

$$\lim_{n \to \infty} r^n = \begin{cases} 0, & -1 < r < 1, \\ 1, & r = 1, \\ \infty, & r > 1, \end{cases}$$

a result that we will use often.

A useful limit theorem

The next theorem enables us to investigate convergence of sequences by examining simpler sequences. It is analogous to Theorem 1.2 of Section 2.1.

Theorem 1.5 *Let*

$$\lim_{n \to \infty} s_n = s \quad \text{and} \quad \lim_{n \to \infty} t_n = t, \tag{3}$$

where s and t are finite. Then

$$\lim_{n \to \infty} (cs_n) = cs \tag{4}$$

if c is a constant;

$$\lim_{n \to \infty} (s_n + t_n) = s + t; \tag{5}$$

$$\lim_{n \to \infty} (s_n - t_n) = s - t; \tag{6}$$

$$\lim_{n \to \infty} (s_n t_n) = st; \tag{7}$$

and

$$\lim_{n \to \infty} \frac{s_n}{t_n} = \frac{s}{t} \tag{8}$$

if t_n is nonzero for all n and $t \neq 0$.

Proof.   We will prove (7) and a special case of (8), and leave the rest to the reader (*Exercises 12–15*). For (7) we write

$$s_n t_n - st = s_n t_n - s t_n + s t_n - st$$
$$= (s_n - s)t_n + s(t_n - t);$$

hence,

$$|s_n t_n - st| \le |s_n - s| \, |t_n| + |s| \, |t_n - t|. \tag{9}$$

Since $\{t_n\}$ converges, it is bounded (Theorem 1.2); hence, there is a number R such that $|t_n| \le R$ for all n, and (9) implies that

$$|s_n t_n - st| \le R|s_n - s| + |s| \, |t_n - t|. \tag{10}$$

Now (3) implies that if $\varepsilon > 0$ there are integers N_1 and N_2 such that

$$|s_n - s| < \varepsilon \qquad \text{if} \quad n \ge N_1 \tag{11}$$

and

$$|t_n - t| < \varepsilon \qquad \text{if} \quad n \ge N_2. \tag{12}$$

If $N = \max(N_1, N_2)$, then (11) and (12) both hold when $n \ge N$, and (10) implies that

$$|s_n t_n - st| \le (R + |s|)\varepsilon \qquad \text{if} \quad n \ge N. \tag{13}$$

This proves (7).

Now consider (8) in the special case where $s_n = 1$ for all n and $t \ne 0$; thus, we want to show that

$$\lim_{n \to \infty} \frac{1}{t_n} = \frac{1}{t}.$$

First observe that since $\lim_{n \to \infty} t_n = t \ne 0$, there is an integer M such that $|t_n| \ge |t|/2$ if $n \ge M$ (*Exercise 16*). If $\varepsilon > 0$, choose N_2 to satisfy (12) and define $\tilde{N} = \max(N_2, M)$. Then

$$\left| \frac{1}{t_n} - \frac{1}{t} \right| = \frac{|t - t_n|}{|t_n| \, |t|} \le \frac{2\varepsilon}{|t|^2} \qquad \text{if} \quad n \ge \tilde{N}; \tag{14}$$

hence, $\lim_{n \to \infty} 1/t_n = 1/t$.

As we saw in Section 2.1 when discussing limits of functions, it is of no consequence that we ended up with multiples of ε in (13) and (14) rather than with ε alone, as apparently required in Definition 1.1.

Example 1.9   To determine the limit of the sequence defined by

$$s_n = \frac{1}{n} \sin \frac{n\pi}{4} + \frac{2(1 + 3/n)}{1 + 1/n},$$

we apply the applicable parts of Theorem 1.5 as follows:

$$\lim_{n \to \infty} s_n = \lim_{n \to \infty} \frac{1}{n} \sin \frac{n\pi}{4} + \frac{2\left[\lim_{n \to \infty} 1 + 3 \lim_{n \to \infty} (1/n) \right]}{\lim_{n \to \infty} 1 + \lim_{n \to \infty} (1/n)}$$

$$= 0 + \frac{2(1 + 3 \cdot 0)}{1 + 0} = 2.$$

Sometimes preliminary manipulations are necessary before applying Theorem 1.5. For example,

$$\lim_{n \to \infty} \left(\frac{(n/2) + \log n}{3n + 4\sqrt{n}} \right) = \lim_{n \to \infty} \frac{\frac{1}{2} + (\log n)/n}{3 + 4n^{-1/2}}$$

$$= \frac{\lim_{n \to \infty} \frac{1}{2} + \lim_{n \to \infty} (\log n)/n}{\lim_{n \to \infty} 3 + 4 \lim_{n \to \infty} n^{-1/2}}$$

$$= \frac{\frac{1}{2} + 0}{3 + 0} \qquad \text{(see Example 1.6)}$$

$$= \frac{1}{6}.$$

Example 1.10   Suppose $-1 < r < 1$ and

$$s_0 = 1, \quad s_1 = 1 + r, \quad s_2 = 1 + r + r^2, \dots,$$
$$s_n = 1 + r + \cdots + r^n.$$

Since

$$s_n - rs_n = (1 + r + \cdots + r^n) - (r + r^2 + \cdots + r^{n+1})$$
$$= 1 - r^{n+1},$$

it follows that

$$s_n = \frac{1 - r^{n+1}}{1 - r}. \tag{15}$$

From Example 1.8, $\lim_{n \to \infty} r^{n+1} = 0$, so (15) and Theorem 1.5 yield

$$\lim_{n \to \infty} (1 + r + \cdots + r^n) = \frac{1}{1 - r} \qquad \text{if} \quad -1 < r < 1.$$

Equations (4), (5), (6), and (7) are valid even if s and t are arbitrary elements in the extended reals, provided their right sides are defined in the extended reals (*Exercise 18*); (8) is valid if s/t is not indeterminate *and* $t \neq 0$ (*Exercise 19*).

Example 1.11   If $-1 < r < 1$, then

$$\lim_{n \to \infty} \frac{r^n}{n!} = \frac{\displaystyle\lim_{n \to \infty} r^n}{\displaystyle\lim_{n \to \infty} n!} = \frac{0}{\infty} = 0,$$

from (8) and Example 1.8; however, if $r > 1$, (8) and Example 1.8 yield

$$\lim_{n \to \infty} \frac{r^n}{n!} = \frac{\displaystyle\lim_{n \to \infty} r^n}{\displaystyle\lim_{n \to \infty} n!} = \frac{\infty}{\infty},$$

an indeterminate form. If $r \leq -1$, then $\lim_{n \to \infty} r^n$ does not exist in the extended reals, so (8) is again not applicable. Theorem 1.4 does not help either, since there is no elementary function f such that $f(n) = r^n/n!$. However, the following argument shows that

$$\lim_{n \to \infty} \frac{r^n}{n!} = 0, \qquad -\infty < r < \infty. \tag{16}$$

There is an integer M such that

$$\frac{|r|}{n} < \frac{1}{2} \qquad \text{if} \quad n \geq M.$$

Let $K = r^m/M!$. Then

$$\left|\frac{r^n}{n!}\right| \leq K \frac{|r|}{M+1} \frac{|r|}{M+2} \cdots \frac{|r|}{n}$$

$$< K \left(\frac{1}{2}\right)^{n-M} \qquad \text{if} \quad n > M.$$

Since $\lim_{n \to \infty} 2^{M-n} K = 0$, this implies (16) [*Exercise 1(b)*].

Limits superior and inferior

Requiring a sequence to converge may be unnecessarily restrictive. Often useful results can be obtained from assumptions on the *limit superior* and *limit inferior* of a sequence, which we consider next.

Theorem 1.6   (a) *If $\{s_n\}$ is bounded above and does not diverge to $-\infty$, then there is a unique real number $\bar{s}$ such that, if $\varepsilon > 0$,*

$$s_n < \bar{s} + \varepsilon \qquad \textit{for large } n \tag{17}$$

and

$$s_n > \bar{s} - \varepsilon \qquad \text{for infinitely many values of } n. \tag{18}$$

(b) *If* $\{s_n\}$ *is bounded below and does not diverge to* ∞*, then there is a unique real number* $\underline{s}$ *such that, if* $\varepsilon > 0$,

$$s_n > \underline{s} - \varepsilon \qquad \text{for large } n \tag{19}$$

and

$$s_n < \underline{s} + \varepsilon \qquad \text{for infinitely many values of } n. \tag{20}$$

Proof of (a). The hypotheses of (a) imply that there are numbers α and β such that $s_n < \beta$ for all n and $s_n > \alpha$ for infinitely many values of n. If we define

$$M_k = \text{l.u.b.}\{s_k, s_{k+1}, \ldots, s_{k+r}, \ldots\},$$

then $\alpha \le M_k \le \beta$, so that $\{M_k\}$ is bounded. Since $\{M_k\}$ is also nonincreasing (why?), it converges, by Theorem 1.3. Let

$$\bar{s} = \lim_{k \to \infty} M_k. \tag{21}$$

If $\varepsilon > 0$, then $M_k < \bar{s} + \varepsilon$ for large k, and since $s_n \le M_k$ for $n \ge k$, $\bar{s}$ satisfies (17). If (18) were false for some positive ε, there would be an integer K such that

$$s_n \le \bar{s} - \varepsilon \qquad \text{if} \quad n \ge K;$$

however, this implies that

$$M_k \le \bar{s} - \varepsilon \qquad \text{if} \quad k \ge K,$$

which contradicts (21). Therefore, $\bar{s}$ as defined by (21) has the required properties. Now we must show that $\bar{s}$ is the only real number with these properties.

If $t < \bar{s}$, the inequality

$$s_n < t + \frac{\bar{s} - t}{2} = \bar{s} - \frac{\bar{s} - t}{2}$$

cannot hold for all large n, because this would contradict (18) with $\varepsilon = (\bar{s} - t)/2$. If $\bar{s} < t$, the inequality

$$s_n \ge t - \frac{t - \bar{s}}{2} = \bar{s} + \frac{t - \bar{s}}{2}$$

cannot hold for infinitely many n, because this would contradict (17) with $\varepsilon = (t - \bar{s})/2$. Therefore, $\bar{s}$ as defined by (21) is the only real number that satisfies (17) and (18) for all positive ε.

This completes the proof of (a). We leave the proof of (b) to the reader (*Exercise 20*).

Definition 1.4   *The numbers $\bar{s}$ and $\underline{s}$ defined in Theorem 1.6 are called the limit superior and limit inferior, respectively, of $\{s_n\}$, and denoted by*

$$\bar{s} = \overline{\lim_{n \to \infty}} \, s_n \quad and \quad \underline{s} = \underline{\lim_{n \to \infty}} \, s_n.$$

We also define

$$\overline{\lim_{n \to \infty}} \, s_n = \infty \qquad if \ \{s_n\} \ is \ not \ bounded \ above;$$

$$\overline{\lim_{n \to \infty}} \, s_n = -\infty \qquad if \ \lim_{n \to \infty} \, s_n = -\infty;$$

$$\underline{\lim_{n \to \infty}} \, s_n = -\infty \qquad if \ \{s_n\} \ is \ not \ bounded \ below;$$

and

$$\underline{\lim_{n \to \infty}} \, s_n = \infty \qquad if \ \lim_{n \to \infty} \, s_n = \infty.$$

Theorem 1.7   *Every sequence $\{s_n\}$ of real numbers has a unique limit superior, $\bar{s}$, and a unique limit inferior, $\underline{s}$, in the extended reals, and*

$$\underline{s} \le \bar{s}. \tag{22}$$

Proof.   The existence and uniqueness of $\bar{s}$ and $\underline{s}$ follow from Theorem 1.6 and Definition 1.4. If $\bar{s}$ and $\underline{s}$ are both finite, then (17) and (19) imply that

$$\underline{s} - \varepsilon < \bar{s} + \varepsilon$$

for every positive ε, which yields (22). If $\underline{s} = -\infty$ or $\bar{s} = \infty$, then (22) is obvious. If $\underline{s} = \infty$ or $\bar{s} = -\infty$, (22) follows immediately from Definition 1.4.

Example 1.12

$$\overline{\lim_{n \to \infty}} \, r^n = \begin{cases} \infty, & |r| > 1, \\ 1, & |r| = 1, \\ 0, & |r| < 1; \end{cases}$$

and

$$\underline{\lim_{n \to \infty}} \, r^n = \begin{cases} \infty, & r > 1, \\ 1, & r = 1, \\ 0, & |r| < 1, \\ -1, & r = -1, \\ -\infty, & r < -1. \end{cases}$$

Also,

$$\overline{\lim_{n \to \infty}} \ n^2 = \underline{\lim_{n \to \infty}} \ n^2 = \infty;$$

$$\overline{\lim_{n \to \infty}} \ (-1)^n \left(1 - \frac{1}{n}\right) = 1, \qquad \underline{\lim_{n \to \infty}} \ (-1)^n \left(1 - \frac{1}{n}\right) = -1;$$

and

$$\overline{\lim_{n \to \infty}} \ [1 + (-1)^n]n^2 = \infty, \qquad \underline{\lim_{n \to \infty}} \ [1 + (-1)^n]n^2 = 0.$$

Theorem 1.8 *If $\{s_n\}$ is a sequence of real numbers, then*

$$\lim_{n \to \infty} s_n = s \tag{23}$$

if and only if

$$\overline{\lim_{n \to \infty}} \ s_n = \underline{\lim_{n \to \infty}} \ s_n = s. \tag{24}$$

Proof. If s is infinite, the equivalence of (23) and (24) follows immediately from their definitions. If $\lim_{n \to \infty} s_n = s$ (finite), then Definition 1.1 implies that (17), (18), (19), and (20) hold with $\overline{s}$ and $\underline{s}$ replaced by s; hence, (24) follows from the uniqueness of $\overline{s}$ and $\underline{s}$. For the converse, suppose $\overline{s} = \underline{s}$ and let s denote their common value. Then (17) and (19) imply that

$$s - \varepsilon < s_n < s + \varepsilon$$

for large n, and (23) follows from Definition 1.1 and the uniqueness of $\lim_{n \to \infty} s_n$ (Theorem 1.1).

Cauchy's convergence criterion

To determine from Definition 1.1 whether a sequence has a limit, it is necessary to guess what the limit is. (This is particularly difficult if the sequence diverges!) To use Theorem 1.8 for this purpose requires finding $\overline{s}$ and $\underline{s}$. The following convergence criterion has neither of these defects.

Theorem 1.9 (Cauchy's Convergence Criterion) *A sequence $\{s_n\}$ of real numbers converges if and only if for every positive ε there is an integer N such that*

$$|s_n - s_m| < \varepsilon \qquad if \quad m, n \geq N. \tag{25}$$

Proof.   Suppose $\lim_{n\to\infty} s_n = s$ and $\varepsilon > 0$. By Definition 1.1 there is an integer N such that

$$|s_r - s| < \frac{\varepsilon}{2} \qquad \text{if} \quad r \geq N.$$

Therefore,

$$|s_n - s_m| = |(s_n - s) + (s - s_m)| \leq |s_n - s| + |s - s_m| < \varepsilon \qquad \text{if} \quad n, m \geq N,$$

which means that the stated condition is necessary for convergence of $\{s_n\}$. To see that it is sufficient, we first observe that it implies that $\{s_n\}$ is bounded (*Exercise 23*), so that $\bar{s}$ and $\underline{s}$ are finite (Theorem 1.6). Now suppose $\varepsilon > 0$ and N satisfies (25). From (17) and (18),

$$|s_n - \bar{s}| < \varepsilon, \tag{26}$$

for some integer $n > N$ and, from (19) and (20),

$$|s_m - \underline{s}| < \varepsilon \tag{27}$$

for some integer $m > N$. Since

$$|\bar{s} - \underline{s}| = |(\bar{s} - s_n) + (s_n - s_m) + (s_m - \underline{s})|$$
$$\leq |\bar{s} - s_n| + |s_n - s_m| + |s_m - \underline{s}|,$$

(25), (26), and (27) imply that

$$|\bar{s} - \underline{s}| < 3\varepsilon.$$

Since ε is an arbitrary positive number, this implies that $\bar{s} = \underline{s}$, and $\{s_n\}$ therefore converges, by Theorem 1.8.

Example 1.13   Let f be differentiable and

$$|f'(x)| \leq r < 1, \qquad -\infty < x < \infty. \tag{28}$$

With x_0 arbitrary, define

$$x_n = f(x_{n-1}), \qquad n \geq 1. \tag{29}$$

We will show that $\{x_n\}$ converges. From (29) and the mean value theorem,

$$x_{n+1} - x_n = f(x_n) - f(x_{n-1})$$
$$= f'(c_n)(x_n - x_{n-1}),$$

where c_n is between x_{n-1} and x_n. This and (28) imply that

$$|x_{n+1} - x_n| \leq r|x_n - x_{n-1}| \qquad \text{if} \quad n \geq 1. \tag{30}$$

The inequality

$$|x_{n+1} - x_n| \leq r^n|x_1 - x_0| \qquad \text{if} \quad n \geq 0, \tag{31}$$

follows by induction from (30). Now, if $n > m$,

$$|x_n - x_m| = |(x_n - x_{n-1}) + (x_{n-1} - x_{n-2}) + \cdots + (x_{m+1} - x_m)|$$
$$\leq |x_n - x_{n-1}| + |x_{n-1} - x_{n-2}| + \cdots + |x_{m+1} - x_m|,$$

and (31) yields

$$|x_n - x_m| \leq |x_1 - x_0| r^m (1 + r + \cdots + r^{n-m-1}). \tag{32}$$

In Example 1.10 we saw that the sequence $\{s_k\}$ defined by

$$s_k = 1 + r + \cdots + r^k$$

converges to $1/(1 - r)$ if $|r| < 1$; moreover, since we have assumed here that $0 < r < 1$, $\{s_k\}$ is nondecreasing, and therefore $s_k < 1/(1 - r)$ for all k. (Why?) Therefore, (32) yields

$$|x_n - x_m| < \frac{|x_1 - x_0|}{1 - r} r^m \qquad \text{if} \quad n > m.$$

Now it follows that

$$|x_n - x_m| < \frac{|x_1 - x_0|}{1 - r} r^N \qquad \text{if} \quad n, m > N,$$

and, since $\lim_{N \to \infty} r^N = 0$, $\{x_n\}$ converges, by Theorem 1.9.

4.1 EXERCISES

1. (a) Show that $\lim_{n \to \infty} s_n = s$ (finite) if and only if $\lim_{n \to \infty} |s_n - s| = 0$.
   (b) Suppose that $|s_n - s| \leq t_n$ for large n and $\lim_{n \to \infty} t_n = 0$. Show that $\lim_{n \to \infty} s_n = s$.

2. Use Definition 1.1 to show that $\{s_n\}$ is convergent:

   (a) $s_n = 2 + \dfrac{1}{n + 1}$ (b) $s_n = \dfrac{\alpha + n}{\beta + n}$ (c) $s_n = \dfrac{1}{n} \sin \dfrac{n\pi}{4}$

3. State necessary and sufficient conditions on a convergent sequence $\{s_n\}$ such that the integer N in Definition 1.1 does not depend upon ε.

* 4. Prove: If $\lim_{n \to \infty} s_n = s$, then $\lim_{n \to \infty} |s_n| = |s|$.

5. Suppose $\{s_n\}$ converges and, for each $\varepsilon > 0$, $|s_n - t_n| < \varepsilon$ for large n. Show that $\lim_{n \to \infty} t_n = \lim_{n \to \infty} s_n$.

6. Complete the proof of Theorem 1.3.

7. Use Theorem 1.3 to show that $\{s_n\}$ converges:

   (a) $s_n = \dfrac{\alpha + n}{\beta + n}$ $(\beta > 0)$ (b) $s_n = \dfrac{n!}{n^n}$

(c) $s_n = \dfrac{r^n}{1 + r^n} \ (r > 0)$ (d) $s_n = \dfrac{(2n)!}{2^{2n}(n!)^2}$

8. Let $y = \text{Tan}^{-1} x$ be the solution of $x = \tan y$ such that $-\pi/2 < y < \pi/2$. Prove: If $x_0 > 0$ and $x_{n+1} = \text{Tan}^{-1} x_n \ (n \geq 0)$, then $\{x_n\}$ converges.

9. Prove: If $\{s_n\}$ is unbounded and monotonic, then either $\lim_{n \to \infty} s_n = \infty$ or $\lim_{n \to \infty} s_n = -\infty$.

10. Prove Theorem 1.4.

11. Use Theorem 1.4 to find $\lim_{n \to \infty} s_n$:

(a) $s_n = \dfrac{\alpha + n}{\beta + n} \ (\beta > 0)$ (b) $s_n = \cos \dfrac{1}{n}$

(c) $s_n = n \sin \dfrac{1}{n}$ (d) $s_n = \log n - n$

(e) $s_n = \dfrac{a^n}{n^r} \ (a, r > 0)$ (f) $s_n = \log(n + 1) - \log(n - 1)$

12. Prove Eq. (4) of Theorem 1.5.

13. Prove Eq. (5) of Theorem 1.5.

14. Obtain Eqs. (6) and (8) of Theorem 1.5 from Eqs. (4), (5), (7), and the special case of (8) ($s_n = s = 1$) proved in the text.

15. Prove Eq. (8) of Theorem 1.5 in the general case, directly from Definition 1.1.

16. Prove: If $\lim_{n \to \infty} t_n = t$, where $0 < |t| < \infty$, then $|t_n| > |t|/2$ for large N.

17. Use Theorem 1.5 to find $\lim_{n \to \infty} s_n$:

(a) $s_n = \dfrac{\alpha + n}{\beta + n} \ (\beta > 0)$ (b) $s_n = \dfrac{r^n}{1 + r^n} \ (r > 0)$

(c) $s_n = n^2 - n$ (d) $s_n = 2^n + \dfrac{1}{n}$

(e) $s_n = \dfrac{n}{n + 1} - \log n$

(f) $s_n = \dfrac{a_0 + a_1 n + \cdots + a_r n^r}{b_0 + b_1 n + \cdots + b_s n^s} \ (a_r > 0, b_s > 0)$

18. Prove: Except when their right sides are indeterminate forms, Eqs. (4), (5), (6), and (7) are also valid if one or both of s and t is infinite.

19. Give an example showing that Eq. (8) may be incorrect if $t = 0$, even if $s \neq 0$ (that is, even though s/t is *not* an indeterminate form).

20. Prove (b) of Theorem 1.6 by a method similar to that used to prove (a).

21. Find $\bar{s}$ and $\underline{s}$ if

(a) $s_n = [(-1)^n + 1]n^2$

(b) $s_n = (1 - r^n)\sin\dfrac{n\pi}{2}$

(c) $s_n = \dfrac{r^{2n}}{1 + r^n} \ (r \neq -1)$

(d) $s_n = n^2 - n$

(e) $s_n = (-1)^n t_n$, where $\lim\limits_{n \to \infty} t_n = t$

22. Suppose $\lim_{n \to \infty} |s_n| = \gamma$ (finite). Show that $\{s_n\}$ diverges unless $\gamma = 0$ or the elements of $\{s_n\}$ have the same sign for large n. (*Hint:* Use Exercise 16.)

23. Prove: The sequence $\{s_n\}$ is bounded if, for some positive ε, there is an integer N such that $|s_n - s_m| < \varepsilon$ whenever $n, m \geq N$.

In Exercises 24–27, assume that $\bar{s}$, $\underline{s}$ (or s), $\bar{t}$, and $\underline{t}$ are elements of the extended reals, and show that the given inequalities or equations hold whenever their right sides are defined (not indeterminate).

24. (a) $\overline{\lim\limits_{n \to \infty}} \ (-s_n) = -\underline{s}$

(b) $\underline{\lim\limits_{n \to \infty}} \ (-s_n) = -\bar{s}$

25. (a) $\overline{\lim\limits_{n \to \infty}} \ (s_n + t_n) \leq \bar{s} + \bar{t}$

(b) $\underline{\lim\limits_{n \to \infty}} \ (s_n + t_n) \geq \underline{s} + \underline{t}$

26. (a) If $s_n \geq 0$, $t_n \geq 0$, then (i) $\overline{\lim\limits_{n \to \infty}} \ s_n t_n \leq \bar{s}\bar{t}$ and (ii) $\underline{\lim\limits_{n \to \infty}} \ s_n t_n \geq \underline{s}\underline{t}$.

(b) If $s_n \leq 0$, $t_n \geq 0$, then (i) $\overline{\lim\limits_{n \to \infty}} \ s_n t_n \leq \bar{s}\underline{t}$ and (ii) $\underline{\lim\limits_{n \to \infty}} \ s_n t_n \geq \underline{s}\bar{t}$.

*27. (a) If $\lim\limits_{n \to \infty} s_n = s > 0$ and $t_n \geq 0$, then (i) $\overline{\lim\limits_{n \to \infty}} \ s_n t_n = s\bar{t}$ and (ii) $\underline{\lim\limits_{n \to \infty}} \ s_n t_n = s\underline{t}$.

(b) If $\lim\limits_{n \to \infty} s_n = s < 0$ and $t_n \geq 0$, then (i) $\overline{\lim\limits_{n \to \infty}} \ s_n t_n = s\underline{t}$ and (ii) $\underline{\lim\limits_{n \to \infty}} \ s_n t_n = s\bar{t}$.

28. Suppose $\{s_n\}$ converges and has only finitely many distinct elements. Show that s_n is constant for large n.

29. Let $t_n = \dfrac{s_1 + s_2 + \cdots + s_n}{n}$, $n = 1, 2, \ldots$.

(a) Prove: If $\lim_{n \to \infty} s_n = s$, then $\lim_{n \to \infty} t_n = s$.
(b) Give an example to show that $\{t_n\}$ may converge even though $\{s_n\}$ does not.

*30. (a) Show that

$$\lim_{n \to \infty} \left(1 - \frac{\alpha}{1}\right)\left(1 - \frac{\alpha}{2}\right) \cdots \left(1 - \frac{\alpha}{n}\right) = 0, \qquad \alpha > 0.$$

(*Hint:* Look at the logarithm of the absolute value of the product.)

(b) Conclude from part (a) that

$$\lim_{n \to \infty} \binom{q}{n} = 0, \qquad q > -1,$$

where $\binom{q}{n}$ is the generalized binomial coefficient of Example 5.3, Section 2.5.

4.2 EARLIER TOPICS REVISITED WITH SEQUENCES

In Chapter 2 we used epsilon–delta definitions and arguments to develop the theory of limits, continuity, and differentiability; thus, f is continuous at x_0 if for each $\varepsilon > 0$ there is a $\delta > 0$ such that $|f(x) - f(x_0)| < \varepsilon$ when $|x - x_0| < \delta$. The same theory can be developed by methods based on sequences. Although we will not carry this out in detail, we will develop it enough to give some examples. First we need another definition about sequences.

Definition 2.1   *A sequence $\{\overline{s}_k\}$ is said to be a subsequence of a sequence $\{s_n\}$ if*

$$\overline{s}_k = s_{n_k}, \qquad k = 0, 1, \ldots,$$

where $\{n_k\}$ is an increasing infinite sequence of integers in the domain of $\{s_n\}$. We denote the subsequence $\{\overline{s}_k\}$ by $\{s_{n_k}\}$.

Note that $\{s_n\}$ is a subsequence of itself, as can be seen by taking $n_k = k$. All other subsequences of $\{s_n\}$ are obtained by deleting elements from $\{s_n\}$ and leaving those remaining in their original relative order.

Example 2.1   If

$$\{s_n\} = \left\{\frac{1}{n}\right\} = \left\{1, \frac{1}{2}, \frac{1}{3}, \ldots, \frac{1}{n}, \ldots\right\},$$

then letting $n_k = 2k$ yields the subsequence

$$\{s_{2k}\} = \left\{\frac{1}{2k}\right\} = \left\{\frac{1}{2}, \frac{1}{4}, \ldots, \frac{1}{2k}, \ldots\right\},$$

and letting $n_k = 2k + 1$ yields

$$\{s_{2k+1}\} = \left\{\frac{1}{2k+1}\right\} = \left\{1, \frac{1}{3}, \ldots, \frac{1}{2k+1}, \ldots\right\}.$$

Since a subsequence $\{s_{n_k}\}$ is again a sequence (with respect to k), we may ask whether it converges.

Example 2.2   The sequence $\{s_n\}$ defined by

$$s_n = (-1)^n \left(1 + \frac{1}{n}\right)$$

does not converge, but it has subsequences which do; for example,

$$\{s_{2k}\} = \left\{1 + \frac{1}{2k}\right\} \quad \text{and} \quad \lim_{k \to \infty} s_{2k} = 1,$$

while

$$\{s_{2k+1}\} = \left\{-1 - \frac{1}{2k+1}\right\} \quad \text{and} \quad \lim_{k \to \infty} s_{2k+1} = -1.$$

It can be shown (*Exercise 1*) that a subsequence $\{s_{n_k}\}$ of $\{s_n\}$ converges to 1 if and only if n_k is even for k sufficiently large ($\geq$ some integer K), or to -1 if and only if n_k is odd for k sufficiently large; otherwise, $\{s_{n_k}\}$ diverges.

The sequence in this example has subsequences which converge to different limits. The next theorem shows that if a sequence converges to a finite limit or diverges to $\pm\infty$, all its subsequences do also.

Theorem 2.1   *If*

$$\lim_{n \to \infty} s_n = s \qquad (-\infty \leq s \leq \infty), \tag{1}$$

then

$$\lim_{k \to \infty} s_{n_k} = s \tag{2}$$

for every subsequence $\{s_{n_k}\}$ of $\{s_n\}$.

Proof.   We consider the case where s is finite and leave the rest to the reader (*Exercise 4*). If (1) holds and $\varepsilon > 0$, there is an integer N such that

$$|s_n - s| < \varepsilon \qquad \text{if} \quad n \geq N.$$

Since $\{n_k\}$ is an increasing sequence there is an integer K such that $n_k \geq N$ if $k \geq K$; therefore,

$$|s_{n_k} - L| < \varepsilon \qquad \text{if} \quad k \geq K,$$

which implies (2).

Theorem 2.2   *If* $\{s_n\}$ *is monotonic and has a subsequence* $\{s_{n_k}\}$ *such that*

$$\lim_{k \to \infty} s_{n_k} = s \qquad (-\infty \le s \le \infty),$$

then

$$\lim_{n \to \infty} s_n = s.$$

Proof.   We consider the case where $\{s_n\}$ is nondecreasing and leave the rest to the reader (*Exercise* 6). Since $\{s_{n_k}\}$ is also nondecreasing in this case, it suffices to show that

$$\mathrm{l.u.b.}\{s_{n_k}\} = \mathrm{l.u.b.}\{s_n\} \tag{3}$$

and invoke Theorem 1.3(a). Since the set of elements of $\{s_{n_k}\}$ is contained in the set of elements of $\{s_n\}$,

$$\mathrm{l.u.b.}\{s_n\} \ge \mathrm{l.u.b.}\{s_{n_k}\}.$$

Since $\{s_n\}$ is nondecreasing, there is for every n an integer n_k such that $s_n \le s_{n_k}$. This implies that

$$\mathrm{l.u.b.}\{s_n\} \le \mathrm{l.u.b.}\{s_{n_k}\}.$$

The last two inequalities imply (3), which completes the proof.

Limit points in terms of sequences

In Section 1.3 we defined *limit point* in terms of neighborhoods: $\bar{x}$ is a limit point of a set S if every neighborhood of $\bar{x}$ contains points of S distinct from $\bar{x}$. The next theorem shows that an equivalent definition can be stated in terms of sequences.

Theorem 2.3   *A point* $\bar{x}$ *is a limit point of a set S if and only if there is a sequence* $\{x_n\}$ *of points in S such that* $\bar{x} \ne x_n$ ($n = 0, 1, \ldots$) *and*

$$\lim_{n \to \infty} x_n = \bar{x}.$$

Proof.   For sufficiency, suppose that the stated condition is satisfied; then for each $\varepsilon > 0$ there is an integer N such that $0 < |x_n - x| < \varepsilon$ if $n \ge N$, and therefore every ε-neighborhood of $\bar{x}$ contains infinitely many points of S. This means that $\bar{x}$ is a limit point of S.

For necessity, let $\bar{x}$ be a limit point of S. Then every interval of the form $(\bar{x} - 1/n, \bar{x} + 1/n)$ contains a point x_n ($\ne \bar{x}$) in S ($n = 1, 2, \ldots$). Since $|x_m - \bar{x}| \le 1/n$ if $m \ge n$, $\lim_{n \to \infty} x_n = \bar{x}$. This completes the proof.

The next theorem will be used to show that continuity can be defined in terms of sequences.

Theorem 2.4   (a) *If* $\{x_n\}$ *is bounded, then it has a convergent subsequence.* (b) *If* $\{x_n\}$ *is unbounded above (below), then it has a subsequence* $\{x_{n_k}\}$ *such that*

$$\lim_{k \to \infty} x_{n_k} = \infty \ (-\infty).$$

Proof.   We prove (a) and leave (b) to the reader (*Exercise 7*). Let S be the set of distinct points that occur as elements of $\{x_n\}$ (for example, if $\{x_n\} = \{(-1)^n\}$, then $S = \{1, -1\}$; if $\{x_n\} = \{1, \frac{1}{2}, 1, \frac{1}{3}, \ldots, 1, 1/n, \ldots\}$, then $S = \{1, \frac{1}{2}, \ldots, 1/n, \ldots\}$). If S contains only finitely many points, then some $\bar{x}$ in S occurs infinitely often in $\{x_n\}$; that is, $\{x_n\}$ has a subsequence $\{x_{n_k}\}$ such that $x_{n_k} = \bar{x} \ (k = 0, 1, \ldots)$. Then $\lim_{k \to \infty} x_{n_k} = \bar{x}$, and we are finished in this case.

If S is infinite, then, since it is bounded (by assumption), the Bolzano–Weierstrass theorem implies that S has a limit point $\bar{x}$. From Theorem 2.3 there is a sequence of points $\{y_j\}$ in S, distinct from $\bar{x}$, such that

$$\lim_{j \to \infty} y_j = \bar{x}. \tag{4}$$

Although each y_j occurs as an element of $\{x_n\}$, $\{y_j\}$ is not necessarily a subsequence of $\{x_n\}$, because if we write

$$y_j = x_{n_j}$$

there is no reason to expect that $\{n_j\}$ is an increasing sequence as required in Definition 2.1. However, it is always possible to pick a subsequence $\{n_{j_k}\}$ of $\{n_j\}$ which is increasing, and then the sequence $\{y_{j_k}\} = \{x_{n_{j_k}}\}$ is a subsequence of both $\{y_j\}$ and $\{x_n\}$. Because of (4) and Theorem 2.1 this subsequence converges to $\bar{x}$. This completes the proof.

Continuity in terms of sequences

We now show that continuity can be defined and studied in terms of sequences.

Theorem 2.5   *Let f be defined on a closed interval $[a, b]$ containing $\bar{x}$. Then f is continuous at $\bar{x}$ (continuous from the right if $\bar{x} = a$, from the left if $\bar{x} = b$) if and only if*

$$\lim_{n \to \infty} f(x_n) = f(\bar{x}) \tag{5}$$

whenever $\{x_n\}$ is a sequence of points in $[a, b]$ such that

$$\lim_{n \to \infty} x_n = \bar{x}. \tag{6}$$

Proof. Assume that $a < \bar{x} < b$; only minor changes in the proof are needed if $\bar{x} = a$ or $\bar{x} = b$. First suppose f is continuous at $\bar{x}$ and $\{x_n\}$ is a sequence of points in $[a, b]$ satisfying (6). If $\varepsilon > 0$, there is a $\delta > 0$ such that

$$|f(x) - f(\bar{x})| < \varepsilon \qquad \text{if} \quad |x - \bar{x}| < \delta.$$

Because of (6) there is an integer N such that $|x_n - \bar{x}| < \delta$, and therefore $|f(x_n) - f(\bar{x})| < \varepsilon$, if $n \geq N$. This implies (5), which shows that the stated condition is necessary.

If f is discontinuous at $\bar{x}$, there is an $\varepsilon_0 > 0$ such that

$$|f(x_n) - f(\bar{x})| \geq \varepsilon_0, \qquad n = 1, 2, \ldots$$

for some x_n satisfying

$$|x_n - \bar{x}| < \frac{1}{n}.$$

The sequence $\{x_n\}$ therefore satisfies (6) but not (5); hence, the given condition cannot hold if f is discontinuous at $\bar{x}$. This proves sufficiency.

Armed with the theorems we have proved so far in this section, we could develop the theory of continuous functions by means of definitions and proofs based on sequences and subsequences. We give one example—a new proof of Theorem 2.4, Section 2.2—and leave others for exercises.

Theorem 2.6   *If f is continuous on a closed interval $[a, b]$, then f is bounded on $[a, b]$.*

Proof.   If f is not bounded on $[a, b]$, there is for each positive integer n a point x_n in $[a, b]$ such that

$$|f(x_n)| > n. \tag{7}$$

Since $\{x_n\}$ is bounded, it has a convergent subsequence $\{x_{n_k}\}$ [Theorem 2.4(a)]. If

$$\bar{x} = \lim_{k \to \infty} x_{n_k},$$

then $\bar{x}$ is a limit point of $[a, b]$, and so $\bar{x} \in [a, b]$. If f is continuous on $[a, b]$, then

$$\lim_{k \to \infty} f(x_{n_k}) = f(\bar{x})$$

by Theorem 2.5, and so

$$\lim_{k \to \infty} |f(x_{n_k})| = |f(\bar{x})|$$

(*Exercise 4, Section 4.1*). But this contradicts (7), which implies that $|f(x_{n_k})| > n_k$, and so

$$\lim_{k \to \infty} |f(x_{n_k})| = \infty.$$

Therefore, f cannot be both continuous and unbounded on $[a, b]$. This completes the proof.

4.2   EXERCISES

1. Let $s_n = (-1)^n(1 + 1/n)$. Show that $\lim_{k \to \infty} s_{n_k} = 1$ if and only if n_k is even for large k, $\lim_{k \to \infty} s_{n_k} = -1$ if and only if n_k is odd for large k, and $\{s_{n_k}\}$ diverges otherwise.

2. Find all elements L in the extended reals that are limits of some subsequence of $\{s_n\}$ and, for each such L, choose a subsequence $\{s_{n_k}\}$ such that $\lim_{k \to \infty} s_{n_k} = L$:

    (a) $s_n = (-1)^n n$ (b) $s_n = \left(1 + \dfrac{1}{n}\right)\cos\dfrac{n\pi}{2}$ (c) $s_n = \left(1 - \dfrac{1}{n^2}\right)\sin\dfrac{n\pi}{2}$

    (d) $s_n = \dfrac{1}{n}$ (e) $s_n = [(-1)^n + 1]n^2$

    (f) $s_n = \dfrac{n+1}{n+2}\left(\sin\dfrac{n\pi}{4} + \cos\dfrac{n\pi}{4}\right)$

3. Construct a sequence $\{s_n\}$ with the following property, or show that none exists: for each positive integer m, $\{s_n\}$ has a subsequence converging to m.

4. Complete the proof of Theorem 2.1.

5. Prove: If $\lim_{n \to \infty} s_n = s$ and $\{s_n\}$ has a subsequence $\{s_{n_k}\}$ such that $(-1)^k s_{n_k} \geq 0$, then $s = 0$.

6. Complete the proof of Theorem 2.2.

7. Prove Theorem 2.4(b).

8. Suppose $\{s_n\}$ is bounded and all convergent subsequences of $\{s_n\}$ converge to the same limit. Show that $\{s_n\}$ is convergent. Give an example showing that the conclusion need not hold if $\{s_n\}$ is unbounded.

9. (a) Let f be defined on a deleted neighborhood N of $\bar{x}$. Show that $\lim_{x \to \bar{x}} f(x) = L$ if and only if $\lim_{n \to \infty} f(x_n) = L$ whenever $\{x_n\}$ is a

sequence of points in N such that $\lim_{n\to\infty} x_n = \bar{x}$. (*Hint:* See the proof of Theorem 2.5.)

   (b) State a result like the one in part (a) for one-sided limits.

10. Give a proof based on sequences for Theorem 2.5, Section 2.2. (*Hint:* Use Theorems 1.3, 2.1, 2.4, and 2.5.)

11. Give a proof based on sequences for Theorem 2.7, Section 2.2.

12. Suppose f is defined on a deleted neighborhood N of $\bar{x}$ and $\{f(x_n)\}$ approaches a limit whenever $\{x_n\}$ is a sequence of points in N and $\lim_{n\to\infty} x_n = \bar{x}$. Show that if $\{x_n\}$ and $\{y_n\}$ are two such sequences, then $\lim_{n\to\infty} f(x_n) = \lim_{n\to\infty} f(y_n)$. Infer from this and Exercise 9 that $\lim_{x\to\bar{x}} f(x)$ exists.

13. Prove: If f is defined in a neighborhood N of $\bar{x}$, then g is differentiable at $\bar{x}$ if and only if

$$\lim_{n\to\infty} \frac{f(x_n) - f(\bar{x})}{x_n - \bar{x}}$$

exists whenever $\{x_n\}$ is a sequence of points in N such that $x_n \neq \bar{x}$ and $\lim_{n\to\infty} x_n = \bar{x}$. (*Hint:* Use Exercise 12.)

4.3 INFINITE SERIES OF CONSTANTS

The theory of sequences developed in the last two sections can be combined with the familiar notion of a finite sum to produce the theory of infinite series. We begin the study of infinite series in this section.

Definition 3.1   *If $\{a_n\}_k^\infty$ is an infinite sequence of real numbers, the symbol*

$$\sum_{n=k}^{\infty} a_n$$

is called an infinite series, and a_n is called the nth term of the series. We say that $\sum_{n=k}^{\infty} a_n$ converges to the sum A, and write

$$\sum_{n=k}^{\infty} a_n = A,$$

if the sequence $\{A_n\}_k^\infty$ defined by

$$A_n = a_k + a_{k+1} + \cdots + a_n, \qquad n \geq k,$$

converges to A. The finite sum A_n is called the nth partial sum of $\sum_{n=k}^{\infty} a_n$. If $\{A_n\}_k^\infty$ diverges, we say that $\sum_{n=k}^{\infty} a_n$ diverges; in particular, if $\lim_{n\to\infty} A_n = \infty$

or $-\infty$, *we say that* $\sum_{n=k}^{\infty} a_n$ *diverges to* ∞ *or* $-\infty$, *and write*

$$\sum_{n=k}^{\infty} a_n = \infty \quad or \quad \sum_{n=k}^{\infty} a_n = -\infty.$$

A divergent infinite series that does not diverge to $\pm\infty$ *is said to oscillate, or be oscillatory.*

We will usually refer to infinite series more briefly as *series*.

Example 3.1   Consider the series

$$\sum_{n=0}^{\infty} r^n, \quad -1 < r < 1.$$

Here $a_n = r^n$ $(n \geq 0)$ and

$$A_n = 1 + r + r^2 + \cdots + r^n = \frac{1 - r^{n+1}}{1 - r}, \tag{1}$$

which converges to $1/(1 - r)$ as $n \to \infty$ (Example 1.10); thus, we write

$$\sum_{n=0}^{\infty} r^n = \frac{1}{1 - r}, \quad -1 < r < 1.$$

If $|r| > 1$, then (1) is still valid, but $\sum_{n=0}^{\infty} r^n$ diverges; if $r > 1$, then

$$\sum_{n=0}^{\infty} r^n = \infty, \tag{2}$$

while if $r < -1$, $\sum_{n=0}^{\infty} r^n$ oscillates, since its partial sums alternate in sign and their magnitudes become arbitrarily large for large n. If $r = -1$, then $A_{2m+1} = 0$ and $A_{2m} = 1$ $(m = 0, 1, \ldots)$, while if $r = 1$, $A_n = n + 1$; in both cases the series diverges, and (2) holds if $r = 1$.

The series $\sum_{n=0}^{\infty} r^n$ is called the *geometric series with ratio r*.

An infinite series can be viewed as a generalization of a finite sum

$$A = \sum_{n=k}^{N} a_n = a_k + a_{k+1} + \cdots + a_N$$

by thinking of the finite sequence $\{a_k, a_{k+1}, \ldots, a_N\}$ as being extended to an infinite sequence $\{a_n\}_k^{\infty}$ with $a_n = 0$ for $n > N$. Then the partial sums of $\sum_{n=k}^{\infty} a_n$ are

$$A_n = a_k + a_{k+1} + \cdots + a_n, \quad k \leq n < N,$$

and

$$A_n = A, \qquad n \geq N;$$

that is, the elements of $\{A_n\}_k^\infty$ equal the finite sum A for $n \geq k$. Therefore, surely $\lim_{n \to \infty} A_n = A$.

The next two theorems can be proved by applying Theorems 1.1 and 1.5 to the partial sums of the series in question (*Exercises 1 and 2*).

Theorem 3.1   *The sum of a convergent series is unique.*

Theorem 3.2   *Let*

$$\sum_{n=k}^{\infty} a_n = A \quad and \quad \sum_{n=k}^{\infty} b_n = B,$$

where A and B are finite. Then

$$\sum_{n=k}^{\infty} (ca_n) = cA$$

if c is a constant,

$$\sum_{n=k}^{\infty} (a_n + b_n) = A + B,$$

and

$$\sum_{n=k}^{\infty} (a_n - b_n) = A - B.$$

These relations also hold if one or both of A and B is infinite, provided the right sides are not indeterminate.

Dropping finitely many terms from a series does not alter convergence or divergence, although it does change the sum of a convergent series if the terms dropped have a nonzero sum. For example, suppose we drop the first k terms of a series $\sum_{n=0}^{\infty} a_n$, and consider the new series $\sum_{n=k}^{\infty} a_n$. Denote the partial sums of the two series by

$$A_n = a_0 + a_1 + \cdots + a_n, \qquad n \geq 0,$$

and

$$A_n' = a_k + a_{k+1} + \cdots + a_n, \qquad n \geq k.$$

Since

$$A_n = (a_0 + a_1 + \cdots + a_{k-1}) + A_n', \qquad n \geq k,$$

it follows that $A = \lim_{n \to \infty} A_n$ exists (in the extended reals) if and only if $A' = \lim_{n \to \infty} A'_n$ does, and in this case

$$A = (a_0 + a_1 + \cdots + a_{k-1}) + A'.$$

From this follows an important principle.

Lemma 3.1 *Suppose the terms of $\sum_{n=k}^{\infty} a_n$ satisfy, for n sufficiently large (that is, for $n \geq$ some integer N), some condition that implies convergence (or divergence) of an infinite series. Then $\sum_{n=k}^{\infty} a_n$ converges (or diverges).*

Example 3.2 Consider the alternating series test, which we will establish later as a special case of a more general test: the series $\sum_{n=k}^{\infty} a_n$ converges if $(-1)^n a_n > 0$, $|a_{n+1}| < |a_n|$, and $\lim_{n \to \infty} a_n = 0$. The terms of

$$\sum_{n=1}^{\infty} \frac{16 + (-2)^n}{n 2^n}$$

do not satisfy these conditions for all $n \geq 1$, but they do satisfy them for sufficiently large n (*Exercise 3*); hence, Lemma 3.1 implies that the series converges.

We will soon give several conditions concerning convergence of a series $\sum_{n=k}^{\infty} a_n$ with nonnegative terms. According to Lemma 3.1 these results apply to series that have at most finitely many negative terms, as long as a_n is nonnegative and satisfies the conditions for n sufficiently large.

When we are interested only in whether $\sum_{n=k}^{\infty} a_n$ converges or diverges, and not in its sum, we will simply say "$\sum a_n$ converges" or "$\sum a_n$ diverges." Lemma 3.1 justifies this convention, subject to the understanding that $\sum a_n$ stands for $\sum_{n=k}^{\infty} a_n$, where k is an integer such that a_n is defined for $n \geq k$. [For example,

$$\sum^{\infty} \frac{1}{(n-6)^2} \quad \text{stands for} \quad \sum_{n=k}^{\infty} \frac{1}{(n-6)^2},$$

where $k \geq 7$.] We write $\sum a_n = \infty \ (-\infty)$ if $\sum a_n$ diverges to $\infty \ (-\infty)$. Finally, let us agree that $\sum_{n=k}^{\infty} a_n$ and $\sum_{n=k-j}^{\infty} a_{n+j}$—where we have obtained the second expression by shifting the index in the first—both represent the same series.

Cauchy's convergence criterion for series

The Cauchy convergence criterion for sequences (Theorem 1.9) yields a useful criterion for convergence of series.

Theorem 3.3   *A series $\sum a_n$ converges if and only if for every $\varepsilon > 0$ there is an integer N such that*

$$\left| a_n + a_{n+1} + \cdots + a_m \right| < \varepsilon \qquad if \quad m \geq n \geq N. \tag{3}$$

Proof.   In terms of the partial sums $\{A_n\}$ of $\sum a_n$,

$$a_n + a_{n+1} + \cdots + a_m = A_m - A_{n-1}.$$

Therefore (3) can be written as

$$\left| A_m - A_{n-1} \right| < \varepsilon \qquad if \quad m \geq n \geq N.$$

Since $\sum a_n$ converges if and only if $\{A_n\}$ converges, the conclusion follows from Theorem 1.9.

Intuitively Theorem 3.3 means that $\sum a_n$ converges if and only if arbitrarily long sums of the form

$$a_n + a_{n+1} + \cdots + a_m, \quad m \geq n,$$

can be made as small as we please by picking n large enough.

Corollary 3.1   *If $\sum a_n$ converges then $\lim_{n \to \infty} a_n = 0$.*

Proof.   If $\sum a_n$ converges and $\varepsilon > 0$, there is an integer N that satisfies (3). By Letting $m = n$ in (3), we conclude that

$$\left| a_n \right| < \varepsilon \qquad if \quad n \geq N,$$

which implies that $\lim_{n \to \infty} a_n = 0$.

It must be emphasized that Corollary 3.1 gives a *necessary* condition for convergence; that is, $\sum a_n$ cannot converge unless $\lim_{n \to \infty} a_n = 0$. The condition is *not sufficient;* $\sum a_n$ may diverge even if $\lim_{n \to \infty} a_n = 0$. We will see examples of this below.

We leave the proof of the following corollary of Theorem 3.3 to the reader (*Exercise 6*).

Corollary 3.2   *If $\sum a_n$ converges, then for each $\varepsilon > 0$ there is an integer K such that*

$$\left| \sum_{n=k}^{\infty} a_n \right| < \varepsilon \qquad if \quad k \geq K.$$

That is,

$$\lim_{k \to \infty} \sum_{n=k}^{\infty} a_n = 0.$$

Example 3.3 Consider the geometric series $\sum r^n$ of Example 3.1. If $|r| \geq 1$, then $\{r^n\}$ does not converge to zero, and therefore $\sum r^n$ diverges, as we saw in Example 3.1. If $|r| < 1$ and $m \geq n$, then the partial sums A_m and A_n satisfy the inequality

$$\begin{aligned}
|A_m - A_n| &= |r^{n+1} + r^{n+2} + \cdots + r^m| \\
&\leq |r|^{n+1}(1 + |r| + \cdots + |r|^{m-n-1}) \\
&= |r|^{n+1} \frac{1 - |r|^{m-n}}{1 - |r|} < \frac{|r|^{n+1}}{1 - |r|}.
\end{aligned} \tag{4}$$

Therefore, if $\varepsilon > 0$, we can choose N so that

$$\frac{|r|^{N+1}}{1 - |r|} < \varepsilon,$$

and then (4) implies that

$$|A_m - A_n| < \varepsilon \qquad \text{if} \quad m \geq n \geq N.$$

Now Theorem 3.3 implies that $\sum r^n$ converges if $|r| < 1$, as we saw in Example 3.1. To illustrate Corollary 3.2 we observe that if $|r| < 1$, then

$$\left| \sum_{n=k}^{\infty} r^n \right| = \left| r^k \sum_{n=k}^{\infty} r^{n-k} \right| = \left| r^k \sum_{n=0}^{\infty} r^n \right| = \frac{|r|^k}{1 - r}.$$

Therefore, if

$$\frac{|r|^K}{1 - r} < \varepsilon,$$

then

$$\left| \sum_{n=k}^{\infty} r^n \right| < \varepsilon \qquad \text{if} \quad k \geq K.$$

Series of nonnegative terms

The theory of series $\sum a_n$ whose terms are nonnegative for sufficiently large n is simpler than the general theory, since such a series either converges to a finite limit or diverges to ∞, as the next theorem shows.

Theorem 3.4 *If* $a_n \geq 0$ *for* $n \geq k$, *then* $\sum a_n$ *converges if its partial sums are bounded, or diverges to* ∞ *if they are not. These are the only possibilities, and, in either case,*

$$\sum_{n=k}^{\infty} a_n = \text{l.u.b.}\{A_n\}_k^{\infty},$$

where

$$A_n = a_k + a_{k+1} + \cdots + a_n, \qquad n \geq k.$$

Proof. Since $A_n = A_{n-1} + a_n$ and $a_n \geq 0$ $(n \geq k)$, the sequence $\{A_n\}$ is non-decreasing, and so the conclusion follows from Theorem 1.3(a) and Definition 3.1.

If $a_n \geq 0$ for sufficiently large n, we will write $\sum a_n < \infty$ if $\sum a_n$ converges. This convention is based on Theorem 3.4, which says that such a series diverges only if $\sum a_n = \infty$. It does not apply to series with infinitely many negative terms, because such series may diverge without diverging to ∞; for example, the series $\sum_{n=0}^{\infty} (-1)^n$ oscillates, since its partial sums are alternately 1 and 0.

Theorem 3.5 (Comparison Test) *Suppose*

$$0 \leq a_r \leq b_r, \qquad r \geq k. \tag{5}$$

Then:

    (a) $\sum a_r < \infty$ *if* $\sum b_r < \infty$.
    (b) $\sum b_r = \infty$ *if* $\sum a_r = \infty$.

Proof of (a). If

$$A_n = a_k + a_{k+1} + \cdots + a_n \quad \text{and} \quad B_n = b_k + b_{k+1} + \cdots + b_n, \qquad n \geq k,$$

then, from (5),

$$A_n \leq B_n. \tag{6}$$

If $\sum b_n < \infty$, then $\{B_n\}$ is bounded above (Theorem 3.4), and (6) implies that $\{A_n\}$ is also; therefore, $\sum a_n < \infty$ (again by Theorem 3.4). This proves (a), and (b) follows from (a) (*Exercise 9*).

Example 3.4 Since

$$\frac{r^n}{n} < r^n, \qquad n \geq 1,$$

and $\sum r^n < \infty$ if $0 < r < 1$, the series $\sum r^n/n$ converges if $0 < r < 1$, by Theorem 3.5(a). Comparing these two series is inconclusive if $r > 1$, since it does not help to know that the terms of $\sum r^n/n$ are smaller than those of the then divergent $\sum r^n$. If $r < 0$, the comparison test does not apply, since the series then have infinitely many negative terms.

Example 3.5   Since

$$r^n < nr^n$$

and $\sum r^n = \infty$ if $r \geq 1$, Theorem 3.5 implies that $\sum nr^n = \infty$ if $r \geq 1$. Comparing these two series is inconclusive if $0 < r < 1$, since it does not help to know that the terms of $\sum nr^n$ are larger than those of the then convergent $\sum r^n$.

The comparison test is useful if we have a collection of series with non-negative terms whose convergence properties are known. We will now use the comparison test as a theoretical tool and to build up such a collection.

Theorem 3.6 (Integral Test)   *Let f be positive, nonincreasing, and locally integrable on $[k, \infty)$ and*

$$c_n = f(n), \qquad n = k, k+1, \ldots \tag{7}$$

Then

$$\sum c_n < \infty \tag{8}$$

if and only if

$$\int_k^\infty f(x)\, dx < \infty. \tag{9}$$

Proof.   We first observe that (9) holds if and only if

$$\sum_{n=k}^\infty \int_n^{n+1} f(x)\, dx < \infty \tag{10}$$

(*Exercise 11*), so it is enough to show that (8) holds if and only if (10) does. From (7) and the assumption that f is nonincreasing,

$$c_{n+1} = f(n+1) \leq f(x) \leq f(n) = c_n, \qquad n \leq x \leq n+1, n \geq k.$$

Therefore,

$$c_{n+1} = \int_n^{n+1} c_{n+1}\, dx \leq \int_n^{n+1} f(x)\, dx \leq \int_n^{n+1} c_n\, dx = c_n, \qquad n \geq k$$

(Lemma 3.1, Section 3.3). From the first inequality and Theorem 3.5(a) [with $a_n = c_{n+1}$ and $b_n = \int_n^{n+1} f(x)\, dx$], (10) implies that $\sum c_{n+1} < \infty$, which is equivalent to (8). From the second inequality and Theorem 3.5(a) [with $a_n = \int_n^{n+1} f(x) \, dx$ and $b_n = c_n$], (8) implies (10). This completes the proof.

Example 3.6   The integral test implies that the series

$$\sum \frac{1}{n^p}, \quad \sum \frac{1}{n(\log n)^p}, \quad \text{and} \quad \sum \frac{1}{n \log n[\log(\log n)]^p}$$

converge if $p > 1$ and diverge if $0 < p \le 1$, because the same is true of the integrals

$$\int_a^\infty \frac{dx}{x^p}, \quad \int_a^\infty \frac{dx}{x(\log x)^p}, \quad \text{and} \quad \int_a^\infty \frac{dx}{x \log x[\log(\log x)]^p}$$

if a is sufficiently large. (See Example 4.16 and *Exercise 14, Section 3.4*.) The three series diverge if $p \le 0$; the first by Corollary 3.1, the second by comparison with the divergent $\sum 1/n$, and the third by comparison with the divergent $\sum 1/(n \log n)$. (The divergence of the last two series for $p \le 0$ also follows from the integral test, but the divergence of the first does not. Why not?) These results can be generalized: if

$$L_0(x) = x \quad \text{and} \quad L_k(x) = \log[L_{k-1}(x)], \qquad k = 1, 2, \ldots,$$

then

$$\sum \frac{1}{L_0(n)L_1(n) \cdots L_k(n)[L_{k+1}(n)]^p}$$

converges if and only if $p > 1$ (*Exercise 13*).

This example provides an infinite family of series with known convergence properties that can be used as standards for the comparison test.

Except for the series of Example 3.6 the integral test is of limited practical value, since convergence or divergence of most of the series to which it can be applied can be determined by simpler tests not requiring integration. However, the method used to prove Theorem 3.6 is often useful for estimating the rapidity of convergence or divergence of a series. This idea is developed in *Exercises 15* and *16*.

Example 3.7   The series

$$\sum_{\infty}^{\infty} \frac{1}{(n^2 + n)^q} \tag{11}$$

converges if $q > \frac{1}{2}$, by comparison with the convergent series $\sum 1/n^{2q}$, since

$$\frac{1}{(n^2 + n)^q} < \frac{1}{n^{2q}}, \qquad n \ge 1.$$

This comparison is inconclusive if $q \leq \frac{1}{2}$, since then

$$\sum \frac{1}{n^{2q}} = \infty$$

and it does not help to know that the terms of (11) are smaller than those of a divergent series. However, we can use the comparison test here, after a little trickery; we observe that

$$\sum_{n=k-1}^{\infty} \frac{1}{(n+1)^{2q}} = \sum_{n=k}^{\infty} \frac{1}{n^{2q}} = \infty, \qquad q \leq \frac{1}{2},$$

and

$$\frac{1}{(n+1)^{2q}} < \frac{1}{(n^2+n)^q}.$$

This implies that

$$\sum \frac{1}{(n^2+n)^q} = \infty, \qquad q \leq \frac{1}{2},$$

because of Theorem 3.5(b).

The next theorem is often applicable where Theorem 3.6 is not. It does not require the kind of trickery that we used in Example 3.7.

Theorem 3.7 *Suppose $a_n \geq 0$ and $b_n > 0$ for $n \geq k$. Then:*
  (a) $\sum a_n < \infty$ if $\sum b_n < \infty$ and $\overline{\lim}_{n \to \infty} a_n/b_n < \infty$.
  (b) $\sum a_n = \infty$ if $\sum b_n = \infty$ and $\underline{\lim}_{n \to \infty} a_n/b_n > 0$.

Proof. We prove (a) and leave (b) to the reader (*Exercise 18*).
      If $\overline{\lim}_{n \to \infty} a_n/b_n < \infty$, then $\{a_n/b_n\}$ is bounded and so there is a constant M such that

$$a_n \leq Mb_n, \qquad n \geq k.$$

Since $\sum b_n < \infty$, Theorem 3.2 implies that $\sum (Mb_n) < \infty$, and now Theorem 3.5(a) implies that $\sum a_n < \infty$.

Example 3.8 Let

$$\sum b_n = \sum \frac{1}{n^{p+q}} \quad \text{and} \quad \sum a_n = \sum \frac{2 + \sin n\pi/6}{(n+1)^p(n-1)^q}.$$

Then

$$\frac{a_n}{b_n} = \frac{2 + \sin n\pi/6}{(1 + 1/n)^p(1 - 1/n)^q},$$

and so

$$\overline{\lim_{n \to \infty}} \frac{a_n}{b_n} = 3 \quad \text{and} \quad \underline{\lim_{n \to \infty}} \frac{a_n}{b_n} = 1.$$

Since $\sum b_n < \infty$ if and only if $p + q > 1$, the same is true of $\sum a_n$, by Theorem 3.7.

The following corollary of Theorem 3.7 is often useful, although it does not apply to the series of Example 3.8.

Corollary 3.3   *If $a_n \geq 0$ and $b_n > 0$ for $n \geq k$, and*

$$\lim_{n \to \infty} \frac{a_n}{b_n} = L,$$

where $0 < L < \infty$, then $\sum a_n$ and $\sum b_n$ converge or diverge together.

Example 3.9   With this corollary we can avoid the trickery used in the second part of Example 3.7, since

$$\lim_{n \to \infty} \frac{1}{(n^2 + n)^q} \bigg/ \frac{1}{n^{2q}} = \lim_{n \to \infty} \frac{1}{(1 + 1/n)^q} = 1,$$

and so

$$\sum \frac{1}{(n^2 + n)^q} \quad \text{and} \quad \sum \frac{1}{n^{2q}}$$

converge or diverge together.

The ratio test

It is sometimes possible to determine whether a series with positive terms converges by comparing the ratios of successive terms with the corresponding ratios of a series known to converge or diverge.

Theorem 3.8   *Suppose $a_n > 0$, $b_n > 0$, and*

$$\frac{a_{n+1}}{a_n} \leq \frac{b_{n+1}}{b_n}.$$ (12)

Then:

(a) $\sum a_n < \infty$ *if* $\sum b_n < \infty$.
(b) $\sum b_n = \infty$ *if* $\sum a_n = \infty$.

Proof of (a). Rewriting (12) as

$$\frac{a_{n+1}}{b_{n+1}} \le \frac{a_n}{b_n},$$

we see that $\{a_n/b_n\}$ is nonincreasing; therefore, $\overline{\lim}_{n\to\infty} a_n/b_n < \infty$, and (a) follows from Theorem 3.7(a).

We leave it to the reader to show that (a) implies (b) (*Exercise 20*).

We will use this theorem to obtain two other widely applicable tests: the ratio test and Raabe's test.

Theorem 3.9 **(Ratio Test)** *Suppose* $a_n > 0$ *for* $n \ge k$. *Then:*

(a) $\sum a_n < \infty$ *if* $\overline{\lim}_{n\to\infty} a_{n+1}/a_n < 1$.
(b) $\sum a_n = \infty$ *if* $\underline{\lim}_{n\to\infty} a_{n+1}/a_n > 1$.

If

$$\underline{\lim_{n\to\infty}} \frac{a_{n+1}}{a_n} \le 1 \le \overline{\lim_{n\to\infty}} \frac{a_{n+1}}{a_n}, \tag{13}$$

then the test is inconclusive; that is, $\sum a_n$ *may converge or diverge.*

Proof. We prove (a) and leave (b) to the reader (*Exercise 22*). If

$$\overline{\lim_{n\to\infty}} \frac{a_{n+1}}{a_n} < 1,$$

then there is a number r such that $0 < r < 1$ and

$$\frac{a_{n+1}}{a_n} < r$$

for n sufficiently large. This can be rewritten as

$$\frac{a_{n+1}}{a_n} < \frac{r^{n+1}}{r^n}$$

and since $\sum r^n < \infty$, (a) follows from Theorem 3.8(a) with $b_n = r^n$.

To see that no conclusion can be drawn if (13) holds, consider

$$\sum a_n = \sum \frac{1}{n^p}.$$

This series converges if $p > 1$ and diverges if $p \leq 1$; however,

$$\overline{\lim_{n \to \infty}} \frac{a_{n+1}}{a_n} = \underline{\lim_{n \to \infty}} \frac{a_{n+1}}{a_n} = 1$$

for every p.

Example 3.10   If

$$\sum a_n = \sum \left(2 + \sin \frac{n\pi}{2} \right) r^n,$$

then

$$\frac{a_{n+1}}{a_n} = r \frac{\left(2 + \sin \dfrac{(n+1)\pi}{2} \right)}{\left(2 + \sin \dfrac{n\pi}{2} \right)},$$

which assumes the values $3r/2$, $2r/3$, $r/2$, and $2r$, each infinitely many times; hence,

$$\overline{\lim_{n \to \infty}} \frac{a_{n+1}}{a_n} = 2r \quad \text{and} \quad \underline{\lim_{n \to \infty}} \frac{a_{n+1}}{a_n} = \frac{r}{2}.$$

Therefore, $\sum a_n$ converges if $0 < r < \frac{1}{2}$ and diverges if $r > 2$; Theorem 3.9 is inconclusive if $\frac{1}{2} \leq r \leq 2$.

The following corollary of Theorem 3.9 is the familiar ratio test usually studied in the first calculus course.

Corollary 3.4   *Suppose $a_n > 0$ $(n \geq k)$ and*

$$\lim_{n \to \infty} \frac{a_{n+1}}{a_n} = L.$$

Then:

(a) $\sum a_n < \infty$ *if $L < 1$.*
(b) $\sum a_n = \infty$ *if $L > 1$.*
The test is inconclusive if $L = 1$.

Example 3.11   The series $\sum a_n = \sum nr^{n-1}$ converges if $0 < r < 1$ and diverges if $r > 1$, since

$$\frac{a_{n+1}}{a_n} = \frac{(n+1)r^n}{nr^{n-1}} = \left(1 + \frac{1}{n} \right) r,$$

and so

$$\lim_{n \to \infty} \frac{a_{n+1}}{a_n} = r.$$

Corollary 3.4 is inconclusive if $r = 1$, but then Corollary 3.1 implies that the series diverges.

Theorem 3.9 does not imply that $\sum a_n < \infty$ if merely

$$\frac{a_{n+1}}{a_n} < 1 \tag{14}$$

for large n (students often erroneously conclude that it does!), since this could occur with $\lim_{n \to \infty} a_{n+1}/a_n = 1$, in which case Theorem 3.9 is inconclusive. However, the next theorem shows that $\sum a_n < \infty$ if (14) is replaced by the stronger condition that

$$\frac{a_{n+1}}{a_n} \leq 1 - \frac{p}{n}$$

for some $p > 1$ and large n. It also shows that $\sum a_n = \infty$ if

$$\frac{a_{n+1}}{a_n} \geq 1 - \frac{q}{n}$$

for some $q < 1$ and large n.

Theorem 3.10 (Raabe's Test) *Suppose $a_n > 0$ for large n and let*

$$M = \overline{\lim_{n \to \infty}} \, n\left(\frac{a_{n+1}}{a_n} - 1\right) \quad \text{and} \quad m = \underline{\lim_{n \to \infty}} \, n\left(\frac{a_{n+1}}{a_n} - 1\right).$$

Then:
   (a) $\sum a_n < \infty$ if $M < -1$.
   (b) $\sum a_n = \infty$ if $m > -1$.
The test is inconclusive if $m \leq -1 \leq M$.

Proof. (a) We need the inequality

$$\frac{1}{(1 + x)^p} > 1 - px, \qquad x > 0, \quad p > 0. \tag{15}$$

This follows from Taylor's theorem (Theorem 5.3, Section 2.5), which implies that

$$\frac{1}{(1 + x)^p} = 1 - px + \frac{1}{2} \frac{p(p + 1)}{(1 + c)^{p+2}} x^2$$

where $0 < c < x$. (Verify.) Since the last term is positive if $p > 0$, (15) follows.

Now suppose $M < -p < -1$. Then there is an integer k such that

$$n\left(\frac{a_{n+1}}{a_n} - 1\right) < -p, \qquad n \geq k,$$

and therefore

$$\frac{a_{n+1}}{a_n} < 1 - \frac{p}{n}, \qquad n \geq k.$$

Hence,

$$\frac{a_{n+1}}{a_n} < \frac{1}{(1 + 1/n)^p}, \qquad n \geq k,$$

as can be seen by letting $x = 1/n$ in (15). From this,

$$\frac{a_{n+1}}{a_n} < \frac{1}{(n + 1)^p}\bigg/\frac{1}{n^p}, \qquad n \geq k,$$

and so $\sum a_n < \infty$ by Theorem 3.8(a), since $\sum 1/n^p < \infty$ if $p > 1$.
  (b) Here we need the inequality

$$(1 - x)^q < 1 - qx, \qquad 0 < x < 1, \quad 0 < q < 1. \tag{16}$$

This also follows from Taylor's theorem, which implies that

$$(1 - x)^q = 1 - qx + q(q - 1)(1 - c)^{q-2}\frac{x^2}{2},$$

where $0 < c < x$.
  Now suppose $-1 < -q < m$. Then there is an integer k such that

$$n\left(\frac{a_{n+1}}{a_n} - 1\right) > -q, \qquad n \geq k,$$

and therefore

$$\frac{a_{n+1}}{a_n} \geq 1 - \frac{q}{n}, \qquad n \geq k.$$

If $q \leq 0$, then $\sum a_n = \infty$, by Corollary 3.1; hence, we may assume that $0 < q < 1$, in which case the last inequality implies that

$$\frac{a_{n+1}}{a_n} > \left(1 - \frac{1}{n}\right)^q, \qquad n \geq k,$$

as can be seen by setting $x = 1/n$ in (16). Hence,

$$\frac{a_{n+1}}{a_n} > \frac{1}{n^q}\bigg/\frac{1}{(n - 1)^q}, \qquad n \geq k,$$

and so $\sum a_n = \infty$, by Theorem 3.8(b), since $\sum 1/n^q = \infty$ if $q < 1$.

Example 3.12   If

$$\sum a_n = \sum \frac{n!}{\alpha(\alpha + 1)(\alpha + 2) \cdots (\alpha + n - 1)}, \qquad \alpha > 0,$$

then

$$\lim_{n \to \infty} \frac{a_{n+1}}{a_n} = \lim_{n \to \infty} \frac{n + 1}{\alpha + n} = 1,$$

so the ratio test is inconclusive. However,

$$\lim_{n \to \infty} n \left(\frac{a_{n+1}}{a_n} - 1 \right) = \lim_{n \to \infty} n \left(\frac{n + 1}{\alpha + n} - 1 \right)$$

$$= \lim_{n \to \infty} \frac{n(1 - \alpha)}{\alpha + n} = 1 - \alpha,$$

and therefore Raabe's test implies that $\sum a_n < \infty$ if $\alpha > 2$ and $\sum a_n = \infty$ if $0 < \alpha < 2$. Raabe's test is inconclusive if $\alpha = 2$, but then the series becomes

$$\sum \frac{n!}{(n + 1)!} = \sum \frac{1}{n + 1},$$

which we know is divergent.

Example 3.13   Consider the series $\sum a_n$, where

$$a_{2m} = \frac{(m!)^2}{\alpha(\alpha + 1) \cdots (\alpha + m)\beta(\beta + 1) \cdots (\beta + m)}$$

and

$$a_{2m+1} = \frac{(m!)^2(m + 1)}{\alpha(\alpha + 1) \cdots (\alpha + m)\beta(\beta + 1) \cdots (\beta + m + 1)},$$

with $0 < \alpha < \beta$. Since

$$2m \left(\frac{a_{2m+1}}{a_{2m}} - 1 \right) = 2m \left(\frac{m + 1}{\beta + m + 1} - 1 \right) = \frac{-2m\beta}{\beta + m + 1}$$

and

$$(2m + 1) \left(\frac{a_{2m+2}}{a_{2m+1}} - 1 \right) = (2m + 1) \left(\frac{m + 1}{\alpha + m + 1} - 1 \right) = \frac{-(2m + 1)\alpha}{\alpha + m + 1},$$

we have

$$\overline{\lim_{n \to \infty}} \left(\frac{a_{n+1}}{a_n} - 1 \right) = -2\alpha \quad \text{and} \quad \underline{\lim_{n \to \infty}} \left(\frac{a_{n+1}}{a_n} - 1 \right) = -2\beta.$$

Raabe's test implies that $\sum a_n < \infty$ if $\alpha > \frac{1}{2}$ and $\sum a_n = \infty$ if $\beta < \frac{1}{2}$; it is inconclusive if $0 < \alpha \leq \frac{1}{2} \leq \beta$.

The next theorem, which will be useful when we study power series (Section 4.5), concludes our study of series with nonnegative terms.

Theorem 3.11 (Cauchy's Root Test)   *If $a_n \geq 0$ for $n \geq k$, then:*
(a) $\sum a_n < \infty$ *if* $\overline{\lim}_{n \to \infty} a_n^{1/n} < 1.$
(b) $\sum a_n = \infty$ *if* $\overline{\lim}_{n \to \infty} a_n^{1/n} > 1.$
The test is inconclusive if $\overline{\lim}_{n \to \infty} a_n^{1/n} = 1.$

Proof.   (a) If $\lim_{n \to \infty} a_n^{1/n} < 1$, there is a number r such that $0 < r < 1$ and $a_n^{1/n} < r$ for n sufficiently large. Therefore, $a_n < r^n$ for large n and Theorem 3.5(a) implies that $\sum a_n < \infty$, since $\sum r^n < \infty$.
    (b) If $\overline{\lim}_{n \to \infty} a_n^{1/n} > 1$, then $a_n^{1/n} > 1$ for infinitely many values of n. This means that $\{a_n\}$ cannot converge to zero, and therefore $\sum a_n = \infty$, by Corollary 3.1.

Example 3.14   If

$$\sum a_n = \sum \left(2 + \sin \frac{n\pi}{4} \right)^n r^n$$

then

$$\overline{\lim}_{n \to \infty} a_n^{1/n} = \overline{\lim}_{n \to \infty} \left(2 + \sin \frac{n\pi}{4} \right) r = 3r,$$

and so $\sum a_n < \infty$ if $r < \frac{1}{3}$ and $\sum a_n = \infty$ if $r > \frac{1}{3}$. The test is inconclusive if $r = \frac{1}{3}$, but then $|a_{8m+2}| = 1$ $(m = 0, 1, \ldots)$, and so $\sum a_n = \infty$, by Corollary 3.1.

Example 3.15   Cauchy's root test is inconclusive if

$$\sum a_n = \sum \frac{1}{n^p},$$

because then

$$\overline{\lim}_{n \to \infty} a_n^{1/n} = \lim_{n \to \infty} \left(\frac{1}{n^p} \right)^{1/n} = \lim_{n \to \infty} \exp\left(-\frac{p}{n} \log n \right) = 1$$

for all p. Of course, we know that $\sum 1/n^p < \infty$ if $p > 1$ and $\sum 1/n^p = \infty$ if $p \leq 1$.

Absolute and conditional convergence

We now drop the assumption that the terms of $\sum a_n$ are nonnegative for large n. In this case $\sum a_n$ may converge in two quite different ways. The first is defined as follows.

Definition 3.2 *A series $\sum a_n$ is said to converge absolutely, or be absolutely convergent, if $\sum |a_n| < \infty$.*

Example 3.16 A convergent series $\sum a_n$ of nonnegative terms is absolutely convergent, since $\sum a_n$ and $\sum |a_n|$ are the same. More generally, any convergent series whose terms are of the same sign for sufficiently large n converges absolutely (*Exercise 27*).

Example 3.17 Consider the series

$$\sum \frac{\sin n\theta}{n^p}, \tag{17}$$

where θ is arbitrary and $p > 1$. Since

$$\left| \frac{\sin n\theta}{n^p} \right| \le \left| \frac{1}{n^p} \right|$$

and $\sum 1/n^p < \infty$ if $p > 1$, Theorem 3.5(a) implies that

$$\sum \left| \frac{\sin n\theta}{n^p} \right| < \infty, \qquad p > 1.$$

Therefore, the series (17) converges absolutely if $p > 1$.

Example 3.18 If $0 < p < 1$, the series

$$\sum \frac{(-1)^n}{n^p}$$

does not converge absolutely, since

$$\sum \left| \frac{(-1)^n}{n^p} \right| = \sum \frac{1}{n^p} = \infty.$$

However, it does converge, by the alternating series test, which is proved below.

Any test for convergence of a series with nonnegative terms can be used to test an arbitrary series $\sum a_n$ for absolute convergence by applying it to $\sum |a_n|$. We used the comparison test this way in Examples 3.17 and 3.18.

Example 3.19 To test the series

$$\sum a_n = \sum (-1)^n \frac{n!}{\alpha(\alpha+1)\cdots(\alpha+n-1)}, \qquad \alpha > 0,$$

for absolute convergence, we apply Raabe's test to

$$\sum |a_n| = \sum \frac{n!}{\alpha(\alpha + 1) \cdots (\alpha + n - 1)}.$$

From Example 3.12, $\sum |a_n| < \infty$ if $\alpha > 2$ and $\sum |a_n| = \infty$ if $\alpha < 2$; therefore, $\sum a_n$ converges absolutely if $\alpha > 2$, but not if $\alpha < 2$. Notice that this does not imply that $\sum a_n$ diverges if $\alpha < 2$.

The proof of the next theorem is analogous to that of Theorem 4.4, Section 3.4. We leave it to the reader (*Exercise 28*).

Theorem 3.12   *If $\sum a_n$ converges absolutely, then it converges.*

For example, Theorem 3.12 implies that

$$\sum \frac{\sin n\theta}{n^p}$$

converges if $p > 1$, since it then converges absolutely (Example 3.17).

The converse of Theorem 3.12 is false; we saw in Example 3.18 that a series may converge without converging absolutely. We say then that it converges *conditionally*, or is *conditionally convergent*; thus, $\sum (-1)^n/n^p$ converges conditionally if $0 < p \le 1$.

Dirichlet's test for series

Except for Theorem 3.3 and Corollary 3.1, the convergence tests we have studied so far apply only to series whose terms have the same sign for large n. The following theorem does not require this. It is analogous to Dirichlet's test for improper integrals (Theorem 4.5, Section 3.4).

Theorem 3.13 (Dirichlet's Test for Series)   *The series $\sum_{n=k}^{\infty} a_n b_n$ converges if* $\lim_{n \to \infty} a_n = 0$,

$$\sum |a_{n+1} - a_n| < \infty, \tag{18}$$

and

$$|b_k + b_{k+1} + \cdots + b_n| \le M, \qquad n = k, k+1, \ldots \tag{19}$$

for some constant M.

Proof.   The proof parallels that of Dirichlet's test for integrals. Define

$$B_n = b_k + b_{k+1} + \cdots + b_n, \qquad n = k, k+1, \ldots,$$

and consider the partial sums

$$S_n = a_k b_k + a_{k+1} b_{k+1} + \cdots + a_n b_n, \qquad n \ge k. \tag{20}$$

By substituting

$$b_k = B_k \quad \text{and} \quad b_n = B_n - B_{n-1}, \qquad n \ge k+1,$$

into (20) we obtain

$$S_n = a_k B_k + a_{k+1}(B_{k+1} - B_k) + \cdots + a_n(B_n - B_{n-1}),$$

which can be rewritten as

$$\begin{aligned}
S_n = {} & (a_k - a_{k+1})B_k + (a_{k+1} - a_{k+2})B_{k+1} + \cdots \\
& + (a_{n-1} - a_n)B_{n-1} + a_n B_n.
\end{aligned} \tag{21}$$

[The procedure that led from (20) to (21) is called *summation by parts;* it is analogous to integration by parts.] Now (21) can be viewed as

$$S_n = T_{n-1} + a_n B_n, \tag{22}$$

where

$$T_{n-1} = (a_k - a_{k+1})B_k + (a_{k+1} - a_{k+2})B_{k+1} + \cdots + (a_{n-1} - a_n)B_{n-1};$$

that is, $\{T_n\}$ is the sequence of partial sums of the series

$$\sum_{j=k}^{\infty} (a_j - a_{j+1})B_j. \tag{23}$$

Since

$$|(a_j - a_{j+1})B_j| \le M|a_j - a_{j+1}|$$

from (19), the comparison test and (18) imply that the series (23) converges absolutely. Theorem 3.12 now implies that $\{T_n\}$ converges. Let $T = \lim_{n \to \infty} T_n$. Since $\{B_n\}$ is bounded and $\lim_{n \to \infty} a_n = 0$, we infer from (22) that

$$\lim_{n \to \infty} S_n = \lim_{n \to \infty} T_{n-1} + \lim_{n \to \infty} a_n B_n = T + 0 = T.$$

Therefore, $\sum a_n b_n$ converges. This completes the proof.

Example 3.20   To apply Dirichlet's test to

$$\sum_{n=2}^{\infty} \frac{\sin n\theta}{n + (-1)^n}, \qquad \theta \ne 2k\pi,$$

we take

$$a_n = \frac{1}{n + (-1)^n} \quad \text{and} \quad b_n = \sin n\theta.$$

(We have deliberately picked an example where $\{a_n\}$ is not nonincreasing; as we will see below, the hypotheses of Dirichlet's theorem can be stated more simply if it is.) Then $\lim_{n \to \infty} a_n = 0$ and

$$|a_{n+1} - a_n| < \frac{3}{n(n-1)}$$

(verify), so that

$$\sum |a_{n+1} - a_n| < \infty.$$

Now

$$B_n = \sin 2\theta + \sin 3\theta + \cdots + \sin n\theta.$$

To show that $\{B_n\}$ is bounded, we use the trigonometric identity

$$\sin r\theta = \frac{\cos(r - \frac{1}{2})\theta - \cos(r + \frac{1}{2})\theta}{2\sin(\theta/2)}, \quad \theta \neq 2k\pi,$$

to write

$$B_n = \frac{(\cos \frac{3}{2}\theta - \cos \frac{5}{2}\theta) + (\cos \frac{5}{2}\theta - \cos \frac{7}{2}\theta) + \cdots + [\cos(n - \frac{1}{2})\theta - \cos(n + \frac{1}{2})\theta]}{2\sin(\theta/2)}$$

$$= \frac{\cos \frac{3}{2}\theta - \cos(n + \frac{1}{2})\theta}{2\sin(\theta/2)},$$

which implies that

$$|B_n| \leq \left| \frac{1}{\sin(\theta/2)} \right|, \quad n = 2, 3, \ldots.$$

Since $\{a_n\}$ and $\{b_n\}$ satisfy the hypotheses of Dirichlet's theorem, $\sum a_n b_n$ converges.

Dirichlet's test takes a simpler form if $\{a_n\}$ is nonincreasing, as follows.

Corollary 3.5   *The series $\sum a_n b_n$ converges if $a_{n+1} \leq a_n$, $\lim_{n \to \infty} a_n = 0$, and*

$$|b_k + b_{k+1} + \cdots + b_n| \leq M, \quad n = k, k+1, \ldots$$

for some constant M.

Proof.   If $a_{n+1} \leq a_n$, then

$$\sum_{n=k}^{m} |a_{n+1} - a_n| = \sum_{n=k}^{m} (a_n - a_{n+1}) = a_k - a_{m+1}$$

and, since $\lim_{m \to \infty} a_{m+1} = 0$,

$$\sum_{n=k}^{\infty} |a_{n+1} - a_n| = a_k < \infty.$$

Therefore, the hypotheses of Theorem 3.13 are satisfied, and so $\sum a_n b_n$ converges.

Example 3.21   The series

$$\sum \frac{\sin n\theta}{n^p},$$

which we know is convergent if $p > 1$ (Example 3.17), also converges if $0 < p \le 1$. This follows from Corollary 3.5 with $a_n = 1/n^p$ and $b_n = \sin n\theta$ (see Example 3.20).

The alternating series test, which should not be new to a reader who has completed an elementary calculus course, follows easily from Corollary 3.5.

Corollary 3.6 (Alternating Series Test)   *The series* $\sum (-1)^n a_n$ *converges if* $0 \le a_{n+1} \le a_n$ *and* $\lim_{n \to \infty} a_n = 0$.

Proof.   Let $b_n = (-1)^n$; then $\{|B_n|\}$ is a sequence of zeros and ones, and therefore bounded. The conclusion now follows from Corollary 3.5.

Grouping terms in a series

The terms of a finite sum can be grouped by inserting parentheses in any way whatever; for example,

$$(1 + 7) + (6 + 5) + 4 = (1 + 7 + 6) + (5 + 4) = (1 + 7) + (6 + 5 + 4).$$

According to the next theorem, the same is true of a convergent infinite series, and of one that diverges to $\pm \infty$.

Theorem 3.14   *Suppose* $\sum_{n=k}^{\infty} a_n = A$ $(-\infty \le A \le \infty)$ *and let* $\{n_j\}_1^{\infty}$ *be an increasing sequence of integers, with* $n_1 \ge k$. *Define*

$$b_1 = a_k + \cdots + a_{n_1},$$
$$b_2 = a_{1+n_1} + \cdots + a_{n_2},$$
$$\vdots$$
$$b_r = a_{1+n_{r-1}} + \cdots + a_{n_r}.$$

Then

$$\sum_{j=1}^{\infty} b_{n_j} = A.$$

Proof. If T_r is the rth partial sum of $\sum_{j=1}^{\infty} b_{n_j}$ and $\{A_n\}$ is the nth partial sum of $\sum_{s=k}^{\infty} a_s$, then

$$T_r = b_1 + b_2 + \cdots + b_r$$
$$= (a_1 + \cdots + a_{n_1}) + (a_{1+n_1} + \cdots + a_{n_2}) + \cdots + (a_{1+n_{r-1}} + \cdots + a_{n_r})$$
$$= A_{n_r}.$$

Thus, $\{T_r\}$ is a subsequence of $\{A_n\}$, and so $\lim_{r \to \infty} T_r = \lim_{n \to \infty} A_n = A$, by Theorem 2.1. This completes the proof.

Example 3.22 If $\sum_{n=0}^{\infty} (-1)^n a_n$ satisfies the hypotheses of the alternating series test and converges to the sum S, Theorem 3.14 enables us to write

$$S = \sum_{n=0}^{k} (-1)^n a_n + (-1)^{k+1} \sum_{j=1}^{\infty} (a_{k+2j-1} - a_{k+2j})$$

and

$$S = \sum_{n=0}^{k} (-1)^n a_n + (-1)^{k+1} \left[a_{k+1} - \sum_{j=1}^{\infty} (a_{k+2j} - a_{k+2j+1}) \right].$$

Since $0 \le a_{n+1} \le a_n$, these two equations imply that $S - S_k$, the error in approximating S by the kth partial sum of the series, is between 0 and $(-1)^{k+1} a_{k+1}$.

Example 3.23 Introducing parentheses in some divergent series can yield seemingly contradictory results. For example, it is tempting to write

$$\sum_{n=1}^{\infty} (-1)^{n+1} = (1 - 1) + (1 - 1) + \cdots = 0 + 0 + \cdots$$

and conclude that $\sum_{n=1}^{\infty} (-1)^n = 0$, but equally tempting to write

$$\sum_{n=1}^{\infty} (-1)^{n+1} = 1 - (1 - 1) - (1 - 1) - \cdots$$
$$= 1 - 0 - 0 - \cdots$$

and conclude that $\sum_{n=1}^{\infty} (-1)^{n+1} = 1$. Of course, there is no contradiction here, since Theorem 3.14 does not apply to this series, and neither of these operations is legitimate.

Rearrangement of series

A finite sum is not changed by rearranging its terms; thus,

$$1 + 3 + 7 = 1 + 7 + 3 = 3 + 1 + 7 = 3 + 7 + 1 = 7 + 1 + 3 = 7 + 3 + 1.$$

This is not true of all infinite series. Let us say that $\sum a_n'$ is a *rearrangement* of $\sum a_n$ if the two series have the same terms, written in different orders. Since the partial sums of the two series may form entirely different sequences, there is no apparent reason to expect them to exhibit the same convergence properties, and in general they do not.

We are interested in what happens if we rearrange the terms of a convergent series. We will see that every rearrangement of an absolutely convergent series has the same sum, but that conditionally convergent series fail—spectacularly—to have this property.

Theorem 3.15 *If* $\sum_{n=1}^{\infty} a_n'$ *is a rearrangement of an absolutely convergent series* $\sum_{n=1}^{\infty} a_n$, *then* $\sum_{n=1}^{\infty} a_n'$ *also converges absolutely, and to the same sum.*

Proof. Let

$$\bar{A}_n = |a_1| + |a_2| + \cdots + |a_n| \quad \text{and} \quad \bar{A}_n' = |a_1'| + |a_2'| + \cdots + |a_n'|$$

For each $n \geq 1$ there is an integer M_n such that $a_1'\ a_2', \ldots, a_n'$ are included among $a_1, a_2, \ldots, a_{M_n}$, and therefore $\bar{A}_n' \leq \bar{A}_{M_n}$. Since $\{\bar{A}_n\}$ is bounded, so is $\{\bar{A}_n'\}$, and therefore $\sum |a_n'| < \infty$ (Theorem 3.4).

Now let

$$A_n = a_1 + a_2 + \cdots + a_n, \qquad A_n' = a_1' + a_2' + \cdots + a_n',$$

$$A = \sum_{n=1}^{\infty} a_n, \quad \text{and} \quad A' = \sum_{n=1}^{\infty} a_n'.$$

We must show that $A = A'$. From Cauchy's convergence criterion and the absolute convergence of $\sum a_n$, there is for each $\varepsilon > 0$ an integer N such that

$$|a_{N+1}| + |a_{N+2}| + \cdots + |a_{N+k}| < \varepsilon, \qquad k = 1, 2, \ldots.$$

Choose N_1 so that $a_1, a_2, \ldots, a_N$ are included among $a_1', a_2', \ldots, a_{N_1}'$. If $n \geq N_1$, then A_n and A_n' both include the terms $a_1, a_2, \ldots, a_N$, which cancel on subtraction; thus, $|A_n - A_n'|$ is dominated by the sum of the absolute values of finitely many terms from $\sum a_n$ with subscripts greater than N. Since every such sum is less than ε,

$$|A_n - A_n'| < \varepsilon \qquad \text{if} \quad n \geq N_1.$$

Therefore, $\lim_{n \to \infty} (A_n - A_n') = 0$ and $A = A'$. This completes the proof.

To investigate the consequences of rearranging a conditionally convergent series, we need the next theorem, which is itself important.

Theorem 3.16 *If* $P = \{a_{n_i}\}_1^{\infty}$ *and* $Q = \{a_{m_j}\}_1^{\infty}$ *are the subsequences of all nonnegative and negative terms in a conditionally convergent series* $\sum a_n$, *then*

$$\sum_{i=1}^{\infty} a_{n_i} = \infty \quad \text{and} \quad \sum_{j=1}^{\infty} a_{m_j} = -\infty.$$

We leave the details of the proof to the reader (*Exercise 40*). The essential idea is this: If both these series converge, then $\sum a_n$ converges absolutely, while if one converges and the other diverges, then $\sum a_n$ diverges to ∞ or $-\infty$. Hence, both must diverge.

The next theorem implies that a conditionally convergent series can be rearranged to produce a series that converges to any given number, diverges to $\pm \infty$, or oscillates.

Theorem 3.17   *The terms of a conditionally convergent series $\sum_{n=1}^{\infty} a_n$ can be rearranged so as to form a series $\sum_{n=1}^{\infty} a'_n$ such that*

$$\overline{\lim_{n \to \infty}} \, A'_n = v \quad and \quad \underline{\lim_{n \to \infty}} \, A'_n = \mu, \tag{24}$$

where

$$A'_n = a'_1 + a'_2 + \cdots + a'_n$$

and μ and v are arbitrarily given elements of the extended reals, with $\mu \leq v$,

---

Proof.   We consider the case where μ and v are finite and leave the other cases to the reader (*Exercise 41*). For convenience we rename the elements of the subsequences P and Q of Theorem 3.16 so that $P = \{\alpha_i\}_1^{\infty}$ and $Q = \{-\beta_j\}_1^{\infty}$. We construct the sequence $\{a'_n\}_1^{\infty}$ in the form

$$\{\alpha_1, \ldots, \alpha_{m_1}, -\beta_1, \ldots, -\beta_{n_1}, \alpha_{1+m_1}, \ldots, \alpha_{m_2}, -\beta_{1+n_1}, \ldots, -\beta_{n_2}, \ldots\},$$

with segments chosen alternately from P and Q so that:
(a) m_1 is the smallest integer such that

$$\sum_{i=1}^{m_1} \alpha_i \geq v;$$

(b) n_1 is the smallest integer such that

$$\sum_{i=1}^{m_1} \alpha_i - \sum_{j=1}^{n_1} \beta_j \leq \mu;$$

and, if $k \geq 2$,
(c) m_k is the smallest integer greater than m_{k-1} such that

$$\sum_{i=1}^{m_k} \alpha_i - \sum_{j=1}^{n_{k-1}} \beta_j \geq v;$$

(d) n_k is the smallest integer greater than n_{k-1} such that

$$\sum_{i=1}^{m_k} \alpha_i - \sum_{j=1}^{n_k} \beta_j \leq \mu.$$

That this construction is possible follows from Theorem 3.16: since $\sum \alpha_i = \sum \beta_j = \infty$, it is possible to choose m_k and n_k so that

$$\sum_{i=m_{k-1}}^{m_k} \alpha_i \quad \text{and} \quad \sum_{j=n_{k-1}}^{n_k} \beta_j$$

are as large as we please, no matter how large m_{k-1} and n_{k-1} are. To see that $\sum_{n=1}^{\infty} a_n'$ satisfies (24), define

$$M_k = (m_1 + m_2 + \cdots + m_k) + (n_1 + n_2 + \cdots + n_{k-1}),$$
$$N_k = (m_1 + m_2 + \cdots + m_k) + (n_1 + n_2 + \cdots + n_k),$$
$$k \geq 2.$$

Since m_k and n_k are the smallest integers with the properties specified above, it follows that

$$v \leq A'_{M_k} < v + \alpha_{m_k}, \qquad k \geq 2, \tag{25}$$

and

$$\mu - \beta_{n_k} < A'_{N_k} \leq \mu, \qquad k \geq 2. \tag{26}$$

Because of our construction,

$$A'_{N_k} \leq A'_n \leq A'_{M_k}, \qquad M_k \leq n \leq N_k, \tag{27}$$

and

$$A'_{N_k} \leq A'_n \leq A'_{M_{k+1}}, \qquad N_k \leq n \leq M_{k+1}. \tag{28}$$

From the first inequality of (25), $A'_n \geq v$ for infinitely many values of n; however, since $\lim_{i \to \infty} \alpha_i = 0$ (why?), the second inequalities in (25), (27), and (28) imply that $A'_n > v + \varepsilon$ for only finitely many values of n, if $\varepsilon > 0$. This implies that $\overline{\lim}_{n \to \infty} A'_n = v$. Since $\lim_{j \to \infty} \beta_j = 0$ also, a similar argument based on the second inequality of (26) and the first inequalities of (26), (27), and (28) implies that $\underline{\lim}_{n \to \infty} A'_n = \mu$. This completes the proof.

Multiplication of series

The product of two finite sums can be written as another finite sum; for example,

$$\begin{aligned}
(a_0 + a_1 + a_2)(b_0 + b_1 + b_2) = \ & a_0 b_0 + a_0 b_1 + a_0 b_2 \\
& + a_1 b_0 + a_1 b_1 + a_1 b_2 \\
& + a_2 b_0 + a_2 b_1 + a_2 b_2,
\end{aligned}$$

where the sum on the right contains each product $a_i b_j$ ($i, j = 0, 1, 2$) exactly once. These products can be rearranged arbitrarily without changing their sum. The corresponding situation for series is more complicated.

Given two series

$$\sum_{n=0}^{\infty} a_n \quad \text{and} \quad \sum_{n=0}^{\infty} b_n$$

(because of applications in Section 4.5 it is convenient here to start the summation index at zero), we can arrange all possible products $a_i b_j$ ($i, j \geq 0$) in a two-dimensional array:

$$
\begin{array}{ccccc}
a_0 b_0 & a_0 b_1 & a_0 b_2 & a_0 b_3 & \cdots \\
a_1 b_0 & a_1 b_1 & a_1 b_2 & a_1 b_3 & \cdots \\
a_2 b_0 & a_2 b_1 & a_2 b_2 & a_2 b_3 & \cdots \\
a_3 b_0 & a_3 b_1 & a_3 b_2 & a_3 b_3 & \cdots \\
\vdots & \vdots & \vdots & \vdots &
\end{array}
\tag{29}
$$

where the subscript on a is constant in each row and the subscript on b is constant in each column. Any sensible definition of the product

$$\left(\sum_{n=0}^{\infty} a_n \right) \left(\sum_{n=0}^{\infty} b_n \right)$$

clearly must involve every product in this array exactly once; thus, we might define the product of the two series to be the series $\sum_{n=0}^{\infty} p_n$, where $\{p_n\}$ is a sequence obtained by ordering the products in (29) according to some method which chooses every product exactly once. One way to do this is indicated by

$$
\begin{array}{ccccc}
a_0 b_0 \rightarrow a_0 b_1 & & a_0 b_2 \rightarrow a_0 b_3 & \cdots \\
\downarrow & & \uparrow & \downarrow \\
a_1 b_0 \leftarrow a_1 b_1 & & a_1 b_2 & a_1 b_3 & \cdots \\
\downarrow & & \uparrow & \downarrow \\
a_2 b_0 \rightarrow a_2 b_1 \rightarrow a_2 b_2 & & a_2 b_3 & \cdots \\
& & & \downarrow \\
a_3 b_0 \leftarrow a_3 b_1 \leftarrow a_3 b_2 \leftarrow a_3 b_3 & \cdots \\
\downarrow \\
\vdots \qquad \vdots \qquad \vdots \qquad \vdots
\end{array}
\tag{30}
$$

and another by

$$
\begin{array}{ccccc}
a_0 b_0 \rightarrow a_0 b_1 & a_0 b_2 \rightarrow a_0 b_3 & a_0 b_4 & \cdots \\
a_1 b_0 & a_1 b_1 & a_1 b_2 & a_1 b_3 & \cdots \\
a_2 b_0 & a_2 b_1 & a_2 b_2 & a_2 b_3 & \cdots \\
a_3 b_0 & a_3 b_1 & a_3 b_2 & a_3 b_3 & \cdots \\
a_4 b_0 & \vdots & \vdots & \vdots
\end{array}
\tag{31}
$$

There are infinitely many others, and to each corresponds a series that we might consider to be the product of the given series. This raises an interesting question:
If

$$\sum_{n=0}^{\infty} a_n = A \quad \text{and} \quad \sum_{n=0}^{\infty} b_n = B,$$

where A and B are finite, does every product series $\sum_{n=0}^{\infty} p_n$ constructed by ordering the products in (29) converge to AB?

The next theorem tells us when the answer to this question is in the affirmative.

Theorem 3.18 *Let*

$$\sum_{n=0}^{\infty} a_n = A \quad and \quad \sum_{n=0}^{\infty} b_n = B,$$

where A and B are finite and at least one element of each series is nonzero. Then $\sum_{n=0}^{\infty} p_n = AB$ for every sequence $\{p_n\}$ obtained by ordering the products in (29) if and only if $\sum a_n$ and $\sum b_n$ converge absolutely. Moreover, in this case $\sum p_n$ converges absolutely.

Proof. First let $\{p_n\}$ be the sequence obtained by arranging the products $\{a_i b_j\}$ according to the scheme indicated in (30), and define

$$\begin{aligned}
A_n &= a_0 + a_1 + \cdots + a_n, & \bar{A}_n &= |a_0| + |a_1| + \cdots + |a_n|, \\
B_n &= b_0 + b_1 + \cdots + b_n, & \bar{B}_n &= |b_0| + |b_1| + \cdots + |b_n|, \\
P_n &= p_0 + p_1 + \cdots + p_n, & \bar{P}_n &= |p_0| + |p_1| + \cdots + |p_n|.
\end{aligned}$$

From (30) we see that

$$P_0 = A_0 B_0, \qquad P_3 = A_1 B_1, \qquad P_8 = A_2 B_2,$$

and in general

$$P_{m^2 - 1} = A_m B_m. \tag{32}$$

Similarly,

$$\bar{P}_{m^2 - 1} = \bar{A}_m \bar{B}_m. \tag{33}$$

If $\sum |a_n| < \infty$ and $\sum |b_n| < \infty$, then $\{\bar{A}_m \bar{B}_m\}$ is bounded and, since

$$\bar{P}_m \leq \bar{P}_{m^2 - 1}, \qquad m \geq 2,$$

(33) implies that $\{\bar{P}_m\}$ is bounded. Therefore, $\sum |p_n| < \infty$. This means that

$\sum p_n$ converges. Now

$$\sum_{n=0}^{\infty} p_n = \lim_{n \to \infty} P_n \qquad \text{(by definition)}$$

$$= \lim_{m \to \infty} P_{m^2 - 1} \qquad \text{(by Theorem 2.1)}$$

$$= \lim_{m \to \infty} A_m B_m \qquad \text{[from (32)]}$$

$$= \left(\lim_{m \to \infty} A_m \right) \left(\lim_{m \to \infty} B_m \right) \qquad \text{(by Theorem 1.5)}$$

$$= AB.$$

Since any other ordering of the products in (29) produces a series $\sum_{n=0}^{\infty} p_n'$ which is a rearrangement of the absolutely convergent series $\sum_{n=0}^{\infty} p_n$, Theorem 3.15 implies that $\sum |p_n'| < \infty$ for every such ordering, and that $\sum_{n=0}^{\infty} p_n' = AB$. This shows that the stated condition is sufficient.

For necessity, again let $\sum_{n=0}^{\infty} p_n$ be obtained from the ordering indicated in (30), and suppose that it and all its rearrangements converge to AB. Then $\sum p_n$ must converge absolutely, because of Theorem 3.17; therefore, $\{\bar{P}_{m^2 - 1}\}$ is bounded, and (33) implies that $\{\bar{A}_m\}$ and $\{\bar{B}_m\}$ are bounded. Therefore, $\sum |a_n| < \infty$ and $\sum |b_n| < \infty$. (It is here that we need the assumption that neither $\sum a_n$ nor $\sum b_n$ consists entirely of zeros. Why?) This completes the proof.

A useful definition of the product of two series is due to Cauchy. This definition is particularly important in connection with power series, as we will see in Section 4.5.

Definition 3.3   *The Cauchy product of* $\sum_{n=0}^{\infty} a_n$ *and* $\sum_{n=0}^{\infty} b_n$ *is* $\sum_{n=0}^{\infty} c_n$, *where*

$$c_n = a_0 b_n + a_1 b_{n-1} + \cdots + a_{n-1} b_1 + a_n b_0. \qquad (34)$$

Thus, c_n *is the sum of all products* $a_i b_j$ *where* $i \geq 0$, $j \geq 0$, *and* $i + j = n$; *it can be written as*

$$c_n = \sum_{r=0}^{n} a_r b_{n-r} = \sum_{r=0}^{n} b_r a_{n-r}.$$

Henceforth, $(\sum_{n=0}^{\infty} a_n)(\sum_{n=0}^{\infty} b_n)$ should be interpreted as the Cauchy product. Notice that

$$\left(\sum_{n=0}^{\infty} a_n \right) \left(\sum_{n=0}^{\infty} b_n \right) = \left(\sum_{n=0}^{\infty} b_n \right) \left(\sum_{n=0}^{\infty} a_n \right)$$

and that the Cauchy product of two series is defined even if one or both diverge.

In the case where both converge it is natural to inquire about the relationship between the product of their sums and the sum of the Cauchy product. Theorem 3.18 yields a partial answer to this question, as follows.

Theorem 3.19   *If $\sum_{n=0}^{\infty} a_n$ and $\sum_{n=0}^{\infty} b_n$ converge absolutely to sums A and B, then their Cauchy product converges absolutely to AB.*

Proof.   Let C_n be the nth partial sum of the Cauchy product; that is,

$$C_n = c_0 + c_1 + \cdots + c_n$$

[see (34)]. Let $\sum_{n=0}^{\infty} p_s$ be the series obtained by ordering the products $\{a_i b_j\}$ according to the scheme indicated in (31), and define P_n to be its nth partial sum; thus,

$$P_n = p_0 + p_1 + \cdots + p_n.$$

Inspection of (31) shows that c_n is the sum of the $n + 1$ terms connected by the diagonal arrows, and therefore $C_n = P_{m_n}$, where

$$m_n = 1 + 2 + \cdots + (n + 1) - 1 = \frac{n(n + 3)}{2}.$$

From Theorem 3.18, $\lim_{n \to \infty} P_{m_n} = AB$, and so $\lim_{n \to \infty} C_n = AB$. To see that $\sum |c_n| < \infty$, we observe that

$$\sum_{r=0}^{n} |c_r| \le \sum_{s=0}^{m_n} |p_s|$$

and recall that $\sum |p_s| < \infty$, from Theorem 3.18. This completes the proof.

Example 3.24   Consider the Cauchy product of $\sum_{n=0}^{\infty} r^n$ with itself; here $a_n = b_n = r^n$ and

$$c_n = r^0 r^n + r^1 r^{n-1} + \cdots + r^{n-1} r^1 + r^n r^0 = (n + 1) r^n,$$

so

$$\left(\sum_{n=0}^{\infty} r^n \right)^2 = \sum_{n=0}^{\infty} (n + 1) r^n.$$

Since

$$\sum_{n=0}^{\infty} r^n = \frac{1}{1 - r}, \qquad |r| < 1,$$

and the convergence is absolute, Theorem 3.19 implies that

$$\sum_{n=0}^{\infty} (n + 1) r^n = \frac{1}{(1 - r)^2}, \qquad |r| < 1.$$

Example 3.25   If

$$\sum_{n=0}^{\infty} a_n = \sum_{n=0}^{\infty} \frac{\alpha^n}{n!} \quad \text{and} \quad \sum_{n=0}^{\infty} b_n = \sum_{n=0}^{\infty} \frac{\beta^n}{n!},$$

then

$$c_n = \sum_{r=0}^{n} \frac{\alpha^{n-r}\beta^r}{(n-r)!r!} = \frac{1}{n!} \sum_{r=0}^{n} \binom{n}{r} \alpha^{n-r}\beta^r = \frac{(\alpha+\beta)^n}{n!};$$

thus,

$$\left(\sum_{n=0}^{\infty} \frac{\alpha^n}{n!} \right) \left(\sum_{n=0}^{\infty} \frac{\beta^n}{n!} \right) = \sum_{n=0}^{\infty} \frac{(\alpha+\beta)^n}{n!}. \tag{35}$$

The series $\sum_{n=0}^{\infty} x^n/n!$ converges absolutely for all x to the sum e^x, as the reader probably recalls from earlier courses. Thus, Theorem 3.19 and (35) imply that

$$e^{\alpha}e^{\beta} = e^{\alpha+\beta},$$

a familiar result.

The Cauchy product of two series may converge under conditions less stringent than those of Theorem 3.19. If one series converges absolutely and the other conditionally, their Cauchy product converges (*Exercise 46*). If $\sum a_n$ and $\sum b_n$ and their Cauchy product all converge, then the sum of the Cauchy product equals the product of the sums of $\sum a_n$ and $\sum b_n$ (*Exercise 33, Section 4.5*). However, the next example shows that the Cauchy product of two conditionally convergent series may diverge.

Example 3.26   If

$$a_n = b_n = \frac{(-1)^{n+1}}{\sqrt{n+1}}$$

then $\sum_{n=0}^{\infty} a_n$ and $\sum_{n=0}^{\infty} b_n$ converge conditionally. The general term of their Cauchy product is

$$c_n = \sum_{r=0}^{n} \frac{(-1)^{r+1}}{\sqrt{r+1}} \frac{(-1)^{n-r+1}}{\sqrt{n-r+1}} = (-1)^n \sum_{r=0}^{n} \frac{1}{\sqrt{r+1}} \frac{1}{\sqrt{n-r+1}}$$

and so

$$|c_n| \geq \sum_{r=0}^{n} \frac{1}{\sqrt{n+1}} \frac{1}{\sqrt{n+1}} = \frac{n+1}{n+1} = 1.$$

Therefore, the Cauchy product diverges, by Corollary 3.1.

4.3   EXERCISES

1. Prove Theorem 3.1.

2. Prove Theorem 3.2.

3. Show that the terms of

$$\sum \frac{16 + (-2)^n}{n2^n}$$

satisfy the conditions of the alternating series test for large n.

4. (a) Prove: If $a_n = b_n$ except for finitely many values of n, then $\sum a_n$ and $\sum b_n$ converge or diverge together.
   (b) Let $b_{n_k} = a_k$ for some increasing sequence $\{n_k\}_1^\infty$ of positive integers, and $b_n = 0$ if n is any other positive integer. Show that

$$\sum_{n=1}^{\infty} b_n \quad \text{and} \quad \sum_{k=1}^{\infty} a_k$$

diverge or converge together, and that in the latter case they have the same sum. (That is, the convergence properties of a series $\sum_{k=1}^{\infty} a_k$ are not changed by inserting zeros between its terms.)

5. (a) Prove: If $\sum a_n$ converges, then

$$\lim_{n \to \infty} (a_n + a_{n+1} + \cdots + a_{n+r}) = 0, \qquad r = 0, 1, \ldots.$$

   (b) Does the result in part (a) imply that $\sum a_n$ converges? Give a reason for your answer.

6. Prove Corollary 3.2.

7. (a) Verify Corollary 3.2 for the convergent series $\sum 1/n^p$ ($p > 1$). (*Hint:* See the proof of Theorem 3.6.)
   (b) Verify Corollary 3.2 for the convergent series $\sum (-1)^n/n$.

8. Prove: If $0 \leq b_n \leq a_n \leq b_{n+1}$, then $\sum a_n$ and $\sum b_n$ converge or diverge together.

9. Show that (a) of Theorem 3.4 implies (b).

10. Determine convergence or divergence:

(a) $\displaystyle \sum \frac{\sqrt{n^2 - 1}}{\sqrt{n^5 + 1}}$

(b) $\displaystyle \sum \frac{1}{n^2[1 + \frac{1}{2}\sin(n\pi/4)]}$

(c) $\displaystyle \sum \frac{(1 - e^{-n})\log n}{n}$

(d) $\displaystyle \sum \cos \frac{\pi}{n^2}$

(e) $\sum \sin \dfrac{\pi}{n^2}$

(f) $\sum \dfrac{1}{n} \tan \dfrac{\pi}{n}$

(g) $\sum \dfrac{1}{n} \cot \dfrac{\pi}{n}$

(h) $\sum \dfrac{\log n}{n^2}$

11. Suppose $f(x) \geq 0$ for $x \geq k$. Prove that $\int_k^\infty f(x)\, dx < \infty$ if and only if

$$\sum_{n=k}^{\infty} \int_n^{n+1} f(x)\, dx < \infty.$$

(*Hint:* Use Theorem 3.4 and Theorem 4.1, Section 3.4.)

12. Use the integral test to find all values of p for which each series converges:

(a) $\sum \dfrac{n}{(n^2 - 1)^p}$

(b) $\sum \dfrac{n^2}{(n^3 + 4)^p}$

(c) $\sum \dfrac{\sinh n}{(\cosh n)^p}$

13. Let L_n be the nth iterated logarithm. Show that

$$\sum \frac{1}{L_0(n)L_1(n) \cdots L_k(n)[L_{k+1}(n)]^p}$$

converges if and only if $p > 1$. (*Hint:* See Exercise 14, Section 3.4.)

14. Suppose g, g', and $(g')^2 - gg''$ are all positive on $[R, \infty)$. Show that

$$\sum \frac{g'(n)}{g(n)} < \infty$$

if and only if $\lim_{x \to \infty} g(x) < \infty$.

15. Let

$$S(p) = \sum_{n=1}^{\infty} \frac{1}{n^p}, \qquad p > 1.$$

Show that

$$\frac{1}{(p-1)(N+1)^{p-1}} < S(p) - \sum_{n=1}^{N} \frac{1}{n^p} < \frac{1}{(p-1)N^{p-1}}.$$

(*Hint:* See the proof of Theorem 3.6.)

16. Suppose f is positive, decreasing, and locally integrable on $[1, \infty)$, and let

$$a_n = \sum_{k=1}^{n} f(k) - \int_1^n f(x)\, dx.$$

(a) Show that $\{a_n\}$ is nonincreasing and nonnegative, and that

$$0 < \lim_{n \to \infty} a_n < f(1).$$

(b) Deduce from part (a) that

$$\gamma = \lim_{n \to \infty} \left(1 + \frac{1}{2} + \frac{1}{3} + \cdots + \frac{1}{n} - \log n \right)$$

exists and $0 < \gamma < 1$. (γ is *Euler's constant;* its value is approximately 0.577.)

17. Determine convergence or divergence:

(a) $\sum \dfrac{2 + \sin n\theta}{n^2 + \sin n\theta}$

(b) $\sum \dfrac{n + 1}{n} r^n \ (r > 0)$

(c) $\sum e^{-n\rho} \cosh n\rho \ (\rho > 0)$

(d) $\sum \dfrac{n + \log n}{n^2 (\log n)^2}$

(e) $\sum \dfrac{n + \log n}{n^2 \log n}$

(f) $\sum \dfrac{(1 + 1/n)^n}{2^n}$

18. Prove Theorem 3.7(b).

19. Let L_n be the nth iterated logarithm. Prove that

$$\sum \frac{1}{[L_0(n)]^{q_0 + 1} [L_1(n)]^{q_1 + 1} \cdots [L_m(n)]^{q_m + 1}}$$

converges if and only if there is at least one nonzero number in $\{q_0, q_1, \dots, q_m\}$, and the first such is positive. (*Hint:* See Exercise 13 and Exercise 56(b), Section 2.4.)

20. Prove that (a) of Theorem 3.8 implies (b).

21. Determine convergence or divergence:

(a) $\sum \dfrac{2 + \sin^2(n\pi/4)}{3^n}$

(b) $\sum \dfrac{n(n + 1)}{4^n}$

(c) $\sum \dfrac{3 - \sin(n\pi/2)}{n(n + 1)}$

(d) $\sum \dfrac{n + (-1)^n}{n(n + 1)} \dfrac{1}{2^n + \cos(n\pi/2)}$

22. Prove Theorem 3.9(b).

23. Determine convergence or divergence, with $r > 0$:

(a) $\sum \dfrac{n!}{r^n}$

(b) $\sum n^p r^n$

(c) $\sum \dfrac{r^n}{n!}$

(d) $\sum \dfrac{r^{2n + 1}}{(2n + 1)!}$

(e) $\sum \dfrac{r^{2n}}{(2n)!}$

24. Determine convergence or divergence:

(a) $\sum \dfrac{(2n)!}{2^{2n}(n!)^2}$

(b) $\sum \dfrac{(3n)!}{3^{3n}n!(n+1)!(n+3)!}$

(c) $\sum \dfrac{2^n n!}{5 \cdot 7 \cdots (2n+3)}$

(d) $\sum \dfrac{\alpha(\alpha+1)\cdots(\alpha+n-1)}{\beta(\beta+1)\cdots(\beta+n-1)} \ (\alpha, \beta > 0)$

25. Determine convergence or divergence:

(a) $\sum \dfrac{n^n[2+(-1)^n]}{2^n}$

(b) $\sum \left(\dfrac{1+\sin 3\theta}{3}\right)^n$

(c) $\sum (n+1)\left[\dfrac{1+\sin(n\pi/6)}{2}\right]^n$

(d) $\sum \left(1-\dfrac{1}{n}\right)^{n^2}$

26. Give counterexamples showing that the following statements are false unless it is assumed that the terms of the series are nonnegative for all n sufficiently large.

   (a) $\sum a_n$ converges if its partial sums are bounded.
   (b) If $b_n \neq 0 \ (n \geq k)$ and $\lim_{n \to \infty} a_n/b_n = L$, where $0 < L < \infty$, then $\sum a_n$ and $\sum b_n$ converge or diverge together.
   (c) If $a_n \neq 0$ and $\overline{\lim}_{n \to \infty} a_{n+1}/a_n < 1$, then $\sum a_n$ converges.
   (d) If $a_n \neq 0$ and $\overline{\lim}_{n \to \infty} n[(a_{n+1}/a_n)-1] < -1$, then $\sum a_n$ converges.

27. Prove: If the terms of the convergent series $\sum a_n$ have the same sign for $n \geq k$, then $\sum a_n$ converges absolutely.

28. Prove Theorem 3.12.

29. Show that the following series converge absolutely.

(a) $\sum (-1)^n \dfrac{1}{n(\log n)^2}$

(b) $\sum \dfrac{\sin n\theta}{2^n}$

(c) $\sum (-1)^n \dfrac{1}{\sqrt{n}} \sin \dfrac{\pi}{n}$

(d) $\sum \dfrac{\cos n\theta}{\sqrt{n^3-1}}$

30. Show that the following series converge.

(a) $\sum \dfrac{n \sin n\theta}{n^2+(-1)^n} \ (-\infty < \theta < \infty)$

(b) $\sum \dfrac{\cos n\theta}{n} \ (\theta \neq 2k\pi, \ k = \text{integer})$

31. Determine whether the following series are absolutely convergent, conditionally convergent, or divergent.

(a) $\sum \dfrac{b_n}{\sqrt{n}} \ (b_{4m} = b_{4m+1} = 1, b_{4m+2} = b_{4m+3} = -1)$

(b) $\displaystyle\sum \frac{1}{n} \sin \frac{n\pi}{6}$ (c) $\displaystyle\sum \frac{1}{n^2} \cos \frac{n\pi}{7}$

(d) $\displaystyle\sum \frac{1 \cdot 3 \cdot 5 \cdots (2n+1)}{4 \cdot 6 \cdot 8 \cdots (2n+4)} \sin n\theta$

32. Let g be a rational function (ratio of two polynomials). Show that $\sum g(n)r^n$ converges absolutely if $|r| < 1$ and diverges if $|r| > 1$. Discuss the possibilities for $|r| = 1$.

33. Prove: If $\sum a_n^2 < \infty$ and $\sum b_n^2 < \infty$, then $\sum a_n b_n$ converges absolutely.

34. (a) Prove: If $\sum a_n$ converges and $\sum a_n^2 = \infty$, then $\sum a_n$ converges conditionally.
    (b) Give an example of a series with the properties described in part (a).

35. Suppose $0 \le a_{n+1} < a_n$, and

$$\lim_{n \to \infty} \frac{b_1 + b_2 + \cdots + b_n}{w_n} > 0,$$

where $\{w_n\}$ is a sequence of positive numbers such that

$$\sum w_n(a_n - a_{n+1}) = \infty.$$

Show that $\sum a_n b_n = \infty$. (*Hint*: Use summation by parts.)

*36. (a) Prove: If $0 < 2\varepsilon < \theta < \pi - 2\varepsilon$, then

$$\lim_{n \to \infty} \frac{|\sin \theta| + |\sin 2\theta| + \cdots + |\sin n\theta|}{n} \ge \frac{\sin \varepsilon}{2}.$$

[*Hint*: Show that $|\sin n\theta| > \sin \varepsilon$ at least "half the time"; more precisely, show that if $|\sin m\theta| \le \sin \varepsilon$ for some integer m, then $|\sin(m+1)\theta| > \sin \varepsilon$.]
    (b) Show that

$$\sum \frac{\sin n\theta}{n^p}$$

converges conditionally if $0 < p \le 1$ and $\theta \ne k\pi$ (k = integer). (*Hint*: Use Exercise 35 and see Example 3.21.)

37. Show that

$$\sum_{n=1}^{\infty} \frac{(-1)^{n+1}}{n} = \frac{1}{2} \sum_{n=1}^{\infty} \frac{1}{n(2n-1)}.$$

38. Let $b_{3m+1} = 1$, $b_{3m+2} = -2$, and $b_{3m+3} = 1$ ($m = 0, 1, 2, \ldots$). Show that

$$\sum_{n=1}^{\infty} \frac{b_n}{n} = \frac{2}{3} \sum_{m=0}^{\infty} \frac{1}{(m+1)(3m+1)(3m+2)}.$$

39. Let $\sum a'_n$ be obtained by rearranging finitely many terms of a convergent series $\sum a_n$. Show that the two series have the same sum.

40. Fill in the details of the proof of Theorem 3.16.

41. Prove Theorem 3.17 for the case where (a) μ is finite and $v = \infty$; (b) $\mu = -\infty$ and $v = \infty$; (c) $\mu = v = \infty$.

42. Prove: If every rearrangement of a series converges, then they all converge to the same sum.

43. Give necessary and sufficient conditions on a divergent series so that it has a convergent rearrangement.

44. A series diverges *unconditionally* to ∞ if every rearrangement of it diverges to ∞. State necessary and sufficient conditions for a series to have this property.

45. Suppose f and g have derivatives of all orders at 0, and let $h = fg$. Show formally that

$$\left[\sum_{n=0}^{\infty} \frac{f^{(n)}(0)}{n!} x^n \right] \left[\sum_{n=0}^{\infty} \frac{g^{(n)}(0)}{n!} x^n \right] = \sum_{n=0}^{\infty} \frac{h^{(n)}(0)}{n!} x^n$$

in the sense of the Cauchy product. (*Hint:* See Exercise 12, Section 2.3.)

46. Prove: If $\sum |a_n| < \infty$ and $\sum b_n$ converges (perhaps conditionally), with $\sum_{n=0}^{\infty} a_n = A$ and $\sum_{n=0}^{\infty} b_n = B$, then the Cauchy product

$$\sum_{n=0}^{\infty} c_n = \left(\sum_{n=0}^{\infty} a_n \right) \left(\sum_{n=0}^{\infty} b_n \right)$$

converges to AB. [*Hint:* Let $\{A_n\}$, $\{B_n\}$, and $\{C_n\}$ be the partial sums of the series. Show that

$$C_n - A_n B = \sum_{r=0}^{n} a_r (B_{n-r} - B)$$

and apply Theorem 3.3 to $\sum |a_n|$.]

4.4   SEQUENCES AND SERIES OF FUNCTIONS

Until now we have considered sequences and series of constants. Now we turn our attention to sequences and series of real-valued functions defined on subsets of the reals.

Throughout this section "subset" means "nonempty subset."

If $F_k, F_{k+1}, \ldots, F_n, \ldots$ are real-valued functions defined on a subset D of the reals, we say that $\{F_n\}$ is an *infinite sequence* (or simply a *sequence*) of *functions on* D. If the sequence of values $\{F_n(x)\}$ converges for each x in some subset S of D, then $\{F_n\}$ defines a limit function on S. The formal definition is as follows.

Definition 4.1   *Suppose $\{F_n\}$ is a sequence of functions on D and the sequence of values $\{F_n(x)\}$ converges for each x in some subset S of D. Then we say that $\{F_n\}$ converges pointwise on S to the limit function F, defined by*

$$F(x) = \lim_{n \to \infty} F_n(x), \qquad x \in S.$$

Example 4.1   The functions

$$F_n(x) = \left(1 - \frac{nx}{n+1}\right)^{n/2}$$

define a sequence on $D = (-\infty, 1]$, and

$$\lim_{n \to \infty} F_n(x) = \begin{cases} \infty, & x < 0, \\ 1, & x = 0, \\ 0, & 0 < x \le 1. \end{cases}$$

Therefore, $\{F_n\}$ converges pointwise on $S = [0, 1]$ to the limit function F defined by

$$F(x) = \begin{cases} 1, & x = 0, \\ 0, & 0 < x \le 1. \end{cases}$$

Example 4.2   Consider the functions

$$F_n(x) = x^n e^{-nx}, \qquad x > 0, \quad n = 1, 2, \ldots$$

(Figure 4.1). Equating the derivative

$$F_n'(x) = nx^{n-1}e^{-nx}(1-x)$$

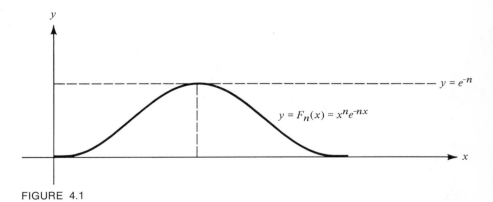

FIGURE 4.1

to zero shows that the maximum value of $F_n(x)$ on $[0, \infty)$ is e^{-n}, attained at $x = 1$. Therefore,

$$|F_n(x)| \le e^{-n}, \qquad x \ge 0,$$

and so $\lim_{n \to \infty} F_n(x) = 0$ for all $x \ge 0$. The limit function in this case is identically zero on $[0, \infty)$.

Example 4.3   Let F_n be defined on $(-\infty, \infty)$ by

$$F_n(x) = \begin{cases} 0, & x < -\dfrac{2}{n}, \\[2mm] -n(2 + nx), & -\dfrac{2}{n} \le x < -\dfrac{1}{n}, \\[2mm] n^2 x, & -\dfrac{1}{n} \le x < \dfrac{1}{n}, \\[2mm] n(2 - nx), & \dfrac{1}{n} \le x < \dfrac{2}{n}, \\[2mm] 0, & x \ge \dfrac{2}{n} \end{cases}$$

(Figure 4.2). Since $F_n(0) = 0$ for all n, it is clear that $\lim_{n \to \infty} F_n(0) = 0$. If $x \ne 0$, then $F_n(x) = 0$ if $n \ge 2/|x|$. Therefore,

$$\lim_{n \to \infty} F_n(x) = 0, \qquad -\infty < x < \infty,$$

so the limit function is identically zero on $(-\infty, \infty)$.

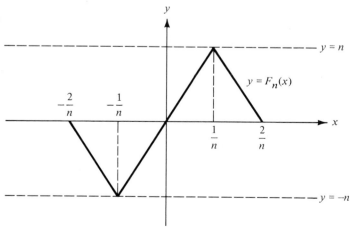

FIGURE 4.2

Example 4.4 For each positive integer n, let S_n be the set of numbers of the form $x = p/q$, where p and q are integers with no common factors and $1 \le q \le n$. Define

$$F_n(x) = \begin{cases} 1, & x \in S_n, \\ 0, & x \notin S_n. \end{cases}$$

If x is irrational, then $x \notin S_n$ for any n, so $F_n(x) = 0$ for $n = 1, 2, \ldots$. If x is rational, then $x \in S_n$, and $F_n(x) = 1$, for all sufficiently large n. Therefore,

$$\lim_{n \to \infty} F_n(x) = F(x) = \begin{cases} 1 & \text{if } x \text{ is rational,} \\ 0 & \text{if } x \text{ is irrational,} \end{cases}$$

Uniform convergence

The pointwise limit of a sequence of functions may differ radically from the functions in the sequence. Thus, in Example 4.1 each F_n is continuous on $[0, 1]$, but F is not. In Example 4.3 the graph of each F_n has two triangular spikes whose height increases with n, while the graph of F (the x axis) has none. In Example 4.4 each F_n is integrable, while F is nonintegrable on every finite interval (*Exercise 3*). There is nothing in Definition 4.1 to preclude these apparent anomalies; although the definition implies that for each x_0 in S, $F_n(x_0)$ approximates $F(x_0)$ if n is sufficiently large, it does not imply that any particular F_n approximates F well over *all* of S. To formulate a definition that does, it is convenient to introduce the notation

$$\|g\|_S = \underset{x \in S}{\text{l.u.b.}} \, |g(x)|$$

and to state the following lemma, whose proof is left to the reader (*Exercise 4*).

Lemma 4.1 *If g and h are defined on S, then*

$$\|g + h\|_S \le \|g\|_S + \|h\|_S$$

and

$$\|gh\|_S \le \|g\|_S \, \|h\|_S.$$

Definition 4.2 *A sequence $\{F_n\}$ of functions defined on a set S is said to converge uniformly to the limit function F on S if*

$$\lim_{n \to \infty} \|F_n - F\|_S = 0.$$

Thus, $\{F_n\}$ converges uniformly to F on S if for each $\varepsilon > 0$ there is an integer N such that

$$\|F_n - F\|_S < \varepsilon \qquad \text{if } n \ge N. \tag{1}$$

If $S = [a, b]$ and F is the function with graph shown in Figure 4.3, (1) implies that the graph of

$$y = F_n(x), \qquad a \le x \le b,$$

lies in the shaded band

$$F(x) - \varepsilon < y < F(x) + \varepsilon, \qquad a \le x \le b,$$

if $n \ge N$.

From Definition 4.2, if $\{F_n\}$ converges uniformly on S, then it converges uniformly on any subset of S (*Exercise 6*).

Example 4.5   The sequence defined by

$$F_n(x) = x^n e^{-nx}, \qquad n = 1, 2, \dots$$

converges uniformly to $F \equiv 0$ (that is, to the identically zero function) on $S = [0, \infty)$, since we saw in Example 4.2 that

$$\|F_n - F\|_S = \|F_n\|_S = e^{-n},$$

and so

$$\|F_n - F\|_S < \varepsilon$$

if $n > -\log \varepsilon$. For these values of n the graph

$$y = F_n(x), \qquad 0 \le x < \infty,$$

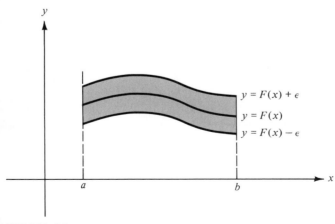

FIGURE 4.3

lies in the strip

$$-\varepsilon \le y \le \varepsilon, \qquad x \ge 0$$

(Figure 4.4).

The next theorem yields alternative definitions of pointwise and uniform convergence.

Theorem 4.1 *Let $\{F_n\}$ be defined on S. Then:*

(a) $\{F_n\}$ converges pointwise to F on S if and only if there is for each $\varepsilon > 0$ and $x_0 \in S$ an integer N, which may depend on x_0 as well as ε, such that

$$\left|F_n(x_0) - F(x_0)\right| < \varepsilon \qquad if \quad n \ge N.$$

(b) $\{F_n\}$ converges uniformly to F on S if and only if there is for each $\varepsilon > 0$ an integer N, which depends only on ε and not on any particular x in S, such that

$$\left|F_n(x) - F(x)\right| < \varepsilon \qquad for\ all\ x\ in\ S\ if\ n \ge N.$$

This theorem follows immediately from Definitions 4.1 and 4.2. We leave its proof to the reader (*Exercise 8*). The next theorem follows immediately from this one.

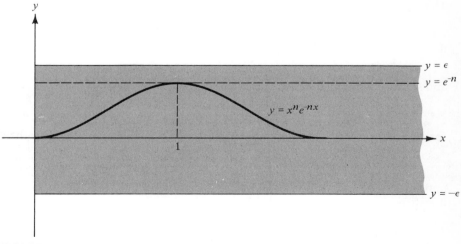

FIGURE 4.4

Theorem 4.2   *If $\{F_n\}$ converges uniformly to F on S, then it converges pointwise to F on S. The converse is false; pointwise convergence does not imply uniform convergence.*

Example 4.6   The sequence $\{F_n\}$ of Example 4.3 converges pointwise to $F \equiv 0$ on $(-\infty, \infty)$, but not uniformly, since

$$\|F_n - F\|_{(-\infty,\infty)} = F_n\left(\frac{1}{n}\right) = \left|F_n\left(\frac{-1}{n}\right)\right| = n,$$

and so

$$\lim_{n \to \infty} \|F_n - F\|_{(-\infty,\infty)} = \infty.$$

However, the convergence is uniform on

$$S_\rho = (-\infty, \rho] \cup [\rho, \infty)$$

for any $\rho > 0$, since

$$\|F_n - F\|_{S_\rho} = 0 \qquad \text{if} \quad n > \frac{2}{\rho}.$$

Example 4.7   If $F_n(x) = x^n$, then $\{F_n\}$ converges pointwise on $S = [0, 1]$ to

$$F(x) = \begin{cases} 1, & x = 1, \\ 0, & 0 \le x < 1. \end{cases}$$

The convergence is not uniform on S, since if $0 < \varepsilon < 1$, then

$$|F_n(x) - F(x)| > 1 - \varepsilon$$

if

$$(1 - \varepsilon)^{1/n} < x < 1,$$

and so

$$\|F_n - F\|_S = 1$$

for all n. However, the convergence is uniform on $[0, \rho]$ if $0 < \rho < 1$, since then

$$\|F_n - F\|_{[0,\rho]} = \rho^n$$

and $\lim_{n \to \infty} \rho^n = 0$. Another way to say the same thing: $\{F_n\}$ converges uniformly on every closed subset of $[0, 1)$.

The next theorem enables us to test a sequence for uniform convergence without guessing what the limit function might be. It is analogous to Cauchy's convergence criterion for sequences of constants (Theorem 1.9).

Theorem 4.3 (Cauchy's Uniform Convergence Criterion) *A sequence $\{F_n\}$ converges uniformly on a set S if and only if for each $\varepsilon > 0$ there is an integer N such that*

$$\|F_n - F_m\|_S < \varepsilon \qquad \text{if} \quad n, m \geq N. \tag{2}$$

Proof. For necessity, suppose $\{F_n\}$ converges uniformly to F on S; then if $\varepsilon > 0$, there is an integer N such that

$$\|F_k - F\|_S < \frac{\varepsilon}{2} \qquad \text{if} \quad k \geq N,$$

and therefore

$$\begin{aligned}
\|F_n - F_m\|_S &= \|(F_n - F) + (F - F_m)\|_S \\
&\leq \|F_n - F\|_S + \|F - F_m\|_S \text{ (Lemma 4.1)} \\
&< \frac{\varepsilon}{2} + \frac{\varepsilon}{2} = \varepsilon \qquad \text{if} \quad m, n \geq N.
\end{aligned}$$

For sufficiency, we first observe that (2) implies that

$$|F_n(x) - F_m(x)| < \varepsilon \qquad \text{if} \quad n, m \geq N,$$

for any fixed x in S. Therefore, Cauchy's convergence criterion for sequences of constants (Theorem 1.9) implies that $\{F_n(x)\}$ converges for each x in S; that is, $\{F_n\}$ converges pointwise to a limit function F on S. To see that the convergence is uniform, we write

$$\begin{aligned}
|F_m(x) - F(x)| &= |[F_m(x) - F_n(x)] + [F_n(x) - F(x)]| \\
&\leq |F_m(x) - F_n(x)| + |F_n(x) - F(x)| \\
&\leq \|F_m - F_n\|_S + |F_n(x) - F(x)|.
\end{aligned}$$

This and (2) imply that

$$|F_m(x) - F(x)| < \varepsilon + |F_n(x) - F(x)| \qquad \text{if} \quad n, m \geq N, \tag{3}$$

Since $\lim_{n \to \infty} F_n(x) = F(x)$,

$$|F_n(x) - F(x)| < \varepsilon$$

for some $n \geq N$, and so (3) implies that

$$|F_m(x) - F(x)| < 2\varepsilon \qquad \text{if} \quad m \geq N.$$

But this inequality holds for all x in S, and so

$$\|F_m - F\|_S \leq 2\varepsilon \qquad \text{if} \quad m \geq N.$$

Since ε is an arbitrary positive number, this implies that $\{F_n\}$ converges uniformly to F on S. This completes the proof.

The next example is similar to Example 1.13.

Example 4.8   Suppose g is differentiable on $S = (-\infty, \infty)$ and

$$|g'(x)| \le r < 1, \qquad -\infty < x < \infty. \tag{4}$$

Let F_0 be bounded on S and

$$F_n(x) = g(F_{n-1}(x)), \qquad n \ge 1.$$

We will show that $\{F_n\}$ converges uniformly on S. For $n \ge 1$,

$$F_{n+1}(x) - F_n(x) = g(F_n(x)) - g(F_{n-1}(x)) = g'(c)[F_n(x) - F_{n-1}(x)]$$

for some c, by the mean value theorem. From this and (4),

$$|F_{n+1}(x) - F_n(x)| \le r|F_n(x) - F_{n-1}(x)|, \qquad -\infty < x < \infty,$$

which implies that

$$\|F_{n+1} - F_n\|_S \le r\|F_n - F_{n-1}\|_S.$$

By induction this implies that

$$\|F_{n+1} - F_n\|_S \le r^n\|F_1 - F_0\|_S. \tag{5}$$

(How do we know that $\|F_1 - F_0\|_S < \infty$?) If $n > m$,

$$\|F_n - F_m\|_S = \|(F_n - F_{n-1}) + (F_{n-1} - F_{n-2}) + \cdots + (F_{m+1} - F_m)\|_S$$
$$\le \|F_n - F_{n-1}\|_S + \|F_{n-1} - F_{n-2}\|_S + \cdots + \|F_{m+1} - F_m\|_S,$$

from Lemma 4.1. Now (5) implies that

$$\|F_n - F_m\|_S \le \|F_1 - F_0\|_S(1 + r + r^2 + \cdots + r^{n-m-1})r^m$$

$$< \|F_1 - F_0\|_S \frac{r^m}{1 - r}.$$

Therefore, if

$$\|F_1 - F_0\|_S \frac{r^N}{1 - r} < \varepsilon,$$

then $\|F_n - F_m\|_S < \varepsilon$ if $n, m \ge N$. According to Theorem 4.3 this implies that $\{F_n\}$ converges uniformly on S.

Properties preserved by uniform convergence

We now study to what extent certain properties of the functions of a uniformly convergent sequence are inherited by the limit function. The first of these is continuity.

Theorem 4.4   *If $\{F_n\}$ converges uniformly to F on S and each F_n is continuous at a point x_0 in S, then so is F. Similar statements hold for continuity from the right and left.*

Proof.   If $x \in S$ and F_n is any function in the sequence, then

$$|F(x) - F(x_0)| \le |F(x) - F_n(x)| + |F_n(x) - F_n(x_0)| + |F_n(x_0) - F(x_0)|$$
$$\le |F_n(x) - F_n(x_0)| + 2\|F_n - F\|_S. \tag{6}$$

Suppose $\varepsilon > 0$; since $\{F_n\}$ converges uniformly to F on S, we can choose n so that $\|F_n - F\|_S < \varepsilon$. For this fixed value of n, (6) implies that

$$|F(x) - F(x_0)| < |F_n(x) - F_n(x_0)| + 2\varepsilon, \qquad x \in S. \tag{7}$$

Since F_n is continuous at x_0, there is a $\delta > 0$ such that

$$|F_n(x) - F_n(x_0)| < \varepsilon \qquad \text{if} \quad |x - x_0| < \delta,$$

and so, from (7),

$$|F(x) - F(x_0)| < 3\varepsilon \qquad \text{if} \quad |x - x_0| < \delta.$$

Therefore, F is continuous at x_0, as we wished to prove. The same argument with only trivial changes applies to the assertion on continuity from the right and left.

Corollary 4.1   *If $\{F_n\}$ converges uniformly to F on S and each F_n is continuous on S, then so is F; that is, the uniform limit of continuous functions is continuous.*

Now we consider the question of integrability of the uniform limit of integrable functions.

Theorem 4.5   *Suppose $\{F_n\}$ converges uniformly to F on $S = [a, b]$ and F and each F_n are integrable on $[a, b]$. Then*

$$\int_a^b F(x)\, dx = \lim_{n \to \infty} \int_a^b F_n(x)\, dx. \tag{8}$$

Proof.   Since

$$\left| \int_a^b F_n(x)\, dx - \int_a^b F(x)\, dx \right| \le \int_a^b |F_n(x) - F(x)|\, dx$$

$$\le (b - a)\|F_n - F\|_S$$

and $\lim_{n \to \infty} \|F_n - F\|_S = 0$, the conclusion follows.

In particular, this theorem implies that (8) holds if each F_n is continuous on $[a, b]$, because then F is continuous, and therefore integrable, on $[a, b]$.

The hypotheses of Theorem 4.5 are stronger than necessary. We state the next theorem so that the reader will be better informed on this subject. Its proof is inaccessible to one who has skipped Section 3.5, and quite involved in any case; therefore, we omit it.

Theorem 4.6 *Suppose $\{F_n\}$ converges pointwise to F and each F_n is integrable on $[a, b]$. Then:*

(a) If the convergence is uniform, then F is integrable on $[a, b]$, and (8) holds.

(b) If the sequence $\{\|F_n\|_{[a,b]}\}$ is bounded and F is integrable on $[a, b]$, then (8) holds.

Part (a) of this theorem shows that it is not necessary to assume in Theorem 4.5 that F is integrable on $[a, b]$, since this follows from the uniform convergence. Part (b) is known as the *bounded convergence theorem*. Neither of its assumptions can be omitted: in Example 4.3, where $\{\|F_n\|_{[0,1]}\}$ is unbounded while F is integrable on $[0, 1]$,

$$\int_0^1 F_n(x)\, dx = 1 \qquad (n = 1, 2, \ldots) \quad \text{but} \quad \int_0^1 F(x)\, dx = 0;$$

in Example 4.4, where $\|F_n\|_{[a,b]} = 1$ for every finite interval $[a, b]$, F_n is integrable for all $n \geq 1$, and F is nonintegrable on every interval (*Exercise 3*).

After Theorems 4.4 and 4.5 it may seem reasonable to expect that if a sequence $\{F_n\}$ of differentiable functions converges uniformly to F on S, then $F' = \lim_{n\to\infty} F_n'$ on S. The next example shows that this need not be so.

Example 4.9 The sequence $\{F_n\}$ defined by

$$F_n(x) = x^n \sin \frac{1}{x^{n-1}}$$

converges uniformly to $F \equiv 0$ on $[r_1, r_2]$ if $0 < r_1 < r_2 < 1$ [or, equivalently, on every compact subset of $(0, 1)$]. However,

$$F_n'(x) = nx^{n-1} \sin \frac{1}{x^{n-1}} - (n-1) \cos \frac{1}{x^{n-1}},$$

so $\{F_n'(x)\}$ does not converge for any x in $(0, 1)$.

Theorem 4.7 *Suppose F_n' is continuous for all $n \geq 1$ and $\{F_n'\}$ converges uniformly on $[a, b]$. Suppose also that the sequence of values $\{F_n(x_0)\}$ converges for some x_0 in $[a, b]$. Then $\{F_n\}$ converges uniformly on $[a, b]$ to a differentiable limit* *function F, and*

$$F'(x) = \lim_{n \to \infty} F_n'(x), \qquad a < x < b, \tag{9}$$

while

$$F'_+(a) = \lim_{n \to \infty} F_n'(a) \quad and \quad F'_-(b) = \lim_{n \to \infty} F_n'(b). \tag{10}$$

Proof.   Since F_n' is continuous on $[a, b]$, we can write

$$F_n(x) = F_n(x_0) + \int_{x_0}^x F_n'(t)\, dt, \qquad a \le x \le b \tag{11}$$

(Theorem 3.11, Section 3.3). Now let

$$L = \lim_{n \to \infty} F_n(x_0)$$

and

$$G(x) = \lim_{n \to \infty} F_n'(x).$$

Since F_n' is continuous and $\{F_n'\}$ converges uniformly to G on $[a, b]$, G is continuous on $[a, b]$ (Corollary 4.1); therefore, (11) and the remark following the proof of Theorem 4.5 imply that $\{F_n\}$ converges pointwise on $[a, b]$ to the limit function

$$F(x) = L + \int_{x_0}^x G(t)\, dt. \tag{12}$$

The convergence is actually uniform, since subtracting (11) from (12) yields

$$|F(x) - F_n(x)| \le |L - F_n(x_0)| + \left| \int_{x_0}^x |G(t) - F_n'(t)|\, dt \right|$$

$$\le |L - F_n(x_0)| + |x - x_0|\, \|G - F_n'\|_{[a,b]}$$

and so

$$\|F - F_n\|_{[a,b]} \le |L - F_n(x_0)| + (b - a)\|G - F_n'\|_{[a,b]},$$

where the right side approaches zero as $n \to \infty$.

Since G is continuous on $[a, b]$, (12) and Theorem 3.10 of Section 3.3 imply (9) and (10). This completes the proof.

Infinite series of functions

In Section 4.3 we defined the sum of an infinite series of constants as the limit of the sequence of partial sums. The same definition can be applied to series of functions, as follows.

Definition 4.3   *If* $\{f_j\}_k^\infty$ *is a sequence of real-valued functions defined on a set D of reals, then* $\sum_{j=k}^\infty f_j$ *is said to be an infinite series (or simply a series) of functions on D. The partial sums of* $\sum_{j=k}^\infty f_j$ *are defined by*

$$F_n = \sum_{j=k}^n f_j, \qquad n \geq k.$$

If $\{F_n\}_k^\infty$ *converges pointwise to a function F on a subset S of D, we say that* $\sum_{j=k}^\infty f_j$ *converges pointwise to the sum F on S, and write*

$$F = \sum_{j=k}^\infty f_j, \qquad x \in S.$$

If $\{F_n\}$ *converges uniformly to F on S, we say that* $\sum_{j=k}^\infty f_j$ *converges uniformly to F on S.*

Example 4.10   The functions

$$f_j(x) = x^j, \qquad j = 0, 1, \ldots,$$

define the infinite series

$$\sum_{j=0}^\infty x^j$$

on $D = (-\infty, \infty)$. Its nth partial sum is

$$F_n(x) = 1 + x + x^2 + \cdots + x^n,$$

or, in closed form,

$$F_n(x) = \begin{cases} \dfrac{1 - x^{n+1}}{1 - x}, & x \neq 1, \\[2mm] n + 1, & x = 1 \end{cases}$$

(see Example 1.10). We have seen earlier that $\{F_n\}$ converges pointwise to

$$F(x) = \frac{1}{1 - x}$$

if $|x| < 1$ and diverges if $|x| \geq 1$; hence, we write

$$\sum_{j=0}^\infty x^j = \frac{1}{1 - x}, \qquad -1 < x < 1.$$

Since the difference

$$F(x) - F_n(x) = \frac{x^{n+1}}{1 - x}$$

can be made arbitrarily large by taking x close to 1,

$$\|F - F_n\|_{(-1,1)} = \infty,$$

and so the convergence is not uniform on $(-1, 1)$. Neither is it uniform on any interval $(-1, r]$ with $-1 < r < 1$, since

$$\|F - F_n\|_{(-1,r]} \ge \tfrac{1}{2}$$

for every n on every such interval. (Why?) The series does converge uniformly on any interval $[-r, r]$ with $0 < r < 1$, since

$$\|F - F_n\|_{[-r,r]} = \frac{r^{n+1}}{1 - r}$$

and $\lim_{n \to \infty} r^n = 0$. Put another way: the series converges uniformly on closed subsets of $(-1, 1)$.

As for series of constants, the convergence—pointwise or uniform—of a series of functions is not changed by altering or omitting finitely many terms. This justifies adopting the convention we used for series of constants: when interested only in whether a series of functions converges, and not in its sum, we will omit the limits on the summation sign and write simply $\sum f_n$.

Tests for uniform convergence of series

Theorem 4.3 is easily converted to a theorem on uniform convergence of series, as follows.

Theorem 4.8 (Cauchy's Criterion for Uniform Convergence of Series)   *A series $\sum f_n$ converges uniformly on a set S if and only if for each $\varepsilon > 0$, there is an integer N such that*

$$\|f_n + f_{n+1} + \cdots + f_m\|_S < \varepsilon \qquad \text{if} \quad m \ge n \ge N. \tag{13}$$

Proof.   Apply Theorem 4.3 to the partial sums of $\sum f_n$, observing that

$$f_n + f_{n+1} + \cdots + f_m = F_m - F_{n-1}.$$

The following corollary provides a useful necessary—but *not* sufficient—condition for uniform convergence of series. It is analogous to Corollary 3.1, and its proof is similar (*Exercise 19*).

Corollary 4.2   *If $\sum f_n$ converges uniformly on S, then $\lim_{n \to \infty} \|f_n\|_S = 0$.*

Theorem 4.8 leads immediately to the following important test for uniform convergence of series.

Theorem 4.9 (Weierstrass' Test)   *The series $\sum f_n$ converges uniformly on S if*

$$\|f_n\|_S \le M_n, \qquad n \ge k, \tag{14}$$

where $\sum M_n < \infty$.

Proof.   From Cauchy's convergence criterion for series of constants, there is for each $\varepsilon > 0$ an integer N such that

$$M_n + M_{n+1} + \cdots + M_m < \varepsilon \qquad \text{if} \quad m \ge n \ge N,$$

which, because of (14), implies that

$$\|f_n\|_S + \|f_{n+1}\|_S + \cdots + \|f_m\|_S < \varepsilon \qquad \text{if} \quad m, n \ge N.$$

Hence, N satisfies (13), and $\sum f_n$ converges uniformly on S, by Theorem 4.8.

Example 4.11   Taking $M_n = 1/n^2$ and recalling that

$$\sum \frac{1}{n^2} < \infty,$$

we see that

$$\sum \frac{1}{x^2 + n^2} \quad \text{and} \quad \sum \frac{\sin nx}{n^2}$$

converge uniformly on $(-\infty, \infty)$.

Example 4.12   The series

$$\sum f_n(x) = \sum \left(\frac{x}{1 + x} \right)^n$$

converges uniformly on any set S such that

$$\left| \frac{x}{1 + x} \right| \le r < 1, \qquad x \in S, \tag{15}$$

because if S is such a set, then

$$\|f_n\|_S \le r^n$$

and Weierstrass' test applies, with

$$\sum M_n = \sum r^n < \infty.$$

Since (15) is equivalent to

$$\frac{-r}{1+r} \le x \le \frac{r}{1-r}, \qquad x \in S,$$

this means that the series converges uniformly on any compact subset of $(-\frac{1}{2}, \infty)$. (Why?) Because of Corollary 4.2 it does not converge uniformly on $S = (-\frac{1}{2}, b)$ with $b < \infty$ or on $S = [a, \infty)$ with $a > -\frac{1}{2}$, because in these cases $\|f_n\|_S = 1$ for all n.

Weierstrass' test is very important, but applicable only to series which actually exhibit a stronger kind of convergence than we have considered so far. We say that $\sum f_n$ converges absolutely on S if $\sum |f_n|$ converges pointwise on S, and absolutely-uniformly on S if $\sum |f_n|$ converges uniformly on S. We leave it to the reader (*Exercise* 24) to verify that our proof of Theorem 4.9 actually shows that $\sum f_n$ converges absolutely-uniformly on S, and to show that if a series converges absolutely-uniformly on S then it converges uniformly there.

The next theorem applies to series that converge uniformly, but perhaps not absolutely-uniformly, on a set S.

Theorem 4.10 (Dirichlet's Test for Uniform Convergence) *The series* $\sum_{n=k}^{\infty} f_n g_n$ *converges uniformly on* S *if:*

(a) $\{f_n\}$ *converges uniformly to zero on* S.

(b) *There is a constant* M *such that*

$$\|g_k + g_{k+1} + \cdots + g_n\|_S \le M, \qquad n \ge k. \tag{16}$$

(c) $\sum (f_{n+1} - f_n)$ *converges absolutely-uniformly on* S.

Proof. The proof is similar to that of Theorem 3.13. Let

$$G_n = g_k + g_{k+1} + \cdots + g_n$$

and consider the partial sums of $\sum_{n=k}^{\infty} f_n g_n$:

$$H_n = f_k g_k + f_{k+1} g_{k+1} + \cdots + f_n g_n. \tag{17}$$

By substituting

$$g_k = G_k \quad \text{and} \quad g_n = G_n - G_{n-1} \qquad (n > k)$$

into (17) we obtain

$$H_n = f_k G_k + f_{k+1}(G_{k+1} - G_k) + \cdots + f_n(G_n - G_{n-1}),$$

which we rewrite as

$$H_n = (f_k - f_{k+1})G_k + (f_{k+1} - f_{k+2})G_{k+1} + \cdots + (f_{n-1} - f_n)G_{n-1} + f_n G_n,$$

or

$$H_n = J_{n-1} + f_n G_n, \tag{18}$$

where

$$J_{n-1} = (f_k - f_{k+1})G_k + (f_{k+1} - f_{k+2})G_{k+1} + \cdots + (f_{n-1} - f_n)G_{n-1}. \tag{19}$$

That is, $\{J_n\}$ is the sequence of partial sums of the series

$$\sum_{j=k}^{\infty} (f_j - f_{j+1})G_j. \tag{20}$$

We now use Cauchy's convergence criterion to show that this series converges uniformly on S. From (16) and the definition of G_j,

$$\left| \sum_{j=n}^{m} [f_j(x) - f_{j+1}(x)]G_j(x) \right| \le M \sum_{j=n}^{m} |f_j(x) - f_{j+1}(x)|, \qquad x \in S,$$

and so

$$\left\| \sum_{j=n}^{m} (f_j - f_{j+1})G_j \right\|_S \le M \left\| \sum_{j=n}^{m} |f_j - f_{j+1}| \right\|_S.$$

Since $\sum (f_j - f_{j+1})$ converges absolutely-uniformly on S, Theorem 4.8 implies that the right side of the last inequality can be made arbitrarily small by choosing n large and $m \ge n$; of course, the same is then true of the left, so Theorem 4.8 implies that (20) converges uniformly on S.

We have now shown that $\{J_n\}$ as defined in (19) converges uniformly to a limit function J on S. Returning to (18) we see that

$$H_n - J = J_{n-1} - J + f_n G_n.$$

Hence, from Lemma 4.1 and (16),

$$\|H_n - J\|_S \le \|J_{n-1} - J\|_S + \|f_n\|_S \|G_n\|_S$$
$$= \|J_{n-1} - J\|_S + M\|f_n\|_S.$$

Since $\{J_{n-1} - J\}$ and $\{f_n\}$ converge uniformly to zero on S, it now follows that $\lim_{n \to \infty} \|H_n - J\|_S = 0$. Thus, we have shown that $\{H_n\}$ converges uniformly on S, which completes the proof.

Corollary 4.3   *The series $\sum_{n=k}^{\infty} f_n g_n$ converges uniformly on S if*

$$f_{n+1}(x) \le f_n(x), \qquad x \in S, \quad n \ge k,$$

and (a) and (b) of Theorem 4.10 hold.

The proof is similar to that of Corollary 3.5. We leave it to the reader (*Exercise 25*).

Example 4.13   Consider the series

$$\sum_{n=1}^{\infty} \frac{\sin nx}{n}$$

with $f_n = 1/n$ (constant), $g_n(x) = \sin nx$, and

$$G_n(x) = \sin x + \sin 2x + \cdots + \sin nx.$$

Proceeding as in Example 3.20, we find that

$$|G_n(x)| \le \frac{1}{|\sin(x/2)|}, \qquad n \ge 1, \quad x \ne 2k\pi \quad (k = \text{integer}).$$

Therefore, $\{\|G_n\|_S\}$ is bounded and the series converges uniformly on any set S on which $\sin x/2$ is bounded away from zero. For example, if $0 < \delta < \pi$, then

$$\left|\sin \frac{x}{2}\right| \ge \sin \frac{\delta}{2}$$

if x is at least δ away from any multiple of 2π; hence, the series converges uniformly on

$$S = \bigcup_{k=-\infty}^{\infty} [2k\pi + \delta, 2(k + 1)\pi - \delta].$$

Since

$$\sum \left|\frac{\sin nx}{n}\right| = \infty, \qquad x \ne 2k\pi$$

(Exercise 36(b), Section 4.3), this result cannot be obtained from Weierstrass' test.

Example 4.14   The series

$$\sum_{n=1}^{\infty} \frac{(-1)^n}{n + x^2}$$

satisfies the hypotheses of Corollary 4.3 on $(-\infty, \infty)$, with

$$f_n(x) = \frac{1}{n + x^2}, \quad g_n = (-1)^n, \quad G_{2m} = 0, \quad \text{and} \quad G_{2m+1} = -1.$$

Therefore, it converges uniformly on $(-\infty, \infty)$. This result cannot be obtained by Weierstrass' test, since

$$\sum \frac{1}{n + x^2} = \infty$$

for all x.

Continuity, differentiability, and integrability of series

We can obtain information on the continuity, differentiability, and integrability of infinite series by applying Theorems 4.4–4.9 to their partial sums. We will state the theorems and give some examples while leaving the proofs to the reader.

Theorem 4.11   If $\sum_{n=k}^{\infty} f_n$ converges uniformly to F on S and each f_n is continuous at a point x_0 in S, then so is F. Similar statements hold for continuity from the right and left.

Proof.   Exercise 27.

Example 4.15   In Example 4.12 we saw that the series

$$F(x) = \sum_{n=0}^{\infty} \left(\frac{x}{1+x} \right)^n$$

converges uniformly on every compact subset of $(-\frac{1}{2}, \infty)$. Since its terms are continuous on every such subset, Theorem 4.1 implies that F is also. In fact, we can state a stronger result: F is continuous on $(-\frac{1}{2}, \infty)$, since every point in $(-\frac{1}{2}, \infty)$ lies in a compact subinterval of $(-\frac{1}{2}, \infty)$.

The same argument and the results of Example 4.13 show that the function

$$G(x) = \sum_{n=1}^{\infty} \frac{\sin nx}{n}$$

is continuous except perhaps at $x_k = 2k\pi$ (k = integer).

From Example 4.14, the function

$$H(x) = \sum_{n=1}^{\infty} (-1)^n \frac{1}{n+x^2}$$

is continuous for all x.

The next theorem gives conditions that permit the interchange of summation and integration of infinite series. It follows from Theorem 4.5. We leave it to the reader to formulate an analog of Theorem 4.6 for series (*Exercise 30*).

Theorem 4.12   If $\sum_{n=k}^{\infty} f_n$ converges uniformly to F on $[a, b]$ and F and each f_n are integrable on $[a, b]$, then

$$\int_a^b F(x)\, dx = \sum_{n=k}^{\infty} \int_a^b f_n(x)\, dx.$$

Example 4.16 From Example 4.10,

$$\frac{1}{1-x} = \sum_{n=0}^{\infty} x^n, \qquad -1 < x < 1.$$

Here the series converges uniformly and the limit function is integrable on any closed interval $[a, b]$ in $(-1, 1)$; hence,

$$\int_a^b \frac{dx}{1-x} = \sum_{n=0}^{\infty} \int_a^b x^n \, dx,$$

and so

$$\log(1-a) - \log(1-b) = \sum_{n=0}^{\infty} \frac{b^{n+1} - a^{n+1}}{n+1}.$$

Letting $a = 0$ and $b = x$, we find that

$$\log(1-x) = -\sum_{n=0}^{\infty} \frac{x^{n+1}}{n+1}, \qquad -1 < x < 1.$$

The next theorem gives conditions that permit the interchange of summation and differentiation of infinite series.

Theorem 4.13 *Suppose f_n is continuously differentiable on $[a, b]$ for each $n \geq k$, $\sum f_n(x_0)$ converges for some x_0 in $[a, b]$, and $\sum_{n=k}^{\infty} f'_n$ converges uniformly on $[a, b]$. Then $\sum_{n=k}^{\infty} f_n$ converges uniformly on $[a, b]$ to a differentiable function F, and*

$$F'(x) = \sum_{n=k}^{\infty} f'_n(x), \qquad a < x < b,$$

while

$$F'_+(a) = \sum_{n=k}^{\infty} f'_n(a) \quad and \quad F'_-(b) = \sum_{n=k}^{\infty} f'_n(b).$$

Proof. *Exercise 32.*

Example 4.17 The series

$$\sum_{n=1}^{\infty} (-1)^n \frac{1}{n} \cos \frac{x}{n} \tag{21}$$

converges at $x_0 = 0$, and the series of everywhere continuous derivatives,

$$\sum_{n=1}^{\infty} (-1)^{n+1} \frac{1}{n^2} \sin \frac{x}{n} \tag{22}$$

converges uniformly on $(-\infty, \infty)$, by Weierstrass' test. Therefore, (21) converges uniformly on $(-\infty, \infty)$, and (22) is its derivative, by Theorem 4.13.

Example 4.18   The series

$$E(x) = \sum_{n=0}^{\infty} \frac{x^n}{n!} = 1 + x + \frac{x^2}{2!} + \frac{x^3}{3!} + \cdots \tag{23}$$

converges uniformly on every interval $[-r, r]$ by Weierstrass' test, because

$$\frac{|x|^n}{n!} \le \frac{r^n}{n!}, \qquad |x| \le r,$$

and

$$\sum \frac{r^n}{n!} < \infty$$

for all r, by the ratio test. Differentiating the right side of (23) term by term yields the series

$$\sum_{n=1}^{\infty} \frac{x^{n-1}}{(n-1)!} = \sum_{n=0}^{\infty} \frac{x^n}{n!},$$

which is the same as (23). Therefore, the differentiated series is also uniformly convergent on $[-r, r]$ for every r, and so the term-by-term differentiation is legitimate and

$$E'(x) = E(x), \qquad -\infty < x < \infty.$$

Example 4.19   Failure to verify that the given series converges at some point can lead to erroneous conclusions. For example, differentiating

$$\sum_{n=1}^{\infty} \cos \frac{x}{n} \tag{24}$$

term by term yields

$$-\sum_{n=1}^{\infty} \frac{1}{n} \sin \frac{x}{n},$$

which converges uniformly on $[-r, r]$ for every r, since

$$\left| \frac{1}{n} \sin \frac{x}{n} \right| \le \frac{|x|}{n^2} \qquad (\textit{Exercise 21, Section 2.3})$$

$$\le \frac{r}{n^2} \qquad \text{if } |x| \le r,$$

and $\sum 1/n^2 < \infty$. We cannot conclude from this that (24) converges uniformly on $[-r, r]$; in fact, it diverges for every x. (Why?)

4.4   EXERCISES

1. Find the set S on which $\{F_n\}$ converges pointwise, and find the limit function:

   (a) $F_n(x) = x^n(1 - x^2)$

   (b) $F_n(x) = nx^n(1 - x^2)$

   (c) $F_n(x) = x^n(1 - x^n)$

   (d) $F_n(x) = \sin\left(1 + \dfrac{1}{n}\right)x$

   (e) $F_n(x) = \dfrac{1 + x^n}{1 + x^{2n}}$

   (f) $F_n(x) = n \sin \dfrac{x}{n}$

   (g) $F_n(x) = n^2\left(1 - \cos\dfrac{x}{n}\right)$

   (h) $F_n(x) = nxe^{-nx^2}$

   (i) $F_n(x) = \dfrac{(x + n)^2}{x^2 + n^2}$

2. Prove: If $\{F_n\}$ converges to F on $[a, b]$ and F_n is nondecreasing for each n, then F is nondecreasing.

3. Show that the functions $\{F_n\}$ of Example 4.4 are integrable and F is non-integrable on every finite interval.

4. Prove Lemma 4.1.

5. Show that $\{F_n\}$ converges uniformly on closed subsets of S, and find the limit function:

   (a) $F_n(x) = x^n \sin nx$, $S = (-1, 1)$

   (b) $F_n(x) = \dfrac{1}{1 + x^{2n}}$, $S = \{x \mid x \neq \pm 1\}$

   (c) $F_n(x) = \dfrac{n^2 \sin x}{1 + n^2 x}$, $S = (0, \infty)$ (*Hint:* See Exercise 21, Section 2.3.)

   In each case show that $\{F_n\}$ does not converge uniformly on S.

6. (a) Show that if $\{F_n\}$ converges uniformly on S, then $\{F_n\}$ converges uniformly on every subset of S.

   (b) Show that if $\{F_n\}$ converges uniformly on $S_1, S_2, \ldots, S_m$, then $\{F_n\}$ converges uniformly on $\bigcup_{k=1}^{m} S_k$.

   (c) Show by example that $\{F_n\}$ may converge uniformly on each of an infinite sequence of sets $S_1, S_2, \ldots$, but fail to converge uniformly on $\bigcup_{k=1}^{\infty} S_k$.

7. Describe the sets on which the sequences of Exercise 1 converge uniformly.

8. Prove Theorem 4.1.

9. Suppose $\{F_n\}$ converges pointwise on $[a, b]$, and for each x in $[a, b]$ there is an open interval I_x such that $\{F_n\}$ converges uniformly on $I_x \cap [a, b]$. Show that $\{F_n\}$ converges uniformly on $[a, b]$.

10. Prove: If $\{F_n\}$ converges uniformly to F on S, then $\lim_{n \to \infty} \|F_n\|_S = \|F\|_S$ if each F_n is bounded on S. (What if they are not?)

11. A sequence $\{F_n\}$ is *uniformly bounded* on S if there is a constant M such that $|F_n(x)| \le M$ for all x in S and $n \ge k$. Prove: If $\{F_n\}$ converges uniformly to F on S, then F is bounded on S if and only if $\{F_n\}$ is uniformly bounded on S.

12. Prove: If $\{F_n\}$ and $\{G_n\}$ converge uniformly to F and G on S, then $\{F_n + G_n\}$ converges uniformly to $F + G$ on S.

13. (a) If $\{F_n\}$ and $\{G_n\}$ converge uniformly to bounded functions F and G on S, then $\{F_n G_n\}$ converges uniformly to FG on S.
    (b) Give an example showing that the conclusion of part (a) may fail to hold if F or G is unbounded on S.

14. (a) Suppose $\{F_n\}$ converges uniformly to F on (a, b) and $\lim_{x \to b-} F_n(x)$ exists (finite) for every n. Show that

$$\lim_{x \to b-} F(x) = \lim_{n \to \infty} \left[\lim_{x \to b-} F_n(x) \right] \qquad \text{(finite)}.$$

(b) State similar results for limits from the right and two-sided limits.

15. Prove: If each F_n is continuous on $[a, b]$ and $\{F_n\}$ converges uniformly on (a, b), then $\{F_n\}$ converges uniformly on $[a, b]$. (*Hint*: See Exercise 14.)

16. Find:

(a) $\lim_{n \to \infty} \int_1^4 \dfrac{n}{x} \sin \dfrac{x}{n} \, dx$ (b) $\lim_{n \to \infty} \int_0^2 \dfrac{dx}{1 + x^{2n}}$

(c) $\lim_{n \to \infty} \int_0^1 nx e^{-nx^2} \, dx$ (d) $\lim_{n \to \infty} \int_0^1 \left(1 + \dfrac{x}{n}\right)^n dx$

17. Prove (without using Theorem 4.6): If each F_n is integrable and $\{F_n\}$ converges uniformly on $[a, b]$, then $\lim_{n \to \infty} \int_a^b F_n(x) \, dx$ exists.

18. Prove (without using Theorem 4.6): If each F_n is nondecreasing and $\{F_n\}$ converges uniformly to F on $[a, b]$, then

$$\lim_{n \to \infty} \int_a^b F_n(x) \, dx = \int_a^b F(x) \, dx.$$

19. Prove Corollary 4.2.

20. Use Weierstrass' test to determine sets on which the following series converge absolutely-uniformly.

(a) $\sum \dfrac{1}{n^{1/2}}\left(\dfrac{x}{1+x}\right)^n$ 　　(b) $\sum \dfrac{1}{n^{3/2}}\left(\dfrac{x}{1+x}\right)^n$

(c) $\sum nx^n(1-x)^n$ 　　(d) $\sum \dfrac{1}{n(x^2+n)}$

(e) $\sum \dfrac{1}{n^x}$ 　　(f) $\sum \dfrac{(1-x^2)^n}{(1+x^2)^n}\sin nx$

21. Show that if $\sum |a_n| < \infty$, then $\sum a_n \cos nx$ and $\sum a_n \sin nx$ define continuous functions on $(-\infty, \infty)$.

22. (a) Give an example showing that the following "comparison test" is invalid: If $\sum f_n$ converges uniformly on S and $\|g_n\|_S \le \|f_n\|_S$, then $\sum g_n$ converges uniformly on S.
    (b) This "comparison test" can be corrected by adding one word to its hypothesis and conclusion. What is the word?

23. (a) Explain the difference between the following statements: (i) $\sum f_n$ converges absolutely and uniformly on S; (ii) $\sum f_n$ converges absolutely-uniformly on S.
    (b) Show that if $\sum f_n$ converges absolutely-uniformly on S, then it converges uniformly on S.

*24. Show that the hypotheses of Weierstrass' test imply that $\sum f_n$ converges absolutely-uniformly on S.

25. Prove Corollary 4.3.

26. Use Dirichlet's test to discuss the uniform convergence of

$$\sum \frac{\sin nx}{1+nx}.$$

27. Prove Theorem 4.11.

28. Suppose $\{a_n\}_1^\infty$ is monotonic and $\lim_{n\to\infty} a_n = 0$. Show that

$$\sum_{n=1}^\infty a_n \sin nx \quad \text{and} \quad \sum_{n=1}^\infty a_n \cos nx$$

define functions continuous for all $x \ne 2k\pi$ ($k = $ integer).

29. Prove Theorem 4.12.

30. Formulate an analog of Theorem 4.6 for series.

31. In Section 4.5 we will see that

$$e^{-x^2} = \sum_{n=0}^\infty (-1)^n \frac{x^{2n}}{n!} \quad \text{and} \quad \sin x = \sum_{n=0}^\infty (-1)^n \frac{x^{2n+1}}{(2n+1)!}$$

for all x, and that in both cases the convergence is uniform on every finite interval. Find series that converge to

(a) $F(x) = \int_0^x e^{-t^2}\, dt$ and   (b) $G(x) = \int_0^x \frac{\sin t}{t}\, dt$

for all x.

32. Prove Theorem 4.13.

33. Show from Example 4.17 that $\sum_{n=1}^{\infty} (-1)^n \sin(x/n)$ converges uniformly on $(-\infty, \infty)$.

34. Prove: If $0 < a_{n+1} < a_n$ and $\sum a_n^k < \infty$ for some positive integer k, then $\sum (-1)^n \sin a_n x$ converges uniformly on $(-\infty, \infty)$.

*35. For $n = 2, 3, 4, \ldots$, define

$$
f_n(x) = \begin{cases}
n^4(x - n + 1/n^3), & n - 1/n^3 \le x \le n, \\
-n^4(x - n - 1/n^3), & n \le x \le n + 1/n^3, \\
0, & |x - n| > 1/n^3,
\end{cases}
$$

and let $F(x) = \sum_{n=2}^{\infty} f_n(x)$. Show that $\int_0^{\infty} F(x)\, dx < \infty$, and conclude that absolute convergence of an improper integral $\int_0^{\infty} F(x)\, dx$ does not imply that $\lim_{x \to \infty} F(x) = 0$, even if F is continuous on $[0, \infty)$.

4.5 POWER SERIES

We now consider a class of series sufficiently general to be interesting, but sufficiently specialized to have strong and easily understood properties.

Definition 5.1   *An infinite series of the form*

$$
\sum_{n=0}^{\infty} a_n(x - x_0)^n,
\tag{1}
$$

where x_0 and $a_0, a_1, \ldots$ are constants, is called a power series in $x - x_0$.

The following theorem summarizes the convergence properties of power series.

Theorem 5.1   *In connection with the power series (1), define R in the extended reals by*

$$
\frac{1}{R} = \overline{\lim_{n \to \infty}} |a_n|^{1/n}.
\tag{2}
$$

In particular, $R = 0$ if $\overline{\lim}_{n\to\infty} |a_n|^{1/n} = \infty$, and $R = \infty$ if $\overline{\lim}_{n\to\infty} |a_n|^{1/n} = 0$. Then the power series converges

(a) only for $x = x_0$ if $R = 0$;

(b) for all x if $R = \infty$, and absolutely-uniformly in every bounded set;

(c) for x in $(x_0 - R, x_0 + R)$ if $0 < R < \infty$, and absolutely-uniformly in every closed subset of this interval. The series diverges if $|x - x_0| > R$. No general statement can be made concerning convergence at the end points $x = x_0 + R$ and $x = x_0 - R$; the series may converge absolutely or conditionally at both, converge conditionally at one and diverge at the other, or diverge at both.

Proof. If

$$\sum |a_n| r^n < \infty \tag{3}$$

for some $r > 0$, then $\sum a_n(x - x_0)^n$ converges absolutely-uniformly in $[x_0 - r, x_0 + r]$ by Weierstrass' test and *Exercise 24, Section 4.4*. From Cauchy's root test, (3) holds if

$$\overline{\lim_{n\to\infty}} \, (|a_n| r^n)^{1/n} < 1,$$

which is equivalent to

$$r \, \overline{\lim_{n\to\infty}} \, |a_n|^{1/n} < 1 \quad (\text{*Exercise 27, Section 4.1*}).$$

From (2), this can be rewritten as $r < R$, which proves the assertions about convergence in (b) and (c).

If $0 \le R < \infty$ and $|x - x_0| > R$, then

$$\frac{1}{R} > \frac{1}{|x - x_0|},$$

so (2) implies that

$$|a_n|^{1/n} \ge \frac{1}{|x - x_0|} \quad \text{and therefore} \quad |a_n(x - x_0)^n| \ge 1$$

for infinitely many values of n; therefore, $\sum a_n(x - x_0)^n$ diverges. This proves the assertion about divergence in (c) and shows that the series diverges for all $x \ne x_0$ if $R = 0$. That it converges for $x = x_0$ in any case is obvious, and so (a) follows.

To prove the assertions in (c) about the possibilities at $x = x_0 + R$ and $x = x_0 - R$ requires examples, which follow. (Also, see *Exercise 1*.)

The number R defined by (2) is the *radius of convergence* of $\sum a_n(x - x_0)^n$. If $R > 0$, the *open* interval $(x_0 - R, x_0 + R)$—or $(-\infty, \infty)$ if $R = \infty$—is the *interval of convergence* of the series. Theorem 5.1 says that a power series with a nonzero radius of convergence converges absolutely-uniformly in every compact subset of its interval of convergence and diverges at every point in the exterior of this interval. On this last we can make a stronger statement: Not only does $\sum a_n(x - x_0)^n$ diverge if $|x - x_0| > R$, but the sequence $\{a_n(x - x_0)^n\}$ is unbounded in this case (*Exercise 3*).

Example 5.1   For the series

$$\sum \frac{\sin n\pi/6}{2^n}(x - 1)^n$$

we have

$$\overline{\lim_{n \to \infty}} |a_n|^{1/n} = \overline{\lim_{n \to \infty}} \left(\frac{|\sin n\pi/6|}{2^n}\right)^{1/n}$$

$$= \frac{1}{2} \overline{\lim_{n \to \infty}} (|\sin n\pi/6|)^{1/n} \qquad (\text{Exercise 27, Section 4.1})$$

$$= \frac{1}{2}(1) = \frac{1}{2}.$$

Therefore, $R = 2$ and Theorem 5.1 implies that the series converges absolutely-uniformly in closed subintervals of $(-1, 3)$ and diverges if $x < -1$ or $x > 3$. Theorem 5.1 does not tell us what happens when $x = -1$ or $x = 3$, but we can see that the series diverges in both these cases, since its general term does not approach zero.

Example 5.2   For the series

$$\sum \frac{x^n}{n}$$

we have

$$\overline{\lim_{n \to \infty}} |a_n|^{1/n} = \overline{\lim_{n \to \infty}} \left(\frac{1}{n}\right)^{1/n} = \overline{\lim_{n \to \infty}} \exp\left(\frac{1}{n} \log \frac{1}{n}\right) = e^0 = 1.$$

Therefore, $R = 1$ and the series converges absolutely-uniformly in closed subintervals of $(-1, 1)$ and diverges if $|x| > 1$. For $x = -1$ the series becomes $\sum (-1)^n/n$, which converges conditionally, and at $x = 1$ it becomes $\sum 1/n$, which diverges.

The next theorem provides an expression for R which, if it is applicable, is usually easier to use than (2).

Theorem 5.2   *The radius of convergence of $\sum a_n(x - x_0)^n$ satisfies*

$$\frac{1}{R} = \lim_{n \to \infty} \left|\frac{a_{n+1}}{a_n}\right|$$

if the limit exists in the extended reals.

Proof.   From Theorem 5.1 it suffices to show that if

$$L = \lim_{n \to \infty} \left| \frac{a_{n+1}}{a_n} \right| \tag{4}$$

exists in the extended reals, then

$$L = \overline{\lim_{n \to \infty}} \, |a_n|^{1/n}. \tag{5}$$

We will show that this is so if $0 < L < \infty$ and leave the cases where $L = 0$ or $L = \infty$ to the reader (*Exercise 7*).

If (4) holds with $0 < L < \infty$ and $0 < \varepsilon < L$, then there is an integer N such that

$$L - \varepsilon < \left| \frac{a_{m+1}}{a_m} \right| < L + \varepsilon \qquad \text{if} \quad m \geq N,$$

and so

$$|a_m|(L - \varepsilon) < |a_{m+1}| < |a_m|(L + \varepsilon) \qquad \text{if} \quad m \geq N.$$

By induction this implies that

$$|a_N|(L - \varepsilon)^{n-N} < |a_n| < |a_N|(L + \varepsilon)^{n-N} \qquad \text{if} \quad n \geq N,$$

and so

$$K_1^{1/n}(L - \varepsilon) < |a_n|^{1/n} < K_2^{1/n}(L + \varepsilon), \tag{6}$$

where

$$K_1 = |a_N|(L - \varepsilon)^{-N} \quad \text{and} \quad K_2 = |a_N|(L + \varepsilon)^{-N}.$$

Now, $\lim_{n \to \infty} K^{1/n} = 1$ if K is any positive number; therefore, from (6),

$$L - \varepsilon \leq \underline{\lim_{n \to \infty}} \, |a_n|^{1/n} \leq \overline{\lim_{n \to \infty}} \, |a_n|^{1/n} \leq L + \varepsilon.$$

Since ε is an arbitrary positive number, we conclude from this that

$$\lim_{n \to \infty} |a_n|^{1/n} = L,$$

which is stronger than (5). This completes the proof.

Example 5.3   For the power series

$$\sum \frac{x^n}{n!}$$

we have

$$\lim_{n \to \infty} \left| \frac{a_{n+1}}{a_n} \right| = \lim_{n \to \infty} \frac{n!}{(n+1)!} = \lim_{n \to \infty} \frac{1}{n+1} = 0.$$

Therefore, $R = \infty$; that is, the series converges for all x, and absolutely-uniformly in every bounded set.

Example 5.4   For the power series

$$\sum n! \, x^n$$

we have

$$\lim_{n \to \infty} \left| \frac{a_{n+1}}{a_n} \right| = \lim_{n \to \infty} \frac{(n+1)!}{n!} = \lim_{n \to \infty} (n+1) = \infty.$$

Therefore $R = 0$ and the series converges only if $x = 0$.

Example 5.5   Theorem 5.2 does not apply directly to

$$\sum \frac{(-1)^n}{4^n n^p} x^{2n} \qquad (p = \text{constant}), \tag{7}$$

which has infinitely many zero coefficients (of odd powers of x). However, by setting $y = x^2$ we obtain the series

$$\sum \frac{(-1)^n}{4^n n^p} y^n, \tag{8}$$

which has nonzero coefficients for which

$$\lim_{n \to \infty} \left| \frac{a_{n+1}}{a_n} \right| = \lim_{n \to \infty} \frac{4^n n^p}{4^{n+1}(n+1)^p} = \frac{1}{4} \lim_{n \to \infty} \left(1 + \frac{1}{n} \right)^{-p} = \frac{1}{4}.$$

Therefore, (8) converges if $|y| < 4$ and diverges if $|y| > 4$. Setting $y = x^2$, we conclude that (7) converges if $|x| < 2$ and diverges if $|x| > 2$. At $x = \pm 2$, (7) becomes $\sum (-1)^n / n^p$, which diverges if $p \le 0$, converges conditionally if $0 < p \le 1$, and converges absolutely if $p > 1$.

Properties of functions defined by power series

We now study the properties of functions defined by power series. Henceforth, we consider only power series with nonzero radii of convergence.

Theorem 5.3   *A power series*

$$f(x) = \sum_{n=0}^{\infty} a_n (x - x_0)^n \tag{9}$$

with positive radius of convergence R is continuous and differentiable in its interval *of convergence, and its derivative can be obtained by differentiating term by term:*

$$f'(x) = \sum_{n=1}^{\infty} na_n(x - x_0)^{n-1} = \sum_{n=0}^{\infty} (n + 1)a_{n+1}(x - x_0)^n. \tag{10}$$

This series also has radius of convergence R.

Proof.   Since differentiability implies continuity, we prove only the former. The power series in (10) has the same radius of convergence as $\sum na_n(x - x_0)^n$. (Why?) Call it R_1. Then Theorem 5.1 implies that

$$\frac{1}{R_1} = \overline{\lim_{n \to \infty}} \, (n|a_n|)^{1/n} = \overline{\lim_{n \to \infty}} \, n^{1/n}|a_n|^{1/n}$$

$$= \left(\lim_{n \to \infty} n^{1/n} \right) \left(\overline{\lim_{n \to \infty}} \, |a_n|^{1/n} \right) \text{(see Exercise 27, Section 4.1)}$$

$$= \left[\lim_{n \to \infty} \exp\left(\frac{\log n}{n} \right) \right] \left(\overline{\lim_{n \to \infty}} \, |a_n|^{1/n} \right) = \frac{e^0}{R} = \frac{1}{R}.$$

Therefore, $R_1 = R$. This means that the power series in (10) converges uniformly in every interval $[x_0 - r, \, x_0 + r]$ such that $0 < r < R$, and Theorem 4.13 implies that it equals $f'(x)$ for all x in $(x_0 - R, \, x_0 + R)$.

Theorem 5.3 can be strengthened as follows.

Theorem 5.4   *A power series (8) has derivatives of all orders in its interval of convergence, and they can be obtained by repeated term-by-term differentiation; thus,*

$$f^{(k)}(x) = \sum_{n=k}^{\infty} n(n - 1) \cdots (n - k + 1)a_n(x - x_0)^{n-k}. \tag{11}$$

The radius of convergence of each of these series is R, the radius of convergence of (9).

Proof.   The proof is by induction. Theorem 5.3 shows that the result holds for $k = 1$. Now suppose it holds for some $k \geq 1$. By shifting the index of summation we can rewrite (11) as

$$f^{(k)}(x) = \sum_{n=0}^{\infty} (n+k)(n+k-1) \cdots (n+1)a_{n+k}(x-x_0)^n, \qquad |x - x_0| < R.$$

Defining

$$b_n = (n + k)(n + k - 1) \cdots (n + 1)a_{n+k}, \tag{12}$$

we rewrite this as

$$f^{(k)}(x) = \sum_{n=0}^{\infty} b_n(x - x_0)^n, \qquad |x - x_0| < R.$$

Theorem 5.3 says that we can differentiate this series term by term to obtain

$$f^{(k+1)}(x) = \sum_{n=1}^{\infty} nb_n(x - x_0)^{n-1}, \qquad |x - x_0| < R.$$

Substituting from (12) for b_n yields

$$f^{(k+1)}(x) = \sum_{n=1}^{\infty} (n + k)(n + k - 1) \cdots (n + 1)na_{n+k}(x - x_0)^{n-1},$$

$$|x - x_0| < R.$$

Shifting the summation index yields

$$f^{(k+1)}(x) = \sum_{n=k+1}^{\infty} n(n - 1) \cdots (n - k)a_n(x - x_0)^{n-k-1}, \qquad |x - x_0| < R,$$

which is of the form (11) with k replaced by $k + 1$. This completes the induction.

Example 5.6   In Example 4.10 we saw that

$$\frac{1}{1 - x} = \sum_{n=0}^{\infty} x^n, \qquad |x| < 1.$$

Differentiating yields

$$\frac{1}{(1 - x)^2} = \sum_{n=1}^{\infty} nx^{n-1} = \sum_{n=0}^{\infty} (n + 1)x^n, \qquad |x| < 1.$$

Repeated differentiation yields

$$\frac{k!}{(1 - x)^{k+1}} = \sum_{n=k}^{\infty} n(n - 1) \cdots (n - k + 1)x^{n-k}$$

$$= \sum_{n=0}^{\infty} (n + k)(n + k - 1) \cdots (n + 1)x^n, \qquad |x| < 1.$$

Example 5.7   By the method of Example 5.5 it can be shown that the series

$$S(x) = \sum_{n=0}^{\infty} (-1)^n \frac{x^{2n+1}}{(2n + 1)!} \quad \text{and} \quad C(x) = \sum_{n=0}^{\infty} (-1)^n \frac{x^{2n}}{(2n)!}$$

converge for all x. Differentiating yields

$$S'(x) = \sum_{n=0}^{\infty} (-1)^n \frac{x^{2n}}{(2n)!} = C(x)$$

and

$$C'(x) = \sum_{n=1}^{\infty} (-1)^n \frac{x^{2n-1}}{(2n-1)!} = -\sum_{n=0}^{\infty} (-1)^n \frac{x^{2n+1}}{(2n+1)!} = -S(x).$$

These results should not surprise the reader who recalls that

$$S(x) = \sin x \quad \text{and} \quad C(x) = \cos x.$$

(We will soon prove this.)

Theorem 5.4 has two important corollaries.

Corollary 5.1   *If*

$$f(x) = \sum_{n=0}^{\infty} a_n(x - x_0)^n, \qquad |x - x_0| < R,$$

then

$$a_n = \frac{f^{(n)}(x_0)}{n!}.$$

Proof.   Setting $x = x_0$ in (11) yields

$$f^{(k)}(x_0) = k! \, a_k.$$

Corollary 5.2   (Uniqueness of Power Series)   *If*

$$\sum_{n=0}^{\infty} a_n(x - x_0)^n = \sum_{n=0}^{\infty} b_n(x - x_0)^n \tag{13}$$

for all x in some interval $(x_0 - r, x_0 + r)$, then

$$a_n = b_n, \qquad n = 0, 1, \ldots. \tag{14}$$

Proof.   Let

$$f(x) = \sum_{n=0}^{\infty} a_n(x - x_0)^n \quad \text{and} \quad g(x) = \sum_{n=0}^{\infty} b_n(x - x_0)^n.$$

From Corollary 5.1,

$$a_n = \frac{f^{(n)}(x_0)}{n!} \quad \text{and} \quad b_n = \frac{g^{(n)}(x_0)}{n!}. \tag{15}$$

From (13), $f = g$ in $(x_0 - r, x_0 + r)$, and therefore

$$f^{(n)}(x_0) = g^{(n)}(x_0), \qquad n = 0, 1, \ldots .$$

This and (15) imply (14).

The next theorem follows from Theorems 4.12 and 5.1. We leave its proof to the reader (*Exercise 15*).

Theorem 5.5 *If a and b are in the interval of convergence of*

$$f(x) = \sum_{n=0}^{\infty} a_n(x - x_0)^n,$$

then

$$\int_a^b f(x)\, dx = \sum_{n=0}^{\infty} \frac{a_n}{n+1}\left[(b - x_0)^{n+1} - (a - x_0)^{n+1}\right];$$

that is, a power series may be integrated term by term between any two points in its interval of convergence.

For an application of this theorem, see Example 4.16.

Taylor's series

So far we have asked for what values of x a given power series converges, and what are the properties of its sum. Now we ask a related question: What properties guarantee that a given function f can be represented as the sum of a convergent power series in $x - x_0$? A partial answer to this question is provided by what we already know: Theorem 5.4 tells us that f must have derivatives of all orders in some neighborhood of x_0, and Corollaries 5.1 and 5.2 tell us that the only power series in $x - x_0$ that can possibly converge to f in such a neighborhood is

$$\sum_{n=0}^{\infty} \frac{f^{(n)}(x_0)}{n!}(x - x_0)^n. \tag{16}$$

This is called the *Taylor series of f about* x_0 (also, the *Maclaurin series of f*, if $x_0 = 0$). Its mth partial sum is the Taylor polynomial

$$T_m(x) = \sum_{n=0}^{m} \frac{f^{(n)}(x_0)}{n!}(x - x_0)^n,$$

defined in Section 2.5.

The Taylor series of an infinitely differentiable function f may converge to a sum different from f. For example, the function

$$f(x) = \begin{cases} e^{-1/x^2}, & x \neq 0, \\ 0, & x = 0, \end{cases}$$

is infinitely differentiable on $(-\infty, \infty)$ and $f^{(n)}(0) = 0$ for $n = 0, 1, \cdots$ (*Exercise 1, Section 2.5*), and so its Maclaurin series is identically zero.

An answer to our question is provided by Theorem 5.3, Section 2.5, which implies that if f is infinitely differentiable on (a, b) and x and x_0 are in (a, b), then, for every integer $n \geq 0$,

$$f(x) - T_n(x) = \frac{f^{(n+1)}(c_n)}{(n+1)!}(x - x_0)^{n+1}, \tag{17}$$

where c_n is between x and x_0. This implies that

$$f(x) = \sum_{n=0}^{\infty} \frac{f^{(n)}(x_0)}{n!}(x - x_0)^n$$

for an x in (a, b) if and only if

$$\lim_{n \to \infty} \frac{f^{(n+1)}(c_n)}{n!}(x - x_0)^n = 0.$$

It is not always easy to check this condition, because the sequence $\{c_n\}$ is usually not precisely known, or even uniquely defined; however, the next theorem is sufficiently general to be useful.

Theorem 5.6 *Suppose f is infinitely differentiable on an interval I and*

$$\lim_{n \to \infty} \frac{r^n}{n!} \|f^{(n)}\|_I = 0. \tag{18}$$

Then, if $x_0 \in I^0$, the Taylor series (16) converges uniformly to f on

$$I_r = I \cap [x_0 - r, x_0 + r].$$

Proof. From (17),

$$\|f - T_n\|_{I_r} \leq \frac{r^{n+1}}{(n+1)!} \|f^{(n+1)}\|_{I_r} \leq \frac{r^{n+1}}{(n+1)!} \|f^{(n+1)}\|_I,$$

and so the conclusion follows from (18).

Example 5.8 If $f(x) = \sin x$ then $\|f^{(k)}\|_{(-\infty,\infty)} = 1, k = 0, 1, \ldots$. Since

$$\lim_{n \to \infty} \frac{r^n}{n!} = 0, \qquad 0 < r < \infty,$$

(Example 1.11), (18) holds for all r. Since

$$f^{(2m)}(0) = 0 \quad \text{and} \quad f^{(2m+1)}(0) = (-1)^m, \qquad m = 0, 1, \ldots,$$

we see from Theorem 5.6, with $I = (-\infty, \infty)$, $x_0 = 0$, and r arbitrary, that

$$\sin x = \sum_{n=0}^{\infty} (-1)^n \frac{x^{2n+1}}{(2n+1)!}, \qquad -\infty < x < \infty,$$

and the convergence is uniform on bounded sets.

A similar argument shows that

$$\cos x = \sum_{n=0}^{\infty} (-1)^n \frac{x^{2n}}{(2n)!}, \qquad -\infty < x < \infty,$$

with uniform convergence on bounded sets.

Example 5.9   If $f(x) = e^x$, then $f^{(k)}(x) = e^x$ and $\|f^{(k)}\|_I = e^r$, $k = 0, 1, \ldots$ if $I = [-r, r]$. Since

$$\lim_{n \to \infty} \frac{r^n}{n!} e^r = 0,$$

we conclude as in Example 5.8 that

$$e^x = \sum_{n=0}^{\infty} \frac{x^n}{n!}, \qquad -\infty < x < \infty,$$

and that the convergence is uniform on bounded sets.

Example 5.10   If $f(x) = (1 + x)^q$, then

$$\frac{f^{(n)}(x)}{n!} = \binom{q}{n}(1 + x)^{q-n} \tag{19}$$

(Example 5.3, Section 2.5). The Maclaurin series

$$\sum_{n=0}^{\infty} \binom{q}{n} x^n \tag{20}$$

is called *the binomial series*. We saw in Example 5.3, Section 2.5, that it equals $(1 + x)^q$ for all x if q is a nonnegative integer. If q is not a nonnegative integer, then

$$\binom{q}{n} \neq 0, \qquad n = 0, 1, \ldots,$$

and Theorem 5.2 implies that its radius of convergence is 1, since

$$\lim_{n \to \infty} \left| \binom{q}{n+1} \middle/ \binom{q}{n} \right| = \lim_{n \to \infty} \left| \frac{q-n}{n+1} \right| = 1.$$

Theorem 5.6 implies that (20) converges to $(1 + x)^q$ on $[0, 1)$, since, from (19),

$$\frac{\|f^{(n)}\|_{[0,1]}}{n!} \le [\max(1, 2^q)] \left|\binom{q}{n}\right|, \qquad n \ge 0,$$

and so, if $0 < r < 1$,

$$\varlimsup_{n \to \infty} \frac{r^n}{n!} \|f^{(n)}\|_{[0,1]} \le [\max(1, 2^q)] \lim_{n \to \infty} \left|\binom{q}{n}\right| r^n = 0,$$

where the last equality follows from the absolute convergence of (20) on $(-1, 1)$.

We cannot prove in this way that (20) converges to $(1 + x)^q$ on $(-1, 0)$; this requires a form of the remainder in Taylor's theorem that we have not studied in this book, or a different kind of proof altogether (*Exercise 20*). The correct result is that

$$(1 + x)^q = \sum_{n=0}^{\infty} \binom{q}{n} x^n, \qquad -1 < x < 1, \tag{21}$$

for all q, and, as we said earlier, the identity holds for all x if q is a nonnegative integer.

Arithmetic operations with power series

We now consider addition and multiplication of power series, and division of one by another.

We leave the proof of the first theorem to the reader (*Exercise 21*).

Theorem 5.7   *If*

$$f(x) = \sum_{n=0}^{\infty} a_n(x - x_0)^n, \qquad |x - x_0| < R_1, \tag{22}$$

$$g(x) = \sum_{n=0}^{\infty} b_n(x - x_0)^n, \qquad |x - x_0| < R_2, \tag{23}$$

and α and β are constants, then

$$\alpha f(x) + \beta g(x) = \sum_{n=0}^{\infty} (\alpha a_n + \beta b_n)(x - x_0)^n, \qquad |x - x_0| < R,$$

where $R \ge \min\{R_1, R_2\}$.

Theorem 5.8   *If f and g are given by (22) and (23), then*

$$f(x)g(x) = \sum_{n=0}^{\infty} c_n(x - x_0)^n, \tag{24}$$

where

$$c_n = \sum_{r=0}^{n} a_r b_{n-r} = \sum_{r=0}^{n} a_{n-r} b_r$$

and the radius of convergence of (24) is $\geq \min\{R_1, R_2\}$.

Proof. Suppose $R_1 \leq R_2$. Since the series (22) and (23) converge absolutely to $f(x)$ and $g(x)$ if $|x - x_0| < R_1$, their Cauchy product converges to $f(x)g(x)$ if $|x - x_0| < R_1$, by Theorem 3.19. The nth term of this product is

$$\sum_{r=0}^{n} a_r(x - x_0)^r b_{n-r}(x - x_0)^{n-r} = \left(\sum_{r=0}^{n} a_r b_{n-r} \right)(x - x_0)^n = c_n(x - x_0)^n,$$

which completes the proof.

Example 5.11 If

$$f(x) = \frac{1}{1 - x} = \sum_{n=0}^{\infty} x^n, \qquad |x| < 1,$$

and

$$g(x) = \sum_{n=0}^{\infty} b_n x^n, \qquad |x| < R,$$

then

$$\frac{g(x)}{1 - x} = \sum_{n=0}^{\infty} s_n x^n, \qquad |x| < \min\{1, R\},$$

where

$$s_n = (1)b_0 + (1)b_1 + \cdots + (1)b_n$$
$$= b_0 + b_1 + \cdots + b_n.$$

Example 5.12 From Example 5.10,

$$(1 + x)^p = \sum_{n=0}^{\infty} \binom{p}{n} x^n, \qquad |x| < 1,$$

and

$$(1 + x)^q = \sum_{n=0}^{\infty} \binom{q}{n} x^n, \qquad |x| < 1.$$

Since

$$(1 + x)^p (1 + x)^q = (1 + x)^{p+q} = \sum_{n=0}^{\infty} \binom{p + q}{n} x^n,$$

while the Cauchy product is $\sum_{n=0}^{\infty} c_n x^n$, with

$$c_n = \sum_{r=0}^{n} \binom{p}{r}\binom{q}{n-r},$$

Corollary 5.2 implies that

$$c_n = \binom{p+q}{n}.$$

This yields the identity

$$\binom{p+q}{n} = \sum_{r=0}^{n} \binom{p}{r}\binom{q}{n-r},$$

valid for all p and q.

The quotient

$$f(x) = \frac{h(x)}{g(x)} \tag{25}$$

of two power series

$$h(x) = \sum_{n=0}^{\infty} c_n(x - x_0)^n, \qquad |x - x_0| < R_1,$$

and

$$g(x) = \sum_{n=0}^{\infty} b_n(x - x_0)^n, \qquad |x - x_0| < R_2,$$

can be represented as a power series

$$f(x) = \sum_{n=0}^{\infty} a_n(x - x_0)^n \tag{26}$$

with positive radius of convergence, provided

$$b_0 = g(x_0) \neq 0.$$

This is surely plausible. Since $g(x_0) \neq 0$ and g is continuous near x_0, the denominator of (25) differs from zero on an interval about x_0, and so f has derivatives of all orders on this interval, because g and h do; however, the proof that its Taylor series about x_0 converges to f near x_0 requires the use of the theory of functions of a complex variable. Therefore, we omit it. However, it is straightforward to compute the coefficients in (26) if we accept the validity of the expansion; since

$$f(x)g(x) = h(x),$$

Theorem 5.8 implies that

$$\sum_{r=0}^{n} a_r b_{n-r} = c_n, \qquad n = 0, 1, \ldots.$$

Solving these equations successively yields

$$a_0 = \frac{c_0}{b_0},$$

$$a_n = \frac{1}{b_0}\left(c_n - \sum_{r=0}^{n-1} b_{n-r} a_r\right), \qquad n \geq 1.$$

It is not worthwhile to memorize these formulas. Rather, it is usually better to view the procedure as follows: Multiply the series f (with unknown coefficients) and g according to the procedure of Theorem 5.8, equate the resulting coefficients with those of h, and solve the resulting equations successively for $a_0, a_1, \ldots$.

Example 5.13   Suppose we wish to find the coefficients in the Maclaurin series

$$\tan x = a_0 + a_1 x + a_2 x^2 + \cdots.$$

We first observe that since $\tan x$ is an odd function, its derivatives of even order vanish at $x_0 = 0$, and so $a_{2m} = 0$, $m = 0, 1, \ldots$; therefore,

$$\tan x = a_1 x + a_3 x^3 + a_5 x^5 + \cdots.$$

Since

$$\tan x = \frac{\sin x}{\cos x},$$

it follows from Example 5.8 that

$$a_1 x + a_3 x^3 + a_5 x^5 + \cdots = \frac{x - \dfrac{x^3}{6} + \dfrac{x^5}{120} + \cdots}{1 - \dfrac{x^2}{2} + \dfrac{x^4}{24} + \cdots}$$

and so

$$(a_1 x + a_3 x^3 + a_5 x^5 + \cdots)\left(1 - \frac{x^2}{2} + \frac{x^4}{24} + \cdots\right) = x - \frac{x^3}{6} + \frac{x^5}{120} + \cdots,$$

or, according to Theorem 5.8,

$$a_1 x + \left(a_3 - \frac{a_1}{2}\right)x^3 + \left(a_5 - \frac{a_3}{2} + \frac{a_1}{24}\right)x^5 + \cdots = x - \frac{x^3}{6} + \frac{x^5}{120} + \cdots.$$

From Corollary 5.2, coefficients of like powers of x on the two sides of this equation must be equal; hence,

$$a_1 = 1,$$

$$a_3 - \frac{a_1}{2} = -\frac{1}{6},$$

$$a_5 - \frac{a_3}{2} + \frac{a_1}{24} = \frac{1}{120},$$

and so

$$a_1 = 1, \quad a_3 = -\frac{1}{6} + \frac{1}{2}(1) = \frac{1}{3}, \quad a_5 = \frac{1}{120} + \frac{1}{2}\left(\frac{1}{3}\right) - \frac{1}{24}(1) = \frac{2}{15}.$$

Therefore,

$$\tan x = x + \frac{x^3}{3} + \frac{2}{15}x^5 + \cdots.$$

Example 5.14 To find the reciprocal of the power series,

$$g(x) = 1 + e^x = 2 + \sum_{n=1}^{\infty} \frac{x^n}{n!},$$

we let $h = 1$ in (25). If

$$\frac{1}{g(x)} = \sum_{n=0}^{\infty} a_n x^n,$$

then

$$1 = (a_0 + a_1 x + a_2 x^2 + a_3 x^3 + \cdots)\left(2 + x + \frac{x^2}{2} + \frac{x^3}{6} + \cdots\right)$$

$$= 2a_0 + (a_0 + 2a_1)x + \left(\frac{a_0}{2} + a_1 + 2a_2\right)x^2$$

$$+ \left(\frac{a_0}{6} + \frac{a_1}{2} + a_2 + 2a_3\right)x^3 + \cdots.$$

From Corollary 5.2,

$$2a_0 = 1,$$

$$a_0 + 2a_1 = 0,$$

$$\frac{a_0}{2} + a_1 + 2a_2 = 0,$$

$$\frac{a_0}{6} + \frac{a_1}{2} + a_2 + 2a_3 = 0,$$

Solving these equations successively yields

$$a_0 = \frac{1}{2},$$

$$a_1 = -\frac{a_0}{2} = -\frac{1}{4},$$

$$a_2 = -\frac{1}{2}\left(\frac{a_0}{2} + a_1\right) = -\frac{1}{2}\left(\frac{1}{4} - \frac{1}{4}\right) = 0,$$

$$a_3 = -\frac{1}{2}\left(\frac{a_0}{6} + \frac{a_1}{2} + a_2\right) = -\frac{1}{2}\left(\frac{1}{12} - \frac{1}{8} + 0\right) = \frac{1}{48},$$

and so

$$\frac{1}{1 + e^x} = \frac{1}{2} - \frac{x}{4} + \frac{x^3}{48} + \cdots.$$

Example 5.15   To find the reciprocal of

$$g(x) = e^x = \sum_{n=0}^{\infty} \frac{x^n}{n!}, \qquad (27)$$

we again let $h = 1$ in (25). If

$$(e^x)^{-1} = \sum_{n=0}^{\infty} a_n x^n,$$

then

$$1 = \left(\sum_{n=0}^{\infty} a_n x^n\right)\left(\sum_{n=0}^{\infty} \frac{x^n}{n!}\right) = \sum_{n=0}^{\infty} c_n x^n,$$

where

$$c_n = \sum_{r=0}^{n} \frac{a_r}{(n-r)!}.$$

From Corollary 5.2, $c_0 = a_0 = 1$ and $c_n = 0$ if $n \geq 1$; hence,

$$a_n = -\sum_{r=0}^{n-1} \frac{a_r}{(n-r)!}, \qquad n \geq 1. \qquad (28)$$

Solving these equations successively for $a_0, a_1, \ldots$ yields

$$a_1 = -\frac{1}{1!}(1) = -1,$$

$$a_2 = -\left[\frac{1}{2!}(1) + \frac{1}{1!}(-1)\right] = \frac{1}{2}$$

$$a_3 = -\left[\frac{1}{3!}(1) + \frac{1}{2!}(-1) + \frac{1}{1!}\left(\frac{1}{2}\right)\right] = -\frac{1}{6}$$

$$a_4 = -\left[\frac{1}{4!}(1) + \frac{1}{3!}(-1) + \frac{1}{2!}\left(\frac{1}{2}\right) + \frac{1}{1!}\left(-\frac{1}{6}\right)\right] = \frac{1}{24}.$$

From this we see that

$$a_k = \frac{(-1)^k}{k!}$$

for $0 \le k \le 4$, and are led to conjecture that this holds for all k. To prove this by induction, we assume that it is so for $0 \le k \le n - 1$ and compute from (28):

$$a_n = -\sum_{r=0}^{n-1} \frac{1}{(n-r)!} \frac{(-1)^r}{r!}$$

$$= -\frac{1}{n!} \sum_{r=0}^{n-1} (-1)^r \binom{n}{r} \qquad [\textit{Exercise 17(a), Section 1.2}]$$

$$= \frac{(-1)^n}{n!} \qquad [\textit{Exercise 17(b), Section 1.2}].$$

Thus, we have shown that

$$(e^x)^{-1} = \sum_{n=0}^{\infty} (-1)^n \frac{x^n}{n!}.$$

Since this is precisely the series that results if x is replaced by $-x$ in (27), we have verified a fundamental property of the exponential function: that

$$(e^x)^{-1} = e^{-x}.$$

This also follows from Example 3.25.

Abel's theorem

From Theorem 5.3 we know that a function f defined by a convergent power series,

$$f(x) = \sum_{n=0}^{\infty} a_n(x - x_0)^n, \qquad |x - x_0| < R, \tag{29}$$

is continuous in the open interval $(x_0 - R, x_0 + R)$. The next theorem concerns the behavior of f as x approaches an end point of the interval of convergence.

Theorem 5.9 (Abel's Theorem)   *Let f be defined by a power series (29) with finite radius of convergence R. Then:*

(a) If $\sum_{n=0}^{\infty} a_n R^n$ converges, then

$$\lim_{x \to (x_0 + R)-} f(x) = \sum_{n=0}^{\infty} a_n R^n;$$

(b) If $\sum_{n=0}^{\infty} (-1)^n a_n R^n$ converges, then

$$\lim_{x \to (x_0 - R)+} f(x) = \sum_{n=0}^{\infty} (-1)^n a_n R^n.$$

Proof.   It is convenient to consider a simpler problem first. Let

$$g(y) = \sum_{n=0}^{\infty} b_n y^n$$

and

$$\sum_{n=0}^{\infty} b_n = s \qquad \text{(finite)}.$$

We will show that

$$\lim_{y \to 1-} g(y) = s. \tag{30}$$

From Example 5.11,

$$g(y) = (1 - y) \sum_{n=0}^{\infty} s_n y^n, \tag{31}$$

where

$$s_n = b_0 + b_1 + \cdots + b_n.$$

Since

$$\frac{1}{1 - y} = \sum_{n=0}^{\infty} y^n, \qquad |y| < 1,$$

we can write

$$s = (1 - y) \sum_{n=0}^{\infty} s y^n, \qquad |y| < 1.$$

Subtracting this from (31) yields

$$g(y) - s = (1 - y) \sum_{n=0}^{\infty} (s_n - s) y^n, \qquad |y| < 1.$$

If $\varepsilon > 0$, choose N so that

$$|s_n - s| < \varepsilon \qquad \text{if} \quad n \geq N + 1.$$

Then, if $0 < y < 1$,

$$|g(y) - s| \leq (1 - y) \sum_{n=0}^{N} |s_n - s| y^n + (1 - y) \sum_{n=N+1}^{\infty} |s_n - s| y^n$$

$$< (1 - y) \sum_{n=0}^{N} |s_n - s| y^n + (1 - y)\varepsilon y^{N+1} \sum_{n=0}^{\infty} y^n$$

$$< (1 - y) \sum_{n=0}^{N} |s_n - s| + \varepsilon.$$

Therefore,

$$|g(y) - s| < 2\varepsilon$$

if

$$(1 - y) \sum_{n=0}^{N} |s_n - s| < \varepsilon.$$

This proves (30).

To obtain (a) from this, let $b_n = a_n R^n$ and $g(y) = f(x_0 + Ry)$; to obtain (b), let $b_n = (-1)^n a_n R^n$ and $g(y) = f(x_0 - Ry)$. We leave the details to the reader (*Exercise 29*).

Example 5.16   The series

$$f(x) = \frac{1}{1 + x} = \sum_{n=0}^{\infty} (-1)^n x^n$$

diverges at $x = 1$, while $\lim_{x \to 1-} f(x) = \frac{1}{2}$. This shows that the converse of Abel's theorem is false. Integrating the series term by term yields

$$\log(1 + x) = \sum_{n=0}^{\infty} (-1)^n \frac{x^{n+1}}{n + 1}, \qquad |x| < 1,$$

where the power series converges at $x = 1$, and Abel's theorem implies that

$$\log 2 = \sum_{n=0}^{\infty} \frac{(-1)^{n+1}}{n + 1}.$$

Example 5.17   If $q \geq 0$, the binomial series (20) converges absolutely for $x = \pm 1$. This is obvious if q is a nonnegative integer, and it follows from Raabe's test for other positive values of q, since

$$\left| \frac{a_{n+1}}{a_n} \right| = \left| \left(\frac{q}{n + 1} \right) \Big/ \left(\frac{q}{n} \right) \right| = \frac{n - q}{n + 1}, \qquad n > q,$$

and

$$\lim_{n\to\infty} n\left(\left|\frac{a_{n+1}}{a_n}\right| - 1\right) = \lim_{n\to\infty} n\left(\frac{n-q}{n+1} - 1\right)$$

$$= \lim_{n\to\infty} \frac{n}{n+1}(-q-1) = -q-1.$$

Therefore, Abel's theorem and (21) imply that

$$\sum_{n=0}^{\infty} \binom{q}{n} = 2^q \quad \text{and} \quad \sum_{n=0}^{\infty} (-1)^n \binom{q}{n} = 0, \qquad q \geq 0.$$

4.5   EXERCISES

1. The possibilities listed in Theorem 5.1(c) for behavior of a power series at the end points of its interval of convergence do not include absolute convergence at one end point and conditional convergence or divergence at the other. Why can't these occur?

2. Find the radius of convergence:

   (a) $\sum \left(\frac{n+1}{n}\right)^{n^2} [2 + (-1)^n]^n x^n$ (b) $\sum 2^{\sqrt{n}}(x-1)^n$

   (c) $\sum \left(2 + \sin\frac{n\pi}{6}\right)^n (x+2)^n$ (d) $\sum n^{\sqrt{n}} x^n$

   (e) $\sum \left(\frac{x}{n}\right)^n$

3. (a) Prove: If $\{a_n r^n\}$ is bounded and $|x_1 - x_0| < r$, then $\sum a_n(x_1 - x_0)^n$ converges.
   (b) Prove: If $\sum a_n(x - x_0)^n$ has radius of convergence R and $|x_1 - x_0| > R$, then $\{a_n(x_1 - x_0)^n\}$ is unbounded.

4. Prove: If g is a rational function defined for all nonnegative integers, then $\sum a_n x^n$ and $\sum a_n g(n) x^n$ have the same radius of convergence. [*Hint:* Use Exercise 27(a), Section 4.1.]

5. Suppose $f(x) = \sum a_n(x - x_0)^n$ has radius of convergence R and $0 < r < R_1 < R$. Show that there is an integer K such that

$$\left| f(x) - \sum_{n=0}^{k} a_n(x - x_0)^n \right| \leq \left(\frac{r}{R_1}\right)^{k+1} \frac{R_1}{R_1 - r}$$

if $|x - x_0| \leq r$ and $k \geq K$.

6. Suppose k is a positive integer and

$$f(x) = \sum_{n=0}^{\infty} a_n x^n$$

has radius of convergence R. Show that the series

$$g(x) = f(x^k) = \sum_{n=0}^{\infty} a_n x^{kn}$$

has radius of convergence $R^{1/k}$.

7. Complete the proof of Theorem 5.2 by showing that (a) $R = 0$ if $\lim_{n\to\infty} |a_{n+1}|/|a_n| = \infty$; (b) $R = \infty$ if $\lim_{n\to\infty} |a_{n+1}|/|a_n| = 0$.

8. Find the radius of convergence:

(a) $\sum (\log n) x^n$

(b) $\sum 2^n n^p (x + 1)^n$

(c) $\sum (-1)^n \binom{2n}{n} x^n$

(d) $\sum (-1)^n \dfrac{n^2 + 1}{n 4^n} (x - 1)^n$

(e) $\sum \dfrac{n^n}{n!} (x + 2)^n$

(f) $\sum \dfrac{\alpha(\alpha + 1) \cdots (\alpha + n - 1)}{\beta(\beta + 1) \cdots (\beta + n - 1)} x^n$

9. Suppose $a_n \neq 0$ for n sufficiently large. Show that

$$\varliminf_{n\to\infty} \left| \frac{a_{n+1}}{a_n} \right| \leq \varliminf_{n\to\infty} |a_n|^{1/n} \quad \text{and} \quad \varlimsup_{n\to\infty} |a_n|^{1/n} \leq \varlimsup_{n\to\infty} \left| \frac{a_{n+1}}{a_n} \right|.$$

Show that this implies Theorem 5.2.

10. Given that

$$\frac{1}{1 - x} = \sum_{n=0}^{\infty} x^n, \qquad |x| < 1,$$

use Theorem 5.3 to express $\sum_{n=0}^{\infty} n^2 x^n$ in closed form.

11. The function

$$J_p(x) = \sum_{n=0}^{\infty} \frac{(-1)^n}{n!(n + p)!} \left(\frac{x}{2} \right)^{p + 2n} \qquad (p = \text{integer} \geq 0)$$

is the *Bessel function of order p*. Show that:
(a) $J_0' = -J_1$
(b) $J_p' = \frac{1}{2}(J_{p-1} - J_{p+1})$, $p \geq 1$
(c) $x^2 J_p'' + x J_p' + (x^2 - p^2) J_p = 0$

12. Given that the power series $f(x) = \sum_{n=0}^{\infty} a_n x^n$ satisfies

$$f'(x) = -2xf(x), \qquad f(0) = 1, \qquad f'(0) = 0,$$

find $\{a_n\}$. Do you recognize f?

13. Let

$$f(x) = \sum_{n=0}^{\infty} a_n x^n, \qquad |x| < R,$$

and $g(x) = f(x^k)$, where k is a positive integer. Show that

$$g^{(r)}(0) = 0 \quad \text{if} \quad r \ne kn \quad \text{and} \quad g^{(kn)}(0) = \frac{(kn)!}{n!} f^{(n)}(0), \qquad n = 0, 1, \ldots.$$

14. Let

$$f(x) = \sum_{n=0}^{\infty} a_n (x - x_0)^n, \qquad |x - x_0| < R,$$

and $f(t_n) = 0$, where $t_n \ne 0$ and $\lim_{n \to \infty} t_n = x_0$. Show that $f(x) \equiv 0 \, (|x - x_0| < R)$. (*Hint:* Rolle's theorem helps here.)

15. Prove Theorem 5.5.

16. Express

$$\int_1^x \frac{\log t}{t - 1} \, dt$$

as a power series in $x - 1$, and find its radius of convergence.

17. By substituting $-x^2$ for x in the geometric series, we obtain

$$\frac{1}{1 + x^2} = \sum_{n=0}^{\infty} (-1)^n x^{2n}, \qquad |x| < 1.$$

Use this to express $f(x) = \text{Tan}^{-1} x \, [f(0) = 0]$ as a power series in x. Then evaluate all derivatives of f at $x_0 = 0$, and find a series of constants which converges to $\pi/6$.

18. Prove: If

$$f(x) = \sum_{n=0}^{\infty} a_n (x - x_0)^n, \qquad |x - x_0| < R,$$

and F is an antiderivative of f on $(x_0 - R, x_0 + R)$, then

$$F(x) = C + \sum_{n=0}^{\infty} \frac{a_n}{n + 1} (x - x_0)^{n+1}, \qquad |x - x_0| < R,$$

where C is a constant.

19. Suppose some derivative of f can be represented by a power series in $x - x_0$ in an interval about x_0. Show that f and all its derivatives can also.

20. Verify Eq. (21) (Example 5.10) by showing that

$$(1 + x)^{-q} \sum_{n=0}^{\infty} \binom{q}{n} x^n = 1, \qquad |x| < 1.$$

(*Hint:* Differentiate.)

21. Prove Theorem 5.7.

22. Find the Maclaurin series of $\cosh x$ and $\sinh x$ from the definition in Eq. (16), and also by applying Theorem 5.7 to the Maclaurin series for e^x and e^{-x}.

23. Give an example where the radius of convergence of the product of two power series is greater than the smaller of the radii of convergence of the factors.

24. Use Theorem 5.8 to find the first four nonzero terms in the Maclaurin series for

(a) $e^x \sin x$ (b) $\dfrac{e^{-x}}{1 + x^2}$ (c) $\dfrac{\cos x}{1 + x^6}$ (d) $(\sin x) \log(1 + x)$

25. Deduce the identity

$$2 \sin x \cos x = \sin 2x$$

from the Maclaurin series for $\sin x$, $\cos x$, and $\sin 2x$.

26. (a) Given that

$$(1 - 2xt + x^2)^{-1/2} = \sum_{n=0}^{\infty} P_n(t) x^n, \qquad |x| < 1, \tag{A}$$

if $-1 < t < 1$, show that $P_0(t) = 1$, $P_1(t) = t$, and

$$P_{n+1}(t) = \frac{2n + 1}{n + 1} t P_n(t) - \frac{n}{n + 1} P_{n-1}(t), \qquad n \geq 1.$$

[*Hint:* First differentiate (A) with respect to x.]

(b) Show from part (a) that P_n is a polynomial of degree n. It is the nth *Legendre polynomial*, and $(1 - 2xt + x^2)^{-1/2}$ is the *generating function* of the sequence $\{P_n\}$.

27. Define (if necessary) each of the following functions so as to be continuous at $x_0 = 0$, and find the first four terms of its Maclaurin series.

(a) $\dfrac{xe^x}{\sin x}$ (b) $\dfrac{\cos x}{1 + x + x^2}$ (c) $\sec x$

(d) $x \csc x$ (e) $\dfrac{\sin 2x}{\sin x}$

28. Let $a_0 = a_1 = 5$ and

$$a_{n+1} = a_n + 6a_{n-1}, \qquad n \geq 1.$$

   (a) Express

$$F(x) = \sum_{n=0}^{\infty} a_n x^n$$

   in closed form.
   (b) Write F as the difference of two geometric series and find an explicit formula for a_n.

29. Show that the general form of Theorem 5.9 follows from the special case proved in the text. (See the hint following that proof.)

30. Starting from the Maclaurin series

$$\log(1 - x) = -\sum_{n=0}^{\infty} \frac{x^{n+1}}{n+1}, \qquad |x| < 1,$$

   use Abel's theorem to evaluate

$$\sum_{n=0}^{\infty} \frac{1}{(n+1)(n+2)}.$$

31. In Example 5.17 we saw that

$$\sum_{n=0}^{\infty} \binom{q}{n} = 2^q, \qquad q \geq 0.$$

   Show that this also holds for $-1 < q < 0$, but not for $q \leq -1$. (*Hint:* See Exercise 30, Section 4.1.)

32. (a) Prove: If $\sum_{n=0}^{\infty} b_n$ converges, then the series

$$g(x) = \sum_{n=0}^{\infty} b_n x^n$$

   converges uniformly on $[0, 1]$. (*Hint:* If $\varepsilon > 0$, there is an integer N such that

$$|a_n + a_{n+1} + \cdots + a_m| < \varepsilon \quad \text{if} \quad n, m \geq N.$$

   Use summation by parts to show that then

$$|a_n x^n + a_{n+1} x^{n+1} + \cdots + a_m x^m| < 2\varepsilon \quad \text{if} \quad 0 \leq x < 1, \quad n, m \geq N.)$$

   This is also known as *Abel's theorem*.
   (b) Show that part (a) implies the restricted form of Theorem 5.9 (concerning g) proved in the text.

*33. Use Exercise 32 to show that if $\sum_{n=0}^{\infty} a_n$, $\sum_{n=0}^{\infty} b_n$, and their Cauchy product $\sum_{n=0}^{\infty} c_n$ all converge, then

$$\left(\sum_{n=0}^{\infty} a_n \right) \left(\sum_{n=0}^{\infty} b_n \right) = \sum_{n=0}^{\infty} c_n.$$

34. Prove: If

$$g(x) = \sum_{n=0}^{\infty} b_n x^n, \qquad |x| < 1,$$

and $b_n \geq 0$, then

$$\sum_{n=0}^{\infty} b_n = \lim_{x \to 1^-} g(x) \qquad \text{(finite or infinite).}$$

35. Use the binomial series and the relation

$$\frac{d}{dx} (\sin^{-1} x) = (1 - x^2)^{-1/2}$$

to obtain the Maclaurin series for $\sin^{-1} x$ $(\sin^{-1} 0 = 0)$. Deduce from this series and Exercise 34 that

$$\sum_{n=0}^{\infty} \binom{2n}{n} \frac{1}{2^{2n}(2n + 1)} = \frac{\pi}{2}.$$

real-valued functions of several variables

5.1 STRUCTURE OF $\mathcal{R}^n$

The rest of this book deals with functions defined on subsets of the real n-dimensional space $\mathcal{R}^n$, which consists of all ordered n-tuples $\mathbf{X} = (x_1, x_2, \ldots, x_n)$ of real numbers, called the *coordinates*, or *components*, of $\mathbf{X}$. In this section we introduce an algebraic structure for $\mathcal{R}^n$, and also consider its *topological* properties, that is, those properties that can be described in terms of a special class of subsets—the neighborhoods—of $\mathcal{R}^n$. In Section 1.3 we studied the topological properties of $\mathcal{R}^1$ (which we will continue to denote simply as $\mathcal{R}$). Most of the definitions and proofs there were stated in terms of neighborhoods in $\mathcal{R}$. We will see that they carry over to $\mathcal{R}^n$ if the concept of neighborhood in $\mathcal{R}^n$ is suitably defined.

Elements of $\mathcal{R}$ have dual interpretations: geometric, as points on the real line, and algebraic, as elements of the field of real numbers. We assume that the reader is familiar with the geometric interpretation of elements of $\mathcal{R}^2$ and $\mathcal{R}^3$ as the coordinates—with respect to rectangular coordinate systems—of points in a plane and three-dimensional space, respectively. Although $\mathcal{R}^n$ cannot be visualized geometrically if $n \geq 4$, geometric ideas from $\mathcal{R}, \mathcal{R}^2$, and $\mathcal{R}^3$ often help us to interpret the properties of $\mathcal{R}^n$ for arbitrary n.

As we said in Section 1.3, the idea of neighborhood is always associated with some definition of "closeness" of points. The following definition imposes an algebraic structure on $\mathscr{R}^n$, in terms of which the distance between two points can be defined in a natural way. In addition, this algebraic structure will be useful later for other purposes.

Definition 1.1   *The vector sum of elements*

$$\mathbf{X} = (x_1, x_2, \ldots, x_n) \quad and \quad \mathbf{Y} = (y_1, y_2, \ldots, y_n)$$

is the element $\mathbf{X} + \mathbf{Y}$ *of* $\mathscr{R}^n$ *given by*

$$\mathbf{X} + \mathbf{Y} = (x_1 + y_1, x_2 + y_2, \ldots, x_n + y_n). \tag{1}$$

If a is a real number, the scalar multiple of $\mathbf{X}$ *by a is the element* $a\mathbf{X}$ *of* $\mathscr{R}^n$ *given by*

$$a\mathbf{X} = (ax_1, ax_2, \ldots, ax_n). \tag{2}$$

Notice that "$+$" has two distinct meanings in (1); on the left it stands for the newly defined addition of elements in $\mathscr{R}^n$; on the right, for ordinary addition of real numbers. However, this can never lead to confusion, since the meaning of "$+$" can always be discerned from the symbols on either side of it. A similar comment applies to the use of juxtaposition to indicate scalar multiplication on the left of (2) and ordinary multiplication on the right.

Example 1.1   In $\mathscr{R}^4$ let

$$\mathbf{X} = (1, -2, 6, 5) \quad and \quad \mathbf{Y} = (3, -5, 4, \tfrac{1}{2}).$$

Then

$$\mathbf{X} + \mathbf{Y} = (4, -7, 10, \tfrac{11}{2})$$

and

$$6\mathbf{X} = (6, -12, 36, 30).$$

We leave the proof of the following theorem to the reader (*Exercise 2*).

Theorem 1.1   *If* $\mathbf{X}$, $\mathbf{Y}$, *and* $\mathbf{Z}$ *are elements of* $\mathscr{R}^n$ *and a and b are real numbers, then:*

(a) $\mathbf{X} + \mathbf{Y} = \mathbf{Y} + \mathbf{X}$. (*Vector addition is commutative.*)
(b) $(\mathbf{X} + \mathbf{Y}) + \mathbf{Z} = \mathbf{X} + (\mathbf{Y} + \mathbf{Z})$. (*Vector addition is associative.*)
(c) *There is a unique vector* $\mathbf{0}$, *called the zero vector, such that*

$$\mathbf{X} + \mathbf{0} = \mathbf{X}$$

for all $\mathbf{X}$ *in* $\mathscr{R}^n$.

(d) *For each* $\mathbf{X}$ *in* $\mathscr{R}^n$ *there is a unique vector* $-\mathbf{X}$ *such that*

$\mathbf{X} + (-\mathbf{X}) = \mathbf{0}$.

(e) $a(b\mathbf{X}) = (ab)\mathbf{X}$.
(f) $(a + b)\mathbf{X} = a\mathbf{X} + b\mathbf{X}$.
(g) $a(\mathbf{X} + \mathbf{Y}) = a\mathbf{X} + a\mathbf{Y}$.
(h) $1\mathbf{X} = \mathbf{X}$.

It is clear that $\mathbf{0} = (0, 0, \ldots, 0)$ and $-\mathbf{X} = (-x_1, -x_2, \ldots, -x_n)$ if $\mathbf{X} = (x_1, x_2, \ldots, x_n)$. As when dealing with real numbers, we will write $\mathbf{X} + (-\mathbf{Y})$ simply as $\mathbf{X} - \mathbf{Y}$. The point $\mathbf{0}$ is called the *origin* of $\mathscr{R}^n$.

A nonempty set $V = \{\mathbf{X}, \mathbf{Y}, \mathbf{Z}, \ldots\}$, together with rules such as (1), associating a unique element of V with every ordered pair of its elements, and (2), associating a unique element of V with every real number and element of V, is said to be a *vector space* if it has the properties listed in Theorem 1.1. The elements of a vector space are called *vectors*. When we wish to emphasize that we are regarding an element of $\mathscr{R}^n$ as part of this algebraic structure, we will speak of it as a vector; otherwise, we will speak of it as a point.

Length, distance, and inner product

Definition 1.2   *The length of the vector* $\mathbf{X} = (x_1, x_2, \ldots, x_n)$ *is defined to be*

$|\mathbf{X}| = (x_1^2 + x_2^2 + \cdots + x_n^2)^{1/2}$.

The distance between points $\mathbf{X}$ *and* $\mathbf{Y}$ *is defined to be* $|\mathbf{X} - \mathbf{Y}|$; *in particular,* $|\mathbf{X}|$ *is the distance between* $\mathbf{X}$ *and the origin, and* $\mathbf{X}$ *is said to be a unit vector if* $|\mathbf{X}| = 1$.

For $n = 1$ this definition of length reduces to the familiar absolute value, and the distance between two points is the length of the interval having them as end points; for $n = 2$ and $n = 3$, the length and distance of Definition 1.1 reduce to the familiar definitions for the plane and three-dimensional space.

Example 1.2   The lengths of the vectors

$$\mathbf{X} = (1, -2, 6, 5) \quad \text{and} \quad \mathbf{Y} = (3, -5, 4, \tfrac{1}{2})$$

are

$$|\mathbf{X}| = [1^2 + (-2)^2 + 6^2 + 5^2]^{1/2} = \sqrt{66}$$

and

$$|\mathbf{Y}| = [3^2 + (-5)^2 + 4^2 + (\tfrac{1}{2})^2]^{1/2} = \frac{\sqrt{201}}{2}.$$

The distance between **X** and **Y** is

$$|\mathbf{X} - \mathbf{Y}| = [(1 - 3)^2 + (-2 + 5)^2 + (6 - 4)^2 + (5 - \tfrac{1}{2})^2]^{1/2} = \frac{\sqrt{149}}{2}.$$

The next definition and lemma will enable us to establish the triangle inequality, an important property of length and distance in $\mathscr{R}^n$.

Definition 1.3   *The inner product* $\mathbf{X} \cdot \mathbf{Y}$ *of* $\mathbf{X} = (x_1, x_2, \ldots, x_n)$ *and* $\mathbf{Y} = (y_1, y_2, \ldots, y_n)$ *is defined by*

$$\mathbf{X} \cdot \mathbf{Y} = x_1 y_1 + x_2 y_2 + \cdots + x_n y_n.$$

Lemma 1.1 (Schwarz's Inequality)   *If* $\mathbf{X}$ *and* $\mathbf{Y}$ *are any two vectors in* $\mathscr{R}^n$, *then*

$$|\mathbf{X} \cdot \mathbf{Y}| \le |\mathbf{X}|\,|\mathbf{Y}| \tag{3}$$

and equality holds if and only if one of the vectors is a scalar multiple of the other.

Proof.   If t is any real number, then

$$0 \le \sum_{i=1}^{n} (x_i - t y_i)^2$$

$$= \sum_{i=1}^{n} x_i^2 - 2t \sum_{i=1}^{n} x_i y_i + t^2 \sum_{i=1}^{n} y_i^2 \tag{4}$$

$$= |\mathbf{X}|^2 - 2(\mathbf{X} \cdot \mathbf{Y})t + t^2 |\mathbf{Y}|^2.$$

If $\mathbf{Y} \ne 0$, the last expression is a second-degree polynomial p in t; from the quadratic formula, its roots are

$$r = \frac{(\mathbf{X} \cdot \mathbf{Y}) \pm \sqrt{(\mathbf{X} \cdot \mathbf{Y})^2 - |\mathbf{X}|^2|\mathbf{Y}|^2}}{|\mathbf{Y}|^2}.$$

Therefore,

$$(\mathbf{X} \cdot \mathbf{Y})^2 \le |\mathbf{X}|^2|\mathbf{Y}|^2, \tag{5}$$

because, if not, p would have two distinct real roots and therefore be negative between them (Figure 1.1), contradicting the inequality (4). Taking square roots in (5) yields (3) if $\mathbf{Y} \ne 0$. If $\mathbf{Y} = 0$, then (3) certainly holds; therefore, (3) holds in any case.

We leave it to the reader to show that equality holds in (3) if and only if one of the vectors is a multiple of the other (*Exercise 6*).

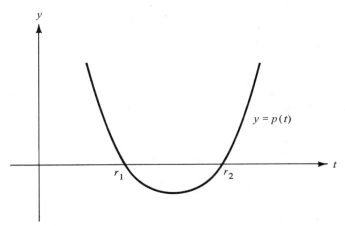

$$y = p(t)$$

FIGURE 1.1

Theorem 1.2 (Triangle Inequality) *If* **X** *and* **Y** *are any two vectors in* $\mathcal{R}^n$, *then*

$$|\mathbf{X} + \mathbf{Y}| \le |\mathbf{X}| + |\mathbf{Y}|, \tag{6}$$

with equality if and only if one of the vectors is a nonnegative multiple of the other.

Proof. By definition,

$$|\mathbf{X} + \mathbf{Y}|^2 = \sum_{i=1}^{n} (x_i + y_i)^2 = \sum_{i=1}^{n} x_i^2 + 2 \sum_{i=1}^{n} x_i y_i + \sum_{i=1}^{n} y_i^2$$

$$= |\mathbf{X}|^2 + 2(\mathbf{X} \cdot \mathbf{Y}) + |\mathbf{Y}|^2$$

$$\le |\mathbf{X}|^2 + 2|\mathbf{X}|\,|\mathbf{Y}| + |\mathbf{Y}|^2 \qquad \text{(by Schwarz's inequality)}$$

$$= (|\mathbf{X}| + |\mathbf{Y}|)^2.$$

Hence,

$$|\mathbf{X} + \mathbf{Y}|^2 \le (|\mathbf{X}| + |\mathbf{Y}|)^2$$

and (6) follows on taking square roots. We leave it to the reader to establish the stated condition for equality in (6) (*Exercise 7*).

Corollary 1.1 *If* **X**, **Y**, *and* **Z** *are three points in* $\mathcal{R}^n$, *then*

$$|\mathbf{X} - \mathbf{Z}| \le |\mathbf{X} - \mathbf{Y}| + |\mathbf{Y} - \mathbf{Z}|.$$

Proof.   Write

$$\mathbf{X} - \mathbf{Z} = (\mathbf{X} - \mathbf{Y}) + (\mathbf{Y} - \mathbf{Z})$$

and apply Theorem 1.2.

Corollary 1.2   *If* $\mathbf{X}$ *and* $\mathbf{Y}$ *are any two vectors in* $\mathscr{R}^n$, *then*

$$|\mathbf{X} - \mathbf{Y}| \geq \big||\mathbf{X}| - |\mathbf{Y}|\big|.$$

Proof.   Since

$$\mathbf{X} = \mathbf{Y} + (\mathbf{X} - \mathbf{Y}),$$

Theorem 1.2 implies that

$$|\mathbf{X}| \leq |\mathbf{Y}| + |\mathbf{X} - \mathbf{Y}|$$

or

$$|\mathbf{X}| - |\mathbf{Y}| \leq |\mathbf{X} - \mathbf{Y}|.$$

Interchanging $\mathbf{X}$ and $\mathbf{Y}$ here yields

$$|\mathbf{Y}| - |\mathbf{X}| \leq |\mathbf{Y} - \mathbf{X}|,$$

and the last two inequalities imply the stated result.

Example 1.3   The angle between two nonzero vectors $\mathbf{X} = (x_1, x_2, x_3)$ and $\mathbf{Y} = (y_1, y_2, y_3)$ in $\mathscr{R}^3$ is defined to be the angle between the directed line segments from the origin to the points $\mathbf{X}$ and $\mathbf{Y}$ (Figure 1.2). Applying the law of cosines to the triangle in Figure 1.2 yields

$$|\mathbf{X} - \mathbf{Y}|^2 = |\mathbf{X}|^2 + |\mathbf{Y}|^2 - 2|\mathbf{X}|\,|\mathbf{Y}|\cos\theta. \tag{7}$$

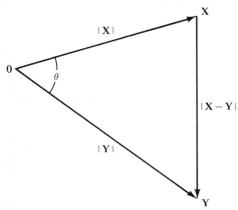

FIGURE 1.2

However,

$$\begin{aligned}
|\mathbf{X} - \mathbf{Y}|^2 &= (x_1 - y_1)^2 + (x_2 - y_2)^2 + (x_3 - y_3)^2 \\
&= (x_1^2 + x_2^2 + x_3^2) + (y_1^2 + y_2^2 + y_3^2) - 2(x_1 y_1 + x_2 y_2 + x_3 y_3) \\
&= |\mathbf{X}|^2 + |\mathbf{Y}|^2 - 2\mathbf{X} \cdot \mathbf{Y}.
\end{aligned}$$

Comparing this with (7) yields

$$\mathbf{X} \cdot \mathbf{Y} = |\mathbf{X}| \, |\mathbf{Y}| \cos \theta.$$

Since $|\cos \theta| \le 1$, this verifies Schwarz's inequality in $\mathscr{R}^3$.

Example 1.4   Connecting the points $\mathbf{0}, \mathbf{X}, \mathbf{Y}$, and $\mathbf{X} + \mathbf{Y}$ in $\mathscr{R}^2$ or $\mathscr{R}^3$ (Figure 1.3) produces a parallelogram with sides of length $|\mathbf{X}|$ and $|\mathbf{Y}|$ and a diagonal of length $|\mathbf{X} + \mathbf{Y}|$. Thus, there is a triangle with sides $|\mathbf{X}|$, $|\mathbf{Y}|$, and $|\mathbf{X} + \mathbf{Y}|$. From this we see geometrically that

$$|\mathbf{X} + \mathbf{Y}| \le |\mathbf{X}| + |\mathbf{Y}|$$

in $\mathscr{R}^2$ or $\mathscr{R}^3$, since the length of one side of a triangle cannot exceed the sum of the lengths of the other two. This verifies (6) for $\mathscr{R}^2$ and $\mathscr{R}^3$, and indicates why it is called the triangle inequality.

The next theorem lists properties of length, distance, and inner product that follow directly from Definitions 1.2 and 1.3. We leave its proof to the reader (*Exercise* 8).

Theorem 1.3   *If* $\mathbf{X}, \mathbf{Y}$, *and* $\mathbf{Z}$ *are elements of* $\mathscr{R}^n$ *and* a *is a scalar, then:*

(*a*) $|a\mathbf{X}| = |a| \, |\mathbf{X}|$.
(*b*) $|\mathbf{X}| \ge 0$, *with equality if and only if* $\mathbf{X} = \mathbf{0}$.
(*c*) $|\mathbf{X} - \mathbf{Y}| \ge 0$, *with equality if and only if* $\mathbf{X} = \mathbf{Y}$.

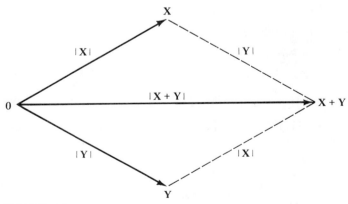

FIGURE 1.3

(d) $\mathbf{X} \cdot \mathbf{Y} = \mathbf{Y} \cdot \mathbf{X}$.
(e) $\mathbf{X} \cdot (\mathbf{Y} + \mathbf{Z}) = \mathbf{X} \cdot \mathbf{Y} + \mathbf{X} \cdot \mathbf{Z}$.
(f) $(c\mathbf{X}) \cdot \mathbf{Y} = \mathbf{X} \cdot (c\mathbf{Y}) = c(\mathbf{X} \cdot \mathbf{Y})$.

Line segments in $\mathscr{R}^n$

The equation of a line through a point $\mathbf{X}_0 = (x_0, y_0, z_0)$ in $\mathscr{R}^3$ can be written parametrically as

$$x = x_0 + u_1 t, \quad y = y_0 + u_2 t, \quad z = z_0 + u_3 t, \qquad -\infty < t < \infty,$$

where u_1, u_2, and u_3 are not all zero. We write this in vector form as

$$\mathbf{X} = \mathbf{X}_0 + t\mathbf{U}, \qquad -\infty < t < \infty, \tag{8}$$

with $\mathbf{U} = (u_1, u_2, u_3)$, and say that the line is *through* $\mathbf{X}_0$ *in the direction of* $\mathbf{U}$.

There are many ways to represent a given line parametrically. For example,

$$\mathbf{X} = \mathbf{X}_0 + s\mathbf{V}, \qquad -\infty < s < \infty, \tag{9}$$

represents the same line as (8) if and only if $\mathbf{V} = a\mathbf{U}$ for some nonzero real number a. If this is so, then the line is traversed in the same direction as s and t vary from $-\infty$ to ∞ in (8) and (9) if $a > 0$; if $a < 0$, in opposite directions.

To write the parametric equation of a line through two points $\mathbf{X}_0$ and $\mathbf{X}_1$ in $\mathscr{R}^3$, we take $\mathbf{U} = \mathbf{X}_1 - \mathbf{X}_0$ in (8), which yields

$$\mathbf{X} = \mathbf{X}_0 + t(\mathbf{X}_1 - \mathbf{X}_0), \qquad -\infty < t < \infty.$$

This can also be written as

$$\mathbf{X} = t\mathbf{X}_1 + (1 - t)\mathbf{X}_0, \qquad -\infty < t < \infty.$$

The line segment from $\mathbf{X}_0$ to $\mathbf{X}_1$ consists of those points for which $0 \le t \le 1$ in this equation.

Example 1.5   The line L defined by

$$x = -1 + 2t, \quad y = 3 - 4t, \quad z = -1, \, -\infty < t < \infty,$$

which can be rewritten as

$$\mathbf{X} = (-1, 3, -1) + t(2, -4, 0), \qquad -\infty < t < \infty, \tag{10}$$

is through $\mathbf{X}_0 = (-1, 3, -1)$ in the direction of $\mathbf{U} = (2, -4, 0)$. The same line can be represented by

$$\mathbf{X} = (-1, 3, -1) + s(1, -2, 0), \qquad -\infty < s < \infty, \tag{11}$$

or by

$$\mathbf{X} = (-1, 3, -1) + \tau(-4, 8, 0), \qquad -\infty < \tau < \infty. \tag{12}$$

Since

$$(1, -2, 0) = \tfrac{1}{2}(2, -4, 0),$$

L is traversed in the same direction as t and s vary from $-\infty$ to ∞ in (10) and (11); however, since

$$(-4, 8, 0) = -2(2, -4, 0),$$

L is traversed in opposite directions as t and τ vary from $-\infty$ to ∞ in (10) and (12).

Setting $t = 1$ in (10), we see that $\mathbf{X}_1 = (1, -1, -1)$ is also on L. The line segment from $\mathbf{X}_0$ to $\mathbf{X}_1$ consists of all points of the form

$$\mathbf{X} = t(1, -1, -1) + (1 - t)(-1, 3, -1), \qquad 0 \leq t \leq 1.$$

These familiar notions in $\mathscr{R}^3$ (which have familiar analogs in $\mathscr{R}$ and $\mathscr{R}^2$) can be generalized to $\mathscr{R}^n$, as follows.

Definition 1.4 *Suppose $\mathbf{X}_0$ and $\mathbf{U}$ are in $\mathscr{R}^n$ and $\mathbf{U} \neq \mathbf{0}$. Then the line through $\mathbf{X}_0$ in the direction of $\mathbf{U}$ is the set of all points in $\mathscr{R}^n$ of the form*

$$\mathbf{X} = \mathbf{X}_0 + t\mathbf{U}, \qquad -\infty < t < \infty.$$

A set of points of the form

$$\mathbf{X} = \mathbf{X}_0 + t\mathbf{U}, \qquad t_1 \leq t \leq t_2,$$

is called a line segment; in particular, the line segment from $\mathbf{X}_0$ to $\mathbf{X}_1$ is the set of points of the form

$$\mathbf{X} = \mathbf{X}_0 + t(\mathbf{X}_1 - \mathbf{X}_0) = t\mathbf{X}_1 + (1 - t)\mathbf{X}_0, \qquad 0 \leq t \leq 1.$$

A given line can be oriented in two opposite directions, since

$$\mathbf{X} = \mathbf{X}_0 + t\mathbf{U}, \qquad -\infty < t < \infty,$$

and

$$\mathbf{X} = \mathbf{X}_0 - t\mathbf{U}, \qquad -\infty < t < \infty,$$

represent the same set of points. When we wish to emphasize that we have chosen a specific orientation, we will speak of a *directed* line.

Neighborhoods and open sets in $\mathscr{R}$

Having defined distance in $\mathscr{R}^n$, we are now able to say what we mean by a neighborhood of a point in $\mathscr{R}^n$.

Definition 1.5   *If* $\varepsilon > 0$, *the* ε-*neighborhood of a point* $\mathbf{X}_0$ *in* $\mathscr{R}^n$ *is the set* $N_\varepsilon(\mathbf{X}_0)$ *of points* $\mathbf{X}$ *in* $\mathscr{R}^n$ *such that*

$$|\mathbf{X} - \mathbf{X}_0| < \varepsilon.$$

An ε-neighborhood of a point $\mathbf{X}_0$ in $\mathscr{R}^2$ is the inside—but not the circumference—of the circle of radius ε about $\mathbf{X}_0$; in $\mathscr{R}^3$ it is the inside—but not the surface—of the sphere of radius ε about $\mathbf{X}_0$.

In Section 1.3 we stated several other definitions in terms of ε-neighborhoods: neighborhood, interior point, interior of a set, open set, closed set, limit point, boundary point, boundary of a set, closure of a set, isolated point, exterior point, and exterior of a set. Since these definitions are the same for $\mathscr{R}^n$ as for $\mathscr{R}$, we will not repeat them. We advise the reader to read them again in Section 1.3, substituting $\mathscr{R}^n$ for $\mathscr{R}$ and $\mathbf{X}_0$ for x_0.

Example 1.6   Let S be the set of points in $\mathscr{R}^2$ in the square bounded by the lines $x = \pm 1$, $y = \pm 1$, except for the origin and the points on the vertical lines $x = \pm 1$ (Figure 1.4); thus,

$$S = \{(x, y) \mid (x, y) \neq (0, 0), \, -1 < x < 1, \, -1 \leq y \leq 1\}.$$

Then every point of S except those on the lines $y = \pm 1$ is an interior point, so

$$S^0 = \{(x, y) \mid (x, y) \neq (0, 0), \, -1 < x, y < 1\}.$$

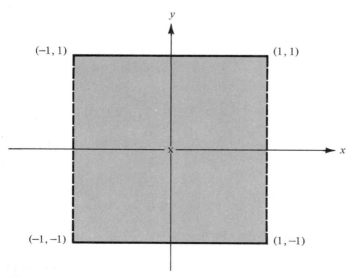

FIGURE 1.4

S is a deleted neighborhood of $(0, 0)$, and is neither open nor closed. The closure of S is

$$\bar{S} = \{(x, y) \mid -1 \le x, y \le 1\}$$

and every point of S is a limit point of S. The origin and the perimeter of S form ∂S, the boundary of S. The exterior of S consists of all points (x, y) such that $|x| > 1$ or $|y| > 1$. The origin is an isolated point of S^c.

Example 1.7   If $\mathbf{X}_0$ is a point in $\mathscr{R}^n$ and r is a positive number, the *open n-sphere of radius r about* $\mathbf{X}_0$ is the set

$$S_r(\mathbf{X}_0) = \{\mathbf{X} \mid |\mathbf{X} - \mathbf{X}_0| < r\}.$$

(Thus, ε-neighborhoods are open n-spheres, and conversely.) If $\mathbf{X}_1$ is in $S_r(\mathbf{X}_0)$ and

$$|\mathbf{X} - \mathbf{X}_1| < \varepsilon = r - |\mathbf{X} - \mathbf{X}_0|,$$

then $\mathbf{X}$ is in $S_r(\mathbf{X}_0)$. (The situation is depicted in Figure 1.5 for $n = 2$.) Thus, $S_r(\mathbf{X}_0)$ contains an ε-neighborhood of each of its points, and is therefore open. We leave it to the reader (*Exercise 15*) to show that the closure of $S_r(\mathbf{X}_0)$ is the *closed n-sphere of radius r about* $\mathbf{X}_0$, defined by

$$\bar{S}_r(\mathbf{X}_0) = \{\mathbf{X} \mid |\mathbf{X} - \mathbf{X}_0| \le r\}.$$

Open and closed n-spheres are generalizations to $\mathscr{R}^n$ of open and closed intervals.

The following lemma will be useful later in this section, when we consider connected sets.

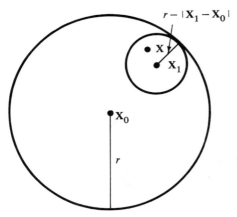

FIGURE 1.5

Lemma 1.2   *If* X_1 *and* X_2 *are in an open n-sphere of radius* ε *about* X_0 *(or, equivalently, in an* ε*-neighborhood of* X_0*), then so is every point on the line segment from* X_1 *to* X_2.

Proof.   If

$$|X_1 - X_0| < \varepsilon, \qquad |X_2 - X_0| < \varepsilon,$$

and

$$X = tX_2 + (1 - t)X_1, \qquad 0 < t < 1,$$

then

$$
\begin{aligned}
|X - X_0| &= |tX_2 + (1 - t)X_1 - tX_0 - (1 - t)X_0| \\
&= |t(X_2 - X_0) + (1 - t)(X_1 - X_0)| \\
&\le t|X_2 - X_0| + (1 - t)|X_1 - X_0| \\
&< t\varepsilon + (1 - t)\varepsilon = \varepsilon.
\end{aligned}
$$

This completes the proof.

The proofs given in Section 1.3 for Theorem 3.1 (the union of open sets is open, the intersection of closed sets closed) and Theorem 3.2 and its Corollary 3.1 (a set is closed if and only if it contains all its limit points) are also valid in $\mathscr{R}^n$. The reader should reread them now. The Heine–Borel theorem also holds in $\mathscr{R}^n$, but the proof in Section 1.3 is valid only for $n = 1$. To prove it for general n we need some preliminary definitions and results which are of interest in their own right.

Definition 1.6   *A sequence of points* $\{X_r\}$ *in* $\mathscr{R}^n$ *is said to converge to the limit* $\bar{X}$ *if*

$$\lim_{r \to \infty} |X_r - \bar{X}| = 0.$$

In this case we write

$$\lim_{r \to \infty} X_r = \bar{X}.$$

The next two theorems follow quickly from this, the definition of distance in $\mathscr{R}^n$, and what we already know about convergence in $\mathscr{R}$. We leave the proofs to the reader (*Exercises 18 and 19*).

Theorem 1.4   *Let*

$$\bar{X} = (\bar{x}_1, \bar{x}_2, \ldots, \bar{x}_n) \quad and \quad X_k = (x_{1k}, x_{2k}, \ldots, x_{nk}), \qquad k = 1, 2, \ldots.$$

Then $\lim_{k \to \infty} \mathbf{X}_k = \bar{\mathbf{X}}$ *if and only if*

$$\lim_{k \to \infty} x_{ik} = \bar{x}_i, \qquad 1 \le i \le n;$$

that is, a sequence $\{\mathbf{X}_k\}$ *of points in* $\mathcal{R}^n$ *converges to a limit* $\bar{\mathbf{X}}$ *if and only if the sequences of components of* $\{\mathbf{X}_k\}$ *converge to the respective components of* $\bar{\mathbf{X}}$.

Theorem 1.5 (Cauchy's Convergence Criterion)   *A sequence* $\{\mathbf{X}_r\}$ *of points in* $\mathcal{R}^n$ *converges if and only if for each* $\varepsilon > 0$ *there is an integer* K *such that*

$$|\mathbf{X}_r - \mathbf{X}_s| < \varepsilon \qquad \text{if} \quad r, s \ge K.$$

The next definition generalizes the definition of diameter of a circle or sphere.

Definition 1.7   *The diameter* $d(S)$ *of a nonempty set* S *is the least upper bound of the distances* $|\mathbf{X} - \mathbf{Y}|$, *where* $\mathbf{X}$ *and* $\mathbf{Y}$ *are in* S. *If* $d(S) < \infty$, S *is said to be bounded; if* $d(S) = \infty$, *it is unbounded.*

Theorem 1.6 (Principle of Nested Sets)   *If* $S_1, S_2, \dots$ *are closed nonempty sets in* $\mathcal{R}^n$ *such that*

$$S_1 \supset S_2 \supset \cdots \supset S_r \supset \cdots \tag{13}$$

and

$$\lim_{r \to \infty} d(S_r) = 0, \tag{14}$$

then the intersection

$$I = \bigcap_{r=1}^{\infty} S_r$$

contains exactly one point.

Proof.   The conclusion is obvious if S_m is finite for some m (why?), so we will assume that every S_r has infinitely many elements. In this case we can choose a sequence of distinct points $\{\mathbf{X}_r\}$ such that $\mathbf{X}_r \in S_r$ $(r = 1, 2, \dots)$. Because of (13), $\mathbf{X}_r \in S_k$ if $r \ge k$, and so

$$|\mathbf{X}_r - \mathbf{X}_s| < d(S_k) \qquad \text{if} \quad r, s \ge k.$$

Now (14) and Theorem 1.5 imply that $\{\mathbf{X}_r\}$ converges to a limit $\bar{\mathbf{X}}$. Since $\bar{\mathbf{X}}$ is a limit point of every S_k and every S_k is closed, $\bar{\mathbf{X}}$ is in every S_k (Corollary 3.1, Section 1.3); therefore $\bar{\mathbf{X}} \in I$, and we now know that $I \ne \varnothing$. Moreover, $\bar{\mathbf{X}}$ is the only point in I, since if $\mathbf{Y} \in I$, then

$$|\bar{\mathbf{X}} - \mathbf{Y}| \le d(S_k), \quad k = 1, 2, \dots,$$

and therefore $|\mathbf{X} - \mathbf{Y}| = 0$, from (14). We have now shown that I contains exactly one point, as was to be proved.

We can now prove the Heine–Borel theorem in $\mathscr{R}^n$. This theorem concerns *compact* sets; as in $\mathscr{R}$, a compact set in $\mathscr{R}^n$ is one that is closed and bounded.

Theorem 1.7 (Heine–Borel Theorem)   *If $\mathscr{H}$ is an open covering for a compact subset S of $\mathscr{R}^n$ (that is, $\mathscr{H}$ is a family of open sets such that $S \subset \cup \{H \,|\, H \in \mathscr{H}\}$), then S can be covered by finitely many sets from $\mathscr{H}$.*

Proof.   The proof is by contradiction. We present it first for $n = 2$, so that the method can be visualized. Suppose there is a covering $\mathscr{H}$ for S from which it is impossible to select a finite subcovering. Since S is bounded, it is contained in a closed square

$$T = \{(x, y) \,|\, a_1 \le x \le a_1 + L, a_2 \le x \le a_2 + L\}$$

with sides of length L (Figure 1.6). Bisecting the sides of T as shown by the dashed lines in Figure 1.6 leads to four closed squares, $T^{(1)}$, $T^{(2)}$, $T^{(3)}$, and $T^{(4)}$, with sides of length $L/2$. Let

$$S^{(i)} = S \cap T^{(i)}, \qquad 1 \le i \le 4.$$

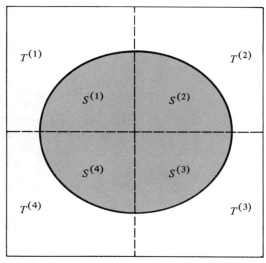

FIGURE 1.6

Each $S^{(i)}$, being the intersection of closed sets, is closed, and

$$S = \bigcup_{i=1}^{4} S^{(i)}.$$

Moreover, $\mathscr{H}$ covers each $S^{(i)}$, but at least one $S^{(i)}$ cannot be covered by any finite subfamily of $\mathscr{H}$, since if all the $S^{(i)}$ could be, then so could S. Let S_1 be a set with this property, chosen from $S^{(1)}$, $S^{(2)}$, $S^{(3)}$, and $S^{(4)}$. We are now back to the situation we started from—a compact set S_1 covered by $\mathscr{H}$, but not by any finite subfamily of $\mathscr{H}$—except that S_1 is contained in a square T_1 with sides of length $L/2$ instead of L. Bisecting the sides of T_1 and repeating the argument, we obtain a subset S_2 of S_1 which has the same properties as S, except that it is contained in a square with sides of length $L/4$. Continuing in this way leads to a sequence of nonempty closed sets S_0 $(=S)$, $S_1, S_2, \ldots$, such that $S_k \supset S_{k+1}$ and $d(S_k) \le L/2^{k-1/2}$ $(k = 0, 1, \ldots)$. From Theorem 1.6 there is a point $\bar{\mathbf{X}}$ in $\bigcap_{k=1}^{\infty} S_k$. Since $\bar{\mathbf{X}} \in S$, there is an open set H in $\mathscr{H}$ which contains $\bar{\mathbf{X}}$, and this H must also contain some ε-neighborhood of $\bar{\mathbf{X}}$. Since every element $\mathbf{X}$ in S_k satisfies the inequality

$$|\mathbf{X} - \bar{\mathbf{X}}| \le \frac{L}{2^{k-1/2}},$$

it follows that $S_k \subset H$ for k sufficiently large. This contradicts our assumption on $\mathscr{H}$, which led us to believe that no S_k could be covered by a finite number of sets from $\mathscr{H}$. Consequently, this assumption must be false: $\mathscr{H}$ must have a finite subfamily which covers S. This completes the proof for $n = 2$.

The idea of the proof is the same for $n > 2$. The counterpart of the square T is the "hypercube" with sides of length L,

$$T = \{(x_1, x_2, \ldots, x_n) \,|\, a_i \le x_i \le a_i + L, \, i = 1, 2, \ldots, n\}.$$

Halving the intervals of variation of the n coordinates $x_1, x_2, \ldots, x_n$ divides T into 2^n closed hypercubes with sides of length $L/2$:

$$T^{(i)} = \{(x_1, x_2, \ldots, x_n) \,|\, b_i \le x_i \le b_i + L/2, \, i = 1, 2, \ldots, n\},$$

where $b_i = a_i$ or $b_i = a_i + L/2$. If no finite subfamily of $\mathscr{H}$ covers S, then at least one of these smaller hypercubes must contain a subset of S which is not covered by any finite subfamily of S. The proof proceeds as for $n = 2$.

The Bolzano–Weierstrass theorem (every bounded infinite set has at least one limit point) is true in $\mathscr{R}^n$, and its proof is the same as in $\mathscr{R}$; see Theorem 3.4, Section 1.3.

Connected sets and regions

Although it is legitimate to consider functions defined on arbitrary domains, we restricted our study of functions of one variable mainly to functions defined on intervals. There are good reasons for this. If we wish to raise questions of continuity and differentiability at every point of the domain D of a function f, then every point of D must be a limit point of D^0. Intervals have this property. Moreover, the definition of $\int_a^b f(x)\, dx$ is obviously applicable only if f is defined on $[a, b]$.

Although we could sensibly consider questions of continuity and differentiability of functions defined on the union of disjoint intervals, it is not fruitful to do so, since many important results simply do not hold for such domains. For example, the intermediate value theorem (Theorem 2.6, Section 2.2; see also *Exercise 25, Section 2.2*) says that if f is continuous on an interval I and $f(x_1) < \mu < f(x_2)$ for some x_1 and x_2 in I, then $f(\bar{x}) = \mu$ for some $\bar{x}$ in I. Theorem 3.9, Section 2.3 says that f is constant on an interval I if $f' \equiv 0$ on I. Neither of these results holds if I is the union of disjoint intervals rather than a single interval; thus, if f is defined on $I = (0, 1) \cup (2, 3)$ by

$$f(x) = \begin{cases} 1, & 0 < x < 1, \\ 0, & 2 < x < 3, \end{cases}$$

then f is continuous on I, but does not assume any value between 0 and 1, and $f' \equiv 0$ on I, but f is not constant.

It is not difficult to see why these results fail to hold for this function: its domain consists of two disconnected pieces. It would be more sensible to regard f as two entirely different functions, one defined on $(0, 1)$ and the other on $(2, 3)$. The two results mentioned are valid for each of these functions.

As we will see when we study functions defined on subsets of $\mathscr{R}^n$, considerations like those just cited as making it natural to consider functions defined on intervals in $\mathscr{R}$ lead us to single out a preferred class of subsets as domains of functions of n variables. These subsets are called *regions*. To define this term, we first need the following definition.

Definition 1.8   *A subset S of $\mathscr{R}^n$ is said to be connected if it is impossible to represent S as the union of two disjoint nonempty sets such that neither contains a limit point of the other; that is, if S cannot be expressed as $S = A \cup B$, where*

$$A \neq \varnothing, \quad B \neq \varnothing, \quad \bar{A} \cap B = \varnothing, \quad \text{and} \quad A \cap \bar{B} = \varnothing. \tag{15}$$

If S can be expressed in this way, then S is said to be disconnected.

Example 1.8   The empty set and singleton sets are connected, because they cannot be represented as the union of two disjoint nonempty sets.

Example 1.9   The space $\mathscr{R}^n$ is connected, because if $\mathscr{R}^n = A \cup B$ with $\bar{A} \cap B = \varnothing$ and $A \cap \bar{B} = \varnothing$, then $\bar{A} \subset A$ and $\bar{B} \subset B$; that is, A and B are both closed, and therefore both open. The only nonempty subset of $\mathscr{R}^n$ that is both open and closed is $\mathscr{R}^n$ itself (*Exercise 23*); hence one of A and B is $\mathscr{R}^n$ and the other is empty.

If $\mathbf{X}_1, \mathbf{X}_2, \ldots, \mathbf{X}_k$ are points in $\mathscr{R}^n$ and L_i is the line segment from $\mathbf{X}_i$ to $\mathbf{X}_{i+1}$ ($i = 1, 2, \ldots, k-1$), we say that $L_1, L_2, \ldots, L_{k-1}$ form a *polygonal path* from $\mathbf{X}_1$ to $\mathbf{X}_k$, and that $\mathbf{X}_1$ and $\mathbf{X}_k$ are *connected* by the polygonal path. For example, Figure 1.7 shows a polygonal path in $\mathscr{R}^2$ connecting $(0, 0)$ to $(3, 3)$. A set S is polygonally connected if every pair of points in S can be connected by a polygonal path lying entirely in S.

Theorem 1.8   *An open set S in $\mathscr{R}^n$ is connected if and only if it is polygonally connected.*

Proof.   For sufficiency we will show that if S is disconnected, then it is not polygonally connected. Let $S = A \cup B$, where A and B satisfy (15). Suppose $\mathbf{X}_1 \in A$ and $\mathbf{X}_2 \in B$, and assume that there is a polygonal path in S connecting $\mathbf{X}_1$ to $\mathbf{X}_2$. Then some line segment L in this path must contain a point $\mathbf{Y}_1$ in A and a point $\mathbf{Y}_2$ in B. The line segment

$$\mathbf{X} = t\mathbf{Y}_2 + (1 - t)\mathbf{Y}_1, \qquad 0 \le t \le 1,$$

is part of L and therefore in S. Now define

$$\rho = \text{l.u.b.}\{\tau \mid t\mathbf{Y}_2 + (1 - t)\mathbf{Y}_1 \in A, 0 \le t \le \tau \le 1\},$$

and let

$$\mathbf{X}_\rho = \rho\mathbf{Y}_2 + (1 - \rho)\mathbf{Y}_1.$$

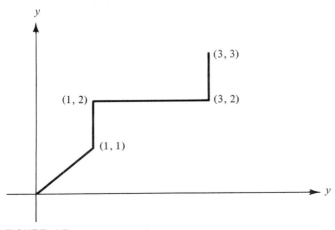

FIGURE 1.7

Then $\mathbf{X}_\rho \in \bar{A} \cap \bar{B}$. However, since $\mathbf{X}_\rho \in A \cup B$ and $\bar{A} \cap B = A \cap \bar{B} = \varnothing$, this is impossible. Because of this contradiction, the assumption that there is a polygonal path in S from $\mathbf{X}_1$ to $\mathbf{X}_2$ must be false. This completes the sufficiency half of the proof.

For necessity, suppose S is a connected open set and $\mathbf{X}_0 \in S$. Let A be the set consisting of $\mathbf{X}_0$ and the points in S that can be connected to $\mathbf{X}_0$ by polygonal paths in S; let B be the other points in S. If $\mathbf{Y}_0 \in S$, there is an ε-neighborhood $N_\varepsilon(\mathbf{Y}_0)$ which is contained in S (because S is open). Any point $\mathbf{Y}_1$ in $N_\varepsilon(\mathbf{Y}_0)$ can be connected to $\mathbf{Y}_0$ by the line segment

$$\mathbf{X} = t\mathbf{Y}_1 + (1 - t)\mathbf{Y}_0, \qquad 0 \le t \le 1,$$

which lies in $N_\varepsilon(\mathbf{Y}_0)$ (Lemma 1.2), and so in S. This implies that $\mathbf{Y}_0$ can be connected to $\mathbf{X}_0$ by a polygonal path in S if and only if every element in $N_\varepsilon(\mathbf{Y}_0)$ can also. Thus, $N_\varepsilon(\mathbf{Y}_0) \subset A$ if $\mathbf{Y}_0 \in A$, and $N_\varepsilon(\mathbf{Y}_0) \subset B$ if $\mathbf{Y}_0 \in B$. Therefore, A and B are open. Since $A \cap B = \varnothing$, this implies that $A \cap \bar{B} = \bar{A} \cap B = \varnothing$ (*Exercise 16*). Since A is nonempty ($\mathbf{X}_0 \in A$), it now follows that $B = \varnothing$, since if $B \neq \varnothing$, S would be disconnected (Definition 1.8); therefore, $A = S$, which completes the proof.

In the proof of sufficiency we did not use the assumption that S is open; in fact, we actually proved that any polygonally connected set—open or not—is connected. The converse is false. A set (not open) may be connected but not polygonally connected (*Exercise 31*).

Our study of functions on $\mathscr{R}^n$ will deal mostly with functions whose domains are *regions*, defined next.

Definition 1.9   *A region S in $\mathscr{R}^n$ is the union of an open connected set with some, all, or none of its boundary; thus, S^0 is connected and every point of S is a limit point of S^0.*

Example 1.10   The intervals are the only regions in $\mathscr{R}$ (*Exercise 33*). The n-sphere $S_r(\mathbf{X}_0)$ (Example 1.7) is a region in $\mathscr{R}^n$, as is its closure $\bar{S}_r(\mathbf{X}_0)$. The set

$$S = \{(x, y) \mid x^2 + y^2 \le 1 \quad \text{or} \quad x^2 + y^2 \ge 4\}$$

[Figure 1.8(a)] is not a region in $\mathscr{R}^2$, since it is not connected. The set S_1 obtained by adding the line segment

$$L_1: \quad \mathbf{X} = t(0, 2) + (1 - t)(0, 1), \qquad 0 < t < 1,$$

to S [Figure 1.8(b)], although connected, is not a region, since points on the line segment are not limit points of S_1^0. The set S_2 obtained by adding to S_1 the

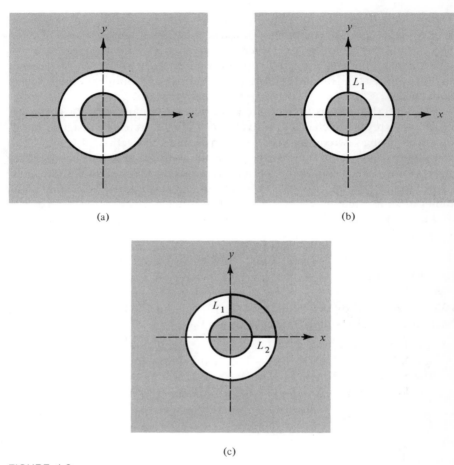

(a)

(b)

(c)

FIGURE 1.8

points bounded by the circles $x^2 + y^2 = 1$ and $x^2 + y^2 = 4$ and the line segments L_1 and

$$L_2: \quad X = t(2, 0) + (1 - t)(1, 0), \qquad 0 < t < 1$$

[Figure 1.8(c)], is a region.

More about sequences in $\mathscr{R}^n$

From Definition 1.6 a sequence $\{X_k\}$ of points in $\mathscr{R}^n$ converges to a limit $\bar{X}$ if and only if for every $\varepsilon > 0$ there is an integer K such that

$$|X_k - \bar{X}| < \varepsilon \qquad \text{if} \quad k \geq K.$$

The $\mathscr{R}^n$ definitions of divergence, boundedness, subsequence, and sums, differences, and constant multiples of sequences are analogous to those given in Sections 4.1 and 4.2 for the case where $n = 1$. Since $\mathscr{R}^n$ is not ordered for $n > 1$, monotonicity, limit inferior and superior of sequences in $\mathscr{R}^n$, and divergence to $\pm\infty$ are left undefined for this case. Products and quotients are also left undefined.

Several theorems from Sections 4.1 and 4.2 remain valid for sequences in $\mathscr{R}^n$, with proofs unchanged, provided $|\ \ |$ is interpreted as distance in $\mathscr{R}^n$. (A trivial change is required: the subscript n, used in Sections 4.1 and 4.2 to identify the elements of the sequence, must be replaced here by k, since n here stands for the dimension of the space.) These include Theorems 1.1 (uniqueness of the limit), 1.2 (boundedness of a convergent sequence), parts of 1.5 (concerning limits of sums, differences, and constant multiples of convergent sequences), and a slightly modified 2.1 (every subsequence of a convergent sequence converges to the limit of the sequence).

5.1 EXERCISES

(With $\mathscr{R}$ replaced by $\mathscr{R}^n$, Exercises 5–8, 10–13, and 16–19—except 18(e)—of Section 1.3 are also suitable for this section.)

1. Find $a\mathbf{X} + b\mathbf{Y}$:
   (a) $\mathbf{X} = (1, 2, -3, 1)$, $\mathbf{Y} = (0, -1, 2, 0)$, $a = 3, b = 6$
   (b) $\mathbf{X} = (1, -1, 2)$, $\mathbf{Y} = (0, -1, 3)$, $a = -1, b = 2$
   (c) $\mathbf{X} = (\frac{1}{2}, \frac{3}{2}, \frac{1}{4}, \frac{1}{6})$, $\mathbf{Y} = (-\frac{1}{2}, 1, 5, \frac{1}{3})$, $a = \frac{1}{2}, b = \frac{1}{6}$

2. Prove Theorem 1.1.

3. Find $|\mathbf{X}|$
   (a) $(1, 2, -3, 1)$ (b) $(\frac{1}{2}, \frac{1}{3}, \frac{1}{4}, \frac{1}{6})$
   (c) $(1, 2, -1, 3, 4)$ (d) $(0, 1, 0, -1, 0, -1)$

4. Find $|\mathbf{X} - \mathbf{Y}|$:
   (a) $\mathbf{X} = (3, 4, 5, -4)$, $\mathbf{Y} = (2, 0, -1, 2)$
   (b) $\mathbf{X} = (-\frac{1}{2}, \frac{1}{2}, \frac{1}{4}, -\frac{1}{4})$, $\mathbf{Y} = (\frac{1}{3}, -\frac{1}{6}, \frac{1}{6}, -\frac{1}{3})$
   (c) $\mathbf{X} = (0, 0, 0)$, $\mathbf{Y} = (2, -1, 2)$
   (d) $\mathbf{X} = (3, -1, 4, 0, -1)$, $\mathbf{Y} = (2, 0, 1, -4, 1)$

5. Find $\mathbf{X} \cdot \mathbf{Y}$:
   (a) $\mathbf{X} = (3, 4, 5, -4)$, $\mathbf{Y} = (3, 0, 3, 3)$
   (b) $\mathbf{X} = (\frac{1}{6}, \frac{11}{12}, \frac{9}{8}, \frac{5}{2})$, $\mathbf{Y} = (-\frac{1}{2}, \frac{1}{2}, \frac{1}{4}, -\frac{1}{4})$
   (c) $\mathbf{X} = (1, 2, -3, 1, 4)$, $\mathbf{X} = (1, 2, -1, 3, 4)$

6. Show that $|\mathbf{X} \cdot \mathbf{Y}| = |\mathbf{X}|\,|\mathbf{Y}|$ if and only if one of $\mathbf{X}$ and $\mathbf{Y}$ is a scalar multiple of the other.

7. Show that $|X + Y| = |X| + |Y|$ if and only if one of X and Y is a nonnegative scalar multiple of the other. (*Hint:* See Exercise 6.)

8. Prove Theorem 1.3.

9. Find a parametric equation of the line through X_0 in the direction of U:
   (a) $X_0 = (1, 2, -3, 1)$, $U = (3, 4, 5, -4)$
   (b) $X_0 = (2, 0, -1, 2, 4)$, $U = (-1, 0, 1, 3, 2)$
   (c) $X_0 = (-\frac{1}{2}, \frac{1}{2}, \frac{1}{4}, -\frac{1}{4})$, $U = (\frac{1}{3}, -\frac{1}{6}, \frac{1}{6}, -\frac{1}{3})$

10. Complete the sentence: The equations
    $$X = X_0 + tU, \qquad -\infty < t < \infty,$$
    and
    $$X = X_1 + sV, \qquad -\infty < s < \infty,$$
    where $U \neq 0$ and $V \neq 0$, represent the same line in $\mathcal{R}^n$ if and only if . . . .

11. Find the equation of the line segment from X_0 to X_1:
    (a) $X_0 = (1, -3, 4, 2)$, $X_1 = (2, 0, -1, 5)$
    (b) $X_0 = (3, 1, -2, 1, 4)$, $X_1 = (2, 0, -1, 4, -3)$
    (c) $X_0 = (1, 2, -1)$, $X_1 = (0, -1, -1)$

12. Find l.u.b.$\{\varepsilon | N_\varepsilon(X_0) \subset S\}$:
    (a) $X_0 = (1, 2, -1, 3)$; $S =$ the open 4-sphere of radius 7 about $(0, 3, -2, 2)$
    (b) $X_0 = (1, 2, -1, 3)$; $S = \{(x_1, x_2, x_3, x_4) | |x_i| \leq 5, 1 \leq i \leq 4\}$
    (c) $X_0 = (3, \frac{5}{2})$; $S =$ the closed triangle with vertices $(2, 0)$, $(2, 2)$, and $(4, 4)$.

13. Find: (i) ∂S; (ii)$\bar{S}$; (iii) S^0; (iv) exterior of S:
    (a) $S = \{(x_1, x_2, x_3, x_4) | |x_i| < 3, i = 1, 2, 3\}$
    (b) $S = \{(x, y, 1) | x^2 + y^2 \leq 1\}$

14. Describe the following sets as open, closed, or neither:
    (a) $S = \{(x_1, x_2, x_3, x_4) | |x_1| > 0, x_2 < 1, x_3 \neq -2\}$
    (b) $S = \{(x_1, x_2, x_3, x_4) | x_1 = 1, x_3 \neq -4\}$
    (c) $S = \{(x_1, x_2, x_3, x_4) | x_1 = 1, -3 \leq x_2 \leq 1, x_4 = -5\}$

15. Show that the closure of the open n-sphere
    $$S_r(X_0) = \{X | |X - X_0| < r\}$$
    is
    $$\bar{S}_r(X_0) = \{X | |X - X_0| \leq r\}.$$

16. Prove: If A and B are open and $A \cap B = \emptyset$, then $A \cap \bar{B} = \bar{A} \cap B = \emptyset$.

17. Show that if $\lim_{r \to \infty} X_r$ exists, then it is unique.

18. Prove Theorem 1.4.

19. Prove Theorem 1.5.

20. Find $\lim_{r \to \infty} \mathbf{X}_r$:
    (a) $\mathbf{X}_r = (r \sin \pi/r, \cos \pi/r, e^{-r})$
    (b) $\mathbf{X}_r = (1 - 1/r^2, \log(r + 1)/(r + 2), (1 + 1/r)^r)$

21. Find $d(S)$:
    (a) $S = \{(x, y, z) \mid |x| \le 2, |y| \le 1, |z - 2| \le 2\}$

    (b) $S = \left\{(x, y) \left| \frac{(x - 1)^2}{9} + \frac{(y - 2)^2}{4} = 1 \right. \right\}$

    (c) $S = $ the triangle in $\mathscr{R}^2$ with vertices $(2, 0)$, $(2, 2)$, and $(4, 4)$
    (d) $S = \{(x_1, x_2, \ldots, x_n) \mid |x_i| \le L, i = 1, 2, \ldots, n\}$
    (e) $S = \{(x, y, z) \mid x \ne 0, |y| \le 1, z > 2\}$

22. Prove that $d(S) = d(\bar{S})$ for any set S in $\mathscr{R}^n$.

23. Prove: If a nonempty subset S of $\mathscr{R}^n$ is both open and closed, then $S = \mathscr{R}^n$.

24. Use the Bolzano–Weierstrass theorem to show that if $S_1, S_2, \ldots$ is an infinite sequence of nonempty compact sets and $S_1 \supset S_2 \supset \cdots \supset S_n \supset \cdots$, then $\bigcap_{n=1}^{\infty} S_n$ is nonempty. Show that the conclusion does not follow if the sets are assumed to be closed rather than compact.

25. Suppose a sequence $U_1, U_2, \ldots$ of open sets covers a compact set S. Without using the Heine–Borel theorem, show that $S \subset \bigcup_{m=1}^{N} U_m$ for some N. [*Hint:* Apply Exercise 24 to the sets $S_n = S \cap (\bigcup_{m=1}^{n} U_m)^c$.]
    This is a seemingly restricted version of the Heine–Borel theorem, valid for the case where the covering family $\mathscr{H}$ is denumerable; however, it can be shown that there is no loss of generality in assuming that this is so.

26. The *distance from a point* $\mathbf{X}_0$ *to a set* S is defined by

    $$\text{dist}(\mathbf{X}_0, S) = \text{g.l.b.}\{|\mathbf{X} - \mathbf{X}_0| \mid \mathbf{X} \in S\}.$$

    (a) Prove: If S is closed and $\mathbf{X}_0 \in \mathscr{R}^n$, there is a point $\mathbf{X}$ in S such that

    $$|\mathbf{X} - \mathbf{X}_0| = \text{dist}(\mathbf{X}_0, S).$$

    (*Hint:* Use Exercise 24.)
    (b) Conclude from part (a) that if S is closed and $\mathbf{X}_0 \notin S$, then $\text{dist}(\mathbf{X}_0, S) > 0$.
    (c) Show that the conclusions of parts (a) and (b) may fail to hold if S is not closed.

*27. The *distance between two sets* S *and* T is defined by

    $$\text{dist}(S, T) = \text{g.l.b.}\{|\mathbf{X} - \mathbf{Y}| \mid \mathbf{X} \in S, \mathbf{Y} \in T.\}$$

    (a) Prove: If S and T are closed there are points $\mathbf{X}$ in S and $\mathbf{Y}$ in T such that

    $$|\mathbf{X} - \mathbf{Y}| = \text{dist}(S, T).$$

    (*Hint:* Use Exercises 24 and 26.)

(b) Conclude from part (a) that if S and T are closed and $S \cap T = \varnothing$, then dist$(S, T) > 0$.

(c) Show that the conclusions of parts (a) and (b) may fail to hold if S or T is not closed.

*28. (a) Prove: If a compact set S is contained in an open set U, there is a positive number r such that the set

$$S_r = \{\mathbf{X} \mid \text{dist}(\mathbf{X}, S) \le r\}$$

is contained in U. (You will need Exercise 26 here.)

(b) Show that S_r is compact.

*29. Let D_1 and D_2 be compact subsets of $\mathscr{R}^n$. Show that

$$D = \{(\mathbf{X}, \mathbf{Y}) \mid \mathbf{X} \in D_1, \ \mathbf{Y} \in D_2\}$$

is a compact subset of $\mathscr{R}^{2n}$.

30. Prove: If S is open and $S = A \cup B$, where $A \cap B = \bar{A} \cap B = A \cap \bar{B} = \varnothing$, then A and B are open.

31. Give an example of a connected set in $\mathscr{R}^2$ that is not polygonally connected.

*32. Prove that a region is connected.

33. Show that the intervals are the only regions in $\mathscr{R}$.

*34. Prove: A bounded sequence in $\mathscr{R}^n$ has a convergent subsequence. (*Hint:* Use Theorem 1.4 and Theorems 2.1 and 2.4(a), Section 4.2.)

35. Define "$\lim_{k \to \infty} \mathbf{X}_k = \infty$" if $\{\mathbf{X}_k\}$ is a sequence in $\mathscr{R}^n$, with $n \ge 2$.

5.2   CONTINUOUS REAL-VALUED FUNCTIONS OF n VARIABLES

We now study real-valued functions of n variables. We will denote the domain of a function f by D_f, and its value at a point $\mathbf{X} = (x_1, x_2, \ldots, x_n)$ by $f(\mathbf{X})$ or $f(x_1, x_2, \ldots, x_n)$. We continue the convention adopted in Section 2.1 for functions of one variable: If a function is defined by a formula such as

$$f(\mathbf{X}) = (1 - x_1^2 - x_2^2 - \cdots - x_n^2)^{1/2} \tag{1}$$

or

$$g(\mathbf{X}) = (1 - x_1^2 - x_2^2 - \cdots - x_n^2)^{-1} \tag{2}$$

without specification of its domain, it is to be understood that its domain is the largest subset of $\mathscr{R}^n$ for which the formula defines a unique real number. Thus, in the absence of any other stipulation, the domain of f in (1) is the closed n-sphere $\{\mathbf{X} \mid \|\mathbf{X}\| \le 1\}$, while the domain of g in (2) is the set $\{\mathbf{X} \mid \|\mathbf{X}\| \ne 1\}$.

The main objective of this section is to study limits and continuity of functions of n variables. The proofs of many of the theorems here are similar to the proofs of their counterparts in Sections 2.1 and 2.2. We leave most of them to the reader.

Definition 2.1   *We say that* $f(\mathbf{X})$ *approaches the limit* L *as* $\mathbf{X}$ *approaches* $\mathbf{X}_0$, *and write*

$$\lim_{\mathbf{X} \to \mathbf{X}_0} f(\mathbf{X}) = L,$$

if $\mathbf{X}_0$ *is a limit point of* D_f *and for every* $\varepsilon > 0$ *there is a* $\delta > 0$ *such that*

$$|f(\mathbf{X}) - L| < \varepsilon$$

for all $\mathbf{X}$ *in* D_f *that satisfy the inequality*

$$0 < |\mathbf{X} - \mathbf{X}_0| < \delta$$

Example 2.1   If

$$g(x, y) = 1 - x^2 - 2y^2,$$

then

$$\lim_{(x,y) \to (x_0, y_0)} g(x, y) = 1 - x_0^2 - 2y_0^2 \tag{3}$$

for every (x_0, y_0). To see this, we write

$$
\begin{aligned}
|g(x, y) - (1 - x_0^2 - 2y_0^2)| &= |(1 - x^2 - 2y^2) - (1 - x_0^2 - 2y_0^2)| \\
&= |x_0^2 - x^2 + 2(y_0^2 - y^2)| \\
&= |(x_0 + x)(x_0 - x) + 2(y_0 + y)(y_0 - y)| \quad (4) \\
&\le [(x_0 + x)^2 + 4(y_0 + y)^2]^{1/2} \\
&\quad \times [(x - x_0)^2 + (y - y_0)^2]^{1/2}
\end{aligned}
$$

(by Schwarz's inequality). If $|\mathbf{X} - \mathbf{X}_0| < 1$, then $|x| < |x_0| + 1$ and $|y| < |y_0| + 1$; this and (4) imply that

$$|g(x, y) - (1 - x_0^2 - 2y_0^2)| < K|\mathbf{X} - \mathbf{X}_0| \quad \text{if} \quad |\mathbf{X} - \mathbf{X}_0| < 1,$$

where

$$K = [(2|x_0| + 1)^2 + 4(2|y_0| + 1)^2]^{1/2}.$$

Therefore, if $\varepsilon > 0$ and

$$|\mathbf{X} - \mathbf{X}_0| < \delta = \min\{1, \varepsilon/K\},$$

then
$$\left| g(x, y) - (1 - x_0^2 - 2y_0^2) \right| < \varepsilon.$$

This proves (3).

Notice that Definition 2.1 does not require that f be defined at $\mathbf{X}_0$, or even on a deleted neighborhood of $\mathbf{X}_0$.

Example 2.2 The function
$$h(x, y) = \frac{\sin \sqrt{1 - x^2 - 2y^2}}{\sqrt{1 - x^2 - 2y^2}}$$

is defined on the interior of the region bounded by the ellipse
$$x^2 + 2y^2 = 1$$

[Figure 2.1(a)]; it is not defined at any point of the ellipse itself, nor on any deleted neighborhood of such a point. Nevertheless,

$$\lim_{(x,y)\to(x_0,y_0)} h(x, y) = 1 \qquad (5)$$

if

$$x_0^2 + 2y_0^2 = 1. \qquad (6)$$

To see this, we recall that

$$\lim_{r\to 0} \frac{\sin r}{r} = 1;$$

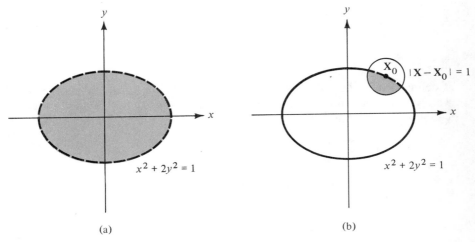

(a)   (b)

FIGURE 2.1

therefore, if $\varepsilon > 0$, there is a $\delta_1 > 0$ such that

$$\left| \frac{\sin u}{u} - 1 \right| < \varepsilon \qquad \text{if} \quad 0 < |u| < \delta_1. \tag{7}$$

From Example 2.1 [specifically, from (3)],

$$\lim_{(x,y)\to(x_0,y_0)} (1 - x^2 - 2y^2) = 0$$

if (6) holds, and so there is a $\delta > 0$ such that

$$0 < (1 - x^2 - 2y^2) < \delta_1^2,$$

and therefore

$$0 < u = \sqrt{1 - x^2 - 2y^2} < \delta_1, \tag{8}$$

if $\mathbf{X}$ is in the interior of the ellipse and $|\mathbf{X} - \mathbf{X}_0| < \delta$, that is, if $\mathbf{X}$ is in the shaded region of Figure 2.1(b). From (7) and (8),

$$|h(x, y) - 1| < \varepsilon$$

if $\mathbf{X}$ is in this region. This proves (5).

We leave the proof of the following theorem to the reader (*Exercise 2*).

Theorem 2.1   *If* $\lim_{\mathbf{X}\to\mathbf{X}_0} f(\mathbf{X})$ *exists, then it is unique.*

When investigating whether a function has a limit at a point $\mathbf{X}_0$, no restriction can be made on the way in which $\mathbf{X}$ approaches $\mathbf{X}_0$, except that $\mathbf{X}$ must be in D_f. The next example shows that incorrect restrictions can lead to incorrect conclusions.

Example 2.3   The function

$$f(x, y) = \frac{xy}{x^2 + y^2}$$

is defined everywhere in $\mathcal{R}^2$ except at $(0, 0)$. Does $\lim_{(x,y)\to(0,0)} f(x, y)$ exist? If we try to answer this question by letting (x, y) approach $(0, 0)$ along the line $y = x$, we see the functional values

$$f(x, x) = \frac{x^2}{2x^2} = \frac{1}{2}$$

and conclude that the limit is $\frac{1}{2}$; if we let (x, y) approach $(0, 0)$ along the line $y = -x$, we see the functional values

$$f(x, -x) = \frac{-x^2}{2x^2} = -\frac{1}{2},$$

and conclude that it equals $-\frac{1}{2}$. Theorem 2.1 implies that these two conclusions cannot both be correct; in fact, they are both incorrect. What we have shown is that

$$\lim_{x \to 0} f(x, x) = \frac{1}{2} \quad \text{and} \quad \lim_{x \to 0} f(x, -x) = -\frac{1}{2}.$$

Since $\lim_{x \to 0} f(x, x)$ and $\lim_{x \to 0} f(x, -x)$ must both equal $\lim_{(x,y) \to (0,0)} f(x, y)$ if the latter exists [*Exercise 3(a)*], we must conclude that the latter does not exist.

The sum, difference, and product of functions of *n* variables are defined in the same way as they are for functions of one variable (Definition 1.1, Section 2.1), and the proof of the next theorem is the same as that of Theorem 1.2, Section 2.1. We leave it to the reader (*Exercise 4*).

Theorem 2.2   *Suppose f and g are defined on a set D*, $\mathbf{X}_0$ *is a limit point of D, and*

$$\lim_{\mathbf{X} \to \mathbf{X}_0} f(\mathbf{X}) = L_1, \qquad \lim_{\mathbf{X} \to \mathbf{X}_0} g(\mathbf{X}) = L_2.$$

Then

$$\lim_{\mathbf{X} \to \mathbf{X}_0} (f + g)(\mathbf{X}) = L_1 + L_2; \tag{9}$$

$$\lim_{\mathbf{X} \to \mathbf{X}_0} (f - g)(\mathbf{X}) = L_1 - L_2; \tag{10}$$

$$\lim_{\mathbf{X} \to \mathbf{X}_0} (fg)(\mathbf{X}) = L_1 L_2; \tag{11}$$

and, if $L_2 \neq 0$,

$$\lim_{\mathbf{X} \to \mathbf{X}_0} \left(\frac{f}{g} \right)(\mathbf{X}) = \frac{L_1}{L_2}. \tag{12}$$

Infinite limits and limits as $|\mathbf{X}| \to \infty$

Definition 2.2   *We say that* $f(\mathbf{X})$ *approaches* ∞ *as* $\mathbf{X}$ *approaches* $\mathbf{X}_0$, *and write*

$$\lim_{\mathbf{X} \to \mathbf{X}_0} f(\mathbf{X}) = \infty,$$

if $\mathbf{X}_0$ *is a limit point of* D_f *and for every real number M there is a* $\delta > 0$ *such that*

$$|f(\mathbf{X})| > M \quad \text{whenever} \quad 0 < |\mathbf{X} - \mathbf{X}_0| < \delta \quad \text{and} \quad \mathbf{X} \in D_f.$$

We say that

$$\lim_{\mathbf{X} \to \mathbf{X}_0} f(\mathbf{X}) = -\infty$$

if $\lim_{\mathbf{X} \to \mathbf{X}_0} (-f)(\mathbf{X}) = \infty$.

Example 2.4   If

$$f(\mathbf{X}) = (1 - x_1^2 - x_2^2 - \cdots - x_n^2)^{-1/2},$$

then

$$\lim_{\mathbf{X} \to \mathbf{X}_0} f(\mathbf{X}) = \infty$$

if $|\mathbf{X}_0| = 1$, because in this case we can write

$$f(\mathbf{X}) = \frac{1}{|\mathbf{X} - \mathbf{X}_0|},$$

and so

$$f(\mathbf{X}) > M \qquad \text{if} \quad 0 < |\mathbf{X} - \mathbf{X}_0| < \delta = \frac{1}{M}.$$

Example 2.5   If

$$f(x, y) = \frac{1}{x + 2y + 1},$$

then $\lim_{(x,y) \to (1, -1)} f(x, y)$ does not exist (why not?), but

$$\lim_{(x,y) \to (1, -1)} |f(x, y)| = \infty.$$

To see this, we observe that

$$|x + 2y + 1| = |(x - 1) + 2(y + 1)|$$
$$\leq \sqrt{5}|\mathbf{X} - \mathbf{X}_0| \qquad \text{(by Schwarz's inequality),}$$

where $\mathbf{X}_0 = (1, -1)$, and so

$$|f(x, y)| \geq \frac{1}{|x + 2y + 1|} \geq \frac{1}{\sqrt{5}|\mathbf{X} - \mathbf{X}_0|}.$$

Therefore,

$$|f(x, y)| > M \qquad \text{if} \quad 0 < |\mathbf{X} - \mathbf{X}_0| < \frac{1}{M\sqrt{5}}.$$

Example 2.6   The function

$$f(x, y, z) = \frac{\left| \sin\left(\dfrac{1}{x^2 + y^2 + z^2}\right) \right|}{x^2 + y^2 + z^2}$$

assumes arbitrarily large values in every neighborhood of $(0, 0, 0)$; for example, if $\mathbf{X}_k = (x_k, y_k, z_k)$, where

$$x_k = y_k = z_k = \frac{1}{\sqrt{3(k + \frac{1}{2})\pi}},$$

then

$$f(\mathbf{X}_k) = (k + \tfrac{1}{2})\pi.$$

However, this does not imply that $\lim_{\mathbf{X} \to \mathbf{0}} f(\mathbf{X}) = \infty$, since, for example, every neighborhood of $(0, 0, 0)$ also contains points

$$\bar{\mathbf{X}}_k = \left(\frac{1}{\sqrt{3k\pi}}, \frac{1}{\sqrt{3k\pi}}, \frac{1}{\sqrt{3k\pi}} \right)$$

for which $f(\bar{\mathbf{X}}_k) = 0$.

Definition 2.3   *If D_f is unbounded, we say that*

$$\lim_{|\mathbf{X}| \to \infty} f(\mathbf{X}) = L \qquad (\textit{finite})$$

if for every $\varepsilon > 0$ there is a number R such that

$$|f(\mathbf{X}) - L| < \varepsilon \qquad \textit{whenever } |\mathbf{X}| \geq R \textit{ and } \mathbf{X} \in D_f.$$

Example 2.7   If

$$f(x, y, z) = \cos\left(\frac{1}{x^2 + 2y^2 + z^2} \right),$$

then

$$\lim_{|\mathbf{X}| \to \infty} f(\mathbf{X}) = 1. \tag{13}$$

To see this, we recall that the continuity of $\cos u$ at $u = 0$ implies that for each $\varepsilon > 0$ there is a $\delta > 0$ such that

$$|\cos u - 1| < \varepsilon \quad \text{if} \quad |u| < \delta.$$

Since

$$\frac{1}{x^2 + 2y^2 + z^2} \leq \frac{1}{|\mathbf{X}|^2},$$

it follows that

$$\frac{1}{x^2 + 2y^2 + z^2} < \delta,$$

and therefore

$$|f(\mathbf{X}) - 1| < \varepsilon,$$

if $|\mathbf{X}| > 1/\sqrt{\delta}$. This proves (13).

Example 2.8   Consider the function defined *only* on the domain

$$D = \{(x, y)\,|\,0 \le y \le ax\}, \qquad 0 < a < 1$$

(Figure 2.2), by

$$f(x, y) = \frac{1}{x - y}.$$

In D,

$$x - y \ge x(1 - a) \tag{14}$$

and

$$|\mathbf{X}|^2 = x^2 + y^2 \le x^2(1 + a^2),$$

so that

$$x \ge \frac{|\mathbf{X}|}{\sqrt{1 + a^2}}.$$

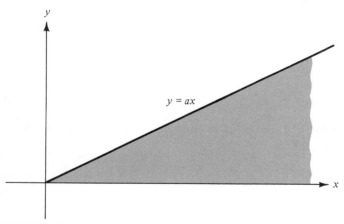

y

$y = ax$

x

FIGURE 2.2

This and (14) imply that

$$x - y \geq \frac{1 - a}{\sqrt{1 + a^2}} |\mathbf{X}|, \qquad \mathbf{X} \in D,$$

and so

$$|f(x, y)| \leq \frac{\sqrt{1 + a^2}}{1 - a} \frac{1}{|\mathbf{X}|}, \qquad \mathbf{X} \in D.$$

Therefore,

$$|f(x, y)| < \varepsilon$$

if $\mathbf{X} \in D$ and

$$|\mathbf{X}| > \frac{\sqrt{1 + a^2}}{1 - a} \frac{1}{\varepsilon}.$$

This implies that $\lim_{|\mathbf{X}| \to \infty} f(x, y) = 0$.

We leave it to the reader to define

$$\lim_{|\mathbf{X}| \to \infty} f(\mathbf{X}) = \infty \quad \text{and} \quad \lim_{|\mathbf{X}| \to \infty} f(\mathbf{X}) = -\infty$$

(*Exercise 7*).

We will continue the convention adopted in Section 2.1: "$\lim_{\mathbf{X} \to \mathbf{X}_0} f(\mathbf{X})$ exists" means that $\lim_{\mathbf{X} \to \mathbf{X}_0} f(\mathbf{X}) = L$, where L is finite; to leave open the possibility that $L = \pm \infty$, we will say that "$\lim_{\mathbf{X} \to \mathbf{X}_0} f(\mathbf{X})$ exists in the extended reals." A similar convention applies to limits as $|\mathbf{X}| \to \infty$.

Theorem 2.2 remains valid if "$\lim_{\mathbf{X} \to \mathbf{X}_0}$" is replaced by "$\lim_{|\mathbf{X}| \to \infty}$" provided the set D is unbounded (*Exercise 10*). Moreover, (9), (10), and (11) are valid in either version of Theorem 2.2 if either or both of L_1 and L_2 is infinite, provided their right sides are not indeterminate, and (12) remains valid if $L_2 \neq 0$ and L_1/L_2 is not indeterminate.

Continuity

We now define continuity for functions of n variables. The definition is, of course, quite similar to the definition for functions of one variable.

Definition 2.4   *A function f defined on a domain D in $\mathscr{R}^n$ is said to be continuous at a point $\mathbf{X}_0$ in D if*

$$\lim_{\mathbf{X} \to \mathbf{X}_0} f(\mathbf{X}) = f(\mathbf{X}_0).$$

The next theorem follows immediately from this and Definition 2.1. We leave the proof to the reader (*Exercise 11*).

Theorem 2.3   *Suppose* X_0 *is in—and a limit point of—D_f. Then f is continuous at X_0 if and only if for each $\varepsilon > 0$ there is a $\delta > 0$ such that*

$$|f(X) - f(X_0)| < \varepsilon$$

whenever

$$|X - X_0| < \delta \quad and \quad X \in D_f.$$

In applying this theorem when $X_0 \in D_f^0$, we will usually omit "and $X \in D_f$," it being understood that $S_\delta(X_0) \subset D_f$.

We will say that *f* is *continuous on S* if *f* is continuous at every point of *S*.

Example 2.9   From Example 2.1 we now see that the function

$$f(x, y) = 1 - x^2 - 2y^2$$

is continuous on $\mathscr{R}^2$.

Example 2.10   If we extend the definition of *h* in Example 2.2 so that

$$h(x, y) = \begin{cases} \dfrac{\sin\sqrt{1 - x^2 - 2y^2}}{\sqrt{1 - x^2 - 2y^2}}, & x^2 + 2y^2 < 1 \\[2mm] 1, & x^2 + 2y^2 = 1, \end{cases}$$

then it follows from Example 2.2 that *h* is continuous on the ellipse

$$x^2 + 2y^2 = 1.$$

We will see below (Example 2.13) that *h* is also continuous on the interior of the ellipse.

Example 2.11   It is impossible to define the function

$$f(x, y) = \frac{xy}{x^2 + y^2}$$

so as to be continuous at the origin, since we saw in Example 2.3 that

$$\lim_{(x,y)\to(0,0)} f(x, y)$$

does not exist.

The next theorem follows from Theorem 2.2 (*Exercise 12*); it is analogous to Theorem 2.2, Section 2.2, and, like the latter, it permits us to investigate continuity of a given function by regarding it as the result of addition, sub-traction, multiplication, and division of simpler functions.

Theorem 2.4 *If f and g are continuous on a set S in $\mathscr{R}^n$, then so are $f + g$, $f - g$, and fg; also, f/g is continuous at each $\mathbf{X}_0$ in S such that $g(\mathbf{X}_0) \neq 0$.*

Vector-valued functions and composite functions

Suppose $g_1, g_2, \ldots, g_n$ are real-valued functions defined on a subset T of $\mathscr{R}^m$, and define the *vector-valued function* $\mathbf{G}$ on T by

$$\mathbf{G}(\mathbf{U}) = (g_1(\mathbf{U}), g_2(\mathbf{U}), \ldots, g_n(\mathbf{U})), \qquad \mathbf{U} \in T.$$

Then $g_1, g_2, \ldots, g_n$ are the *component functions* of $\mathbf{G} = (g_1, g_2, \ldots, g_n)$. We say that

$$\lim_{\mathbf{U} \to \mathbf{U}_0} \mathbf{G}(\mathbf{U}) = \mathbf{L} = (L_1, L_2, \ldots, L_n)$$

if

$$\lim_{\mathbf{U} \to \mathbf{U}_0} g_i(\mathbf{U}) = L_i, \qquad 1 \leq i \leq n,$$

and that $\mathbf{G}$ is *continuous at* $\mathbf{U}_0$ if $g_1, g_2, \ldots, g_n$ are each continuous at $\mathbf{U}_0$.

The next theorem follows immediately from this definition. We leave its proof to the reader (*Exercise 13*).

Theorem 2.5 *For a vector-valued function $\mathbf{G}$,*

$$\lim_{\mathbf{U} \to \mathbf{U}_0} \mathbf{G}(\mathbf{U}) = \mathbf{L}$$

if and only if for each $\varepsilon > 0$ there is a $\delta > 0$ such that

$$|\mathbf{G}(\mathbf{U}) - \mathbf{L}| < \varepsilon \quad \text{whenever} \quad 0 < |\mathbf{U} - \mathbf{U}_0| < \delta \quad \text{and} \quad \mathbf{U} \in T.$$

Similarly, $\mathbf{G}$ is continuous at $\mathbf{U}_0$ if and only if for each $\varepsilon > 0$ there is a $\delta > 0$ such that

$$|\mathbf{G}(\mathbf{U}) - \mathbf{G}(\mathbf{U}_0)| < \varepsilon \quad \text{whenever} \quad |\mathbf{U} - \mathbf{U}_0| < \delta \quad \text{and} \quad \mathbf{U} \in T.$$

The next theorem is analogous to Theorem 2.3, Section 2.2.

Theorem 2.6 *Let f be defined on a subset of $\mathscr{R}^n$ and let the vector-valued function $\mathbf{G} = (g_1, g_2, \ldots, g_n)$ be defined on a domain $D_{\mathbf{G}}$ in $\mathscr{R}^m$. Let the set*

$$T = \{\mathbf{U} \mid \mathbf{U} \in D_{\mathbf{G}} \quad \text{and} \quad \mathbf{G}(\mathbf{U}) \in D_f\}$$

(*Figure 2.3*) *be nonempty, and define the composite function*

$$h = f \cdot \mathbf{G}$$

on T by

$$h(\mathbf{U}) = f(\mathbf{G}(\mathbf{U})), \qquad \mathbf{U} \in T.$$

Now suppose $\mathbf{U}_0$ is in T and a limit point of T, $\mathbf{G}$ is continuous at $\mathbf{U}_0$, and f is continuous at $\mathbf{X}_0 = \mathbf{G}(\mathbf{U}_0)$. Then h is continuous at $\mathbf{U}_0$.

Proof.   Suppose $\varepsilon > 0$. Since f is continuous at $\mathbf{X}_0 = \mathbf{G}(\mathbf{U}_0)$, there is an $\varepsilon_1 > 0$. such that

$$\left| f(\mathbf{X}) - f(\mathbf{G}(\mathbf{U}_0)) \right| < \varepsilon \tag{15}$$

if

$$\left| \mathbf{X} - \mathbf{G}(\mathbf{U}_0) \right| < \varepsilon_1 \quad \text{and} \quad \mathbf{X} \in D_f. \tag{16}$$

Since $\mathbf{G}$ is continuous at $\mathbf{U}_0$, there is a $\delta > 0$ such that

$$\left| \mathbf{G}(\mathbf{U}) - \mathbf{G}(\mathbf{U}_0) \right| < \varepsilon_1 \qquad \text{if} \quad \left| \mathbf{U} - \mathbf{U}_0 \right| < \delta \quad \text{and} \quad \mathbf{U} \in D_{\mathbf{G}}.$$

By taking $\mathbf{X} = \mathbf{G}(\mathbf{U})$ in (15) and (16), we see that

$$\left| h(\mathbf{U}) - h(\mathbf{U}_0) \right| = \left| f(\mathbf{G}(\mathbf{U})) - f(\mathbf{G}(\mathbf{U}_0)) \right| < \varepsilon$$

if

$$\left| \mathbf{U} - \mathbf{U}_0 \right| < \delta \quad \text{and} \quad \mathbf{U} \in T.$$

This completes the proof.

Example 2.12   If

$$f(s) = \sqrt{s}$$

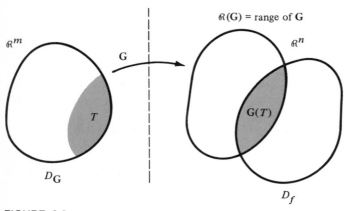

FIGURE 2.3

and

$$g(x, y) = 1 - x^2 - 2y^2,$$

then $D_f = [0, \infty)$, $D_g = \mathscr{R}^2$, and

$$T = \{(x, y) \mid x^2 + 2y^2 \le 1\}.$$

Theorem 2.3 and the results of Example 2.1 imply that g is continuous on $\mathscr{R}^2$. [We can obtain the same result by observing that the functions $p_1(x, y) = x$ and $p_2(x, y) = y$ are continuous on $\mathscr{R}^2$ and applying Theorem 2.4.] Since f is continuous on D_f, the function

$$h(x, y) = f(g(x, y)) = \sqrt{1 - x^2 - 2y^2}$$

is continuous on T.

Example 2.13   If

$$g(x, y) = \sqrt{1 - x^2 - 2y^2}$$

and

$$f(s) = \begin{cases} \dfrac{\sin s}{s}, & s \ne 0, \\ 1, & s = 0, \end{cases}$$

then $D_f = (-\infty, \infty)$ and

$$D_g = T = \{(x, y) \mid x^2 + 2y^2 \le 1\}.$$

In Example 2.12 we saw that g (we called it h there) is continuous on T. Since f is continuous on D_f, the composite function $h = f \circ g$ defined by

$$h(x, y) = \begin{cases} \dfrac{\sin \sqrt{1 - x^2 - 2y^2}}{\sqrt{1 - x^2 - 2y^2}}, & x^2 + 2y^2 < 1, \\ 1, & x^2 + 2y^2 = 1, \end{cases}$$

is continuous on T. This implies the result of Example 2.2.

Bounded functions

The definitions of *bounded above*, *bounded below*, and *bounded* on a set S are the same for functions of n variables as for functions of one variable, as are the definitions of *least upper bound* and *greatest lower bound* of a function on a set S (Section 2.2). The proofs of the next two theorems are similar to those of Theorems 2.4 and 2.5 of Section 2.2 (*Exercises 17 and 18*).

Theorem 2.7   *If f is continuous on a compact set S in $\mathscr{R}^n$, then f is bounded on S.*

Theorem 2.8   *Let f be continuous on a compact set S in $\mathscr{R}^n$ and*

$$\alpha = \underset{\mathbf{X} \in S}{\text{g.l.b.}}\ f(\mathbf{X}), \qquad \beta = \underset{\mathbf{X} \in S}{\text{l.u.b.}}\ f(\mathbf{X}).$$

Then

$$f(\bar{\mathbf{X}}) = \alpha \quad and \quad f(\bar{\bar{\mathbf{X}}}) = \beta$$

for some $\bar{\mathbf{X}}$ and $\bar{\bar{\mathbf{X}}}$ in S.

The next theorem generalizes Theorem 2.6 of Section 2.2.

Theorem 2.9 (Intermediate Value Theorem)   *Let f be continuous on a region S in $\mathscr{R}^n$. Suppose $\mathbf{A}$ and $\mathbf{B}$ are in S and*

$$f(\mathbf{A}) < u < f(\mathbf{B}).$$

Then $f(\mathbf{C}) = u$ for some $\mathbf{C}$ in S.

Proof.   If there is no such $\mathbf{C}$ then $S = R \cup T$, where

$$R = \{\mathbf{X} \,|\, \mathbf{X} \in S \text{ and } f(\mathbf{X}) < u\}$$

and

$$T = \{\mathbf{X} \,|\, \mathbf{X} \in S \text{ and } f(\mathbf{X}) > u\}.$$

If $\mathbf{X}_0 \in R$, the continuity of f implies that there is a $\delta > 0$ such that $f(\mathbf{X}) < u$ if $|\mathbf{X} - \mathbf{X}_0| < \delta$ and $\mathbf{X} \in S$. This means that $\mathbf{X}_0 \notin \bar{T}$. Therefore, $R \cap \bar{T} = \varnothing$. Similarly, $\bar{R} \cap T = \varnothing$. Therefore, S is disconnected (Definition 1.8), which contradicts the assumption that S is a region (*Exercise 32, Section 5.1*). Hence, we conclude that $f(\mathbf{C}) = u$ for some $\mathbf{C}$ in S.

Uniform continuity

The definition of uniform continuity is the same for functions of n variables as for functions of one variable: f is uniformly continuous on a subset S of its domain in $\mathscr{R}^n$ if for every $\varepsilon > 0$ there is a $\delta > 0$ such that

$$|f(\mathbf{X}) - f(\mathbf{X}')| < \varepsilon$$

whenever $|\mathbf{X} - \mathbf{X}'| < \delta$ and $\mathbf{X}, \mathbf{X}' \in S$. We emphasize again that δ must depend only on ε and S, and not on the particular points $\mathbf{X}$ and $\mathbf{X}'$.

The proof of the next theorem is analogous to that of Theorem 2.7, Section 2.2; we leave it to the reader (*Exercise 19*).

Theorem 2.10   *If f is continuous on a compact set S in $\mathscr{R}^n$, then f is uniformly continuous on S.*

5.2   EXERCISES

The following exercises have analogs for $\mathscr{R}^n$: 5, 8, 9, 10, 11, 25, 27, 28, 29, and 32 of Section 2.1; 8, 9, 10, 15, 16, 20, 25, 29, and 30 of Section 2.2.

1. Find $\lim_{\mathbf{X} \to \mathbf{X}_0} f(\mathbf{X})$ and justify your answer with an ε–δ argument, as required by Definition 2.1:

   (a) $f(\mathbf{X}) = 3x + 4y + z - 2$, $\mathbf{X}_0 = (1, 2, 1)$

   (b) $f(\mathbf{X}) = \dfrac{x^3 - y^3}{x - y}$, $\mathbf{X}_0 = (1, 1)$

   (c) $f(\mathbf{X}) = \dfrac{\sin(x + 4y + 2z)}{x + 4y + 2z}$, $\mathbf{X}_0 = (-2, 1, -1)$

   (d) $f(\mathbf{X}) = (x^2 + y^2) \log(x^2 + y^2)^{1/2}$, $\mathbf{X}_0 = (0, 0)$

   (e) $f(\mathbf{X}) = \dfrac{\sin(x - y)}{\sqrt{x - y}}$, $\mathbf{X}_0 = (2, 2)$

   (f) $f(\mathbf{X}) = \dfrac{1}{|\mathbf{X}|} e^{-1/|\mathbf{X}|}$, $\mathbf{X}_0 = \mathbf{0}$

2. Prove Theorem 2.1.

3. If $\lim_{x \to x_0} g(x) = y_0$ and $\lim_{x \to x_0} f(x, g(x)) = L$, we say that $f(x, y)$ *approaches L as (x, y) approaches (x_0, y_0) along the curve $y = g(x)$.*
    (a) Prove: If $\lim_{(x,y) \to (x_0,y_0)} f(x, y) = L$, then $f(x, y)$ approaches L as (x, y) approaches (x_0, y_0) along any curve $y = g(x)$ through (x_0, y_0).
    (b) We saw in Example 2.3 that if

   $$f(x, y) = \frac{xy}{x^2 + y^2}$$

   then $\lim_{(x,y) \to (0,0)} f(x, y)$ does not exist. Show, however, that $f(x, y)$ approaches a value L_a as (x, y) approaches $(0, 0)$ along any curve $y = g(x)$ which passes through $(0, 0)$ with slope a. Find L_a.
    (c) Show that the function

   $$g(x, y) = \frac{x^3 y^4}{(x^2 + y^6)^3}$$

   approaches 0 as (x, y) approaches $(0, 0)$ along a curve as described in (b), but that $\lim_{(x,y) \to (0,0)} g(x, y)$ does not exist.

4. Prove Theorem 2.2.

5. Decide directly from Definition 2.2 whether $\lim_{\mathbf{X}\to\mathbf{X}_0} f(\mathbf{X}) = \pm\infty$:

   (a) $f(\mathbf{X}) = \dfrac{|\sin(x + 2y + 4z)|}{(x + 2y + 4z)^2}$, $\mathbf{X}_0 = (2, -1, 0)$

   (b) $f(\mathbf{X}) = \dfrac{1}{\sqrt{x - y}}$, $\mathbf{X}_0 = (0, 0)$

   (c) $f(\mathbf{X}) = \dfrac{\sin 1/x}{\sqrt{x - y}}$, $\mathbf{X}_0 = (0, 0)$

   (d) $f(\mathbf{X}) = \dfrac{4y^2 - x^2}{(x - 2y)^2}$, $\mathbf{X}_0 = (2, 1)$

   (e) $f(\mathbf{X}) = \dfrac{\sin(x + 2y + 4z)}{(x + 2y + 4z)^2}$, $\mathbf{X}_0 = (2, -1, 0)$

6. Find $\lim_{|\mathbf{X}|\to\infty} f(\mathbf{X})$, if it exists:

   (a) $f(\mathbf{X}) = \dfrac{\log(x^2 + 2y^2 + 4z^2)}{x^2 + y^2 + z^2}$ (b) $f(\mathbf{X}) = \dfrac{\sin(x^2 + y^2)}{\sqrt{x^2 + y^2}}$

   (c) $f(\mathbf{X}) = e^{-(x+y)^2}$ (d) $f(\mathbf{X}) = e^{-x^2 - y^2}$

   (e) $f(\mathbf{X}) = \begin{cases} \dfrac{\sin(x^2 - y^2)}{x^2 - y^2}, & x \neq \pm y, \\ \\ 1, & x = \pm y \end{cases}$

7. Define: (a) $\lim_{|\mathbf{X}|\to\infty} f(\mathbf{X}) = \infty$; (b) $\lim_{|\mathbf{X}|\to\infty} f(\mathbf{X}) = -\infty$.

8. Let

$$f(\mathbf{X}) = \frac{|x_1|^{a_1}|x_2|^{a_2}\cdots|x_n|^{a_n}}{|\mathbf{X}|^b}.$$

   For what nonnegative values of $b, a_1, a_2, \ldots, a_n$ does $\lim_{\mathbf{X}\to 0} f(\mathbf{X})$ (a) exist? (b) exist in the extended reals?

9. Let

$$g(\mathbf{X}) = \frac{(x^2 + y^4)^3}{1 + x^6 y^4}.$$

   Show that $\lim_{x\to\infty} g(x, ax) = \infty$ for any real number a. Does

$$\lim_{|\mathbf{X}|\to\infty} g(\mathbf{X}) = \infty?$$

10. Theorem 2.2 remains valid with "$\lim_{\mathbf{X}\to\mathbf{X}_0}$" replaced by "$\lim_{|\mathbf{X}|\to\infty}$." How would its proof have to be changed to show this?

11. Prove Theorem 2.3.

12. Prove Theorem 2.4.

13. Prove Theorem 2.5.

14. For each f in Exercise 1, find the largest set S on which f is continuous, or can be defined so as to be continuous.

15. Repeat Exercise 14 for the functions in Exercise 6.

16. Give an example of a function f on $\mathscr{R}^2$ such that f is not continuous at $(0, 0)$, but $f(0, y)$ is a continuous function of y on $(-\infty, \infty)$ and $f(x, 0)$ is a continuous function of x on $(-\infty, \infty)$.

17. Prove Theorem 2.7. (*Hint:* See the proof of Theorem 2.4, Section 2.2.)

18. Prove Theorem 2.8. (*Hint:* See the proof of Theorem 2.5, Section 2.2.)

19. Prove Theorem 2.10. (*Hint:* See the proof of Theorem 2.7, Section 2.2.)

*20. Suppose $\bar{\mathbf{X}} \in D_f \subset \mathscr{R}^n$ and $\bar{\mathbf{X}}$ is a limit point of D_f. Show that f is continuous at $\bar{\mathbf{X}}$ if and only if $\lim_{k \to \infty} f(\mathbf{X}_k) = f(\bar{\mathbf{X}})$ whenever $\{\mathbf{X}_k\}$ is a sequence of points in D_f such that $\lim_{k \to \infty} \mathbf{X}_k = \bar{\mathbf{X}}$. (*Hint:* See the proof of Theorem 2.5, Section 4.2.)

5.3 PARTIAL DERIVATIVES AND THE DIFFERENTIAL

To say that a function of one variable has a derivative at x_0 is the same as to say that it is differentiable at x_0. The situation is not so simple for functions of more than one variable. First, there is no specific number that can be called *the* derivative of f at a point $\mathbf{X}_0$ in $\mathscr{R}^n$; indeed, there are infinitely many numbers—the directional derivatives of f at $\mathbf{X}_0$, defined below—analogous to the derivative of a function of one variable. Second, we will see that the existence of directional derivatives at $\mathbf{X}_0$ does not imply that f is differentiable at $\mathbf{X}_0$, if differentiability at $\mathbf{X}_0$ is to imply—as it does for functions of one variable—that $f(\mathbf{X}) - f(\mathbf{X}_0)$ can be approximated well near $\mathbf{X}_0$ by a simple linear function, or even that f is continuous at $\mathbf{X}_0$.

We now define directional and partial derivatives of functions of several variables. However, we will still have occasion to refer to derivatives of functions of one variable. We will call them "ordinary" derivatives when we wish to distinguish between them and the partial derivatives we are about to define.

Definition 3.1 *Let* $\mathbf{\Phi}$ *be a unit vector and* $\mathbf{X}$ *a point in* $\mathscr{R}^n$. *The directional derivative of* f *at* $\mathbf{X}$ *in the direction of* $\mathbf{\Phi}$ *is defined by*

$$\frac{\partial f(\mathbf{X})}{\partial \mathbf{\Phi}} = \lim_{t \to 0} \frac{f(\mathbf{X} + t\mathbf{\Phi}) - f(\mathbf{X})}{t}$$

if the limit exists. That is, $\partial f(\mathbf{X})/\partial \mathbf{\Phi}$ is the ordinary derivative of the function

$$h(t) = f(\mathbf{X} + t\mathbf{\Phi})$$

at $t = 0$, if the latter exists.

Example 3.1 Let $\mathbf{\Phi} = (\phi_1, \phi_2, \phi_3)$ and

$$f(x, y, z) = 3xyz + 2x^2 + z^2.$$

Then

$$h(t) = f(x + t\phi_1, y + t\phi_2, z + t\phi_3),$$
$$= 3(x + t\phi_1)(y + t\phi_2)(z + t\phi_3) + 2(x + t\phi_1)^2 + (z + t\phi_3)^2$$

and

$$h'(t) = 3\phi_1(y + t\phi_2)(z + t\phi_3) + 3\phi_2(x + t\phi_1)(z + t\phi_3)$$
$$+ 3\phi_3(x + t\phi_1)(y + t\phi_2) + 4\phi_1(x + t\phi_1) + 2\phi_3(z + t\phi_3).$$

Therefore,

$$\frac{\partial f(\mathbf{X})}{\partial \mathbf{\Phi}} = h'(0) = (3yz + 4x)\phi_1 + 3xz\phi_2 + (3xy + 2z)\phi_3. \tag{1}$$

The directional derivatives that we are most interested in are those in the directions of the unit vectors

$$\mathbf{E}_1 = (1, 0, \ldots, 0), \quad \mathbf{E}_2 = (0, 1, 0, \ldots, 0), \quad \ldots, \quad \mathbf{E}_n = (0, \ldots, 0, 1).$$

(All components of $\mathbf{E}_i$ are zero except for the ith, which is 1.) Since $\mathbf{X}$ and $\mathbf{X} + t\mathbf{E}_i$ differ only in the ith coordinate, $\partial f(\mathbf{X})/\partial \mathbf{E}_i$ is called the *partial derivative of f, with respect to x_i, at $\mathbf{X}$.* It is also denoted by $\partial f(\mathbf{X})/\partial x_i$ or $f_{x_i}(\mathbf{X})$; thus,

$$\frac{\partial f(\mathbf{X})}{\partial x_1} = f_{x_1}(\mathbf{X}) = \lim_{t \to 0} \frac{f(x_1 + t, x_2, \ldots, x_n) - f(x_1, x_2, \ldots, x_n)}{t},$$

$$\frac{\partial f(\mathbf{X})}{\partial x_i} = f_{x_i}(\mathbf{X}) = \lim_{t \to 0} \frac{\left(\begin{array}{l} f(x_1, \ldots, x_{i-1}, x_i + t, x_{i+1}, \ldots, x_n) \\ \quad - f(x_1, \ldots, x_{i-1}, x_i, x_{i+1}, \ldots, x_n) \end{array}\right)}{t},$$

$$2 \le i \le n,$$

$$\frac{\partial f(\mathbf{X})}{\partial x_n} = f_{x_n}(\mathbf{X}) = \lim_{t \to 0} \frac{f(x_1, \ldots, x_{n-1}, x_n + t) - f(x_1, \ldots, x_{n-1}, x_n)}{t},$$

if the limits exist.

If we write $\mathbf{X} = (x, y)$, then we denote the partial derivatives accordingly; thus,

$$\frac{\partial f(x, y)}{\partial x} = f_x(x, y) = \lim_{h \to 0} \frac{f(x + h, y) - f(x, y)}{h}$$

and

$$\frac{\partial f(x, y)}{\partial y} = f_y(x, y) = \lim_{h \to 0} \frac{f(x, y + h) - f(x, y)}{h}.$$

It can be seen from these definitions that to compute $f_{x_i}(\mathbf{X})$ we simply differentiate f with respect to x_i according to the rules for ordinary differentiation, while treating the other variables as constants.

Example 3.2   Let

$$f(x, y, z) = 3xyz + 2x^2 + z^2 \tag{2}$$

as in Example 3.1. Taking $\boldsymbol{\Phi} = \mathbf{E}_1$ (that is, setting $\phi_1 = 1$ and $\phi_2 = \phi_3 = 0$) in (1), we find that

$$\frac{\partial f(\mathbf{X})}{\partial x} = \frac{\partial f(\mathbf{X})}{\partial \mathbf{E}_1} = 3yz + 4x,$$

which is the result obtained by regarding y and z as constants in (2) and taking the ordinary derivative with respect to x. Similarly,

$$\frac{\partial f(\mathbf{X})}{\partial y} = \frac{\partial f(\mathbf{X})}{\partial \mathbf{E}_2} = 3xz$$

and

$$\frac{\partial f(\mathbf{X})}{\partial z} = \frac{\partial f(\mathbf{X})}{\partial \mathbf{E}_3} = 3xy + 2z.$$

The next theorem follows from the rule just given for calculating partial derivatives (*Exercise 6*).

Theorem 3.1   *If $f_{x_i}(\mathbf{X})$ and $g_{x_i}(\mathbf{X})$ exist, then*

$$\frac{\partial(f + g)(\mathbf{X})}{\partial x_i} = f_{x_i}(\mathbf{X}) + g_{x_i}(\mathbf{X}),$$

$$\frac{\partial(fg)(\mathbf{X})}{\partial x_i} = f_{x_i}(\mathbf{X})g(\mathbf{X}) + f(\mathbf{X})g_{x_i}(\mathbf{X}),$$

and, if $g(\mathbf{X}) \neq 0$,

$$\frac{\partial(f/g)(\mathbf{X})}{\partial x_i} = \frac{g(\mathbf{X})f_{x_i}(\mathbf{X}) - f(\mathbf{X})g_{x_i}(\mathbf{X})}{[g(\mathbf{X})]^2}.$$

If $f_{x_i}(\mathbf{X})$ exists at every point of a set D, then it defines a function, f_{x_i}, on D. If this function has a partial derivative with respect to x_j on a subset of D, we denote it by

$$\frac{\partial}{\partial x_j}\left(\frac{\partial f}{\partial x_i}\right) = \frac{\partial^2 f}{\partial x_j \, \partial x_i} = f_{x_i x_j}.$$

Similarly,

$$\frac{\partial}{\partial x_k}\left(\frac{\partial^2 f}{\partial x_j \partial x_i}\right) = \frac{\partial^3 f}{\partial x_k \partial x_j \partial x_i} = f_{x_i x_j x_k}.$$

The function obtained by differentiating f successively with respect to x_{i_1}, $x_{i_2}, \ldots, x_{i_r}$ is denoted by

$$\frac{\partial^r f}{\partial x_{i_r} \partial x_{i_{r-1}} \cdots \partial x_{i_1}} = f_{x_{i_1} \cdots x_{i_{r-1}} x_{i_r}};$$

it is said to be an *rth-order partial derivative of f*.

Example 3.3   The function

$$f(x, y) = 3x^2 y^3 + xy$$

has partial derivatives everywhere. Its first-order partial derivatives are

$$f_x(x, y) = 6xy^3 + y, \qquad f_y(x, y) = 9x^2 y^2 + x.$$

Its second-order partial derivatives are

$$f_{xx}(x, y) = 6y^3, \qquad\qquad f_{yy}(x, y) = 18x^2 y,$$
$$f_{xy}(x, y) = 18xy^2 + 1, \qquad f_{yx}(x, y) = 18xy^2 + 1.$$

There are eight third-order partial derivatives. Some examples are

$$f_{xxy}(x, y) = 18y^2, \quad f_{xyx}(x, y) = 18y^2, \quad f_{yxx}(x, y) = 18y^2.$$

Example 3.4   Let us compute $f_{xx}(0, 0), f_{yy}(0, 0), f_{xy}(0, 0)$, and $f_{yx}(0, 0)$ if

$$f(x, y) = \begin{cases} \dfrac{(x^2 y + xy^2)\sin(x - y)}{x^2 + y^2}, & (x, y) \neq (0, 0), \\[2mm] 0, & (x, y) = (0, 0). \end{cases}$$

The ordinary rules for differentiation, applied separately to x and y, yield

$$f_x(x, y) = \frac{(2xy + y^2)\sin(x - y) + (x^2 y + xy^2)\cos(x - y)}{x^2 + y^2}$$

$$- \frac{2x(x^2 y + xy^2)\sin(x - y)}{(x^2 + y^2)^2}, \qquad (x, y) \neq (0, 0),$$

(3)

and

$$f_y(x, y) = \frac{(x^2 + 2xy)\sin(x - y) - (x^2 y + xy^2)\cos(x - y)}{x^2 + y^2}$$

$$- \frac{2y(x^2 y + xy^2)\sin(x - y)}{(x^2 + y^2)^2}, \qquad (x, y) \neq (0, 0).$$

(4)

These formulas do not apply if $(x, y) = (0, 0)$, so we find $f_x(0, 0)$ and $f_y(0, 0)$ from their definitions as difference quotients:

$$f_x(0, 0) = \lim_{x \to 0} \frac{f(x, 0) - f(0, 0)}{x} = \lim_{x \to 0} \frac{0 - 0}{x} = 0,$$

$$f_y(0, 0) = \lim_{y \to 0} \frac{f(0, y) - f(0, 0)}{y} = \lim_{y \to 0} \frac{0 - 0}{y} = 0.$$

Setting $y = 0$ in (3) and (4) yields

$$f_x(x, 0) = 0, \quad f_y(x, 0) = \sin x, \qquad x \neq 0,$$

and so

$$f_{xx}(0, 0) = \lim_{x \to 0} \frac{f_x(x, 0) - f_x(0, 0)}{x} = \lim_{x \to 0} \frac{0 - 0}{x} = 0,$$

$$f_{yx}(0, 0) = \lim_{x \to 0} \frac{f_y(x, 0) - f_y(0, 0)}{x} = \lim_{x \to 0} \frac{\sin x - 0}{x} = 1.$$

Setting $x = 0$ in (3) and (4) yields

$$f_x(0, y) = -\sin y, \quad f_y(0, y) = 0, \qquad y \neq 0,$$

and so

$$f_{xy}(0, 0) = \lim_{y \to 0} \frac{f_x(0, y) - f_x(0, 0)}{y} = \lim_{y \to 0} \frac{-\sin y - 0}{y} = -1,$$

$$f_{yy}(0, 0) = \lim_{y \to 0} \frac{f_y(0, y) - f_y(0, 0)}{y} = \lim_{y \to 0} \frac{0 - 0}{y} = 0.$$

This example shows that $f_{xy}(X_0)$ and $f_{yx}(X_0)$ may differ. However, the next theorem shows that they are equal if f satisfies a fairly mild condition.

Theorem 3.2 *Suppose f, f_x, f_y, and f_{xy} exist in a neighborhood N of (x_0, y_0), and f_{xy} is continuous at (x_0, y_0). Then $f_{yx}(x_0, y_0)$ exists and*

$$f_{yx}(x_0, y_0) = f_{xy}(x_0, y_0). \tag{5}$$

Proof. Suppose $\varepsilon > 0$. Choose $\delta > 0$ so that the open square

$$S_\delta = \{(x, y) \,|\, |x - x_0| < \delta, |y - y_0| < \delta\}$$

is in N and

$$|f_{xy}(\hat{x}, \hat{y}) - f_{xy}(x_0, y_0)| < \varepsilon \quad \text{if} \quad (\hat{x}, \hat{y}) \in S_\delta. \tag{6}$$

This is possible because of the continuity of f_{xy} at (x_0, y_0) The function

$$A(h, k) = f(x_0 + h, y_0 + k) - f(x_0 + h, y_0)$$
$$- f(x_0, y_0 + k) + f(x_0, y_0) \tag{7}$$

is defined if $-\delta < h, k < \delta$; moreover,

$$A(h, k) = \phi(x_0 + h) - \phi(x_0), \tag{8}$$

where

$$\phi(x) = f(x, y_0 + k) - f(x, y_0).$$

Since

$$\phi'(x) = f_x(x, y_0 + k) - f_x(x, y_0), \qquad |x - x_0| < \delta,$$

(8) and the mean value theorem imply that

$$A(h, k) = [f_x(\hat{x}, y_0 + k) - f_x(\hat{x}, y_0)]h, \tag{9}$$

where $\hat{x}$ is between x_0 and $x_0 + h$. The mean value theorem, applied to $f_x(\hat{x}, y)$ (where $\hat{x}$ is regarded as constant), also implies that

$$f_x(\hat{x}, y_0 + k) - f_x(\hat{x}, y_0) = f_{xy}(\hat{x}, \hat{y})k,$$

where $\hat{y}$ is between y_0 and $y_0 + k$. From this and (9),

$$A(h, k) = f_{xy}(\hat{x}, \hat{y})hk.$$

Now (6) implies that

$$\left| \frac{A(h, k)}{hk} - f_{xy}(x_0, y_0) \right| = |f_{xy}(\hat{x}, \hat{y}) - f_{xy}(x_0, y_0)| < \varepsilon$$

$$\text{if} \quad 0 < |h|, |k| < \delta, \tag{10}$$

Since (7) implies that

$$\lim_{k \to 0} \frac{A(h, k)}{hk} = \lim_{k \to 0} \frac{f(x_0 + h, y_0 + k) - f(x_0 + h, y_0)}{hk}$$

$$- \lim_{k \to 0} \frac{f(x_0, y_0 + k) - f(x_0, y_0)}{hk}$$

$$= \frac{f_y(x_0 + h, y_0) - f_y(x_0, y_0)}{h},$$

it follows from (10) that

$$\left| \frac{f_y(x_0 + h, y_0) - f_y(x_0, y_0)}{h} - f_{xy}(x_0, y_0) \right| \le \varepsilon \qquad \text{if} \quad 0 < |h| < \delta.$$

This proves (5).

The next theorem follows from this. We leave its proof to the reader (*Exercises 11 and 12*).

Theorem 3.3 *Suppose f and all its partial derivatives of order $\leq r$ exist in a neighborhood of $\mathbf{X}_0$ in $\mathscr{R}^n$, and are continuous at $\mathbf{X}_0$. Then*

$$f_{x_{i_1} x_{i_2} \ldots x_{i_r}}(\mathbf{X}_0) = f_{x_{j_1} x_{j_2} \ldots x_{j_r}}(\mathbf{X}_0) \tag{11}$$

if each of the variables x_k appears the same number of times in $\{x_{i_1}, x_{i_2}, \ldots, x_{i_r}\}$ and $\{x_{j_1}, x_{j_2}, \ldots, x_{j_r}\}$. If this number is r_k, we denote the common value of the two sides of (11) by

$$\frac{\partial^r f(\mathbf{X}_0)}{\partial x_1^{r_1} \, \partial x_2^{r_2} \cdots \partial x_n^{r_n}}, \tag{12}$$

it being understood that

$$0 \leq r_k \leq r, \qquad 1 \leq k \leq n, \tag{13}$$

and

$$r_1 + r_2 + \cdots + r_n = r. \tag{14}$$

This theorem says, for example, that if f satisfies its hypotheses with $k = 4$ at a point $\mathbf{X}_0$ in $\mathscr{R}^n$ ($n \geq 2$), then

$$f_{xxyy}(\mathbf{X}_0) = f_{xyxy}(\mathbf{X}_0) = f_{xyyx}(\mathbf{X}_0) = f_{yyxx}(\mathbf{X}_0) = f_{yxyx}(\mathbf{X}_0) = f_{yxxy}(\mathbf{X}_0),$$

and their common value is denoted by

$$\frac{\partial^4 f(\mathbf{X}_0)}{\partial x^2 \, \partial y^2}.$$

It can be shown (*Exercise 13*) that the number of partial derivatives (11) that can be given in the form (12) for a fixed ordered n-tuple $(r_1, r_2, \ldots, r_n)$ satisfying (13) and (14) equals the *multinomial coefficient*

$$\frac{r!}{r_1! r_2! \cdots r_n!}.$$

Differentiable functions of several variables

A function of several variables may have first-order partial derivatives at a point but fail to be continuous there. For example, if

$$f(x, y) = \begin{cases} \dfrac{xy}{x^2 + y^2}, & (x, y) \neq (0, 0), \\[2mm] 0, & (x, y) = (0, 0), \end{cases} \tag{15}$$

then

$$f_x(0, 0) = f_y(0, 0) = 0,$$

but f is not continuous at $(0, 0)$. (See Examples 2.3 and 2.11.) Therefore, if differentiability of a function of several variables is to be a stronger property than continuity—as it is for functions of one variable—its definition must require more than the existence of first partial derivatives. *Exercise 1 of Section 2.3* characterizes differentiability of a function f of one variable in a way that suggests the proper generalization: f is differentiable at x_0 if and only if

$$\lim_{x \to x_0} \frac{f(x) - f(x_0) - m(x - x_0)}{x - x_0} = 0$$

for some constant m, in which case $m = f'(x_0)$.

The generalization to functions of n variables is as follows.

Definition 3.2   *A function f is said to be differentiable at an interior point* $\mathbf{X}_0 = (c_1, c_2, \ldots, c_n)$ *of its domain if*

$$\lim_{\mathbf{X} \to \mathbf{X}_0} \frac{f(\mathbf{X}) - f(\mathbf{X}_0) - \sum\limits_{i=1}^{n} m_i(x_i - c_i)}{|\mathbf{X} - \mathbf{X}_0|} = 0 \tag{16}$$

for some constants $m_1, m_2, \ldots, m_n$.

Example 3.5   Let

$$f(x, y) = x^2 + 2xy.$$

We will show that f is differentiable at any point (x_0, y_0), as follows:

$$\begin{aligned}
f(x, y) - f(x_0, y_0) &= x^2 + 2xy - x_0^2 - 2x_0 y_0 \\
&= x^2 - x_0^2 + 2(xy - x_0 y_0) \\
&= (x - x_0)(x + x_0) + 2(xy - x_0 y) + 2(x_0 y - x_0 y_0) \\
&= (x + x_0 + 2y)(x - x_0) + 2x_0(y - y_0) \\
&= 2(x_0 + y_0)(x - x_0) + 2x_0(y - y_0) \\
&\quad + (x - x_0 + 2y - 2y_0)(x - x_0).
\end{aligned}$$

If

$$m_1 = 2(x_0 + y_0) \quad \text{and} \quad m_2 = 2x_0 \tag{17}$$

then

$$\frac{|f(x, y) - f(x_0, y_0) - m_1(x - x_0) - m_2(y - y_0)|}{|\mathbf{X} - \mathbf{X}_0|}$$

$$= \frac{|x - x_0| \, |(x - x_0) + 2(y - y_0)|}{|\mathbf{X} - \mathbf{X}_0|} \le \sqrt{5} |\mathbf{X} - \mathbf{X}_0| \quad \text{(Schwarz's inequality)}.$$

This implies that

$$\lim_{\mathbf{X} \to \mathbf{X}_0} \frac{f(x, y) - f(x_0, y_0) - m_1(x - x_0) - m_2(y - y_0)}{|\mathbf{X} - \mathbf{X}_0|} = 0,$$

and therefore f is differentiable at (x_0, y_0).

From (17), $m_1 = f_x(x_0, y_0)$ and $m_2 = f_y(x_0, y_0)$ in this example. The next theorem shows that this is no accident.

Theorem 3.4 *If f is differentiable at $\mathbf{X}_0 = (c_1, c_2, \ldots, c_n)$, then $f_{x_1}(\mathbf{X}_0)$, $f_{x_2}(\mathbf{X}_0), \ldots, f_{x_n}(\mathbf{X}_0)$ exist, and*

$$f_{x_i}(\mathbf{X}_0) = m_i, \qquad i = 1, 2, \ldots, n, \tag{18}$$

where $m_1, m_2, \ldots, m_n$ are the constants in (16).

Proof. If we let $\mathbf{X} = \mathbf{X}_0 + t\mathbf{E}_i$, so that $x_i = c_i + t$, $x_j = c_j$ ($j \neq i$), and $|\mathbf{X} - \mathbf{X}_0| = |t|$, then (16) and the differentiability of f at $\mathbf{X}_0$ imply that

$$\lim_{t \to 0} \frac{f(\mathbf{X}_0 + t\mathbf{E}_i) - f(\mathbf{X}_0) - m_i t}{t} = 0.$$

Hence,

$$\lim_{t \to 0} \frac{f(\mathbf{X}_0 + t\mathbf{E}_i) - f(\mathbf{X}_0)}{t} = m_i,$$

which proves (18), since the limit on the left is $f_{x_i}(\mathbf{X}_0)$, by definition.

A function L of the form

$$L(\mathbf{X}) = m_1 x_1 + m_2 x_2 + \cdots + m_n x_n, \tag{19}$$

where $m_1, m_2, \ldots, m_n$ are constants, is said to be *linear*. From Definition 3.2, f is differentiable at $\mathbf{X}_0$ if and only if there is a linear function L such that $f(\mathbf{X}) - f(\mathbf{X}_0)$ can be approximated so well near $\mathbf{X}_0$ by

$$L(\mathbf{X}) - L(\mathbf{X}_0) = L(\mathbf{X} - \mathbf{X}_0)$$

that

$$f(\mathbf{X}) - f(\mathbf{X}_0) = L(\mathbf{X} - \mathbf{X}_0) + \varepsilon(\mathbf{X})(|\mathbf{X} - \mathbf{X}_0|), \tag{20}$$

where

$$\lim_{\mathbf{X} \to \mathbf{X}_0} \varepsilon(\mathbf{X}) = 0. \tag{21}$$

Theorem 3.5 *If f is differentiable at $X_0 = (c_1, c_2, \ldots, c_n)$, then it is continuous there.*

Proof. From (19) and Schwarz's inequality,

$$|L(X - X_0)| \le M|X - X_0|,$$

where

$$M = (m_1^2 + m_2^2 + \cdots + m_n^2)^{1/2}.$$

This and (20) imply that

$$|f(X) - f(X_0)| \le (M + |\varepsilon(X)|)|X - X_0|,$$

which, with (21), implies that f is continuous at X_0.

This theorem implies that the function f defined by (15) is not differentiable at (0, 0), since it is not continuous there. However, $f_x(0, 0)$ and $f_y(0, 0)$ exist, and so the converse of Theorem 3.4 is false; that is, a function may have partial derivatives at a point without being differentiable there.

The differential

Theorem 3.4 implies that if f is differentiable at X_0, then there is only one linear function L that satisfies (20) and (21):

$$L(X) = f_{x_1}(X_0)x_1 + f_{x_2}(X_0)x_2 + \cdots + f_{x_n}(X_0)x_n.$$

This function is called the *differential of f at X_0.* We will denote it by $d_{X_0}f$, and its value by $(d_{X_0}f)(X)$; thus,

$$(d_{X_0}f)(X) = f_{x_1}(X_0)x_1 + f_{x_2}(X_0)x_2 + \cdots + f_{x_n}(X_0)x_n. \tag{22}$$

In terms of the differential, (16) can be rewritten as

$$\lim_{X \to X_0} \frac{f(X) - f(X_0) - (d_{X_0}f)(X - X_0)}{|X - X_0|} = 0.$$

For convenience in writing $d_{X_0}f$—and to conform with older and by now standard notation—we introduce the function dx_i, defined by

$$dx_i(X) = x_i;$$

that is, dx_i is the function whose value at a point in $\mathscr{R}^n$ is the ith coordinate of the point. It is the differential of the function $g_i(X) = x_i$. From (22) we see that

$$d_{X_0}f = f_{x_1}(X_0)\, dx_1 + f_{x_2}(X_0)\, dx_2 + \cdots + f_{x_n}(X_0)\, dx_n. \tag{23}$$

If we write $\mathbf{X} = (x, y, \ldots,)$, then we write

$$d_{\mathbf{X}_0} f = f_x(\mathbf{X}_0) \, dx + f_y(\mathbf{X}_0) \, dy + \cdots,$$

where $dx, dy, \ldots$ are the functions defined by

$$dx(\mathbf{X}) = x, \quad dy(\mathbf{X}) = y, \quad \ldots$$

When it is not necessary to emphasize the specific point $\mathbf{X}_0$, (23) can be written more simply as

$$df = f_{x_1} \, dx_1 + f_{x_2} \, dx_2 + \cdots + f_{x_n} \, dx_n.$$

When dealing with a specific function at an arbitrary point of its domain, we may use the hybrid notation

$$df = f_{x_1}(\mathbf{X}) \, dx_1 + f_{x_2}(\mathbf{X}) \, dx_2 + \cdots + f_{x_n}(\mathbf{X}) \, dx_n.$$

Example 3.6   We saw in Example 3.5 that the function

$$f(x, y) = x^2 + 2xy$$

is differentiable at every $\mathbf{X}$ in $\mathscr{R}^n$, with differential

$$df = (2x + 2y) \, dx + 2x \, dy.$$

To find $d_{\mathbf{X}_0} f$ with $\mathbf{X}_0 = (1, 2)$ we set $x_0 = 1$ and $y_0 = 2$; thus,

$$d_{\mathbf{X}_0} f = 6 \, dx + 2 \, dy$$

and

$$(d_{\mathbf{X}_0} f)(\mathbf{X} - \mathbf{X}_0) = 6(x - 1) + 2(y - 2).$$

Since $f(1, 2) = 5$, the differentiability of f at $(1, 2)$ implies that

$$\lim_{(x,y) \to (1,2)} \frac{f(x, y) - 5 - 6(x - 1) - 2(y - 2)}{\sqrt{(x - 1)^2 + (y - 2)^2}} = 0.$$

Example 3.7   The differential of a function $f = f(x)$ of one variable is given by

$$d_{x_0} f = f'(x_0) \, dx,$$

where dx is the identity function; that is,

$$dx(t) = t.$$

For example, if

$$f(x) = 3x^2 + 5x^3,$$

then

$$df = (6x + 15x^2)\, dx.$$

If $x_0 = -1$, then

$$d_{x_0} f = 9\, dx, \qquad (d_{x_0} f)(x - x_0) = 9(x + 1),$$

and, since $f(-1) = -2$,

$$\lim_{x \to -1} \frac{f(x) + 2 - 9(x + 1)}{x + 1} = 0.$$

Unfortunately the notation for the differential is so complicated that it obscures the simplicity of the concept. The peculiar symbols df, dx, dy, etc., were introduced in the early stages of the development of calculus to represent very small ("infinitesimal") increments in the variables; however, in modern usage they are not quantities at all, but linear functions. This meaning of the symbol dx differs from its meaning in $\int_a^b f(x)\, dx$, where it serves merely to identify the variable of integration; indeed, some authors omit it in the latter context and write simply $\int_a^b f$.

We leave the proof of the next theorem to the reader (*Exercise 17*).

Theorem 3.6   *If f and g are differentiable at $\mathbf{X}_0$, then so are $f + g$ and fg; the same is true of f/g if $g(\mathbf{X}_0) \neq 0$. The differentials are given by*

$$d_{\mathbf{X}_0}(f + g) = d_{\mathbf{X}_0} f + d_{\mathbf{X}_0} g,$$
$$d_{\mathbf{X}_0}(fg) = f(\mathbf{X}_0)\, d_{\mathbf{X}_0} g + g(\mathbf{X}_0)\, d_{\mathbf{X}_0} f,$$

and

$$d_{\mathbf{X}_0}\left(\frac{f}{g}\right) = \frac{g(\mathbf{X}_0)\, d_{\mathbf{X}_0} f - f(\mathbf{X}_0)\, d_{\mathbf{X}_0} g}{[g(\mathbf{X}_0)]^2}.$$

The next theorem provides a widely applicable sufficient condition for differentiability.

Theorem 3.7   *Suppose $f_{x_1}, f_{x_2}, \ldots, f_{x_n}$ exist in a neighborhood of $\mathbf{X}_0 = (c_1, c_2, \ldots, c_n)$ and are continuous at $\mathbf{X}_0$. Then f is differentiable at $\mathbf{X}_0$.*

Proof.   Suppose $\varepsilon > 0$. Our assumptions imply that there is a $\delta > 0$ such that $f_{x_1}, f_{x_2}, \ldots, f_{x_n}$ are defined in the n-sphere

$$S_\delta(\mathbf{X}_0) = \{\mathbf{X} \,|\, |\mathbf{X} - \mathbf{X}_0| < \delta\}$$

and

$$|f_{x_j}(\mathbf{X}) - f_{x_j}(\mathbf{X}_0)| < \varepsilon \qquad \text{if} \quad |\mathbf{X} - \mathbf{X}_0| < \delta. \tag{24}$$

Define

$$\mathbf{X}_j = (x_1, \ldots, x_j, c_{j+1}, \ldots, c_n), \quad 1 \le j \le n - 1,$$

and

$$\mathbf{X}_n = (x_1, x_2, \ldots, x_n) = \mathbf{X}.$$

Thus, $\mathbf{X}_j$ differs from $\mathbf{X}_{j-1}$ ($j = 1, 2, \ldots, n$) in the jth component only, and the line segment from $\mathbf{X}_{j-1}$ to $\mathbf{X}_j$ is in $S_\delta(\mathbf{X}_0)$ if $\mathbf{X}$ is, as we now assume. Now write

$$f(\mathbf{X}) - f(\mathbf{X}_0) = f(\mathbf{X}_n) - f(\mathbf{X}_0) = \sum_{j=1}^{n} [f(\mathbf{X}_j) - f(\mathbf{X}_{j-1})] \tag{25}$$

and consider the auxiliary functions

$$\begin{aligned}
g_1(t) &= f(t, c_2, \ldots, c_n), \\
g_j(t) &= f(x_1, \ldots, x_{j-1}, t, c_{j+1}, \ldots, c_n), \quad 2 \le j \le n - 1, \\
g_n(t) &= f(x_1, \ldots, x_{n-1}, t),
\end{aligned} \tag{26}$$

where, in each case, all variables except t are temporarily regarded as constants. Since

$$f(\mathbf{X}_j) - f(\mathbf{X}_{j-1}) = g_j(x_j) - g_j(c_j),$$

the mean value theorem implies that

$$f(\mathbf{X}_j) - f(\mathbf{X}_{j-1}) = g_j'(\tau_j)(x_j - c_j),$$

where τ_j is between x_j and c_j. From (26),

$$g_j'(\tau_j) = f_{x_j}(\hat{\mathbf{X}}_j),$$

where $\hat{\mathbf{X}}_j$ is on the line segment from $\mathbf{X}_{j-1}$ to $\mathbf{X}_j$; therefore,

$$f(\mathbf{X}_j) - f(\mathbf{X}_{j-1}) = f_{x_j}(\hat{\mathbf{X}}_j)(x_j - c_j),$$

and (25) implies that

$$\begin{aligned}
f(\mathbf{X}) - f(\mathbf{X}_0) &= \sum_{j=1}^{n} f_{x_j}(\hat{\mathbf{X}}_j)(x_j - c_j) \\
&= \sum_{j=1}^{n} f_{x_j}(\mathbf{X}_0)(x_j - c_j) + \sum_{j=1}^{n} [f_{x_j}(\hat{\mathbf{X}}_j) - f_{x_j}(\mathbf{X}_0)](x_j - c_j).
\end{aligned}$$

From this and (24),

$$\left| f(\mathbf{X}) - f(\mathbf{X}_0) - \sum_{j=1}^{n} f_{x_j}(\mathbf{X}_0)(x_j - c_j) \right| \le \varepsilon \sum_{j=1}^{n} |x_j - c_j| \le n\varepsilon |\mathbf{X} - \mathbf{X}_0|,$$

which implies that f is differentiable at $\mathbf{X}_0$. This completes the proof.

We say that f is *continuously differentiable* on a subset S of $\mathscr{R}^n$ if f_{x_1}, $f_{x_2}, \ldots, f_{x_n}$ are continuous on S. Theorem 3.7 implies that such a function is differentiable at each $\mathbf{X}_0$ in S.

Example 3.8   If

$$f(x, y) = \frac{x^2 + y^2}{x - y},$$

then

$$f_x(x, y) = \frac{2x}{x - y} - \frac{x^2 + y^2}{(x - y)^2} \quad \text{and} \quad f_y(x, y) = \frac{2y}{x - y} + \frac{x^2 + y^2}{(x - y)^2}.$$

Since f_x and f_y are continuous on

$$S = \{(x, y) \mid x \neq y\},$$

f is continuously differentiable on S.

Example 3.9   The conditions of Theorem 3.7 are not necessary for differentiability; that is, a function may be differentiable at a point even if its first partial derivatives are not continuous there. For example, let

$$f(x, y) = \begin{cases} (x - y)^2 \sin \dfrac{1}{x - y}, & x \neq y, \\ 0, & x = y. \end{cases}$$

Then

$$f_x(x, y) = 2(x - y) \sin \frac{1}{x - y} - \cos \frac{1}{x - y}, \qquad x \neq y,$$

and

$$f_x(x, x) = \lim_{h \to 0} \frac{f(x + h, x) - f(x, x)}{h} = \lim_{h \to 0} \frac{h^2 \sin(1/h)}{h} = 0,$$

so that f_x exists for all (x, y), but is not continuous on the line $y = x$. The same is true of f_y, since

$$f_y(x, y) = -2(x - y) \sin \frac{1}{x - y} + \cos \frac{1}{x - y}, \qquad x \neq y,$$

and

$$f_y(x, x) = \lim_{h \to 0} \frac{f(x, x + h) - f(x, x)}{h} = -\lim_{h \to 0} \frac{h^2 \sin(1/h)}{h} = 0.$$

Now,

$$\frac{f(x, y) - f(0, 0) - f_x(0, 0)x - f_y(0, 0)y}{\sqrt{x^2 + y^2}} = \begin{cases} \dfrac{(x - y)^2}{\sqrt{x^2 + y^2}} \sin \dfrac{1}{x - y}, & x \ne y, \\ \\ 0, & x = y, \end{cases}$$

and

$$\left| \frac{(x - y)^2}{\sqrt{x^2 + y^2}} \sin \frac{1}{x - y} \right| \le \frac{2(x^2 + y^2)}{\sqrt{x^2 + y^2}} = 2\sqrt{x^2 + y^2}, \qquad x \ne y;$$

therefore,

$$\lim_{(x,y)\to(0,0)} \frac{f(x, y) - f(0, 0) - f_x(0, 0)x - f_y(0, 0)y}{\sqrt{x^2 + y^2}} = 0,$$

and so f is differentiable at $(0, 0)$ even though f_x and f_y are not continuous there.

Geometric interpretation of differentiability

In Section 2.3 we saw that if a function f of one variable is differentiable at x_0, then the curve $y = f(x)$ has a tangent line

$$y = T(x) = f(x_0) + f'(x_0)(x - x_0)$$

which approximates it so well near x_0 that

$$\lim_{x \to x_0} \frac{f(x) - T(x)}{x - x_0} = 0.$$

Moreover, the tangent line is the "limit" of the secant line through the points $(x_1, f(x_1))$ and $(x_0, f(x_0))$ as x_1 approaches x_0 (Figure 3.2, Section 2.3). Differentiability of a function of n variables has an analogous geometric interpretation. We illustrate for $n = 2$.

If f is defined in a region D in $\mathcal{R}^2$, the set of points (x, y, z) such that

$$z = f(x, y), \qquad (x, y) \in D, \tag{27}$$

is a *surface* in $\mathcal{R}^3$ (Figure 3.1). If f is differentiable at $\mathbf{X}_0 = (x_0, y_0)$, then the plane

$$z = T(x, y) = f(\mathbf{X}_0) + f_x(\mathbf{X}_0)(x - x_0) + f_y(\mathbf{X}_0)(y - y_0) \tag{28}$$

intersects the surface (27) at $(x_0, y_0, f(x_0, y_0))$, and approximates the latter so well near this point that

$$\lim_{(x,y)\to(x_0,y_0)} \frac{f(x, y) - T(x, y)}{\sqrt{(x - x_0)^2 + (y - y_0)^2}} = 0$$

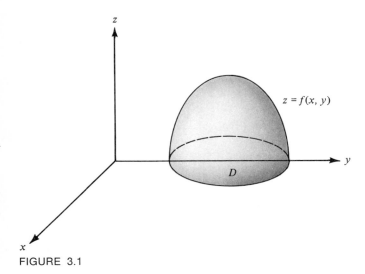

FIGURE 3.1

(Figure 3.2). Moreover, (28) is the only plane in $\mathcal{R}^3$ with these properties (*Exercise 27*). We say that this plane is *tangent to the surface* $z = f(x, y)$ *at the point* $(x_0, y_0, f(x_0, y_0))$. We will now show that it is the "limit" of "secant planes" associated with the surface $z = f(x, y)$, just as a tangent line to a curve $y = f(x)$ in $\mathcal{R}^2$ is the limit of secant lines to the curve (Section 2.3).

Let $\mathbf{X}_i = (x_i, y_i)$ $(i = 1, 2, 3)$. The equation of the "secant plane" through the points $(x_i, y_i, f(x_i, y_i))$ $(i = 1, 2, 3)$ on the surface $z = f(x, y)$ (Figure 3.3) is

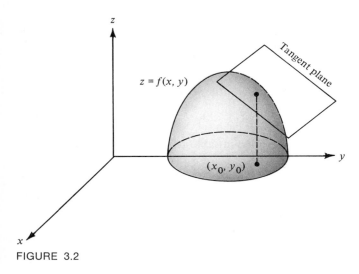

FIGURE 3.2

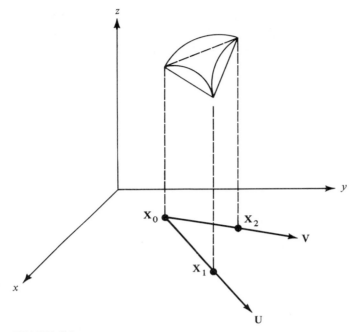

FIGURE 3.3

of the form

$$z = f(\mathbf{X}_0) + A(x - x_0) + B(y - y_0),$$ (29)

where A and B satisfy the system

$$f(\mathbf{X}_1) = f(\mathbf{X}_0) + A(x_1 - x_0) + B(y_1 - y_0),$$
$$f(\mathbf{X}_2) = f(\mathbf{X}_0) + A(x_2 - x_0) + B(y_2 - y_0).$$

Solving for A and B yields

$$A = \frac{[f(\mathbf{X}_1) - f(\mathbf{X}_0)](y_2 - y_0) - [f(\mathbf{X}_2) - f(\mathbf{X}_0)](y_1 - y_0)}{(x_1 - x_0)(y_2 - y_0) - (x_2 - x_0)(y_1 - y_0)}$$ (30)

and

$$B = \frac{[f(\mathbf{X}_2) - f(\mathbf{X}_0)](x_1 - x_0) - [f(\mathbf{X}_1) - f(\mathbf{X}_0)](x_2 - x_0)}{(x_1 - x_0)(y_2 - y_0) - (x_2 - x_0)(y_1 - y_0)}$$ (31)

if

$$(x_1 - x_0)(y_2 - y_0) - (x_2 - x_0)(y_1 - y_0) \neq 0,$$ (32)

which is equivalent to the requirement that $\mathbf{X}_0$, $\mathbf{X}_1$, and $\mathbf{X}_2$ do not lie on a line (*Exercise 25*). If we write

$$\mathbf{X}_1 = \mathbf{X}_0 + t\mathbf{U} \quad \text{and} \quad \mathbf{X}_2 = \mathbf{X}_0 + t\mathbf{V}$$

where $\mathbf{U} = (u_1, u_2)$ and $\mathbf{V} = (v_1, v_2)$ are fixed nonzero vectors (Figure 3.3), then (30), (31), and (32) take the more convenient forms

$$A = \frac{\dfrac{f(\mathbf{X}_0 + t\mathbf{U}) - f(\mathbf{X}_0)}{t} v_2 - \dfrac{f(\mathbf{X}_0 + t\mathbf{V}) - f(\mathbf{X}_0)}{t} u_2}{u_1 v_2 - u_2 v_1}, \tag{33}$$

$$B = \frac{\dfrac{f(\mathbf{X}_0 + t\mathbf{V}) - f(\mathbf{X}_0)}{t} u_1 - \dfrac{f(\mathbf{X}_0 + t\mathbf{U}) - f(\mathbf{X}_0)}{t} v_1}{u_1 v_2 - u_2 v_1}, \tag{34}$$

and

$$u_1 v_2 - u_2 v_1 \neq 0.$$

If f is differentiable at $\mathbf{X}_0$, then

$$f(\mathbf{X}) - f(\mathbf{X}_0) = f_x(\mathbf{X}_0)(x - x_0) + f_y(\mathbf{X}_0)(y - y_0) + \varepsilon(\mathbf{X})|\mathbf{X} - \mathbf{X}_0|, \tag{35}$$

where

$$\lim_{\mathbf{X} \to \mathbf{X}_0} \varepsilon(\mathbf{X}) = 0. \tag{36}$$

Substituting first $\mathbf{X} = \mathbf{X}_0 + t\mathbf{U}$ and then $\mathbf{X} = \mathbf{X}_0 + t\mathbf{V}$ in (35) and dividing by t yields

$$\frac{f(\mathbf{X}_0 + t\mathbf{U}) - f(\mathbf{X}_0)}{t} = f_x(\mathbf{X}_0)u_1 + f_y(\mathbf{X}_0)u_2 + E_1(t)|\mathbf{U}| \tag{37}$$

and

$$\frac{f(\mathbf{X}_0 + t\mathbf{V}) - f(\mathbf{X}_0)}{t} = f_x(\mathbf{X}_0)v_1 + f_y(\mathbf{X}_0)v_2 + E_2(t)|\mathbf{V}|, \tag{38}$$

where

$$E_1(t) = \varepsilon(\mathbf{X}_0 + t\mathbf{U})|t|/t \quad \text{and} \quad E_2(t) = \varepsilon(\mathbf{X}_0 + t\mathbf{V})|t|/t,$$

and so, because of (36),

$$\lim_{t \to 0} E_i(t) = 0, \qquad i = 1, 2. \tag{39}$$

Substituting (37) and (38) into (33) and (34) yields

$$A = f_x(\mathbf{X}_0) + \Delta_1(t), \qquad B = f_y(\mathbf{X}_0) + \Delta_2(t), \tag{40}$$

where

$$\Delta_1(t) = \frac{v_2|\mathbf{U}|E_1(t) - u_2|\mathbf{V}|E_2(t)}{u_1 v_2 - u_2 v_1}$$

and

$$\Delta_2(t) = \frac{u_1|V|E_2(t) - v_1|U|E_1(t)}{u_1 v_2 - u_2 v_1},$$

and so, because of (39),

$$\lim_{t \to 0} \Delta_i(t) = 0, \qquad i = 1, 2. \tag{41}$$

From (29) and (40), the equation of the secant plane is

$$z = f(\mathbf{X}_0) + [f_x(\mathbf{X}_0) + \Delta_1(t)](x - x_0) + [f_y(\mathbf{X}_0) + \Delta_2(t)](y - y_0).$$

Therefore, because of (41), the secant plane "approaches" the tangent plane (28) as t approaches zero.

Maxima and minima

We say that $\mathbf{X}_0$ is a *local extreme point* of f if there is an n-sphere $S_\delta(\mathbf{X}_0)$ (with $\delta > 0$) such that

$$f(\mathbf{X}) - f(\mathbf{X}_0)$$

does not change sign in $S_\delta(\mathbf{X}_0) \cap D_f$. More specifically, $\mathbf{X}_0$ is a *local maximum point* if

$$f(\mathbf{X}) \le f(\mathbf{X}_0)$$

or a *local minimum point* if

$$f(\mathbf{X}) \ge f(\mathbf{X}_0)$$

for all $\mathbf{X}$ in $S_\delta(\mathbf{X}_0) \cap D_f$.

The next theorem is analogous to Theorem 3.4, Section 2.3.

Theorem 3.8   *Suppose f is defined in a neighborhood of $\mathbf{X}_0$ in $\mathscr{R}^n$ and $f_{x_1}(\mathbf{X}_0)$, $f_{x_2}(\mathbf{X}_0), \ldots, f_{x_n}(\mathbf{X}_0)$ exist. Let $\mathbf{X}_0$ be a local extreme point of f. Then*

$$f_{x_i}(\mathbf{X}_0) = 0, \qquad 1 \le i \le n. \tag{42}$$

Proof.   Let

$$\mathbf{E}_1 = (1, 0, \ldots, 0), \quad \mathbf{E}_2 = (0, 1, 0, \ldots, 0), \quad \ldots, \quad \mathbf{E}_n = (0, 0, \ldots, 1),$$

and

$$g_i(t) = f(\mathbf{X}_0 + t\mathbf{E}_i), \qquad 1 \le i \le n.$$

Then g_i is differentiable at $t = 0$, with

$$g_i'(0) = f_{x_i}(\mathbf{X}_0)$$

(Definition 3.1). Since $\mathbf{X}_0$ is a local extreme point of f, $t_0 = 0$ is a local extreme point of g_i. From Theorem 3.4, Section 2.3, this implies that $g_i'(0) = 0$. Now (42) follows, and the proof is complete.

The converse of Theorem 3.8 is false, since (42) may hold at a point $\mathbf{X}_0$ which is not a local extreme point of f; for example, if $\mathbf{X}_0 = (0, 0)$ and

$$f(x, y) = x^3 + y^3.$$

We will say that a point $\mathbf{X}_0$ where (42) holds is a *critical point* of f. Thus, if f is defined in a neighborhood of a local extreme point $\mathbf{X}_0$, then $\mathbf{X}_0$ is a critical point of f; however, a critical point need not be a local extreme point of f.

The use of Theorem 3.8 for finding local extreme points is covered in most elementary calculus courses, so we will not pursue it here; however, we will use the theorem for theoretical purposes and in connection with extreme value problems involving constraints, in Section 6.5.

5.3   EXERCISES

1. Calculate $\partial f(\mathbf{X})/\partial\boldsymbol{\Phi}$:

   (a) $f(x, y) = x^2 + 2xy \cos x$, $\boldsymbol{\Phi} = \left(\dfrac{1}{\sqrt{3}}, -\sqrt{\dfrac{2}{3}}\right)$

   (b) $f(x, y, z) = e^{-x+y^2+2z}$, $\boldsymbol{\Phi} = \left(\dfrac{1}{\sqrt{3}}, -\dfrac{1}{\sqrt{3}}, \dfrac{1}{\sqrt{3}}\right)$

   (c) $f(\mathbf{X}) = |\mathbf{X}|^2$, $\boldsymbol{\Phi} = \left(\dfrac{1}{\sqrt{n}}, \dfrac{1}{\sqrt{n}}, \ldots, \dfrac{1}{\sqrt{n}}\right)$

   (d) $f(x, y, z) = \log(1 + x + y + z)$, $\boldsymbol{\Phi} = (0, 1, 0)$

2. Let

   $$f(x, y) = \begin{cases} \dfrac{xy \sin x}{x^2 + y^2}, & (x, y) \neq (0, 0), \\ 0, & (x, y) = (0, 0), \end{cases}$$

   and let $\boldsymbol{\Phi} = (\phi_1, \phi_2)$ be a unit vector. Find $\partial f(0, 0)/\partial\boldsymbol{\Phi}$.

3. Find $\partial f(\mathbf{X}_0)/\partial\boldsymbol{\Phi}$, where $\boldsymbol{\Phi}$ is the unit vector in the direction of $\mathbf{X}_1 - \mathbf{X}_0$:
   (a) $f(x, y, z) = \sin \pi xyz$; $\mathbf{X}_0 = (1, 1, -2)$, $\mathbf{X}_1 = (3, 2, -1)$
   (b) $f(x, y, z) = e^{-(x^2+y^2+2z)}$; $\mathbf{X}_0 = (1, 0, -1)$, $\mathbf{X}_1 = (2, 0, -1)$

(c) $f(x, y, z) = \log(1 + x + y + z)$; $\mathbf{X}_0 = (1, 0, 1)$, $\mathbf{X}_1 = (3, 0, -1)$
(d) $f(\mathbf{X}) = |\mathbf{X}|^4$; $\mathbf{X}_0 = \mathbf{0}$, $\mathbf{X}_1 = (1, 1, \ldots, 1)$

4. Give a geometrical interpretation of the directional derivative of a function of two variables.

5. Find all first-order partial derivatives:
   (a) $f(x, y, z) = \log(x + y + 2z)$
   (b) $f(x, y, z) = x^2 + 3xyz + 2xy$
   (c) $f(x, y, z) = xe^{yz}$
   (d) $f(x, y, z) = z + \sin x^2 y$

6. Prove Theorem 3.1.

7. Find all second-order partial derivatives of the functions in Exercise 5.

8. Find all second-order partial derivatives at $(0, 0)$:

   (a) $f(x, y) = \begin{cases} \dfrac{xy(x^2 - y^2)}{x^2 + y^2}, & (x, y) \neq (0, 0) \\ 0, & (x, y) = (0, 0) \end{cases}$

   (b) $f(x, y) = \begin{cases} x^2 \tan^{-1} \dfrac{y}{x} - y^2 \tan^{-1} \dfrac{x}{y}, & x \neq 0, y \neq 0 \\ 0, & x = 0 \text{ or } y = 0 \end{cases}$

   (Here $|\tan^{-1} u| < \pi/2$.)

9. Find a function f of two variables such that f_{xy} exists for all (x, y), but f_y exists nowhere.

10. Let u and v be functions of two variables whose second-order partial derivatives are continuous and satisfy

    $$u_x = v_y, \qquad u_y = -v_x$$

    in a region S. Show that

    $$u_{xx} + u_{yy} = v_{xx} + v_{yy} = 0$$

    in S.

11. Let f be a function of $(x_1, x_2, \ldots, x_n)$ $(n \geq 2)$ such that f_{x_i}, f_{x_j}, and $f_{x_i x_j}$ $(i \neq j)$ exist in a neighborhood of $\mathbf{X}_0$ and $f_{x_i x_j}$ is continuous at $\mathbf{X}_0$. Use Theorem 3.2 to prove that $f_{x_j x_i}(\mathbf{X}_0)$ exists and equals $f_{x_i x_j}(\mathbf{X}_0)$.

12. Use Exercise 11 repeatedly to prove Theorem 3.3.

*13. Let $r_1, r_2, \ldots, r_n$ be nonnegative integers such that

    $$r_1 + r_2 + \cdots + r_n = r.$$

(a) Show that

$$(z_1 + z_2 + \cdots + z_n)^r = \sum_r \frac{r!}{r_1! r_2! \cdots r_n!} z_1^{r_1} z_2^{r_2} \cdots z_n^{r_n},$$

where $\sum_r$ denotes summation over all n-tuples $(r_1, r_2, \ldots, r_n)$ that satisfy the stated conditions. (*Hint*: For $r = 2$, this follows from Exercise 17, Section 1.2. Use induction on r.)

(b) Show that there are

$$\frac{r!}{r_1! r_2! \cdots r_n!} \qquad \text{(\textit{multinomial coefficient})}$$

ordered n-tuples of integers $(i_1, i_2, \ldots, i_n)$ which contain r_1 ones, r_2 twos, $\ldots$, and r_n n's.

(c) Let f be a function of $(x_1, x_2, \ldots, x_n)$. Show that there are

$$\frac{r!}{r_1! r_2! \cdots r_n!}$$

partial derivatives $f_{x_{i_1} x_{i_2} \cdots x_{i_r}}$ that involve differentiation r_i times with respect to x_i, for $i = 1, 2, \ldots, n$.

14. Show that the function

$$f(x, y) = \begin{cases} \dfrac{x^2 y}{x^6 + 2y^2}, & (x, y) \neq (0, 0), \\[2mm] 0, & (x, y) = (0, 0), \end{cases}$$

has a directional derivative in every direction at $(0, 0)$, but is not continuous there.

15. Prove: If f_x and f_y are bounded in a neighborhood of (x_0, y_0), then f is continuous at (x_0, y_0).

16. As in Example 3.5, show directly from Definition 3.2 that f is differentiable at $\mathbf{X}_0$:
    (a) $f(x, y) = 2x^2 + 3xy + y^2$, $\mathbf{X}_0 = (1, 2)$
    (b) $f(x, y, z) = 2x^2 + 3x + 4yz$, $\mathbf{X}_0 = (1, 1, 1)$
    (c) $f(\mathbf{X}) = |\mathbf{X}|^2$, $\mathbf{X}_0$ arbitrary

17. Prove Theorem 3.6.

18. (a) Suppose f_x exists in a neighborhood of (x_0, y_0) and is continuous at (x_0, y_0), while f_y merely exists at (x_0, y_0). Show that f is differentiable at (x_0, y_0).
    (b) Generalize part (a) to functions of n variables, so as to obtain a result stronger than Theorem 3.7.

19. Find df and $d_{X_0}f$, and write $(d_{X_0}f)(X - X_0)$:
    (a) $f(x, y) = x^3 + 4xy^2 + 2xy \sin x$, $X_0 = (0, -2)$
    (b) $f(x, y, z) = e^{-(x+y+z)}$, $X_0 = (0, 0, 0)$
    (c) $f(X) = \log(1 + x_1 + 2x_2 + 3x_3 + \cdots + nx_n)$, $X_0 = 0$
    (d) $f(X) = |X|^{2r}$; $X_0 = (1, 1, 1, \ldots, 1)$

20. (a) Suppose f is differentiable at X_0 and $\Phi = (\phi_1, \phi_2, \ldots, \phi_n)$ is a unit vector. Show that

$$\frac{\partial f(X_0)}{\partial \Phi} = f_{x_1}(X_0)\phi_1 + f_{x_2}(X_0)\phi_2 + \cdots + f_{x_n}(X_0)\phi_n.$$

    (b) For what unit vector Φ does $\partial f(X_0)/\partial \Phi$ attain its maximum value?

21. The height of a flagpole is determined by measuring the angle α between the horizontal plane and the line of sight from a point 30 feet from the base of the flagpole to its tip. Suppose the error in pacing off the 30 feet is less than one tenth of a foot and α is found to be 60 degrees, with an error of less than one tenth of a degree. Use the differential to estimate an upper bound for the percentage error in the determination of the height.

22. Let

$$f(x, y) = \begin{cases} x^2 \sin \dfrac{1}{x} + y^2 \sin \dfrac{1}{y}, & x \neq 0 \text{ and } y \neq 0, \\[2mm] x^2 \sin \dfrac{1}{x}, & x \neq 0 \text{ and } y = 0, \\[2mm] y^2 \sin \dfrac{1}{y}, & x = 0 \text{ and } y \neq 0, \\[2mm] 0, & x = y = 0. \end{cases}$$

Show that f is differentiable at $(0, 0)$, but f_x and f_y are not continuous there.

*23. Prove: If f_x and f_y are differentiable at (x_0, y_0), then $f_{xy}(x_0, y_0) = f_{yx}(x_0, y_0)$. [*Hint*: Consider $A(h, k)$ as in the proof of Theorem 3.2.]

*24. (a) Let f_{x_i} and f_{x_j} be differentiable at a point X_0 in $\mathcal{R}^n$. Show from Exercise 23 that

$$f_{x_i x_j}(X_0) = f_{x_j x_i}(X_0).$$

    (b) Use part (a) repeatedly to prove that if all $(r - 1)$st-order partial derivatives of f are differentiable at X_0, then $f_{x_{i_1} x_{i_2} \ldots x_{i_r}}(X_0)$ depends only on the number of differentiations with respect to each variable, and not on the order in which they are performed.

25. Prove that (x_0, y_0), (x_1, y_1), and (x_2, y_2) lie on a line if and only if

$$(x_1 - x_0)(y_2 - y_0) - (x_2 - x_0)(y_1 - y_0) = 0.$$

26. Find the equation of the tangent plane to the surface $z = f(x, y)$ at $(x_0, y_0, z_0) = (x_0, y_0, f(x_0, y_0))$:
    (a) $f(x, y) = x^2 + y^2 - 1, (x_0, y_0) = (1, 2)$
    (b) $f(x, y) = 2x + 3y + 1, (x_0, y_0) = (1, -1)$
    (c) $f(x, y) = xy \sin xy, (x_0, y_0) = (1, \pi/2)$
    (d) $f(x, y) = x^2 - 2y^2 + 3xy, (x_0, y_0) = (2, -1)$

27. Prove: If f is differentiable at (x_0, y_0) and

$$\lim_{(x,y) \to (x_0,y_0)} \frac{f(x, y) - a - b(x - x_0) - c(y - y_0)}{\sqrt{(x - x_0)^2 + (y - y_0)^2}} = 0,$$

then $a = f(x_0, y_0)$, $b = f_x(x_0, y_0)$, and $c = f_y(x_0, y_0)$.

5.4   THE CHAIN RULE AND TAYLOR'S THEOREM

We now consider the problem of differentiating a composite function $h(\mathbf{U}) = f(\mathbf{G}(\mathbf{U}))$, where $\mathbf{G} = (g_1, g_2, \ldots, g_n)$ is a vector-valued function as defined in Section 5.2. We will say that $\mathbf{G}$ is differentiable at $\mathbf{U}_0$ if $g_1, g_2, \ldots, g_n$ are differentiable at $\mathbf{U}_0$.

The following theorem is analogous to Theorem 3.3, Section 2.3.

Theorem 4.1 (The Chain Rule)   *Suppose f is differentiable at $\mathbf{X}_0 = (c_1, c_2, \ldots, c_n)$ in $\mathscr{R}^n$, $\mathbf{G} = (g_1, g_2, \ldots, g_n)$ is differentiable at $\mathbf{U}_0 = (\gamma_1, \gamma_2, \ldots, \gamma_m)$ in $\mathscr{R}^m$, and $\mathbf{X}_0 = \mathbf{G}(\mathbf{U}_0)$. Then the composite function, $h = f \circ \mathbf{G}$, defined by*

$$h(\mathbf{U}) = f(\mathbf{G}(\mathbf{U})), \tag{1}$$

is differentiable at $\mathbf{U}_0$, and

$$d_{\mathbf{U}_0} h = f_{x_1}(\mathbf{X}_0) \, d_{\mathbf{U}_0} g_1 + f_{x_2}(\mathbf{X}_0) \, d_{\mathbf{U}_0} g_2 + \cdots + f_{x_n}(\mathbf{X}_0) \, d_{\mathbf{U}_0} g_n. \tag{2}$$

Proof.   We leave it to the reader to show that $\mathbf{U}_0$ is an interior point of the domain of h (*Exercise 1*), so it is legitimate to ask if h is differentiable at $\mathbf{U}_0$.
Since f is differentiable at $\mathbf{X}_0$,

$$f(\mathbf{X}) - f(\mathbf{X}_0) = \sum_{i=1}^{n} f_{x_i}(\mathbf{X}_0)(x_i - c_i) + \varepsilon(\mathbf{X})|\mathbf{X} - \mathbf{X}_0|, \tag{3}$$

where

$$\lim_{\mathbf{X} \to \mathbf{X}_0} \varepsilon(\mathbf{X}) = 0,$$

and, for convenience, we define $\varepsilon(\mathbf{X}_0) = 0$ so that ε is continuous at $\mathbf{X}_0$. Substituting $\mathbf{X} = \mathbf{G}(\mathbf{U})$ and $\mathbf{X}_0 = \mathbf{G}(\mathbf{U}_0)$ in (3) and recalling (1) yields

$$h(\mathbf{U}) - h(\mathbf{U}_0) = \sum_{i=1}^{n} f_{x_i}(\mathbf{X}_0)[g_i(\mathbf{U}) - g_i(\mathbf{U}_0)]$$
$$+ \varepsilon(\mathbf{G}(\mathbf{U}))|\mathbf{G}(\mathbf{U}) - \mathbf{G}(\mathbf{U}_0)|. \tag{4}$$

Since $g_1, g_2, \ldots, g_n$ are differentiable at $\mathbf{U}_0$,

$$g_i(\mathbf{U}) - g_i(\mathbf{U}_0) = (d_{\mathbf{U}_0} g_i)(\mathbf{U} - \mathbf{U}_0) + \varepsilon_i(\mathbf{U})|\mathbf{U} - \mathbf{U}_0|, \tag{5}$$

where

$$\lim_{\mathbf{U} \to \mathbf{U}_0} \varepsilon_i(\mathbf{U}) = 0, \qquad 1 \le i \le n. \tag{6}$$

Substituting (5) into (4) yields

$$h(\mathbf{U}) - h(\mathbf{U}_0) = \sum_{i=1}^{n} f_{x_i}(\mathbf{X}_0)(d_{\mathbf{U}_0} g_i)(\mathbf{U} - \mathbf{U}_0)$$
$$+ \left(\sum_{i=1}^{n} f_{x_i}(\mathbf{X}_0)\varepsilon_i(\mathbf{U}) \right)|\mathbf{U} - \mathbf{U}_0| \tag{7}$$
$$+ \varepsilon(\mathbf{G}(\mathbf{U}))|\mathbf{G}(\mathbf{U}) - \mathbf{G}(\mathbf{U}_0)|.$$

We leave it to the reader to show from (5) and (6) that

$$|\mathbf{G}(\mathbf{U}) - \mathbf{G}(\mathbf{U}_0)| \le [M + \sigma(\mathbf{U})]|\mathbf{U} - \mathbf{U}_0|,$$

where M is a suitable constant and

$$\lim_{\mathbf{U} \to \mathbf{U}_0} \sigma(\mathbf{U}) = 0.$$

From this and (7),

$$h(\mathbf{U}) - h(\mathbf{U}_0) = \sum_{i=1}^{n} f_{x_i}(\mathbf{X}_0)(d_{\mathbf{U}_0} g_i)(\mathbf{U} - \mathbf{U}_0) + E(\mathbf{U})|\mathbf{U} - \mathbf{U}_0|, \tag{8}$$

where

$$|E(\mathbf{U})| \le \sum_{i=1}^{n} |f_{x_i}(\mathbf{X}_0)| \, |\varepsilon_i(\mathbf{U})| + |\varepsilon(\mathbf{G}(\mathbf{U}))|[M + \sigma(\mathbf{U})] \tag{9}$$

Observing that

$$\lim_{\mathbf{U} \to \mathbf{U}_0} \varepsilon(\mathbf{G}(\mathbf{U})) = \varepsilon(\mathbf{G}(\mathbf{U}_0)) = \varepsilon(\mathbf{X}_0) = 0$$

(Theorem 2.6), we conclude from (6) and (9) that

$$\lim_{\mathbf{U} \to \mathbf{U}_0} E(\mathbf{U}) = 0.$$

This and (8) imply that h is differentiable at $\mathbf{U}_0$ and $d_{\mathbf{U}_0}h$ is given by (2), which completes the proof.

Example 4.1   Let

$$f(x, y, z) = 2x^2 + 4xy + 3yz,$$

$$g_1(u, v) = u^2 + v^2, \quad g_2(u, v) = u^2 - 2v^2, \quad g_3(u, v) = uv,$$

and

$$h(u, v) = f(g_1(u, v), g_2(u, v), g_3(u, v)).$$

Let $\mathbf{U}_0 = (1, -1)$ and

$$\mathbf{X}_0 = (g_1(\mathbf{U}_0), g_2(\mathbf{U}_0), g_3(\mathbf{U}_0)) = (2, -1, -1).$$

Then

$$f_x(\mathbf{X}_0) = 4, \qquad f_y(\mathbf{X}_0) = 5, \qquad f_z(\mathbf{X}_0) = -3,$$

$$\frac{\partial g_1(\mathbf{U}_0)}{\partial u} = 2, \qquad \frac{\partial g_1(\mathbf{U}_0)}{\partial v} = -2,$$

$$\frac{\partial g_2(\mathbf{U}_0)}{\partial u} = 2, \qquad \frac{\partial g_2(\mathbf{U}_0)}{\partial v} = 4,$$

$$\frac{\partial g_3(\mathbf{U}_0)}{\partial u} = -1, \qquad \frac{\partial g_3(\mathbf{U}_0)}{\partial v} = 1.$$

Therefore,

$$d_{\mathbf{U}_0}g_1 = 2\,du - 2\,dv, \quad d_{\mathbf{U}_0}g_2 = 2\,du + 4\,dv, \quad d_{\mathbf{U}_0}g_3 = -du + dv,$$

and, from (2),

$$\begin{aligned}
d_{\mathbf{U}_0}h &= f_x(\mathbf{X}_0)\,d_{\mathbf{U}_0}g_1 + f_y(\mathbf{X}_0)\,d_{\mathbf{U}_0}g_2 + f_z(\mathbf{X}_0)\,d_{\mathbf{U}_0}g_3 \\
&= 4(2\,du - 2\,dv) + 5(2\,du + 4\,dv) - 3(-du + dv) \\
&= 21\,du + 9\,dv.
\end{aligned}$$

Since

$$d_{\mathbf{U}_0}h = h_u(\mathbf{U}_0)\,du + h_v(\mathbf{U}_0)\,dv,$$

we conclude that

$$h_u(\mathbf{U}_0) = 21 \quad \text{and} \quad h_v(\mathbf{U}_0) = 9, \tag{10}$$

which can also be obtained by writing h explicitly in terms of (u, v) and differentiating; thus,

$$\begin{aligned}
h(u, v) &= 2[g_1(u, v)]^2 + 4g_1(u, v)g_2(u, v) + 3g_2(u, v)g_3(u, v) \\
&= 2(u^2 + v^2)^2 + 4(u^2 + v^2)(u^2 - 2v^2) + 3(u^2 - 2v^2)uv \\
&= 6u^4 + 3u^3v - 6uv^3 - 6v^4.
\end{aligned}$$

Hence,

$$h_u(u, v) = 24u^3 + 9u^2v - 6v^3,$$
$$h_v(u, v) = 3u^3 - 18uv^2 - 24v^3,$$

so that $h_u(1, -1) = 21$ and $h_v(1, -1) = 9$, which agrees with (10).

Corollary 4.1 *Under the assumptions of Theorem 4.1,*

$$\frac{\partial h(\mathbf{U_0})}{\partial u_i} = \sum_{j=1}^{n} \frac{\partial f(\mathbf{X_0})}{\partial x_j} \frac{\partial g_j(\mathbf{U_0})}{\partial u_i}, \qquad 1 \le i \le m. \tag{11}$$

Proof. Substituting

$$d_{\mathbf{U_0}} g_i = \frac{\partial g_i(\mathbf{U_0})}{\partial u_1} du_1 + \frac{\partial g_i(\mathbf{U_0})}{\partial u_2} du_2 + \cdots + \frac{\partial g_i(\mathbf{U_0})}{\partial u_m} du_m, \qquad 1 \le i \le n,$$

into (2) and collecting multipliers of $du_1, du_2, \ldots, du_m$ yields

$$d_{\mathbf{U_0}} h = \sum_{i=1}^{m} \left[\sum_{j=1}^{n} \frac{\partial f(\mathbf{X_0})}{\partial x_j} \frac{\partial g_j(\mathbf{U_0})}{\partial u_i} \right] du_i.$$

However, from Theorem 3.4,

$$d_{\mathbf{U_0}} h = \sum_{i=1}^{m} \frac{\partial h(\mathbf{U_0})}{\partial u_i} du_i.$$

Comparing the last two equations yields (11).

When it is not important to emphasize the particular point $\mathbf{X_0}$, we write (11) less formally as

$$\frac{\partial h}{\partial u_i} = \sum_{j=1}^{n} \frac{\partial f}{\partial x_j} \frac{\partial g_j}{\partial u_i}, \qquad 1 \le i \le m, \tag{12}$$

it being understood that in calculating $\partial h(\mathbf{U_0})/\partial u_i$, $\partial g_j/\partial u_i$ is evaluated at $\mathbf{U_0}$ and $\partial f/\partial x_j$ at $\mathbf{X_0} = \mathbf{G}(\mathbf{U_0})$.

The formulas (11) and (12) can also be simplified by replacing the symbol $\mathbf{G}$ with $\mathbf{X} = \mathbf{X(U)}$; then we write

$$h(\mathbf{U}) = f(\mathbf{X(U)})$$

and

$$\frac{\partial h(\mathbf{U_0})}{\partial u_i} = \sum_{j=1}^{n} \frac{\partial f(\mathbf{X_0})}{\partial x_j} \frac{\partial x_j(\mathbf{U_0})}{\partial u_i},$$

or simply

$$\frac{\partial h}{\partial u_i} = \sum_{j=1}^{n} \frac{\partial f}{\partial x_j} \frac{\partial x_j}{\partial u_i}.$$

Example 4.2   Let (r, θ) be polar coordinates in the xy plane; thus,

$$x = r \cos \theta, \qquad y = r \sin \theta.$$

Suppose $f = f(x, y)$ is differentiable on a set S and let

$$h(r, \theta) = f(r \cos \theta, r \sin \theta).$$

If $(r \cos \theta, r \sin \theta) \in S$, (13) implies that

$$\frac{\partial h}{\partial r} = \frac{\partial f}{\partial x} \frac{\partial x}{\partial r} + \frac{\partial f}{\partial y} \frac{\partial y}{\partial r} = \cos \theta \frac{\partial f}{\partial x} + \sin \theta \frac{\partial f}{\partial y} \tag{14}$$

and

$$\frac{\partial h}{\partial \theta} = \frac{\partial f}{\partial x} \frac{\partial x}{\partial \theta} + \frac{\partial f}{\partial y} \frac{\partial y}{\partial \theta} = -r \sin \theta \frac{\partial f}{\partial x} + r \cos \theta \frac{\partial f}{\partial y},$$

where f_x and f_y are evaluated at $(x, y) = (r \cos \theta, r \sin \theta)$.

The proof of Corollary 4.1 suggests a straightforward way to calculate the partial derivatives of a composite function without using (13) explicitly. If $h(\mathbf{U}) = f(\mathbf{X}(\mathbf{U}))$, then Theorem 4.1—in the more casual notation introduced before Example 4.1—implies that

$$dh = f_{x_1} \, dx_1 + f_{x_2} \, dx_2 + \cdots + f_{x_n} \, dx_n, \tag{15}$$

where $dx_1, dx_2, \cdots, dx_n$ must, in turn, be written in terms of the independent differentials (differentials of the independent variables) $du_1, du_2, \ldots, du_m$:

$$dx_i = \frac{\partial x_i}{\partial u_1} \, du_1 + \frac{\partial x_i}{\partial u_2} \, du_2 + \cdots + \frac{\partial x_i}{\partial u_m} \, du_m.$$

Substituting this into (15) and collecting the multipliers of $du_1, du_2, \ldots, du_m$ yields (13).

Example 4.3   If

$$h(r, \theta) = f(x(r, \theta), y(r, \theta), z)$$

then

$$dh = f_x \, dx + f_y \, dy + f_z \, dz.$$

But

$$dx = \frac{\partial x}{\partial r} \, dr + \frac{\partial x}{\partial \theta} \, d\theta \quad \text{and} \quad dy = \frac{\partial y}{\partial r} \, dr + \frac{\partial y}{\partial \theta} \, d\theta;$$

hence

$$dh = f_x\left(\frac{\partial x}{\partial r}\,dr + \frac{\partial x}{\partial \theta}\,d\theta\right) + f_y\left(\frac{\partial y}{\partial r}\,dr + \frac{\partial y}{\partial \theta}\,d\theta\right) + f_z\,dz$$

$$= \left(f_x\frac{\partial x}{\partial r} + f_y\frac{\partial y}{\partial r}\right)dr + \left(f_x\frac{\partial x}{\partial \theta} + f_y\frac{\partial y}{\partial \theta}\right)d\theta + f_z\,dz,$$

and therefore

$$h_r = f_x\frac{\partial x}{\partial r} + f_y\frac{\partial y}{\partial r}, \quad h_\theta = f_x\frac{\partial x}{\partial \theta} + f_y\frac{\partial y}{\partial \theta}, \quad h_z = f_z.$$

Example 4.4 Let

$$h(x) = f(x, y(x, z(x)), z(x)).$$

Then

$$dh = f_x\,dx + f_y\,dy + f_z\,dz, \tag{16}$$
$$dy = y_x\,dx + y_z\,dz, \tag{17}$$

and

$$dz = z'\,dx. \tag{18}$$

Substituting (18) into (17) yields

$$dy = (y_x + y_z z')\,dx$$

and substituting this and (18) into (16) yields

$$dh = [f_x + f_y(y_x + y_z z') + f_z z']\,dx;$$

hence,

$$h' = f_x + f_y(y_x + y_z z') + f_z z'.$$

Here f_x, f_y, and f_z are evaluated at $(x, y(x, z(x)), z(x))$, y_x and y_z at $(x, z(x))$, and z' at x.

Higher derivatives of composite functions

Higher derivatives of composite functions can be computed by repeatedly applying the chain rule. For example, differentiating (13) with respect to u_k yields

$$\frac{\partial^2 h}{\partial u_k\,\partial u_i} = \sum_{j=1}^{n}\frac{\partial}{\partial u_k}\left(\frac{\partial f}{\partial x_j}\frac{\partial x_j}{\partial u_i}\right)$$

$$= \sum_{j=1}^{n}\frac{\partial f}{\partial x_j}\frac{\partial^2 x_j}{\partial u_k\,\partial u_i} + \sum_{j=1}^{n}\frac{\partial x_j}{\partial u_i}\frac{\partial}{\partial u_k}\left(\frac{\partial f}{\partial x_j}\right). \tag{19}$$

Care must be exercised in finding

$$\frac{\partial}{\partial u_k}\left(\frac{\partial f}{\partial x_j}\right),$$

which really stands here for

$$\frac{\partial}{\partial u_k}\left[\frac{\partial f(\mathbf{X}(\mathbf{U}))}{\partial x_j}\right]. \tag{20}$$

The safest procedure is to write temporarily

$$g(\mathbf{X}) = \frac{\partial f(\mathbf{X})}{\partial x_j};$$

then (20) becomes

$$\frac{\partial g(\mathbf{X}(\mathbf{U}))}{\partial u_k} = \sum_{s=1}^{n} \frac{\partial g(\mathbf{X}(\mathbf{U}))}{\partial x_s} \frac{\partial x_s(\mathbf{U})}{\partial u_k}.$$

Since

$$\frac{\partial g}{\partial x_s} = \frac{\partial^2 f}{\partial x_s \, \partial x_j},$$

this yields

$$\frac{\partial}{\partial u_k}\left(\frac{\partial f}{\partial x_k}\right) = \sum_{s=1}^{n} \frac{\partial^2 f}{\partial x_s \, \partial x_j} \frac{\partial x_s}{\partial u_k}.$$

Substituting this into (19) yields

$$\frac{\partial^2 h}{\partial u_k \, \partial u_i} = \sum_{j=1}^{n} \frac{\partial f}{\partial x_j} \frac{\partial^2 x_j}{\partial u_k \, \partial u_i} + \sum_{j=1}^{n} \frac{\partial x_j}{\partial u_i} \sum_{s=1}^{n} \frac{\partial^2 f}{\partial x_s \, \partial x_j} \frac{\partial x_s}{\partial u_k}. \tag{21}$$

To compute $h_{u_i u_k}(\mathbf{U}_0)$ from this formula we evaluate the partial derivatives of $x_1, x_2, \ldots, x_n$ at $\mathbf{U}_0$ and those of f at $\mathbf{X}_0 = \mathbf{X}(\mathbf{U}_0)$. The formula is valid if $x_1, x_2, \ldots, x_n$ and their first partial derivatives are differentiable at $\mathbf{U}_0$ and $f, f_{x_1}, f_{x_2}, \ldots, f_{x_n}$ and their first partial derivatives are differentiable at $\mathbf{X}_0$.

Instead of memorizing (21), the reader should understand how it is derived, and use the method—rather than the formula—when calculating second partial derivatives of composite functions. The same method applies to the calculation of higher derivatives.

Example 4.5 Suppose f_x and f_y in Example 4.2 are differentiable on an open set S in $\mathscr{R}^2$. Differentiating (14) with respect to r yields

$$\frac{\partial^2 h}{\partial r^2} = \cos \theta \, \frac{\partial}{\partial r}\left(\frac{\partial f}{\partial x}\right) + \sin \theta \, \frac{\partial}{\partial r}\left(\frac{\partial f}{\partial y}\right)$$

$$= \cos \theta \left(\frac{\partial^2 f}{\partial x^2} \frac{\partial x}{\partial r} + \frac{\partial^2 f}{\partial y \, \partial x} \frac{\partial y}{\partial r}\right) + \sin \theta \left(\frac{\partial^2 f}{\partial x \, \partial y} \frac{\partial x}{\partial r} + \frac{\partial^2 f}{\partial y^2} \frac{\partial y}{\partial r}\right) \tag{22}$$

if $(x, y) \in S$. Since

$$\frac{\partial x}{\partial r} = \cos\theta, \quad \frac{\partial y}{\partial r} = \sin\theta, \quad \text{and} \quad \frac{\partial^2 f}{\partial x \, \partial y} = \frac{\partial^2 f}{\partial y \, \partial x}$$

if $(x, y) \in S$ (*Exercise 23, Section 5.3*), (22) yields

$$\frac{\partial^2 h}{\partial r^2} = \cos^2\theta \frac{\partial^2 f}{\partial x^2} + 2\sin\theta \cos\theta \frac{\partial^2 f}{\partial x \, \partial y} + \sin^2\theta \frac{\partial^2 f}{\partial y^2}.$$

Differentiating (14) with respect to θ yields

$$\frac{\partial^2 h}{\partial \theta \, \partial r} = -\sin\theta \frac{\partial f}{\partial x} + \cos\theta \frac{\partial f}{\partial y} + \cos\theta \frac{\partial}{\partial \theta}\left(\frac{\partial f}{\partial x}\right) + \sin\theta \frac{\partial}{\partial \theta}\left(\frac{\partial f}{\partial y}\right)$$

$$= -\sin\theta \frac{\partial f}{\partial x} + \cos\theta \frac{\partial f}{\partial y} + \cos\theta \left(\frac{\partial^2 f}{\partial x^2}\frac{\partial x}{\partial \theta} + \frac{\partial^2 f}{\partial y \, \partial x}\frac{\partial y}{\partial \theta}\right)$$

$$+ \sin\theta \left(\frac{\partial^2 f}{\partial x \, \partial y}\frac{\partial x}{\partial \theta} + \frac{\partial^2 f}{\partial y^2}\frac{\partial y}{\partial \theta}\right).$$

Since

$$\frac{\partial x}{\partial \theta} = -r\sin\theta \quad \text{and} \quad \frac{\partial y}{\partial \theta} = r\cos\theta,$$

it follows that

$$\frac{\partial^2 h}{\partial \theta \, \partial r} = -\sin\theta \frac{\partial f}{\partial x} + \cos\theta \frac{\partial f}{\partial y} - r\sin\theta \cos\theta \left(\frac{\partial^2 f}{\partial x^2} - \frac{\partial^2 f}{\partial y^2}\right)$$

$$+ r(\cos^2\theta - \sin^2\theta) \frac{\partial^2 f}{\partial x \, \partial y}.$$

The mean value theorem

For a composite function of the form

$$h(t) = f(x_1(t), x_2(t), \dots, x_n(t)),$$

where t is a real variable, $x_1, x_2, \dots, x_n$ are differentiable at t_0, and f is differentiable at $\mathbf{X}_0 = \mathbf{X}(t_0)$, (11) takes the form

$$h'(t_0) = \sum_{j=1}^{n} f_{x_j}(\mathbf{X}(t_0))x_j'(t_0). \tag{23}$$

This will be useful in the following proof.

Theorem 4.2 (Mean Value Theorem for Functions of Several Variables) *Let f be continuous at* $\mathbf{X}_1 = (a_1, a_2, \ldots, a_n)$ *and* $\mathbf{X}_2 = (b_1, b_2, \ldots, b_n)$, *and differentiable on the line segment L from* $\mathbf{X}_1$ *to* $\mathbf{X}_2$. *Then*

$$f(\mathbf{X}_2) - f(\mathbf{X}_1) = \sum_{i=1}^{n} f_{x_i}(\mathbf{X}_0)(b_i - a_i) = (d_{\mathbf{X}_0} f)(\mathbf{X}_2 - \mathbf{X}_1) \tag{24}$$

for some $\mathbf{X}_0$ *on L, distinct from* $\mathbf{X}_1$ *and* $\mathbf{X}_2$.

Proof.   The equation of L is

$$\mathbf{X} = \mathbf{X}(t) = t\mathbf{X}_2 + (1 - t)\mathbf{X}_1, \qquad 0 \le t \le 1.$$

Our hypotheses imply that the function

$$h(t) = f(\mathbf{X}(t))$$

is continuous on $[0, 1]$ and differentiable on $(0, 1)$. Since

$$x_j(t) = tb_j + (1 - t)a_j,$$

(23) implies that

$$h'(t) = \sum_{i=1}^{n} f_{x_i}(\mathbf{X}(t))(b_i - a_i), \qquad 0 < t < 1.$$

From the mean value theorem for functions of one variable (Theorem 3.8, Section 2.3),

$$h(1) - h(0) = h'(t_0),$$

where $t_0 \in (0, 1)$. Since $h(1) = f(\mathbf{X}_2)$ and $h(0) = f(\mathbf{X}_1)$, this implies (24), with $\mathbf{X}_0 = \mathbf{X}(t_0)$.

Corollary 4.2   *If* $f_{x_1}, f_{x_2}, \ldots, f_{x_n}$ *are identically zero in an open region S of* $\mathscr{R}^n$, *then f is constant in S.*

Proof.   Let $\mathbf{X}_0$ and $\mathbf{X}$ be in S. Since S is an open region, it is polygonally connected (Theorem 1.8). Therefore, there are points

$$\mathbf{X}_0, \mathbf{X}_1, \ldots, \mathbf{X}_n = \mathbf{X}$$

such that the line segment L_i from $\mathbf{X}_{i-1}$ to $\mathbf{X}_i$ is in S, for $i = 1, 2, \ldots, n$. From Theorem 4.2,

$$f(\mathbf{X}_i) - f(\mathbf{X}_{i-1}) = \sum_{i=1}^{n} (d_{\tilde{\mathbf{X}}_i} f)(\mathbf{X}_i - \mathbf{X}_{i-1}),$$

where $\tilde{\mathbf{X}}_i$ is on L_i, and so in S. Therefore,

$$f_{x_1}(\tilde{\mathbf{X}}_i) = f_{x_2}(\tilde{\mathbf{X}}_i) = \cdots = f_{x_n}(\tilde{\mathbf{X}}_i) = 0,$$

which means that $d_{\tilde{\mathbf{X}}_i} f \equiv 0$. Hence,

$$f(\mathbf{X}_0) = f(\mathbf{X}_1) = \cdots = f(\mathbf{X}_n);$$

that is, $f(\mathbf{X}) = f(\mathbf{X}_0)$ for every $\mathbf{X}$ in S. This completes the proof.

Higher differentials and Taylor's theorem

Suppose f is defined in a neighborhood N of $\mathbf{X}_0 = (c_1, c_2, \ldots, c_n)$ and $\mathbf{\Phi} = (\phi_1, \phi_2, \ldots, \phi_n)$ is an arbitrary vector. Then the set of points

$$\mathbf{X} = \mathbf{X}_0 + t\mathbf{\Phi}, \qquad |t| < r,$$

is in N for some $r > 0$, so that the function

$$h(t) = f(\mathbf{X}_0 + t\mathbf{\Phi})$$

is defined for $|t| < r$. From Theorem 4.1 [see also (23)],

$$h'(t) = \sum_{i=1}^{n} f_{x_i}(\mathbf{X}_0 + t\mathbf{\Phi})\phi_i$$

if f is differentiable in N, and

$$h''(t) = \sum_{j=1}^{n} \frac{\partial}{\partial x_j} \sum_{i=1}^{n} \frac{\partial f(\mathbf{X}_0 + t\mathbf{\Phi})}{\partial x_i} \phi_i$$

$$= \sum_{i,j=1}^{n} \frac{\partial^2 f(\mathbf{X}_0 + t\mathbf{\Phi})}{\partial x_j \, \partial x_i} \phi_i \phi_j$$

if $f_{x_1}, f_{x_2}, \ldots, f_{x_n}$ are differentiable in N. Continuing in this way, we see that

$$h^{(r)}(t) = \sum_{i_1, i_2, \ldots, i_r = 1}^{n} \frac{\partial^r f(\mathbf{X}_0 + t\mathbf{\Phi})}{\partial x_{i_r} \, \partial x_{i_{r-1}} \cdots \partial x_{i_1}} \phi_{i_1} \phi_{i_2} \cdots \phi_{i_r} \tag{25}$$

if all partial derivatives of f of order $\leq r - 1$ are differentiable in N.

Definition 4.1 *Suppose $r \geq 1$ and all partial derivatives of f of order $\leq r - 1$ are differentiable in a neighborhood of $\mathbf{X}_0$. Then the rth differential of f at $\mathbf{X}_0$, denoted by $d^r_{\mathbf{X}_0} f$, is defined by*

$$(d^r_{\mathbf{X}_0})(\mathbf{\Phi}) = \sum_{i_1, i_2, \ldots, i_r = 1}^{n} \frac{\partial^r f(\mathbf{X}_0)}{\partial x_{i_r} \, \partial x_{i_{r-1}} \cdots \partial x_{i_1}} \phi_{i_1} \phi_{i_2} \cdots \phi_{i_r}. \tag{26}$$

For convenience we define

$$(d^0_{\mathbf{X}_0} f)(\mathbf{\Phi}) = f(\mathbf{X}_0).$$

Notice that $d^1_{\mathbf{X}_0} f = d_{\mathbf{X}_0} f$.

Under the assumptions of Definition 4.1 the value of

$$\frac{\partial^r f(\mathbf{X}_0)}{\partial x_{i_r} \, \partial x_{i_{r-1}} \cdots \partial x_{i_1}}$$

depends only on the number of times f is differentiated with respect to each variable, and not on the order in which the differentiations are performed (*Exercise 24, Section 5.3*); hence, *Exercise 13, Section 5.3*, implies that (26) can be rewritten as

$$(d^r_{\mathbf{X}_0} f)(\mathbf{\Phi}) = \sum_r \frac{r!}{r_1! r_2! \cdots r_n!} \frac{\partial^r f(\mathbf{X}_0)}{\partial x_1^{r_1} \, \partial x_2^{r_2} \cdots \partial x_n^{r_n}} \phi_1^{r_1} \phi_2^{r_2} \cdots \phi_n^{r_n},$$

where $\sum_r$ indicates summation over all ordered n-tuples $(r_1, r_2, \ldots, r_n)$ of nonnegative integers such that

$$r_1 + r_2 + \cdots + r_n = r.$$

In particular, if $n = 2$,

$$(d^r_{\mathbf{X}_0} f)(\mathbf{\Phi}) = \sum_{j=0}^r \binom{r}{j} \frac{\partial^r f(x_0, y_0)}{\partial x^j \, \partial y^{r-j}} \phi_1^j \phi_2^{r-j}.$$

Example 4.6   Let

$$f(x, y) = \frac{1}{1 + ax + by},$$

where a and b are constants. Then

$$\frac{\partial^r f(x, y)}{\partial x^j \, \partial y^{r-j}} = (-1)^r r! \frac{a^j b^{r-j}}{(1 + ax + by)^{r+1}},$$

and so

$$(d^r_{\mathbf{X}_0} f)(\mathbf{\Phi}) = \frac{(-1)^r r!}{(1 + ax_0 + by_0)^{r+1}} \sum_{j=0}^r \binom{r}{j} a^j b^{r-j} \phi_1^j \phi_2^{r-j}$$

$$= \frac{(-1)^r r!}{(1 + ax_0 + by_0)^{r+1}} (a\phi_1 + b\phi_2)^r$$

if $1 + ax_0 + by_0 \neq 0$.

Example 4.7   Let

$$f(\mathbf{X}) = \exp\left(-\sum_{j=1}^n a_j x_j\right),$$

where $a_1, a_2, \ldots, a_n$ are constants. Then

$$\frac{\partial^r f(\mathbf{X})}{\partial x_1^{r_1} \partial x_2^{r_2} \cdots \partial x_n^{r_n}} = (-1)^r a_1^{r_1} a_2^{r_2} \cdots a_n^{r_n} \exp\left(-\sum_{j=1}^{n} a_j x_j\right).$$

Therefore, if $\mathbf{X}_0 = \mathbf{0}$,

$$(d_{\mathbf{X}_0}^r f)(\mathbf{\Phi}) = (-1)^r \sum_r \frac{r!}{r_1! r_2! \cdots r_n!} a_1^{r_1} a_2^{r_2} \cdots a_n^{r_n} \phi_1^{r_1} \phi_2^{r_2} \cdots \phi_n^{r_n}$$

$$= (-1)^r (a_1 \phi_1 + a_2 \phi_2 + \cdots + a_n \phi_n)^r$$

(*Exercise 13, Section 5.3*).

The next theorem is analogous to Taylor's theorem for functions of one variable (Theorem 5.3, Section 2.5).

Theorem 4.3 (Taylor's Theorem for Functions of Several Variables) *Suppose f and its partial derivatives of order $\leq k$ are differentiable at $\mathbf{X}_0$ and $\mathbf{X}$ in $\mathcal{R}^n$, and on the line segment L connecting them. Then*

$$f(\mathbf{X}) = \sum_{r=0}^{k} \frac{1}{r!} (d_{\mathbf{X}_0}^r f)(\mathbf{X} - \mathbf{X}_0) + \frac{1}{(k+1)!} (d_{\tilde{\mathbf{X}}}^{k+1} f)(\mathbf{X} - \mathbf{X}_0), \tag{27}$$

where $\tilde{\mathbf{X}}$ is on L and distinct from $\mathbf{X}_0$ and $\mathbf{X}$.

Proof. Define

$$h(t) = f(\mathbf{X}_0 + t(\mathbf{X} - \mathbf{X}_0)). \tag{28}$$

With $\mathbf{\Phi} = \mathbf{X} - \mathbf{X}_0$, our assumptions and the discussion preceding Definition 4.1 imply that $h, h', \ldots, h^{(k+1)}$ exist on $[0, 1]$. From Taylor's theorem for functions of one variable,

$$h(1) = \sum_{r=0}^{k} \frac{h^{(r)}(0)}{r!} + \frac{h^{(k+1)}(\tau)}{(k+1)!}, \tag{29}$$

where $0 < \tau < 1$. From (28),

$$h(0) = f(\mathbf{X}_0) \quad \text{and} \quad h(1) = f(\mathbf{X}). \tag{30}$$

From (25) and (26), with $\mathbf{\Phi} = \mathbf{X} - \mathbf{X}_0$,

$$h^{(r)}(0) = (d_{\mathbf{X}_0}^r f)(\mathbf{X} - \mathbf{X}_0), \qquad 1 \leq r \leq k, \tag{31}$$

and

$$h^{(k+1)}(\tau) = (d_{\tilde{\mathbf{X}}}^{k+1} f)(\mathbf{X} - \mathbf{X}_0), \tag{32}$$

where

$$\tilde{\mathbf{X}} = \mathbf{X}_0 + \tau(\mathbf{X} - \mathbf{X}_0)$$

is on L and distinct from $\mathbf{X}_0$ and $\mathbf{X}$. Substituting (30), (31), and (32) into (29) yields (27) and completes the proof.

Example 4.8   With $\mathbf{X}_0 = (0, 0)$, Theorem 4.3 and the results of Example 4.6 imply that if $1 + ax + by > 0$, then

$$\frac{1}{1 + ax + by} = \sum_{r=0}^{k} (-1)^r (ax + by)^r + (-1)^{k+1} \left(\frac{ax + by}{1 + a\tau x + b\tau y} \right)^{k+1}$$

where $0 < \tau < 1$. [Note that τ depends on k as well as (x, y).]

Example 4.9   With $\mathbf{X}_0 = \mathbf{0}$, Theorem 4.3 and the results of Example 4.7 imply that

$$\exp\left(-\sum_{j=1}^{n} a_j x_j \right) = \sum_{r=0}^{k} \frac{(-1)^r}{r!} (a_1 x_1 + a_2 x_2 + \cdots + a_n x_n)^r$$

$$+ \frac{(-1)^{k+1}}{(k+1)!} (a_1 x_1 + a_2 x_2 + \cdots + a_n x_n)^{k+1}$$

$$\times \exp\left[-\tau \left(\sum_{j=1}^{n} a_j x_j \right) \right],$$

where $0 < \tau < 1$.

By analogy with the situation for functions of one variable, we define the kth *Taylor polynomial of f about* $\mathbf{X}_0$ by

$$T_k(\mathbf{X}) = \sum_{r=0}^{k} \frac{1}{r!} (d_{\mathbf{X}_0}^r f)(\mathbf{X} - \mathbf{X}_0)$$

if the differentials exist; then (27) can be rewritten as

$$f(\mathbf{X}) = T_k(\mathbf{X}) + \frac{1}{(k+1)!} (d_{\tilde{\mathbf{X}}}^{k+1} f)(\mathbf{X} - \mathbf{X}_0).$$

A sufficient condition for relative extreme values

The next theorem leads to a useful sufficient condition for local maxima and minima. It is related to the analogous theorem for functions of one variable (Theorem 5.1, Section 2.5). Both that theorem and this one indicate how well the Taylor polynomial for f approximates f. Strictly speaking, however, the present theorem is not a generalization of Theorem 5.1, Section 2.5 (*Exercise 20*).

Theorem 4.4 *Suppose f and its partial derivatives of order $\leq k - 1$ are differentiable in a neighborhood N of $\mathbf{X}_0$ and all of its kth-order partial derivatives are continuous at $\mathbf{X}_0$. Then*

$$\lim_{\mathbf{X} \to \mathbf{X}_0} \frac{f(\mathbf{X}) - T_k(\mathbf{X})}{|\mathbf{X} - \mathbf{X}_0|^k} = 0. \tag{33}$$

Proof. If $\varepsilon > 0$, there is a $\delta > 0$ such that the n-sphere $S_\delta(\mathbf{X}_0)$ is in N and all kth-order partial derivatives of f satisfy

$$\left| \frac{\partial^k f(\tilde{\mathbf{X}})}{\partial x_{i_k} \partial x_{i_{k-1}} \cdots \partial x_{i_1}} - \frac{\partial^k f(\mathbf{X}_0)}{\partial x_{i_k} \partial x_{i_{k-1}} \cdots \partial x_{i_1}} \right| < \varepsilon, \qquad \tilde{\mathbf{X}} \in S_\delta(\mathbf{X}_0). \tag{34}$$

Now suppose $\mathbf{X} \in S_\delta(\mathbf{X}_0)$. From Theorem 4.3 (with k replaced by $k - 1$),

$$f(\mathbf{X}) = T_{k-1}(\mathbf{X}) + \frac{1}{k!}(d_{\tilde{\mathbf{X}}}^k f)(\mathbf{X} - \mathbf{X}_0),$$

where $\tilde{\mathbf{X}}$ is on the line segment from $\mathbf{X}_0$ to $\mathbf{X}$, and therefore in $S_\delta(\mathbf{X}_0)$. This can be rewritten as

$$f(\mathbf{X}) = T_k(\mathbf{X}) + \frac{1}{k!}[(d_{\tilde{\mathbf{X}}}^k f)(\mathbf{X} - \mathbf{X}_0) - (d_{\mathbf{X}_0}^k f)(\mathbf{X} - \mathbf{X}_0)]. \tag{35}$$

But (34) implies that

$$|(d_{\tilde{\mathbf{X}}}^k f)(\mathbf{X} - \mathbf{X}_0) - (d_{\mathbf{X}_0}^k f)(\mathbf{X} - \mathbf{X}_0)| < n^k \varepsilon |\mathbf{X} - \mathbf{X}_0|^k \tag{36}$$

(*Exercise 19*), and therefore (35) yields

$$\frac{|f(\mathbf{X}) - T_k(\mathbf{X})|}{|\mathbf{X} - \mathbf{X}_0|^k} < \frac{n^k \varepsilon}{k!}, \qquad \mathbf{X} \in S_\delta(\mathbf{X}_0).$$

This implies (33) and completes the proof.

Let r be a positive integer and $\mathbf{X}_0 = (c_1, c_2, \ldots, c_n)$. A function of the form

$$p(\mathbf{X}) = \sum_r a_{r_1 r_2 \cdots r_n} (x_1 - c_1)^{r_1} (x_2 - c_2)^{r_2} \cdots (x_n - c_n)^{r_n}, \tag{37}$$

where the coefficients $a_{r_1 r_2 \cdots r_n}$ are constants and the summation is over all n-tuples of nonnegative integers $(r_1, r_2, \ldots, r_n)$ such that

$$r_1 + r_2 + \cdots + r_n = r,$$

is a *homogeneous polynomial of degree r in $\mathbf{X} - \mathbf{X}_0$*, provided at least one of the coefficients is nonzero. For example, if f satisfies the conditions of Definition 4.1, then the function

$$p(\mathbf{X}) = (d_{\mathbf{X}_0}^r f)(\mathbf{X} - \mathbf{X}_0)$$

is such a polynomial, provided at least one of the r-th order mixed partial derivatives of f at $\mathbf{X}_0$ is nonzero.

Clearly, $p(\mathbf{X}_0) = 0$ if p is a homogeneous polynomial of degree $r \geq 1$ in $\mathbf{X} - \mathbf{X}_0$. If $p(\mathbf{X}) \geq 0$ for all $\mathbf{X}$, we say that p is *positive semidefinite*; if $p(\mathbf{X}) > 0$ except when $\mathbf{X} = \mathbf{X}_0$, it is *positive definite*. Similarly, p is *negative semidefinite* if $p(\mathbf{X}) \leq 0$ or *negative definite* if $p(\mathbf{X}) < 0$ for all $\mathbf{X} \neq \mathbf{X}_0$. In all these cases p is *semidefinite*.

With p as in (37),

$$p(-\mathbf{X} + 2\mathbf{X}_0) = (-1)^r p(\mathbf{X}),$$

so p cannot be semidefinite if r is odd.

Example 4.10   The polynomial

$$p(x, y, z) = x^2 + y^2 + z^2 + xy + xz + yz$$

is homogeneous of degree 2 in $\mathbf{X} = (x, y, z)$. It can be rewritten as

$$p(x, y, z) = \tfrac{1}{2}[(x + y)^2 + (y + z)^2 + (z + x)^2].$$

This is clearly nonnegative and $p(\bar{x}, \bar{y}, \bar{z}) = 0$ if and only if

$$\bar{x} + \bar{y} = \bar{y} + \bar{z} = \bar{z} + \bar{x} = 0,$$

which is equivalent to $(\bar{x}, \bar{y}, \bar{z}) = (0, 0, 0)$. Therefore, p is positive definite and $-p$ is negative definite.

The polynomial

$$p_1(x, y, z) = x^2 + y^2 + z^2 + 2xy$$

can be rewritten as

$$p_1(x, y, z) = (x + y)^2 + z^2.$$

Again it is obvious that $p_1(x, y, z)$ cannot be negative, but $p(1, -1, 0) = 0$; hence, p_1 is positive semidefinite and $-p_1$ is negative semidefinite.

The polynomial

$$p_2(x, y, z) = x^2 - y^2 + z^2$$

is not semidefinite, since, for example,

$$p_2(1, 0, 0) > 0 \quad \text{and} \quad p_2(0, 1, 0) < 0.$$

Theorem 3.8 implies that if f is differentiable and attains a local extreme value at $\mathbf{X}_0$, then

$$d_{\mathbf{X}_0} f = 0 \tag{38}$$

[since $f_{x_1}(\mathbf{X}_0) = f_{x_2}(\mathbf{X}_0) = \cdots = f_{x_n}(\mathbf{X}_0) = 0$]. However, the converse is false. The next theorem provides a method for deciding whether a point satisfying (38) is an extreme point. It is related to Theorem 5.2, Section 2.5.

Theorem 4.5 *Suppose f satisfies the hypotheses of Theorem 4.4 with $k \geq 2$, and*

$$d_{\mathbf{X}_0}^r f \equiv 0 \quad (1 \leq r \leq k-1), \qquad d_{\mathbf{X}_0}^k f \not\equiv 0. \tag{39}$$

Then:

(a) $\mathbf{X}_0$ is not a local extreme point of f unless $d_{\mathbf{X}_0}^k f$ is semidefinite as a polynomial in $\mathbf{X} - \mathbf{X}_0$, and therefore $\mathbf{X}_0$ is certainly not a local extreme point of f if k is odd;

(b) $\mathbf{X}_0$ is a local minimum point of f if $d_{\mathbf{X}_0}^k f$ is positive definite, or a local maximum point if $d_{\mathbf{X}_0}^k f$ is negative definite;

(c) if $d_{\mathbf{X}_0}^k f$ is semidefinite, then $\mathbf{X}_0$ may be a local extreme point of f, but it need not be.

Proof.   From (39) and Theorem 4.4,

$$\lim_{\mathbf{X} \to \mathbf{X}_0} \frac{f(\mathbf{X}) - f(\mathbf{X}_0) - (1/k!)(d_{\mathbf{X}_0}^k)(\mathbf{X} - \mathbf{X}_0)}{|\mathbf{X} - \mathbf{X}_0|^k} = 0. \tag{40}$$

If we set $\mathbf{X} = \mathbf{X}_0 + t\mathbf{U}$, where $\mathbf{U}$ is a constant vector, then

$$(d_{\mathbf{X}_0}^k f)(\mathbf{X} - \mathbf{X}_0) = t^k (d_{\mathbf{X}_0}^k f)(\mathbf{U}),$$

so (40) implies that

$$\lim_{t \to 0} \frac{f(\mathbf{X}_0 + t\mathbf{U}) - f(\mathbf{X}_0) - (t^k/k!)(d_{\mathbf{X}_0}^k f)(\mathbf{U})}{t^k} = 0,$$

or

$$\lim_{t \to 0} \frac{f(\mathbf{X}_0 + t\mathbf{U}) - f(\mathbf{X}_0)}{t^k} = \frac{1}{k!}(d_{\mathbf{X}_0}^k f)(\mathbf{U}) \tag{41}$$

for any constant vector $\mathbf{U}$. To prove (a), suppose $d_{\mathbf{X}_0}^k f$ is not semidefinite; then there are vectors $\mathbf{U}_1$ and $\mathbf{U}_2$ such that

$$(d_{\mathbf{X}_0}^k f)(\mathbf{U}_1) > 0 \quad \text{and} \quad (d_{\mathbf{X}_0}^k f)(\mathbf{U}_2) < 0.$$

This and (41) imply that

$$f(\mathbf{X}_0 + t\mathbf{U}_1) > f(\mathbf{X}_0) \quad \text{and} \quad f(\mathbf{X}_0 + t\mathbf{U}_2) < f(\mathbf{X}_0)$$

for t sufficiently small; hence, $\mathbf{X}_0$ is not a local extreme point of f.

To prove (b), first assume that $d_{\mathbf{X}_0}^k f$ is positive definite. Then it can be shown that there is a $\rho > 0$ such that

$$\frac{(d_{\mathbf{X}_0}^k f)(\mathbf{X} - \mathbf{X}_0)}{k!} \geq \rho |\mathbf{X} - \mathbf{X}_0|^k \tag{42}$$

for all $\mathbf{X}$ (Exercise 21). From (40) there is a $\delta > 0$ such that

$$\frac{f(\mathbf{X}) - f(\mathbf{X}_0) - (1/k!)(d_{\mathbf{X}_0}^k f)(\mathbf{X} - \mathbf{X}_0)}{|\mathbf{X} - \mathbf{X}_0|^k} > -\frac{\rho}{2} \quad \text{if} \quad |\mathbf{X} - \mathbf{X}_0| < \delta,$$

and therefore

$$f(\mathbf{X}) - f(\mathbf{X}_0) > \frac{1}{k!}(d_{\mathbf{X}_0}^k)(\mathbf{X} - \mathbf{X}_0) - \frac{\rho}{2}|\mathbf{X} - \mathbf{X}_0|^k \quad \text{if} \quad |\mathbf{X} - \mathbf{X}_0| < \delta.$$

This and (42) imply that

$$f(\mathbf{X}) - f(\mathbf{X}_0) > \frac{\rho}{2}|\mathbf{X} - \mathbf{X}_0|^k \quad \text{if} \quad |\mathbf{X} - \mathbf{X}_0| < \delta,$$

which implies that $\mathbf{X}_0$ is a local minimum point of f. This proves half of (b); we leave the other half to the reader (*Exercise 22*).

To prove (c) merely requires examples; see *Exercise 23*.

Corollary 4.3   *Suppose f, f_x, and f_y are differentiable in a neighborhood of a critical point $\mathbf{X}_0$ of f, and f_{xx}, f_{yy}, and f_{xy} are continuous at $\mathbf{X}_0$. Let*

$$D = f_{xx}(\mathbf{X}_0)f_{yy}(\mathbf{X}_0) - f_{xy}^2(\mathbf{X}_0).$$

Then:

   (a) $\mathbf{X}_0$ is a local extreme point of f if $D > 0$; it is a local minimum point if $f_{xx}(\mathbf{X}_0) > 0$, a local maximum point if $f_{xx}(\mathbf{X}_0) < 0$;
   (b) $\mathbf{X}_0$ is not a local extreme point of f if $D < 0$.

Proof.   Write $\mathbf{X} - \mathbf{X}_0 = (u, v)$ and

$$p(u, v) = (d_{\mathbf{X}_0}^2 f)(u, v) = Au^2 + 2Buv + Cv^2,$$

where $A = f_{xx}(\mathbf{X}_0)$, $B = f_{xy}(\mathbf{X}_0)$, and $C = f_{yy}(\mathbf{X}_0)$, so that

$$D = AC - B^2.$$

If $D > 0$, then $A \neq 0$, and we can write

$$(d_{\mathbf{X}_0}^2 f)(u, v) = A\left(u^2 + \frac{2B}{A}uv + \frac{B^2}{A^2}v^2\right) + \left(C - \frac{B^2}{A}\right)v^2$$

$$= A\left(u + \frac{B}{A}v\right)^2 + \frac{D}{A}v^2.$$

This cannot vanish unless $u = v = 0$; hence, $d_{\mathbf{X}_0}^2 f$ is positive definite if $A > 0$ or negative definite if $A < 0$, and Theorem 4.5(b) implies (a).

If $D < 0$, there are three possibilities:

1. $A \neq 0$; then $p(1, 0) = A$ and $p\left(-\dfrac{B}{A}, 1\right) = \dfrac{D}{A}$;

2. $C \neq 0$; then $p(0, 1) = C$ and $p\left(1, \dfrac{-B}{C}\right) = \dfrac{D}{C}$;

3. $A = C = 0$; then $B \neq 0$ and $p(1, 1) = -p(1, -1) = 2B$.

In each case the two given values of p differ in sign, and so $\mathbf{X}_0$ is not a local extreme point of f, from Theorem 4.5(a).

Example 4.11   If

$$f(x, y) = e^{ax^2 + by^2}$$

then

$$f_x(x, y) = 2axf(x, y), \qquad f_y(x, y) = 2byf(x, y);$$

therefore,

$$f_x(0, 0) = f_y(0, 0) = 0,$$

and $(0, 0)$ is a critical point of f. To apply Corollary 4.3, we calculate

$$f_{xx}(x, y) = (2a + 4a^2x^2)f(x, y),$$
$$f_{yy}(x, y) = (2b + 4b^2y^2)f(x, y),$$
$$f_{xy}(x, y) = 4abxyf(x, y).$$

Therefore,

$$D = f_{xx}(0, 0)f_{yy}(0, 0) - f_{xy}^2(0, 0) = (2a)(2b) - (0)(0) = 4ab.$$

Corollary 4.3 implies that $(0, 0)$ is a local minimum point if a and b are positive, a local maximum if they are negative, and neither if one is positive and the other is negative. Corollary 4.3 does not apply if a or b is zero.

5.4   EXERCISES

In the exercises on the use of the chain rule, assume that the functions satisfy appropriate differentiability conditions.

1. Under the assumptions of Theorem 4.1, show that $\mathbf{U}_0$ is an interior point of the domain of h.

2. Let $h(U) = f(G(U))$ and find $d_{U_0}h$ by Theorem 4.1, and then by writing h explicitly as a function of U:

(a) $f(x, y) = 3x^2 + 4xy^2 + 3x$, $\quad g_1(u, v) = ve^{u+v-1}$, $\quad g_2(u, v) = e^{-u+v-1}$, $(u_0, v_0) = (0, 1)$

(b) $f(x, y, z) = e^{-(x+y+z)}$, $g_1(u, v, w) = \log u - \log v + \log w$, $g_2(u, v, w) = -2 \log u - 3 \log w$, $g_3(u, v, w) = \log u + \log v + 2 \log w$, $(u_0, v_0, w_0) = (1, 1, 1)$

(c) $f(x, y) = (x + y)^2$, $g_1(u, v) = u \cos v$, $g_2(u, v) = u \sin v$, $(u_0, v_0) = (3, \pi/2)$

(d) $f(x, y, z) = x^2 + y^2 + z^2$, $g_1(u, v, w) = u \cos v \sin w$, $g_2(u, v, w) = u \cos v \cos w$; $g_3(u, v, w) = u \sin v$, $(u_0, v_0, w_0) = (4, \pi/3, \pi/6)$

3. Let $h(r, \theta, z) = f(x, y, z)$, where $x = r \cos \theta$ and $y = r \sin \theta$. Find h_r, h_θ, and h_z in terms of f_x, f_y, and f_z.

4. Let $h(r, \theta, \phi) = f(x, y, z)$, where $x = r \sin \phi \cos \theta$, $y = r \sin \phi \sin \theta$, and $z = r \cos \phi$. Find h_r, h_θ, and h_ϕ in terms of f_x, f_y, and f_z.

5. Prove:

(a) If $h(u, v) = f(u^2 + v^2)$, then

$$v \frac{\partial h}{\partial u} - u \frac{\partial h}{\partial v} = 0.$$

(b) If $h(u, v) = f(\sin u + \cos v)$, then

$$\sin v \frac{\partial h}{\partial u} + \cos u \frac{\partial h}{\partial v} = 0.$$

(c) If $h(u, v) = f(u/v)$, then

$$u \frac{\partial h}{\partial u} + v \frac{\partial h}{\partial v} = 0.$$

(d) If $h(u, v) = f(g(u, v), -g(u, v))$, then

$$dh = (f_x - f_y)(g_u \, du + g_v \, dv).$$

6. Find h_y and h_z if

$$h(y, z) = g(x(y, z), y, z, w(y, z)).$$

7. Let $g(x, y, z, w) = \log(1 + 3x + 4y + 2z + w)$, $z(x, y) = x^2 + y^2$, $w(x, y) = 2xy$. $y(x) = -3x$, and $h(x) = g(x, y(x), z(x, y(x)), w(x, y(x)))$. Find $h'(1)$.

8. Suppose that g_1 and g_2 are differentiable and f is continuous for all x. Show that

$$\frac{d}{dx} \int_{g_1(x)}^{g_2(x)} f(u) \, du = f(g_2(x))g_2'(x) - f(g_1(x))g_1'(x).$$

9. We say that f is *homogeneous of degree r* if D_f is open and there is a constant r such that $f(\lambda x, \lambda y) = \lambda^r f(x, y)$ whenever $\lambda > 0$ and (x, y) and $(\lambda x, \lambda y)$ are in D_f.
   (a) Prove: If f is differentiable and homogeneous of degree r, then

   $$x f_x(x, y) + y f_y(x, y) = r f(x, y).$$

   (This is *Euler's theorem for homogeneous functions*.)
   (b) Generalize part (a) to functions of n variables.

10. If $h(r, \theta) = f(r \cos \theta, r \sin \theta)$, show that

    $$f_{xx} + f_{yy} = h_{rr} + \frac{1}{r} h_r + h_{\theta\theta}.$$

11. Let

    $$h(u, v) = f\left(\frac{u}{u^2 + v^2}, \frac{-v}{u^2 + v^2}\right).$$

    Show that $h_{uu} + h_{vv} = 0$ if $f_{xx} + f_{yy} = 0$.

12. Let $h(u, v) = f(a(u, v), b(u, v))$, where $a_u = b_v$ and $a_v = -b_u$. Show that

    $$h_{uu} + h_{vv} = (f_{xx} + f_{yy})(a_u^2 + a_v^2).$$

13. Prove: If

    $$u(x, t) = f(x - ct) + g(x + ct),$$

    then $u_{tt} = c^2 u_{xx}$.

14. Let $h(u, v) = f(u + v, u - v)$. Show that:
    (a) $f_{xx} - f_{yy} = h_{uv}$ (b) $f_{xx} + f_{yy} = \frac{1}{2}(h_{uu} + h_{vv})$

15. Returning to Exercise 4, find h_{rr} and $h_{r\theta}$ in terms of the partial derivatives of f.

16. Let $h_{uv} = 0$ for all (u, v). Show that h is of the form

    $$h(u, v) = F(u) + G(v).$$

    Use this and Exercise 14(a) to show that if $f_{xx} - f_{yy} = 0$ for all (x, y), then

    $$f(x, y) = F(x + y) + G(x - y).$$

17. Prove or give a counterexample: If f is differentiable and $f_x = 0$ in a region D, then $f(x_1, y) = f(x_2, y)$ whenever (x_1, y) and (x_2, y) are in D; that is, $f(x, y)$ depends only on y.

18. Find $T_3(\mathbf{X})$:
    (a) $f(x, y) = e^x \cos y$, $\mathbf{X}_0 = (0, 0)$
    (b) $f(x, y) = e^{-x-y}$, $\mathbf{X}_0 = (0, 0)$

(c) $f(x, y, z) = (x + y + z - 3)^5$, $X_0 = (1, 1, 1)$
(d) $f(x, y, z) = \sin x \sin y \sin z$, $X_0 = (0, 0, 0)$

19. In the proof of Theorem 4.4, show that Eq. (36) is valid.

20. Why is Theorem 4.4 not a generalization of Theorem 5.1, Section 2.5?

21. Suppose p is a homogeneous polynomial of degree r in Y and $p(Y) > 0$ for all nonzero Y in $\mathscr{R}^n$. Show that there is a $\rho > 0$ such that $p(Y) \geq \rho|Y|^r$ for all Y in $\mathscr{R}^n$. (*Hint*: p assumes a minimum on the set $\{Y \mid |Y| = 1\}$.) Use this to establish the inequality in Eq. (42).

22. Complete the proof of Theorem 4.5(b).

23. (a) Show that $(0, 0)$ is critical point of each of the following functions, and that they have positive semidefinite second differentials at $(0, 0)$:

$$p(x, y) = x^2 - 2xy + y^2 + x^4 + y^4;$$
$$q(x, y) = x^2 - 2xy + y^2 - x^4 - y^4.$$

(b) Show that D, as defined in Corollary 4.3, is zero for both p and q.
(c) Show that $(0, 0)$ is a local minimum point of p, but not a local extreme point of q.

24. Classify the critical points:
(a) $f(x, y) = \cos(x^2 + y^2)$ (b) $f(x, y) = e^{x^2 + y^2}$
(c) $f(x, y) = e^{x^4 + y^6}$.   (d) $f(x, y) = \sin(x^2 + y^2)$

vector-valued functions of several variables

6.1 LINEAR TRANSFORMATIONS AND MATRICES

In this and subsequent sections it will often be convenient to write vectors vertically; thus, instead of $\mathbf{X} = (x_1, x_2, \ldots, x_n)$, we will write

$$\mathbf{X} = \begin{bmatrix} x_1 \\ x_2 \\ \vdots \\ x_n \end{bmatrix}$$

in situations where it is more convenient to do so. These arise in connection with matrix operations, defined below.

We have defined vector-valued functions as ordered n-tuples of real-valued functions, in connection with composite functions $h = f \circ \mathbf{G}$, where f is real-valued and $\mathbf{G}$ vector-valued. We now consider vector-valued functions as objects of interest on their own.

If $f_1, f_2, \ldots, f_m$ are real-valued functions defined on a set D in $\mathscr{R}^n$, then

$$
\mathbf{F} = \begin{bmatrix} f_1 \\ f_2 \\ \vdots \\ f_m \end{bmatrix}
$$

assigns to every $\mathbf{X}$ in D an m-vector

$$
\mathbf{F(X)} = \begin{bmatrix} f_1(\mathbf{X}) \\ f_2(\mathbf{X}) \\ \vdots \\ f_m(\mathbf{X}) \end{bmatrix}
$$

We recall that $f_1, f_2, \ldots, f_m$ are called the *component functions*, or simply *components*, of $\mathbf{F}$. To indicate that the domain of $\mathbf{F}$ is in $\mathscr{R}^n$ and its range in $\mathscr{R}^m$, we write

$$
\mathbf{F} \colon \mathscr{R}^n \to \mathscr{R}^m.
$$

We also say that $\mathbf{F}$ is a *transformation* from $\mathscr{R}^n$ to $\mathscr{R}^m$. If $m = 1$, we identify $\mathbf{F}$ with its single component function f_1, and regard it as a real-valued function.

Example 1.1   The transformation $\mathbf{F} \colon \mathscr{R}^2 \to \mathscr{R}^3$ defined by

$$
\mathbf{F}(x, y) = \begin{bmatrix} 2x + 3y \\ -x + 4y \\ x - y \end{bmatrix}
$$

has component functions

$$
f_1(x, y) = 2x + 3y, \quad f_2(x, y) = -x + 4y, \quad f_3(x, y) = x - y.
$$

Linear transformations

The simplest interesting transformations from $\mathscr{R}^n$ to $\mathscr{R}^m$ are the *linear transformations*, defined as follows.

Definition 1.1   *A transformation* $\mathbf{L} \colon \mathscr{R}^n \to \mathscr{R}^m$ *defined on all of* $\mathscr{R}^n$ *is said to be linear if*

$$
\mathbf{L}(\mathbf{X} + \mathbf{Y}) = \mathbf{L}(\mathbf{X}) + \mathbf{L}(\mathbf{Y})
$$

for all $\mathbf{X}$ *and* $\mathbf{Y}$ *in* $\mathscr{R}^n$, *and*

$$\mathbf{L}(a\mathbf{X}) = a\mathbf{L}(\mathbf{X})$$

for all $\mathbf{X}$ *in* $\mathscr{R}^n$ *and real numbers* a.

Theorem 1.1   *A transformation* $\mathbf{L}: \mathscr{R}^n \to \mathscr{R}^m$ *defined on all of* $\mathscr{R}^n$ *is linear if and only if*

$$\mathbf{L}(\mathbf{X}) = \begin{bmatrix} a_{11}x_1 + a_{12}x_2 + \cdots + a_{1n}x_n \\ a_{21}x_1 + a_{22}x_2 + \cdots + a_{2n}x_n \\ \vdots \\ a_{m1}x_1 + a_{m2}x_2 + \cdots + a_{mn}x_n \end{bmatrix}, \tag{1}$$

where the a_{ij}'s *are constants.*

Proof.   It can be seen by induction (*Exercise 1*) that if $\mathbf{L}$ is linear, then

$$\mathbf{L}(a_1\mathbf{X}_1 + a_2\mathbf{X}_2 + \cdots + a_k\mathbf{X}_k) = a_1\mathbf{L}(\mathbf{X}_1) + a_2\mathbf{L}(\mathbf{X}_2) + \cdots + a_k\mathbf{L}(\mathbf{X}_k) \tag{2}$$

for any vectors $\mathbf{X}_1, \mathbf{X}_2, \ldots, \mathbf{X}_k$ and real numbers $a_1, a_2, \ldots, a_k$. Any $\mathbf{X}$ in $\mathscr{R}^n$ can be written as

$$\mathbf{X} = \begin{bmatrix} x_1 \\ x_2 \\ \vdots \\ x_n \end{bmatrix} = x_1 \begin{bmatrix} 1 \\ 0 \\ \vdots \\ 0 \end{bmatrix} + x_2 \begin{bmatrix} 0 \\ 1 \\ \vdots \\ 0 \end{bmatrix} + \cdots + x_n \begin{bmatrix} 0 \\ 0 \\ \vdots \\ 1 \end{bmatrix}$$

$$= x_1\mathbf{E}_1 + x_2\mathbf{E}_2 + \cdots + x_n\mathbf{E}_n.$$

Applying (2) with $k = n$, $\mathbf{X}_i = \mathbf{E}_i$, and $a_i = x_i$ yields

$$\mathbf{L}(\mathbf{X}) = x_1\mathbf{L}(\mathbf{E}_1) + x_2\mathbf{L}(\mathbf{E}_2) + \cdots + x_n\mathbf{L}(\mathbf{E}_n). \tag{3}$$

Now denote

$$\mathbf{L}(\mathbf{E}_j) = \begin{bmatrix} a_{1j} \\ a_{2j} \\ \vdots \\ a_{mj} \end{bmatrix},$$

so that (3) becomes

$$\mathbf{L}(\mathbf{X}) = x_1 \begin{bmatrix} a_{11} \\ a_{21} \\ \vdots \\ a_{m1} \end{bmatrix} + x_2 \begin{bmatrix} a_{12} \\ a_{22} \\ \vdots \\ a_{m2} \end{bmatrix} + \cdots + x_n \begin{bmatrix} a_{1n} \\ a_{2n} \\ \vdots \\ a_{mn} \end{bmatrix},$$

which is equivalent to (1). This proves that if **L** is linear, then it has the form (1). We leave the proof of the converse to the reader (*Exercise 2*).

We call the rectangular array

$$\mathbf{A} = \begin{bmatrix} a_{11} & a_{12} & \cdots & a_{1n} \\ a_{21} & a_{21} & \cdots & a_{2n} \\ \vdots & \vdots & & \vdots \\ a_{m1} & a_{m2} & \cdots & a_{mn} \end{bmatrix} \tag{4}$$

the *matrix* of the linear transformation (1). The number a_{ij} in the ith row and jth column of **A** is called the (i, j)th *element of* **A**. We say that **A** is an $m \times n$ matrix, since it has m rows and n columns. We will sometimes abbreviate (4) as

$$\mathbf{A} = [a_{ij}].$$

Example 1.2   The transformation **F** of Example 1.1 is linear. Its matrix is

$$\begin{bmatrix} 2 & 3 \\ -1 & 4 \\ 1 & -1 \end{bmatrix}.$$

We now recall the matrix operations that we need to study the differential calculus of transformations. *Exercises 7, 8, and 9* provide a review of their properties.

Definition 1.2   (a) *If c is a real number and* $\mathbf{A} = [a_{ij}]$ *is an* $m \times n$ *matrix, then* $c\mathbf{A}$ *is the* $m \times n$ *matrix defined by*

$$c\mathbf{A} = [ca_{ij}];$$

that is, cA is obtained by multiplying every element of **A** *by c.*
     (b) *If* $\mathbf{A} = [a_{ij}]$ *and* $\mathbf{B} = [b_{ij}]$ *are* $m \times n$ *matrices, then their sum is the* $m \times n$ *matrix*

$$\mathbf{A} + \mathbf{B} = [a_{ij} + b_{ij}];$$

that is, the sum of two $m \times n$ *matrices is obtained by adding corresponding elements. The sum of two matrices is not defined unless they have the same number of rows and the same number of columns.*
     (c) *If* $\mathbf{A} = [a_{ij}]$ *is an* $m \times p$ *matrix and* $\mathbf{B} = [b_{ij}]$ *is a* $p \times n$ *matrix, then the product* $\mathbf{C} = \mathbf{AB}$ *is the* $m \times n$ *matrix with*

$$c_{ij} = a_{i1}b_{1j} + a_{i2}b_{2j} + \cdots + a_{ip}b_{pj} = \sum_{k=1}^{p} a_{ik}b_{kj}, \quad 1 \le i \le m, \quad 1 \le j \le n.$$

Thus, the (i, j)th *element of* **AB** *is obtained by multiplying each element in the* *ith row of* **A** *by the corresponding element in the jth column of* **B**, *and adding the products. This definition requires that* **A** *have the same number of columns as* **B** *has rows; otherwise,* **AB** *is undefined.*

Example 1.3   Let

$$A = \begin{bmatrix} 2 & 1 & 2 \\ -1 & 0 & 3 \\ 0 & 1 & 0 \end{bmatrix}, \qquad B = \begin{bmatrix} 0 & 1 & 1 \\ -1 & 0 & 2 \\ 3 & 0 & 1 \end{bmatrix},$$

and

$$C = \begin{bmatrix} 5 & 0 & 1 & 2 \\ 3 & 0 & -3 & 1 \\ 1 & 0 & -1 & 1 \end{bmatrix}.$$

Then

$$2A = \begin{bmatrix} 2(2) & 2(1) & 2(2) \\ 2(-1) & 2(0) & 2(3) \\ 2(0) & 2(1) & 2(0) \end{bmatrix} = \begin{bmatrix} 4 & 2 & 4 \\ -2 & 0 & 6 \\ 0 & 2 & 0 \end{bmatrix}$$

and

$$A + B = \begin{bmatrix} 2+0 & 1+1 & 2+1 \\ -1-1 & 0+0 & 3+2 \\ 0+3 & 1+0 & 0+1 \end{bmatrix} = \begin{bmatrix} 2 & 2 & 3 \\ -2 & 0 & 5 \\ 3 & 1 & 1 \end{bmatrix}.$$

The $(2, 3)$ entry in the product **AC** is obtained by multiplying the elements of the second row of **A** by those of the third column of **C** and adding the products:

$$c_{23} = (-1)(1) + (0)(-3) + (3)(-1) = -4.$$

The full product is given by

$$\begin{bmatrix} 2 & 1 & 2 \\ -1 & 0 & 3 \\ 0 & 1 & 0 \end{bmatrix} \begin{bmatrix} 5 & 0 & 1 & 2 \\ 3 & 0 & -3 & 1 \\ 1 & 0 & -1 & 1 \end{bmatrix} = \begin{bmatrix} 15 & 0 & -3 & 7 \\ -2 & 0 & -4 & 1 \\ 3 & 0 & -3 & 1 \end{bmatrix}.$$

Notice that **A** + **C**, **B** + **C**, **CA**, and **CB** are undefined.

The next theorem shows why Definition 1.2 is appropriate. We leave its proof to the reader (*Exercise 11*).

Theorem 1.2   (a) *If* L_1 *and* L_2 *are linear transformations from* $\mathcal{R}^n$ *to* $\mathcal{R}^m$, *with matrices* A_1 *and* A_2 *respectively, then* $aL_1 + bL_2$ *is a linear transformation from* $\mathcal{R}^n$ *to* $\mathcal{R}^m$, *with matrix* $aA_1 + bA_2$.

(b) *If* $L_1: \mathcal{R}^n \to \mathcal{R}^p$ *and* $L_2: \mathcal{R}^p \to \mathcal{R}^m$ *are linear transformations with matrices* A_1 *and* A_2, *respectively, then the composite function* $L_3 = L_2 \circ L_1$, *defined by*

$$L_3(X) = L_2(L_1(X)),$$

is a linear transformation from $\mathcal{R}^n$ *to* $\mathcal{R}^m$, *with matrix* $A_2 A_1$.

(c) *If we regard the vector*

$$X = \begin{bmatrix} x_1 \\ x_2 \\ \vdots \\ x_n \end{bmatrix}$$

as an $n \times 1$ *matrix, then the linear transformation (1) can be written as*

$$L(X) = AX.$$

Example 1.4   If

$$L_1(X) = \begin{bmatrix} 2x + 3y \\ 3x + 2y \\ -x + y \end{bmatrix} \quad \text{and} \quad L_2(X) = \begin{bmatrix} -x - y \\ 4x + y \\ x \end{bmatrix},$$

then

$$A_1 = \begin{bmatrix} 2 & 3 \\ 3 & 2 \\ -1 & 1 \end{bmatrix} \quad \text{and} \quad A_2 = \begin{bmatrix} -1 & -1 \\ 4 & 1 \\ 1 & 0 \end{bmatrix}.$$

The linear transformation

$$L = 2L_1 + L_2$$

is defined by

$$L(X) = 2L_1(X) + L_2(X)$$

$$= 2\begin{bmatrix} 2x + 3y \\ 3x + 2y \\ -x + y \end{bmatrix} + \begin{bmatrix} -x - y \\ 4x + y \\ x \end{bmatrix}$$

$$= \begin{bmatrix} 3x + 5y \\ 10x + 5y \\ -x + 2y \end{bmatrix};$$

its matrix is

$$\mathbf{A} = \begin{bmatrix} 3 & 5 \\ 10 & 5 \\ -1 & 2 \end{bmatrix} = 2\mathbf{A}_1 + \mathbf{A}_2.$$

Example 1.5 Let

$$\mathbf{L}_1(\mathbf{X}) = \begin{bmatrix} x + 2y \\ 3x + 4y \end{bmatrix} : \mathscr{R}^2 \to \mathscr{R}^2,$$

and

$$\mathbf{L}_2(\mathbf{U}) = \begin{bmatrix} u + v \\ -u - 2v \\ 3u + v \end{bmatrix} : \mathscr{R}^2 \to \mathscr{R}^3.$$

Then $\mathbf{L}_3 = \mathbf{L}_2 \cdot \mathbf{L}_1 : \mathscr{R}^2 \to \mathscr{R}^3$ is given by

$$\mathbf{L}_3(\mathbf{X}) = \mathbf{L}_2(\mathbf{L}_1(\mathbf{X})) = \begin{bmatrix} (x + 2y) + (3x + 4y) \\ -(x + 2y) - 2(3x + 4y) \\ 3(x + 2y) + (3x + 4y) \end{bmatrix}$$

$$= \begin{bmatrix} 4x + 6y \\ -7x - 10y \\ 6x + 10y \end{bmatrix}.$$

The matrices of $\mathbf{L}_1$ and $\mathbf{L}_2$ are

$$\mathbf{A}_1 = \begin{bmatrix} 1 & 2 \\ 3 & 4 \end{bmatrix} \quad \text{and} \quad \mathbf{A}_2 = \begin{bmatrix} 1 & 1 \\ -1 & -2 \\ 3 & 1 \end{bmatrix},$$

respectively. The matrix of $\mathbf{L}_3$ is

$$\mathbf{C} = \begin{bmatrix} 4 & 6 \\ -7 & -10 \\ 6 & 10 \end{bmatrix};$$

verify that $\mathbf{C} = \mathbf{A}_2\mathbf{A}_1$.

Example 1.6 The linear transformations of Example 1.5 can be written as

$$\mathbf{L}_1(\mathbf{X}) = \begin{bmatrix} 1 & 2 \\ 3 & 4 \end{bmatrix}\begin{bmatrix} x \\ y \end{bmatrix}, \quad \mathbf{L}_2(\mathbf{U}) = \begin{bmatrix} 1 & 1 \\ -1 & -2 \\ 3 & 1 \end{bmatrix}\begin{bmatrix} u \\ v \end{bmatrix},$$

and

$$\mathbf{L}_3(\mathbf{X}) = \begin{bmatrix} 4 & 6 \\ -7 & -10 \\ 6 & 10 \end{bmatrix} \begin{bmatrix} x \\ y \end{bmatrix}.$$

A new notation for the differential

If a real-valued function $f \colon \mathscr{R}^n \to \mathscr{R}$ is differentiable at $\mathbf{X}_0$, then

$$d_{\mathbf{X}_0} f = f_{x_1}(\mathbf{X}_0) \, dx_1 + f_{x_2}(\mathbf{X}_0) \, dx_2 + \cdots + f_{x_n}(\mathbf{X}_0) \, dx_n.$$

This can be written as a matrix product

$$d_{\mathbf{X}_0} f = [f_{x_1}(\mathbf{X}_0) \quad f_{x_2}(\mathbf{X}_0) \quad \cdots \quad f_{x_n}(\mathbf{X}_0)] \begin{bmatrix} dx_1 \\ dx_2 \\ \vdots \\ dx_n \end{bmatrix}. \tag{5}$$

We define the *differential matrix of f at $\mathbf{X}_0$* by

$$f'(\mathbf{X}_0) = [f_{x_1}(\mathbf{X}_0) \quad f_{x_2}(\mathbf{X}_0) \quad \cdots \quad f_{x_n}(\mathbf{X}_0)] \tag{6}$$

and the *differential linear transformation* by

$$d\mathbf{X} = \begin{bmatrix} dx_1 \\ dx_2 \\ \vdots \\ dx_n \end{bmatrix}.$$

Then (5) can be rewritten as

$$d_{\mathbf{X}_0} f = f'(\mathbf{X}_0) \, d\mathbf{X}. \tag{7}$$

This is analogous to the corresponding formula for functions of one variable (Example 3.7, Section 5.3), and shows that the differential matrix $f'(\mathbf{X}_0)$ is a natural generalization of the derivative. With this new notation we can express the defining property of the differential in a way similar to the form that applies for $n = 1$:

$$\lim_{\mathbf{X} \to \mathbf{X}_0} \frac{f(\mathbf{X}) - f(\mathbf{X}_0) - f'(\mathbf{X}_0)(\mathbf{X} - \mathbf{X}_0)}{|\mathbf{X} - \mathbf{X}_0|} = 0,$$

where $\mathbf{X}_0 = (c_1, c_2, \ldots, c_n)$ and $f'(\mathbf{X}_0)(\mathbf{X} - \mathbf{X}_0)$ is the matrix product

$$[f_{x_1}(\mathbf{X}_0) \quad f_{x_2}(\mathbf{X}_0) \quad \cdots \quad f_{x_n}(\mathbf{X}_0)] \begin{bmatrix} x_1 - c_1 \\ x_2 - c_2 \\ \vdots \\ x_n - c_n \end{bmatrix}.$$

As before, we omit the $\mathbf{X}_0$ in (6) and (7) when it is not necessary to emphasize the specific point; thus, we write

$$f' = [f_{x_1} \quad f_{x_2} \quad \cdots \quad f_{x_n}]$$

and

$$df = f' \, d\mathbf{X}.$$

Example 1.7 If

$$f(x, y, z) = 4x^2yz^3,$$

then

$$f'(x, y, z) = [8xyz^3 \quad 4x^2z^3 \quad 12x^2yz^2].$$

In particular, if $\mathbf{X}_0 = (1, -1, 2)$, then

$$f'(\mathbf{X}_0) = [-64 \quad 32 \quad -48]$$

and therefore

$$d_{\mathbf{X}_0}f = f'(\mathbf{X}_0) \, d\mathbf{X} = [-64 \quad 32 \quad -48] \begin{bmatrix} dx \\ dy \\ dz \end{bmatrix}$$

$$= -64 \, dx + 32 \, dy - 48 \, dz.$$

The norm of a matrix

We will need the following definition in the next section.

Definition 1.3 *The norm,* $\|\mathbf{A}\|$, *of an* $m \times n$ *matrix* $\mathbf{A}$ *is the smallest number such that*

$$|\mathbf{AX}| \le \|\mathbf{A}\| \, |\mathbf{X}|$$

for all $\mathbf{X}$ *in* $\mathscr{R}^n$.

To justify this definition we must show that $\|\mathbf{A}\|$ exists. The components of $\mathbf{Y} = \mathbf{AX}$ are

$$y_i = a_{i1}x_1 + a_{i2}x_2 + \cdots + a_{in}x_n, \qquad 1 \le i \le m.$$

By Schwarz's inequality,

$$y_i^2 \le (a_{i1}^2 + a_{i2}^2 + \cdots + a_{in}^2)|\mathbf{X}|^2.$$

Summing this over $i = 1, 2, \ldots, m$ yields

$$|\mathbf{Y}|^2 \leq \left(\sum_{i=1}^{m} \sum_{j=1}^{n} a_{ij}^2 \right) |\mathbf{X}|^2;$$

therefore, the set

$$B = \{ K \, | \, |\mathbf{AX}| \leq K|\mathbf{X}| \text{ for all } \mathbf{X} \text{ in } \mathscr{R}^n \}$$

is nonempty. Since B is bounded below by zero, it has a greatest lower bound α. If $\varepsilon > 0$, then $\alpha + \varepsilon$ is in B, because if not, then no number less than $\alpha + \varepsilon$ could be in B, and therefore $\alpha + \varepsilon$ would be a lower bound for B, contradicting the definition of α. Hence,

$$|\mathbf{AX}| \leq (\alpha + \varepsilon)|\mathbf{X}|, \qquad \mathbf{X} \in \mathscr{R}^n.$$

Since ε is an arbitrary positive number, this implies that

$$|\mathbf{AX}| \leq \alpha|\mathbf{X}|, \qquad \mathbf{X} \in \mathscr{R}^n,$$

so that $\alpha \in B$. Since no smaller number is in B, we conclude that $\|\mathbf{A}\| = \alpha$.

Example 1.8   If

$$\mathbf{A} = \begin{bmatrix} 1 & 0 & 1 \\ 0 & 1 & 1 \end{bmatrix},$$

then

$$\sum_{i=1}^{2} \sum_{j=1}^{3} a_{ij}^2 = 4;$$

Therefore, Schwarz's inequality implies that

$$|\mathbf{AX}| \leq 2|\mathbf{X}|, \qquad \mathbf{X} \in \mathscr{R}^2.$$

However, a better inequality is available, since $\|\mathbf{A}\| = \sqrt{3}$ (we must defer the proof of this until Section 6.5), and so

$$|\mathbf{AX}| \leq \sqrt{3}|\mathbf{X}|, \qquad \mathbf{X} \in \mathscr{R}^3,$$

with equality if and only if $\mathbf{X}$ is a multiple of

$$\mathbf{X}_0 = \begin{bmatrix} 1 \\ 1 \\ 2 \end{bmatrix}.$$

(See Example 5.8, Section 6.5.)

Square matrices

Linear transformations from $\mathscr{R}^n$ to itself will be important when we discuss the inverse function theorem in Section 6.3, and change of variables in multiple integrals in Section 7.3. The matrix of such a transformation is *square;* that is, it has the same number of rows and columns. We will review, briefly and without proofs, as much of the theory of square matrices as we need.

We assume that the reader knows the definition of the determinant of a square matrix, and how to evaluate determinants. If

$$\mathbf{A} = \begin{bmatrix} a_{11} & a_{12} & \cdots & a_{1n} \\ a_{21} & a_{22} & \cdots & a_{2n} \\ & & \vdots & \\ a_{n1} & a_{n2} & \cdots & a_{nn} \end{bmatrix},$$

we say that A is *of order n*, and denote its determinant by

$$\det \mathbf{A} = \begin{vmatrix} a_{11} & a_{12} & \cdots & a_{1n} \\ a_{21} & a_{22} & \cdots & a_{2n} \\ & & \vdots & \\ a_{n1} & a_{n2} & \cdots & a_{nn} \end{vmatrix}.$$

The *transpose*, $\mathbf{A}^t$, of a matrix $\mathbf{A}$ (square or not) is the matrix obtained by interchanging the rows and columns of $\mathbf{A}$; thus, if

$$\mathbf{A} = \begin{bmatrix} 1 & 2 & 3 \\ 3 & 1 & 4 \\ 0 & 1 & 2 \end{bmatrix}, \quad \text{then} \quad \mathbf{A}^t = \begin{bmatrix} 1 & 3 & 0 \\ 2 & 1 & 1 \\ 3 & 4 & 2 \end{bmatrix}.$$

A square matrix and its transpose have the same determinant; thus,

$$\det \mathbf{A}^t = \det \mathbf{A}.$$

If $\mathbf{A} = \mathbf{A}^t$, then $\mathbf{A}$ is *symmetric;* for example,

$$\mathbf{A} = \begin{bmatrix} 1 & 2 & 3 \\ 2 & 0 & 1 \\ 3 & 1 & -3 \end{bmatrix}$$

is symmetric.

From linear algebra we know that if $\det \mathbf{A} \neq 0$, and only in this case, there is a unique matrix $\mathbf{A}^{-1}$—the *inverse* of $\mathbf{A}$—such that

$$\mathbf{A}^{-1}\mathbf{A} = \mathbf{A}\mathbf{A}^{-1} = \mathbf{I},$$

where $\mathbf{I}$ is the $n \times n$ *identity* matrix, so called because $\mathbf{IB} = \mathbf{BI} = \mathbf{B}$ if $\mathbf{B}$ is any square matrix of the same order as $\mathbf{I}$. The diagonal elements of $\mathbf{I}$ are ones, while all other elements are zeros; thus, if $n = 3$,

$$I = \begin{bmatrix} 1 & 0 & 0 \\ 0 & 1 & 0 \\ 0 & 0 & 1 \end{bmatrix}.$$

If $\det A \neq 0$, so that A has an inverse, we say that A is *nonsingular;* if $\det A = 0$, so that A does not have an inverse, then A is *singular.*

Theorem 1.3 (Cramer's Rule) *Suppose A is of order n and $\det A \neq 0$. Then for each Y in $\mathcal{R}^n$ there is a unique X in $\mathcal{R}^n$ that satisfies the system of n equations in n unknowns,*

$$y_1 = a_{11}x_1 + a_{12}x_2 + \cdots + a_{1n}x_n$$
$$y_2 = a_{21}x_1 + a_{22}x_2 + \cdots + a_{2n}x_n$$
$$\vdots \qquad\qquad \vdots$$
$$y_n = a_{n1}x_1 + a_{n2}x_2 + \cdots + a_{nn}x_n$$

(or, in matrix form, $Y = AX$). This solution is given by

$$y_i = \frac{D_i}{\det A}, \qquad i \leq 1 \leq n,$$

where D_i is the determinant of the matrix obtained by replacing the ith column of A with Y; thus,

$$D_1 = \begin{vmatrix} y_1 & a_{12} & \cdots & a_{1n} \\ y_2 & a_{22} & \cdots & a_{2n} \\ & & \vdots & \\ y_n & a_{n2} & \cdots & a_{nn} \end{vmatrix}, \quad D_2 = \begin{vmatrix} a_{11} & y_1 & a_{13} & \cdots & a_{1n} \\ a_{21} & y_2 & a_{23} & \cdots & a_{2n} \\ & & \vdots & & \\ a_{n1} & y_n & a_{n3} & \cdots & a_{nn} \end{vmatrix}, \quad \cdots,$$

$$D_n = \begin{vmatrix} a_{11} & \cdots & a_{1,n-1} & y_1 \\ a_{21} & \cdots & a_{2,n-1} & y_2 \\ & \vdots & & \\ a_{n1} & \cdots & a_{n,n-1} & y_n \end{vmatrix}.$$

The proof of this theorem can be found in any linear algebra text. It implies the next theorem (*Exercise 20*).

Theorem 1.4 *Suppose A is of order n and $\det A \neq 0$. Let A_{ij} be the matrix of order $n - 1$ obtained by striking out the ith row and jth column of A. Then the elements of $A^{-1} = [b_{ij}]$ are given by*

$$b_{ij} = (-1)^{i+j} \frac{\det A_{ji}}{\det A}. \tag{8}$$

The determinant of $\mathbf{A}_{ij}$ is the *minor* of a_{ij}, and

$$c_{ij} = (-1)^{i+j} \det \mathbf{A}_{ij}$$

is its *cofactor*. Theorems 1.3 and 1.4 follow from the fact that for any $n \times n$ matrix $\mathbf{A}$,

$$\sum_{k=1}^{n} a_{ik}c_{jk} = \begin{cases} \det \mathbf{A}, & i = j, \\ 0, & i \neq j; \end{cases}$$

similarly,

$$\sum_{k=1}^{n} a_{ki}c_{kj} = \begin{cases} \det \mathbf{A}, & i = j, \\ 0, & i \neq j. \end{cases}$$

In words: The sum of the products of the elements of a row (column) of $\mathbf{A}$ and their cofactors equals $\det \mathbf{A}$; the sum of the products of the elements of a row (column) and the cofactors of the corresponding elements of a different row (column) equals zero.

Example 1.9 The matrix of the system

$$\begin{aligned} 4x + 2y + \quad z &= 1 \\ 3x - \quad y + 2z &= 2 \\ y + 2z &= 0 \end{aligned}$$

is

$$\mathbf{A} = \begin{bmatrix} 4 & 2 & 1 \\ 3 & -1 & 2 \\ 0 & 1 & 2 \end{bmatrix}.$$

Computing $\det \mathbf{A}$ from the cofactors of the elements of the first row yields

$$\det \mathbf{A} = 4 \begin{vmatrix} -1 & 2 \\ 1 & 2 \end{vmatrix} - 2 \begin{vmatrix} 3 & 2 \\ 0 & 2 \end{vmatrix} + 1 \begin{vmatrix} 3 & -1 \\ 0 & 1 \end{vmatrix} \tag{9}$$

$$= 4(-4) - 2(6) + 1(3) = -25.$$

Using Cramer's rule to solve the system yields

$$x = -\frac{1}{25} \begin{vmatrix} 1 & 2 & 1 \\ 2 & -1 & 2 \\ 0 & 1 & 2 \end{vmatrix} = \frac{2}{5}, \qquad y = -\frac{1}{25} \begin{vmatrix} 4 & 1 & 1 \\ 3 & 2 & 2 \\ 0 & 0 & 2 \end{vmatrix} = -\frac{2}{5},$$

and

$$z = -\frac{1}{25} \begin{vmatrix} 4 & 2 & 1 \\ 3 & -1 & 2 \\ 0 & 1 & 0 \end{vmatrix} = \frac{1}{5}.$$

To compute $\mathbf{A}^{-1}$ from Theorem 1.4 we calculate all cofactors of $\mathbf{A}$:

$$c_{11} = \begin{vmatrix} -1 & 2 \\ 1 & 2 \end{vmatrix} = -4, \quad c_{12} = -\begin{vmatrix} 3 & 2 \\ 0 & 2 \end{vmatrix} = -6, \quad c_{13} = \begin{vmatrix} 3 & -1 \\ 0 & 1 \end{vmatrix} = 3,$$

$$c_{21} = -\begin{vmatrix} 2 & 1 \\ 1 & 2 \end{vmatrix} = -3, \quad c_{22} = \begin{vmatrix} 4 & 1 \\ 0 & 2 \end{vmatrix} = 8, \quad c_{23} = -\begin{vmatrix} 4 & 2 \\ 0 & 1 \end{vmatrix} = -4,$$

$$c_{31} = \begin{vmatrix} 2 & 1 \\ -1 & 2 \end{vmatrix} = 5, \quad c_{32} = -\begin{vmatrix} 4 & 1 \\ 3 & 2 \end{vmatrix} = -5, \quad c_{33} = \begin{vmatrix} 4 & 2 \\ 3 & -1 \end{vmatrix} = -10.$$

From this, (8), and (9),

$$\mathbf{A}^{-1} = -\frac{1}{25} \begin{bmatrix} -4 & -3 & 5 \\ -6 & 8 & -5 \\ 3 & -4 & -10 \end{bmatrix}.$$

The system of n equations in n unknowns

$$\begin{aligned} a_{11}x_1 + a_{12}x_2 + \cdots + a_{1n}x_n &= 0 \\ a_{21}x_1 + a_{22}x_2 + \cdots + a_{2n}x_n &= 0 \\ &\vdots \\ a_{n1}x_1 + a_{n2}x_2 + \cdots + a_{nn}x_n &= 0 \end{aligned} \tag{10}$$

(or, in matrix form, $\mathbf{AX} = \mathbf{0}$), with zeros on the right, is said to be *homogeneous*. It is obviously satisfied by $\mathbf{X}_0 = (0, 0, \ldots, 0)$, which is called the *trivial solution*. It may also have *nontrivial solutions* other than $\mathbf{X}_0$. The next theorem gives necessary and sufficient conditions for this.

Theorem 1.5   *The homogeneous system (10) of n equations in n unknowns has nontrivial solutions if and only if* $\det \mathbf{A} = 0$, *or, equivalently,* $\mathbf{A}$ *is singular.*

The following theorems will also be useful later.

Theorem 1.6   *If* $\mathbf{A}_1, \mathbf{A}_2, \ldots, \mathbf{A}_k$ *are n × n matrices, then*

$$\det(\mathbf{A}_1\mathbf{A}_2 \cdots \mathbf{A}_k) = (\det \mathbf{A}_1)(\det \mathbf{A}_2) \cdots (\det \mathbf{A}_k).$$

Theorem 1.7   *If* $\mathbf{A}_1, \mathbf{A}_2, \ldots, \mathbf{A}_k$ *are nonsingular n × n matrices, then so is* $\mathbf{A}_1\mathbf{A}_2 \cdots \mathbf{A}_k$, *and*

$$(\mathbf{A}_1\mathbf{A}_2 \cdots \mathbf{A}_k)^{-1} = \mathbf{A}_k^{-1}\mathbf{A}_{k-1}^{-1} \cdots \mathbf{A}_1^{-1}.$$

6.1 EXERCISES

1. Prove: If $L: \mathcal{R}^n \to \mathcal{R}^m$ is a linear transformation, then
$$L(a_1 X_1 + a_2 X_2 + \cdots + a_k X_k) = a_1 L(X_1) + a_2 L(X_2) + \cdots + a_k L(X_k)$$
if $X_1, X_2, \ldots, X_k$ are in $\mathcal{R}^n$ and $a_1, a_2, \ldots, a_k$ are real numbers.

2. Prove that the transformation L defined by Eq. (1) is linear.

3. Find the matrix of L:

(a) $L(X) = \begin{bmatrix} 3x + 4y + 6z \\ 2x - 4y + 2z \\ 7x + 2y + 3z \end{bmatrix}$ (b) $L(X) = \begin{bmatrix} 2x_1 + 4x_2 \\ 3x_1 - 2x_2 \\ 7x_1 - 4x_2 \\ 6x_1 + x_2 \end{bmatrix}$

4. Find cA:

(a) $c = 4$, $A = \begin{bmatrix} 2 & 2 & 4 & 6 \\ 0 & 0 & 1 & 3 \\ 3 & 4 & 7 & 11 \end{bmatrix}$ (b) $c = -2$, $A = \begin{bmatrix} 1 & 3 & 0 \\ 0 & 1 & 2 \\ 1 & -1 & 3 \end{bmatrix}$

5. Find $A + B$:

(a) $A = \begin{bmatrix} -1 & 2 & 3 \\ 0 & 1 & 4 \\ 0 & -1 & 4 \end{bmatrix}$, $B = \begin{bmatrix} -1 & 0 & 3 \\ 5 & 6 & -7 \\ 0 & -1 & 2 \end{bmatrix}$

(b) $A = \begin{bmatrix} 0 & 5 \\ 3 & 2 \\ 1 & 7 \end{bmatrix}$, $B = \begin{bmatrix} -1 & 2 \\ 0 & 3 \\ 4 & 7 \end{bmatrix}$

6. Find AB:

(a) $A = \begin{bmatrix} -1 & 2 & 3 \\ 0 & 1 & 4 \\ 0 & -1 & 4 \end{bmatrix}$, $B = \begin{bmatrix} -1 & 2 \\ 0 & 3 \\ 4 & 7 \end{bmatrix}$

(b) $A = \begin{bmatrix} 5 & 3 & 2 & 1 \\ 6 & 7 & 4 & 1 \end{bmatrix}$, $B = \begin{bmatrix} 1 \\ 3 \\ 4 \\ 7 \end{bmatrix}$

7. Prove: If A and B are $m \times n$ matrices and r and s are real numbers, then
(a) $r(sA) = (rs)A$; (b) $(r + s)A = rA + sA$; (c) $r(A + B) = rA + rB$.

8. Prove: If **A**, **B**, and **C** are $m \times p$, $p \times q$, and $q \times n$ matrices, respectively, then $(\mathbf{AB})\mathbf{C} = \mathbf{A}(\mathbf{BC})$.

9. Prove: If **A**, **B**, and **C** are $m \times n$ matrices, then $(\mathbf{A} + \mathbf{B}) + \mathbf{C} = \mathbf{A} + (\mathbf{B} + \mathbf{C})$.

10. Suppose **A** + **B** and **AB** are both defined. What can be said about **A** and **B**?

11. (a) Prove Theorem 1.2(a).
    (b) Prove Theorem 1.2(b).
    (c) Prove Theorem 1.2(c).

12. Find the matrix of $a\mathbf{L}_1 + b\mathbf{L}_2$:

(a) $\mathbf{L}_1(x, y, z) = \begin{bmatrix} 3x + 2y + & z \\ x + 4y + 2z \\ 3x - 4y + & z \end{bmatrix}$,

$\mathbf{L}_2(x, y, z) = \begin{bmatrix} -x + y - & z \\ -2x + y + 3z \\ y + & z \end{bmatrix}$, $a = 2, b = -1$

(b) $\mathbf{L}_1(x, y) = \begin{bmatrix} 2x + 3y \\ x - & y \\ 4x + & y \end{bmatrix}$, $\mathbf{L}_2(x, y) = \begin{bmatrix} 3x - y \\ x + y \\ -x - y \end{bmatrix}$, $a = 4, b = 2$

13. Find the matrices of $\mathbf{L}_1 \circ \mathbf{L}_2$ and $\mathbf{L}_2 \circ \mathbf{L}_1$, where $\mathbf{L}_1$ and $\mathbf{L}_2$ are as in Exercise 12(a).

14. Write the transformations of Exercise 12 in the form $\mathbf{L}(\mathbf{X}) = \mathbf{A}\mathbf{X}$.

15. Find f' and $f'(\mathbf{X}_0)$:
    (a) $f(x, y, z) = 3x^2yz$, $\mathbf{X}_0 = (1, -1, 1)$
    (b) $f(x, y) = \sin(x + y)$, $\mathbf{X}_0 = (\pi/4, \pi/4)$
    (c) $f(x, y, z) = xye^{-xz}$, $\mathbf{X}_0 = (1, 2, 0)$
    (d) $f(x, y, z) = \tan(x + 2y + z)$, $\mathbf{X}_0 = (\pi/4, -\pi/8, \pi/4)$
    (e) $f(\mathbf{X}) = |\mathbf{X}|(\mathscr{R}^n \rightarrow \mathscr{R})$, $\mathbf{X}_0 = (1/\sqrt{n}, 1/\sqrt{n}, \ldots, 1/\sqrt{n})$

16. Let $\mathbf{A} = [a_{ij}]$ be an $m \times n$ matrix and

$$\lambda = \max\{|a_{ij}| \,|\, 1 \leq i \leq m, 1 \leq i \leq n\}.$$

Show that $\|\mathbf{A}\| \leq \lambda\sqrt{mn}$.

17. Prove: If **A** has at least one nonzero entry, then $\|\mathbf{A}\| \neq 0$.

18. Prove: $\|\mathbf{A} + \mathbf{B}\| \leq \|\mathbf{A}\| + \|\mathbf{B}\|$.

19. Prove: $\|\mathbf{AB}\| \leq \|\mathbf{A}\| \, \|\mathbf{B}\|$.

20. Show that Theorem 1.3 implies Theorem 1.4.

21. Solve by Cramer's rule:

$$\text{(a)} \quad \begin{aligned} x + y + 2z &= 1 \\ 2x - y + z &= -1 \\ x - 2y - 3z &= 2 \end{aligned} \qquad \text{(b)} \quad \begin{aligned} x + y - z &= 5 \\ 3x - 2y + 2z &= 0 \\ 4x + 2y - 3z &= 14 \end{aligned}$$

$$\text{(c)} \quad \begin{aligned} x + 2y + 3z &= -5 \\ x - z &= -1 \\ x + y + 2z &= -4 \end{aligned} \qquad \text{(d)} \quad \begin{aligned} x - y + z - 2w &= 1 \\ 2x + y - 3z + 3w &= 4 \\ 3x + 2y + w &= 13 \\ 2x + y - z &= 4 \end{aligned}$$

22. Find A^{-1} by the method of Theorem 1.4:

(a) $\begin{bmatrix} 1 & -2 \\ 3 & 4 \end{bmatrix}$

   (b) $\begin{bmatrix} 1 & 2 & 3 \\ 1 & 0 & -1 \\ 1 & 1 & 2 \end{bmatrix}$

(c) $\begin{bmatrix} 4 & 2 & 1 \\ 3 & -1 & 2 \\ 0 & 1 & 2 \end{bmatrix}$

   (d) $\begin{bmatrix} 1 & 0 & 1 \\ 0 & 1 & 1 \\ 1 & 1 & 0 \end{bmatrix}$

(e) $\begin{bmatrix} 1 & 2 & 0 & 0 \\ -2 & 3 & 0 & 0 \\ 0 & 0 & 2 & 3 \\ 0 & 0 & -1 & 2 \end{bmatrix}$

   (f) $\begin{bmatrix} 1 & 1 & 2 & -1 \\ 2 & 2 & -1 & 3 \\ -1 & 4 & 1 & 2 \\ 3 & 1 & 0 & 1 \end{bmatrix}$

23. Show that the rule stated after Theorem 1.4 for expanding a determinant by cofactors implies (a) Theorem 1.3; (b) Theorem 1.4.

*24. For $i, j = 1, 2, \ldots, m$, let $a_{ij} = a_{ij}(X)$ be a real-valued function continuous on a compact set K in $\mathscr{R}^n$. Suppose the $m \times m$ matrix

$$A(X) = [a_{ij}(X)]$$

is nonsingular for each X in K, and define the $m \times m$ matrix

$$B(X, Y) = [b_{ij}(X, Y)]$$

by

$$B(X, Y) = A^{-1}(X)A(Y) - I.$$

Show that for each $\varepsilon > 0$ there is a $\delta > 0$ such that

$$|b_{ij}(X, Y)| < \varepsilon, \qquad i, j = 1, 2, \ldots, m,$$

if $|X - Y| < \delta$. [Hint: Show that b_{ij} is continuous on the set

$$\{(X, Y) \,|\, X \in K, Y \in K\}.$$

Then assume that the conclusion is false and use Exercise 34, Section 5.1, to obtain a contradiction.]

6.2 CONTINUITY AND DIFFERENTIABILITY OF TRANSFORMATIONS

Throughout the rest of this chapter transformations $\mathbf{F}$ and points $\mathbf{X}$ should be considered as written in vertical form when they appear in equations; however, for convenience in printing, we will continue to write $\mathbf{F} = (f_1, f_2, \ldots, f_m)$ and $\mathbf{X} = (x_1, x_2, \ldots, x_n)$ otherwise.

Continuous transformations

In Section 5.2 we defined a vector-valued function (transformation) to be continuous at $\mathbf{X}_0$ if each of its component functions is. We leave it to the reader to show that this implies the following theorem (*Exercise 1*).

Theorem 2.1   *Suppose* $\mathbf{X}_0$ *is in—and a limit point of—the domain of* $\mathbf{F} : \mathscr{R}^n \to \mathscr{R}^m$. *Then* $\mathbf{F}$ *is continuous at* $\mathbf{X}_0$ *if and only if for each* $\varepsilon > 0$ *there is a* $\delta > 0$ *such that*

$$\left| \mathbf{F}(\mathbf{X}) - \mathbf{F}(\mathbf{X}_0) \right| < \varepsilon \tag{1}$$

if $|\mathbf{X} - \mathbf{X}_0| < \delta$ *and* $\mathbf{X} \in D_{\mathbf{F}}$.

This theorem is the same as Theorem 2.3, Section 5.2, except that the "absolute value" in (1) now stands for distance in $\mathscr{R}^m$ rather than $\mathscr{R}$.

If $\mathbf{L}$ is a constant vector, then "$\lim_{\mathbf{X} \to \mathbf{X}_0} \mathbf{F}(\mathbf{X}) = \mathbf{L}$" means that

$$\lim_{\mathbf{X} \to \mathbf{X}_0} \left| \mathbf{F}(\mathbf{X}) - \mathbf{L} \right| = 0.$$

Theorem 2.1 implies that $\mathbf{F}$ is continuous at $\mathbf{X}_0$ if and only if

$$\lim_{\mathbf{X} \to \mathbf{X}_0} \mathbf{F}(\mathbf{X}) = \mathbf{F}(\mathbf{X}_0).$$

Example 2.1   The linear transformation

$$\mathbf{L}(\mathbf{X}) = \begin{bmatrix} x + y + z \\ 2x - 3y + z \\ 2x + y - z \end{bmatrix}$$

is continuous at every $\mathbf{X}_0$ in $\mathscr{R}^3$, since

$$\mathbf{L}(\mathbf{X}) - \mathbf{L}(\mathbf{X}_0) = \mathbf{L}(\mathbf{X} - \mathbf{X}_0) = \begin{bmatrix} (x - x_0) + (y - y_0) + (z - z_0) \\ 2(x - x_0) - 3(y - y_0) + (z - z_0) \\ 2(x - x_0) + (y - y_0) - (z - z_0) \end{bmatrix},$$

and applying Schwarz's inequality to each component yields

$$|\mathbf{L}(\mathbf{X}) - \mathbf{L}(\mathbf{X}_0)|^2 \le (3 + 14 + 6)|\mathbf{X} - \mathbf{X}_0|^2 = 23|\mathbf{X} - \mathbf{X}_0|^2;$$

therefore,

$$|\mathbf{L}(\mathbf{X}) - \mathbf{L}(\mathbf{X}_0)| < \varepsilon \quad \text{if} \quad |\mathbf{X} - \mathbf{X}_0| < \sqrt{\frac{\varepsilon}{23}}.$$

Differentiable transformations

In Section 5.4 we defined a vector-valued function (transformation) to be differentiable at a point if each of its components is. The next theorem characterizes this property in a useful way.

Theorem 2.2   *A transformation* $\mathbf{F} = (f_1, f_2, \ldots, f_m)$ *defined in a neighborhood of* $\mathbf{X}_0 = (c_1, c_2, \ldots, c_n)$ *is differentiable at* $\mathbf{X}_0$ *if and only if there is a constant* $m \times n$ *matrix* $\mathbf{A}$ *such that*

$$\lim_{\mathbf{X} \to \mathbf{X}_0} \frac{\mathbf{F}(\mathbf{X}) - \mathbf{F}(\mathbf{X}_0) - \mathbf{A}(\mathbf{X} - \mathbf{X}_0)}{|\mathbf{X} - \mathbf{X}_0|} = \mathbf{0}. \tag{2}$$

If (2) holds, then $\mathbf{A}$ *is given uniquely by*

$$\mathbf{A} = \left[\frac{\partial f_i(\mathbf{X}_0)}{\partial x_j}\right] = \begin{bmatrix} \dfrac{\partial f_1(\mathbf{X}_0)}{\partial x_1} & \dfrac{\partial f_1(\mathbf{X}_0)}{\partial x_2} & \cdots & \dfrac{\partial f_1(\mathbf{X}_0)}{\partial x_n} \\[2mm] \dfrac{\partial f_2(\mathbf{X}_0)}{\partial x_1} & \dfrac{\partial f_2(\mathbf{X}_0)}{\partial x_2} & \cdots & \dfrac{\partial f_2(\mathbf{X}_0)}{\partial x_n} \\ & & \vdots & \\ \dfrac{\partial f_m(\mathbf{X}_0)}{\partial x_1} & \dfrac{\partial f_m(\mathbf{X}_0)}{\partial x_2} & \cdots & \dfrac{\partial f_m(\mathbf{X}_0)}{\partial x_n} \end{bmatrix}. \tag{3}$$

Proof.   If $\mathbf{F}$ is differentiable at $\mathbf{X}_0$, then so are $f_1, f_2, \ldots, f_m$, by definition. Hence,

$$\lim_{\mathbf{X} \to \mathbf{X}_0} \frac{f_i(\mathbf{X}) - f_i(\mathbf{X}_0) - \displaystyle\sum_{j=1}^{n} \frac{\partial f_i(\mathbf{X}_0)}{\partial x_j}(x_j - c_j)}{|\mathbf{X} - \mathbf{X}_0|} = 0, \qquad 1 \le i \le m. \tag{4}$$

Since the length of the vector in (2) is the square root of the sum of the squares of the quantities in (4) (as $i = 1, 2, \ldots, m$), (4) implies (2), with $\mathbf{A}$ as in (3). This proves necessity.

For sufficiency, let (2) hold, with $\mathbf{A} = [a_{ij}]$. Since each component of the vector in (2) approaches zero as $\mathbf{X}$ approaches $\mathbf{X}_0$, it follows that

$$\lim_{\mathbf{X} \to \mathbf{X}_0} \frac{f_i(\mathbf{X}) - f_i(\mathbf{X}_0) - \sum_{j=1}^{n} a_{ij}(x_j - c_j)}{|\mathbf{X} - \mathbf{X}_0|} = 0, \qquad 1 \le i \le m,$$

so that each f_i is differentiable at $\mathbf{X}_0$, and therefore so is $\mathbf{F}$ (by definition) and, by Theorem 3.4, Section 5.3,

$$a_{ij} = \frac{\partial f_i(\mathbf{X}_0)}{\partial x_j}, \qquad 1 \le i \le m, \quad 1 \le j \le n.$$

Therefore, (3) holds. This completes the proof.

A transformation $\mathbf{T} \colon \mathscr{R}^n \to \mathscr{R}^m$ of the form

$$\mathbf{T}(\mathbf{X}) = \mathbf{U} + \mathbf{A}(\mathbf{X} - \mathbf{X}_0),$$

where $\mathbf{U}$ is a constant vector in $\mathscr{R}^m$, $\mathbf{X}_0$ is a constant vector in $\mathscr{R}^n$, and $\mathbf{A}$ is a constant $m \times n$ matrix, is said to be *affine*. Theorem 2.2 says that if $\mathbf{F}$ is differentiable at $\mathbf{X}_0$, then it can be well approximated by an affine transformation.

Example 2.2 The components of the transformation.

$$\mathbf{F}(\mathbf{X}) = \begin{bmatrix} x^2 + 2xy + z \\ x + 2xz + y \\ x^2 + y^2 + z^2 \end{bmatrix}$$

are differentiable at $\mathbf{X}_0 = (1, 0, 2)$. Evaluating the partial derivatives of the components there yields

$$\mathbf{A} = \begin{bmatrix} 2 & 2 & 1 \\ 5 & 1 & 2 \\ 2 & 0 & 4 \end{bmatrix}.$$

(Verify). Therefore, Theorem 2.2 implies that the affine transformation

$$\mathbf{T}(\mathbf{X}) = \mathbf{F}(\mathbf{X}_0) + \mathbf{A}(\mathbf{X} - \mathbf{X}_0)$$

$$= \begin{bmatrix} 3 \\ 5 \\ 5 \end{bmatrix} + \begin{bmatrix} 2 & 2 & 1 \\ 5 & 1 & 2 \\ 2 & 0 & 4 \end{bmatrix} \begin{bmatrix} x - 1 \\ y \\ z - 2 \end{bmatrix}$$

satisfies

$$\lim_{\mathbf{X} \to \mathbf{X}_0} \frac{\mathbf{F}(\mathbf{X}) - \mathbf{T}(\mathbf{X})}{|\mathbf{X} - \mathbf{X}_0|} = \mathbf{0}.$$

Differential of a transformation

If $\mathbf{F} = (f_1, f_2, \ldots, f_m)$ is differentiable at $\mathbf{X}_0$, we define its differential at $\mathbf{X}_0$ to be the linear transformation

$$d_{\mathbf{X}_0}\mathbf{F} = \begin{bmatrix} d_{\mathbf{X}_0} f_1 \\ d_{\mathbf{X}_0} f_2 \\ \vdots \\ d_{\mathbf{X}_0} f_m \end{bmatrix}. \tag{5}$$

We call the matrix $\mathbf{A}$ in (3) *the differential matrix of* $\mathbf{F}$ *at* $\mathbf{X}_0$, and denote it by $\mathbf{F}'(\mathbf{X}_0)$; thus,

$$\mathbf{F}'(\mathbf{X}_0) = \begin{bmatrix} \dfrac{\partial f_1(\mathbf{X}_0)}{\partial x_1} & \dfrac{\partial f_1(\mathbf{X}_0)}{\partial x_2} & \cdots & \dfrac{\partial f_1(\mathbf{X}_0)}{\partial x_n} \\ \dfrac{\partial f_2(\mathbf{X}_0)}{\partial x_1} & \dfrac{\partial f_2(\mathbf{X}_0)}{\partial x_2} & \cdots & \dfrac{\partial f_2(\mathbf{X}_0)}{\partial x_n} \\ & & \vdots & \\ \dfrac{\partial f_m(\mathbf{X}_0)}{\partial x_1} & \dfrac{\partial f_m(\mathbf{X}_0)}{\partial x_2} & \cdots & \dfrac{\partial f_m(\mathbf{X}_0)}{\partial x_n} \end{bmatrix}. \tag{6}$$

From Theorem 2.2 the differential can be written in terms of the differential matrix as

$$d_{\mathbf{X}_0}\mathbf{F} = \mathbf{F}'(\mathbf{X}_0) \begin{bmatrix} dx_1 \\ dx_2 \\ \vdots \\ dx_n \end{bmatrix}, \tag{7}$$

or more succinctly as

$$d_{\mathbf{X}_0}\mathbf{F} = \mathbf{F}'(\mathbf{X}_0)\, d\mathbf{X},$$

where

$$d\mathbf{X} = \begin{bmatrix} dx_1 \\ dx_2 \\ \vdots \\ dx_n \end{bmatrix},$$

as defined earlier.

When it is not necessary to emphasize the particular point $\mathbf{X}_0$, we write (5) as

$$dF = \begin{bmatrix} df_1 \\ df_2 \\ \vdots \\ df_m \end{bmatrix},$$

(6) as

$$F' = \begin{bmatrix} \dfrac{\partial f_1}{\partial x_1} & \dfrac{\partial f_1}{\partial x_2} & \cdots & \dfrac{\partial f_1}{\partial x_n} \\[2ex] \dfrac{\partial f_2}{\partial x_1} & \dfrac{\partial f_2}{\partial x_2} & \cdots & \dfrac{\partial f_2}{\partial x_n} \\[2ex] & & \vdots & \\[1ex] \dfrac{\partial f_m}{\partial x_1} & \dfrac{\partial f_m}{\partial x_2} & \cdots & \dfrac{\partial f_m}{\partial x_n} \end{bmatrix},$$

and (7) as

$$dF = F' \, d\mathbf{X}.$$

With the differential notation we can rewrite (2) as

$$\lim_{\mathbf{X} \to \mathbf{X}_0} \frac{F(\mathbf{X}) - F(\mathbf{X}_0) - F'(\mathbf{X}_0)(\mathbf{X} - \mathbf{X}_0)}{|\mathbf{X} - \mathbf{X}_0|} = \mathbf{0}.$$

Example 2.3   If F is a linear transformation,

$$F(\mathbf{X}) = \begin{bmatrix} a_{11}x_1 + a_{12}x_2 + \cdots + a_{1n}x_n \\ a_{21}x_1 + a_{22}x_2 + \cdots + a_{2n}x_n \\ \vdots \\ a_{m1}x_1 + a_{m2}x_2 + \cdots + a_{mn}x_n \end{bmatrix} = \mathbf{AX},$$

where $\mathbf{A} = [a_{ij}]$, then

$$F' = \mathbf{A};$$

that is, the differential matrix of a linear transformation is independent of $\mathbf{X}$ and equals the matrix of the transformation. For example, the differential matrix of

$$F(x_1, x_2, x_3) = \begin{bmatrix} 1 & 2 & 3 \\ 2 & 1 & 0 \end{bmatrix} \begin{bmatrix} x_1 \\ x_2 \\ x_3 \end{bmatrix}$$

is

$$F' = \begin{bmatrix} 1 & 2 & 3 \\ 2 & 1 & 0 \end{bmatrix}.$$

If $F(\mathbf{X}) = \mathbf{X}$ (the identity transformation), then, $F' = \mathbf{I}$ (the identity matrix).

Example 2.4 The transformation

$$F(x, y) = \begin{bmatrix} \dfrac{x}{x^2 + y^2} \\ \dfrac{y}{x^2 + y^2} \\ 2xy \end{bmatrix}$$

is differentiable at every point of $\mathscr{R}^2$ except $(0, 0)$, and

$$F'(x, y) = \begin{bmatrix} \dfrac{y^2 - x^2}{(x^2 + y^2)^2} & \dfrac{-2xy}{(x^2 + y^2)^2} \\ \dfrac{-2xy}{(x^2 + y^2)^2} & \dfrac{x^2 - y^2}{(x^2 + y^2)^2} \\ 2y & 2x \end{bmatrix}.$$

In particular,

$$F'(1, 1) = \begin{bmatrix} 0 & -\dfrac{1}{2} \\ -\dfrac{1}{2} & 0 \\ 2 & 2 \end{bmatrix}$$

and so

$$\lim_{(x,y) \to (1,1)} \frac{1}{\sqrt{(x - 1)^2 + (y - 1)^2}} \left(F(x, y) - \begin{bmatrix} \dfrac{1}{2} \\ \dfrac{1}{2} \\ 2 \end{bmatrix} - \begin{bmatrix} 0 & -\dfrac{1}{2} \\ -\dfrac{1}{2} & 0 \\ 2 & 2 \end{bmatrix} \begin{bmatrix} x - 1 \\ y - 1 \end{bmatrix} \right) = \begin{bmatrix} 0 \\ 0 \\ 0 \end{bmatrix}.$$

If $n = m$, the differential matrix is square, and its determinant is called the *Jacobian of* **F**. The standard notation for this determinant is

$$\frac{\partial(f_1, f_2, \ldots, f_n)}{\partial(x_1, x_2, \ldots, x_n)} = \begin{vmatrix} \dfrac{\partial f_1}{\partial x_1} & \dfrac{\partial f_1}{\partial x_2} & \cdots & \dfrac{\partial f_1}{\partial x_n} \\ \dfrac{\partial f_2}{\partial x_1} & \dfrac{\partial f_2}{\partial x_2} & \cdots & \dfrac{\partial f_2}{\partial x_n} \\ \vdots & & & \\ \dfrac{\partial f_n}{\partial x_1} & \dfrac{\partial f_n}{\partial x_2} & \cdots & \dfrac{\partial f_n}{\partial x_n} \end{vmatrix}.$$

We will write the Jacobian of $\mathbf{F}$ more simply as $J\mathbf{F}$, and its value at $\mathbf{X}_0$ as $J\mathbf{F}(\mathbf{X}_0)$.

Since an $n \times n$ matrix is nonsingular if and only if its determinant is nonzero, it follows that if $\mathbf{F}: \mathscr{R}^n \to \mathscr{R}^n$ is differentiable at $\mathbf{X}_0$, then $\mathbf{F}'(\mathbf{X}_0)$ is nonsingular if and only if $J\mathbf{F}(\mathbf{X}_0) \neq 0$. We will soon use this important fact.

Example 2.5 If

$$\mathbf{F}(x, y, z) = \begin{vmatrix} x^2 - 2x & + z \\ x & + 2xy + z^2 \\ x & + y + z \end{vmatrix},$$

then

$$\frac{\partial(f_1, f_2, f_3)}{\partial(x_1, x_2, x_3)} = J\mathbf{F}(\mathbf{X}) = \begin{vmatrix} 2x - 2 & 0 & 1 \\ 1 + 2y & 2x & 2z \\ 1 & 1 & 1 \end{vmatrix}$$

$$= (2x - 2)\begin{vmatrix} 2x & 2z \\ 1 & 1 \end{vmatrix} + \begin{vmatrix} 1 + 2y & 2x \\ 1 & 1 \end{vmatrix}$$

$$= (2x - 2)(2x - 2z) + (1 + 2y - 2x).$$

In particular, $J\mathbf{F}(1, -1, 1) = -3$, so the differential matrix

$$\mathbf{F}'(1, -1, 1) = \begin{bmatrix} 0 & 0 & 1 \\ -1 & 2 & 2 \\ 1 & 1 & 1 \end{bmatrix}$$

is invertible.

Properties of differentiable transformations

We leave the proofs of the next two theorems to the reader (*Exercises 16 and 17*).

Theorem 2.3 *If $\mathbf{F}: \mathscr{R}^n \to \mathscr{R}^m$ is differentiable at $\mathbf{X}_0$, then it is continuous there.*

Theorem 2.4 *If $\mathbf{F} = (f_1, f_2, \ldots, f_m)$ and the partial derivatives*

$$\frac{\partial f_i}{\partial x_j}, \qquad 1 \le i \le m, \quad 1 \le j \le n, \tag{8}$$

are continuous at $\mathbf{X}_0$, then $\mathbf{F}$ is differentiable at $\mathbf{X}_0$.

We say that $\mathbf{F}$ is continuously differentiable on a set S if S is contained in an open set on which the partial derivatives in (8) are continuous. The next three lemmas give properties of continuously differentiable transformations which we will need later.

Lemma 2.1 *Suppose* $\mathbf{F}: \mathscr{R}^n \to \mathscr{R}^m$ *is continuously differentiable on a neighborhood* N *of* $\mathbf{X}_0$. *Then for every* $\varepsilon > 0$, *there is a* $\delta > 0$ *such that*

$$|\mathbf{F}(\mathbf{X}) - \mathbf{F}(\mathbf{Y})| < (\|\mathbf{F}'(\mathbf{X}_0)\| + \varepsilon)|\mathbf{X} - \mathbf{Y}| \quad \text{if} \quad \mathbf{X}, \mathbf{Y} \in S_\delta(\mathbf{X}_0). \tag{9}$$

Proof. Consider the auxiliary function

$$\mathbf{G}(\mathbf{X}) = \mathbf{F}(\mathbf{X}) - \mathbf{F}'(\mathbf{X}_0)\mathbf{X}.$$

The components of $\mathbf{G}$ are

$$g_i(\mathbf{X}) = f_i(\mathbf{X}) - \sum_{j=1}^n \frac{\partial f_i(\mathbf{X}_0)}{\partial x_j} x_j,$$

and therefore

$$\frac{\partial g_i(\mathbf{X})}{\partial x_j} = \frac{\partial f_i(\mathbf{X})}{\partial x_j} - \frac{\partial f_i(\mathbf{X}_0)}{\partial x_j}.$$

Thus, $\partial g_i/\partial x_j$ is continuous on N and vanishes at $\mathbf{X}_0$. Because of this, there is a $\delta > 0$ such that

$$\left| \frac{\partial g_i(\mathbf{X})}{\partial x_j} \right| < \frac{\varepsilon}{\sqrt{mn}} \quad \text{for } 1 \le i \le m, \quad 1 \le j \le n, \quad \text{if} \quad |\mathbf{X} - \mathbf{X}_0| < \delta. \tag{10}$$

Now suppose $\mathbf{X}, \mathbf{Y} \in S_\delta(\mathbf{X}_0)$. By the mean value theorem,

$$g_i(\mathbf{X}) - g_i(\mathbf{Y}) = \sum_{j=1}^n \frac{\partial g_i(\mathbf{X}_i)}{\partial x_j} (x_j - y_j), \tag{11}$$

where $\mathbf{X}_i$ is on the line segment from $\mathbf{X}$ to $\mathbf{Y}$, and so in $S_\delta(\mathbf{X}_0)$. From (10), (11), and Schwarz's inequality,

$$(g_i(\mathbf{X}) - g_i(\mathbf{Y}))^2 \le \left(\sum_{j=1}^n \left[\frac{\partial g_i(\mathbf{X}_i)}{\partial x_j} \right]^2 \right) |\mathbf{X} - \mathbf{Y}|^2 < \frac{\varepsilon^2}{m} |\mathbf{X} - \mathbf{Y}|^2.$$

Summing this from $i = 1$ to $i = m$ and taking square roots yields

$$|\mathbf{G}(\mathbf{X}) - \mathbf{G}(\mathbf{Y})| < \varepsilon|\mathbf{X} - \mathbf{Y}| \quad \text{if} \quad \mathbf{X}, \mathbf{Y} \in S_\delta(\mathbf{X}_0). \tag{12}$$

To complete the proof, we note that

$$\mathbf{F}(\mathbf{X}) - \mathbf{F}(\mathbf{Y}) = \mathbf{G}(\mathbf{X}) - \mathbf{G}(\mathbf{Y}) + \mathbf{F}'(\mathbf{X}_0)(\mathbf{X} - \mathbf{Y}), \tag{13}$$

and so (12) and the triangle inequality imply (9).

Lemma 2.2 *Suppose* **F** *satisfies the hypotheses of Lemma 2.1, $m = n$, and* **F'(X_0)** *is nonsingular. Let*

$$r = \frac{1}{\|(\mathbf{F'(X_0)})^{-1}\|} \tag{14}$$

Then for every $\varepsilon > 0$ there is a $\delta > 0$ such that

$$|\mathbf{F(X)} - \mathbf{F(Y)}| \geq (r - \varepsilon)|\mathbf{X} - \mathbf{Y}| \quad \text{if} \quad \mathbf{X}, \mathbf{Y} \in S_\delta(\mathbf{X_0}). \tag{15}$$

Proof. Since

$$\mathbf{X} - \mathbf{Y} = [\mathbf{F'(X_0)}]^{-1}\mathbf{F'(X_0)(X - Y)},$$

(14) implies that

$$|\mathbf{X} - \mathbf{Y}| \leq \frac{1}{r}|\mathbf{F'(X_0)(X - Y)}|;$$

therefore,

$$|\mathbf{F'(X_0)(X - Y)}| \geq r|\mathbf{X} - \mathbf{Y}|. \tag{16}$$

From (13),

$$|\mathbf{F(X)} - \mathbf{F(Y)}| \geq \big||\mathbf{F'(X_0)(X - Y)}| - |\mathbf{G(X)} - \mathbf{G(Y)}|\big|,$$

and therefore (12) and (16) imply (15).

A stronger conclusion holds if **F** is linear (*Exercise 20*).

Lemma 2.3 *If* **F**: $\mathcal{R}^n \to \mathcal{R}^m$ *is continuously differentiable on an open set containing a compact set D, then there is a constant M such that*

$$|\mathbf{F(Y)} - \mathbf{F(X)}| \leq M|\mathbf{Y} - \mathbf{X}| \quad \text{if} \quad \mathbf{X}, \mathbf{Y} \in D. \tag{17}$$

Proof. On

$$S = \{(\mathbf{X}, \mathbf{Y}) \,|\, \mathbf{X}, \mathbf{Y} \in D\} \subset \mathcal{R}^{2n}$$

define

$$g(\mathbf{X}, \mathbf{Y}) = \begin{cases} \dfrac{|\mathbf{F(Y)} - \mathbf{F(X)} - \mathbf{F'(X)(Y - X)}|}{|\mathbf{Y} - \mathbf{X}|}, & \mathbf{Y} \neq \mathbf{X}, \\[3mm] 0, & \mathbf{Y} = \mathbf{X}. \end{cases}$$

Then g is obviously continuous for all $(\mathbf{X}, \mathbf{Y})$ in S such that $\mathbf{X} \neq \mathbf{Y}$. We now show that

$$\lim_{(\mathbf{X},\mathbf{Y}) \to (\mathbf{X}_0,\mathbf{X}_0)} g(\mathbf{X}, \mathbf{Y}) = 0 = g(\mathbf{X}_0, \mathbf{X}_0); \tag{18}$$

that is, g is also continuous at points $(\mathbf{X}_0, \mathbf{X}_0)$ in S.

Suppose $\varepsilon > 0$ and $\mathbf{X}_0 \in D$. Since the partial derivatives of $f_1, f_2, \ldots, f_m$ are continuous on an open set containing D, there is a $\delta > 0$ such that

$$\left| \frac{\partial f_i(\mathbf{X})}{\partial x_j} - \frac{\partial f_i(\mathbf{Y})}{\partial x_j} \right| < \frac{\varepsilon}{\sqrt{mn}} \qquad \text{if} \quad \mathbf{X}, \mathbf{Y} \in S_\delta(\mathbf{X}_0), \quad 1 \le i \le m, \quad 1 \le j \le n.$$

$$\tag{19}$$

[Note that $\partial f_i / \partial x_j$ is uniformly continuous on $\overline{S_\delta(\mathbf{X}_0)}$ for δ sufficiently small, from Theorem 2.10, Section 5.2.] Applying the mean value theorem to $f_1, f_2, \ldots, f_m$, we find that if $\mathbf{X}, \mathbf{Y} \in S_\delta(\mathbf{X}_0)$, then

$$f_i(\mathbf{X}) - f_i(\mathbf{Y}) = \sum_{j=1}^{n} \frac{\partial f_i(\mathbf{X}_i)}{\partial x_j}(x_j - y_j),$$

where $\mathbf{X}_i$ is on the line segment from $\mathbf{X}$ to $\mathbf{Y}$. From this,

$$\left[f_i(\mathbf{X}) - f_i(\mathbf{Y}) - \sum_{j=1}^{n} \frac{\partial f_i(\mathbf{X})}{\partial x_j}(x_j - y_j) \right]^2$$

$$= \left[\sum_{j=1}^{n} \left[\frac{\partial f_i(\mathbf{X}_i)}{\partial x_j} - \frac{\partial f_i(\mathbf{X})}{\partial x_j} \right](x_j - y_j) \right]^2$$

$$\le |\mathbf{X} - \mathbf{Y}|^2 \sum_{j=1}^{n} \left[\frac{\partial f_i(\mathbf{X}_i)}{\partial x_j} - \frac{\partial f_i(\mathbf{X})}{\partial x_j} \right]^2 \qquad \text{(by Schwarz's inequality)}$$

$$< \frac{\varepsilon^2}{m} |\mathbf{X} - \mathbf{Y}|^2 \qquad \text{[by (19)].}$$

Summing from $i = 1$ to $i = m$ and taking square roots yields

$$|\mathbf{F}(\mathbf{Y}) - \mathbf{F}(\mathbf{X}) - \mathbf{F}'(\mathbf{X})(\mathbf{Y} - \mathbf{X})| < \varepsilon |\mathbf{Y} - \mathbf{X}| \quad \text{if} \quad \mathbf{X}, \mathbf{Y} \in S_\delta(\mathbf{X}_0).$$

This implies (18) and completes the proof that g is continuous on S. Since D is compact, so is S (*Exercise 29, Section 5.1*), and therefore g is bounded on S (Theorem 2.7, Section 5.2); thus, for some M_1,

$$|\mathbf{F}(\mathbf{Y}) - \mathbf{F}(\mathbf{X}) - \mathbf{F}'(\mathbf{X})(\mathbf{Y} - \mathbf{X})| \le M_1 |\mathbf{X} - \mathbf{Y}| \text{ if } \mathbf{X}, \mathbf{Y} \in D.$$

But

$$|\mathbf{F}(\mathbf{Y}) - \mathbf{F}(\mathbf{X})| \le |\mathbf{F}(\mathbf{Y}) - \mathbf{F}(\mathbf{X}) - \mathbf{F}'(\mathbf{X})(\mathbf{Y} - \mathbf{X})| + |\mathbf{F}'(\mathbf{X})(\mathbf{Y} - \mathbf{X})|$$

$$\le (M_1 + \|\mathbf{F}'(\mathbf{X})\|)|\mathbf{Y} - \mathbf{X}|. \tag{20}$$

Since

$$\|\mathbf{F}'(\mathbf{X})\| \le \left(\sum_{i=1}^{m} \sum_{j=1}^{n} \left[\frac{\partial f_i(\mathbf{X})}{\partial x_j} \right]^2 \right)^{1/2}$$

and the partial derivatives $\{\partial f_i/\partial x_j\}$ are bounded on D, it follows that $\|\mathbf{F}'(\mathbf{X})\|$ is bounded on D; if

$$\|\mathbf{F}'(\mathbf{X})\| \le M_2, \qquad \mathbf{X} \in D,$$

then (20) implies (17), with $M = M_1 + M_2$.

The chain rule for transformations

With differential matrices we can write the chain rule for transformations in a form analogous to that of the chain rule for real-valued functions of one variable (Theorem 3.3, Section 2.3).

Theorem 2.5   *Suppose* $\mathbf{F} \colon \mathscr{R}^n \to \mathscr{R}^m$ *is differentiable at* $\mathbf{X}_0$, $\mathbf{G} \colon \mathscr{R}^k \to \mathscr{R}^n$ *is differentiable at* $\mathbf{U}_0$, *and* $\mathbf{X}_0 = \mathbf{G}(\mathbf{U}_0)$. *Then the composite function* $\mathbf{H} = \mathbf{F} \circ \mathbf{G} \colon \mathscr{R}^k \to \mathscr{R}^m$, *defined by*

$$\mathbf{H}(\mathbf{U}) = \mathbf{F}(\mathbf{G}(\mathbf{U})),$$

is differentiable at $\mathbf{U}_0$,

$$\mathbf{H}'(\mathbf{U}_0) = \mathbf{F}'(\mathbf{G}(\mathbf{U}_0))\mathbf{G}'(\mathbf{U}_0), \tag{21}$$

and

$$d_{\mathbf{U}_0}\mathbf{H} = d_{\mathbf{X}_0}\mathbf{F} \circ d_{\mathbf{U}_0}\mathbf{G}, \tag{22}$$

where $\circ$ *stands for composition.*

Proof.   The components of $\mathbf{H}$ are $h_1, h_2, \ldots, h_m$, where

$$h_i(\mathbf{U}) = f_i(\mathbf{G}(\mathbf{U})).$$

Applying Theorem 4.1, Section 5.4, to h_i yields

$$d_{\mathbf{U}_0}h_i = \sum_{j=1}^{n} \frac{\partial f_i(\mathbf{X}_0)}{\partial x_j} d_{\mathbf{U}_0}g_j, \qquad 1 \le i \le m. \tag{23}$$

Since

$$d_{\mathbf{U}_0}\mathbf{H} = \begin{bmatrix} d_{\mathbf{U}_0}h_1 \\ d_{\mathbf{U}_0}h_2 \\ \vdots \\ d_{\mathbf{U}_0}h_m \end{bmatrix} \quad \text{and} \quad d_{\mathbf{U}_0}\mathbf{G} = \begin{bmatrix} d_{\mathbf{U}_0}g_1 \\ d_{\mathbf{U}_0}g_2 \\ \vdots \\ d_{\mathbf{U}_0}g_n \end{bmatrix},$$

the m equations in (23) can be written in matrix form as

$$d_{\mathbf{U}_0}\mathbf{H} = \mathbf{F}'(\mathbf{X}_0)\, d_{\mathbf{U}_0}\mathbf{G} = \mathbf{F}'(\mathbf{G}(\mathbf{U}_0))\, d_{\mathbf{U}_0}\mathbf{G}. \tag{24}$$

But

$$d_{\mathbf{U}_0}\mathbf{G} = \mathbf{G}'(\mathbf{U}_0)\, d\mathbf{U},$$

where

$$d\mathbf{U} = \begin{bmatrix} du_1 \\ du_2 \\ \vdots \\ du_k \end{bmatrix},$$

so that (24) can be rewritten as

$$d_{\mathbf{U}_0}\mathbf{H} = \mathbf{F}'(\mathbf{G}(\mathbf{U}_0))\mathbf{G}'(\mathbf{U}_0)\, d\mathbf{U}.$$

On the other hand,

$$d_{\mathbf{U}_0}\mathbf{H} = \mathbf{H}'(\mathbf{U}_0)\, d\mathbf{U}.$$

Comparing the last two equations yields (21). We leave it to the reader to show that (21) implies (22) (*Exercise 21*).

Example 2.6 Let $\mathbf{U}_0 = (1, -1)$,

$$\mathbf{G}(\mathbf{U}) = \mathbf{G}(u, v) = \begin{bmatrix} \sqrt{u} \\ \sqrt{u^2 + 3v^2} \\ \sqrt{v + 2} \end{bmatrix}, \qquad \mathbf{F}(\mathbf{X}) = \mathbf{F}(x, y, z) = \begin{bmatrix} x^2 + y^2 + 2z^2 \\ x^2 - y^2 \end{bmatrix},$$

and

$$\mathbf{H}(\mathbf{U}) = \mathbf{F}(\mathbf{G}(\mathbf{U})).$$

Since $\mathbf{G}$ is differentiable at $\mathbf{U}_0 = (1, -1)$ and $\mathbf{F}$ is differentiable at $\mathbf{X}_0 = \mathbf{G}(\mathbf{U}_0) = (1, 2, 1)$, Theorem 2.5 implies that $\mathbf{H}$ is differentiable at $(1, -1)$. To find $\mathbf{H}'(1, -1)$ from (21), we first find that

$$\mathbf{G}'(\mathbf{U}) = \begin{bmatrix} \dfrac{1}{2\sqrt{u}} & 0 \\[2ex] \dfrac{u}{\sqrt{u^2 + 3v^2}} & \dfrac{3v}{\sqrt{u^2 + 3v^2}} \\[2ex] 0 & \dfrac{1}{2\sqrt{v + 2}} \end{bmatrix}$$

and

$$\mathbf{F}'(\mathbf{X}) = \begin{bmatrix} 2x & 2y & 4z \\ 2x & -2y & 0 \end{bmatrix};$$

then, from (21),

$$\mathbf{H}'(1, -1) = \mathbf{F}'(1, 2, 1)\mathbf{G}'(1, -1)$$

$$= \begin{bmatrix} 2 & 4 & 4 \\ 2 & -4 & 0 \end{bmatrix} \begin{bmatrix} \frac{1}{2} & 0 \\ \frac{1}{2} & -\frac{3}{2} \\ 0 & \frac{1}{2} \end{bmatrix} = \begin{bmatrix} 3 & -4 \\ -1 & 6 \end{bmatrix}.$$

We can check this by expressing **H** directly in terms of (u, v),

$$\mathbf{H}(u, v) = \begin{bmatrix} (\sqrt{u})^2 + (\sqrt{u^2 + 3v^2})^2 + 2(\sqrt{v + 2})^2 \\ (\sqrt{u})^2 - (\sqrt{u^2 + 3v^2})^2 \end{bmatrix}$$

$$= \begin{bmatrix} u + u^2 + 3v^2 + 2v + 4 \\ u - u^2 - 3v^2 \end{bmatrix},$$

and differentiating to obtain

$$\mathbf{H}'(u, v) = \begin{bmatrix} 1 + 2u & 6v + 2 \\ 1 - 2u & -6v \end{bmatrix},$$

which yields

$$\mathbf{H}'(1, -1) = \begin{bmatrix} 3 & -4 \\ -1 & 6 \end{bmatrix},$$

as we saw before.

6.2   EXERCISES

1. Show that the following definitions are equivalent.
   (A) $\mathbf{F} = (f_1, f_2, \ldots, f_m)$ is continuous at $\mathbf{X}_0$ if $f_1, f_2, \ldots, f_m$ are.
   (B) $\mathbf{F}$ is continuous at $\mathbf{X}_0$ if for each $\varepsilon > 0$, there is a $\delta > 0$ such that $|\mathbf{F}(\mathbf{X}) - \mathbf{F}(\mathbf{X}_0)| < \varepsilon$ whenever $|\mathbf{X} - \mathbf{X}_0| < \delta$ and $\mathbf{X} \in D_{\mathbf{F}}$.

2. Prove that Eq. (2) holds if

   (a) $\mathbf{F}(\mathbf{X}) = \begin{bmatrix} 3x + 4y \\ 2x - y \\ x + y \end{bmatrix}$, $\mathbf{X}_0 = (x_0, y_0, z_0)$

   (b) $\mathbf{F}(\mathbf{X}) = \begin{bmatrix} 2x^2 + xy + 1 \\ xy \\ x^2 + y^2 \end{bmatrix}$, $\mathbf{X}_0 = (1, -1)$

(c) $\mathbf{F}(\mathbf{X}) = \begin{bmatrix} \sin(x + y) \\ \sin(y + z) \\ \sin(x + z) \end{bmatrix}$, $\mathbf{X}_0 = (\pi/4, 0, \pi/4)$

3. Suppose $\mathbf{F} \colon \mathscr{R}^n \to \mathscr{R}^m$ and $h \colon \mathscr{R}^n \to \mathscr{R}$ have the same domain and are continuous at $\mathbf{X}_0$. Show that the product $h\mathbf{F} = (hf_1, hf_2, \ldots, hf_m)$ is continuous at $\mathbf{X}_0$.

4. Suppose $\mathbf{F}$ and $\mathbf{G}$ are transformations from $\mathscr{R}^n$ to $\mathscr{R}^m$, with common domain D. Show that if $\mathbf{F}$ and $\mathbf{G}$ are continuous at $\mathbf{X}_0$, then so are $\mathbf{F} + \mathbf{G}$ and $\mathbf{F} - \mathbf{G}$.

5. Suppose $\mathbf{F} \colon \mathscr{R}^n \to \mathscr{R}^m$ is defined in a neighborhood of $\mathbf{X}_0$ and continuous at $\mathbf{X}_0$, $\mathbf{G} \colon \mathscr{R}^k \to \mathscr{R}^n$ is defined in a neighborhood of $\mathbf{U}_0$ and continuous at $\mathbf{U}_0$, and $\mathbf{X}_0 = \mathbf{G}(\mathbf{U}_0)$. Prove that the composite function $\mathbf{H} = \mathbf{F} \circ \mathbf{G}$ is continuous at $\mathbf{U}_0$.

6. Prove: If $\mathbf{F} \colon \mathscr{R}^n \to \mathscr{R}^m$ is continuous on a set S, then $|\mathbf{F}|$ is continuous on S.

7. Prove: If $\mathbf{F} \colon \mathscr{R}^n \to \mathscr{R}^m$ is continuous on a compact set S, then $|\mathbf{F}|$ is bounded on S, and there are points $\mathbf{X}_0$ and $\mathbf{X}_1$ in S such that

$$|\mathbf{F}(\mathbf{X}_0)| \leq |\mathbf{F}(\mathbf{X})| \leq |\mathbf{F}(\mathbf{X}_1)|, \qquad \mathbf{X} \in S;$$

that is, $|\mathbf{F}|$ attains its greatest lower bound and least upper bound on S. (*Hint:* Use Exercise 6.)

8. Prove that a linear transformation is continuous.

* 9. Let $\mathbf{A}$ be an $m \times n$ matrix.
   (a) Use Exercises 7 and 8 to show that the quantities

$$M(\mathbf{A}) = \max \left\{ \frac{|\mathbf{AX}|}{|\mathbf{X}|} \,\middle|\, \mathbf{X} \neq \mathbf{0} \right\},$$

   and

$$m(\mathbf{A}) = \min \left\{ \frac{|\mathbf{AX}|}{|\mathbf{X}|} \,\middle|\, \mathbf{X} \neq \mathbf{0} \right\}$$

   exist. [*Hint:* Consider the function $\mathbf{L}(\mathbf{Y}) = \mathbf{AY}$ on $S = \{\mathbf{Y} \,|\, |\mathbf{Y}| = 1\}$.]
   (b) Show that $M(\mathbf{A}) = \|\mathbf{A}\|$.
   (c) Prove: If $n > m$ or $n = m$ and $\mathbf{A}$ is singular, then $m(\mathbf{A}) = 0$. (This requires a result from linear algebra on the existence of nontrivial solutions of $\mathbf{AX} = \mathbf{0}$.)
   (d) Prove: If $n = m$ and $\mathbf{A}$ is nonsingular, then

$$m(\mathbf{A})M(\mathbf{A}^{-1}) = m(\mathbf{A}^{-1})M(\mathbf{A}) = 1.$$

10. We say that $\mathbf{F} \colon \mathscr{R}^n \to \mathscr{R}^m$ is uniformly continuous on S if each of its components is. Prove: If $\mathbf{F}$ is uniformly continuous on S, then for each $\varepsilon > 0$ there is a $\delta > 0$ such that

$$|\mathbf{F}(\mathbf{X}) - \mathbf{F}(\mathbf{Y})| < \varepsilon \quad \text{if} \quad |\mathbf{X} - \mathbf{Y}| < \delta \quad \text{and} \quad \mathbf{X}, \mathbf{Y} \in S.$$

11. Show that if $\mathbf{F}$ is continuous on $\mathscr{R}^n$ and $\mathbf{F}(\mathbf{X} + \mathbf{Y}) = \mathbf{F}(\mathbf{X}) + \mathbf{F}(\mathbf{Y})$ for all $\mathbf{X}$ and $\mathbf{Y}$ in $\mathscr{R}^n$, then $\mathbf{F}$ is linear. (*Hint:* The rational numbers are dense in the reals.)

12. Find an affine transformation $\mathbf{G}$ such that

$$\lim_{\mathbf{X} \to \mathbf{X}_0} \frac{\mathbf{F}(\mathbf{X}) - \mathbf{G}(\mathbf{X})}{\mathbf{X} - \mathbf{X}_0} = 0$$

if:

(a) $\mathbf{F}(x, y, z) = \begin{bmatrix} x^2 + y + 2z \\ \cos(x + y + z) \\ e^{xyz} \end{bmatrix}$, $\mathbf{X}_0 = (1, -1, 0)$

(b) $\mathbf{F}(x, y) = \begin{bmatrix} e^x \cos y \\ e^x \sin y \end{bmatrix}$, $\mathbf{X}_0 = (0, \pi/2)$

(c) $\mathbf{F}(x, y, z) = \begin{bmatrix} x^2 - y^2 \\ y^2 - z^2 \\ z^2 - x^2 \end{bmatrix}$, $\mathbf{X}_0 = (1, 1, 1)$

13. Find $\mathbf{F}'$ and $J\mathbf{F}$ for the transformations in Exercise 12.

14. Find $\mathbf{F}'$:

(a) $\mathbf{F}(x, y, z) = \begin{bmatrix} (x + y + z)e^x \\ (x^2 + y^2)e^{-x} \end{bmatrix}$

(b) $\mathbf{F}(x) = \begin{bmatrix} g_1(x) \\ g_2(x) \\ \vdots \\ g_n(x) \end{bmatrix}$ (c) $\mathbf{F}(x, y, z) = \begin{bmatrix} e^x \sin yz \\ e^y \sin xz \\ e^z \sin xy \end{bmatrix}$

15. Prove: If $\mathbf{G}_1$ and $\mathbf{G}_2$ are affine transformations and

$$\lim_{\mathbf{X} \to \mathbf{X}_0} \frac{\mathbf{G}_1(\mathbf{X}) - \mathbf{G}_2(\mathbf{X})}{|\mathbf{X} - \mathbf{X}_0|} = 0,$$

then $\mathbf{G}_1 = \mathbf{G}_2$.

16. Prove Theorem 2.3.

17. Prove Theorem 2.4. (*Hint:* Use Theorem 3.7, Section 5.3.)

18. Show that if $\mathbf{F}: \mathscr{R}^n \to \mathscr{R}^m$ is differentiable at $\mathbf{X}_0$ and $\varepsilon > 0$, then there is a $\delta > 0$ such that

$$|\mathbf{F}(\mathbf{X}) - \mathbf{F}(\mathbf{X}_0)| \le (\|\mathbf{F}'(\mathbf{X}_0)\| + \varepsilon)|\mathbf{X} - \mathbf{X}_0| \qquad \text{if} \quad |\mathbf{X} - \mathbf{X}_0| < \delta.$$

Compare this with Lemma 2.1.

*19. Suppose $\mathbf{F}: \mathscr{R}^n \to \mathscr{R}^n$ is differentiable at $\mathbf{X}_0$ and $\mathbf{F}'(\mathbf{X}_0)$ is nonsingular. Let

$$r = \frac{1}{\|[\mathbf{F}'(\mathbf{X}_0)]^{-1}\|}$$

and suppose $\varepsilon > 0$. Show that there is a $\delta > 0$ such that

$$|\mathbf{F}(\mathbf{X}) - \mathbf{F}(\mathbf{X}_0)| \ge (r - \varepsilon)|\mathbf{X} - \mathbf{X}_0| \qquad \text{if} \quad |\mathbf{X} - \mathbf{X}_0| < \delta.$$

Compare this with Lemma 2.2.

20. Prove: If $\mathbf{L}: \mathscr{R}^n \to \mathscr{R}^m$ is defined by $\mathbf{L}(\mathbf{X}) = \mathbf{A}\mathbf{X}$, where $\mathbf{A}$ is nonsingular, then

$$|\mathbf{L}(\mathbf{X}) - \mathbf{L}(\mathbf{Y})| \ge \frac{1}{\|\mathbf{A}^{-1}\|}|\mathbf{X} - \mathbf{Y}|$$

for all $\mathbf{X}$ and $\mathbf{Y}$ in $\mathscr{R}^n$.

21. Complete the proof of Theorem 2.5 by showing that Eq. (21) implies Eq. (22).

22. Use Theorem 2.5 to find $\mathbf{H}'(\mathbf{U}_0)$, where $\mathbf{H}(\mathbf{U}) = \mathbf{F}(\mathbf{G}(\mathbf{U}))$. Check your results by expressing $\mathbf{H}$ directly in terms of $\mathbf{U}$ and differentiating.

(a) $\mathbf{F}(x, y, z) = \begin{bmatrix} x^2 + y^2 \\ z \\ x^2 + y^2 \end{bmatrix}$, $\mathbf{G}(u, v, w) = \begin{bmatrix} w \cos u \sin v \\ w \sin u \sin v \\ w \cos v \end{bmatrix}$, $\mathbf{U}_0 = (\pi/2, \pi/2, 2)$

(b) $\mathbf{F}(x, y) = \begin{bmatrix} x^2 - y^2 \\ y \\ x \end{bmatrix}$, $\mathbf{G}(u, v) = \begin{bmatrix} v \cos u \\ v \sin u \end{bmatrix}$, $\mathbf{U}_0 = (3, \pi/4)$

(c) $\mathbf{F}(x, y, z) = \begin{bmatrix} 3x + 4y + 2z + 6 \\ 4x - 2y + z - 1 \\ -x + y + z - 2 \end{bmatrix}$, $\mathbf{G}(u, v) = \begin{bmatrix} u - v \\ u + v \\ u - 2v \end{bmatrix}$, $\mathbf{U}_0$ arbitrary

(d) $\mathbf{F}(x, y) = \begin{bmatrix} x + y \\ x - y \end{bmatrix}$, $\mathbf{G}(u, v, w) = \begin{bmatrix} 2u - v + w \\ e^{u^2 - v^2} \end{bmatrix}$, $\mathbf{U}_0 = (1, 1, -2)$

(e) $\mathbf{F}(x, y) = \begin{bmatrix} x^2 + y^2 \\ x^2 - y^2 \end{bmatrix}$, $\mathbf{G}(u, v) = \begin{bmatrix} e^u \cos v \\ e^u \sin v \end{bmatrix}$, $\mathbf{U}_0 = (0, 0)$

(f) $F(x, y) = \begin{bmatrix} x + 2y \\ x - y^2 \\ x^2 + y \end{bmatrix}$, $G(u, v) = \begin{bmatrix} u + 2v \\ 2u - v^2 \end{bmatrix}$, $U_0 = (1, -2)$

23. Suppose $\mathbf{F}$ and $\mathbf{G}$ are continuously differentiable on $\mathscr{R}^n$, with values in $\mathscr{R}^n$, and let $\mathbf{H} = \mathbf{F} \circ \mathbf{G}$. Show that

$$\frac{\partial(h_1, h_2, \ldots, h_n)}{\partial(u_1, u_2, \ldots, u_n)} = \frac{\partial(f_1, f_2, \ldots, f_n)}{\partial(x_1, x_2, \ldots, x_n)} \frac{\partial(g_1, g_2, \ldots, g_n)}{\partial(u_1, u_2, \ldots, u_n)}.$$

Where are these Jacobians to be evaluated?

24. Suppose $\mathbf{F}: \mathscr{R}^n \to \mathscr{R}^m$ and $\bar{\mathbf{X}}$ is a limit point of $D_{\mathbf{F}}$ contained in $D_{\mathbf{F}}$. Show that $\mathbf{F}$ is continuous at $\bar{\mathbf{X}}$ if and only if $\lim_{k \to \infty} \mathbf{F}(\mathbf{X}_k) = \mathbf{F}(\bar{\mathbf{X}})$ whenever $\{\mathbf{X}_k\}$ is a sequence of points in $D_{\mathbf{F}}$ such that $\lim_{k \to \infty} \mathbf{X}_k = \bar{\mathbf{X}}$. (*Hint:* See Exercise 20, Section 5.2.)

*25. Suppose $\mathbf{F}: \mathscr{R}^n \to \mathscr{R}^m$ is continuous on a compact subset S of $\mathscr{R}^n$. Show that $\mathbf{F}(S)$ is a compact subset of $\mathscr{R}^m$.

*26. Find $\mathbf{F}'$ and $J\mathbf{F}$:

(a) $F(r, \theta) = \begin{bmatrix} r \cos \theta \\ r \sin \theta \end{bmatrix}$ (b) $F(r, \theta, \phi) = \begin{bmatrix} r \cos \theta \cos \phi \\ r \sin \theta \cos \phi \\ r \sin \phi \end{bmatrix}$

(c) $F(r, \theta, z) = \begin{bmatrix} r \cos \theta \\ r \sin \theta \\ z \end{bmatrix}$

6.3   THE INVERSE FUNCTION THEOREM

So far our discussion of transformations has dealt mainly with properties that could just as well be defined and studied by considering the component functions individually. Now we turn to questions involving a transformation as a whole, which cannot be studied by regarding it simply as a collection of component functions.

In this section we restrict our attention to transformations from $\mathscr{R}^n$ to itself, that is, to transformations whose ranges and domains are subsets of $\mathscr{R}^n$. It is useful to interpret such transformations geometrically, as follows. If $\mathbf{F} = (f_1, f_2, \ldots, f_n)$, we can think of the components of

$$\mathbf{F}(\mathbf{X}) = (f_1(\mathbf{X}), f_2(\mathbf{X}), \ldots, f_n(\mathbf{X}))$$

as the coordinates of a point $\mathbf{U} = \mathbf{U}(\mathbf{X})$ in another "copy" of $\mathscr{R}^n$. We may write the transformation as $\mathbf{U} = \mathbf{F}(\mathbf{X})$, or in terms of its components as

$$u_1 = f_1(\mathbf{X}), \quad u_2 = f_2(\mathbf{X}), \quad \ldots, \quad u_n = f_n(\mathbf{X}).$$

We say that **F** *maps* **X** *to* **U**, and that **U** *is the image of* **X** *under* **F**. On occasion we will also write $\partial u_i/\partial x_j$ to mean $\partial f_i/\partial x_j$. If $S \subset D_\mathbf{F}$, then the set

$$\mathbf{F}(S) = \{\mathbf{Y} \mid \mathbf{Y} = \mathbf{F}(\mathbf{X}), \mathbf{X} \in S\}$$

is the *image of S under* **F**.

Example 3.1 If

$$\begin{bmatrix} u \\ v \end{bmatrix} = \mathbf{F}(x, y) = \begin{bmatrix} x^2 + y^2 \\ x^2 - y^2 \end{bmatrix},$$

then

$$u = f_1(x, y) = x^2 + y^2, \qquad v = f_2(x, y) = x^2 - y^2,$$

and

$$u_x(x, y) = \frac{\partial f_1(x, y)}{\partial x} = 2x, \qquad u_y(x, y) = \frac{\partial f_1(x, y)}{\partial y} = 2y,$$

$$v_x(x, y) = \frac{\partial f_2(x, y)}{\partial x} = 2x, \qquad v_y(x, y) = \frac{\partial f_2(x, y)}{\partial y} = -2y.$$

To find **F** $(\mathscr{R}^2)$ we observe that

$$u + v = 2x^2, \qquad u - v = 2y^2,$$

and so

$$\mathbf{F}(\mathscr{R}^2) \subset T = \{(u, v) \mid u + v \geq 0, u - v \geq 0\},$$

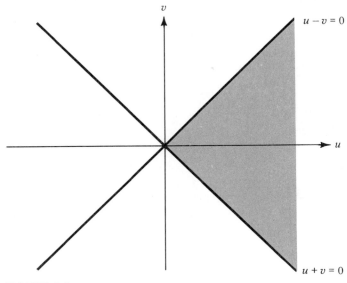

FIGURE 3.1

which is the part of the *uv* plane shaded in Figure 3.1. If $(u, v) \in T$, then

$$\mathbf{F}\left(\frac{\sqrt{u+v}}{2}, \frac{\sqrt{u-v}}{2}\right) = \begin{bmatrix} u \\ v \end{bmatrix},$$

and so $\mathbf{F}(\mathscr{R}^2) = T$.

Invertible transformations

A transformation $\mathbf{F}$ is *one-to-one*, or *invertible*, if $\mathbf{F}(\mathbf{X}_1)$ and $\mathbf{F}(\mathbf{X}_2)$ are distinct whenever $\mathbf{X}_1$ and $\mathbf{X}_2$ are distinct points of $D_{\mathbf{F}}$. In this case we can define a function $\mathbf{G}$ on

$$R(\mathbf{F}) = \{\mathbf{Y} \mid \mathbf{Y} = \mathbf{F}(\mathbf{X}) \text{ for some } \mathbf{X} \in D_{\mathbf{F}}\}$$

$[R(\mathbf{F})$ is the range of $\mathbf{F}]$, by defining $\mathbf{G}(\mathbf{Y})$ to be the unique point in $D_{\mathbf{F}}$ such that $\mathbf{F}(\mathbf{X}) = \mathbf{Y}$. Then

$$D_{\mathbf{G}} = R(\mathbf{F}) \quad \text{and} \quad R(\mathbf{G}) = D_{\mathbf{F}}.$$

Moreover, $\mathbf{G}$ is one-to-one,

$$\mathbf{G}(\mathbf{F}(\mathbf{X})) = \mathbf{X}, \qquad \mathbf{X} \in D_{\mathbf{F}},$$

and

$$\mathbf{F}(\mathbf{G}(\mathbf{Y})) = \mathbf{Y}, \qquad \mathbf{Y} \in D_{\mathbf{G}}.$$

We say that $\mathbf{G}$ is the *inverse* of $\mathbf{F}$, and occasionally write $\mathbf{G} = \mathbf{F}^{-1}$. The relation between $\mathbf{F}$ and $\mathbf{G}$ is symmetric; that is, $\mathbf{F}$ is also the inverse of $\mathbf{G}$, and we may write $\mathbf{F} = \mathbf{G}^{-1}$.

Example 3.2   The linear transformation

$$\begin{bmatrix} u \\ v \end{bmatrix} = \mathbf{L}(x, y) = \begin{bmatrix} x - y \\ x + y \end{bmatrix} \tag{1}$$

maps (x, y) onto (u, v), where

$$\begin{aligned} u &= x - y \\ v &= x + y. \end{aligned} \tag{2}$$

It is one-to-one and $R(\mathbf{L}) = \mathscr{R}^2$, since for each (u, v) in $\mathscr{R}^2$ there is exactly one (x, y) such that $\mathbf{L}(x, y) = (u, v)$. This is so because the system (2) can be solved uniquely for (x, y) in terms of (u, v):

$$\begin{aligned} x &= \tfrac{1}{2}(u + v), \\ y &= \tfrac{1}{2}(-u + v). \end{aligned} \tag{3}$$

Thus,

$$\mathbf{L}^{-1}(u, v) = \begin{bmatrix} \frac{1}{2}(u + v) \\ \frac{1}{2}(-u + v) \end{bmatrix}.$$

Example 3.3   The linear transformation

$$\begin{bmatrix} u \\ v \end{bmatrix} = \mathbf{L}_1(x, y) = \begin{bmatrix} x + y \\ 2x + 2y \end{bmatrix}$$

maps (x, y) onto (u, v), where

$$\begin{aligned} u &= x + y, \\ v &= 2x + 2y. \end{aligned} \tag{4}$$

It is obviously not one-to-one, since every point on the line

$$x + y = c \text{ (constant)}$$

is mapped onto the single point $(c, 2c)$; hence, $\mathbf{L}_1$ does not have an inverse.

The crucial difference between the transformations of Examples 3.2 and 3.3 is that the matrix of $\mathbf{L}$ is nonsingular while that of $\mathbf{L}_1$ is singular. Thus, $\mathbf{L}$[see (1)] can be written as

$$\begin{bmatrix} u \\ v \end{bmatrix} = \begin{bmatrix} 1 & -1 \\ 1 & 1 \end{bmatrix}\begin{bmatrix} x \\ y \end{bmatrix}, \tag{5}$$

where the matrix has the inverse

$$\begin{bmatrix} \frac{1}{2} & \frac{1}{2} \\ -\frac{1}{2} & \frac{1}{2} \end{bmatrix}.$$

(Verify.) Multiplying both sides of (5) by this matrix yields

$$\begin{bmatrix} \frac{1}{2} & \frac{1}{2} \\ -\frac{1}{2} & \frac{1}{2} \end{bmatrix}\begin{bmatrix} u \\ v \end{bmatrix} = \begin{bmatrix} x \\ y \end{bmatrix},$$

which is equivalent to (3).

The matrix of $\mathbf{L}_1$,

$$\begin{bmatrix} 1 & 1 \\ 2 & 2 \end{bmatrix},$$

is singular and therefore (4) cannot be solved uniquely for (x, y) in terms of (u, v); in fact, it cannot be solved at all unless $v = 2u$.

The following theorem settles the question of invertibility of linear transformations from $\mathscr{R}^n$ to $\mathscr{R}^n$. We omit the proof (*Exercise 3*).

Theorem 3.1   *The linear transformation*

$$\mathbf{U} = \mathbf{L}(\mathbf{X}) = \mathbf{A}\mathbf{X} \qquad (\mathcal{R}^n \to \mathcal{R}^n)$$

is invertible if and only if $\mathbf{A}$ *is nonsingular, in which case* $R(\mathbf{L}) = \mathcal{R}^n$ *and*

$$\mathbf{L}^{-1}(\mathbf{U}) = \mathbf{A}^{-1}\mathbf{U}.$$

Polar coordinates

We will need polar coordinates in some of the following examples, so we now review them.

The coordinates of any point (x, y) can be written in infinitely many ways as

$$x = r \cos \theta, \qquad y = r \sin \theta, \tag{6}$$

where

$$r^2 = x^2 + y^2$$

and, if $r > 0$, θ is the angle from the x axis to the line segment from $(0, 0)$ to (x, y), measured counterclockwise (Figure 3.2).

For each $(x, y) \neq (0, 0)$ there are infinitely many values of θ, differing by integral multiples of 2π, that satisfy (6). If θ is any of these, we say it is an *argument* of (x, y), and write

$$\theta = \arg(x, y).$$

By itself this does not define a function; however, if ϕ is an arbitrary fixed number, the statement

$$\theta = \arg(x, y), \qquad \phi \le \theta < \phi + 2\pi$$

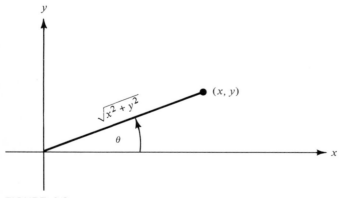

FIGURE 3.2

does, since every half-open interval $[\phi, \phi + 2\pi)$ contains exactly one argument of (x, y).

We do not define $\arg(0, 0)$, since (6) places no restriction on θ if $(x, y) = (0, 0)$ and, therefore, $r = 0$.

The transformation

$$\begin{bmatrix} r \\ \theta \end{bmatrix} = \mathbf{G}(x, y) = \begin{bmatrix} \sqrt{x^2 + y^2} \\ \arg(x, y) \end{bmatrix}, \qquad \phi \leq \arg(x, y) < \phi + 2\pi,$$

is defined and one-to-one on

$$D_{\mathbf{G}} = \{(x, y) \,|\, (x, y) \neq (0, 0)\}$$

and its range is

$$R(\mathbf{G}) = \{(r, \theta) \,|\, r > 0, \phi \leq \theta < \phi + 2k\pi\}.$$

For example, if $\phi = 0$, then

$$\mathbf{G}(1, 1) = \begin{bmatrix} \sqrt{2} \\ \dfrac{\pi}{4} \end{bmatrix},$$

since $\pi/4$ is the unique argument of $(1, 1)$ in $[0, 2\pi)$; if $\phi = \pi$, then

$$\mathbf{G}(1, 1) = \begin{bmatrix} \sqrt{2} \\ \dfrac{9\pi}{4} \end{bmatrix},$$

since $9\pi/4$ is the unique argument of $(1, 1)$ in $[\pi, 3\pi)$.

If $\arg(x_0, y_0) = \phi$, then (x_0, y_0) is on the half-line shown in Figure 3.3; $\mathbf{G}$ is not continuous at (x_0, y_0), since every neighborhood of (x_0, y_0) contains

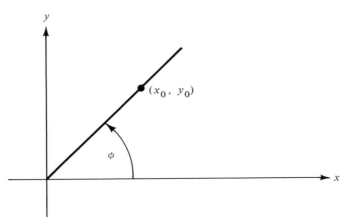

FIGURE 3.3

points (x, y) for which the second component of $G(x, y)$ is arbitrarily close to $\phi + 2\pi$, while the second component of $G(x_0, y_0)$ is ϕ. We will show later, however, that G is continuous—in fact, continuously differentiable—on the plane with this half-line deleted.

Local invertibility

A transformation F may fail to be one-to-one, but be one-to-one on a subset S of D_F. By this we mean that $F(X_1)$ and $F(X_2)$ are distinct whenever X_1 and X_2 are distinct points of S. In this case F is not invertible, but if F_S is defined on S by

$$F_S(X) = F(X), \qquad X \in S,$$

and left undefined for $X \notin S$, then F_S is invertible. We say that F_S is the *restriction of F to S*, and that F_S^{-1} is *the inverse of F restricted to S*. The domain of F_S^{-1} is $F(S)$.

If F is one-to-one on a neighborhood of X_0, we say that F is *locally invertible* at X_0. If this is true for every X_0 in a set S, then F is *locally invertible on S*.

Example 3.4   The transformation

$$\begin{bmatrix} u \\ v \end{bmatrix} = F(x, y) = \begin{bmatrix} x^2 - y^2 \\ 2xy \end{bmatrix} \tag{7}$$

is not one-to-one, since

$$F(-x, -y) = F(x, y). \tag{8}$$

It is one-to-one on S if and only if S does not contain any pair of distinct points of the form (x_0, y_0) and $(-x_0, -y_0)$; (8) implies the necessity of this condition, and its sufficiency follows from the fact that if

$$F(x_1, y_1) = F(x_0, y_0), \tag{9}$$

then

$$(x_1, y_1) = (x_0, y_0) \quad \text{or} \quad (x_1, y_1) = (-x_0, -y_0). \tag{10}$$

To see this, suppose (9) holds; then

$$x_1^2 - y_1^2 = x_0^2 - y_0^2 \tag{11}$$

and

$$2x_1 y_1 = 2x_0 y_0. \tag{12}$$

Squaring (11) yields

$$x_1^4 - 2x_1^2 y_1^2 + y_1^4 = x_0^4 - 2x_0^2 y_0^2 + y_0^4;$$

this and (12) imply that

$$x_1^4 + y_1^4 = x_0^4 + y_0^4,$$

or

$$x_1^4 - x_0^4 = y_0^4 - y_1^4. \tag{13}$$

From (11),

$$x_1^2 - x_0^2 = y_1^2 - y_0^2. \tag{14}$$

Factoring (13) yields

$$(x_1^2 - x_0^2)(x_1^2 + x_0^2) = (y_0^2 - y_1^2)(y_0^2 + y_1^2).$$

If either side of (14) is nonzero, we can cancel here to obtain

$$x_1^2 + x_0^2 = -y_0^2 - y_1^2,$$

which implies that $x_0 = x_1 = y_0 = y_1 = 0$, and so (10) holds in this case. On the other hand, if both sides of (14) vanish, then

$$x_1 = \pm x_0, \qquad y_1 = \pm y_0.$$

From (12) the same sign must be chosen in these equalities, which proves that (8) implies (10) in this case also.

   We now see, for example, that $\mathbf{F}$ is one-to-one on every set S of the form

$$S = \{(x, y) \,|\, ax + by > 0\},$$

where a and b are constants, not both zero. Geometrically, S is an open half-plane; that is, the set of points on one side of—but not on—the line

$$ax + by = 0$$

(Figure 3.4). Therefore, $\mathbf{F}$ is locally invertible at every $\mathbf{X}_0 \neq (0, 0)$, since every such point lies in a half-plane of this form; $\mathbf{F}$ is not locally invertible at $(0, 0)$. (Why not?) Thus, $\mathbf{F}$ is locally invertible on the entire plane with $(0, 0)$ removed.

   It is instructive to find $\mathbf{F}_S^{-1}$ for a specific choice of S. Suppose S is the open right half-plane:

$$S = \{(x, y) \,|\, x > 0\}. \tag{15}$$

Then $\mathbf{F}(S)$ is the entire uv plane except for the nonpositive u axis. To see this, note that every point in S can be written in polar coordinates as

$$x = r \cos \theta, \quad y = r \sin \theta, \quad r > 0, \quad -\frac{\pi}{2} < \theta < \frac{\pi}{2};$$

therefore, from (7), $\mathbf{F}(x, y)$ has coordinates (u, v), where

$$u = x^2 - y^2 = r^2(\cos^2 \theta - \sin^2 \theta) = r^2 \cos 2\theta,$$
$$v = 2xy = 2r^2 \cos \theta \sin \theta = r^2 \sin 2\theta.$$

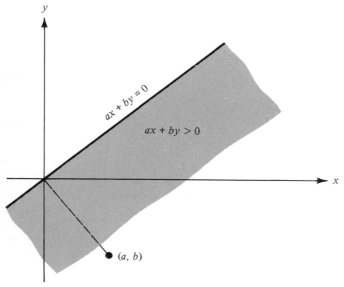

FIGURE 3.4

Every point in the (u, v) plane can be written in polar coordinates as

$$u = \rho \cos \alpha, \qquad v = \rho \sin \alpha,$$

where either $\rho = 0$ or

$$\rho = \sqrt{u^2 + v^2} > 0, \qquad -\pi \le \alpha < \pi,$$

and the points for which $\rho = 0$ or $\alpha = -\pi$ are of the form $(u, 0)$, with $u \le 0$ (Figure 3.5). If $(u, v) = \mathbf{F}(x, y)$ for some (x, y) in S, (15) implies that $\rho > 0$ and

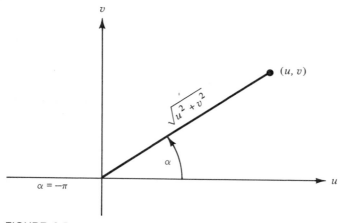

FIGURE 3.5

$-\pi < \alpha < \pi$; conversely, any point in the uv plane with polar coordinates (ρ, α) satisfying these conditions is the image under $\mathbf{F}$ of the point

$$(x, y) = (\rho^{1/2} \cos \alpha/2, \rho^{1/2} \sin \alpha/2) \in S.$$

Thus,

$$\mathbf{F}_S^{-1}(u, v) = \begin{bmatrix} (u^2 + v^2)^{1/4} \cos[\tfrac{1}{2} \arg(u, v)] \\ (u^2 + v^2)^{1/4} \sin[\tfrac{1}{2} \arg(u, v)] \end{bmatrix}, \qquad -\pi < \arg(u, v) < \pi.$$

Because of (8), $\mathbf{F}$ also maps the open left half-plane

$$S_1 = \{(x, y) \,|\, x < 0\}$$

onto $F(S)$, and

$$\mathbf{F}_{S_1}^{-1}(u, v) = \begin{bmatrix} (u^2 + v^2)^{1/4} \cos[\tfrac{1}{2} \arg(u, v)] \\ (u^2 + v^2)^{1/4} \sin[\tfrac{1}{2} \arg(u, v)] \end{bmatrix}, \qquad \pi < \arg(u, v) < 3\pi,$$

$$= -\mathbf{F}_S^{-1}(u, v).$$

Example 3.5   The transformation

$$\begin{bmatrix} u \\ v \end{bmatrix} = \mathbf{F}(x, y) = \begin{bmatrix} e^x \cos y \\ e^x \sin y \end{bmatrix}$$

is not one-to-one, since

$$\mathbf{F}(x, y + 2k\pi) = \mathbf{F}(x, y) \tag{16}$$

if k is any integer. It is one-to-one on a set S if and only if S does not contain any pair of points (x_0, y_0) and $(x_0, y_0 + 2k\pi)$, where k is a nonzero integer. This condition is necessary because of (16); we leave it to the reader to show that it is sufficient (*Exercise 9*).

This implies, for example, that $\mathbf{F}$ is one-to-one on

$$S_\phi = \{(x, y) \,|\, -\infty < x < \infty, \phi \leq y < \phi + 2\pi\}, \tag{17}$$

where ϕ is arbitrary. Geometrically, S_ϕ is the infinite strip bounded by the lines $y = \phi$ and $y = \phi + 2\pi$; the lower boundary is in S_ϕ, but the upper is not (Figure 3.6). Since every point is in the interior of some such strip, $\mathbf{F}$ is locally invertible on the entire plane.

The range of $\mathbf{F}_{S_\phi}$ is the entire uv plane except for the origin, since any $(u, v) \neq (0, 0)$ can be written uniquely as

$$\begin{bmatrix} u \\ v \end{bmatrix} = \begin{bmatrix} \rho \cos \alpha \\ \rho \sin \alpha \end{bmatrix},$$

where

$$\rho > 0, \qquad \phi \leq \alpha < \phi + 2\pi,$$

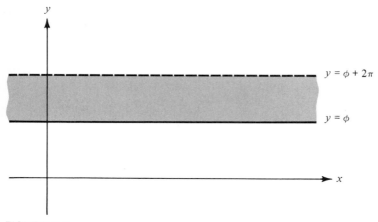

FIGURE 3.6

and is therefore the image under $\mathbf{F}$ of

$$(x, y) = (\log \rho, \alpha) \in S.$$

The origin is not in $R(\mathbf{F})$, since

$$|\mathbf{F}(x, y)|^2 = (e^x \cos y)^2 + (e^x \sin y)^2 = e^{2x} \neq 0.$$

The inverse of $\mathbf{F}$ restricted to S_ϕ is

$$\mathbf{F}_{S_\phi}^{-1}(u, v) = \begin{bmatrix} \log(u^2 + v^2)^{1/2} \\ \arg(u, v) \end{bmatrix}, \qquad \phi \leq \arg(u, v) < \phi + 2\pi;$$

its domain is the entire (u, v) plane except for $(0, 0)$.

Regular transformations

The question of invertibility of an arbitrary transformation $\mathbf{F}: \mathscr{R}^n \to \mathscr{R}^n$ is too general to have a useful answer. However, there is a useful and easily applicable sufficient condition which implies that one-to-one restrictions of continuously differentiable transformations have continuously differentiable inverses.

To motivate our study of this question, let us first assume that $\mathbf{F}$ is the linear transformation

$$\mathbf{F}(\mathbf{X}) = \mathbf{A}\mathbf{X} = \begin{bmatrix} a_{11} & a_{12} & \cdots & a_{1n} \\ a_{21} & a_{22} & \cdots & a_{2n} \\ \vdots & & & \vdots \\ a_{n1} & a_{n2} & \cdots & a_{nn} \end{bmatrix} \begin{bmatrix} x_1 \\ x_2 \\ \vdots \\ x_n \end{bmatrix}.$$

From Theorem 3.1, $\mathbf{F}$ is invertible if and only if $\mathbf{A}$ is nonsingular, in which case $R(\mathbf{F}) = \mathscr{R}^n$ and

$$\mathbf{F}^{-1}(\mathbf{Y}) = \mathbf{A}^{-1}\mathbf{Y}.$$

Since $\mathbf{A}$ and $\mathbf{A}^{-1}$ are the differential matrices of $\mathbf{F}$ and $\mathbf{F}^{-1}$, respectively, we can say that a linear transformation is invertible if and only if its differential matrix $\mathbf{F}'$ is nonsingular, in which case the differential matrix of $\mathbf{F}^{-1}$ is given by

$$(\mathbf{F}^{-1})' = (\mathbf{F}')^{-1}.$$

[Notice that $(\mathbf{F}')^{-1}$ is the inverse of the $n \times n$ matrix $\mathbf{F}'$, and not the inverse of a function; indeed, $\mathbf{F}'$ is not a function from $\mathscr{R}^n$ to $\mathscr{R}^n$, which is the only kind for which we have defined function inverses.]

Because of this it is tempting to conjecture that if $\mathbf{F} \colon \mathscr{R}^n \to \mathscr{R}^n$ is continuously differentiable and $\mathbf{F}'(\mathbf{X})$ is nonsingular—or, equivalently, $J\mathbf{F}(\mathbf{X}) \neq 0$—for $\mathbf{X}$ in a set S, then $\mathbf{F}$ is one-to-one on S. However, this is false; for example, if

$$\mathbf{F}(x, y) = \begin{bmatrix} e^x \cos y \\ e^x \sin y \end{bmatrix},$$

then

$$J\mathbf{F}(x, y) = \begin{vmatrix} e^x \cos y & -e^x \sin y \\ e^x \sin y & e^x \cos y \end{vmatrix} = e^{2x} \neq 0, \tag{18}$$

but $\mathbf{F}$ is not one-to-one on $\mathscr{R}^2$. (See Example 3.5.) The best that can be said in general is that if $\mathbf{F}$ is continuously differentiable and $J\mathbf{F}(\mathbf{X}) \neq 0$ in an open set T, then $\mathbf{F}$ is locally invertible on T and the local inverses are continuously differentiable. This is part of the inverse function theorem, which we will prove presently. Before that we need the following definition.

Definition 3.1   *A transformation* $\mathbf{F} \colon \mathscr{R}^n \to \mathscr{R}^n$ *is said to be regular on an open set S if $\mathbf{F}$ is one-to-one and continuously differentiable on S and $J\mathbf{F}(\mathbf{X}) \neq 0$ if $\mathbf{X} \in S$.*

Example 3.6   The transformations of Examples 3.1–3.5 are continuously differentiable throughout $\mathscr{R}^2$.

If

$$\mathbf{F}(x, y) = \begin{bmatrix} x - y \\ x + y \end{bmatrix}$$

(Example 3.2), then

$$J\mathbf{F}(x, y) = \begin{vmatrix} 1 & -1 \\ 1 & 1 \end{vmatrix} = 2$$

and $\mathbf{F}$ is one-to-one on $\mathscr{R}^2$; hence, $\mathbf{F}$ is regular on $\mathscr{R}^2$.
  If

$$\mathbf{F}(x, y) = \begin{bmatrix} x + y \\ 2x + 2y \end{bmatrix}$$

(Example 3.3), then

$$J\mathbf{F}(x, y) = \begin{vmatrix} 1 & 1 \\ 2 & 2 \end{vmatrix} = 0,$$

so $\mathbf{F}$ is not regular on any subset of $\mathscr{R}^2$.
  If

$$\mathbf{F}(x, y) = \begin{bmatrix} x^2 - y^2 \\ 2xy \end{bmatrix}$$

(Example 3.4), then

$$J\mathbf{F}(x, y) = \begin{vmatrix} 2x & -2y \\ 2y & 2x \end{vmatrix} = 2(x^2 + y^2),$$

so $\mathbf{F}$ is regular on any open set S on which it is one-to-one, provided $(0, 0) \notin S$. For example, $\mathbf{F}$ is regular on the open half-plane $\{(x, y)\,|\,x > 0\}$, since we saw in Example 3.4 that $\mathbf{F}$ is one-to-one on this half-plane.
  If

$$\mathbf{F}(x, y) = \begin{bmatrix} e^x \cos y \\ e^x \cos y \end{bmatrix}$$

(Example 3.5), then $J\mathbf{F}(x, y) = e^{2x}$ [see (18)], and so $\mathbf{F}$ is regular on any open set on which it is one-to-one. The interior of S_ϕ [defined by (17)] is an example of such a set.

Theorem 3.2 *Suppose* $\mathbf{F}: \mathscr{R}^n \to \mathscr{R}^n$ *is regular on an open set S and let* $\mathbf{G} = \mathbf{F}_S^{-1}$. *Then $\mathbf{F}(S)$ is open, $\mathbf{G}$ is continuously differentiable on $\mathbf{F}(S)$, and*

$$\mathbf{G}'(\mathbf{Y}) = [\mathbf{F}'(\mathbf{X})]^{-1}, \qquad \text{where } \mathbf{Y} = \mathbf{F}(\mathbf{X}).$$

(The reader will find it helpful to work through this proof for the case where $n = 2$. See *Exercise 12.*)

Proof.   To see that $\mathbf{F}(S)$ is open, we will show that if $\mathbf{X}_0 \in S$, then a neighborhood of $\mathbf{F}(\mathbf{X}_0)$ is in $\mathbf{F}(S)$. Since S is open, there is a $\rho > 0$ such that the closed n-sphere $\overline{S_\rho(\mathbf{X}_0)}$ is contained in S. Let B be the boundary of $\overline{S_\rho(\mathbf{X}_0)}$; thus,

$$B = \{\mathbf{X}\,|\,|\mathbf{X} - \mathbf{X}_0| = \rho\}.$$

The function

$$\sigma(\mathbf{X}) = |\mathbf{F}(\mathbf{X}) - \mathbf{F}(\mathbf{X}_0)|$$

is continuous on S, and therefore on B, which is compact; hence, by Theorem 2.8, Section 5.2, there is a point $\mathbf{X}_1$ on B where $\sigma(\mathbf{X})$ attains its minimum value, say m, on B. Moreover, $m \neq 0$, since $\mathbf{X}_1 \neq \mathbf{X}_0$ and $\mathbf{F}$ is one-to-one on S. Therefore,

$$|\mathbf{F}(\mathbf{X}) - \mathbf{F}(\mathbf{X}_0)| \geq m > 0 \quad \text{if} \quad |\mathbf{X} - \mathbf{X}_0| = \rho$$

We will now show that if

$$|\mathbf{Y} - \mathbf{F}(\mathbf{X}_0)| < \frac{m}{2}, \tag{19}$$

then

$$\mathbf{F}(\bar{\mathbf{X}}) = \mathbf{Y}$$

for some $\bar{\mathbf{X}}$ in $S_\rho(\mathbf{X}_0)$. To this end let $\mathbf{Y}$ be a fixed point that satisfies (19) and consider the function

$$\sigma_1(\mathbf{X}) = |\mathbf{Y} - \mathbf{F}(\mathbf{X})|^2,$$

which is continuous on S and therefore attains a minimum value μ on the compact set $\overline{S_\rho(\mathbf{X}_0)}$; that is, there is an $\bar{\mathbf{X}}$ in $\overline{S_\rho(\mathbf{X}_0)}$ such that

$$\sigma_1(\mathbf{X}) \geq \sigma_1(\bar{\mathbf{X}}) = \mu, \qquad \mathbf{X} \in \overline{S_\rho(\mathbf{X}_0)}.$$

Setting $\mathbf{X} = \mathbf{X}_0$, we conclude from this and (19) that

$$\mu \leq \sigma_1(\mathbf{X}_0) < \frac{m^2}{4}.$$

Therefore, since $\sigma_1(\mathbf{X}) \geq m^2/4$ if $\mathbf{X}$ is on B, the boundary of $\overline{S_\rho(\mathbf{X}_0)}$, the minimum of $\sigma_1(\mathbf{X})$ on $\overline{S_\rho(\mathbf{X}_0)}$ must be attained at a point $\bar{\mathbf{X}}$ in the open n-sphere $S_\rho(\mathbf{X}_0)$. Now, $\sigma_1(\mathbf{X})$ can be written as

$$\sigma_1(\mathbf{X}) = \sum_{j=1}^{n} [y_j - f_j(\mathbf{X})]^2,$$

so it is differentiable on $S_\rho(\mathbf{X}_0)$, and therefore its first partial derivatives vanish at the local minimum point $\bar{\mathbf{X}}$ (Theorem 3.8, Section 5.3); thus,

$$\sum_{j=1}^{n} \frac{\partial f_j(\bar{\mathbf{X}})}{\partial x_i} [y_j - f_j(\bar{\mathbf{X}})] = 0, \qquad 1 \leq i \leq n,$$

or, in matrix form,

$$F'(\bar{\mathbf{X}})[\mathbf{Y} - \mathbf{F}(\bar{\mathbf{X}})] = 0.$$

Since $F'(\bar{\mathbf{X}})$ is invertible, this implies that $\mathbf{Y} = \mathbf{F}(\bar{\mathbf{X}})$. Thus, we have shown that every $\mathbf{Y}$ that satisfies (19) is in $\mathbf{F}(S)$. This proves that $\mathbf{F}(S)$ is open.

Now we must show that $\mathbf{G}$ is differentiable on $\mathbf{F}(S)$. Since

$$\mathbf{G}(\mathbf{F}(\mathbf{X})) = \mathbf{X}, \qquad \mathbf{X} \in S,$$

the chain rule (Theorem 2.5) implies that

$$\mathbf{G}'(\mathbf{Y}_0)\mathbf{F}'(\mathbf{X}_0) = \mathbf{I}$$

(see Example 2.3), where $\mathbf{Y}_0 = \mathbf{F}(\mathbf{X}_0)$, *if* $\mathbf{G}$ is differentiable at $\mathbf{Y}_0$. Therefore, if $\mathbf{G}$ is differentiable at $\mathbf{Y}_0$ its differential matrix must be

$$\mathbf{G}'(\mathbf{Y}_0) = [\mathbf{F}'(\mathbf{X}_0)]^{-1}, \tag{20}$$

and so, if we are to show that $\mathbf{G}$ is differentiable at $\mathbf{Y}_0$, we must show that

$$\lim_{\mathbf{Y} \to \mathbf{Y}_0} \mathbf{H}(\mathbf{Y}) = \mathbf{0}.$$

where

$$\mathbf{H}(\mathbf{Y}) = \frac{\mathbf{G}(\mathbf{Y}) - \mathbf{G}(\mathbf{Y}_0) - [\mathbf{F}'(\mathbf{X}_0)]^{-1}(\mathbf{Y} - \mathbf{Y}_0)}{|\mathbf{Y} - \mathbf{Y}_0|} \quad \text{if} \quad \mathbf{Y} \ne \mathbf{Y}_0. \tag{21}$$

By Lemma 2.2 there is a $\lambda > 0$ and an open neighborhood N of $\mathbf{X}_0$ such that $N \subset S$ and

$$|\mathbf{F}(\mathbf{X}) - \mathbf{F}(\mathbf{X}_0)| \ge \lambda|\mathbf{X} - \mathbf{X}_0| \quad \text{if} \quad \mathbf{X} \in N. \tag{22}$$

(*Exercise 19, Section 6.2* also implies this.) Since $\mathbf{F}$ satisfies the hypotheses of the present theorem on N, the first part of this proof shows that $\mathbf{F}(N)$ is an open set containing $\mathbf{Y}_0 = \mathbf{F}(\mathbf{X}_0)$. Therefore, there is a $\delta > 0$ such that $\mathbf{X} = \mathbf{G}(\mathbf{Y})$ is in N if $\mathbf{Y} \in S_\delta(\mathbf{Y}_0)$. Setting $\mathbf{X} = \mathbf{G}(\mathbf{Y})$ and $\mathbf{X}_0 = \mathbf{G}(\mathbf{Y}_0)$ in (22) yields

$$|\mathbf{F}(\mathbf{G}(\mathbf{Y})) - \mathbf{F}(\mathbf{G}(\mathbf{Y}_0))| \ge \lambda|\mathbf{G}(\mathbf{Y}) - \mathbf{G}(\mathbf{Y}_0)| \quad \text{if} \quad \mathbf{Y} \in S_\delta(\mathbf{Y}_0).$$

Since $\mathbf{F}(\mathbf{G}(\mathbf{Y})) = \mathbf{Y}$, this can be rewritten as

$$|\mathbf{G}(\mathbf{Y}) - \mathbf{G}(\mathbf{Y}_0)| \le \frac{1}{\lambda}|\mathbf{Y} - \mathbf{Y}_0| \quad \text{if} \quad \mathbf{Y} \in S_\delta(\mathbf{Y}_0), \tag{23}$$

which means that $\mathbf{G}$ is continuous at $\mathbf{Y}_0$.

We can rewite (21) as

$$\mathbf{H}(\mathbf{Y}) = -\frac{|\mathbf{G}(\mathbf{Y}) - \mathbf{G}(\mathbf{Y}_0)|}{|\mathbf{Y} - \mathbf{Y}_0|}[\mathbf{F}'(\mathbf{X}_0)]^{-1}\mathbf{H}_1(\mathbf{G}(\mathbf{Y})), \qquad \mathbf{Y} \ne \mathbf{Y}_0, \tag{24}$$

where

$$\mathbf{H}_1(\mathbf{X}) = \begin{cases} \dfrac{\mathbf{F}(\mathbf{X}) - \mathbf{F}(\mathbf{X}_0) - \mathbf{F}'(\mathbf{X}_0)(\mathbf{X} - \mathbf{X}_0)}{|\mathbf{X} - \mathbf{X}_0|}, & \mathbf{X} \ne \mathbf{X}_0, \\ \\ 0, & \mathbf{X} = \mathbf{X}_0. \end{cases} \tag{25}$$

From (23) and (24),

$$|\mathbf{H}(\mathbf{Y})| \le \frac{1}{\lambda} \|[\mathbf{F}'(\mathbf{X}_0)]^{-1}\| \; \|\mathbf{H}_1(\mathbf{G}(\mathbf{Y}))\| \quad \text{if} \quad 0 < |\mathbf{Y} - \mathbf{Y}_0| < \delta. \tag{26}$$

Since $\mathbf{G}$ is continuous at $\mathbf{Y}_0$ and $\mathbf{H}_1$ is continuous at $\mathbf{X}_0 = \mathbf{G}(\mathbf{Y}_0)$ [because of (25) and the differentiability of $\mathbf{F}$ at $\mathbf{X}_0$], it can be seen by applying Theorem 2.6, Section 5.2, to the components of $\mathbf{H}_1$ that

$$\lim_{\mathbf{Y} \to \mathbf{Y}_0} \mathbf{H}_1(\mathbf{G}(\mathbf{Y})) = \mathbf{H}_1(\mathbf{G}(\mathbf{Y}_0)) = \mathbf{H}_1(\mathbf{X}_0) = \mathbf{0}.$$

This and (26) imply that

$$\lim_{\mathbf{Y} \to \mathbf{Y}_0} \mathbf{H}(\mathbf{Y}) = \mathbf{0}.$$

This proves that $\mathbf{G}$ is differentiable at $\mathbf{Y}_0$, and therefore throughout $\mathbf{F}(N)$, since $\mathbf{Y}_0$ is an arbitrary point of $\mathbf{F}(N)$.

Dropping the zero subscript in (20), we have

$$\mathbf{G}'(\mathbf{Y}) = [\mathbf{F}'(\mathbf{X})]^{-1}, \qquad \mathbf{Y} \in \mathbf{F}(\mathbf{N}).$$

To see that $\mathbf{G}$ is *continuously* differentiable on $\mathbf{F}(N)$, we observe that, by Theorem 1.4, each element of $\mathbf{G}'(\mathbf{Y})$ [that is, each partial derivative $\partial g_i(\mathbf{Y})/\partial y_j, 1 \le i, j \le n$] can be written as the ratio—with nonzero denominator—of determinants with elements of the form

$$\frac{\partial f_r(\mathbf{G}(\mathbf{Y}))}{\partial x_s}; \tag{27}$$

since $\partial f_r/\partial x_s$ is continuous on N and $\mathbf{G}$ is continuous on $\mathbf{F}(N)$, Theorem 2.6, Section 5.2, implies that (27) is continuous on $\mathbf{F}(N)$. Since a determinant is a continuous function of its elements, it now follows that the entries of $\mathbf{G}'(\mathbf{Y})$ are continuous on $\mathbf{F}(N)$. This completes the proof.

Branches of the inverse

If $\mathbf{F}$ is regular on an open set S, we say that $\mathbf{F}_S^{-1}$ is a *branch of* $\mathbf{F}^{-1}$. This is a convenient terminology but is not meant to imply that $\mathbf{F}$ actually has an inverse. Because of Theorem 3.2, this definition implies that a branch of $\mathbf{F}^{-1}$, if it exists, must be continuously differentiable on an open set.

By definition it is possible to define a branch of $\mathbf{F}^{-1}$ on a set $T \subset R(\mathbf{F})$ if and only if $T = \mathbf{F}(S)$, where $\mathbf{F}$ is regular on S. There may be open subsets of $R(\mathbf{F})$ which do not have this property, and therefore no branch of $\mathbf{F}^{-1}$ can be defined on them. It is also possible that $T = \mathbf{F}(S_1) = \mathbf{F}(S_2)$, where S_1 and S_2

are distinct subsets of $D_{\mathbf{F}}$; in this case more than one branch of $\mathbf{F}^{-1}$ is defined on T. Thus, we saw in Example 3.4 that two branches of $\mathbf{F}^{-1}$ may be defined on a set T; in Example 3.5 infinitely many branches of $\mathbf{F}^{-1}$ are defined on the same set.

It is also useful to define branches of the argument function. To do this, we think of the relationship between polar and rectangular coordinates in terms of the transformation

$$\begin{bmatrix} x \\ y \end{bmatrix} = \mathbf{F}(r, \theta) = \begin{bmatrix} r \cos \theta \\ r \sin \theta \end{bmatrix}, \tag{28}$$

where for the moment we regard r and θ as rectangular coordinates of a point in an $r\theta$ plane. Let S be an open subset of the right half of this plane (that is, $S \subset \{(r, \theta) \,|\, r > 0\}$) which does not contain any pair of points (r, θ) and $(r, \theta + 2k\pi)$, where k is a nonzero integer. This means that $\mathbf{F}$ is one-to-one on S. It is also continuously differentiable, with

$$\mathbf{F}'(r, \theta) = \begin{bmatrix} \cos \theta & -r \sin \theta \\ \sin \theta & r \cos \theta \end{bmatrix} \tag{29}$$

and

$$J\mathbf{F}(r, \theta) = r > 0, \qquad (r, \theta) \in S; \tag{30}$$

hence $\mathbf{F}$ is regular on S. Now let $T = \mathbf{F}(S)$, the set of points in the xy plane whose polar coordinates are in S. Theorem 3.2 states that T is open and $\mathbf{F}_S$ has a continuously differentiable inverse (which we denote by $\mathbf{G}$, rather than $\mathbf{F}_S^{-1}$, for typographical reasons),

$$\begin{bmatrix} r \\ \theta \end{bmatrix} = \mathbf{G}(x, y) = \begin{bmatrix} \sqrt{x^2 + y^2} \\ \arg_S(x, y) \end{bmatrix}, \qquad (x, y) \in T,$$

where $\arg_S(x, y)$ is the unique value of $\arg(x, y)$ such that

$$(r, \theta) = (\sqrt{x^2 + y^2}, \arg_S(x, y)) \in S.$$

We say that $\arg_S(x, y)$ is a *branch of the argument defined on T*. Theorem 3.2 also implies that

$$\mathbf{G}'(x, y) = [\mathbf{F}'(r, \theta)]^{-1} = \begin{bmatrix} \cos \theta & \sin \theta \\ \dfrac{-\sin \theta}{r} & \dfrac{\cos \theta}{r} \end{bmatrix} \qquad [\text{see } (29)]$$

$$= \begin{bmatrix} \dfrac{x}{\sqrt{x^2 + y^2}} & \dfrac{y}{\sqrt{x^2 + y^2}} \\ \dfrac{-y}{x^2 + y^2} & \dfrac{x}{x^2 + y^2} \end{bmatrix} \qquad [\text{see } (28)],$$

so that

$$\frac{\partial \, \arg_S(x, \, y)}{\partial x} = \frac{-y}{x^2 + y^2}, \qquad \frac{\partial \, \arg_S(x, \, y)}{\partial y} = \frac{x}{x^2 + y^2}. \tag{31}$$

A branch of arg$(x, \, y)$ can be defined on an open set T of the xy plane if and only if the polar coordinates of the points in T form an open subset of the $r\theta$ plane which does not intersect the θ axis or contain any two points of the form $(r, \, \theta)$ and $(r, \, \theta + 2k\pi)$, where k is a nonzero integer. Obviously no subset containing the origin $(x, \, y) = (0, \, 0)$ has this property, nor does any deleted neighborhood of the origin (*Exercise 17*), so there are open sets on which no branch of the argument can be defined; however, if one branch can be defined on T, then so can infinitely many others (*Exercise 16*). All branches of arg$(x, \, y)$ have the same partial derivatives, given in (31).

Example 3.7   The set

$$T = \{(x, \, y) \, | \, (x, \, y) \neq (x, \, 0) \text{ with } x \geq 0\},$$

which is the entire xy plane with the nonnegative x axis deleted, can be written as $T = \mathbf{F}(S_k)$, where $\mathbf{F}$ is as in (28), k is an integer, and

$$S_k = \{(r, \, \theta) \, | \, r > 0, \, 2k\pi < \theta < 2(k + 1)\pi\}.$$

For each integer k we can define a branch $\arg_{S_k}(x, \, y)$ of the argument in S_k, by taking $\arg_{S_k}(x, \, y)$ to be the value of arg$(x, \, y)$ that satisfies

$$2k\pi < \arg_{S_k}(x, \, y) < 2(k + 1)\pi.$$

Each of these branches is continuously differentiable in T, with derivatives as given in (31), and

$$\arg_{S_k}(x, \, y) - \arg_{S_j}(x, \, y) = 2(k - j)\pi, \qquad (x, \, y) \in T.$$

Example 3.8   Returning to the transformation

$$\begin{bmatrix} u \\ v \end{bmatrix} = \mathbf{F}(x, \, y) = \begin{bmatrix} x^2 - y^2 \\ 2xy \end{bmatrix},$$

we now see from Example 3.4 that a branch $\mathbf{G}$ of $\mathbf{F}^{-1}$ can be defined on any subset T of the uv plane on which a branch of arg$(u, \, v)$ can be defined, and $\mathbf{G}$ has the form

$$\begin{bmatrix} x \\ y \end{bmatrix} = \mathbf{G}(u, \, v) = \begin{bmatrix} (u^2 + v^2)^{1/4} \cos[\frac{1}{2} \arg(u, \, v)] \\ (u^2 + v^2)^{1/4} \sin[\frac{1}{2} \arg(u, \, v)] \end{bmatrix}, \qquad (u, \, v) \in T, \tag{32}$$

where arg$(u, \, v)$ is a branch of the argument defined on T. If $\mathbf{G}_1$ and $\mathbf{G}_2$ are branches of $\mathbf{F}^{-1}$ defined on the same set T, then $\mathbf{G}_1 = \pm \mathbf{G}_2$. (Why?)

From Theorem 3.2,

$$\mathbf{G}'(u, v) = [\mathbf{F}'(x, y)]^{-1} = \begin{bmatrix} 2x & -2y \\ 2y & 2x \end{bmatrix}^{-1}$$

$$= \frac{1}{2(x^2 + y^2)} \begin{bmatrix} x & y \\ -y & x \end{bmatrix}.$$

Substituting for x and y in terms of u and v from (32), we find that

$$\frac{\partial x}{\partial u} = \frac{\partial y}{\partial v} = \frac{x}{2(x^2 + y^2)} = \tfrac{1}{2}(u^2 + v^2)^{-1/4} \cos[\tfrac{1}{2} \arg(u, v)] \tag{33}$$

and

$$\frac{\partial x}{\partial v} = -\frac{\partial y}{\partial u} = \frac{y}{2(x^2 + y^2)} = \tfrac{1}{2}(u^2 + v^2)^{-1/4} \sin[\tfrac{1}{2} \arg(u, v)]. \tag{34}$$

It is essential, of course, that the same branch of the argument be used here and in (32).

We leave it to the reader (*Exercise 20*) to verify that (33) and (34) can also be obtained by differentiating (32) directly.

Example 3.9   If

$$\begin{bmatrix} u \\ v \end{bmatrix} = \mathbf{F}(x, y) = \begin{bmatrix} e^x \cos y \\ e^x \sin y \end{bmatrix}$$

(Example 3.5), we can also define a branch $\mathbf{G}$ of $\mathbf{F}^{-1}$ on any subset T of the uv plane on which a branch of $\arg(u, v)$ can be defined, and $\mathbf{G}$ has the form

$$\begin{bmatrix} x \\ y \end{bmatrix} = \mathbf{G}(u, v) = \begin{bmatrix} \log(u^2 + v^2)^{1/2} \\ \arg(u, v) \end{bmatrix}. \tag{35}$$

From Theorem 3.2,

$$\mathbf{G}'(u, v) = [\mathbf{F}'(x, y)]^{-1} = \begin{bmatrix} e^x \cos y & -e^x \sin y \\ e^x \sin y & e^x \cos y \end{bmatrix}^{-1}$$

$$= \begin{bmatrix} e^{-x} \cos y & e^{-x} \sin y \\ -e^{-x} \sin y & e^{-x} \cos y \end{bmatrix}.$$

Substituting for x and y in terms of u and v from (35), we find that

$$\frac{\partial x}{\partial u} = \frac{\partial y}{\partial v} = \frac{\cos[\arg(u, v)]}{(u^2 + v^2)^{1/2}}, \qquad \frac{\partial x}{\partial v} = -\frac{\partial y}{\partial u} = \frac{\sin[\arg(u, v)]}{(u^2 + v^2)^{1/2}}.$$

Again we must use the same branch of $\arg(u, v)$ here and in (35). Since the branches of the argument differ by integral multiples of 2π, (35) implies that if G_1 and G_2 are branches of F^{-1}, both defined on T, then

$$G_1(u, v) - G_2(u, v) = \begin{bmatrix} 0 \\ 2k\pi \end{bmatrix} \qquad (k = \text{integer}).$$

The inverse function theorem

Examples 3.4 and 3.5 show that a continuously differentiable function F may fail to have an inverse on a set S even if $J F(X) \neq 0$ on S. However, the next theorem shows that in this case F is locally invertible on S.

Theorem 3.3 (Inverse Function Theorem)   *Let $F: \mathscr{R}^n \to \mathscr{R}^n$ be continuously differentiable and suppose $J F(X) \neq 0$ on S. Then if $X_0 \in S$, there is an open neighborhood N of X_0 on which F is regular. Moreover, $F(N)$ is open and $G = F_N^{-1}$ is continuously differentiable on $F(N)$, with*

$$G'(Y) = [F'(X)]^{-1} \quad [\text{where } Y = F(X)], \qquad Y \in F(N).$$

Proof.   Lemma 2.2 implies that there is an open neighborhood N of X_0 on which F is one-to-one. The rest of the conclusions then follow from applying Theorem 3.2 to F on N.

Corollary 3.1   *Suppose F is continuously differentiable in an open neighborhood N_0 of X_0 and $J F(X_0) \neq 0$. Then there is a neighborhood N of X_0 on which the conclusions of Theorem 3.3 hold.*

Proof.   *Exercise 25.*

Example 3.10   Let $X_0 = (1, 2, 1)$ and

$$\begin{bmatrix} u \\ v \\ w \end{bmatrix} = F(x, y, z) = \begin{bmatrix} x + y + (z - 1)^2 + 1 \\ y + z + (x - 1)^2 - 1 \\ z + x + (y - 2)^2 + 3 \end{bmatrix}.$$

Then

$$F'(x, y, z) = \begin{bmatrix} 1 & 1 & 2z - 2 \\ 2x - 2 & 1 & 1 \\ 1 & 2y - 4 & 1 \end{bmatrix},$$

so that

$$JF(X_0) = \begin{vmatrix} 1 & 1 & 0 \\ 0 & 1 & 1 \\ 1 & 0 & 1 \end{vmatrix} = 2.$$

In this case it is difficult to describe N or find $G = F_N^{-1}$ explicitly; however, we know that $F(N)$ is a neighborhood of $U_0 = F(X_0) = (4, 2, 5)$, that $G(U_0) = X_0 = (1, 2, 1)$, and that

$$G'(U_0) = [F'(X_0)]^{-1} = \begin{bmatrix} 1 & 1 & 0 \\ 0 & 1 & 1 \\ 1 & 0 & 1 \end{bmatrix}^{-1}$$

$$= \frac{1}{2}\begin{bmatrix} 1 & -1 & 1 \\ 1 & 1 & -1 \\ -1 & 1 & 1 \end{bmatrix}.$$

Therefore,

$$G(U) = \begin{bmatrix} 1 \\ 2 \\ 1 \end{bmatrix} + \frac{1}{2}\begin{bmatrix} 1 & -1 & 1 \\ 1 & 1 & -1 \\ -1 & 1 & 1 \end{bmatrix}\begin{bmatrix} u - 4 \\ v - 2 \\ w - 5 \end{bmatrix} + E(U),$$

where

$$\lim_{U \to (4,2,5)} \frac{E(U)}{\sqrt{(u - 4)^2 + (v - 2)^2 + (w - 5)^2}} = 0;$$

thus, we have approximated G near $U_0 = (4, 2, 5)$ by an affine transformation.

Theorem 3.3 and (30) imply that the transformation (28) is locally invertible on $S = \{(r, \theta) \,|\, r > 0\}$, which means that it is possible to define a branch of $\arg(x, y)$ in a neighborhood of any point $(x_0, y_0) \ne (0, 0)$. (In fact, infinitely many branches can be so defined; see *Exercise 16*.) It also implies, as we have already seen, that the transformation of Example 3.4 is locally invertible everywhere except at $(0, 0)$, where its Jacobian vanishes, and the transformation of Example 3.5 is locally invertible everywhere.

6.3   EXERCISES

1. Prove or disprove:
    (a) If $F: \mathscr{R}^n \to \mathscr{R}^n$ is one-to-one, then so are its component functions.
    (b) If the component functions of $F: \mathscr{R}^n \to \mathscr{R}^n$ are one-to-one, then so is F.

2. Prove: If $\mathbf{F}$ is invertible, then $\mathbf{F}^{-1}$ is unique.

3. Prove Theorem 3.1.

4. Prove: The linear transformation $\mathbf{L}(\mathbf{X}) = \mathbf{AX}$ cannot be one-to-one on any open set if $\mathbf{A}$ is singular. (*Hint:* There is a nonzero vector $\mathbf{X}_0$ such that $\mathbf{AX}_0 = \mathbf{0}$.)

5. Let

$$\mathbf{G}(x, y) = \begin{bmatrix} \sqrt{x^2 + y^2} \\ \arg(x, y) \end{bmatrix}, \qquad \pi/2 \leq \arg(x, y) < 5\pi/2.$$

Find:
(a) $\mathbf{G}(0, 1)$ (b) $\mathbf{G}(1, 0)$ (c) $\mathbf{G}(-1, 0)$
(d) $\mathbf{G}(2, 2)$ (e) $\mathbf{G}(-1, 1)$

6. Same as Exercise 5, except that $-2\pi \leq \arg(x, y) < 0$.

7. (a) Prove: If $f: \mathscr{R} \to \mathscr{R}$ is continuous and locally invertible on (a, b), then f is invertible on (a, b).
   (b) Give an example showing that the continuity assumption is needed in part (a).

8. Let

$$\mathbf{F}(x, y) = \begin{bmatrix} x^2 - y^2 \\ 2xy \end{bmatrix}$$

(Example 3.4) and

$$S = \{(x, y) \mid ax + by > 0\} \qquad (a^2 + b^2 \neq 0).$$

Find $\mathbf{F}(S)$ and $\mathbf{F}_S^{-1}$. If

$$S_1 = \{(x, y) \mid ax + by < 0\},$$

show that $\mathbf{F}(S_1) = \mathbf{F}(S)$ and $\mathbf{F}_{S_1}^{-1} = -\mathbf{F}_S^{-1}$.

9. Show that the transformation

$$\begin{bmatrix} u \\ v \end{bmatrix} = \mathbf{F}(x, y) = \begin{bmatrix} e^x \cos y \\ e^x \sin y \end{bmatrix}$$

(Example 3.5) is one-to-one on any set S which does not contain any pair of points (x_0, y_0) and $(x_0, y_0 + 2k\pi)$, where k is a nonzero integer.

10. Suppose $\mathbf{F}: \mathscr{R}^n \to \mathscr{R}^n$ is continuous and invertible on a compact set S. Show that $\mathbf{F}_S^{-1}$ is continuous. [*Hint:* If $\mathbf{F}_S^{-1}$ is not continuous at $\overline{\mathbf{Y}}$ in $\mathbf{F}(S)$, then

there is an $\varepsilon_0 > 0$ and a sequence $\{Y_k\}$ in $F(S)$ such that $\lim_{k \to \alpha} Y_k = \bar{Y}$ while $|F_S^{-1}(Y_k) - F_S^{-1}(\bar{Y})| \geq \varepsilon_0, \ k = 1, 2, \ldots$. Use Exercise 34, Section 5.1 to obtain a contradiction.]

11. Find F^{-1} and $(F^{-1})'$:

(a) $\begin{bmatrix} u \\ v \end{bmatrix} = F(x, y) = \begin{bmatrix} 4x + 2y \\ -3x + y \end{bmatrix}$

(b) $\begin{bmatrix} u \\ v \\ w \end{bmatrix} = F(x, y, z) = \begin{bmatrix} -x + y + 2z \\ 3x + y - 4z \\ -x - y + 2z \end{bmatrix}$

12. Suppose the function

$$F(x, y) = \begin{bmatrix} u(x, y) \\ v(x, y) \end{bmatrix}$$

is regular on an open set $S \subset \mathcal{R}^2$.
(a) Show that $F(S)$ is open.
(b) Prove that $G = F^{-1}$ is differentiable on $F(S)$, and that $G' = (F')^{-1}$.
(c) Show that G is continuously differentiable on $F(S)$.
(*Hint*: Follow the proof of Theorem 3.2.)

13. In addition to the assumptions of Theorem 3.2, suppose all qth-order $(q > 1)$ partial derivatives of the components of F are continuous on S. Show that all qth-order partial derivatives of F_S^{-1} are continuous on $F(S)$.

14. If

$$\begin{bmatrix} u \\ v \end{bmatrix} = F(x, y) = \begin{bmatrix} x^2 + y^2 \\ x^2 - y^2 \end{bmatrix}$$

(Example 3.1) find the four branches of F^{-1} defined on

$$T_1 = \{(u, v) \mid u + v > 0, u - v > 0\},$$

and find their differential matrices by differentiating the various branches directly with respect to u and v, as well as by Eq. (20).

15. Suppose A is an invertible $n \times n$ matrix and

$$U = F(X) = A \begin{bmatrix} x_1^2 \\ x_2^2 \\ \vdots \\ x_n^2 \end{bmatrix}.$$

Describe the sets on which F is regular, find $R(F)$, and discuss the various branches of F^{-1}. Use Eq. (20) to find the differential matrix of any branch of F^{-1}.

16. Prove: If one branch of $\arg(x, y)$ can be defined on a set T, then so can infinitely many others.

*17. Let $\theta(x, y)$ be a branch of $\arg(x, y)$ defined on an open set S.
   (a) Show that $\theta(x, y)$ cannot assume a local extreme value at any point of S.
   (b) Prove: If the line segment from (x_0, y_0) to (ax_0, ay_0) is in S, then $\theta(ax_0, ay_0) = \theta(x_0, y_0)$.
   (c) Show that S cannot contain a subset of the form

$$A = \{(x, y) \mid 0 < r_1 \leq \sqrt{x^2 + y^2} \leq r_2\}.$$

   (d) Show that no branch of $\arg(x, y)$ can be defined on a deleted neighborhood of the origin.

18. Graph the various branches of f^{-1} if:
   (a) $f(x) = \sin x$ (b) $f(x) = \cos x$ (c) $f(x) = \tan x$

19. Obtain Eq. (31) formally by differentiating:

   (a) $\arg(x, y) = \cos^{-1} \dfrac{x}{\sqrt{x^2 + y^2}}$ (b) $\arg(x, y) = \sin^{-1} \dfrac{y}{\sqrt{x^2 + y^2}}$

   (c) $\arg(x, y) = \tan^{-1} \dfrac{y}{x}$

   Where do these formulas come from? What is the disadvantage of using any one of them to define $\arg(x, y)$?

20. For the transformation $\mathbf{F}$ of Example 3.4, find a branch $\mathbf{G}$ of $\mathbf{F}^{-1}$ defined on $T = \{(u, v) \mid au + bv > 0\}$. Find $\mathbf{G}'$ by direct differentiation with respect to u and v, and by means of Eq. (20).

21. Same as Exercise 20, for $\mathbf{F}$ as defined in Example 3.5.

22. Let us say that a transformation

$$\mathbf{F}(x, y) = \begin{bmatrix} u(x, y) \\ v(x, y) \end{bmatrix}$$

   is *analytic* on a set S if it is continuously differentiable and

$$u_x = v_y, \qquad u_y = -v_x$$

   on S. Prove: If $\mathbf{F}$ is analytic and regular on S then $\mathbf{F}_S^{-1}$ is analytic on $\mathbf{F}(S)$.

23. Prove: If $\mathbf{U} = \mathbf{F}(\mathbf{X})$ and $\mathbf{X} = \mathbf{G}(\mathbf{U})$ are inverse functions, then

$$\frac{\partial(u_1, u_2, \ldots, u_n)}{\partial(x_1, x_2, \ldots, x_n)} \frac{\partial(x_1, x_2, \ldots, x_n)}{\partial(u_1, u_2, \ldots, u_n)} = 1.$$

   Where are the Jacobians to be evaluated?

24. Give an example of a transformation $\mathbf{F}\colon \mathscr{R}^n \to \mathscr{R}^n$ which is invertible, but not regular, on $\mathscr{R}^n$.

25. Prove Corollary 3.1.

26. Find an affine transformation $\mathbf{A}$ which so well approximates the branch $\mathbf{G}$ of $\mathbf{F}^{-1}$ defined near $\mathbf{U}_0 = \mathbf{F}(\mathbf{X}_0)$ that

$$\lim_{\mathbf{U}\to\mathbf{U}_0} \frac{\mathbf{G}(\mathbf{U}) - \mathbf{A}(\mathbf{U})}{|\mathbf{U} - \mathbf{U}_0|} = \mathbf{0}.$$

(a) $\begin{bmatrix} u \\ v \end{bmatrix} = \mathbf{F}(x, y) = \begin{bmatrix} x^4 y^5 - 4x \\ x^3 y^2 - 3y \end{bmatrix}$, $\mathbf{X}_0 = (1, -1)$

(b) $\begin{bmatrix} u \\ v \end{bmatrix} = \mathbf{F}(x, y) = \begin{bmatrix} x^2 y + xy \\ 2xy + xy^2 \end{bmatrix}$, $\mathbf{X}_0 = (1, 1)$

(c) $\begin{bmatrix} u \\ v \\ w \end{bmatrix} = \mathbf{F}(x, y, z) = \begin{bmatrix} 2x^2 y + x^3 + z \\ x^3 + yz \\ x + y + z \end{bmatrix}$, $\mathbf{X}_0 = (0, 1, 1)$

(d) $\begin{bmatrix} u \\ v \\ w \end{bmatrix} = \mathbf{F}(x, y, z) = \begin{bmatrix} x \cos y \cos z \\ x \sin y \cos z \\ x \sin z \end{bmatrix}$, $\mathbf{X}_0 = (1, \pi/2, \pi)$

27. If $\mathbf{F}$ is defined by

$$\begin{bmatrix} x \\ y \\ z \end{bmatrix} = \mathbf{F}(r, \theta, \phi) = \begin{bmatrix} r \cos \theta \cos \phi \\ r \sin \theta \cos \phi \\ r \sin \phi \end{bmatrix}$$

and $\mathbf{G}$ is a branch of $\mathbf{F}^{-1}$, find $\mathbf{G}'$ in terms of r, θ, and ϕ. [*Hint*: See Exercise 26(b), Section 6.2.]

28. If $\mathbf{F}$ is defined by

$$\begin{bmatrix} x \\ y \\ z \end{bmatrix} = \mathbf{F}(r, \theta, z) = \begin{bmatrix} r \cos \theta \\ r \sin \theta \\ z \end{bmatrix}$$

and $\mathbf{G}$ is a branch of $\mathbf{F}^{-1}$, find $\mathbf{G}'$ on terms of r, θ, and z. [*Hint*: See Exercise 26(c), Section 6.2.]

*29. Suppose $\mathbf{F}\colon \mathscr{R}^n \to \mathscr{R}^m$ is regular on an open set S and T is a closed subset of S. Show that $\mathbf{F}(\partial T) = \partial(\mathbf{F}(T))$. [*Hint*: Use Exercise 25, Section 6.2, and Theorem 3.2 to show that $\partial(\mathbf{F}(T)) \subset \mathbf{F}(\partial T)$. Then apply this result with $\mathbf{F}$ and T replaced by $\mathbf{F}^{-1}$ and $\mathbf{F}(T)$ to show that $\mathbf{F}(\partial T) \subset \partial(\mathbf{F}(T))$.]

6.4 THE IMPLICIT FUNCTION THEOREM

In this section we consider transformations from $\mathscr{R}^{n+m}$ to $\mathscr{R}^m$. It will be convenient to denote points in $\mathscr{R}^{n+m}$ by

$$(\mathbf{X}, \mathbf{U}) = (x_1, x_2, \ldots, x_n, u_1, u_2, \ldots, u_m).$$

To lead into the problem that we are interested in, we first ask whether the linear system of m equations in $m + n$ variables.

$$
\begin{aligned}
a_{11}x_1 + a_{12}x_2 + \cdots + a_{1n}x_n + b_{11}u_1 + b_{12}u_2 + \cdots + b_{1m}u_m &= 0 \\
a_{21}x_1 + a_{22}x_2 + \cdots + a_{2n}x_n + b_{21}u_1 + b_{22}u_2 + \cdots + b_{2m}u_m &= 0 \\
&\vdots \\
a_{m1}x_1 + a_{m2}x_2 + \cdots + a_{mn}x_n + b_{m1}u_1 + b_{m2}u_2 + \cdots + b_{mm}u_m &= 0
\end{aligned}
\tag{1}
$$

determines $u_1, u_2, \ldots, u_m$ uniquely in terms of $x_1, x_2, \ldots, x_n$. Rewriting it in matrix form as

$$\mathbf{AX} + \mathbf{BU} = \mathbf{0},$$

where

$$
\mathbf{A} = \begin{bmatrix} a_{11} & a_{12} & \cdots & a_{1n} \\ a_{21} & a_{22} & \cdots & a_{2n} \\ & & \vdots & \\ a_{m1} & a_{m2} & \cdots & a_{mn} \end{bmatrix}, \quad
\mathbf{B} = \begin{bmatrix} b_{11} & b_{12} & \cdots & b_{1m} \\ b_{21} & b_{22} & \cdots & b_{2m} \\ & & \vdots & \\ b_{m1} & b_{m2} & \cdots & b_{mm} \end{bmatrix},
$$

$$
\mathbf{X} = \begin{bmatrix} x_1 \\ x_2 \\ \vdots \\ x_n \end{bmatrix}, \quad \text{and} \quad \mathbf{U} = \begin{bmatrix} u_1 \\ u_2 \\ \vdots \\ u_m \end{bmatrix},
$$

we see that (1) can be solved uniquely for $\mathbf{U}$ in terms of $\mathbf{X}$ if the square matrix $\mathbf{B}$ is nonsingular, and the solution is

$$\mathbf{U} = -\mathbf{B}^{-1}\mathbf{AX}.$$

For our purposes it is convenient to restate this: If

$$\mathbf{F}(\mathbf{X}, \mathbf{U}) = \mathbf{AX} + \mathbf{BU}, \tag{2}$$

where $\mathbf{B}$ is nonsingular, then the system

$$\mathbf{F}(\mathbf{X}, \mathbf{U}) = \mathbf{0}$$

determines $\mathbf{U}$ as a function of $\mathbf{X}$, for all $\mathbf{X}$ in $\mathscr{R}^n$.

Notice that $\mathbf{F}$ in (2) is a linear transformation. If $\mathbf{F}$ is a more general transformation from $\mathscr{R}^{n+m}$ to $\mathscr{R}^m$, we can still ask whether the system

$$\mathbf{F}(\mathbf{X}, \mathbf{U}) = \mathbf{0},$$

or, in terms of components,

$$f_1(x_1, x_2, \ldots, x_n, u_1, u_2, \ldots, u_m) = 0$$
$$f_2(x_1, x_2, \ldots, x_n, u_1, u_2, \ldots, u_m) = 0$$
$$\vdots$$
$$f_m(x_1, x_2, \ldots, x_n, u_1, u_2, \ldots, u_m) = 0,$$

can be solved for $\mathbf{U}$ in terms of $\mathbf{X}$. However, the situation is now more complicated, even if $m = 1$. For example, suppose $m = 1$ and

$$f(x, y, u) = 1 - x^2 - y^2 - u^2.$$

If $x^2 + y^2 > 1$, then no value of u satisfies

$$f(x, y, u) = 0; \tag{3}$$

however, infinitely many functions $u = u(x, y)$ satisfy (3) on the set

$$S = \{(x, y) \mid x^2 + y^2 \le 1\}.$$

They are of the form

$$u(x, y) = \varepsilon(x, y)\sqrt{1 - x^2 - y^2},$$

where $\varepsilon(x, y)$ can be chosen arbitrarily, for each (x, y) in S, to be 1 or -1. We can narrow the choice of functions to two by requiring that u be continuous on S; then

$$u(x, y) = \sqrt{1 - x^2 - y^2} \tag{4}$$

or

$$u(x, y) = -\sqrt{1 - x^2 - y^2}.$$

We can define a unique continuous solution u of (3) by specifying its value at a single interior point of S. For example, if we require that

$$u\left(\frac{1}{\sqrt{3}}, \frac{1}{\sqrt{3}}\right) = \frac{1}{\sqrt{3}},$$

then u must be as defined by (4).

The question of whether an arbitrary system

$$\mathbf{F}(\mathbf{X}, \mathbf{U}) = \mathbf{0}$$

determines U as a function of X is too general to have a satisfactory answer. However, there is a theorem—the implicit function theorem—which answers this question affirmatively in an important special case. To facilitate the statement of this theorem, we partition the differential matrix of F: $\mathscr{R}^{n+m} \rightarrow \mathscr{R}^m$:

$$
F' = \begin{bmatrix}
\dfrac{\partial f_1}{\partial x_1} & \dfrac{\partial f_2}{\partial x_2} & \cdots & \dfrac{\partial f_1}{\partial x_n} & \bigg| & \dfrac{\partial f_1}{\partial u_1} & \dfrac{\partial f_1}{\partial u_2} & \cdots & \dfrac{\partial f_1}{\partial u_m} \\[2mm]
\dfrac{\partial f_2}{\partial x_1} & \dfrac{\partial f_2}{\partial x_2} & \cdots & \dfrac{\partial f_2}{\partial x_n} & \bigg| & \dfrac{\partial f_2}{\partial u_1} & \dfrac{\partial f_2}{\partial u_2} & \cdots & \dfrac{\partial f_2}{\partial u_m} \\[2mm]
& & & & \vdots & & & & \\[2mm]
\dfrac{\partial f_m}{\partial x_1} & \dfrac{\partial f_m}{\partial x_2} & \cdots & \dfrac{\partial f_m}{\partial x_n} & \bigg| & \dfrac{\partial f_m}{\partial u_1} & \dfrac{\partial f_m}{\partial u_2} & \cdots & \dfrac{\partial f_m}{\partial u_m}
\end{bmatrix},
\tag{5}
$$

or

$$F' = [F_X, F_U],$$

where F_X is the submatrix to the left of the dashed line in (5) and F_U is to the right.

For the linear transformation (2), $F_X = A$ and $F_U = B$, and we have seen that the system $F(X, U) = 0$ defines U as a function of X for all X in $\mathscr{R}^n$ if F_U is nonsingular. The next theorem shows that a related result holds for more general transformations.

Theorem 4.1 (Implicit Function Theorem)   *Suppose* F: $\mathscr{R}^{n+m} \rightarrow \mathscr{R}^m$ *is continuously differentiable on an open set S of $\mathscr{R}^{n+m}$ containing (X_0, U_0). Let $F(X_0, U_0) = 0$ and suppose $F_U(X_0, U_0)$ is nonsingular. Then there is a neighborhood M of (X_0, U_0), contained in S, and a neighborhood N of X_0 in $\mathscr{R}^n$ on which a unique transformation G: $\mathscr{R}^n \rightarrow \mathscr{R}^m$ is defined, such that*

$$(X, G(X)) \in M, \qquad X \in N,$$
$$G(X_0) = U_0,
\tag{6}$$

and

$$F(X, G(X)) = 0, \qquad X \in N.
\tag{7}$$

Moreover, G is continuously differentiable on N and

$$G'(X) = -[F_U(X, G(X))]^{-1}F_X(X, G(X)).
\tag{8}$$

(The reader may find it instructive to work through this proof for specific values of m and n, as in *Exercises 3 and 4*.)

Proof.   Define $\Phi: \mathscr{R}^{n+m} \to \mathscr{R}^{n+m}$ by

$$\Phi(\mathbf{X}, \mathbf{U}) = \begin{bmatrix} x_1 \\ x_2 \\ \vdots \\ x_n \\ f_1(\mathbf{X}, \mathbf{U}) \\ f_2(\mathbf{X}, \mathbf{U}) \\ \vdots \\ f_m(\mathbf{X}, \mathbf{U}) \end{bmatrix}, \tag{9}$$

or, in "horizontal" notation, by

$$\Phi(\mathbf{X}, \mathbf{U}) = (\mathbf{X}, \mathbf{F}(\mathbf{X}, \mathbf{U})); \tag{10}$$

then Φ is continuously differentiable on S and, since $\mathbf{F}(\mathbf{X}_0, \mathbf{U}_0) = \mathbf{0}$,

$$\Phi(\mathbf{X}_0, \mathbf{U}_0) = (\mathbf{X}_0, \mathbf{0}). \tag{11}$$

Its differential matrix is

$$\Phi' = \begin{bmatrix} 1 & 0 & \cdots & 0 & 0 & 0 & \cdots & 0 \\ 0 & 1 & \cdots & 0 & 0 & 0 & \cdots & 0 \\ & & & \vdots & & & & \\ 0 & 0 & \cdots & 1 & 0 & 0 & \cdots & 0 \\ \dfrac{\partial f_1}{\partial x_1} & \dfrac{\partial f_1}{\partial x_2} & \cdots & \dfrac{\partial f_1}{\partial x_n} & \dfrac{\partial f_1}{\partial u_1} & \dfrac{\partial f_1}{\partial u_2} & \cdots & \dfrac{\partial f_1}{\partial u_m} \\ \dfrac{\partial f_2}{\partial x_1} & \dfrac{\partial f_2}{\partial x_2} & \cdots & \dfrac{\partial f_2}{\partial x_n} & \dfrac{\partial f_2}{\partial u_1} & \dfrac{\partial f_2}{\partial u_2} & \cdots & \dfrac{\partial f_2}{\partial u_m} \\ & & & \vdots & & & & \\ \dfrac{\partial f_m}{\partial x_1} & \dfrac{\partial f_m}{\partial x_2} & \cdots & \dfrac{\partial f_m}{\partial x_n} & \dfrac{\partial f_m}{\partial u_1} & \dfrac{\partial f_m}{\partial u_2} & \cdots & \dfrac{\partial f_m}{\partial u_m} \end{bmatrix} = \begin{bmatrix} \mathbf{I} & \mathbf{0} \\ \mathbf{F_X} & \mathbf{F_U} \end{bmatrix},$$

where $\mathbf{I}$ is the $n \times n$ identity matrix, $\mathbf{0}$ is the $m \times n$ matrix with all zero elements, and $\mathbf{F_X}$ and $\mathbf{F_U}$ are as shown in (5). By expanding det Φ' and the determinants that evolve from it in terms of the cofactors of their first rows it can be shown in n steps that

$$J\Phi = \det \Phi' = \begin{vmatrix} \dfrac{\partial f_1}{\partial u_1} & \dfrac{\partial f_1}{\partial u_2} & \cdots & \dfrac{\partial f_1}{\partial u_m} \\ \dfrac{\partial f_2}{\partial u_1} & \dfrac{\partial f_2}{\partial u_2} & \cdots & \dfrac{\partial f_2}{\partial u_m} \\ & & \vdots & \\ \dfrac{\partial f_m}{\partial u_1} & \dfrac{\partial f_m}{\partial u_2} & \cdots & \dfrac{\partial f_m}{\partial u_m} \end{vmatrix} = \det \mathbf{F_U}.$$

In particular,

$$J\mathbf{\Phi}(\mathbf{X}_0, \mathbf{U}_0) = \det \mathbf{F}_U(\mathbf{X}_0, \mathbf{U}_0) \neq 0.$$

Since $\mathbf{\Phi}$ is continuously differentiable on S, Corollary 3.1 implies that $\mathbf{\Phi}$ is regular on some open neighborhood M of $(\mathbf{X}_0, \mathbf{U}_0)$, and that $\tilde{M} = \mathbf{\Phi}(M)$ is open.

Because of the form of $\mathbf{\Phi}$ [see (9) or (10)] it is appropriate to denote points of $\tilde{M}$ by $(\mathbf{X}, \mathbf{V})$, where $\mathbf{V} \in \mathscr{R}^m$. The inverse function theorem also implies that there is a continuously differentiable transformation $\mathbf{\Gamma} = \mathbf{\Gamma}(\mathbf{X}, \mathbf{V})$, defined on $\tilde{M}$ with values in M, which is inverse to $\mathbf{\Phi}$; from (10), this means that

$$\mathbf{\Gamma}(\mathbf{X}, \mathbf{F}(\mathbf{X}, \mathbf{U})) = (\mathbf{X}, \mathbf{U}), \qquad (\mathbf{X}, \mathbf{U}) \in M.$$

Since every $(\mathbf{X}, \mathbf{V})$ in $\tilde{M}$ is of the form $(\mathbf{X}, \mathbf{F}(\mathbf{X}, \mathbf{U}))$ for some $(\mathbf{X}, \mathbf{U})$ in M, this implies that the component representation of $\mathbf{\Gamma}$ is of the form

$$\mathbf{\Gamma}(\mathbf{X}, \mathbf{V}) = \begin{bmatrix} x_1 \\ x_2 \\ \vdots \\ x_n \\ h_1(\mathbf{X}, \mathbf{V}) \\ h_2(\mathbf{X}, \mathbf{V}) \\ \vdots \\ h_m(\mathbf{X}, \mathbf{V}) \end{bmatrix},$$

or, in "horizontal" notation

$$\mathbf{\Gamma}(\mathbf{X}, \mathbf{V}) = (\mathbf{X}, \mathbf{H}(\mathbf{X}, \mathbf{V})), \tag{12}$$

where $\mathbf{H}: \mathscr{R}^{n+m} \to \mathscr{R}^m$ is continuously differentiable on $\tilde{M}$. Since $\mathbf{\Gamma} = \mathbf{\Phi}^{-1}$, (11) implies that

$$\mathbf{\Gamma}(\mathbf{X}_0, \mathbf{0}) = (\mathbf{X}_0, \mathbf{U}_0),$$

and therefore, from (12),

$$\mathbf{H}(\mathbf{X}_0, \mathbf{0}) = \mathbf{U}_0, \tag{13}$$

because $\mathbf{\Gamma}$ is one-to-one on $\tilde{M}$. We will need this below.

Again since $\mathbf{\Gamma} = \mathbf{\Phi}^{-1}$, (10) and (12) imply that

$$(\mathbf{X}, \mathbf{V}) = \mathbf{\Phi}(\mathbf{\Gamma}(\mathbf{X}, \mathbf{V})) = \mathbf{\Phi}(\mathbf{X}, \mathbf{H}(\mathbf{X}, \mathbf{V}))$$
$$= (\mathbf{X}, \mathbf{F}(\mathbf{X}, \mathbf{H}(\mathbf{X}, \mathbf{V}))), \qquad (\mathbf{X}, \mathbf{V}) \in \tilde{M},$$

and so

$$\mathbf{F}(\mathbf{X}, \mathbf{H}(\mathbf{X}, \mathbf{V})) = \mathbf{V}, \qquad (\mathbf{X}, \mathbf{V}) \in \tilde{M}. \tag{14}$$

From (11), $(\mathbf{X}_0, \mathbf{0}) \in \tilde{M}$ and, since $\tilde{M}$ is open, there is a neighborhood N of $\mathbf{X}_0$ in $\mathscr{R}^n$ such that $(\mathbf{X}, \mathbf{0}) \in \tilde{M}$ if $\mathbf{X} \in N$ (*Exercise 2*). Therefore, (14) implies (7) if we define

$$\mathbf{G}(\mathbf{X}) = \mathbf{H}(\mathbf{X}, \mathbf{0});$$

moreover, $(\mathbf{X}, \mathbf{G}(\mathbf{X})) \in M$ for all $\mathbf{X}$ in N, and $\mathbf{G}(\mathbf{X}_0) = \mathbf{U}_0$, from (13). To see that $\mathbf{G}$ is the only function with these properties, suppose $\mathbf{G}_1 : \mathscr{R}^n \to \mathscr{R}^m$ also has them; then

$$\Phi(\mathbf{X}, \mathbf{G}_1(\mathbf{X})) = (\mathbf{X}, \mathbf{F}(\mathbf{X}, \mathbf{G}_1(\mathbf{X}))) = (\mathbf{X}, \mathbf{0})$$

and

$$\Phi(\mathbf{X}, \mathbf{G}(\mathbf{X})) = (\mathbf{X}, \mathbf{F}(\mathbf{X}, \mathbf{G}(\mathbf{X}))) = (\mathbf{X}, \mathbf{0})$$

for all $\mathbf{X}$ in N. This implies that $\mathbf{G}(\mathbf{X}) = \mathbf{G}_1(\mathbf{X})$, since Φ is one-to-one on M.
Since the partial derivatives

$$\frac{\partial h_i}{\partial x_j}, \qquad 1 \le i \le m, \quad 1 \le j \le n,$$

are continuous functions of $(\mathbf{X}, \mathbf{V})$ on $\tilde{M}$, they are continuous with respect to $\mathbf{X}$ on the subset $\{(\mathbf{X}, \mathbf{0}) \mid \mathbf{X} \in N\}$ of $\tilde{M}$; therefore, $\mathbf{G}$ is continuously differentiable on N. To verify (8), we write (7) in terms of components; thus

$$f_r(x_1, x_2, \ldots, x_n, g_1(\mathbf{X}), g_2(\mathbf{X}), \ldots, g_m(\mathbf{X})) = 0, \qquad 1 \le r \le m, \quad \mathbf{X} \in N.$$

Since f_r and $g_1, g_2, \ldots, g_m$ are continuously differentiable on their respective domains, the chain rule (Theorem 4.1, Section 5.4) implies that

$$\frac{\partial f_r(\mathbf{X}, \mathbf{G}(\mathbf{X}))}{\partial x_s} + \sum_{j=1}^{m} \frac{\partial f_r(\mathbf{X}, \mathbf{G}(\mathbf{X}))}{\partial u_j} \frac{\partial g_j(\mathbf{X})}{\partial x_s} = 0, \; 1 \le r \le m, \; 1 \le s \le n, \tag{15}$$

or, in matrix form,

$$\mathbf{F}_\mathbf{X}(\mathbf{X}, \mathbf{G}(\mathbf{X})) + \mathbf{F}_\mathbf{U}(\mathbf{X}, \mathbf{G}(\mathbf{X})) \mathbf{G}'(\mathbf{X}) = \mathbf{0}. \tag{16}$$

Since $(\mathbf{X}, \mathbf{G}(\mathbf{X})) \in M$ for all $\mathbf{X}$ in N and $\mathbf{F}_\mathbf{U}(\mathbf{X}, \mathbf{U})$ is nonsingular when $(\mathbf{X}, \mathbf{U}) \in M$, we can multiply (16) on the left by $\mathbf{F}_\mathbf{U}^{-1}(\mathbf{X}, \mathbf{G}(\mathbf{X}))$ to obtain (8). This completes the proof.

The following corollary is the implicit function theorem for $m = 1$.

Corollary 4.1 *Suppose $f : \mathscr{R}^{n+1} \to \mathscr{R}$ is continuously differentiable on an open set containing $(\mathbf{X}_0, u_0)$, with*

$$f(\mathbf{X}_0, u_0) = 0$$

and

$$f_u(\mathbf{X}_0, u_0) \ne 0.$$

Then there is a neighborhood M of $(\mathbf{X}_0, u_0)$, *contained in S, and a neighborhood N of* $\mathbf{X}_0$ *in* $\mathscr{R}^n$ *on which is defined a unique function* $g: \mathscr{R}^n \to \mathscr{R}$ *such that*

$$(\mathbf{X}, g(\mathbf{X})) \in M, \qquad \mathbf{X} \in N,$$

$$g(\mathbf{X}_0) = \mathbf{U}_0,$$

and

$$f(\mathbf{X}, g(\mathbf{X})) = 0, \qquad \mathbf{X} \in N.$$

Furthermore, g is continuously differentiable on N, and

$$g_{x_i}(\mathbf{X}) = -\frac{f_{x_i}(\mathbf{X}, g(\mathbf{X}))}{f_u(\mathbf{X}, g(\mathbf{X}))}, \qquad 1 \le i \le n.$$

Under the assumptions of Theorem 4.1 we also say that (6) and (7) *determine* $\mathbf{U} = \mathbf{U}(\mathbf{X})$ *near* $\mathbf{X}_0$, and rewrite them as

$$\mathbf{U}(\mathbf{X}_0) = \mathbf{U}_0$$

and

$$\mathbf{F}(\mathbf{X}, \mathbf{U}(\mathbf{X})) = \mathbf{0}, \qquad \mathbf{X} \in N,$$

thus eliminating the symbol $\mathbf{G}$. A similar convention applies to Corollary 4.1.

Example 4.1   Let

$$f(x, y, u) = 1 - x^2 - y^2 - u^2$$

and $(x_0, y_0, u_0) = (\frac{1}{2}, -\frac{1}{2}, 1/\sqrt{2})$; then

$$f_x(x, y, u) = -2x, \quad f_y(x, y, u) = -2y, \quad f_u(x, y, u) = -2u.$$

Since f is continuously differentiable everywhere and $f_u(x_0, y_0, u_0) = -\sqrt{2} \ne 0$, Corollary 4.1 implies that the conditions

$$1 - x^2 - y^2 - u^2 = 0, \qquad u\left(\frac{1}{2}, -\frac{1}{2}\right) = \frac{1}{\sqrt{2}},$$

determine $u = u(x, y)$ near $(x_0, y_0) = (\frac{1}{2}, -\frac{1}{2})$ so that

$$u_x(x, y) = -\frac{f_x(x, y, u(x, y))}{f_u(x, y, u(x, y))} = \frac{-2x}{2u(x, y)}, \tag{17}$$

and

$$u_y(x, y) = -\frac{f_y(x, y, u(x, y))}{f_u(x, y, u(x, y))} = \frac{-2y}{2u(x, y)}. \tag{18}$$

It is not necessary to memorize formulas like (17) and (18), since they can be obtained by applying the chain rule to the identity

$$f(x, y, u(x, y)) = 0$$

if f and u are differentiable.

Example 4.2   Let

$$f(x, y, z) = x^3 y^2 z^2 + 3xy^4 z^4 - 3x^6 y^6 z^7 + 12x - 13 \tag{19}$$

and $(x_0, y_0, z_0) = (1, -1, 1)$, so that

$$f(x_0, y_0, z_0) = 0.$$

Then

$$f_x(x, y, z) = 3x^2 y^2 z^2 + 3y^4 z^4 - 18x^5 y^6 z^7 + 12,$$
$$f_y(x, y, z) = 2x^3 yz^2 + 12xy^3 z^4 - 18x^6 y^5 z^7,$$
$$f_z(x, y, z) = 2x^3 y^2 z + 12xy^4 z^3 - 21x^6 y^6 z^6.$$

Since $f_z(1, -1, 1) = -7 \neq 0$, Corollary 4.1 implies that the conditions

$$z(1, -1) = 1 \tag{20}$$

and

$$f(x, y, z(x, y)) = 0 \tag{21}$$

determine z as a continuously differentiable function of (x, y) near $(1, -1)$.

If we try to solve (19) for z, we see very clearly that Theorem 4.1 and Corollary 4.1 are *existence* theorems; that is, they tell us that there is a function $z = z(x, y)$ that satisfies (20) and (21), but not how to find it. In this case there is no convenient formula for the function, although its partial derivatives can be expressed conveniently in terms of x, y, and $z(x, y)$:

$$z_x(x, y) = -\frac{f_x(x, y, z(x, y))}{f_z(x, y, z(x, y))}, \qquad z_y(x, y) = -\frac{f_y(x, y, z(x, y))}{f_z(x, y, z(x, y))}.$$

In particular, since $z(1, -1) = 1$,

$$z_x(1, -1) = -\frac{0}{-7} = 0, \qquad z_y(1, -1) = \frac{4}{-7} = -\frac{4}{7}.$$

Example 4.3   Let $m = 2$, $n = 3$, and

$$\mathbf{F}(\mathbf{X}, \mathbf{G}) = \begin{bmatrix} 2x^2 + y^2 + z^2 + u^2 - v^2 \\ x^2 + z^2 + 2u - v \end{bmatrix}.$$

If $\mathbf{X}_0 = (1, -1, 1)$ and $\mathbf{U}_0 = (0, 2)$, then $\mathbf{F}(\mathbf{X}_0, \mathbf{G}_0) = \mathbf{0}$; moreover,

$$\mathbf{F}_\mathbf{U}(\mathbf{X}, \mathbf{U}) = \begin{bmatrix} 2u & -2v \\ 2 & -1 \end{bmatrix}$$

and so

$$\det \mathbf{F}_\mathbf{U}(\mathbf{X}_0, \mathbf{U}_0) = \begin{vmatrix} 0 & -4 \\ 2 & -1 \end{vmatrix} = 8 \neq 0.$$

Hence, the conditions

$$\mathbf{F}(\mathbf{X}, \mathbf{U}) = \mathbf{0} \quad \text{and} \quad \mathbf{U}(1, -1, 1) = (0, 2)$$

determine $\mathbf{U} = \mathbf{U}(\mathbf{X})$ near $\mathbf{X}_0$. Although it is difficult to find $\mathbf{U}(\mathbf{X})$ explicitly, we can approximate it near $\mathbf{X}_0$ by an affine transformation. Thus, from (8),

$$\mathbf{U}'(\mathbf{X}_0) = -[\mathbf{F}_\mathbf{U}(\mathbf{X}_0, \mathbf{U}(\mathbf{X}_0))]^{-1} \mathbf{F}_\mathbf{X}(\mathbf{X}_0, \mathbf{U}(\mathbf{X}_0)) \tag{22}$$

$$= -\begin{bmatrix} 0 & -4 \\ 2 & -1 \end{bmatrix}^{-1} \begin{bmatrix} 4 & -2 & 2 \\ 2 & 0 & 2 \end{bmatrix}$$

$$= -\frac{1}{8}\begin{bmatrix} -1 & 4 \\ -2 & 0 \end{bmatrix}\begin{bmatrix} 4 & -2 & 2 \\ 2 & 0 & 2 \end{bmatrix}$$

$$= -\frac{1}{8}\begin{bmatrix} 4 & 2 & 6 \\ -8 & 4 & -4 \end{bmatrix};$$

therefore,

$$\lim_{\mathbf{X} \to (1, -1, 1)} \frac{\begin{bmatrix} u(x, y) \\ v(x, y) \end{bmatrix} - \begin{bmatrix} 0 \\ 2 \end{bmatrix} + \frac{1}{8}\begin{bmatrix} 4 & 2 & 6 \\ -8 & 4 & -4 \end{bmatrix}\begin{bmatrix} x - 1 \\ y + 1 \\ z - 1 \end{bmatrix}}{[(x-1)^2 + (y+1)^2 + (z-1)^2]^{1/2}} = \begin{bmatrix} 0 \\ 0 \end{bmatrix}.$$

Again, it is not necessary to memorize (22), since the partial derivatives of an implicitly defined function can be obtained from the chain rule and Cramer's rule, as in the next example.

Example 4.4 Let $u = u(x, y)$ and $v = v(x, y)$ be differentiable and satisfy

$$\begin{aligned} x^2 + 2y^2 + 3z^2 + u^2 + v &= 6 \\ 2x^3 + 4y^2 + 2z^2 + u + v^2 &= 9 \end{aligned} \tag{23}$$

and

$$u(1, -1, 0) = -1, \qquad v(1, -1, 0) = 2. \tag{24}$$

To find u_x and v_x we differentiate (23) with respect to x, obtaining

$$2x + 2uu_x + v_x = 0$$
$$6x^2 + u_x + 2vv_x = 0.$$

Therefore,

$$\begin{bmatrix} 2u & 1 \\ 1 & 2v \end{bmatrix} \begin{bmatrix} u_x \\ v_x \end{bmatrix} = - \begin{bmatrix} 2x \\ 6x^2 \end{bmatrix},$$

and Cramer's rule yields

$$u_x = - \frac{\begin{vmatrix} 2x & 1 \\ 6x^2 & 2v \end{vmatrix}}{\begin{vmatrix} 2u & 1 \\ 1 & 2v \end{vmatrix}} = \frac{6x^2 - 4xv}{4uv - 1}$$

and

$$v_x = - \frac{\begin{vmatrix} 2u & 2x \\ 1 & 6x^2 \end{vmatrix}}{\begin{vmatrix} 2u & 1 \\ 1 & 2v \end{vmatrix}} = \frac{2x - 12x^2u}{4uv - 1}$$

if $4uv \neq 1$; in particular, from (24),

$$u_x(1, -1, 0) = \frac{-2}{-9} = \frac{2}{9}, \qquad u_y(1, -1, 0) = \frac{14}{-9} = -\frac{14}{9}.$$

Jacobians

It is convenient to extend the notation introduced earlier (Section 6.2) for the Jacobian of a transformation $\mathbf{F}: \mathscr{R}^m \to \mathscr{R}^m$. If $f_1, f_2, \ldots, f_m$ are real-valued functions of k variables, $k \geq m$, and $\xi_1, \xi_2, \ldots, \xi_m$ are any m of the variables, then we denote the determinant

$$\begin{vmatrix} \dfrac{\partial f_1}{\partial \xi_1} & \dfrac{\partial f_1}{\partial \xi_2} & \cdots & \dfrac{\partial f_1}{\partial \xi_m} \\[2ex] \dfrac{\partial f_2}{\partial \xi_1} & \dfrac{\partial f_2}{\partial \xi_2} & \cdots & \dfrac{\partial f_2}{\partial \xi_m} \\[2ex] & & \vdots & \\[2ex] \dfrac{\partial f_m}{\partial \xi_1} & \dfrac{\partial f_m}{\partial \xi_2} & \cdots & \dfrac{\partial f_m}{\partial \xi_m} \end{vmatrix},$$

which we call the *Jacobian of* $f_1, f_2, \ldots, f_m$ *with respect to* $\xi_1, \xi_2, \ldots, \xi_m$, by

$$\frac{\partial(f_1, f_2, \ldots, f_m)}{\partial(\xi_1, \xi_2, \ldots, \xi_m)};$$

for its value at a point **P** we write

$$\frac{\partial(f_1, f_2, \ldots, f_m)}{\partial(\xi_1, \xi_2, \ldots, \xi_m)}\bigg|_{\mathbf{P}}.$$

Throughout this section $f_1, f_2, \ldots, f_m$ denote the components of a transformation $\mathbf{F}: \mathscr{R}^{n+m} \to \mathscr{R}^m$.

Example 4.5 If

$$\mathbf{F}(x, y, z) = \begin{bmatrix} 3x^2 + 2xy + z^2 \\ 4x^2 + 2xy^2 + z^3 \end{bmatrix},$$

then

$$\frac{\partial(f_1, f_2)}{\partial(x, y)} = \begin{vmatrix} 6x + 2y & 2x \\ 8x + 2y^2 & 4xy \end{vmatrix},$$

$$\frac{\partial(f_1, f_2)}{\partial(y, z)} = \begin{vmatrix} 2x & 2z \\ 4xy & 3z^2 \end{vmatrix},$$

and

$$\frac{\partial(f_1, f_2)}{\partial(z, x)} = \begin{vmatrix} 2z & 6x + 2y \\ 3z^2 & 8x + 2y^2 \end{vmatrix}.$$

The values of these Jacobians at $\mathbf{X}_0 = (-1, 1, 0)$ are

$$\frac{\partial(f_1, f_2)}{\partial(x, y)}\bigg|_{\mathbf{X}_0} = \begin{vmatrix} -4 & -2 \\ -6 & -4 \end{vmatrix} = 4,$$

$$\frac{\partial(f_1, f_2)}{\partial(y, z)}\bigg|_{\mathbf{X}_0} = \begin{vmatrix} -2 & 0 \\ -4 & 0 \end{vmatrix} = 0,$$

and

$$\frac{\partial(f_1, f_2)}{\partial(z, x)}\bigg|_{\mathbf{X}_0} = \begin{vmatrix} 0 & -4 \\ 0 & -6 \end{vmatrix} = 0.$$

The requirement in Theorem 4.1 that $\mathbf{F_U}(\mathbf{X}_0, \mathbf{U}_0)$ be nonsingular is equivalent to the requirement that

$$\frac{\partial(f_1, f_2, \ldots, f_m)}{\partial(u_1, u_2, \ldots, u_m)}\bigg|_{(\mathbf{X}_0, \mathbf{U}_0)} \neq 0.$$

So far we have considered only the problem of solving a continuously differentiable system

$$F(X, U) = 0 \ (F: \mathscr{R}^{n+m} \to \mathscr{R}^m) \tag{25}$$

for the last m variables, $u_1, u_2, \ldots, u_m$, in terms of the first n, $x_1, x_2, \ldots, x_n$. This was merely for convenience; (25) can be solved near (X_0, U_0) for any m of the variables in terms of the other n, provided only that the Jacobian of $f_1, f_2, \ldots, f_m$ with respect to the m variables is nonzero at (X_0, U_0). This can be seen by renaming the variables and applying Theorem 4.1. The following theorem provides a convenient formula for the first partial derivatives of the functions so determined.

Theorem 4.2 *Suppose* $F = (f_1, f_2, \ldots, f_m)$ *is a differentiable function of $n + m$ variables* $x_1, x_2, \ldots, x_{n+m}$ $(n \geq 1)$. *Let* $\xi_1, \xi_2, \ldots, \xi_m$ *be m distinct variables chosen from* $x_1, x_2, \ldots, x_{n+m}$, *and let* $\eta_1, \eta_2, \ldots, \eta_n$ *be the rest. Suppose the system*

$$F = 0$$

determines $\xi_1, \xi_2, \ldots, \xi_m$ *as differentiable functions of* $\eta_1, \eta_2, \ldots, \eta_n$ *in some open subset S of the* $(\eta_1, \eta_2, \ldots, \eta_m)$ *space. Then*

$$\frac{\partial \xi_i}{\partial \eta_j} = -\frac{\dfrac{\partial(f_1, f_2, \ldots, f_i, \ldots, f_m)}{\partial(\xi_1, \xi_2, \ldots, \eta_j, \ldots, \xi_m)}}{\dfrac{\partial(f_1, f_2, \ldots, f_i, \ldots, f_m)}{\partial(\xi_1, \xi_2, \ldots, \xi_i, \ldots, \xi_m)}}, \qquad 1 \leq i \leq m, \quad 1 \leq j \leq n, \tag{26}$$

provided the denominator is nonzero.

Notice that the determinant in the numerator on the right of (26) is obtained by replacing the ith column of the determinant in the denominator, which is

$$\begin{bmatrix} \dfrac{\partial f_1}{\partial \xi_i} \\[2mm] \dfrac{\partial f_2}{\partial \xi_i} \\[2mm] \vdots \\[2mm] \dfrac{\partial f_m}{\partial \xi_i} \end{bmatrix}, \quad \text{by} \quad \begin{bmatrix} \dfrac{\partial f_1}{\partial \eta_j} \\[2mm] \dfrac{\partial f_2}{\partial \eta_j} \\[2mm] \vdots \\[2mm] \dfrac{\partial f_m}{\partial \eta_j} \end{bmatrix}.$$

The next example proves a special case of Theorem 4.2. Other special cases are considered in *Exercises 16, 17, and 18.*

Example 4.6   Let

$$\mathbf{F}(x, y, z) = \begin{bmatrix} f(x, y, z) \\ g(x, y, z) \end{bmatrix}$$

be continuously differentiable in a neighborhood of (x_0, y_0, z_0), where $\mathbf{F}(x_0, y_0, z_0) = \mathbf{0}$, and suppose

$$\frac{\partial(f, g)}{\partial(x, z)}\bigg|_{(x_0, y_0, z_0)} \neq 0. \tag{27}$$

Then Theorem 4.1 (see also *Exercise 17*) implies that the conditions

$$f(x, y, z) = 0, \qquad g(x, y, z) = 0, \tag{28}$$

and

$$x(y_0) = x_0, \qquad z(y_0) = z_0,$$

determine x and z as continuously differentiable functions of y near y_0. Differentiating (28) with respect to y and regarding x and z as functions of y yields

$$f_x x' + f_y + f_z z' = 0$$
$$g_x x' + g_y + g_z z' = 0.$$

Rewriting this as

$$f_x x' + f_z z' = -f_y$$
$$g_x x' + g_z z' = -g_y,$$

and solving for x' and z' by Cramer's rule yields

$$x' = \frac{\begin{vmatrix} -f_y & f_z \\ -g_y & g_z \end{vmatrix}}{\begin{vmatrix} f_x & f_z \\ g_x & g_z \end{vmatrix}} = -\frac{\dfrac{\partial(f, g)}{\partial(y, z)}}{\dfrac{\partial(f, g)}{\partial(x, z)}} \tag{29}$$

and

$$z' = \frac{\begin{vmatrix} f_x & -f_y \\ g_x & -g_y \end{vmatrix}}{\begin{vmatrix} f_x & f_z \\ g_x & g_z \end{vmatrix}} = -\frac{\dfrac{\partial(f, g)}{\partial(x, y)}}{\dfrac{\partial(f, g)}{\partial(x, z)}}; \tag{30}$$

(27) implies that $\partial(f, g)/\partial(x, z)$ is nonzero if y is sufficiently near y_0.

Example 4.7   Let $\mathbf{X}_0 = (1, 1, 2)$ and

$$\mathbf{F}(x, y, z) = \begin{bmatrix} f(x, y, z) \\ g(x, y, z) \end{bmatrix} = \begin{bmatrix} 6x & + 6y & + 4z^3 - 44 \\ -x^2 - & y^2 & + 8z & - 14 \end{bmatrix};$$

then $\mathbf{F}(\mathbf{X}_0) = \mathbf{0}$,

$$\frac{\partial(f, g)}{\partial(x, z)} = \begin{vmatrix} 6 & 12z^2 \\ -2x & 8 \end{vmatrix}$$

and

$$\frac{\partial(f, g)}{\partial(x, z)}\bigg|_{(1,1,2)} = \begin{vmatrix} 6 & 48 \\ -2 & 8 \end{vmatrix} = 144 \neq 0.$$

Therefore, Theorem 4.1 implies that the conditions (28) and

$$x(1) = 1, \qquad z(1) = 2, \tag{31}$$

determine x and y as continuously differentiable functions of y near $y_0 = 1$. From (29) and (30),

$$x' = -\frac{\dfrac{\partial(f, g)}{\partial(y, z)}}{\dfrac{\partial(f, g)}{\partial(x, z)}} = -\frac{\begin{vmatrix} 6 & 12z^2 \\ -2y & 8 \end{vmatrix}}{\begin{vmatrix} 6 & 12z^2 \\ -2x & 8 \end{vmatrix}} = -\frac{2 + yz^2}{2 + xz^2}$$

and

$$z' = -\frac{\dfrac{\partial(f, g)}{\partial(x, y)}}{\dfrac{\partial(f, g)}{\partial(x, z)}} = -\frac{\begin{vmatrix} 6 & 6 \\ -2x & -2y \end{vmatrix}}{\begin{vmatrix} 6 & 12z^2 \\ -2x & 8 \end{vmatrix}} = \frac{y - x}{4 + 2xz^2}.$$

These equations hold near $y = 1$; together with (31) they imply that

$$x'(1) = -1, \qquad z'(1) = 0.$$

Example 4.8   Continuing with Example 4.7, Theorem 4.1 implies that the conditions (28) and

$$y(1) = 1, \qquad z(1) = 2,$$

also determine y and z as functions of x near $x_0 = 1$, since

$$\frac{\partial(f, g)}{\partial(y, z)} = \begin{vmatrix} 6 & 12z^2 \\ -2y & 8 \end{vmatrix}$$

and

$$\frac{\partial(f, g)}{\partial(y, z)}\bigg|_{(1,1,2)} = \begin{vmatrix} 6 & 48 \\ -2 & 8 \end{vmatrix} = 144 \neq 0.$$

However, Theorem 4.1 does not imply that the conditions (28) and

$$x(2) = 1, \qquad y(2) = 1,$$

define x and y as functions of z near $z_0 = 2$, since

$$\frac{\partial(f, g)}{\partial(x, y)} = \begin{vmatrix} 6 & 6 \\ -2x & -2y \end{vmatrix}$$

and

$$\frac{\partial(f, g)}{\partial(x, y)}\Big|_{(1,1,2)} = \begin{vmatrix} 6 & 6 \\ -2 & -2 \end{vmatrix} = 0.$$

We close this section by observing that the functions $\xi_1, \xi_2, \ldots, \xi_m$ of Theorem 4.2 have higher derivatives if $f_1, f_2, \ldots, f_m$ do (this is stated more precisely in *Exercise 22*), and they may be obtained by differentiating (26), using the chain rule.

Example 4.9   Suppose u and v are functions of (x, y) which satisfy

$$f(x, y, u, v) = x - u^2 - v^2 + 9 = 0$$
$$g(x, y, u, v) = y - u^2 + v^2 - 10 = 0.$$

Then

$$\frac{\partial(f, g)}{\partial(u, v)} = \begin{vmatrix} -2u & -2v \\ -2u & 2v \end{vmatrix} = -8uv.$$

From Theorem 4.2, if $uv \neq 0$, then

$$u_x = \frac{1}{8uv} \frac{\partial(f, g)}{\partial(x, v)} = \frac{1}{8uv} \begin{vmatrix} 1 & -2v \\ 0 & 2v \end{vmatrix} = \frac{1}{4u},$$

$$u_y = \frac{1}{8uv} \frac{\partial(f, g)}{\partial(y, v)} = \frac{1}{8uv} \begin{vmatrix} 0 & -2v \\ 1 & 2v \end{vmatrix} = \frac{1}{4u},$$

$$v_x = \frac{1}{8uv} \frac{\partial(f, g)}{\partial(u, x)} = \frac{1}{8uv} \begin{vmatrix} -2u & 1 \\ -2u & 0 \end{vmatrix} = \frac{1}{4v},$$

$$v_y = \frac{1}{8uv} \frac{\partial(f, g)}{\partial(u, y)} = \frac{1}{8uv} \begin{vmatrix} -2u & 0 \\ -2u & 1 \end{vmatrix} = -\frac{1}{4v}.$$

These can be differentiated as many times as we wish. For example,

$$u_{xx} = -\frac{u_x}{4u^2} = -\frac{1}{16u^3},$$

$$u_{xy} = -\frac{u_y}{4u^2} = -\frac{1}{16u^3},$$

$$v_{yx} = \frac{v_x}{4v^2} = \frac{1}{16v^2}.$$

6.4   EXERCISES

1. Solve for $\mathbf{U} = (u, \ldots)$ as a function of $\mathbf{X} = (x, \ldots)$:

(a) $\begin{bmatrix} 1 & 1 \\ 1 & -1 \end{bmatrix} \begin{bmatrix} u \\ v \end{bmatrix} + \begin{bmatrix} 1 & -1 \\ 2 & -3 \end{bmatrix} \begin{bmatrix} x \\ y \end{bmatrix} = \begin{bmatrix} 0 \\ 0 \end{bmatrix}$

$u - v + w + 3x + 2y = 0$

(b) $-u + v + w - \quad x + \quad y = 0$

$\quad u + v - w \qquad + \quad y = 0$

(c) $\begin{aligned} 3u + \quad v + y &= \sin x \\ u + 2v + x &= \sin y \end{aligned}$

$2u + 2v + \quad w + 2x + 2y + \quad z = 0$

(d) $\quad u - \quad v + 2w + \quad x - \quad y + 2z = 0$

$3u + 2v - \quad w + 3x + 2y - \quad z = 0$

2. Suppose $\mathbf{X}_0 \in \mathscr{R}^n$ and $\mathbf{U}_0 \in \mathscr{R}^m$. Prove: If N_1 is a neighborhood of $(\mathbf{X}_0, \mathbf{U}_0)$ in $\mathscr{R}^{n+m}$, there is a neighborhood N of $\mathbf{X}_0$ in $\mathscr{R}^n$ such that $(\mathbf{X}, \mathbf{U}_0) \in N_1$ if $\mathbf{X} \in N$.

3. Prove Corollary 4.1 by specializing the proof of Theorem 4.1 to the case where $m = 1$.

4. Write out the proof of Theorem 4.1 for $m = 2$ and $n = 1$.

5. Show that Theorem 4.1 implies Theorem 3.3.

6. Give an example, for arbitrary m and n, where $\mathbf{F} \cdot \mathscr{R}^{n+m} \to \mathscr{R}^m$ satisfies the hypotheses of Theorem 4.1, except that $\mathbf{F}_\mathbf{U}(\mathbf{X}_0, \mathbf{U}_0)$ is singular, and the conditions

$$\mathbf{F}(\mathbf{X}, \mathbf{U}) = \mathbf{0}, \qquad \mathbf{U}(\mathbf{X}_0) = \mathbf{U}_0,$$

(a) determine $\mathbf{U}$ as a continuously differentiable function of $\mathbf{X}$ near $\mathbf{X}_0$;
(b) determine $\mathbf{U}$ as a function of $\mathbf{X}$ near $\mathbf{X}_0$, but $\mathbf{U}$ is not differentiable at $\mathbf{X}_0$;
(c) do not determine $\mathbf{U}$ as a function of $\mathbf{X}$ near $\mathbf{X}_0$.

7. Let $u = u(x, y)$ be determined near $(1, 1)$ by

$$x^2 yu + 2xy^2 u^3 - 3x^3 y^3 u^5 = 0, \qquad u(1, 1) = 1.$$

Find $u_x(1, 1)$ and $y_y(1, 1)$.

8. Let $u = u(x, y, z)$ be determined near $(1, 1, 1)$ by

$$x^2 y^5 z^2 u^5 + 2xy^2 u^3 - 3x^3 z^2 u = 0, \qquad u(1, 1, 1) = 1.$$

Find $u_x(1, 1, 1)$, $u_y(1, 1, 1)$, and $u_z(1, 1, 1)$.

9. Find $u(x_0, y_0)$, $u_x(x_0, y_0)$, and $u_y(x_0, y_0)$ if:
   (a) $2x^2 + y^2 + ue^u = 6$, $(x_0, y_0) = (1, 2)$
   (b) $u(x + 1) + x(y + 2) + y(u - 2) = 0$, $(x_0, y_0) = (-1, -2)$
   (c) $1 - e^u \sin(x + y) = 0$, $(x_0, y_0) = (\pi/4, \pi/4)$
   (d) $x \log u + y \log x + u \log y = 0$, $(x_0, y_0) = (1, 1)$

10. (a) Find $u(x_0, y_0)$, $u_x(x_0, y_0)$, and $u_y(x_0, y_0)$ for all continuously differentiable functions u that satisfy

$$2x^2y^4 - 3uxy^3 + u^2x^4y^3 = 0$$

near $(x_0, y_0) = (1, 1)$.
   (b) Repeat part (a) for

$$\cos u \cos x + \sin u \sin y = 0, \qquad (x_0, y_0) = (0, \pi).$$

11. Suppose $\mathbf{U} = (u, v)$ is continuously differentiable with respect to (x, y) and satisfies

$$\begin{aligned}
x^2 + 4y^2 + z^2 - 2u^2 + v^2 &= -4 \\
(x + z)^2 \quad + u - v &= -3
\end{aligned} \tag{A}$$

and

$$u(1, \tfrac{1}{2}, -1) = -2, \qquad v(1, \tfrac{1}{2}, -1) = 1.$$

Find $\mathbf{U}'(1, \tfrac{1}{2}, -1)$.

12. Let u and v be continuously differentiable and satisfy

$$\begin{aligned}
u + 2u^2 + v^2 + x^2 + 2v - x &= 0 \\
xuv + e^u \sin(v + x) &= 0
\end{aligned}$$

and $u(0) = v(0) = 0$. Find $u'(0)$ and $v'(0)$.

13. Let $\mathbf{U} = (u, v, w)$ be continuously differentiable with respect to (x, y) and satisfy

$$\begin{aligned}
x^2y + xy^2 + u^2 - (v + w)^2 &= -3 \\
e^{x+y} - u - v - w &= -2 \\
(x + y)^2 + u + v + w^2 &= 3
\end{aligned} \tag{A}$$

and $\mathbf{U}(1, -1) = (1, 2, 0)$. Find $\mathbf{U}'(1, -1)$.

14. Two continuously differentiable transformations $\mathbf{U} = (u, v)$ satisfy

$$\begin{aligned}
xyu - 4yu + 9xv &= 0 \\
2xy - 3y^2 - v^2 &= 0
\end{aligned}$$

near $(x_0, y_0) = (1, 1)$. Find the value of each and its differential matrix at $(1, 1)$.

15. Suppose u, v, and w are continuously differentiable functions of (x, y, z) which satisfy

$$e^x \cos y + e^z \cos u + e^v \cos w + x = 3$$
$$e^x \sin y + e^z \sin u + e^v \cos w \quad\quad = 1$$
$$e^x \tan y + e^z \tan u + e^v \tan w + z = 0$$

near $(x_0, y_0, z_0) = (0, 0, 0)$, and $u(0, 0, 0) = v(0, 0, 0) = w(0, 0, 0) = 0$. Find $u_x(0, 0, 0)$, $v_x(0, 0, 0)$, and $w_x(0, 0, 0)$ by the chain rule and Cramer's rule.

16. Under the assumptions of Theorem 4.1, show that

$$\frac{\partial g_i}{\partial x_j} = -\frac{\dfrac{\partial(f_1, f_2, \ldots, f_i, \ldots, f_m)}{\partial(u_1, u_2, \ldots, x_j, \ldots, u_m)}}{\dfrac{\partial(f_1, f_2, \ldots, f_i, \ldots, f_m)}{\partial(u_1, u_2, \ldots, u_i, \ldots, u_m)}}$$

by using Cramer's rule to solve the system in Eq. (15).

17. Let $\mathbf{F} = (f, g)$ be continuously differentiable in a neighborhood of $\mathbf{X}_0 = (x_0, y_0, z_0)$, $\mathbf{F}(\mathbf{X}_0) = \mathbf{0}$, and

$$\frac{\partial(f, g)}{\partial(x, z)}\bigg|_{\mathbf{X}_0} \neq 0.$$

Show how Theorem 4.1 implies that the conditions

$$\mathbf{F}(x, y, z) = \mathbf{0}, \qquad x(y_0) = x_0, \qquad z(y_0) = z_0,$$

determine x and z as continuously differentiable functions of y near y_0.

18. Let $\mathbf{F} = (f, g, h)$ be continuously differentiable in a neighborhood of $\mathbf{P}_0 = (x_0, y_0, z_0, u_0, v_0)$, $\mathbf{F}(\mathbf{P}_0) = \mathbf{0}$, and

$$\frac{\partial(f, g, h)}{\partial(y, z, u)}\bigg|_{\mathbf{P}_0} \neq 0.$$

Show how Theorem 4.1 implies that the conditions

$$\mathbf{F}(x, y, z, u, v) = \mathbf{0}, \quad y(x_0, v_0) = y_0, \quad z(x_0, v_0) = z_0, \quad u(x_0, v_0) = u_0,$$

determine y, z, and u as continuously differentiable functions of (x, v) near (x_0, v_0), and use Cramer's rule to express their first partial derivatives as ratios of Jacobians.

19. Decide which pairs of the variables x, y, z, u, and v are determined as functions of the others by the system

$$x + 2y + 3z + \quad u + 6v = 0$$
$$2x + 4y + \quad z + 2u + 2v = 0,$$

and solve for them.

20. Show that the conditions (A) in Exercise 11 and

$$y(1, -1, -2) = \tfrac{1}{2}, \qquad v(1, -1, -2) = 1,$$

determine y and v as functions of (x, z, u) near $(x_0, z_0, u_0) = (1, -1, -2)$, and find $y_x(1, -1, -2)$ and $v_u(1, -1, -2)$.

21. Let x, u, and v be continuously differentiable functions of (w, y) which satisfy conditions (A) of Exercise 13 and

$$x(0, -1) = 1, \qquad u(0, -1) = 1, \qquad v(0, -1) = 2.$$

Find their first partial derivatives with respect to y and w at $(0, -1)$.

22. In addition to the assumptions of Theorem 4.1, suppose that $\mathbf{F}$ has all partial derivatives of order $\le q$ in S. Show that $\mathbf{G}$ has all partial derivatives of order $\le q$ in N. (*Hint:* Use Theorem 4.2 and induction on q.)

23. Calculate all first and second partial derivatives at $(x_0, y_0) = (1, 1)$ of the functions u and v that satisfy

$$x^2 + y^2 + u^2 + v^2 = 3$$
$$x + y + u + v = 3,$$
$$u(1, 1) = 0, \qquad v(1, 1) = 1.$$

24. Calculate all first and second partial derivatives at $(x_0, y_0) = (1, -1)$ of the functions that satisfy

$$u^2 - v^2 = x - y - 2$$
$$2uv = x + y - 2,$$
$$u(1, -1) = -1, \qquad v(1, -1) = 1.$$

25. Suppose $f_1, f_2, \ldots, f_n$ are continuously differentiable in a region S in $\mathscr{R}^n$, ϕ is continuously differentiable in a region T of $\mathscr{R}^n$,

$$(f_1(\mathbf{X}), f_2(\mathbf{X}), \ldots, f_n(\mathbf{X})) \in T,$$
$$\phi(f_1(\mathbf{X}), f_2(\mathbf{X}), \ldots, f_n(\mathbf{X})) = 0, \qquad \mathbf{X} \in S,$$

and

$$\sum_{j=1}^{n} \phi_{u_j}^2(\mathbf{U}) > 0, \qquad \mathbf{U} \in T.$$

Show that

$$\frac{\partial(f_1, f_2, \ldots, f_n)}{\partial(x_1, x_2, \ldots, x_n)} = 0, \qquad \mathbf{X} \in S.$$

6.5   THE METHOD OF LAGRANGE MULTIPLIERS

Suppose f and $g_1, g_2, \ldots, g_n$ are real-valued functions defined on an open set D of $\mathscr{R}^n$ and

$$g_1(\mathbf{X}) = g_2(\mathbf{X}) = \cdots = g_m(\mathbf{X}) = 0 \tag{1}$$

if and only if $\mathbf{X}$ is in a nonempty subset D_1 of D. If $\mathbf{X}_0 \in D_1$ and there is a neighborhood N of $\mathbf{X}_0$ such that

$$f(\mathbf{X}) \le f(\mathbf{X}_0) \tag{2}$$

for every $\mathbf{X}$ in $N \cap D_1$, then we say that $\mathbf{X}_0$ is *a local maximum point of f subject to the constraints* (1). If (2) is replaced by

$$f(\mathbf{X}) \ge f(\mathbf{X}_0), \tag{3}$$

then "maximum" is replaced by "minimum." A local maximum or minimum of f subject to (2) is also called a *local extreme point of f subject to* (2). More briefly, we also speak of *constrained* local maximum, minimum, or extreme points. If (2)—or (3)—holds for all $\mathbf{X}$ in D_1, we omit the adjective "local."

Example 5.1   To find the point in the plane

$$3x + 4y + z = 1 \tag{4}$$

closest to $(-1, 1, 1)$, we minimize

$$f(x, y, z) = (x + 1)^2 + (y - 1)^2 + (z - 1)^2$$

subject to the constraint

$$g(x, y, z) = 3x + 4y + z - 1 = 0.$$

Thus, D_1 is the plane defined by (4); if $(x, y, z) \in D_1$, then

$$z = 1 - 3x - 4y, \tag{5}$$

and we can minimize f subject to (4) by finding the ordinary (unconstrained) minimum of the composite function

$$h(x, y) = f(x, y, z(x, y)) = (x + 1)^2 + (y - 1)^2 + (3x + 4y)^2.$$

Since

$$h_x(x, y) = 2(x + 1) + 6(3x + 4y) = 20x + 24y + 2,$$

and

$$h_y(x, y) = 2(y - 1) + 8(3x + 4y) = 24x + 34y - 2,$$

a minimum point (x_0, y_0) of h must satisfy

$$20x_0 + 24y_0 = -2$$
$$24x_0 + 34y_0 = 2$$

(Theorem 3.8, Section 5.3). This system has the unique solution

$$x_0 = -\frac{29}{26}, \qquad y_0 = \frac{22}{26}, \tag{6}$$

and, since it is geometrically evident that f has a constrained minimum, these must be its x and y coordinates. To obtain its z coordinate we substitute (6) into (5):

$$z = 1 - 3x_0 - 4y_0 = 1 - 3\left(-\frac{29}{26}\right) - 4\left(\frac{22}{26}\right) = \frac{25}{26}.$$

Thus, the point on the plane (4) closest to $(-1, 1, 1)$ is

$$(x_0, y_0, z_0) = \left(-\frac{29}{26}, \frac{22}{26}, \frac{25}{26}\right);$$

the minimum distance from $(-1, 1, 1)$ to the plane is

$$\sqrt{f(x_0, y_0, z_0)} = \left[\left(-\frac{3}{26}\right)^2 + \left(-\frac{4}{26}\right)^2 + \left(-\frac{1}{26}\right)^2\right]^{1/2} = \frac{1}{\sqrt{26}}.$$

Example 5.2   To find the extreme value of

$$f(x, y, z, w) = x^2 + y^2 + z^2 + w^2$$

subject to the constraints

$$x + y + z + \ w = 10$$
$$x - y + z + 3w = \ 6,$$

we can solve for x and y, obtaining

$$x = x(z, w) = 8 - z - 2w$$
$$y = y(z, w) = 2 + w, \tag{7}$$

and look for an ordinary (unconstrained) extreme point of the composite function

$$h(z, w) = f(x(z, w), y(z, w), z, w)$$
$$= (8 - z - 2w)^2 + (w + 2)^2 + z^2 + w^2.$$

Since

$$h_z(z, w) = -2(8 - z - 2w) + 2z = 4(z + w - 4)$$

and

$$h_w(z, w) = -4(8 - z - 2w) + 2(w + 2) + 2w = 4(z + 3w - 7),$$

an extreme point (z_0, w_0) must satisfy

$$z_0 + w_0 = 4, \qquad z_0 + 3w_0 = 7.$$

This system has the unique solution

$$z_0 = \frac{5}{2}, \qquad w_0 = \frac{3}{2}. \tag{8}$$

From Corollary 4.3, Section 5.4, (z_0, w_0) is a minimum point of h, since $h_{zz} = 4 > 0$ and

$$h_{zz} h_{ww} - h_{zw}^2 = 32 > 0.$$

Substituting (8) into (7) yields

$$x_0 = \frac{5}{2}, \qquad y_0 = \frac{7}{2};$$

thus, the minimum value of f subject to (7) is

$$f\left(\frac{5}{2}, \frac{7}{2}, \frac{5}{2}, \frac{3}{2}\right) = \left(\frac{5}{2}\right)^2 + \left(\frac{7}{2}\right)^2 + \left(\frac{5}{2}\right)^2 + \left(\frac{3}{2}\right)^2 = 27.$$

It was convenient to reduce the constrained extremum problems of Examples 5.1 and 5.2 to ordinary (unconstrained) extremum problems, because it was easy to solve the constraint equations for some of the variables in terms of the others. Often this is not the case, but fortunately the *method of Lagrange multipliers* makes it unnecessary. The following theorem is the basis for this method.

Theorem 5.1 *Let f and $g_1, g_2, \ldots, g_m$ be continuously differentiable real-valued functions on an open set D of $\mathcal{R}^n$, where $n > m \geq 1$, and suppose $\mathbf{X}_0 = (c_1, c_2, \ldots, c_n)$ is in D and an extreme point of f subject to the constraints*

$$g_i(\mathbf{X}) = 0, \qquad 1 \leq i \leq m. \tag{9}$$

Suppose also that at least one of the Jacobians

$$\left. \frac{\partial(g_1, g_2, \ldots, g_m)}{\partial(x_{i_1}, x_{i_2}, \ldots, x_{i_m})} \right|_{\mathbf{X}_0} \qquad (1 \leq i_1 < i_2 < \cdots < i_m \leq n) \tag{10}$$

is nonzero. Then there are constants $\lambda_1, \lambda_2, \ldots, \lambda_m$ such that

$$\frac{\partial f(\mathbf{X}_0)}{\partial x_i} + \sum_{j=1}^{m} \lambda_j \frac{\partial g_j(\mathbf{X}_0)}{\partial x_i} = 0, \qquad 1 \leq i \leq n. \tag{11}$$

Proof. For convenience we assume that

$$\left. \frac{\partial(g_1, g_2, \ldots, g_m)}{\partial(x_1, x_2, \ldots, x_m)} \right|_{\mathbf{X}_0} \neq 0 \tag{12}$$

and write

$$\mathbf{Y} = (x_{m+1}, x_{m+2}, \ldots, x_n).$$

[A proof based on the assumption that some other Jacobian in (10) is nonzero, can be obtained from this by simply renaming the variables.] Then the implicit function theorem implies that (9) and the conditions

$$x_i(\mathbf{Y}_0) = c_i, \qquad 1 \le i \le m,$$

determine $x_1, x_2, \ldots, x_m$ as continuously differentiable functions of $\mathbf{Y}$ in a neighborhood N of $\mathbf{Y}_0 = (c_{m+1}, c_{m+2}, \ldots, c_n)$. Now let r be a fixed integer, $m+1 \le r \le n$. From the chain rule,

$$
\begin{aligned}
\text{(a)} \quad & \sum_{j=1}^{m} \frac{\partial g_i(\mathbf{X}_0)}{\partial x_j} \frac{\partial x_j(\mathbf{Y}_0)}{\partial x_r} + \frac{\partial g_i(\mathbf{X}_0)}{\partial x_r} = 0, \qquad 1 \le i \le m, \\[2mm]
\text{(b)} \quad & \sum_{j=1}^{m} \frac{\partial f(\mathbf{X}_0)}{\partial x_j} \frac{\partial x_j(\mathbf{Y}_0)}{\partial x_r} + \frac{\partial f(\mathbf{X}_0)}{\partial x_r} = 0,
\end{aligned}
\tag{13}
$$

where (13a) follows from the identities

$$g_i(x_1(\mathbf{Y}), x_2(\mathbf{Y}), \ldots, x_m(\mathbf{Y}), \mathbf{Y}) = 0, \qquad \mathbf{Y} \in N, \quad 1 \le i \le m$$

[see (9)], and (13b) from the fact that if f has a constrained extremum at $\mathbf{X}_0$, then the composite function

$$h(\mathbf{Y}) = f(x_1(\mathbf{Y}), x_2(\mathbf{Y}), \ldots, x_m(\mathbf{Y}), \mathbf{Y})$$

has an ordinary extremum at $\mathbf{Y}_0$, and so $h_{x_r}(\mathbf{Y}_0) = 0$, $\qquad 1 \le r \le m$.
From (13), the system

$$
\begin{bmatrix}
\dfrac{\partial g_1(\mathbf{X}_0)}{\partial x_1} & \dfrac{\partial g_1(\mathbf{X}_0)}{\partial x_2} & \cdots & \dfrac{\partial g_1(\mathbf{X}_0)}{\partial x_m} & \dfrac{\partial g_1(\mathbf{X}_0)}{\partial x_r} \\[3mm]
\dfrac{\partial g_2(\mathbf{X}_0)}{\partial x_1} & \dfrac{\partial g_2(\mathbf{X}_0)}{\partial x_2} & \cdots & \dfrac{\partial g_2(\mathbf{X}_0)}{\partial x_m} & \dfrac{\partial g_2(\mathbf{X}_0)}{\partial x_r} \\[3mm]
& & \vdots & & \\[2mm]
\dfrac{\partial g_m(\mathbf{X}_0)}{\partial x_1} & \dfrac{\partial g_m(\mathbf{X}_0)}{\partial x_2} & \cdots & \dfrac{\partial g_m(\mathbf{X}_0)}{\partial x_m} & \dfrac{\partial g_m(\mathbf{X}_0)}{\partial x_r} \\[3mm]
\dfrac{\partial f(\mathbf{X}_0)}{\partial x_1} & \dfrac{\partial f(\mathbf{X}_0)}{\partial x_2} & \cdots & \dfrac{\partial f(\mathbf{X}_0)}{\partial x_m} & \dfrac{\partial f(\mathbf{X}_0)}{\partial x_r}
\end{bmatrix}
\begin{bmatrix}
\alpha_{1r} \\[1mm] \alpha_{2r} \\[1mm] \vdots \\[1mm] \alpha_{mr} \\[1mm] \alpha_{m+1,r}
\end{bmatrix}
=
\begin{bmatrix}
0 \\[1mm] 0 \\[1mm] \vdots \\[1mm] 0 \\[1mm] 0
\end{bmatrix}
$$

has the nontrivial solution

$$\alpha_{1r} = \frac{\partial x_1(\mathbf{Y}_0)}{\partial x_r}, \quad \alpha_{2r} = \frac{\partial x_2(\mathbf{Y}_0)}{\partial x_r}, \quad \ldots, \quad \alpha_{mr} = \frac{\partial x_m(\mathbf{Y}_0)}{\partial x_r}, \quad \alpha_{m+1,r} = 1;$$

therefore its matrix is singular (Theorem 1.5). Since the transpose of a singular matrix is singular, the system

$$
\begin{bmatrix}
\dfrac{\partial g_1(\mathbf{X}_0)}{\partial x_1} & \dfrac{\partial g_2(\mathbf{X}_0)}{\partial x_1} & \cdots & \dfrac{\partial g_m(\mathbf{X}_0)}{\partial x_1} & \dfrac{\partial f(\mathbf{X}_0)}{\partial x_1} \\[2ex]
\dfrac{\partial g_1(\mathbf{X}_0)}{\partial x_2} & \dfrac{\partial g_2(\mathbf{X}_0)}{\partial x_2} & \cdots & \dfrac{\partial g_m \mathbf{X}_0)}{\partial x_2} & \dfrac{\partial f(\mathbf{X}_0)}{\partial x_2} \\[2ex]
& & \vdots & & \\[2ex]
\dfrac{\partial g_1(\mathbf{X}_0)}{\partial x_m} & \dfrac{\partial g_2(\mathbf{X}_0)}{\partial x_m} & \cdots & \dfrac{\partial g_m(\mathbf{X}_0)}{\partial x_m} & \dfrac{\partial f(\mathbf{X}_0)}{\partial x_m} \\[2ex]
\dfrac{\partial g_1(\mathbf{X}_0)}{\partial x_r} & \dfrac{\partial g_2(\mathbf{X}_0)}{\partial x_r} & \cdots & \dfrac{\partial g_m(\mathbf{X}_0)}{\partial x_r} & \dfrac{\partial f(\mathbf{X}_0)}{\partial x_r}
\end{bmatrix}
\begin{bmatrix}
\lambda_{1r} \\[1ex] \lambda_{2r} \\[1ex] \vdots \\[1ex] \lambda_{mr} \\[1ex] \lambda_{m+1,\,r}
\end{bmatrix}
=
\begin{bmatrix}
0 \\ 0 \\ \vdots \\ 0 \\ 0
\end{bmatrix}
\tag{14}
$$

also has a nontrivial solution (Theorem 1.5); moreover, since (12) implies that the system

$$
\begin{bmatrix}
\dfrac{\partial g_1(\mathbf{X}_0)}{\partial x_1} & \dfrac{\partial g_2(\mathbf{X}_0)}{\partial x_1} & \cdots & \dfrac{\partial g_m(\mathbf{X}_0)}{\partial x_1} \\[2ex]
\dfrac{\partial g_1(\mathbf{X}_0)}{\partial x_2} & \dfrac{\partial g_2(\mathbf{X}_0)}{\partial x_2} & \cdots & \dfrac{\partial g_m(\mathbf{X}_0)}{\partial x_2} \\[2ex]
& & \vdots & \\[2ex]
\dfrac{\partial g_1(\mathbf{X}_0)}{\partial x_m} & \dfrac{\partial g_2(\mathbf{X}_0)}{\partial x_m} & \cdots & \dfrac{\partial g_m(\mathbf{X}_0)}{\partial x_m}
\end{bmatrix}
\begin{bmatrix}
\xi_1 \\[1ex] \xi_2 \\[1ex] \vdots \\[1ex] \xi_m
\end{bmatrix}
=
\begin{bmatrix}
0 \\ 0 \\ \vdots \\ 0
\end{bmatrix}
\tag{15}
$$

has only the trivial solution, it follows that $\lambda_{m+1,r} \neq 0$. (Why?) Therefore, we may assume that $\lambda_{m+1,r} = 1$. Thus, (14) implies that

$$
\frac{\partial f(\mathbf{X}_0)}{\partial x_i} + \sum_{j=1}^{m} \lambda_{jr} \frac{\partial g_j(\mathbf{X}_0)}{\partial x_i} = 0 \qquad \text{if } 1 \le i \le m \text{ or } i = r.
\tag{16}
$$

If $n = m + 1$, this proves (11), with $\lambda_j = \lambda_{jr}$, $1 \le j \le n$. Suppose $n > m + 1$, $m + 1 \le s \le n$, and $s \neq r$. By the argument just given, there are constants $\lambda_{1s}, \lambda_{2s}, \ldots, \lambda_{ns}$ such that

$$
\frac{\partial f(\mathbf{X}_0)}{\partial x_i} + \sum_{j=1}^{m} \lambda_{js} \frac{\partial g_j(\mathbf{X}_0)}{\partial x_i} = 0 \qquad \text{if } 1 \le i \le m \quad \text{or} \quad i = s.
$$

Subtracting the first m of these equations from the first m of (16) yields

$$
\sum_{j=1}^{m} (\lambda_{jr} - \lambda_{js}) \frac{\partial g_j(\mathbf{X}_0)}{\partial x_i} = 0, \qquad 1 \le i \le m;
$$

that is, $(\xi_1, \xi_2, \ldots, \xi_m)$, defined by $\xi_i = \lambda_{ir} - \lambda_{is}$ $(1 \le i \le m)$, is a solution of (15).

Since (15) has only the trivial solution,

$$\lambda_{jr} = \lambda_{js}, \qquad 1 \le j \le n, \quad m+1 \le r, \quad s \le n;$$

hence, λ_{jr} is independent of r. This proves (11).

Theorem 5.1 implies that if $\mathbf{X}_0$ is an extreme point of f subject to constraints

$$g_i(\mathbf{X}) = 0, \qquad 1 \le i \le m, \tag{17}$$

then $\mathbf{X}_0$ is a critical point of

$$F = f + \sum_{j=1}^{m} \lambda_j g_j \tag{18}$$

for some constants $\lambda_1, \lambda_2, \ldots, \lambda_m$. This provides a method for finding constrained extreme points of f subject to (17):

1. Find the critical points of the auxiliary function (18), treating $\lambda_1, \lambda_2, \ldots, \lambda_m$ as fixed but unspecified constants; these are obtained by equating the first partial derivatives of F to zero, and relating the extreme points of F to $\lambda_1, \lambda_2, \ldots, \lambda_m$.
2. Determine $\lambda_1, \lambda_2, \ldots, \lambda_m$ so that the critical points obtained in (a) satisfy the constraints (17). Since $F(\mathbf{X}_0) = f(\mathbf{X}_0)$ if $\mathbf{X}_0$ is such a point, we say that $\mathbf{X}_0$ is a *critical point of f subject to the constraints* (17), or, more briefly, a *constrained critical point of f*. (We will see below that it is not always necessary to actually find the values of $\lambda_1, \lambda_2, \ldots, \lambda_m$.)
3. Determine which constrained critical points of f are constrained extreme points. Each of the latter must be one of the former, but the converse is not true.

The parameters $\lambda_1, \lambda_2, \ldots, \lambda_m$ are called *Lagrange multipliers*, and the procedure is *the method of Lagrange multipliers*.

Example 5.3   To minimize

$$f(x, y, z) = (x + 1)^2 + (y - 1)^2 + (z - 1)^2$$

subject to

$$3x + 4y + z = 1 \tag{19}$$

(this is the problem considered in Example 5.1), we form the auxiliary function

$$F(x, y, z) = (x + 1)^2 + (y - 1)^2 + (z - 1)^2 + \lambda(3x + 4y + z - 1);$$

then

$$F_x(x, y, z) = 2(x + 1) + 3\lambda$$
$$F_y(x, y, z) = 2(y - 1) + 4\lambda$$
$$F_z(x, y, z) = 2(z - 1) + \lambda.$$

Equating these partial derivatives to zero and solving for x_0, y_0, and z_0 yields

$$x_0 = -1 - \frac{3\lambda}{2}, \quad y_0 = 1 - 2\lambda, \quad z_0 = 1 - \frac{\lambda}{2}. \tag{20}$$

To determine λ so that (x_0, y_0, z_0) satisfies the constraints, we substitute this into (19):

$$3\left(-1 - \frac{3\lambda}{2}\right) + 4(1 - 2\lambda) + \left(1 - \frac{\lambda}{2}\right) = 1.$$

This yields $\lambda = \frac{1}{13}$, and substituting this into (20) shows that

$$(x_0, y_0, z_0) = \left(-\frac{29}{26}, \frac{22}{26}, \frac{25}{26}\right)$$

is the only constrained critical point of f that satisfies (19). As in Example 5.1, we argue on geometrical grounds that f must have a constrained minimum, and so this must be it.

In this example we could just as well have taken the auxiliary function to be

$$F_1(x, y, z) = (x + 1)^2 + (y - 1)^2 + (z - 1)^2 + \lambda(3x + 4y + z),$$

which differs from F only by a constant, and therefore has the same critical points. We could also have replace λ by -2λ, and considered the auxiliary function

$$F_2(x, y, z) = (x + 1)^2 + (y - 1)^2 + (z - 1)^2 - 2\lambda(3x + 4y + z).$$

In general it is legitimate to add an arbitrary constant to an auxiliary function and replace the Lagrange multipliers $\lambda_1, \lambda_2, \ldots, \lambda_m$ by $c_1\lambda_1, c_2\lambda_2, \ldots, c_m\lambda_m$, where $c_1, c_2, \ldots, c_m$ are any convenient nonzero constants. This allows us to dispense with "nuisance constants."

Example 5.4   To minimize

$$f(x, y, z, w) = x^2 + y^2 + z^2 + w^2$$

subject to the constraints

$$\begin{aligned} x + y + z + w &= 10 \\ x - y + z + 3w &= 6 \end{aligned} \tag{21}$$

(this is the problem considered in Example 5.2), we form the auxiliary function

$$F(x, y, z, w) = x^2 + y^2 + z^2 + w^2 - 2\lambda(x + y + z + w) - 2\mu(x - y + z + 3w);$$

then

$$F_x(x, y, z, w) = 2x - 2\lambda - 2\mu$$
$$F_y(x, y, z, w) = 2y - 2\lambda + 2\mu$$
$$F_z(x, y, z, w) = 2z - 2\lambda - 2\mu$$
$$F_w(x, y, z, w) = 2w - 2\lambda - 6\mu.$$

Equating these to zero yields

$$x_0 = \lambda + \mu, \quad y_0 = \lambda - \mu, \quad z_0 = \lambda + \mu, \quad w_0 = \lambda + 3\mu. \qquad (22)$$

To determine λ and μ so that (x_0, y_0, z_0, w_0) satisfies the constraints, we sub-stitute (22) into (21):

$$(\lambda + \mu) + (\lambda - \mu) + (\lambda + \mu) + (\lambda + 3\mu) = 10$$
$$(\lambda + \mu) - (\lambda - \mu) + (\lambda + \mu) + (3\lambda + 9\mu) = 6,$$

or

$$4\lambda + 4\mu = 10$$
$$4\lambda + 12\mu = 6.$$

Solving this system yields

$$\lambda = 3, \qquad \mu = -\frac{1}{2},$$

and so, from (22),

$$(x_0, y_0, z_0, w_0) = \left(\frac{5}{2}, \frac{7}{2}, \frac{5}{2}, \frac{3}{2}\right)$$

is the only critical point of f subject to (21).

This agrees with the result of Example 5.2; however, there we were able to ascertain that $f(x_0, y_0, z_0, w_0)$ is the minimum of f subject to (21) by applying Corollary 4.3, Section 5.4, a sufficient condition for ordinary extrema, to the composite function h. We do not wish to develop sufficient conditions for use in conjunction with the method of Lagrange multipliers; rather, we will restrict our attention to problems for which the nature of the constrained critical points can be determined on physical, geometric, or intuitive grounds. Thus, in Example 5.4, $f(x, y, z, w)$ is the square of the distance from (x, y, z, w) to the origin, and it can be shown that among the points that satisfy (21) there is one closest to the origin; (x_0, y_0, z_0, w_0) is that point.

Example 5.5   We will show that

$$x^{1/p}y^{1/q} \le \frac{x}{p} + \frac{y}{q}, \qquad x, y \ge 0, \qquad (23)$$

if

$$\frac{1}{p} + \frac{1}{q} = 1, \qquad p > 0, \quad q > 0.$$

To do this we first find the maximum of

$$f(x, y) = x^{1/p} y^{1/q}$$

subject to the constraint

$$\frac{x}{p} + \frac{y}{q} = \sigma, \qquad x \geq 0, \quad y \geq 0, \tag{24}$$

where σ is a fixed positive number. Since f is continuous it must assume a maximum at some point (x_0, y_0) on the line segment (24), and (x_0, y_0) cannot be an end point of the segment, since $f(p\sigma, 0) = f(0, q\sigma) = 0$; therefore, (x_0, y_0) is in the open first quadrant $\{(x, y) | x > 0, y > 0\}$, which can be taken to be the set D of Theorem 5.1. Consequently, (x_0, y_0) is a critical point of the auxiliary function

$$F(x, y) = x^{1/p} y^{1/q} - \lambda \left(\frac{x}{p} + \frac{y}{q} \right)$$

for some λ. Therefore,

$$0 = F_x(x_0, y_0) = \frac{1}{px_0} f(x_0, y_0) - \frac{\lambda}{p}$$

$$0 = F_y(x_0, y_0) = \frac{1}{qy_0} f(x_0, y_0) - \frac{\lambda}{q}.$$

These equations imply that $x_0 = y_0$ and so, from (24), $x_0 = y_0 = \sigma$. Therefore,

$$f(x, y) \leq f(\sigma, \sigma) = \sigma^{1/p} \sigma^{1/q} = \sigma,$$

which, because of (24), implies (23).

This result can be generalized (*Exercise 27*). It can also be used to generalize the Schwarz inequality (*Exercise 28*).

Example 5.6   The distance between two curves in $\mathcal{R}^2$ is the minimum value of $|\mathbf{X}_1 - \mathbf{X}_2|$, where $\mathbf{X}_1$ is on one curve and $\mathbf{X}_2$ is on the other. To find the distance between the ellipse

$$x^2 + 2y^2 = 1$$

and the line

$$x + y = 4$$

(Figure 5.1), we minimize

$$d^2 = (x_1 - x_2)^2 + (y_1 - y_2)^2$$

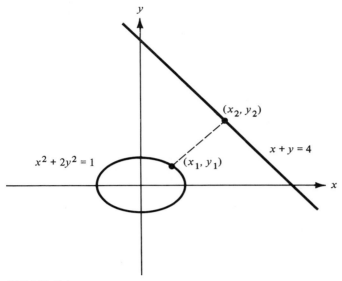

FIGURE 5.1

subject to the constraints

$$x_1^2 + 2y_1^2 = 1, \tag{25}$$
$$x_2 + y_2 = 4. \tag{26}$$

Introducing the auxiliary function

$$f(x_1, y_1, x_2, y_2) = (x_1 - x_2)^2 + (y_1 - y_2)^2 + \lambda(x_1^2 + 2y_1^2) + \mu(x_2 + y_2)$$

and equating its first partial derivatives to zero yields

$$2(x_1 - x_2) + 2\lambda x_1 = 0 \tag{27}$$
$$2(y_1 - y_2) + 4\lambda y_1 = 0 \tag{28}$$
$$-2(x_1 - x_2) + \mu = 0 \tag{29}$$
$$-2(y_1 - y_2) + \mu = 0 \tag{30}$$

From (27) and (29),

$$\mu = -2\lambda x_1,$$

and, from (28) and (30),

$$\mu = -4\lambda y_1.$$

Therefore, $\lambda x_1 = 2\lambda y_1$, and, since (27) and (28) imply that $\lambda \neq 0$ (the curves do not intersect), we conclude that

$$x_1 = 2y_1. \tag{31}$$

Substituting this in (25) yields $y_1 = \pm 1/\sqrt{6}$; since it is obvious from Figure 5.1 that (x_1, y_1) is in the first quadrant, $y_1 = 1/\sqrt{6}$. From this and (31),

$$(x_1, y_1) = \left(\frac{2}{\sqrt{6}}, \frac{1}{\sqrt{6}} \right). \tag{32}$$

Together with (29) and (30), this implies that

$$x_2 - y_2 = x_1 - y_1 = \frac{1}{\sqrt{6}}.$$

This and (26) form a pair of simultaneous equations for (x_2, y_2) which can be solve to yield

$$(x_2, y_2) = \left(2 + \frac{1}{2\sqrt{6}}, 2 - \frac{1}{2\sqrt{6}} \right).$$

From this and (32) the distance between the curves is

$$\left[\left(2 + \frac{1}{2\sqrt{6}} - \frac{2}{\sqrt{6}} \right)^2 + \left(2 - \frac{1}{2\sqrt{6}} - \frac{1}{\sqrt{6}} \right)^2 \right]^{1/2} = \sqrt{2} \left(2 - \frac{3}{2\sqrt{6}} \right).$$

Constrained extrema of quadratic forms

An *eigenvalue* of a square matrix $\mathbf{A} = [a_{ij}]$ is a number λ such that the system

$$\mathbf{AX} = \lambda \mathbf{X}, \tag{33}$$

or, equivalently,

$$(\mathbf{A} - \lambda \mathbf{I})\mathbf{X} = \mathbf{0},$$

has a solution $\mathbf{X} \neq \mathbf{0}$. Such a solution is called an *eigenvector* of $\mathbf{A}$. From Theorem 1.5, λ is an eigenvalue of $\mathbf{A}$ if and only if
$$\det(\mathbf{A} - \lambda \mathbf{I}) = 0.$$

It can be shown that if $\mathbf{A}$ is symmetric ($a_{ij} = a_{ji}$, $1 \leq i, j \leq n$), then

$$\det(\mathbf{A} - \lambda I) = (-1)^n (\lambda - \lambda_1)(\lambda - \lambda_2) \cdots (\lambda - \lambda_n),$$

where $\lambda_1, \lambda_2, \ldots, \lambda_n$ are real numbers.

Now suppose $\mathbf{A}$ is symmetric. To find the maximum or minimum of the quadratic form

$$Q(\mathbf{X}) = \sum_{i,j=1}^{n} a_{ij} x_i x_j$$

subject to the constraint

$$\sum_{i=1}^{n} x_i^2 = 1,$$ (34)

we form the auxiliary function

$$F(\mathbf{X}) = Q(\mathbf{X}) - \lambda \sum_{i=1}^{n} x_i^2$$

and equate its first partial derivatives to zero:

$$F_{x_i}(\mathbf{X}) = 2 \sum_{j=1}^{n} a_{ij} x_j - 2\lambda x_i = 0, \qquad 1 \le i \le n.$$

This implies that a vector $\mathbf{X}$ is a constrained critical point of Q subject to (34) if and only if it is a unit vector which satisfies (33) for some λ; that is, if and only if λ is an eigenvalue and $\mathbf{X}$ an associated eigenvector of $\mathbf{A}$ with unit length. If $\mathbf{X}$ satisfies (33) and (34), then

$$Q(\mathbf{X}) = \sum_{i=1}^{n} \left(\sum_{j=1}^{n} a_{ij} x_j \right) x_i = \sum_{i=1}^{n} (\lambda x_i) x_i$$

$$= \lambda \sum_{i=1}^{n} x_i^2 = \lambda;$$

therefore, the largest and smallest eigenvalues of A are the maximum and minimum values of Q subject to (34).

It can be shown that an eigenvector of unit length associated with an intermediate (neither the largest nor smallest) eigenvalue of $\mathbf{A}$ is not a local extreme value of Q subject to (34), although it is a constrained critical point.

Example 5.7   Let

$$Q(\mathbf{X}) = x^2 + y^2 + 2z^2 - 2xy + 4xz + 4yz.$$

The matrix of Q is

$$\mathbf{A} = \begin{bmatrix} 1 & -1 & 2 \\ -1 & 1 & 2 \\ 2 & 2 & 2 \end{bmatrix}$$

and

$$\det(\mathbf{A} - \lambda \mathbf{I}) = \begin{vmatrix} 1-\lambda & -1 \cdot & 2 \\ -1 & 1-\lambda & 2 \\ 2 & 2 & 2-\lambda \end{vmatrix}$$

$$= -(\lambda + 2)(\lambda - 2)(\lambda - 4),$$

and so

$$\lambda_1 = 4, \qquad \lambda_2 = 2, \qquad \lambda_3 = -2$$

are the eigenvalues of $\mathbf{A}$. Hence, $\lambda_1 = 4$ and $\lambda_3 = -2$ are the maximum and minimum of Q subject to the constraint

$$x^2 + y^2 + z^2 = 1. \tag{35}$$

To find the points (x_0, y_0, z_0) at which Q attains its constrained maximum, we first find an eigenvector of $\mathbf{A}$ corresponding to $\lambda_1 = 4$. To do this we find a nontrivial solution of the system

$$(\mathbf{A} - 4\mathbf{I}) \begin{bmatrix} x \\ y \\ z \end{bmatrix} = \begin{bmatrix} -3 & -1 & 2 \\ -1 & -3 & 2 \\ 2 & 2 & -2 \end{bmatrix} \begin{bmatrix} x \\ y \\ z \end{bmatrix} = \begin{bmatrix} 0 \\ 0 \\ 0 \end{bmatrix}.$$

All such solutions are multiples of

$$\begin{bmatrix} x \\ y \\ z \end{bmatrix} = \begin{bmatrix} 1 \\ 1 \\ 2 \end{bmatrix};$$

normalizing this so as to satisfy (35) yields

$$(x_0, y_0, z_0) = \pm \left(\frac{1}{\sqrt{6}}, \frac{1}{\sqrt{6}}, \frac{2}{\sqrt{6}} \right).$$

Example 5.8   In Section 6.1 we defined the norm, $\|\mathbf{A}\|$, of an $m \times n$ matrix $\mathbf{A}$ to be the smallest number such that

$$|\mathbf{AX}| \le \|\mathbf{A}\| \, |\mathbf{X}|$$

for all $\mathbf{X}$ in $\mathcal{R}^n$. This implies that $\|\mathbf{A}\|^2$ is the maximum of

$$Q(\mathbf{X}) = |\mathbf{AX}|^2$$

subject to the constraint

$$\sum_{i=1}^{n} x_i^2 = 1$$

(*Exercise 9, Section 6.2*). Now

$$Q(\mathbf{X}) = \sum_{1=1}^{m} \left(\sum_{j=1}^{n} a_{ij} x_j \right)^2,$$

which can be multiplied out and rewritten as

$$Q(\mathbf{X}) = \sum_{i,j=1}^{n} b_{ij} x_i x_j$$

where $\mathbf{B} = [b_{ij}]$ is a symmetric $n \times n$ matrix; thus, $\|\mathbf{A}\|^2 = \lambda_M$, where λ_M is the largest eigenvalue of $\mathbf{B}$, and

$$\|\mathbf{AX}\| = \|\mathbf{A}\|\,\|\mathbf{X}\|$$

if and only if $\mathbf{X} = \mathbf{0}$ or $\mathbf{X}$ is an eigenvector of $\mathbf{B}$ corresponding to λ_M.

For example, let

$$\mathbf{A} = \begin{bmatrix} 1 & 0 & 1 \\ 0 & 1 & 1 \end{bmatrix};$$

then

$$\mathbf{A} \begin{bmatrix} x_1 \\ x_2 \\ x_3 \end{bmatrix} = \begin{bmatrix} x_1 + x_3 \\ x_2 + x_3 \end{bmatrix},$$

and so

$$Q(\mathbf{X}) = |\mathbf{AX}|^2 = (x_1 + x_3)^2 + (x_2 + x_3)^2$$
$$= x_1^2 + x_2^2 + 2x_3^2 + 2x_1 x_3 + 2x_2 x_3,$$

which is of the form

$$Q(\mathbf{X}) = \sum_{i,j=1}^{3} b_{ij} x_i x_j$$

with

$$\mathbf{B} = [b_{ij}] = \begin{bmatrix} 1 & 0 & 1 \\ 0 & 1 & 1 \\ 1 & 1 & 2 \end{bmatrix}.$$

Now,

$$\det(\mathbf{B} - \lambda\mathbf{I}) = \begin{vmatrix} 1 - \lambda & 0 & 1 \\ 0 & 1 - \lambda & 1 \\ 1 & 1 & 2 - \lambda \end{vmatrix}$$
$$= -\lambda(\lambda - 1)(\lambda - 3).$$

Since 3 is the largest eigenvalue of $\mathbf{B}$,

$$\|\mathbf{A}\| = \sqrt{3},$$

and a nonzero vector $\mathbf{X}$ satisfies

$$|\mathbf{AX}| = \sqrt{3}|\mathbf{X}|$$

if and only if $\mathbf{X}$ is an eigenvector of $\mathbf{B}$ corresponding to $\lambda = 3$. These eigenvectors are the nonzero multiples of

$$\mathbf{X}_0 = \begin{bmatrix} 1 \\ 1 \\ 2 \end{bmatrix}.$$

The next theorem provides further information on the relationship between the eigenvalues of a symmetric matrix and constrained extrema of its quadratic form. We omit the proof.

Theorem 5.2   *Suppose*

$$\lambda_1 \geq \lambda_2 \geq \cdots \geq \lambda_n$$

are the eigenvalues of an $n \times n$ symmetric matrix $\mathbf{A}$. For some integer r, $1 \leq r < n$, let

$$\mathbf{X}_i = \begin{bmatrix} c_{i1} \\ c_{i2} \\ \vdots \\ c_{in} \end{bmatrix}, \qquad 1 \leq i \leq r,$$

be eigenvectors of $\mathbf{A}$ corresponding to $\lambda_1, \lambda_2, \ldots, \lambda_r$. Then λ_{r+1} is the maximum value of the quadratic form

$$Q(\mathbf{X}) = \sum_{i,j=1}^{n} a_{ij} x_i x_j$$

subject to the constraints

$$\sum_{j=1}^{n} x_j^2 = 1, \qquad \sum_{j=1}^{n} c_{ij} x_j = 0, \qquad 1 \leq i \leq r,$$

and Q attains this constrained maximum only for unit eigenvectors of $\mathbf{A}$ corresponding to the eigenvalue λ_{r+1}.

A problem in data smoothing

In physical and statistical problems one is often presented with a sequence of observations

$$u_r = f_r + \varepsilon_r, \qquad r = 0, \pm 1, \pm 2, \ldots,$$

where $\{f_r\}$ is the sequence of values of some observable at equally spaced values of the independent variable and ε_r is a random error of observation.

Suppose it is known that

$$f_r = a_0 + a_1 r + \cdots + a_{2k} r^{2k},$$

where k is known, but $a_0, a_1, \ldots, a_{2k}$ are not. That is, f_r is an unknown polynomial of degree $\leq 2k$. The problem is to "smooth" the data, that is, to perform an operation that will decrease the effect of the measurement error. The method we will consider is called *least-squares smoothing*; it consists of operating on the data $\{u_r\}$ to produce "smoothed" data $\{v_r\}$ by means of the formula

$$v_r = \sum_{j=-n}^{n} w_j u_{r-j},$$

where $n \geq k$ and $w_{-n}, \ldots, w_0, \ldots, w_n$ (the *weighting coefficients*) are chosen so that

$$\sum_{j=-n}^{n} w_j^2 = \text{minimum} \tag{36}$$

subject to the requirement that $v_r = f_r$ if $\varepsilon_r = 0$ $(r = 0, \pm 1, \pm 2, \ldots)$, or, equivalently, that

$$f_r = \sum_{j=-n}^{n} w_j f_{r-j}, \qquad r = 0, \pm 1, \pm 2, \ldots,$$

if f_r is any polynomial of degree $\leq 2k$. This holds if and only if

$$\sum_{j=-n}^{n} w_j j^s = \begin{cases} 1, & s = 0, \\ 0, & 1 \leq s \leq 2k \end{cases} \tag{37}$$

(*Exercise 21*). Thus, to find the weighting coefficients we must minimize (36) subject to (37). To do this we form the auxiliary function

$$F(W) = \sum_{j=-n}^{n} w_j^2 - 2 \sum_{s=0}^{2k} \lambda_s \sum_{j=-n}^{n} w_j j^s$$

and equate its first partial derivatives to zero. This shows that

$$w_j = \sum_{s=0}^{2k} \lambda_s j^s, \qquad -n \leq j \leq n; \tag{38}$$

that is, the weighting coefficients are themselves values of a certain polynomial of degree $\leq 2k$. The multipliers $\lambda_0, \lambda_1, \ldots, \lambda_{2k}$ are found by substituting (38) into (37). This leads to the system

$$\sum_{s=0}^{2k} \sigma_{r+s} \lambda_s = \begin{cases} 1, & r = 0, \\ 0, & 1 \leq r \leq 2k, \end{cases}$$

where

$$\sigma_i = \sum_{j=-n}^{n} j^i.$$

After solving this system for $\lambda_0, \lambda_1, \ldots, \lambda_{2k}$ we can compute the weighting coefficients from (38).

The critical reader will question whether the weighting coefficients obtained in this way actually minimize $\sum_{j=-n}^{n} w_j^2$. *Exercise 22* indicates a proof that this is so.

6.5   EXERCISES

Unless there are other instructions use the method of Lagrange multipliers in the following exercises.

1. Find the point on the plane $2x + 3y + z = 7$ closest to $(1, -2, 3)$ (a) without using Lagrange multipliers; (b) using Lagrange multipliers.

2. Find the minimum of

$$f(x, y, z, w) = x^2 + 2y^2 + z^2 + 2w^2$$

subject to

$$x + y + z + 3w = 1$$
$$x + y + 2z + w = 2$$

(a) without using Lagrange multipliers; (b) using Lagrange multipliers.

3. Work through the proof of Theorem 5.1 with (a) $m = 1$ and $n = 2$; (b) $m = 2$ and $n = 3$.

4. Find the extreme values of

$$f(x, y, z) = 2x + 3y + z$$

subject to

$$x^2 + 2y^2 + 3z^2 = 1.$$

5. The sides and bottom of a rectangular box have total area A. Find the largest volume the box can have.

6. The sides, top, and bottom of a rectangular box have total area A. Find the largest volume the box can have.

7. A rectangular box with no top is to have volume V. Find the dimensions that minimize its surface area.

8. Suppose a, b, and c are positive. Find the dimensions of the rectangular box of greatest volume with three faces in the coordinate planes and a vertex in the plane

$$\frac{x}{a} + \frac{y}{b} + \frac{z}{c} = 1.$$

9. Two vertices of a triangle are $(-a, 0)$ and $(a, 0)$, and the third is on the ellipse

$$\frac{x^2}{a^2} + \frac{y^2}{b^2} = 1 \qquad (a, b > 0).$$

What is the largest area the triangle can have?

10. Show that the triangle with the greatest possible area for a given perimeter is equilateral. [Hint: The area of a triangle with sides x, y, and z and perimeter s is

$$A = [s(s - x)(s - y)(s - z)]^{1/2}.]$$

11. Derive the formula for the distance from (x_0, y_0, z_0) to the plane

$$ax + by + cz = \sigma.$$

12. Let $X_i = (x_i, y_i, z_i)$, $1 \le i \le n$. Find the point in the plane

$$ax + by + cz = \sigma$$

for which $\sum_{i=1}^{n} |X - X_i|^2$ is a minimum.

13. Find the distance between parabola $y = 1 + x^2$ and the line $x + y = -1$.

14. Find the distance between the ellipsoid

$$3x^2 + 9y^2 + 6z^2 = 10$$

and the plane

$$3x + 3y + 6z = 70.$$

15. Suppose f, ϕ, and ψ are continuously differentiable on $\mathscr{R}^n$ and let X and Y be points in $\mathscr{R}^n$. Show that if $(\bar{X}, \bar{Y})$ is an extreme point of $g(X, Y) = f(X - Y)$ subject to the constraints $\phi(X) = \psi(Y) = 0$, then one of the following must be true: (i) $\bar{U} = \bar{X} - \bar{Y}$ is an unconstrained critical point of $f = f(U)$; (ii) $\bar{X}$ is a critical point of ϕ; (iii) $\bar{Y}$ is a critical point of ψ; (iv) $\phi'(\bar{X}) = k\psi'(\bar{Y})$, where $k \ne 0$.

16. Use the results of Exercises 11 and 15 to find the distance between the ellipse

$$x^2 + y^2 + z^2 + xy + xz + yz = 52$$

and the plane

$$2x + 3y + 7z = 70.$$

17. Find the extreme values of the quadratic form

$$Q(x, y, z) = 2xy + 2xz + 2yz$$

subject to

$$x^2 + y^2 + z^2 = 1.$$

18. Find the extreme values of the quadratic form

$$Q(x, y, z) = 3x^2 + 2y^2 + 3z^2 + 2xz$$

subject to

$$x^2 + y^2 + z^2 = 1.$$

19. Find the extreme values of

$$f(x, y) = x^2 + 8xy + 4y^2$$

subject to

$$x^2 + 2xy + 4y^2 = 1.$$

20. Find $\|A\|$ and $m(A)$ (see Exercise 9, Section 6.2), and nonzero vectors X_1 and X_2 such that $|AX_1| = \|A\| \, |X_1|$ and $|AX_2| = m(A)|X_1|$ if

(a) $A = \begin{bmatrix} 1 & 2 \\ 2 & 1 \end{bmatrix}$ (b) $A = \begin{bmatrix} 1 & 1 & 1 & 1 \\ 1 & -1 & -1 & 1 \end{bmatrix}$

21. Show that

$$f_r = \sum_{j=-n}^{n} w_j f_{r-j},$$

whenever f_r is a polynomial in r of degree $\leq 2k$, if and only if the coefficients $\{w_j\}$ satisfy Eq. (37).

22. Suppose

$$\sum_{j=-n}^{n} w_j j^r = \sum_{j=-n}^{n} u_j j^r, \qquad 0 \leq r \leq 2k,$$

and w_j is a polynomial of degree $\leq 2k$:

$$w_j = \sum_{s=0}^{2k} \lambda_s j^s.$$

Show that

$$\sum_{j=-n}^{n} w_j^2 \leq \sum_{j=-n}^{n} u_j^2$$

with equality if and only if $u_j = w_j$, $-n \leq j \leq n$. [*Hint:* Write $u_j = (u_j - w_j) + w_j$.]

23. Find $x_1, x_2, \ldots, x_n$ so as to minimize

$$Q(X) = \sum_{r=1}^{n} x_r^2$$

subject to

$$\sum_{r=1}^{n} x_r = 1, \qquad \sum_{r=1}^{n} r x_r = 0.$$

Assume that $n \geq 2$.

24. Find the extreme values of

$$f(x, y, z, w) = xw - yz$$

subject to

$$x^2 + 2y^2 = 4, \qquad 2z^2 + w^2 = 9.$$

25. Let a, b, c, and d be positive. Find the maximum value, and the points for which it is attained, for

$$f(x, y, z, w) = xw - yz$$

subject to

$$ax^2 + by^2 = 1, \qquad cz^2 + dw^2 = 1,$$

if (a) $ad > bc$; (b) $ad = bc$.

26. Prove Schwarz's inequality by maximizing $\sum_{i=1}^{n} a_i b_i$ subject to

$$\sum_{i=1}^{n} b_i^2 = \sigma^2.$$

27. Let $x_1, x_2, \ldots, x_m, r_1, r_2, \ldots, r_m$ be positive and

$$r_1 + r_2 + \cdots + r_m = r.$$

Show that

$$(x_1^{r_1} x_2^{r_2} \cdots x_m^{r_m})^{1/r} \leq \frac{r_1 x_1 + r_2 x_2 + \cdots + r_m x_m}{r}$$

and give necessary and sufficient conditions for equality. (*Hint:* Maximize $x_1^{r_1} x_2^{r_2} \cdots x_m^{r_m}$ subject to $\sum_{j=1}^{m} r_j x_j = \sigma$, $x_1 > 0$, $x_2 > 0, \ldots, x_n > 0$.) State carefully why the constrained critical point so obtained is a constrained maximum point.

28. Let $A = [a_{ij}]$ be an $m \times n$ matrix, suppose $p_1, p_2, \ldots, p_m$ are positive and

$$\sum_{j=1}^{m} \frac{1}{p_j} = 1,$$

and define

$$\sigma_i = \sum_{j=1}^{m} |a_{ij}|^{p_i}, \qquad 1 \leq i \leq m.$$

Use Exercise 27 to show that

$$\left| \sum_{j=1}^{n} a_{1j} a_{2j} \cdots a_{mj} \right| \leq \sigma_1^{1/p_1} \sigma_2^{1/p_2} \cdots \sigma_m^{1/p_m}.$$

(With $m = 2$ this is *Hölder's inequality*, which reduces to Schwarz's inequality if $p_1 = p_2 = 2$.)

integrals of functions of several variables

7.1 DEFINITION AND EXISTENCE OF THE MULTIPLE INTEGRAL

We now consider the Riemann integral of a real-valued function f defined on a subset of $\mathscr{R}^n$, where $n \geq 2$. Much of this development is analogous to that given earlier for $n = 1$, but there is an important difference: For $n = 1$ we considered integrals over closed intervals only, where for $n > 1$ we must consider more complicated regions of integration. To defer complications due to geometry, we first consider integrals over rectangles in $\mathscr{R}^n$, which we now define.

Integrals over rectangles

If $S_1, S_2, \ldots, S_n$ are subsets of $\mathscr{R}$, then their *Cartesian product*,

$$S_1 \times S_2 \times \cdots \times S_n,$$

is the set of all points $(x_1, x_2, \ldots, x_n)$ in $\mathscr{R}^n$ such that $x_1 \in S_1, x_2 \in S_2, \ldots, x_n \in S_n$. For example, the Cartesian product of two closed intervals,

$$[a_1, b_1] \times [a_2, b_2] = \{(x, y) \mid a_1 \leq x \leq b_1, a_2 \leq y \leq b_2\},$$

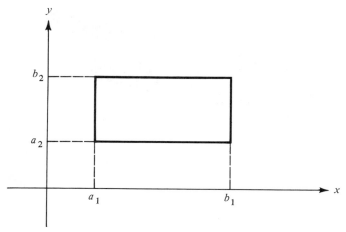

FIGURE 1.1

is a rectangle in $\mathscr{R}^2$, with sides parallel to the x and y axes (Figure 1.1); the Cartesian product of three closed intervals,

$$[a_1, b_1] \times [a_2, b_2] \times [a_3, b_3]$$
$$= \{(x, y, z) \mid a_1 \leq x \leq b_1, a_2 \leq y \leq b_2, a_3 \leq z \leq b_3\}$$

is a rectangular parallelepiped in $\mathscr{R}^3$ with faces parallel to the coordinate axes (Figure 1.2).

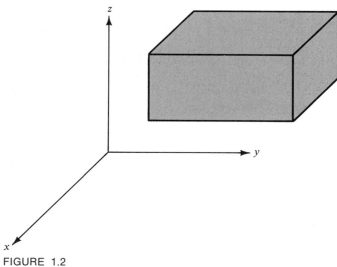

FIGURE 1.2

Definition 1.1 *A coordinate rectangle in $\mathscr{R}^n$ is the Cartesian product of n closed intervals; that is,*

$$R = I_1 \times I_2 \times \cdots \times I_n,$$

where

$$I_r = [a_r, b_r], \qquad r = 1, 2, \ldots, n.$$

The content of R is defined to be

$$V(R) = (b_1 - a_1)(b_2 - a_2) \cdots (b_n - a_n).$$

The numbers $b_1 - a_1, b_2 - a_2, \ldots, b_n - a_n$ are called the edge lengths of R. If they are equal, then R is called a coordinate cube. If $a_r = b_r$ for some r, we say that R is degenerate; otherwise, it is nondegenerate.

If $n = 1, 2$, or 3, then $V(R)$ is, respectively, the length of an interval, the area of a rectangle, or the volume of a rectangular parallelepiped. Henceforth, "rectangle" or "cube" will always mean "coordinate rectangle" or "coordinate cube" unless it is stated otherwise.

Now suppose

$$P_r: a_r = a_{r0} < a_{r1} < \cdots < a_{rm_r} = b_r \qquad (r = 1, 2, \ldots, n)$$

is a partition of the subinterval $I_r = [a_r, b_r]$; then the set of all rectangles in $\mathscr{R}^n$ that can be written as

$$J_1 \times J_2 \times \cdots \times J_n,$$

where J_r is a subinterval of I_r from the partition P_r, is called a *partition* of $R = I_1 \times I_2 \times \cdots \times I_n$. We denote this partition of R by

$$\mathbf{P} = P_1 \times P_2 \times \cdots \times P_n \tag{1}$$

and define its norm to be the maximum of the norms of $P_1, P_2, \ldots, P_n$ (see Section 3.1); thus,

$$\|\mathbf{P}\| = \max\{\|P_1\|, \|P_2\|, \ldots, \|P_n\|\}.$$

Put another way, $\|\mathbf{P}\|$ is the largest of the edge lengths of all the subrectangles in $\mathbf{P}$.

Geometrically, a rectangle in $\mathscr{R}^2$ is partitioned by drawing horizontal and vertical lines through it (Figure 1.3); in $\mathscr{R}^3$, by drawing planes through it parallel to the coordinate axes. Partitioning divides a rectangle R into finitely many subrectangles which we can number in arbitrary order as $R_1, R_2, \ldots, R_k$; sometimes it is convenient to write

$$\mathbf{P} = \{R_1, R_2, \ldots, R_k\}$$

rather than (1).

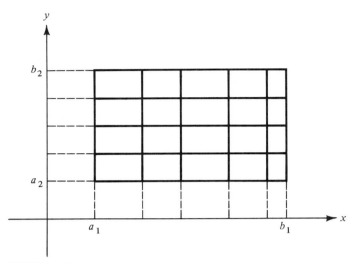

FIGURE 1.3

If $\mathbf{P} = P_1 \times P_2 \times \cdots \times P_n$ and $\mathbf{P}' = P'_1 \times P'_2 \times \cdots \times P'_n$ are partitions of the same rectangle, then $\mathbf{P}'$ is a *refinement* of $\mathbf{P}$ if P'_i is a refinement of P_i for $i = 1, 2, \ldots, n$ (see Section 3.1).

Suppose f is a real-valued function defined on a rectangle R in $\mathscr{R}^n$, $\mathbf{P} = \{R_1, R_2, \ldots, R_k\}$ is a partition of R, and $\mathbf{X}_j$ is an arbitrary point in R_j. Then

$$\sigma = \sum_{j=1}^{k} f(\mathbf{X}_j)V(R_j)$$

is called a *Riemann sum of f over* $\mathbf{P}$. Since $\mathbf{X}_j$ can be chosen arbitrarily in R_j, there are infinitely many Riemann sums for a given function f over any partition $\mathbf{P}$ of R.

The following definition is similar to Definition 1.1, Section 3.1.

Definition 1.2 *A function f is Riemann integrable over a rectangle R in $\mathscr{R}^n$ if there is a number L with the following property: For every $\varepsilon > 0$ there is a $\delta > 0$ such that every Riemann sum of f over any partition $\mathbf{P} = \{R_1, R_2, \ldots, R_k\}$ of R satisfies the inequality*

$$\left| \sum_{j=1}^{k} f(\mathbf{X}_j)V(R_j) - L \right| < \varepsilon$$

provided only that $\mathbf{X}_j \in R_j$ ($j = 1, 2, \ldots, k$) and $\|\mathbf{P}\| < \delta$. The number L is called *the Riemann integral of f over R, and is denoted by*

$$\int_R f(\mathbf{X}) \, d\mathbf{X}.$$

If R is degenerate, we define $\int_R f(\mathbf{X}) \, d\mathbf{X} = 0$ for any f defined on R.

The integral is also written as

$$\int_R f(x, y) \, d(x, y) \qquad (n = 2),$$

$$\int_R f(x, y, z) \, d(x, y, z) \qquad (n = 3),$$

or

$$\int_R f(x_1, x_2, \ldots, x_n) \, d(x_1, x_2, \ldots, x_n) \qquad \text{(arbitrary } n\text{)}.$$

Here $d\mathbf{X}$ does not stand for the differential of $\mathbf{X}$, as defined in Section 6.2; it merely identifies $x_1, x_2, \ldots, x_n$—the components of $\mathbf{X}$—as the variables of integration. To avoid this minor inconsistency, some authors write simply $\int_R f$ rather than $\int_R f(\mathbf{X}) \, d\mathbf{X}$.

As in the case where $n = 1$, we will say simply "integrable" or "integral" when we mean "Riemann integrable" or "Riemann integral." If $n \geq 2$, we call the integral of Definition 1.2 a *multiple integral;* for $n = 2$ and $n = 3$ we also speak of *double* and *triple integrals*, respectively. When we wish to distinguish between multiple integrals and the integral we studied in Chapter 3 ($n = 1$), we will call the latter an *ordinary* integral.

Example 1.1 Let

$$R = [a, b] \times [c, d]$$

and

$$f(x, y) = x + y.$$

Let P_1 and P_2 be partitions of $[a, b]$ and $[c, d]$; thus,

$$P_1 : a = x_0 < x_1 < \cdots < x_r = b$$

and

$$P_2 : c = y_0 < y_1 < \cdots < y_s = d.$$

A typical Riemann sum of f over $\mathbf{P} = P_1 \times P_2$ is of the form

$$\sigma = \sum_{i=1}^{r} \sum_{j=1}^{s} (\xi_{ij} + \eta_{ij})(x_i - x_{i-1})(y_j - y_{j-1}), \tag{2}$$

where

$$x_{i-1} \le \xi_{ij} \le x_i \quad \text{and} \quad y_{j-1} \le \eta_{ij} \le y_j. \tag{3}$$

The midpoints of $[x_{i-1}, x_i]$ and $[y_{j-1}, y_j]$ are

$$\bar{x}_i = \frac{x_i + x_{i-1}}{2} \quad \text{and} \quad \bar{y}_j = \frac{y_j + y_{j-1}}{2}, \tag{4}$$

and (3) implies that

$$|\xi_{ij} - \bar{x}_i| \le \frac{x_i - x_{i-1}}{2} \le \frac{\|P_1\|}{2} \le \frac{\|P\|}{2} \tag{5}$$

and

$$|\eta_{ij} - \bar{y}_j| \le \frac{y_j - y_{j-1}}{2} \le \frac{\|P_2\|}{2} \le \frac{\|P\|}{2}. \tag{6}$$

Now we rewrite (2) as

$$\sigma = \sum_{i=1}^{r} \sum_{j=1}^{s} (\bar{x}_i + \bar{y}_j)(x_i - x_{i-1})(y_j - y_{j-1})$$

$$+ \sum_{i=1}^{r} \sum_{j=1}^{s} [(\xi_{ij} - \bar{x}_i) + (\eta_{ij} - \bar{y}_j)](x_i - x_{i-1})(y_j - y_{j-1}). \tag{7}$$

To find $\int_R f(x, y)\, d(x, y)$ from this, we recall that

$$\sum_{i=1}^{r} (x_i - x_{i-1}) = b - a, \qquad \sum_{j=1}^{s} (y_j - y_{j-1}) = d - c \tag{8}$$

(see Example 1.1, Section 3.1), and

$$\sum_{i=1}^{r} (x_i^2 - x_{i-1}^2) = b^2 - a^2, \qquad \sum_{j=1}^{s} (y_j^2 - y_{j-1}^2) = d^2 - c^2 \tag{9}$$

(see Example 1.2, Section 3.1).

Because of (5) and (6) the absolute value of the second sum in (7) does not exceed

$$\|P\| \sum_{i=1}^{r} \sum_{j=1}^{s} (x_i - x_{i-1})(y_j - y_{j-1}) = \|P\| \left[\sum_{i=1}^{r} (x_i - x_{i-1}) \right]\left[\sum_{j=1}^{s} (y_j - y_{j-1}) \right]$$

$$= \|P\|(b - a)(d - c)$$

[see (8)], and so (7) implies that

$$\left| \sigma - \sum_{i=1}^{r} \sum_{j=1}^{s} (\bar{x}_i + \bar{y}_j)(x_i - x_{i-1})(y_j - y_{j-1}) \right| \le \|P\|(b - a)(d - c). \tag{10}$$

Now

$$\sum_{i=1}^{r} \sum_{j=1}^{s} \bar{x}_i(x_i - x_{i-1})(y_j - y_{j-1})$$

$$= \left[\sum_{i=1}^{r} \bar{x}_i(x_i - x_{i-1}) \right] \left[\sum_{j=1}^{s} (y_j - y_{j-1}) \right]$$

$$= (d - c) \sum_{i=1}^{r} \bar{x}_i(x_i - x_{i-1}) \qquad [\text{from (8)}] \quad (11)$$

$$= \frac{d - c}{2} \sum_{i=1}^{r} (x_i^2 - x_{i-1}^2) \qquad [\text{from (4)}]$$

$$= \frac{d - c}{2}(b^2 - a^2) \qquad\qquad [\text{from (9)}].$$

Similarly,

$$\sum_{i=1}^{r} \sum_{j=1}^{s} \bar{y}_j(x_i - x_{i-1})(y_j - y_{j-1}) = \frac{b - a}{2}(d^2 - c^2). \qquad (12)$$

Therefore, (10) can be written as

$$\left| \sigma - \frac{d - c}{2}(b^2 - a^2) - \frac{b - a}{2}(d^2 - c^2) \right| \leq \|\mathbf{P}\|(b - a)(d - c).$$

Since the right side can be made as small as we wish by choosing $\|\mathbf{P}\|$ sufficiently small, it follows that

$$\int_R (x + y)\, d(x, y) = \tfrac{1}{2}[(d - c)(b^2 - a^2) + (b - a)(d^2 - c^2)].$$

Upper and lower integrals

The development of the integral over a rectangle in $\mathcal{R}^n$ $(n \geq 2)$ is similar to that given in Chapter 3 for $n = 1$. We therefore leave to the exercises proofs that are merely repetitions of the proofs given earlier for $n = 1$.

Theorem 1.1   *If f is unbounded on the nondegenerate rectangle R in $\mathcal{R}^n$, then f is not integrable over R.*

Proof.   *Exercise 4.*

Because of Theorem 1.1 we need consider only bounded functions in connection with Definition 1.2. As in the case where $n = 1$, it is now convenient to define the upper and lower integrals of a bounded function over a rectangle.

Definition 1.3   *Suppose f is bounded on a rectangle R in $\mathscr{R}^n$, let $\mathbf{P} = \{R_1, R_2, \ldots, R_k\}$ be a partition of R, and*

$$M_j = \text{l.u.b. } f(\mathbf{X}), \qquad m_j = \text{g.l.b. } f(\mathbf{X}).$$
$$\quad\; \mathbf{X} \in R_j \qquad\qquad\qquad \mathbf{X} \in R_j$$

The upper sum of f over $\mathbf{P}$ is defined to be

$$S(\mathbf{P}) = \sum_{j=1}^{k} M_j V(R)$$

and the upper integral of f over R, denoted by $\overline{\int_R} f(\mathbf{X}) \, d\mathbf{X}$, is the greatest lower bound of all such upper sums. The lower sum of f over $\mathbf{P}$ is defined to be

$$s(\mathbf{P}) = \sum_{j=1}^{k} m_j V(R),$$

and the lower integral of f over R, denoted by $\underline{\int_R} f(\mathbf{X}) \, d\mathbf{X}$, is the least upper bound of all such lower sums.

If

$$m \le f(\mathbf{X}) \le M \qquad \text{for } \mathbf{X} \text{ in } R,$$

then

$$mV(R) \le s(\mathbf{P}) \le S(\mathbf{P}) \le MV(R);$$

therefore, $\overline{\int_R} f(\mathbf{X}) \, d\mathbf{X}$ and $\underline{\int_R} f(\mathbf{X}) \, d\mathbf{X}$ exist, are unique, and satisfy the inequalities

$$mV(R) \le \overline{\int_R} f(\mathbf{X}) \, d\mathbf{X} \le MV(R)$$

and

$$mV(R) \le \underline{\int_R} f(\mathbf{X}) \, d\mathbf{X} \le MV(R).$$

The upper and lower integrals are also written as

$$\overline{\int_R} f(x, y) \, d(x, y) \quad \text{and} \quad \underline{\int_R} f(x, y) \, d(x, y) \qquad (n = 2),$$

$$\overline{\int_R} f(x, y, z) \, d(x, y, z) \quad \text{and} \quad \underline{\int_R} f(x, y, z) \, d(x, y, z) \qquad (n = 3),$$

or

$$\overline{\int_R} f(x_1, x_2, \ldots, x_n) \, d(x_1, x_2, \ldots, x_n)$$

and

$$\underline{\int_R} f(x_1, x_2, \ldots, x_n) \, d(x_1, x_2, \ldots, x_n) \qquad (n \text{ arbitrary}).$$

Example 1.2 As in Example 1.1 let $R = [a, b] \times [c, d]$ and

$$f(x, y) = x + y,$$

and let P_1 and P_2 be partitions of $[a, b]$ and $[c, d]$; thus,

$$P_1 : a = x_0 < x_1 < \cdots < x_r = b$$

and

$$P_2 : c = y_0 < y_1 < \cdots < y_s = d.$$

The maximum and minimum value of $f(x, y)$ on the rectangle $[x_{i-1}, x_i] \times [y_{j-1}, y_j]$ are $x_i + y_j$ and $x_{i-1} + y_{j-1}$, respectively; therefore,

$$S(\mathbf{P}) = \sum_{i=1}^{r} \sum_{j=1}^{s} (x_i + y_j)(x_i - x_{i-1})(y_j - y_{j-1}) \tag{13}$$

and

$$s(\mathbf{P}) = \sum_{i=1}^{r} \sum_{j=1}^{s} (x_{i-1} + y_{j-1})(x_i - x_{i-1})(y_j - y_{j-1}). \tag{14}$$

By substituting

$$x_i = \frac{(x_i + x_{i-1})}{2} + \frac{(x_i - x_{i-1})}{2} \quad \text{and} \quad y_j = \frac{(y_j + y_{j-1})}{2} + \frac{(y_j - y_{j-1})}{2}$$

into (13), we find that

$$
\begin{aligned}
S(\mathbf{P}) = {} & \frac{1}{2} \sum_{i=1}^{r} \sum_{j=1}^{s} (x_i^2 - x_{i-1}^2)(y_j - y_{j-1}) \\
& + \frac{1}{2} \sum_{i=1}^{r} \sum_{j=1}^{s} (x_i - x_{i-1})^2 (y_j - y_{j-1}) \\
& + \frac{1}{2} \sum_{i=1}^{r} \sum_{j=1}^{s} (x_i - x_{i-1})(y_j^2 - y_{j-1}^2) \\
& + \frac{1}{2} \sum_{i=1}^{r} \sum_{j=1}^{s} (x_i - x_{i-1})(y_j - y_{j-1})^2 \\
= {} & \tfrac{1}{2}(\Sigma_1 + \Sigma_2 + \Sigma_3 + \Sigma_4).
\end{aligned}
\tag{15}
$$

Here $\tfrac{1}{2}\Sigma_1$ and $\tfrac{1}{2}\Sigma_2$ are as in (11) and (12),

$$0 < \Sigma_2 \le \|P_1\| \sum_{i=1}^{r} \sum_{j=1}^{s} (x_i - x_{i-1})(y_j - y_{j-1}) = \|P_1\|(b - a)(d - c)$$

and

$$0 < \Sigma_4 \le \|P_2\| \sum_{i=1}^{r} \sum_{j=1}^{s} (x_i - x_{i-1})(y_j - y_{j-1}) = \|P_2\|(b - a)(d - c).$$

From (11), (12), (15), and the last two inequalities,

$$I < S(\mathbf{P}) < I + \|\mathbf{P}\|(b - a)(d - c),$$

where

$$I = \frac{(b - a)(d^2 - c^2) + (b^2 - a^2)(d - c)}{2}.$$

From this we see that

$$\overline{\int_R}(x + y)\,d(x, y) = I.$$

After substituting

$$x_{i-1} = \frac{(x_i + x_{i-1})}{2} - \frac{(x_i - x_{i-1})}{2} \quad \text{and} \quad y_j = \frac{(y_j + y_{j-1})}{2} - \frac{(y_j - y_{j-1})}{2}$$

into (14), a similar argument shows that

$$I - \|\mathbf{P}\|(b - a)(d - c) < s(\mathbf{P}) < I,$$

so

$$\underline{\int_R}(x + y)\,d(x, y) = I.$$

We now prove an analog of Lemma 2.1, Section 3.2.

Lemma 1.1 *Suppose $|f(\mathbf{X})| \leq K$ if $\mathbf{X}$ is in the rectangle*

$$R = [a_1, b_1] \times [a_2, b_2] \times \cdots \times [a_n, b_n].$$

Let $\mathbf{P} = P_1 \times P_2 \times \cdots \times P_n$ and $\mathbf{P}' = P_1' \times P_2' \times \cdots \times P_n'$ be partitions of R, where P_j' is obtained by adding r_j partition points to P_j ($j = 1, 2, \ldots, n$). Then

$$S(\mathbf{P}) - 2KV(R)\left(\sum_{j=1}^{n} \frac{r_j}{b_j - a_j}\right)\|\mathbf{P}\| \leq S(\mathbf{P}') \leq S(\mathbf{P}) \qquad (16)$$

and

$$s(\mathbf{P}) \leq s(\mathbf{P}') \leq s(\mathbf{P}) + 2KV(R)\left(\sum_{j=1}^{n} \frac{r_j}{b_j - a_j}\right)\|\mathbf{P}\|. \qquad (17)$$

Proof. It suffices to prove (16), since (17) follows on applying (16) to $-f$ (*Exercise 6*). We first assume that P_1' is obtained by adding one point to P_1, and $P_j' = P_j$ for $j = 2, \ldots, n$. (We choose to refine P_1—rather than some other P_j—for notational convenience only.) If P_i is defined by

$$P_i: a_i = a_{i0} < a_{i1} < \cdots < a_{im_i} = b_i, \qquad 1 \leq i \leq n,$$

then a typical subrectangle of $\mathbf{P}$ is of the form

$$R_{j_1 j_2 \cdots j_n} = [a_{1,j_1}, a_{1,j_1+1}] \times [a_{2,j_2}, a_{2,j_2+1}] \times \cdots \times [a_{n,j_n}, a_{n,j_n+1}]. \quad (18)$$

Let c be the additional point introduced into P_1 to obtain P'_1, and suppose

$$a_{1k} < c < a_{1,k+1}.$$

If $j_1 \neq k$, then $R_{j_1 j_2 \cdots j_n}$, as defined in (18), is common to $\mathbf{P}$ and $\mathbf{P}'$, and therefore the terms associated with it in $S(\mathbf{P}')$ and $S(\mathbf{P})$ cancel in the difference $S(\mathbf{P}) - S(\mathbf{P}')$. To analyze the terms that do not cancel, we introduce some notation.

Let

$$R^{(1)}_{k j_2 \cdots j_n} = [a_{1k}, c] \times [a_{2,j_2}, a_{2,j_2+1}] \times \cdots \times [a_{n,j_n}, a_{n,j_n+1}],$$

$$R^{(2)}_{k j_2 \cdots j_n} = [c, a_{1,k+1}] \times [a_{2,j_2}, a_{2,j_2+1}] \times \cdots \times [a_{n,j_n}, a_{n,j_n+1}],$$

$$M_{k j_2 \cdots j_n} = \text{l.u.b.} \{ f(\mathbf{X}) \mid \mathbf{X} \in R_{k j_2 \cdots j_n} \} \quad (19)$$

and

$$M^{(i)}_{k j_2 \cdots j_n} = \text{l.u.b.} \{ f(\mathbf{X}) \mid \mathbf{X} \in R^{(i)}_{k j_2 \cdots j_n} \}, \qquad i = 1, 2; \quad (20)$$

then $S(\mathbf{P}) - S(\mathbf{P}')$ is the sum of terms of the form

$$[M_{k j_2 \cdots j_n}(a_{1,k+1} - a_{1k}) - M^{(1)}_{k j_2 \cdots j_n}(c - a_{1k}) - M^{(2)}_{k j_2 \cdots j_n}(a_{1,k+1} - c)]$$
$$\times (a_{2,j_2+1} - a_{2,j_2}) \cdots (a_{n,j_n+1} - a_{n,j_n}). \quad (21)$$

The terms within the brackets can be rewritten as

$$(M_{k j_2 \cdots j_n} - M^{(1)}_{k j_2 \cdots j_n})(c - a_{1k}) + (M_{k j_2 \cdots j_n} - M^{(2)}_{k j_2 \cdots j_n})(a_{1,k+1} - c), \quad (22)$$

which is nonnegative, because of (19) and (20). Therefore,

$$S(\mathbf{P}') \leq S(\mathbf{P}). \quad (23)$$

Moreover, the quantity in (22) is not greater than $2K(a_{1,k+1} - a_{1k})$, so (21) implies that the general surviving term in $S(\mathbf{P}) - S(\mathbf{P}')$ is not greater than

$$2K\|\mathbf{P}\|(a_{2,j_2+1} - a_{2,j_2}) \cdots (a_{n,j_n+1} - a_{n,j_n}).$$

The sum of these terms, as $j_2, \ldots, j_n$ assume all possible values ($1 \leq j_k \leq m_k - 1$, $k = 2, 3, \ldots, n$) is

$$2K\|\mathbf{P}\|(b_2 - a_2) \cdots (b_n - a_n) = \frac{2K\|\mathbf{P}\|V(R)}{b_1 - a_1}.$$

This implies that

$$S(\mathbf{P}) \leq S(\mathbf{P}') + \frac{2K\|\mathbf{P}\|V(R)}{b_1 - a_1}.$$

This and (23) imply (16) for $r_1 = 1$ and $r_2 = \cdots = r_n = 0$. Repeating this argument finitely many times (as in the proof of Lemma 2.1, Section 3.2) yields (16) for the general case.

Lemma 1.1 implies the following theorems and lemma, whose proofs are analogous to the proofs of their counterparts in Section 3.2.

Theorem 1.2   *If f is bounded on a rectangle R, then*

$$\underline{\int_R} f(\mathbf{X}) \, d\mathbf{X} \le \overline{\int_R} f(\mathbf{X}) \, d\mathbf{X}.$$

Proof.   *Exercise 7.*

Theorem 1.3   *If f is integrable on a rectangle R, then*

$$\underline{\int_R} f(\mathbf{X}) \, d\mathbf{X} = \int_R f(\mathbf{X}) \, d\mathbf{X} = \overline{\int_R} f(\mathbf{X}) \, d\mathbf{X}.$$

Proof.   *Exercise 8.*

Lemma 1.2   *If f is bounded on a rectangle R, then there is for each $\varepsilon > 0$ a $\delta > 0$ such that*

$$S(\mathbf{P}) < \overline{\int_R} f(\mathbf{X}) \, d\mathbf{X} + \varepsilon$$

and

$$\underline{\int_R} f(\mathbf{X}) \, d\mathbf{X} - \varepsilon < s(\mathbf{P})$$

if $\|\mathbf{P}\| < \delta$.

Proof.   *Exercise 9.*

Theorem 1.4   *If f is bounded on a rectangle R and*

$$\underline{\int_R} f(\mathbf{X}) \, d\mathbf{X} = \overline{\int_R} f(\mathbf{X}) \, d\mathbf{X} = L,$$

then f is integrable on R and

$$\int_R f(\mathbf{X}) \, d\mathbf{X} = L.$$

Proof. *Exercise 10.*

Theorems 1.3 and 1.4 imply that a bounded function f is integrable over a rectangle R if and only if

$$\underline{\int_R} f(\mathbf{X})\, d\mathbf{X} = \overline{\int_R} f(\mathbf{X})\, d\mathbf{X}.$$

Theorem 1.5 *If f is bounded on a rectangle R, then $\int_R f(\mathbf{X})\, d\mathbf{X}$ exists if and only if for every $\varepsilon > 0$, there is a partition $\mathbf{P}$ of R such that*

$$S(\mathbf{P}) - s(\mathbf{P}) < \varepsilon.$$

Proof. *Exercise 11.*

Theorem 1.5 provides a useful criterion for integrability. The next theorem is an immediate consequence.

Theorem 1.6 *If f is continuous on a rectangle R, then f is integrable on R.*

Proof. *Exercise 12.*

Sets with zero content

The next definition will enable us to establish integrability for a large class of functions.

Definition 1.4 *A subset S of $\mathscr{R}^n$ has zero content if for each $\varepsilon > 0$ there is a set of rectangles $T_1, T_2, \ldots, T_m$ such that*

$$E \subset \bigcup_{j=1}^{m} T_j \tag{24}$$

and

$$\sum_{j=1}^{m} V(T_j) < \varepsilon. \tag{25}$$

Example 1.3 Since the empty set is contained in every rectangle, it has zero content. If E consists of finitely many points $\mathbf{X}_1, \mathbf{X}_2, \ldots, \mathbf{X}_m$, then $\mathbf{X}_j$ can be enclosed in a rectangle T_j such that

$$V(T_j) < \frac{\varepsilon}{m}, \qquad j = 1, 2, \ldots, m;$$

then (24) and (25) hold, and so E has zero content.

Example 1.4   Any bounded set E with only finitely many limit points has zero content. To see this we first observe that if E has no limit points, then it must be finite, by the Bolzano–Weierstrass theorem, and therefore have zero content, by Example 1.3. If E has limit points $X_1, X_2, \ldots, X_m$, let $R_1, R_2, \ldots, R_m$ be rectangles such that $X_i \in R_i^0$ and

$$V(R_i) < \frac{\varepsilon}{2m}, \qquad i = 1, 2, \ldots, m. \tag{26}$$

The set of points of E that are not in $\bigcup_{j=1}^{m} R_j$ has no limit points (why?) and, being bounded, must be finite (again by the Bolzano–Weierstrass theorem); hence, it can be covered by rectangles $R_1', R_2', \ldots, R_p'$ with

$$V(R_j') < \frac{\varepsilon}{2p}, \qquad j = 1, 2, \ldots, p. \tag{27}$$

Now,

$$E \subset \bigcup_{i=1}^{m} R_i \bigcup_{j=1}^{p} R_j'$$

and, from (26) and (27),

$$\sum_{i=1}^{m} V(R_i) + \sum_{j=1}^{p} V(R_j') < \varepsilon.$$

Example 1.5   If f is continuous on $[a, b]$, then the curve

$$y = f(x), \qquad a \le x \le b \tag{28}$$

(that is, the set $\{(x, y) \mid y = f(x), a \le x \le b\}$), has zero content in $\mathscr{R}^2$. To see this, suppose $\varepsilon > 0$ and choose $\delta > 0$ such that

$$|f(x) - f(x')| < \varepsilon \qquad \text{if } x, x' \in [a, b] \quad \text{and} \quad |x - x'| < \delta; \tag{29}$$

this is possible because f is uniformly continuous on $[a, b]$ (Theorem 2.7, Section 2.2). Now partition $[a, b]$ with

$$P: a = x_0 < x_1 < \cdots < x_r = b,$$

suppose $\|P\| < \delta$, and suppose

$$x_{i-1} \le \xi_i \le x_i, \qquad i = 1, 2, \ldots, n.$$

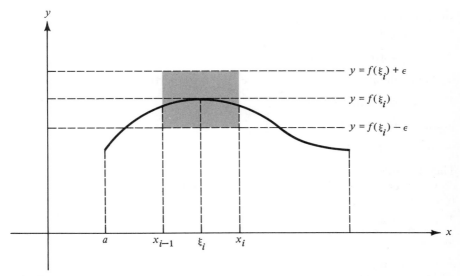

FIGURE 1.4

Then, from (29),

$$|f(x) - f(\xi_i)| < \varepsilon \qquad \text{if } x_{i-1} \le x \le x_i.$$

This means that every point on the curve (28) above the interval $[x_{i-1}, x_i]$ is in a rectangle with area $2\varepsilon(x_i - x_{i-1})$ (Figure 1.4). Since the total area of these rectangles is $2\varepsilon(b - a)$, this proves that the curve has zero content.

The next lemma follows immediately from Definition 1.4 (*Exercise 20*).

Lemma 1.3 *The union of finitely many sets with zero content has zero content.*

Theorem 1.7 *Suppose f is bounded on a rectangle*

$$R = [a_1, b_1] \times [a_2, b_2] \times \cdots \times [a_n, b_n] \qquad (30)$$

and continuous except on a subset E of R with zero content. Then f is integrable on R.

Proof. Suppose $\varepsilon > 0$ and $T_1, T_2, \ldots, T_m$ are rectangles satisfying (24) and (25). We may assume that they are contained in R, since if not their intersections with R would be contained in R, and still satisfy (24) and (25). We may also assume that

$$R^0 \cap E \subset \bigcup_{j=1}^{m} T_j^0, \qquad (31)$$

since if this were not so we could simply enlarge $T_1, T_2, \ldots, T_m$ slightly, while maintaining (25). Now suppose

$$T_j = [a_{1j}, b_{1j}] \times [a_{2j}, b_{2j}] \times \cdots \times [a_{n_j}, b_{n_j}], \qquad j = 1, 2, \ldots, m,$$

let P_{i0} be the partition of $[a_i, b_i]$ [see (30)] with partition points

$$a_i, b_i, a_{i1}, b_{i1}, a_{i2}, b_{i2}, \ldots, a_{im}, b_{im}$$

(these are not in order), and let

$$\mathbf{P}_0 = P_{10} \times P_{20} \times \cdots \times P_{n0}.$$

Then $\mathbf{P}_0$ consists of $T_1, T_2, \ldots, T_m$ and other rectangles $T'_1, T'_2, \ldots, T'_k$ which do not intersect E. [We need (31) to be sure that $T'_i \cap E = \varnothing, i = 1, 2, \ldots, k.$] If we let

$$B = \bigcup_{j=1}^{m} T_j \quad \text{and} \quad C = \bigcup_{i=1}^{k} T'_i,$$

then f is continuous on the compact set C and $R = B \cup C$.

If $\mathbf{P} = \{R_1, R_2, \ldots, R_k\}$ is a refinement of $\mathbf{P}_0$, then every subrectangle of $\mathbf{P}$ is contained entirely in B or entirely in C; therefore, we can write

$$S(\mathbf{P}) - s(\mathbf{P}) = \sum_1 (M_j - m_j)V(R_j) + \sum_2 (M_j - m_j)V(R_j), \tag{32}$$

where $\sum_1$ and $\sum_2$ are summations over values of j for which $R_j \subset B$ and $R_j \subset C$, respectively. Now suppose

$$|f(\mathbf{X})| \le M \qquad \text{for } \mathbf{X} \text{ in } R;$$

then

$$\sum_1 (M_j - m_j)V(R_j) \le 2M \sum_1 V(R_j) = 2M \sum_{j=1}^{m} V(T_j) < 2M\varepsilon. \tag{33}$$

Since f is uniformly continuous on C (Theorem 2.10, Section 5.2), there is a $\delta > 0$ such that $M_j - m_j < \varepsilon$ if $\|\mathbf{P}\| < \delta$ and $R_j \subset C$; hence,

$$\sum_2 (M_j - m_j)V(R_j) < \varepsilon \sum_2 V(R_j) \le \varepsilon V(R).$$

This, (32), and (33) imply that

$$S(\mathbf{P}) - s(\mathbf{P}) < [2M + V(R)]\varepsilon$$

if $\|\mathbf{P}\| < \delta$ and $\mathbf{P}$ is a refinement of $\mathbf{P}_0$. Theorem 1.5 now implies that $\int_R f(\mathbf{X}) \, d\mathbf{X}$ exists, and the proof is complete.

Example 1.6   If

$$f(x, y) = \begin{cases} x + y, & 0 \le x < y \le 1, \\ 5, & 0 \le y \le x \le 1, \end{cases}$$

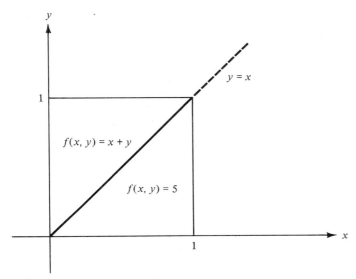

FIGURE 1.5

then f is continuous on $R = [0, 1] \times [0, 1]$ except on the line segment

$$y = x, \qquad 0 \le x \le 1$$

(Figure 1.5). Since the line segment has zero content (Example 1.5), f is integrable over R.

Integrals over more general subsets of $\mathscr{R}^n$

We can now define the integral of a bounded function over more general subsets of $\mathscr{R}^n$.

Definition 1.5 *Suppose f is bounded on a bounded subset S of $\mathscr{R}^n$, and let f_S be defined on all of $\mathscr{R}^n$ by*

$$f_S(\mathbf{X}) = \begin{cases} f(\mathbf{X}), & \mathbf{X} \in S, \\ 0, & \mathbf{X} \notin S. \end{cases} \tag{34}$$

Then the integral of f over S is defined by

$$\int_S f(\mathbf{X}) \, d\mathbf{X} = \int_R f_S(\mathbf{X}) \, d\mathbf{X},$$

where R is any rectangle containing S, provided $\int_R f_S(\mathbf{X}) \, d\mathbf{X}$ exists.

To see that this definition makes sense, we must show that if R_1 and R_2 are two rectangles containing S and $\int_{R_1} f_s(\mathbf{X}) \, d\mathbf{X}$ exists, then so does $\int_{R_2} f_s(\mathbf{X}) \, d\mathbf{X}$, and the two integrals are equal. The proof of this is sketched in *Exercise 21*.

If the integral $\int_S d\mathbf{X}$ (whose integrand is $f \equiv 1$) exists, we call it the *content* (also, *area* if $n = 2$ or *volume* if $n = 3$) of S, and denote it by $V(S)$. We will discuss this further in the examples in the next section, and look more critically at the concept in Section 7.3.

Theorem 1.8   *Suppose f is bounded on a bounded set S, and continuous except on a subset E of S with zero content. Suppose also that ∂S has zero content. Then f is integrable over S.*

Proof.   Let f_s be as in (34). Since a discontinuity of f_s is either a discontinuity of f or a point of ∂S, the set of discontinuities of f_s is the union of two sets of zero content, and therefore of zero content (Lemma 1.3); therefore, f_s is integrable over any rectangle containing S (from Theorem 1.7), and consequently on S.

Differentiable surfaces

Differentiable surfaces, defined as follows, form an important class of sets of zero content in $\mathcal{R}^n$.

Definition 1.6   *A differentiable surface S in $\mathcal{R}^n$ ($n > 1$) is the image of a compact subset D of $\mathcal{R}^m$, where $m < n$, under a continuously differentiable transformation $\mathbf{G} \colon \mathcal{R}^m \to \mathcal{R}^n$. If $m = 1$, S is also called a differentiable curve.*

Example 1.7   The circle

$$\{(x, y) \,|\, x^2 + y^2 = 9\}$$

is a differentiable curve in $\mathcal{R}^2$, since it is the image of $D = [0, 2\pi]$ under the continuously differentiable transformation $\mathbf{G} \colon \mathcal{R} \to \mathcal{R}^2$ defined by

$$\mathbf{X} = \mathbf{G}(\theta) = \begin{bmatrix} 3 \cos \theta \\ 3 \sin \theta \end{bmatrix}.$$

Example 1.8   The sphere

$$\{(x, y, z) \,|\, x^2 + y^2 + z^2 = 4\}$$

is a differentiable surface in $\mathscr{R}^3$, since it is the image of

$$D = \{(\theta, \phi) \mid 0 \le \theta \le 2\pi, \ -\pi/2 \le \phi \le \pi/2\}$$

under the continuously differentiable transformation $\mathbf{G} \colon \mathscr{R}^2 \to \mathscr{R}^3$ defined by

$$\mathbf{X} = \mathbf{G}(\theta, \phi) = \begin{bmatrix} 2 \cos \theta \cos \phi \\ 2 \sin \theta \cos \phi \\ 2 \sin \phi \end{bmatrix}.$$

Example 1.9 The set

$$\{(x_1, x_2, x_3, x_4) \mid x_i \ge 0 \ (i = 1, 2, 3, 4), \ x_1 + x_2 \le 1, \ x_3 + x_4 \le 1\}$$

is a differentiable surface in $\mathscr{R}^4$, since it is the image of $D = [0, 1] \times [0, 1]$ under the continuously differentiable transformation $\mathbf{G} \colon \mathscr{R}^2 \to \mathscr{R}^4$ defined by

$$\mathbf{X} = \mathbf{G}(u, v) = \begin{bmatrix} u \\ 1 - u \\ v \\ 1 - v \end{bmatrix}.$$

Theorem 1.9 *A differentiable surface in $\mathscr{R}^n$ has zero content.*

Proof. Let S, D, and $\mathbf{G}$ be as in Definition 1.6. From Lemma 2.3, Section 6.2, there is a constant M such that

$$|\mathbf{G}(\mathbf{X}) - \mathbf{G}(\mathbf{Y})| \le M|\mathbf{X} - \mathbf{Y}| \quad \text{if} \quad \mathbf{X}, \mathbf{Y} \in D. \tag{35}$$

Since D is bounded, it is contained in a cube

$$C = [a_1, b_1] \times [a_2, b_2] \times \cdots \times [a_m, b_m],$$

where

$$b_i - a_i = L, \quad i = 1, 2, \ldots, m.$$

Suppose we partition C into N^m smaller cubes by partitioning each of the intervals $[a_i, b_i]$ into N equal subintervals. Let $R_1, R_2, \ldots, R_k$ be the smaller cubes so produced that contain points of D, and select points $\mathbf{X}_1, \mathbf{X}_2, \ldots, \mathbf{X}_k$ such that $\mathbf{X}_i \in D \cap R_i$ for $i = 1, 2, \ldots, k$. If $\mathbf{Y} \in D \cap R_i$, then

$$|\mathbf{G}(\mathbf{X}_i) - \mathbf{G}(\mathbf{Y})| \le M|\mathbf{X}_i - \mathbf{Y}| \tag{36}$$

[from (35)], and

$$|\mathbf{X}_i - \mathbf{Y}| \le \frac{L\sqrt{m}}{N} \tag{37}$$

(*Exercise 25*); therefore, (36) implies that $\mathbf{G}(\mathbf{Y})$ lies in a cube $\tilde{R}_i$ in $\mathscr{R}^n$, with sides of length $2ML\sqrt{m}/N$, centered at $\mathbf{G}(\mathbf{X}_i)$ [*Exercise 26(b)*]. Now

$$\sum_{i=1}^{k} V(\tilde{R}_i) \le k\left(\frac{2ML\sqrt{m}}{N}\right)^n \le N^m\left(\frac{2ML\sqrt{m}}{N}\right)^n = (2ML\sqrt{m})^n N^{m-n},$$

and, since $n > m$, we can make the sum on the left arbitrarily small by taking N large. Therefore, S has zero content.

Theorems 1.8 and 1.9 imply the following theorem (*Exercise 27*).

Theorem 1.10 *Suppose S is a bounded set in $\mathscr{R}^n$, with boundary consisting of a finite number of differentiable surfaces. Let f be bounded on S, and continuous except on a set of zero content. Then f is integrable on S.*

Example 1.10   Let

$$S = \{(x, y) \,|\, x^2 + y^2 = 1, x \ge 0\};$$

thus, S is bounded by a semicircle and a line segment (Figure 1.6), both differentiable curves in $\mathscr{R}^2$. Let

$$f(x, y) = \begin{cases} (1 - x^2 - y^2)^{1/2}, & (x, y) \in S, y \ge 0, \\ -(1 - x^2 - y^2)^{1/2}, & (x, y) \in S, y < 0. \end{cases}$$

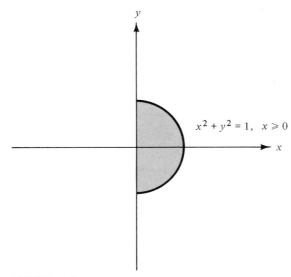

$x^2 + y^2 = 1, \quad x \ge 0$

FIGURE 1.6

Then f is continuous on S except on the line segment

$$y = 0, \qquad 0 \leq x < 1,$$

which has zero content, from Example 1.5; hence, Theorem 1.10 implies that f is integrable over S.

Properties of multiple integrals

We now list some theorems on properties of multiple integrals. For the case where S is a rectangle, the proofs of Theorems 1.11–1.16 are the same as those of their counterparts in Section 3.3. Once established for rectangles, they follow for general bounded regions, by Definition 1.5.

Theorem 1.11   *If f and g are integrable on S, then so is $f + g$, and*

$$\int_S (f + g)(\mathbf{X})\, dX = \int_S f(\mathbf{X})\, d\mathbf{X} + \int_S g(\mathbf{X})\, d\mathbf{X}.$$

Proof.   *Exercise 28.*

Theorem 1.12   *If f is integrable on S and c is a constant, then cf is integrable on S, and*

$$\int_S (cf)(\mathbf{X})\, d\mathbf{X} = c \int_S f(\mathbf{X})\, d\mathbf{X}.$$

Proof.   *Exercise 29.*

Theorem 1.13   *If f and g are integrable on S and $f(\mathbf{X}) \leq g(\mathbf{X})$ for $\mathbf{X}$ in S, then*

$$\int_S f(\mathbf{X})\, d\mathbf{X} \leq \int_S g(\mathbf{X})\, d\mathbf{X}.$$

Proof.   *Exercise 30.*

Theorem 1.14   *Suppose f is integrable on S, $|f(\mathbf{X})| \leq M$ for $\mathbf{X}$ in S, and $V(S)$ exists. Then*

$$\left| \int_S f(\mathbf{X})\, d\mathbf{X} \right| \leq MV(S).$$

Proof.   *Exercise 31.*

Theorem 1.15   *If f and g are integrable over S, then so is the product fg.*

Proof.   *Exercise 32.*

Theorem 1.16   *Suppose u is continuous and v is integrable and nonnegative on a region S. Then*

$$\int_S u(\mathbf{X})v(\mathbf{X})\, d\mathbf{X} = u(\mathbf{X}_0) \int_S v(\mathbf{X})\, d\mathbf{X}$$

for some $\mathbf{X}_0$ in S.

Proof.   *Exercise 33.*

Lemma 1.4   *If S is contained in a bounded set T and f is integrable over S, then f_S as defined in (34) is integrable over T, and*

$$\int_T f_S(\mathbf{X})\, d\mathbf{X} = \int_S f(\mathbf{X})\, d\mathbf{X}.$$

Proof.   Suppose R is a rectangle containing T. Then R also contains S (Figure 1.7), and

$$\int_S f(\mathbf{X})\, d\mathbf{X} = \int_R f_S(\mathbf{X})\, d\mathbf{X} \tag{38}$$

by Definition 1.5. Now let

$$(f_S)_T(\mathbf{X}) = \begin{cases} f_S(\mathbf{X}), & \mathbf{X} \in T, \\ 0, & \mathbf{X} \notin T. \end{cases}$$

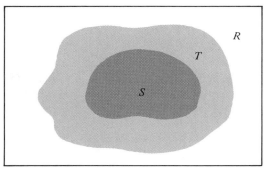

FIGURE 1.7

By Definition 1.5, f_S is integrable over T and

$$\int_T f_S(\mathbf{X}) \, d\mathbf{X} = \int_R (f_S)_T(\mathbf{X}) \, d\mathbf{X} \tag{39}$$

if the integral on the right exists. However, $(f_S)_T = f_S$ if $S \subset T$ (verify); hence, the integral on the right of (39) equals $\int_R f_S(\mathbf{X}) \, d\mathbf{X}$, which exists by assumption [see (38)]. Therefore, f_S is integrable over T and

$$\int_T f_S(\mathbf{X}) \, d\mathbf{X} = \int_R f_S(\mathbf{X}) \, d\mathbf{X} \qquad \text{[from (39)]}$$

$$= \int_S f(\mathbf{X}) \, d\mathbf{X} \qquad \text{[from (38)]}.$$

This completes the proof.

Theorem 1.17 *If f is integrable over disjoint sets S_1 and S_2, then f is integrable over $S = S_1 \cup S_2$ and*

$$\int_S f(\mathbf{X}) \, d\mathbf{X} = \int_{S_1} f(\mathbf{X}) \, d\mathbf{X} + \int_{S_2} f(\mathbf{X}) \, d\mathbf{X}. \tag{40}$$

Proof. Define

$$f_i(\mathbf{X}) = \begin{cases} f(\mathbf{X}), & \mathbf{X} \in S_i, \\ 0, & \mathbf{X} \notin S_i, \end{cases} \qquad i = 1, 2.$$

Then f_i is integrable over S_i, so Lemma 1.4 implies that f_i is integrable over S and, since $(f_i)_S = f_i$,

$$\int_S f_i(\mathbf{X}) \, d\mathbf{X} = \int_{S_i} f_i(\mathbf{X}) \, d\mathbf{X} = \int_{S_i} f(\mathbf{X}) \, d\mathbf{X}, \qquad i = 1, 2.$$

Theorem 1.11 now implies that $f_1 + f_2$ is integrable over S and

$$\int_S (f_1 + f_2)(\mathbf{X}) \, d\mathbf{X} = \int_{S_1} f(\mathbf{X}) \, d\mathbf{X} + \int_{S_2} f(\mathbf{X}) \, d\mathbf{X}. \tag{41}$$

But since $S_1 \cap S_2 = \varnothing$, $f_1 + f_2 = f$; therefore, (41) implies (40). This completes the proof.

We leave it to the reader to prove the following extension of Theorem 1.17 [*Exercise 35(b)*].

Corollary 1.1 *If f is integrable over sets S_1 and S_2 whose intersection has zero content, then f is integrable on $S = S_1 \cup S_2$ and (40) is valid.*

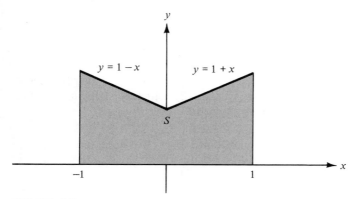

FIGURE 1.8

Example 1.11 Let

$$S = \{(x, y) \mid -1 \leq x \leq 1, 0 \leq y \leq 1 + |x|\}$$

(Figure 1.8),

$$S_1 = \{(x, y) \mid 0 \leq x \leq 1, 0 \leq y \leq 1 + x\},$$

and

$$S_2 = \{(x, y) \mid -1 \leq x \leq 0, 0 \leq y \leq 1 - x\}$$

(Figure 1.9). Then $S = S_1 \cup S_2$ and

$$S_1 \cap S_2 = \{(0, y) \mid 0 \leq y \leq 1\}$$

has zero content. Hence, Corollary 1.1 implies that if f is integrable over S_1 and S_2, then it is also integrable over $S = S_1 \cup S_2$, and that (40) holds.

We will discuss this example further in the next section.

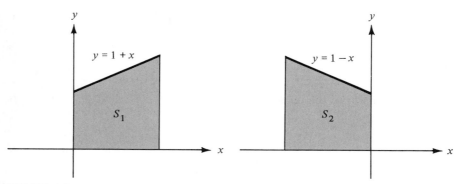

FIGURE 1.9

Lebesgue's existence criterion

A set E in $\mathscr{R}^n$ has *Lebesgue measure zero* if for every $\varepsilon > 0$ there is a finite or infinite sequence of rectangles $T_1, T_2, \ldots$ such that

$$E \subset \bigcup_j T_j$$

and

$$\sum_j V(T_j) < \varepsilon.$$

This definition enables us to state necessary and sufficient conditions for existence of a multiple integral.

Theorem 1.18   *If f is bounded on a rectangle R, then $\int_R f(\mathbf{X}) \, d\mathbf{X}$ exists if and only if the set of discontinuities of f form a set of Lebesgue measure zero.*

We omit the proof of this theorem, which is analogous to that of Theorem 5.1, Section 3.5.

7.1   EXERCISES

1. Evaluate directly from Definition 1.2:
   (a) $\int_R (3x + 2y) \, d(x, y)$; $R = [0, 2] \times [1, 3]$
   (b) $\int_R xy \, d(x, y)$; $R = [0, 1] \times [0, 1]$

2. Suppose $\int_a^b f(x) \, dx$ and $\int_c^d g(y) \, dy$ exist, and let $R = [a, b] \times [c, d]$. Criticize the following "proof" that $\int_R f(x)g(y) \, d(x, y)$ exists and equals $[\int_a^b f(x) \, dx][\int_c^d g(y) \, dy]$.
      "Proof." Let

   $$P_1: a = x_0 < x_1 < \cdots < x_r = b \quad \text{and} \quad P_2: c = y_0 < y_1 < \cdots < y_s = d$$

   be partitions of $[a, b]$ and $[c, d]$, and $\mathbf{P} = P_1 \times P_2$. Then a typical Riemann sum of fg over $\mathbf{P}$ is of the form

   $$\sigma = \sum_{i=1}^{r} \sum_{j=1}^{s} f(\xi_i)g(\eta_j)(x_i - x_{i-1})(y_j - y_{j-1}) = \sigma_1 \sigma_2,$$

   where

   $$\sigma_1 = \sum_{i=1}^{r} f(\xi_i)(x_i - x_{i-1}) \quad \text{and} \quad \sigma_2 = \sum_{j=1}^{s} g(\eta_j)(y_j - y_{j-1})$$

are typical Riemann sums of f over $[a, b]$ and g over $[c, d]$. Since f and g are integrable over these intervals,

$$\left| \sigma_1 - \int_a^b f(x)\, dx \right| \quad \text{and} \quad \left| \sigma_2 - \int_c^d g(y)\, dy \right|$$

can be made arbitrarily small by taking $\|P_1\|$ and $\|P_2\|$ sufficiently small. From this it is straightforward to show that

$$\left| \sigma - \left(\int_a^b f(x)\, dx \right) \left(\int_c^d g(y)\, dy \right) \right|$$

can be made arbitrarily small by taking $\|\mathbf{P}\|$ sufficiently small. This implies the stated result.

3. Suppose $f(x, y) \geq 0$ on $R = [a, b] \times [c, d]$. Justify the interpretation of $\int_R f(x, y)\, d(x, y)$, if it exists, as the volume of the region in $\mathscr{R}^3$ bounded by the surfaces $z = f(x, y)$ and the planes $z = 0$, $x = a$, $x = b$, $y = c$, and $y = d$.

4. Prove Theorem 1.1. (*Hint:* See the proof of Theorem 1.1, Section 3.1.)

5. Suppose

$$f(x, y) = \begin{cases} 0 & \text{if } x \text{ and } y \text{ are rational,} \\ 1 & \text{if } x \text{ is rational and } y \text{ is irrational,} \\ 2 & \text{if } x \text{ is irrational and } y \text{ is rational,} \\ 3 & \text{if } x \text{ and } y \text{ are irrational.} \end{cases}$$

Find

$$\overline{\int_R} f(x, y)\, d(x, y) \quad \text{and} \quad \underline{\int_R} f(x, y)\, d(x, y) \text{ if } R = [a, b] \times [c, d].$$

6. Given that Eq. (16) of Lemma 1.1 is valid in general, show that Eq. (17) is also.

7. Prove Theorem 1.2. (*Hint:* See the proof of Theorem 2.1, Section 3.2.)

8. Prove Theorem 1.3. (*Hint:* See the proof of Theorem 2.2, Section 3.2.)

9. Prove Lemma 1.2. (*Hint:* See the proof of Lemma 2.2, Section 3.2.)

10. Prove Theorem 1.4. (*Hint:* See the proof of Theorem 2.3, Section 3.2.)

11. Prove Theorem 1.5. (*Hint:* See the proof of Theorem 2.4, Section 3.2.)

12. Prove Theorem 1.6. (*Hint:* See the proof of Theorem 2.5, Section 3.2.)

13. Prove: If f is integrable over a rectangle R, then f is integrable over any subrectangle of R. (*Hint:* Use Theorem 1.5; see the proof of Theorem 3.7, Section 3.3.)

14. (a) Suppose f is integrable over a rectangle R and $\mathbf{P} = \{R_1, R_2, \ldots, R_k\}$ is a partition of R. Show that

$$\int_R f(\mathbf{X})\, d\mathbf{X} = \sum_{j=1}^{k} \int_{R_j} f(\mathbf{X})\, d\mathbf{X}.$$

(*Hint:* Use Exercise 13.)

(b) Use (a) to show that if f is continuous on R and $\mathbf{P}$ is a partition of R, then there is a Riemann sum of f over $\mathbf{P}$ which equals $\int_R f(\mathbf{X})\, d\mathbf{X}$.

15. Suppose f is continuously differentiable on a rectangle R. Show that there is a constant M such that

$$\left| \sigma - \int_R f(\mathbf{X})\, d\mathbf{X} \right| \le M \|\mathbf{P}\|$$

if σ is any Riemann sum of f over a partition $\mathbf{P}$ of R. [*Hint:* Use Exercise 14(b) and Theorem 4.2, Section 5.4.]

16. Give an example of a denumerable set in $\mathscr{R}^2$ which does not have zero content.

17. Prove:

(a) If S_1 and S_2 have zero content, so does $S_1 \cup S_2$.

(b) If S_1 has zero content and $S_2 \subset S_1$, then S_2 has zero content.

(c) If S has zero content, then so does $\overline{S}$.

18. Show that a degenerate rectangle has zero content according to Definition 1.4.

19. Suppose f is continuous on a compact set S in $\mathscr{R}^n$. Show that the surface $z = f(\mathbf{X})$ $(\mathbf{X} \in S)$ has zero content in $\mathscr{R}^{n+1}$. (*Hint:* See Example 1.5.)

20. Prove Lemma 1.3.

*21. Suppose R and $\tilde{R}$ are rectangles, $R \subset \tilde{R}$, g is bounded on $\tilde{R}$, and $g(\mathbf{X}) = 0$ if $\mathbf{X} \notin R$.

(a) Show that $\int_{\tilde{R}} g(\mathbf{X})\, d\mathbf{X}$ exists if and only if $\int_R g(\mathbf{X})\, d\mathbf{X}$ exists, and in this case,

$$\int_{\tilde{R}} g(\mathbf{X})\, d\mathbf{X} = \int_R g(\mathbf{X})\, d\mathbf{X}.$$

(b) Use (a) to show that Definition 1.5 is legitimate; that is, the existence and value of $\int_S f(\mathbf{X})\, d\mathbf{X}$ does not depend on the particular rectangle chosen to contain S.

22. Suppose f is defined on S and $f(\mathbf{X}) \ge \rho > 0$ on a subset T of $S \cap \partial S$ which does not have zero content. Show that f is not integrable over S.

23. (a) Suppose h is bounded and $h(\mathbf{X}) = 0$ except on a set of zero content. Show that $\int_S h(\mathbf{X})\, d\mathbf{X} = 0$ for any bounded set S.
    (b) Suppose $\int_S f(\mathbf{X})\, d\mathbf{X}$ exists, g is bounded on S, and $f(\mathbf{X}) = g(\mathbf{X})$ except for $\mathbf{X}$ in a set of zero content. Show that g is integrable over S and

$$\int_S g(\mathbf{X})\, d\mathbf{X} = \int_S f(\mathbf{X})\, d\mathbf{X}.$$

24. Suppose f is integrable over a set T and T_0 is a subset of T such that ∂T_0 has zero content. Show that f is integrable over T_0.

25. Show that if $\mathbf{X}$ and $\mathbf{Y}$ are in

$$R = [a_1, b_1] \times [a_2, b_2] \times \cdots \times [a_m, b_m],$$

then

$$|\mathbf{X} - \mathbf{Y}| \leq \sqrt{m} \max_{1 \leq i \leq m} (b_i - a_i).$$

Use this to establish Eq. (37).

*26. The *center* of a rectangle

$$R = [a_1, b_1] \times [a_2, b_2] \times \cdots \times [a_n, b_n]$$

is the point

$$\mathbf{X}_0 = \left(\frac{a_1 + b_1}{2}, \frac{a_2 + b_2}{2}, \ldots, \frac{a_n + b_n}{2} \right).$$

   (a) Let C be a cube in $\mathscr{R}^n$ with edge length s. Show that if $\mathbf{X}_0$ is the center of C and $\mathbf{X}$ is in C, then

$$|\mathbf{X} - \mathbf{X}_0| \leq \frac{s\sqrt{n}}{2}.$$

   (b) Prove: If $\mathbf{X}$ and $\mathbf{X}_0$ are points in $\mathscr{R}^n$ and $|\mathbf{X} - \mathbf{X}_0| \leq r$, then $\mathbf{X}$ is contained in a cube with edge length $2r$, centered at $\mathbf{X}_0$.

27. Use Theorems 1.8 and 1.9 to prove Theorem 1.10.

28. Prove Theorem 1.11. (*Hint*: See the proof of Theorem 3.1, Section 3.3.)

29. Prove Theorem 1.12.

30. Prove Theorem 1.13. (*Hint*: See the proof of Lemma 3.1, Section 3.3.)

31. Prove Theorem 1.14.

32. Prove Theorem 1.15. (*Hint*: See the proof of Theorem 3.5, Section 3.3.)

33. Prove Theorem 1.16. (*Hint*: See the proof of Theorem 3.6, Section 3.3.)

34. Under the assumptions of Exercise 2, give a correct proof that $\int_R f(x)g(y)\, d(x, y)$ exists and equals $[\int_a^b f(x)\, dx][\int_c^d g(y)\, dy]$.

35. (a) Suppose f is integrable over S and S_0 is obtained by removing a set of zero content from S. Show that f is integrable over S_0 and $\int_{S_0} f(\mathbf{X})\, d\mathbf{X} = \int_S f(\mathbf{X})\, d\mathbf{X}$.
    (b) Prove Corollary 1.1.

36. The set $U_f = \{\mathbf{X} \,|\, f(\mathbf{X}) \neq 0\}$ is called the *support* of f. Prove: If f is continuous on a bounded set S, then $\int f(\mathbf{X})\, d\mathbf{X}$ exists if and only if $U_f \cap \partial S$ has Lebesgue measure zero. (*Hint:* Use Theorem 1.18.)

37. Prove: If $\int_S f(\mathbf{X})\, d\mathbf{X}$ exists and U_f is as defined in Exercise 36, then $U_f \cap \partial S$ has Lebesgue measure zero. (*Hint:* Use Theorem 1.18.)

7.2 EVALUATION OF MULTIPLE INTEGRALS

Except for very simple examples it is impractical to evaluate multiple integrals directly from Definitions 1.2 and 1.5. Fortunately, this can usually be accomplished by evaluating n successive ordinary integrals. To motivate the method, let us first assume that f is continuous on $R = [a, b] \times [c, d]$. Then for each y in $[c, d]$, $f(x, y)$ is continuous with respect to x on $[a, b]$ and so the integral

$$F(y) = \int_a^b f(x, y)\, dx$$

exists. Moreover, the uniform continuity of f on R implies that F is continuous (*Exercise 3*), and therefore integrable, on $[c, d]$. We say that

$$I_1 = \int_c^d F(y)\, dy = \int_c^d \left[\int_a^b f(x, y)\, dx \right] dy$$

is an *iterated* integral of f over R; we will usually write it as

$$I_1 = \int_c^d dy \int_a^b f(x, y)\, dx.$$

Another iterated integral can also be defined by writing

$$G(x) = \int_c^d f(x, y)\, dy, \qquad a \leq x \leq b,$$

and defining

$$I_2 = \int_a^b G(x)\, dx = \int_a^b \left[\int_c^d f(x, y)\, dy \right] dx,$$

which we usually write as

$$I_2 = \int_a^b dx \int_c^d f(x, y)\, dy.$$

Example 2.1 Let

$$f(x, y) = x + y$$

and $R = [0, 1] \times [1, 2]$. Then

$$F(y) = \int_0^1 f(x, y) \, dx = \int_0^1 (x + y) \, dx$$

$$= \left(\frac{x^2}{2} + xy \right) \Big|_{x=0}^{1} = \frac{1}{2} + y,$$

and

$$I_1 = \int_1^2 F(y) \, dy = \int_1^2 \left(\frac{1}{2} + y \right) dy$$

$$= \left(\frac{y}{2} + \frac{y^2}{2} \right) \Big|_1^2 = 3 - 1 = 2.$$

Also,

$$G(x) = \int_1^2 (x + y) \, dy = \left(xy + \frac{y^2}{2} \right) \Big|_{y=1}^{2}$$

$$= (2x + 2) - \left(x + \frac{1}{2} \right) = x + \frac{3}{2},$$

and

$$I_2 = \int_0^1 G(x) \, dx = \int_0^1 \left(x + \frac{3}{2} \right) dx$$

$$= \left(\frac{x^2}{2} + \frac{3x}{2} \right) \Big|_0^1 = 2.$$

In this example $I_1 = I_2$; moreover, on setting $a = 0$, $b = 1$, $c = 1$, and $d = 2$ in Example 1.1 we see that

$$\int_R (x + y) \, d(x, y) = 2,$$

so that the common value of the iterated integrals equals the multiple integral. The following theorem shows that this is no accident.

Theorem 2.1 *Suppose f is integrable over $R = [a, b] \times [c, d]$ and the integral*

$$F(y) = \int_a^b f(x, y) \, dx$$

exists for each y in $[c, d]$. Then F is integrable over $[c, d]$ and

$$\int_c^d F(y) \, dy = \int_R f(x, y) \, d(x, y); \tag{1}$$

that is,

$$\int_c^d dy \int_a^b f(x, y) \, dx = \int_R f(x, y) \, d(x, y). \tag{2}$$

Proof. Let

$$P_1: a = x_0 < x_1 < \cdots < x_r = b$$

and

$$P_2: c = y_0 < y_1 < \cdots < y_s = d$$

be partitions of $[a, b]$ and $[c, d]$, and $\mathbf{P} = P_1 \times P_2$. Suppose

$$y_{j-1} \leq \eta_j \leq y_j, \qquad j = 1, 2, \ldots, s, \tag{3}$$

so that

$$\sigma = \sum_{j=1}^{s} F(\eta_j)(y_j - y_{j-1}) \tag{4}$$

is a typical Riemann sum of F over P_2. Since

$$F(\eta_j) = \int_a^b f(x, \eta_j)\, dx = \sum_{i=1}^{r} \int_{x_{i-1}}^{x} f(x, \eta_j)\, dx,$$

(3) implies that if

$$m_{ij} = \text{g.l.b.}\{f(x, y) \,|\, x_{i-1} \leq x \leq x_i, \, y_{j-1} \leq y \leq y_j\}$$

and

$$M_{ij} = \text{l.u.b.}\{f(x, y) \,|\, x_{i-1} \leq x \leq x_i, \, y_{j-1} \leq y \leq y_j\}$$

then

$$\sum_{i=1}^{r} m_{ij}(x_i - x_{i-1}) \leq F(\eta_j) \leq \sum_{i=1}^{r} M_{ij}(x_i - x_{i-1}).$$

Multiplying this by $y_j - y_{j-1}$ and summing from $j = 1$ to $j = s$, we see from (4) that

$$s(\mathbf{P}) \leq \sigma \leq S(\mathbf{P}), \tag{5}$$

where $s(\mathbf{P})$ and $S(\mathbf{P})$ are the upper and lower sums of f over $\mathbf{P}$. Now let $\hat{s}(P_2)$ and $\hat{S}(P_2)$ be the upper and lower sums of F over P_2; since they are respectively the greatest lower bound and least upper bound of the Riemann sums of F over P_2 (*Exercise 6, Section 3.1*), (5) implies that

$$s(\mathbf{P}) \leq \hat{s}(P_2) \leq \hat{S}(P_2) \leq S(\mathbf{P}). \tag{6}$$

Since f is integrable over R, there is for each $\varepsilon > 0$ a partition $\mathbf{P}$ of R such that $S(\mathbf{P}) - s(\mathbf{P}) < \varepsilon$, from Theorem 1.5; consequently, from (6), there is a partition of $[c, d]$ such that $\hat{S}(P_2) - \hat{s}(P_2) < \varepsilon$, and so F is integrable over $[c, d]$, again from Theorem 1.5 (or Theorem 2.4, Section 3.2).

It remains to verify (2). From the definition of integrability there is for each $\varepsilon > 0$ a $\delta > 0$ such that

$$\left| \int_c^d F(y)\, dy - \sigma \right| < \varepsilon \quad \text{if} \quad \|P_2\| < \delta.$$

This and (5) imply that

$$s(\mathbf{P}) - \varepsilon < \int_c^d F(y)\,dy < S(\mathbf{P}) + \varepsilon \quad \text{if} \quad \|\mathbf{P}\| < \delta,$$

and this implies that

$$\underline{\int_R} f(x,\,y)\,d(x,\,y) - \varepsilon \le \int_c^d F(y)\,dy \le \overline{\int_R} f(x,\,y)\,d(x,\,y) + \varepsilon. \tag{7}$$

Since

$$\underline{\int_R} f(x,\,y)\,d(x,\,y) = \overline{\int_R} f(x,\,y)\,d(x,\,y)$$

(Theorem 1.3) and ε can be made arbitrarily small, (7) implies (1), and therefore (2). This completes the proof.

   If f is continuous on R, then it satisfies the hypotheses of Theorem 2.1 (*Exercise 3*), and therefore (2) is valid in this case.
   If $\int_R f(x,\,y)\,d(x,\,y)$ and

$$\int_c^d f(x,\,y)\,dy, \qquad a \le x \le b,$$

exist, then by interchanging x and y in Theorem 2.1 we see that

$$\int_a^b dx \int_c^d f(x,\,y)\,dy = \int_R f(x,\,y)\,d(x,\,y).$$

This and (2) yield the following corollary to Theorem 2.1.

Corollary 2.1   *If f is integrable over $[a,\,b] \times [c,\,d]$, then*

$$\int_a^b dx \int_c^d f(x,\,y)\,dy = \int_c^d dy \int_a^b f(x,\,y)\,dx,$$

provided $\int_c^d f(x,\,y)\,dy$ exists for $a \le x \le b$ and $\int_a^b f(x,\,y)\,dx$ exists for $c \le y \le d$. In particular, these hypotheses hold if f is continuous on $[a,\,b] \times [c,\,d]$.

Example 2.2   The function

$$f(x,\,y) = 3xy + 2x$$

is continuous everywhere, and therefore (2) holds for every rectangle R. Let $R = [0,\,1] \times [1,\,2]$; then (2) yields

$$\int_R (x + y)\,d(x,\,y) = \int_1^2 dy \int_0^1 (x + y)\,dx = \int_1^2 \left[\left(\frac{x^2}{2} + xy \right) \Big|_{x=0}^1 \right] dy$$

$$= \int_1^2 \left(\frac{1}{2} + y \right) dy = \left(\frac{y}{2} + \frac{y^2}{2} \right) \Big|_1^2 = 2.$$

Since f also satisfies the hypotheses of Theorem 2.1 with x and y interchanged, we can calculate the double integral from the iterated integral in which the integrations are performed in the opposite order; thus,

$$\int_R (x + y) \, d(x, y) = \int_0^1 dx \int_1^2 (x + y) \, dy = \int_0^1 \left[\left(xy + \frac{y^2}{2} \right) \Big|_{y=1}^2 \right] dx$$

$$= \int_0^1 \left(x + \frac{3}{2} \right) dx = \left(\frac{x^2}{2} + \frac{3x}{2} \right) \Big|_0^1 = 2.$$

A plausible partial converse of Theorem 2.1 would be that if $\int_c^d dy \int_a^b f(x, y) \, dx$ exists, then so does $\int_R f(x, y) \, d(x, y)$; however, the next example shows that this need not be so.

Example 2.3   Let f be defined on $R = [0, 1] \times [0, 1]$ by

$$f(x, y) = \begin{cases} 2xy & \text{if } y \text{ is rational,} \\ y & \text{if } y \text{ is irrational.} \end{cases}$$

Since f is discontinuous except on the line $x = \frac{1}{2}$, its set of discontinuities does not have Lebesgue measure zero; hence, Theorem 1.18 implies that f is not integrable over R. (See *Exercise 7* for a more elementary proof of this.) However,

$$\int_0^1 f(x, y) \, dx = y, \qquad 0 \le y \le 1,$$

and

$$\int_0^1 dy \int_0^1 f(x, y) \, dx = \int_0^1 y \, dy = \frac{1}{2}.$$

Theorem 2.1 can be extended to multiple integrals over subsets of $\mathcal{R}^n$. The proof of the next theorem is analogous to that of Theorem 2.1. We leave it to the reader (*Exercise 11*).

Theorem 2.2   *Let*

$$R = [a_1, b_1] \times [a_2, b_2] \times \cdots \times [a_n, b_n]$$

and

$$\tilde{R} = [a_1, b_1] \times [a_2, b_2] \times \cdots \times [a_{n-1}, b_{n-1}].$$

Suppose f is integrable over R and the integral

$$F(x_n) = \int_{\tilde{R}} f(x_1, x_2, \ldots, x_{n-1}, x_n) \, d(x_1, x_2, \ldots, x_{n-1})$$

exists for each x_n in $[a_n, b_n]$. Then F is integrable over $[a_n, b_n]$ and

$$\int_R f(\mathbf{X}) \, d\mathbf{X} = \int_{a_n}^{b_n} F(x_n) \, dx_n,$$

which we also write as

$$\int_R f(\mathbf{X}) \, d\mathbf{X} = \int_{a_n}^{b_n} dx_n \int_R f(x_1, x_2, \ldots, x_{n-1}, x_n) \, d(x_1, x_2, \ldots, x_{n-1}).$$

If $n = 3$, this becomes

$$\int_R f(x, y, z) \, d(x, y, z) = \int_{a_3}^{b_3} dz \int_{\bar{R}} f(x, y, z) \, d(x, y). \tag{8}$$

If we also assume that the integral

$$G(y, z) = \int_{a_1}^{b_1} f(x, y, z) \, dx$$

exists for each (y, z) in $[a_2, b_2] \times [a_3, b_3]$, and regard z as fixed, then applying Theorem 2.1 to f as a function of x and y alone yields

$$\int_{\bar{R}} f(x, y, z) \, d(x, y) = \int_{a_2}^{b_2} dy \int_{a_1}^{b_1} f(x, y, z) \, dx, \qquad a_3 \le z \le b_3.$$

Substituting this in (8) yields

$$\int_R f(x, y, z) \, d(x, y, z) = \int_{a_3}^{b_3} dz \int_{a_2}^{b_2} dy \int_{a_1}^{b_1} f(x, y, z) \, dx.$$

This is a special case of the next theorem, which can be proved by repeated application of Theorem 2.1 (*Exercise 13*).

Theorem 2.3   *Let $I_j = [a_j, b_j], j = 1, 2, \ldots, n$, and suppose f is integrable over $R = I_1 \times I_2 \times \cdots \times I_n$. Suppose also that for $j = 1, 2, \ldots, n - 1$ the integral*

$$g_j(x_{j+1}, \ldots, x_n) = \int_{I_1 \times I_2 \cdots \times I_j} f(\mathbf{X}) \, d(x_1, x_2, \ldots, x_j)$$

exists for all

$$(x_{j+1}, \ldots, x_n) \quad \text{in} \quad I_{j+1} \times \cdots \times I_n.$$

Then the iterated integral

$$\int_{a_n}^{b_n} dx_n \int_{a_{n-1}}^{b_{n-1}} dx_{n-1} \cdots \int_{a_2}^{b_2} dx_2 \int_{a_1}^{b_1} f(\mathbf{X}) \, dx_1 \tag{9}$$

exists and equals $\int_R f(\mathbf{X}) \, d\mathbf{X}$.

Theorem 2.4   *If f is integrable over*

$$R = [a_1, b_1] \times [a_2, b_2] \times \cdots \times [a_n, b_n]$$

and the iterated integral (9) exists, then

$$\int_R f(\mathbf{X}) \, d\mathbf{X} = \int_{a_n}^{b_n} dx_n \int_{a_{n-1}}^{b_{n-1}} dx_{n-1} \cdots \int_{a_2}^{b_2} dx_2 \int_{a_1}^{b_1} f(\mathbf{X}) \, dx_1.$$

For $n = 2$ this theorem is weaker than Theorem 2.1. (Why?) We omit its proof for general n.

Example 2.4 Let $R = [0, 1] \times [1, 2] \times [0, 1]$ and

$$f(x, y, z) = x + y + z.$$

Then f is continuous and therefore integrable on R, and

$$g_1(y, z) = \int_0^1 (x + y + z)\, dx = \left(\frac{x^2}{2} + xy + xz\right)\Bigg|_{x=0}^1 = y + z + \frac{1}{2},$$

$$g_2(z) = \int_1^2 g_1(y, z)\, dy = \int_1^2 \left(y + z + \frac{1}{2}\right) dy$$

$$= \left(\frac{y^2}{2} + yz + \frac{y}{2}\right)\Bigg|_{y=1}^2 = 2 + z,$$

and

$$\int_R f(x, y, z)\, d(x, y, z) = \int_0^1 (2 + z)\, dz = \left(2z + \frac{z^2}{2}\right)\Bigg|_0^1 = \frac{5}{2}.$$

The hypotheses of Theorems 2.3 and 2.4 are stated so as to justify successive integrations with respect to x_1, then x_2, then x_3, and so forth. It is legitimate to use other orders of integration if the hypotheses are adjusted accordingly. For example, suppose $\{i_1, i_2, \ldots, i_n\}$ is a permutation of $\{1, 2, \ldots, n\}$ and $\int_R f(\mathbf{X})\, d\mathbf{X}$ exists, along with

$$\int_{I_{i_1} \times I_{i_2} \times \cdots \times I_{i_j}} f(\mathbf{X})\, d(x_{i_1}, x_{i_2}, \ldots, x_{i_j}), \qquad 1 \le j \le n - 1, \tag{10}$$

for each

$$(x_{i_{j+1}}, x_{i_{j+2}}, \ldots, x_{i_n}) \text{ in } I_{i_{j+1}} \times I_{i_{j+2}} \times \cdots \times I_{i_n}. \tag{11}$$

Then by simply renaming the variables we infer from Theorem 2.3 that

$$\int_R f(\mathbf{X})\, d\mathbf{X} = \int_{a_{i_n}}^{b_{i_n}} dx_{i_n} \int_{a_{i_{n-1}}}^{b_{i_{n-1}}} dx_{i_{n-1}} \cdots \int_{a_{i_2}}^{b_{i_2}} dx_{i_2} \int_{a_{i_1}}^{b_{i_1}} f(\mathbf{X})\, dx_{i_1}. \tag{12}$$

A similar reformulation applies to Theorem 2.4.

Since there are $n!$ permutations of $\{1, 2, \ldots, n\}$, there are $n!$ ways of evaluating a multiple integral over a rectangle in $\mathscr{R}^n$, provided the integrand satisfies appropriate hypotheses. In particular, if f is continuous on R and $\{i_1, i_2, \ldots, i_n\}$ is any permutation of $\{1, 2, \ldots, n\}$, then f is continuous with respect to $(x_{i_1}, x_{i_2}, \ldots, x_{i_j})$ on $I_{i_1} \times I_{i_2} \times \cdots \times I_{i_j}$ for each fixed

$$(x_{i_{j+1}}, x_{i_{j+2}}, \ldots, x_{i_n})$$

satisfying (11); therefore, the integrals (10) exist for every permutation of $\{1, 2, \ldots, n\}$ (Theorem 1.6). We summarize this in the next theorem, which now follows from Theorem 2.3.

Theorem 2.5   *If f is continuous on*

$$R = [a_1, b_1] \times [a_2, b_2] \times \cdots \times [a_n, b_n],$$

then $\int_R f(\mathbf{X}) \, d\mathbf{X}$ can be evaluated by iterated integrals in any of the $n!$ ways indicated in (12).

The conclusion of Theorem 2.5 may hold under weaker assumptions on f; that is, continuity is sufficient, but not necessary.

Example 2.5   If f is continuous on $R = [a_1, b_1] \times [a_2, b_2] \times [a_3, b_3]$, then

$$
\begin{aligned}
\int_R f(x, y, z) \, d(x, y, z) &= \int_{a_3}^{b_3} dz \int_{a_2}^{b_2} dy \int_{a_1}^{b_1} f(x, y, z) \, dx \\
&= \int_{a_2}^{b_2} dy \int_{a_3}^{b_3} dz \int_{a_1}^{b_1} f(x, y, z) \, dx \\
&= \int_{a_3}^{b_3} dz \int_{a_1}^{b_1} dx \int_{a_2}^{b_2} f(x, y, z) \, dy \\
&= \int_{a_1}^{b_1} dx \int_{a_3}^{b_3} dz \int_{a_2}^{b_2} f(x, y, z) \, dy \\
&= \int_{a_2}^{b_2} dy \int_{a_1}^{b_1} dx \int_{a_3}^{b_3} f(x, y, z) \, dz \\
&= \int_{a_1}^{b_1} dx \int_{a_2}^{b_2} dy \int_{a_3}^{b_3} f(x, y, z) \, dz.
\end{aligned}
$$

Integrals over more general sets

We now consider the problem of evaluating multiple integrals over more general sets. First suppose f is integrable over

$$S = \{(x, y) \mid u(y) \le x \le v(y), c \le y \le d\} \tag{13}$$

(Figure 2.1). If $u(y) \ge a$ and $v(y) \le b$ for $c \le y \le d$, and

$$f_S(x, y) = \begin{cases} f(x, y), & (x, y) \in S, \\ 0, & (x, y) \notin S, \end{cases} \tag{14}$$

then

$$\int_S f(x, y) \, d(x, y) = \int_R f_S(x, y) \, d(x, y),$$

where $R = [a, b] \times [c, d]$ (Definition 1.5). From Theorem 2.1,

$$\int_R f_S(x, y) \, d(x, y) = \int_c^d dy \int_a^b f_S(x, y) \, dx$$

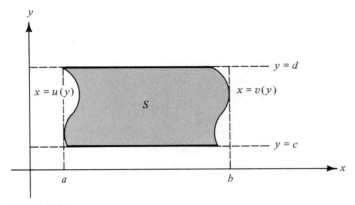

FIGURE 2.1

provided $\int_a^b f_S(x, y)\, dx$ exists for each y in $[c, d]$. From (13) and (14), this integral can be written as

$$\int_{u(y)}^{v(y)} f(x, y)\, dx. \tag{15}$$

Thus, we have proved the following theorem.

Theorem 2.6   *If f is integrable over the set S in (13) and the integral (15) exists for $c \le y \le d$, then*

$$\int_S f(x, y)\, d(x, y) = \int_c^d dy \int_{u(y)}^{v(y)} f(x, y)\, dx. \tag{16}$$

The assumptions of Theorem 2.6 are satisfied if f is continuous on S and u and v are continuous on $[c, d]$ (*Exercise 16*).

Interchanging x and y in Theorem 2.6 shows that if f is integrable over

$$S = \{(x, y)\,|\, u(x) \le y \le v(x), \, a \le x \le b\} \tag{17}$$

(Figure 2.2) and

$$\int_{u(x)}^{v(x)} f(x, y)\, dy$$

exists for $a \le x \le b$, then

$$\int_S f(x, y)\, d(x, y) = \int_a^b dx \int_{u(x)}^{v(x)} f(x, y)\, dy. \tag{18}$$

2.6   Suppose

$$f(x, y) = xy$$

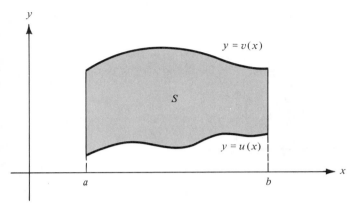

FIGURE 2.2

and S is the region bounded by the curves $x = y^2$ and $x = y$ (Figure 2.3). Since S can be represented in the form (13) as

$$S = \{(x, y) \mid y^2 \leq x \leq y,\ 0 \leq y \leq 1\},$$

(16) yields

$$\int_S xy\, d(x, y) = \int_0^1 dy \int_{y^2}^y xy\, dx,$$

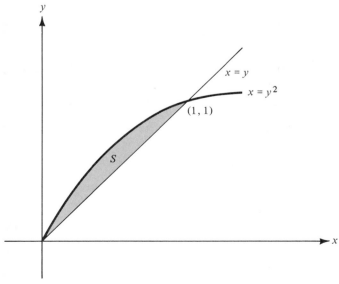

FIGURE 2.3

which, incidentally, can be written as

$$\int_S xy \, d(x, y) = \int_0^1 y \, dy \int_{y^2}^y x \, dx,$$

since y is independent of x. Evaluating the iterated integral yields

$$\int_S xy \, d(x, y) = \int_0^1 \left(\frac{x^2}{2}\Big|_{y^2}^y\right) y \, dy = \frac{1}{2}\int_0^1 (y^3 - y^5) \, dy$$

$$= \frac{1}{2}\left(\frac{y^4}{4} - \frac{y^6}{6}\right)\Big|_0^1 = \frac{1}{24}.$$

In this case we can also represent S in the form (17) as

$$S = \{(x, y)|x \le y \le \sqrt{x}, 0 \le x \le 1\};$$

hence, from (18),

$$\int_S xy \, d(x, y) = \int_0^1 x \, dx \int_x^{\sqrt{x}} y \, dy = \int_0^1 \left(\frac{y^2}{2}\Big|_x^{\sqrt{x}}\right) x \, dx$$

$$= \frac{1}{2}\int_0^1 (x^2 - x^3) \, dx = \frac{1}{2}\left(\frac{x^3}{3} - \frac{x^4}{4}\right)\Big|_0^1 = \frac{1}{24}.$$

***Example* 2.7**   To evaluate

$$\int_S (x + y) \, d(x, y),$$

where

$$S = \{(x, y)| -1 \le x \le 1, 0 \le y \le 1 + |x|\}$$

(Example 1.11; see Figure 1.8), we invoke Corollary 1.1 and write

$$\int_S (x + y) \, d(x, y) = \int_{S_1} (x + y) \, d(x, y) + \int_{S_2} (x + y) \, d(x, y),$$

where

$$S_1 = \{(x, y)|0 \le x \le 1, 0 \le y \le 1 + x\}$$

and

$$S_2 = \{(x, y)| -1 \le x \le 0, 0 \le y \le 1 - x\}$$

(Figure 1.9). From Theorem 2.6,

$$\int_{S_1} (x + y) \, d(x, y) = \int_0^1 dx \int_0^{1+x} (x + y) \, dy = \int_0^1 \left[\frac{(x + y)^2}{2}\Big|_{y=0}^{1+x}\right] dx$$

$$= \frac{1}{2}\int_0^1 [(2x + 1)^2 - x^2] \, dx$$

$$= \frac{1}{2}\left[\frac{(2x + 1)^3}{6} - \frac{x^3}{3}\right]\Big|_0^1 = 2$$

and

$$\int_{S_2} (x + y) \, d(x, y) = \int_{-1}^{0} dx \int_{0}^{1-x} (x + y) \, dy = \int_{-1}^{0} \left[\frac{(x + y)^2}{2} \Big|_{y=0}^{1-x} \right] dx$$

$$= \frac{1}{2} \int_{-1}^{0} (1 - x^2) \, dx = \frac{1}{2} \left(x - \frac{x^3}{3} \right) \Big|_{-1}^{0} = \frac{1}{3}.$$

Therefore,

$$\int_{S} (x + y) \, d(x, y) = 2 + \frac{1}{3} = \frac{7}{3}.$$

Example 2.8   To find the area A of the region bounded by the curves

$$y = x^2 + 1 \quad \text{and} \quad y = 9 - x^2$$

(Figure 2.4), we evaluate

$$A = \int_{S} d(x, y),$$

where

$$S = \{(x, y) \,|\, x^2 + 1 \le y \le 9 - x^2, \ -2 \le x \le 2\}.$$

According to Theorem 2.6,

$$A = \int_{-2}^{2} dx \int_{x^2+1}^{9-x^2} dy = \int_{-2}^{2} [(9 - x^2) - (x^2 + 1)] \, dx$$

$$= \int_{-2}^{2} (8 - 2x^2) \, dx = \left(8x - \frac{2x^3}{3} \right) \Big|_{-2}^{2} = \frac{64}{3}.$$

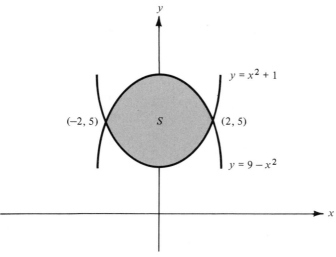

FIGURE 2.4

Theorem 2.6 has an analog for $n > 2$. Suppose f is integrable over a set S of points $\mathbf{X} = (x_1, x_2, \ldots, x_n)$ satisfying the inequalities

$$u_j(x_{j+1}, \ldots, x_n) \le x_j \le v_j(x_{j+1}, \ldots, x_n) \quad \text{if} \quad j = 1, 2, \ldots, n - 1$$

and

$$a_n \le x_n \le b.$$

Then under appropriate additional assumptions it can be shown, by an argument analogous to the one that led to Theorem 2.6, that

$$\int_S f(\mathbf{X})\, d\mathbf{X} = \int_{a_n}^{b_n} dx_n \int_{u_n(x_n)}^{v_n(x_n)} dx_{n-1} \cdots \int_{u_2(x_3,\ldots,x_n)}^{v_2(x_3,\ldots,x_n)} dx_2 \int_{u_1(x_2,\ldots,x_n)}^{v_1(x_2,\ldots,x_n)} f(\mathbf{X})\, dx_1$$

These additional assumptions are tedious to state for general n. The following theorem contains a complete statement for $n = 3$. It can be deduced from Theorem 2.3 (with $n = 3$) as Theorem 2.6 was from Theorem 2.1 (*Exercise 20*).

Theorem 2.7 *Suppose f is integrable over*

$$S = \{(x, y, z) \,|\, u_1(y, z) \le x \le v_1(y, z), u_2(z) \le y \le v_2(z), c \le z \le d\} \quad (19)$$

and let

$$S(z) = \{(x, y) \,|\, u_1(y, z) \le x \le v_1(y, z), u_2(z) \le y \le v_2(z)\}$$

for each z in $[c, d]$. Then

$$\int_S f(x, y, z)\, d(x, y, z) = \int_c^d dz \int_{u_2(z)}^{v_2(z)} dy \int_{u_1(y,z)}^{v_1(y,z)} f(x, y, z)\, dx \quad (20)$$

provided

$$\int_{u_1(y,z)}^{v_1(y,z)} f(x, y, z)\, dx$$

exists for all (y, z) such that

$$c \le z \le d \quad \text{and} \quad u_2(z) \le y \le v_2(z),$$

and

$$\int_{S(z)} f(x, y, z)\, d(x, y)$$

exists for all z in $[c, d]$.

The next theorem, also for $n = 3$, can be deduced from Theorem 2.4 (*Exercise 21*).

Theorem 2.8 *Suppose f is integrable over S as in (19) and the iterated integral on the right of (20) exists. Then (20) is valid.*

Either of the last two theorems implies that (20) holds if f is continuous on S and u_1, u_2, v_1, and v_2 are continuous (*Exercise 22*); an analogous statement is true for general n.

Example 2.9   Suppose f is continuous on the region S in $\mathscr{R}^3$ bounded by the coordinate planes and the plane

$$x + y + 2z = 2$$

(Figure 2.5); thus,

$$S = \{(x, y, z) \mid 0 \le x \le 2 - y - 2z, \, 0 \le y \le 2 - 2z, \, 0 \le z \le 1\}.$$

From Theorem 2.7,

$$\int_S f(x, y, z) \, d(x, y, z) = \int_0^1 dz \int_0^{2-2z} dy \int_0^{2-y-2z} f(x, y, z) \, dx.$$

There are five other iterated integrals that equal the multiple integral. We leave it to the reader to verify that

$$\int_S f(x, y, z) \, d(x, y, z) = \int_0^2 dy \int_0^{1-y/2} dz \int_0^{2-y-2z} f(x, y, z) \, dx$$

$$= \int_0^1 dz \int_0^{2-2z} dx \int_0^{2-x-2z} f(x, y, z) \, dy$$

$$= \int_0^2 dx \int_0^{1-x/2} dz \int_0^{2-x-2z} f(x, y, z) \, dy$$

$$= \int_0^2 dx \int_0^{2-x} dy \int_0^{1-x/2-y/2} f(x, y, z) \, dz$$

$$= \int_0^2 dy \int_0^{2-y} dx \int_0^{1-x/2-y/2} f(x, y, z) \, dz$$

(*Exercise 23*).

Thus far we have viewed the iterated integral as a tool for evaluating multiple integrals. In some problems the iterated integral is itself the object of interest. In this case a result like Theorem 2.6 may be used to evaluate the iterated integral. The procedure is as follows.

1. Express the given iterated integral as a multiple integral, and check to see that the multiple integral exists.
2. Look for another iterated integral, which equals the multiple integral and is easier to evaluate than the given one. The two iterated integrals must be equal, by Theorem 2.6.

This procedure is called *changing the order of integration* of an iterated integral.

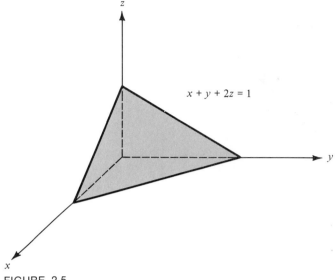

FIGURE 2.5

Example 2.10   The iterated integral

$$I = \int_0^1 dy \int_0^y e^{-(x-1)^2} dx$$

is hard to evaluate because

$$e^{-(x-1)^2}$$

has no elementary indefinite integral. The set of points (x, y) that enter into the integration—which we call the *region of integration*—is

$$S = \{(x, y) \mid 0 \le x \le y, 0 \le y \le 1\}$$

(Figure 2.6). Therefore,

$$I = \int_S e^{-(x-1)^2} d(x, y),\tag{21}$$

and this multiple integral exists because its integrand is continuous. Since S can also be written as

$$S = \{(x, y) \mid x \le y \le 1, 0 \le x \le 1\},$$

Theorem 2.6 implies that

$$\int_S e^{-(x-1)^2} d(x, y) = \int_0^1 e^{-(x-1)^2} dx \int_x^1 dy = -\int_0^1 (x-1)e^{-(x-1)^2} dx$$

$$= \frac{1}{2} e^{-(x-1)^2} \Big|_0^1 = \frac{1}{2}(1 - e^{-1}).$$

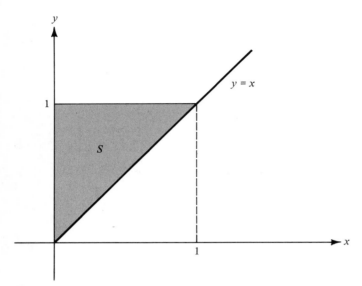

FIGURE 2.6

This and (21) imply that

$$I = \frac{1}{2}(1 - e^{-1}).$$

Example 2.11 Suppose f is continuous on $[a, \infty)$ and y satisfies the simple differential equation

$$y''(x) = f(x), \qquad x > a, \tag{22}$$

with initial conditions

$$y(a) = y'(a) = 0. \tag{23}$$

Integrating (22) and invoking (23) yields

$$y'(x) = \int_a^x f(t)\, dt,$$

and integrating this and invoking (23) again yields

$$y(x) = \int_a^x ds \int_a^s f(t)\, dt,$$

This can be reduced to a single integral as follows. Since the function

$$g(s, t) = f(t)$$

is continuous for all (s, t) such that $t \geq a$, g is integrable over

$$S = \{(s, t) \,|\, a \leq t \leq s, \, a \leq s \leq x\}$$

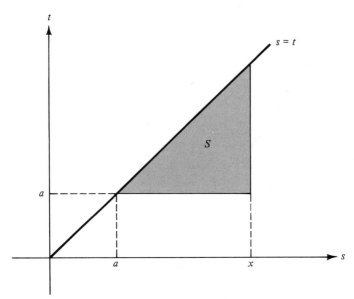

FIGURE 2.7

(Figure 2.7), and Theorem 2.6 implies that

$$\int_S f(t)\, d(s, t) = \int_a^x ds \int_a^s f(t)\, dt = y(x). \tag{24}$$

However, S can also be described as

$$S = \{(s, t)\,|\, t \le s \le x,\ a \le t \le x\},$$

and so Theorem 2.6 implies that

$$\int_S f(t)\, d(s, t) = \int_a^x f(t)\, dt \int_t^x ds = \int_a^x (x - t)f(t)\, dt.$$

Comparing this with (24) yields

$$y(x) = \int_a^x (x - t)f(t)\, dt.$$

7.2 EXERCISES

1. Evaluate:

(a) $\displaystyle\int_0^2 dy \int_{-1}^1 (x + 3y)\, dx$

(b) $\displaystyle\int_1^2 dx \int_0^1 (x^3 + y^4)\, dy$

(c) $\displaystyle\int_{\pi/2}^{2\pi} x\, dx \int_1^2 \sin xy\, dy$

(d) $\displaystyle\int_0^{\log 2} y\, dy \int_0^1 xe^{x^2 y}\, dx$

2. Let $I_j = [a_j, b_j]\, (j = 1, 2, 3)$ and suppose f is integrable on $R = I_1 \times I_2 \times I_3$. Prove:

(a) If the integral

$$G(y, z) = \int_{a_1}^{b_1} f(x, y, z)\, dx$$

exists for $(y, z) \in I_2 \times I_3$, then G is integrable over $I_2 \times I_3$ and

$$\int_R f(x, y, z)\, d(x, y, z) = \int_{I_2 \times I_3} G(y, z)\, d(y, z).$$

(b) If the integral

$$H(z) = \int_{I_1 \times I_2} f(x, y, z)\, d(x, y)$$

exists for $z \in I_3$, then H is integrable over I_3 and

$$\int_R f(x, y, z)\, d(x, y, z) = \int_{a_3}^{b_3} H(z)\, dz.$$

(*Hint for both parts:* See the proof of Theorem 2.1.)

* 3. Prove that if f is continuous on $[a, b] \times [c, d]$, then the function

$$F(y) = \int_a^b f(x, y)\, dx$$

is continuous on $[c, d]$. (*Hint:* Use Theorem 2.10, Section 5.2.)

4. Suppose

$$f(x', y') \geq f(x, y) \qquad \text{if } a \leq x \leq x' \leq b, \quad c \leq y \leq y' \leq d.$$

Show that f satisfies the hypotheses of Theorem 2.1 on $R = [a, b] \times [c, d]$. (*Hint:* See the proof of Theorem 2.6, Section 3.2.)

5. Evaluate by means of iterated integrals:

(a) $\int_R (xy + 1)\, d(x, y);\ R = [0, 1] \times [1, 2]$

(b) $\int_R (2x + 3y)\, d(x, y);\ R = [1, 3] \times [1, 2]$

(c) $\int_R \dfrac{xy}{\sqrt{x^2 + y^2}}\, d(x, y);\ R = [0, 1] \times [0, 1]$

(d) $\int_R x \cos xy \cos 2\pi x\, d(x, y);\ R = [0, \tfrac{1}{4}] \times [0, 2\pi]$

6. Let A be the set of points of the form $(2^{-n}p, 2^{-n}q)$, where p and q are odd integers and n is a nonnegative integer. Let

$$f(x, y) = \begin{cases} 1, & (x, y) \notin A, \\ 0, & (x, y) \in A. \end{cases}$$

Show that f is not integrable over any rectangle $R = [a, b] \times [c, d]$, but

$$\int_a^b dx \int_c^d f(x, y)\, dy = \int_c^d dy \int_a^b f(x, y)\, dx = (b - a)(d - c).$$

(*Hint:* Use Theorem 1.18.)

7. Let

$$f(x, y) = \begin{cases} 2xy & \text{if } y \text{ is rational,} \\ y & \text{if } y \text{ is irrational,} \end{cases}$$

and $R = [0, 1] \times [0, 1]$ (Example 2.3).

(a) Calculate $\overline{\int}_R f(x, y)\, d(x, y)$ and $\underline{\int}_R f(x, y)\, d(x, y)$, and show that f is not integrable over R.

(b) Calculate $\int_0^1 \left[\overline{\int}_0^1 f(x, y)\, dy\right] dx$ and $\int_0^1 \left[\underline{\int}_0^1 f(x, y)\, dy\right] dx$.

8. Let $R = [0, 1] \times [0, 1] \times [0, 1]$, $\tilde{R} = [0, 1] \times [0, 1]$, and

$$f(x, y, z) = \begin{cases} 2xy + 2xz & \text{if } y \text{ and } z \text{ are rational,} \\ y + 2xz & \text{if } y \text{ is irrational and } z \text{ is rational,} \\ 2xy + z & \text{if } y \text{ is rational and } z \text{ is irrational,} \\ y + z & \text{if } y \text{ and } z \text{ are irrational.} \end{cases}$$

Calculate:

(a) $\overline{\int}_R f(x, y, z)\, d(x, y, z)$ and $\underline{\int}_R f(x, y, z)\, d(x, y, z)$

(b) $\overline{\int}_{\tilde{R}} f(x, y, z)\, d(x, y)$ and $\underline{\int}_{\tilde{R}} f(x, y, z)\, d(x, y)$

(c) $\int_0^1 dy \int_0^1 f(x, y, z)\, dx$ and $\int_0^1 dz \int_0^1 dy \int_0^1 f(x, y, z)\, dx$

9. Suppose f is bounded on $R = [a, b] \times [c, d]$. Prove:

(a) $\underline{\int}_R f(x, y)\, d(x, y) \le \int_a^b \left[\underline{\int}_c^d f(x, y)\, dy\right] dx$

$$\le \int_a^b \left[\overline{\int}_c^d f(x, y)\, dy\right] dx \le \overline{\int}_R f(x, y)\, d(x, y)$$

(b) $\underline{\int}_R f(x, y)\, d(x, y) \le \int_a^b \left[\overline{\int}_c^d f(x, y)\, dy\right] dx$

$$\le \overline{\int}_a^b \left[\underline{\int}_c^d f(x, y)\, dy\right] dx \le \overline{\int}_R f(x, y)\, d(x, y)$$

10. Use Exercise 9 to prove the following generalization of Theorem 2.1: If f is integrable over $R = [a, b] \times [c, d]$, then

$$\overline{\int}_c^b f(x, y)\, dy \quad \text{and} \quad \underline{\int}_c^d f(x, y)\, dy$$

are integrable over $[a, b]$, and

$$\int_a^b \left[\int_c^d f(x, y) \, dy \right] dx = \int_a^b \left[\int_c^d f(x, y) \, dy \right] dx = \int_R f(x, y) \, d(x, y).$$

11. Prove Theorem 2.2. (*Hint*: See the proof of Theorem 2.1.)

12. Let $I_1, I_2, \ldots, I_n$ be closed intervals and suppose f is integrable over $R = I_1 \times I_2 \times \cdots \times I_n$. Prove: If $1 \le j \le n - 1$ and the integral

$$g(x_{j+1}, x_{j+2}, \ldots, x_n) = \int_{I_1 \times I_2 \times \cdots \times I_j} f(x_1, x_2, \ldots, x_n) \, d(x_1, x_2, \ldots, x_j)$$

exists for each $(x_{j+1}, x_{j+2}, \ldots, x_n)$ in $I_{j+1} \times I_{j+2} \times \cdots \times I_n$, then

$$\int_{I_{j+1} \times I_{j+2} \times \cdots \times I_n} g(x_{j+1}, x_{j+2}, \ldots, x_n) \, d(x_{j+1}, x_{j+2}, \ldots, x_n)$$

exists and equals $\int_R f(\mathbf{X}) \, d\mathbf{X}$. (*Hint*: See the proof of Theorem 2.1.)

13. Prove Theorem 2.3.

14. Evaluate:

(a) $\int_R (x - 2y + 3z) \, d(x, y, z)$; $R = [-2, 0] \times [2, 5] \times [-3, 2]$

(b) $\int_R e^{-x^2 - y^2} \sin x \sin z \, d(x, y, z)$; $R = [-1, 1] \times [0, 2] \times [0, \pi/2]$

(c) $\int_R (xy + 2xz + yz) \, d(x, y, z)$; $R = [-1, 1] \times [0, 1] \times [-1, 1]$

(d) $\int_R x^2 y^3 z e^{xy^2 z^2} \, d(x, y, z)$; $R = [0, 1] \times [0, 1] \times [0, 1]$

15. Suppose f is defined on $R = [a_1, b_1] \times [a, b_2] \times \cdots \times [a, b_n]$ and

$$f(x_1', x_2', \ldots, x_n') \ge f(x_1, x_2, \ldots, x_n)$$

if $x_1' \ge x_1, x_2' \ge x_2, \ldots, x_n' \ge x_n$. Show that f satisfies the hypotheses of Theorem 2.4. (*Hint*: See Exercise 4.)

16. Show that the assumptions of Theorem 2.6 are satisfied if f is continuous on S and u and v are continuous on $[c, d]$.

17. Evaluate:

(a) $\int_S (2x + y^2) \, d(x, y)$; $S = \{(x, y) \mid 0 \le x \le 9 - y^2, -3 \le y \le 3\}$

(b) $\int_S xy \, d(x, y)$; S is bounded by $y = |x|$ and $y = 6 - x^2$

(c) $\int_S 2xy \, d(x, y)$; S is bounded by $y = x^2$ and $x = y^2$

(d) $\int_S e^x \frac{\sin y}{y} \, d(x, y)$: $S = \{(x, y) \mid \log y \le x \le \log y + \log 2, \pi/2 \le y \le \pi\}$

18. Evaluate $\int_S (x + y)\, d(x, y)$, where S is bounded by $y = x^2$ and $y = 2x$, using iterated integrals of both possible types.

19. Find the area of the set bounded by the given curves:
    (a) $y = x^2 + 9$, $y = x^2 - 9$, $x = -1$, $x = 1$
    (b) $y = x + 2$, $y = 4 - x$, $x = 0$
    (c) $x = y^2 - 4$, $x = 4 - y^2$
    (d) $y = e^{2x}$, $y = -2x$, $x = 0$, $x = 3$

20. Prove Theorem 2.7.

21. Prove Theorem 2.8.

22. Show that the hypotheses of Theorems 2.7 and 2.8 are satisfied if f is continuous on S and u_1, u_2, v_1, and v_2 are continuous.

23. In Example 2.9 verify the last five representations of $\int_S f(x, y, z)\, d(x, y, z)$ as iterated integrals.

24. Let S be the region in $\mathcal{R}^3$ bounded by the coordinate planes and the plane $x + 2y + 3z = 1$. Let f be continuous on S. Set up six iterated integrals that equal $\int_S f(x, y, z)\, d(x, y, z)$.

25. Evaluate:
    (a) $\int_S x\, d(x, y, z)$; S is bounded by the coordinate planes and the plane $3x + y + z = 2$.

    (b) $\int_S y e^z\, d(x, y, z)$; $S = \{(x, y, z) \mid 0 \le x \le 1, 0 \le y \le \sqrt{x}, 0 \le z \le y^2\}$

    (c) $\int_S xyz\, d(x, y, z)$;
    $$S = \{(x, y, z) \mid 0 \le y \le 1, 0 \le x \le \sqrt{1 - y^2}, 0 \le z \le \sqrt{x^2 + y^2}\}$$

    (d) $\int_S yz\, d(x, y, z)$; $S = \{(x, y, z) \mid z^2 \le x \le \sqrt{z}, 0 \le y \le z, 0 \le z \le 1\}$

26. Find the volume of S:
    (a) S is bounded by the surfaces $z = x^2 + y^2$ and $z = 8 - x^2 - y^2$.
    (b) $S = \{(x, y, z) \mid 0 \le z \le x^2 + y^2$, $(x, y, 0)$ is in the triangle with vertices $(0, 1, 0), (0, 0, 0)$, and $(1, 0, 0)\}$
    (c) $S = \{(x, y, z) \mid 0 \le y \le x^2, 0 \le x \le 2, 0 \le z \le y^2\}$
    (d) $S = \{(x, y, z) \mid x \ge 0, y \ge 0, 0 \le z \le 4 - 4x^2 - 4y^2\}$

27. Let $R = [a_1, b_1] \times [a_2, b_2] \times \cdots \times [a_n, b_n]$. Evaluate

    (a) $\int_R (x_1 + x_2 + \cdots + x_n)\, d\mathbf{X}$ (b) $\int_R (x_1^2 + x_2^2 + \cdots + x_n^2)\, d\mathbf{X}$

    (c) $\int_R x_1 x_2 \cdots x_n\, d\mathbf{X}$

28. Assuming that f is continuous, express

$$\int_{1/2}^{1} dy \int_{-\sqrt{1-y^2}}^{\sqrt{1-y^2}} f(x, y)\, dx$$

as an iterated integral with the order of integration reversed.

29. Evaluate $\int_S (x + y)\, d(x, y)$ of Example 2.7 by means of iterated integrals in which the first integration is with respect to x.

30. Reverse the order of integration and evaluate:

$$\int_0^1 x\, dx \int_0^{\sqrt{1-x^2}} \frac{dy}{\sqrt{x^2 + y^2}}$$

31. Suppose f is continuous on $[a, \infty)$,

$$y^{(n)}(t) = f(t), \qquad t \geq a,$$

and

$$y(a) = y'(a) = \cdots = y^{(n-1)}(a) = 0.$$

(a) By repeated integration show that

$$y(t) = \int_0^t dt_n \int_0^{t_n} dt_{n-1} \cdots \int_0^{t_3} dt_2 \int_0^{t_2} f(t_1)\, dt_1.$$

(b) By reversing the order of integration in the iterated integral of part (a), show that

$$y(t) = \frac{1}{(n-1)!} \int_0^x (x - t)^{n-1} f(t)\, dt.$$

(*Hint:* See Example 2.11.)

*32. Let $T_\rho = [0, \rho] \times [0, \rho]$, $\rho > 0$. By calculating

$$I_1(a) = \lim_{\rho \to \infty} \int_{T_\rho} e^{-xy} \sin ax\, d(x, y)$$

in two different ways, show that

$$\int_0^\infty \frac{\sin ax}{x}\, dx = \frac{\pi}{2} \quad \text{if} \quad a > 0.$$

7.3   CHANGE OF VARIABLES IN MULTIPLE INTEGRALS

In Section 3.3 we saw that a change of variables may simplify the evaluation of an ordinary integral. We now consider change of variables in multiple integrals. To do this we must first look more carefully at the notion of the content of a set.

Jordan-measurable sets

In Section 7.1 we defined the content of a set S to be

$$V(S) = \int_S d\mathbf{X} \tag{1}$$

if the integral exists. If R is a rectangle containing S, then (1) can be rewritten as

$$V(S) = \int_R \psi_S(\mathbf{X}) \, d\mathbf{X}, \tag{2}$$

where ψ_S is the characteristic function of S, defined by

$$\psi_S(\mathbf{X}) = \begin{cases} 1, & \mathbf{X} \in S, \\ 0, & \mathbf{X} \notin S. \end{cases}$$

From *Exercise 21, Section 7.1*, the existence and value of $V(S)$ does not depend on the particular choice of the enclosing rectangle R.

We now examine this definition more closely, letting S be an arbitrary bounded set in $\mathscr{R}^n$ and R be a rectangle containing S, and considering the upper and lower integrals of ψ_S over R. Let $\mathbf{P} = \{R_1, R_2, \ldots, R_k\}$ be a partition of R. For each j in $\{1, 2, \ldots, k\}$ there are three possibilities:

1. $R_j \subset S$; then

$$\min_{\mathbf{X} \in R_j} \psi_S(\mathbf{X}) = \max_{\mathbf{X} \in R_j} \psi_S(\mathbf{X}) = 1.$$

2. $R_j \cap S \neq \varnothing$ and $R_j \cap S^c \neq \varnothing$; then

$$\min_{\mathbf{X} \in R_j} \psi_S(\mathbf{X}) = 0 \quad \text{and} \quad \max_{\mathbf{X} \in R_j} \psi_S(\mathbf{X}) = 1.$$

3. $R_j \subset S^c$; then

$$\min_{\mathbf{X} \in R_j} \psi_S(\mathbf{X}) = \max_{\mathbf{X} \in R_j} \psi_S(\mathbf{X}) = 0.$$

If $\sum_1$ and $\sum_2$ denote summation over those j's for which possibilities 1 and 2 hold, respectively, then the upper and lower sums of ψ_S over $\mathbf{P}$ are

$$S(\mathbf{P}) = \sum_1 V(R_j) + \sum_2 V(R_j) \tag{3}$$
$$= \text{total content of the subrectangles in } \mathbf{P} \text{ that intersect } S,$$

and

$$s(\mathbf{P}) = \sum_1 V(R_j) \tag{4}$$
$$= \text{total content of the subrectangles in } \mathbf{P} \text{ that are contained in } S.$$

The upper and lower integrals

$$\overline{\int_R} \psi_S(\mathbf{X}) \, d\mathbf{X} \quad \text{and} \quad \underline{\int_R} \psi_S(\mathbf{X}) \, d\mathbf{X}$$

are called the *outer* and *inner Jordan content*, respectively, of *S*. The integral (2) exists if and only if they are equal, in which case *S* is said to be *Jordan-measurable*, with *Jordan content*—or simply *content*—$V(S)$.

We leave it to the reader (*Exercise 2*) to show that a set has content zero according to Definition 1.4 if and only if it has Jordan content zero.

Theorem 3.1   *A bounded set S is Jordan-measurable if and only if its boundary has zero content.*

Proof.   Let *R* be a rectangle containing *S*. The set of discontinuities of ψ_S is ∂S (*Exercise 9, Section 2.2*); hence, $\int_R \psi_S(\mathbf{X})\, d\mathbf{X}$ exists if $V(\partial S) = 0$, by Theorem 1.8. For the converse, suppose ∂S does not have zero content and let **P** be a partition of *R*; from (3) and (4),

$$S(\mathbf{P}) - s(\mathbf{P}) = \sum_2 V(R_j),$$

which is the total content of the subrectangles in **P** that intersect both *S* and S^c. Since these subrectangles contain ∂S, which does not have zero content, there is an $\varepsilon_0 > 0$ such that

$$S(\mathbf{P}) - s(\mathbf{P}) \geq \varepsilon_0$$

for every partition **P** of *R*. By Theorem 1.5 this implies that ψ_S is not integrable over *R*, and so *S* is not Jordan-measurable.

Theorems 1.8 and 3.1 imply the following corollary.

Corollary 3.1   *If f is continuous on a Jordan-measurable set S, then $\int_S f(\mathbf{X})\, d\mathbf{X}$ exists.*

Lemma 3.1   *Suppose $V(K) = 0$ and ε and ρ are positive numbers. Then there are cubes $C_1, C_2, \ldots, C_r$ with edges of length $\leq \rho$ such that*

$$K \subset \bigcup_{j=1}^{r} C_j \tag{5}$$

and

$$\sum_{j=1}^{r} V(C_j) < \varepsilon. \tag{6}$$

Proof.   Since $V(K) = 0$,

$$\int_C \psi_K(\mathbf{X})\, d\mathbf{X} = 0 \tag{7}$$

if C is any cube containing K. From (7) and the definition of the integral, there is a $\delta > 0$ such that if $\mathbf{P}$ is any partition of C with $\|\mathbf{P}\| \leq \delta$ and σ is any Riemann sum of ψ_K over $\mathbf{P}$, then

$$\sigma \leq \varepsilon. \tag{8}$$

Now suppose $\mathbf{P} = \{C_1, C_2, \ldots, C_k\}$ is a partition of C into cubes, with

$$\|\mathbf{P}\| \leq \min(\rho, \delta), \tag{9}$$

and let $C_1, C_2, \ldots, C_k$ be numbered so that $C_j \cap K \neq \varnothing$ if $1 \leq j \leq r$ (we assume that $K \neq \varnothing$) and $C_j \cap K = \varnothing$ if $r + 1 \leq j \leq k$. Then (5) holds, and a typical Riemann sum of ψ_K over $\mathbf{P}$ is of the form

$$\sigma = \sum_{j=1}^{r} \psi_K(\mathbf{X}_j) V(C_j),$$

with $\mathbf{X}_j \in C_j$, $j = 1, 2, \ldots, r$. In particular, we can choose $\mathbf{X}_j$ from K, so that $\psi_K(\mathbf{X}_j) = 1$ and

$$\sigma = \sum_{j=1}^{r} V(C_j),$$

and (8) and (9) imply that $C_1, C_2, \ldots, C_r$ have the required properties.

Transformations of Jordan-measurable sets

In Section 6.3 we said that a transformation $\mathbf{G} \colon \mathscr{R}^n \to \mathscr{R}^n$ is regular on an open set S if it is one-to-one and continuously differentiable on S and its Jacobian has no zeros in S. For convenience in this section we will also say that $\mathbf{G}$ is regular on an arbitrary (not necessarily open) set S if it is regular on an open set containing S.

To formulate the theorem on change of variables in multiple integrals, we must first consider the question of preservation of Jordan-measurability under a regular transformation.

Lemma 3.2 *Suppose $\mathbf{G} \colon \mathscr{R}^n \to \mathscr{R}^n$ is continuously differentiable on an open set S and K is a closed subset of S with zero content. Then $\mathbf{G}(K)$ has zero content.*

Proof. Since K is a compact subset of the open set S, there is a positive number ρ_1 such that the compact set

$$K_{\rho_1} = \{\mathbf{X} \mid \text{dist}(\mathbf{X}, K) \leq \rho_1\}$$

is contained in S (*Exercise 28, Section 5.1*). From Lemma 2.3, Section 6.2, there is a constant M such that

$$|\mathbf{G}(\mathbf{Y}) - \mathbf{G}(\mathbf{X})| \leq M|\mathbf{Y} - \mathbf{X}| \quad \text{if} \quad \mathbf{X}, \mathbf{Y} \in K_{\rho_1}. \tag{10}$$

Now suppose $\varepsilon > 0$ and choose $\rho < \rho_1/\sqrt{n}$ in Lemma 3.1, so that the covering cubes $C_1, C_2, \ldots, C_r$ satisfying (5) and (6) are contained in K_{ρ_1} (Figure 3.1). If $\mathbf{X}_j$ is the center of C_j, s_j is its edge length, and $\mathbf{X} \in C_j$, then

$$|\mathbf{X} - \mathbf{X}_j| \leq \frac{\sqrt{n}}{2} s_j$$

(*Exercise 26, Section 7.1*), so (10) implies that

$$|\mathbf{G}(\mathbf{X}) - \mathbf{G}(\mathbf{X}_j)| \leq \frac{M\sqrt{n}}{2} s_j,$$

which in turn implies that $\mathbf{G}(C_j)$ is contained in a cube $\tilde{C}_j$ with edge length $M\sqrt{n}s_j$, centered at $\mathbf{G}(\mathbf{X}_j)$ (*Exercise 26, Section 7.1*). Since

$$V(\tilde{C}_j) = (M\sqrt{n})^n s_j^n = (M\sqrt{n})^n V(C_j),$$

we now see that

$$\mathbf{G}(K) \subset \bigcup_{j=1}^{r} \tilde{C}_j$$

and

$$\sum_{j=1}^{r} V(\tilde{C}_j) \leq (M\sqrt{n})^n \sum_{j=1}^{r} V(C_j) < (M\sqrt{n})^n \varepsilon.$$

Since $(M\sqrt{n})^n$ does not depend on ε, it follows that $V(\mathbf{G}(K)) = 0$.

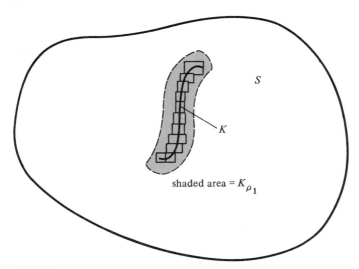

FIGURE 3.1

Theorem 3.2 *Suppose* $\mathbf{G}\colon \mathscr{R}^n \to \mathscr{R}^n$ *is regular on a closed Jordan-measurable set* S. *Then* $\mathbf{G}(S)$ *is closed and Jordan-measurable.*

Proof. We leave to the reader the proof that $\mathbf{G}(S)$ is closed (*Exercise 25, Section 6.2*). Since S is Jordan-measurable, Theorem 3.1 implies that $V(\partial S) = 0$, and therefore $V(\mathbf{G}(\partial S)) = 0$, by Lemma 3.2. But $\mathbf{G}(\partial S) = \partial(\mathbf{G}(S))$ (*Exercise 29, Section 6.3*), so $V(\partial(\mathbf{G}(S))) = 0$, which implies that $\mathbf{G}(S)$ is Jordan-measurable, again by Theorem 3.1.

Change of content under a linear transformation

To motivate and prove the main result of this section—the rule for change of variables in multiple integrals—we need to know how $V(\mathbf{L}(S))$ is related to $V(S)$ if S is a closed Jordan-measurable set and $\mathbf{L}$ is a nonsingular linear transformation. [Theorem 3.2 implies that $\mathbf{L}(S)$ is Jordan-measurable in this case.] The next lemma from linear algebra will help to establish this relationship. We omit its proof.

Lemma 3.3 *A nonsingular* $n \times n$ *matrix* $\mathbf{A}$ *can be written as*

$$\mathbf{A} = \mathbf{E}_k \mathbf{E}_{k-1} \cdots \mathbf{E}_1, \tag{11}$$

where each $\mathbf{E}_i$ *is a matrix which can be obtained from the* $n \times n$ *identity matrix* $\mathbf{I}$ *by one of the following operations:*
  (a) *interchanging two rows of* $\mathbf{I}$;
  (b) *multiplying a row of* $\mathbf{I}$ *by a nonzero constant;*
  (c) *adding a multiple of one row of* $\mathbf{I}$ *to another.*

Matrices of the kind described in this lemma are called *elementary* matrices. The key to the proof of the lemma is that if $\mathbf{E}$ is an elementary $n \times n$ matrix and $\mathbf{A}$ is any $n \times n$ matrix (in fact, $\mathbf{A}$ can be $n \times p$, with $n \neq p$), then $\mathbf{EA}$ equals the matrix obtained by applying to $\mathbf{A}$ the same operation that must be applied to $\mathbf{I}$ to produce $\mathbf{E}$ (*Exercise 7*); also, the inverse of an elementary matrix of type (a), (b), or (c) is an elementary matrix of the same type (*Exercise 8*).

The next example illustrates the procedure for finding the factorization (11).

Example 3.1 The matrix

$$\mathbf{A} = \begin{bmatrix} 0 & 1 & 1 \\ 1 & 0 & 1 \\ 2 & 2 & 0 \end{bmatrix}$$

is nonsingular, since det $\mathbf{A} = 4$. Interchanging the first two rows yields

$$\mathbf{A}_1 = \begin{bmatrix} 1 & 0 & 1 \\ 0 & 1 & 1 \\ 2 & 2 & 0 \end{bmatrix} = \hat{\mathbf{E}}_1 \mathbf{A},$$

where

$$\hat{\mathbf{E}}_1 = \begin{bmatrix} 0 & 1 & 0 \\ 1 & 0 & 0 \\ 0 & 0 & 1 \end{bmatrix}.$$

Subtracting twice the first row of $\mathbf{A}_1$ from the third yields

$$\mathbf{A}_2 = \begin{bmatrix} 1 & 0 & 1 \\ 0 & 1 & 1 \\ 0 & 2 & -2 \end{bmatrix} = \hat{\mathbf{E}}_2 \hat{\mathbf{E}}_1 \mathbf{A},$$

where

$$\hat{\mathbf{E}}_2 = \begin{bmatrix} 1 & 0 & 0 \\ 0 & 1 & 0 \\ -2 & 0 & 1 \end{bmatrix}.$$

Subtracting twice the second row of $\mathbf{A}_2$ from the third yields

$$\mathbf{A}_3 = \begin{bmatrix} 1 & 0 & 1 \\ 0 & 1 & 1 \\ 0 & 0 & -4 \end{bmatrix} = \hat{\mathbf{E}}_3 \hat{\mathbf{E}}_2 \hat{\mathbf{E}}_1 \mathbf{A},$$

where

$$\hat{\mathbf{E}}_3 = \begin{bmatrix} 1 & 0 & 0 \\ 0 & 1 & 0 \\ 0 & -2 & 1 \end{bmatrix}.$$

Multiplying the third row of $\mathbf{A}_3$ by $-\frac{1}{4}$ yields

$$\mathbf{A}_4 = \begin{bmatrix} 1 & 0 & 1 \\ 0 & 1 & 1 \\ 0 & 0 & 1 \end{bmatrix} = \hat{\mathbf{E}}_4 \hat{\mathbf{E}}_3 \hat{\mathbf{E}}_2 \hat{\mathbf{E}}_1 \mathbf{A},$$

where

$$\hat{\mathbf{E}}_4 = \begin{bmatrix} 1 & 0 & 0 \\ 0 & 1 & 0 \\ 0 & 0 & -\frac{1}{4} \end{bmatrix}.$$

Subtracting the third row of $\mathbf{A}_4$ from the first yields

$$\mathbf{A}_5 = \begin{bmatrix} 1 & 0 & 0 \\ 0 & 1 & 1 \\ 0 & 0 & 1 \end{bmatrix} = \hat{\mathbf{E}}_5 \hat{\mathbf{E}}_4 \hat{\mathbf{E}}_3 \hat{\mathbf{E}}_2 \hat{\mathbf{E}}_1 \mathbf{A},$$

where

$$\hat{\mathbf{E}}_5 = \begin{bmatrix} 1 & 0 & -1 \\ 0 & 1 & 0 \\ 0 & 0 & 1 \end{bmatrix}.$$

Finally, subtracting the third row of $\mathbf{A}_5$ from the second yields

$$I = \hat{\mathbf{E}}_6 \hat{\mathbf{E}}_5 \hat{\mathbf{E}}_4 \hat{\mathbf{E}}_3 \hat{\mathbf{E}}_2 \hat{\mathbf{E}}_1 \mathbf{A}, \tag{12}$$

where

$$\hat{\mathbf{E}}_6 = \begin{bmatrix} 1 & 0 & 0 \\ 0 & 1 & -1 \\ 0 & 0 & 1 \end{bmatrix}.$$

From (12) and Theorem 1.7, Section 6.1,

$$\mathbf{A} = (\hat{\mathbf{E}}_6 \hat{\mathbf{E}}_5 \hat{\mathbf{E}}_4 \hat{\mathbf{E}}_3 \hat{\mathbf{E}}_2 \hat{\mathbf{E}}_1)^{-1} = \hat{\mathbf{E}}_1^{-1} \hat{\mathbf{E}}_2^{-1} \hat{\mathbf{E}}_3^{-1} \hat{\mathbf{E}}_4^{-1} \hat{\mathbf{E}}_5^{-1} \hat{\mathbf{E}}_6^{-1}.$$

Therefore,

$$\mathbf{A} = \mathbf{E}_6 \mathbf{E}_5 \mathbf{E}_4 \mathbf{E}_3 \mathbf{E}_2 \mathbf{E}_1,$$

where

$$\mathbf{E}_1 = \hat{\mathbf{E}}_6^{-1} = \begin{bmatrix} 1 & 0 & 0 \\ 0 & 1 & 1 \\ 0 & 0 & 1 \end{bmatrix}, \qquad \mathbf{E}_2 = \hat{\mathbf{E}}_5^{-1} = \begin{bmatrix} 1 & 0 & 1 \\ 0 & 1 & 0 \\ 0 & 0 & 1 \end{bmatrix},$$

$$\mathbf{E}_3 = \hat{\mathbf{E}}_4^{-1} = \begin{bmatrix} 1 & 0 & 0 \\ 0 & 1 & 0 \\ 0 & 0 & -4 \end{bmatrix}, \qquad \mathbf{E}_4 = \hat{\mathbf{E}}_3^{-1} = \begin{bmatrix} 1 & 0 & 0 \\ 0 & 1 & 0 \\ 0 & 2 & 1 \end{bmatrix},$$

$$\mathbf{E}_5 = \hat{\mathbf{E}}_2^{-1} = \begin{bmatrix} 1 & 0 & 0 \\ 0 & 1 & 0 \\ 2 & 0 & 1 \end{bmatrix}, \qquad \mathbf{E}_6 = \hat{\mathbf{E}}_1^{-1} = \begin{bmatrix} 0 & 1 & 0 \\ 1 & 0 & 0 \\ 0 & 0 & 1 \end{bmatrix}.$$

[*Exercise 8(c)*].

Lemma 3.3 and Theorem 1.2(b), Section 6.1, imply that an arbitrary invertible linear transformation $\mathbf{L}: \mathscr{R}^n \to \mathscr{R}^n$, defined by

$$\mathbf{L}(\mathbf{X}) = \mathbf{A}\mathbf{X}, \tag{13}$$

can be written as a composition

$$\mathbf{L} = \mathbf{L}_k \circ \mathbf{L}_{k-1} \circ \cdots \circ \mathbf{L}_1, \tag{14}$$

where

$$\mathbf{L}_i(\mathbf{X}) = \mathbf{E}_i \mathbf{X}.$$

Theorem 3.3   *If S is a closed Jordan-measurable subset of $\mathscr{R}^n$ and $\mathbf{L}: \mathscr{R}^n \to \mathscr{R}^n$ is the invertible linear transformation (13), then*

$$V(\mathbf{L}(S)) = |\det \mathbf{A}| V(S). \tag{15}$$

Proof.   Theorem 3.2 implies that $\mathbf{L}(S)$ is Jordan-measurable. If (15) holds whenever S is a rectangle, then it holds for any closed Jordan-measurable set. To see this, let R be a rectangle containing S and, if $\varepsilon > 0$, let $\mathbf{P}$ be a partition of R such that

$$S(\mathbf{P}) - s(\mathbf{P}) < \varepsilon, \tag{16}$$

where $S(\mathbf{P})$ and $s(\mathbf{P})$ are the upper and lower sums of the characteristic function ψ_S over $\mathbf{P}$. With the notation introduced before (3) and (4),

$$s(\mathbf{P}) = \sum_1 V(R_j) \le V(S) \le \sum_1 V(R_j) + \sum_2 V(R_j) = S(\mathbf{P}), \tag{17}$$

and

$$\sum_1 V(\mathbf{L}(R_j)) \le V(\mathbf{L}(S)) \le \sum_1 V(\mathbf{L}(R_j)) + \sum_2 V(\mathbf{L}(R_j)).$$

If we assume that (15) holds whenever S is a rectangle, then

$$V(\mathbf{L}(R_j)) = |\det \mathbf{A}| V(R_j),$$

so the last equation implies that

$$s(\mathbf{P}) \le \frac{V(\mathbf{L}(S))}{|\det \mathbf{A}|} = S(\mathbf{P}).$$

This, (16), and (17) imply that

$$\left| V(S) - \frac{V(\mathbf{L}(S))}{|\det \mathbf{A}|} \right| < \varepsilon;$$

hence, since ε can be made arbitrarily small, (15) follows for any Jordan-measurable set.

Now suppose $\mathbf{A}$ in (13) is an elementary matrix; that is,

$$\mathbf{L}(\mathbf{X}) = \mathbf{Y} = \mathbf{E}\mathbf{X},$$

and suppose

$$R = [a_1, b_1] \times [a_2, b_2] \times \cdots \times [a_n, b_n] = I_1 \times I_2 \times \cdots \times I_n.$$

Case 1.   If **E** is obtained by interchanging the ith and jth rows of **I**, then

$$y_r = \begin{cases} x_r & \text{if } r \neq i \text{ and } r \neq j; \\ x_j & \text{if } r = i; \\ x_i & \text{if } r = j; \end{cases}$$

hence $L(R)$ is the Cartesian product of $I_1, I_2, \ldots, I_n$ with I_i and I_j interchanged; therefore,

$$V(L(R)) = V(R) = |\det \mathbf{E}| V(R),$$

since $\det \mathbf{E} = -1$ in this case [*Exercise 8(a)*].

    Case 2.   If **E** is obtained by multiplying the rth row of **I** by a, then

$$y_r = \begin{cases} x_r & \text{if } r \neq i; \\ ax_i & \text{if } r = i; \end{cases}$$

and so

$$L(R) = I_1 \times \cdots \times I_{i-1} \times I_i' \times I_{i+1} \times \cdots \times I_n,$$

where I_i' is an interval with length equal to $|a|$ times the length of I_i; hence,

$$V(L(R)) = |a| V(R) = |\det \mathbf{E}| V(R),$$

since $\det \mathbf{E} = a$ in this case [*Exercise 8(a)*].

    Case 3.   If **E** is obtained by adding a times the jth row of **I** to its ith row, then

$$y_r = \begin{cases} x_r & \text{if } r \neq i; \\ x_i + ax_j & \text{if } r = i; \end{cases}$$

hence

$$L(R) = \{(y_1, y_2, \ldots, y_n) \mid a_i + ay_j \leq y_i \leq b_i + ay_j \text{ and } a_r \leq y_r \leq b_r \text{ if } r \neq i\},$$

which is a parallelogram if $n = 2$ and a parallelepiped if $n = 3$ (Figure 3.2). Now

$$V(L(R)) = \int_{L(R)} d\mathbf{Y},$$

which we can evaluate as an iterated integral in which the first integration is with respect to y_i. For example, if $i = 1$, then

$$V(L(R)) = \int_{a_n}^{b_n} dy_n \int_{a_{n-1}}^{b_{n-1}} dy_{n-1} \cdots \int_{a_2}^{b_2} dy_2 \int_{a_1 + ay_j}^{b_1 + ay_j} dy_1. \tag{18}$$

Since

$$\int_{a_1 + ay_j}^{b_1 + ay_j} dx_1 = \int_{a_1}^{b_1} dx_1,$$

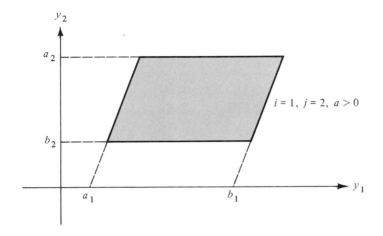

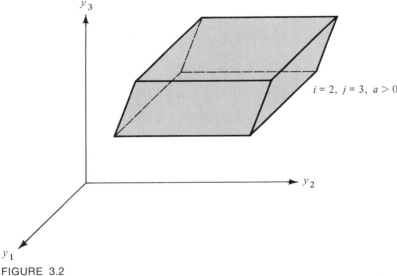

FIGURE 3.2

(18) can be rewritten as

$$V(\mathbf{L}(R)) = \int_{a_n}^{b_n} dy_n \int_{a_{n-1}}^{b_{n-1}} dy_{n-1} \cdots \int_{a_2}^{b_2} dy_2 \int_{a_1}^{b_1} dy_1$$
$$= (b_n - a_n)(b_{n-1} - a_{n-1}) \cdots (b_1 - a_1) = V(R).$$

Hence,

$$V(\mathbf{L}(R)) = |\det \mathbf{E}| V(R),$$

since $\det \mathbf{E} = 1$ in this case [*Exercise 8(a)*].

From what we have shown so far, (15) holds if **A** is an elementary matrix and S is any closed Jordan-measurable set. If **A** is an arbitrary nonsingular matrix, then we can write it as a product of elementary matrices (11) and apply our known result successively to $\mathbf{L}_1, \mathbf{L}_2, \ldots, \mathbf{L}_k$ [see (14)]; this yields

$$V(\mathbf{L}(S)) = |\det \mathbf{E}_k| |\det \mathbf{E}_{k-1}| \cdots |\det \mathbf{E}_1| V(S) = |\det \mathbf{A}| V(S),$$

by Theorem 1.6, Section 6.1. We leave the details to the reader (*Exercise 12*).

Formulation of the rule for change of variables

For the present it is convenient to think of the range and domain of a transformation $\mathbf{G}: \mathcal{R}^n \to \mathcal{R}^n$ as subsets of distinct copies of $\mathcal{R}^n$. We will denote the copy containing $D_\mathbf{G}$ as $\mathcal{U}^n$, and write $\mathbf{G}: \mathcal{U}^n \to \mathcal{R}^n$ and

$$\mathbf{X} = \mathbf{G}(\mathbf{U}).$$

Although this reverses the previous roles of **X** and **U** [we previously wrote $\mathbf{U} = \mathbf{G}(\mathbf{X})$], it is convenient here and should not cause confusion.

If **G** is regular (actually it need only be one-to-one) on a subset S of $\mathcal{U}^n$, then each **X** in $\mathbf{G}(S)$ can be identified by specifying the unique point **U** in S such that $\mathbf{X} = \mathbf{G}(\mathbf{U})$. We say that the components of **U** are the **U**-*coordinates of* **X**. We will give examples later in this section.

We now formulate the rule for change of variables in a multiple integral. Since we are for the present interested only in "discovering" the rule, we will make any assumptions that ease this task, deferring questions of rigor until the proof.

Suppose we wish to evaluate $\int_T f(\mathbf{X}) \, d\mathbf{X}$, where T is the image of a set S under the regular transformation $\mathbf{X} = \mathbf{G}(\mathbf{U})$. For simplicity we take S to be a rectangle and assume that f is continuous on $T = \mathbf{G}(S)$.

Now suppose $\mathbf{P} = \{R_1, R_2, \ldots, R_k\}$ is a partition of S and $T_j = \mathbf{G}(R_j)$ (Figure 3.3); then

$$\int_T f(\mathbf{X}) \, d\mathbf{X} = \sum_{j=1}^k \int_{T_j} f(\mathbf{X}) \, d\mathbf{X}. \tag{19}$$

Since f is continuous there is a point $\mathbf{X}_j$ in T_j such that

$$\int_{T_j} f(\mathbf{X}) \, d\mathbf{X} = f(\mathbf{X}_j) V(T_j)$$

(Theorem 1.16), so (19) can be rewritten as

$$\int_T f(\mathbf{X}) \, d\mathbf{X} = \sum_{j=1}^k f(\mathbf{X}_j) V(T_j). \tag{20}$$

Now we approximate $V(T_j)$. If

$$\mathbf{X}_j = \mathbf{G}(\mathbf{U}_j), \tag{21}$$

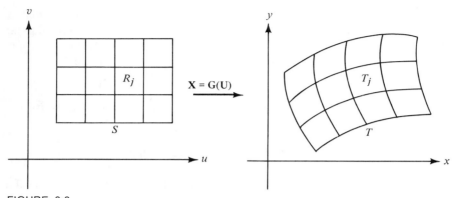

FIGURE 3.3

then $\mathbf{U}_j \in R_j$ and, since $\mathbf{G}$ is differentiable at $\mathbf{U}_j$,

$$\mathbf{G}(\mathbf{U}) \approx \mathbf{G}(\mathbf{U}_j) + \mathbf{G}'(\mathbf{U}_j)(\mathbf{U} - \mathbf{U}_j) \tag{22}$$

(here $\mathbf{G}$ and $\mathbf{U} - \mathbf{U}_j$ are written as column matrices and $\mathbf{G}'$ is a differential matrix), where "$\approx$" means "approximately equal" in a sense that we could make precise if we wished (see Theorem 2.2, Section 6.2).

The affine transformation on the right of (22) can be thought of as the composition of the linear transformation

$$\mathbf{L}(\mathbf{U}) = \mathbf{G}'(\mathbf{U}_j)\mathbf{U} \tag{23}$$

and the "shift"

$$\mathbf{X} = [\mathbf{G}(\mathbf{U}_j) - \mathbf{G}'(\mathbf{U}_j)] + \mathbf{L}(\mathbf{U}),$$

where the vector in parentheses is constant. The shift does not change the content of a set, and the linear transformation (23) multiplies the content by

$$|\det \mathbf{G}'(\mathbf{U}_j)| = |J\mathbf{G}(\mathbf{U}_j)|,$$

where $J\mathbf{G}$ is the Jacobian of $\mathbf{G}$. Here we know that $\mathbf{G}'(\mathbf{U}_j)$ is nonsingular because $\mathbf{G}$ is regular, and therefore Theorem 3.3 implies that

$$V(T_j) \approx |J\mathbf{G}(\mathbf{U}_j)|V(R_j);$$

substituting this and (21) into (20) yields

$$\int_T f(\mathbf{X})\, d\mathbf{X} \approx \sum_{j=1}^{k} f(\mathbf{G}(\mathbf{U}_j))|J\mathbf{G}(\mathbf{U}_j)|V(R_j).$$

But the sum on the right is a Riemann sum for the integral

$$\int_S f(\mathbf{G}(\mathbf{U}))|J\mathbf{G}(\mathbf{U})|\, d\mathbf{U};$$

thus, if we could show that the errors in the various uses of "$\approx$" can be made arbitrarily small by choosing $\|\mathbf{P}\|$ sufficiently small, then we could conclude that

$$\int_T f(\mathbf{X}) \, d\mathbf{X} = \int_S f(\mathbf{G}(\mathbf{U}))|J\mathbf{G}(\mathbf{U})| \, d\mathbf{U}.$$

This can be done, but it is difficult; rather, we will use a different approach, which was published in the *American Mathematical Monthly* [Vol. 61 (1948), pp. 81–85] by J. Schwartz.

The main theorem

We now prove the following form of the rule for change of variable in a multiple integral.

Theorem 3.4   *Suppose* $\mathbf{G}$ $\mathscr{U}^n \to \mathscr{R}^n$ *is regular on a closed Jordan-measurable set S and f is continuous on* $\mathbf{G}(S)$. *Then*

$$\int_{\mathbf{G}(S)} f(\mathbf{X}) \, d\mathbf{X} = \int_S f(\mathbf{G}(\mathbf{U}))|J\mathbf{G}(\mathbf{U})| \, d\mathbf{U}. \tag{24}$$

Before turning to the proof we first observe that both integrals in (24) exist, by Corollary 3.1, since their integrands are continuous; S is Jordan-measurable by assumption, and $\mathbf{G}(S)$ is Jordan-measurable by Theorem 3.2.

For the proof we need the following definition and lemmas.

Definition 3.1   *If* $\mathbf{A} = [a_{ij}]$ *is an* $n \times n$ *matrix, let*

$$|\mathbf{A}| = \max_{1 \le i \le n} \sum_{j=1}^n |a_{ij}|.$$

Lemma 3.4   *Suppose* $\mathbf{G} \colon \mathscr{U}^n \to \mathscr{R}^n$ *is continuously differentiable on a cube* C *in* $\mathscr{U}^n$, *and let* $\mathbf{A}$ *be a nonsingular* $n \times n$ *matrix. Then*

$$V(\mathbf{G}(C)) \le |\det \mathbf{A}| \left\{ \max_{\mathbf{Y} \in C} |\mathbf{A}^{-1}\mathbf{G}'(\mathbf{Y})| \right\}^n V(C). \tag{25}$$

Proof.   Suppose the edges of C have length s, let $\mathbf{U}_0 = (c_1, c_2, \ldots, c_n)$ be its center, and suppose $\mathbf{U} \in C$. If $\mathbf{H} = (h_1, h_2, \ldots, h_n)$ is continuously differentiable on $\mathbf{C}$, then applying the mean value theorem to the components of $\mathbf{H}$ yields

$$h_i(\mathbf{U}) - h_i(\mathbf{U}_0) = \sum_{j=1}^n \frac{\partial h_i(\mathbf{Y}_i)}{\partial u_j}(u_j - c_j),$$

where $\mathbf{Y}_i \in C$; hence,

$$|h_i(\mathbf{U}) - h_i(\mathbf{U}_0)| \leq \frac{s}{2} \max_{\mathbf{Y} \in C} |\mathbf{H}'(\mathbf{Y})|, \qquad 1 \leq i \leq n$$

(see Definition 3.1). This means that $\mathbf{H}(C)$ is contained in a cube with center $\mathbf{X}_0 = \mathbf{H}(\mathbf{U}_0)$ and edges of length

$$s \max_{\mathbf{Y} \in C} |\mathbf{H}'(\mathbf{Y})|;$$

therefore

$$V(\mathbf{H}(C)) \leq \left\{ \max_{\mathbf{Y} \in C} |\mathbf{H}'(\mathbf{Y})| \right\}^n s^n = \left\{ \max_{\mathbf{Y} \in C} |\mathbf{H}'(\mathbf{Y})| \right\}^n V(C). \tag{26}$$

Now let

$$\mathbf{L}(\mathbf{X}) = \mathbf{A}^{-1}\mathbf{X}$$

and set $\mathbf{H} = \mathbf{L} \circ \mathbf{G}$; then

$$\mathbf{H}(C) = \mathbf{L}(\mathbf{G}(C)) \quad \text{and} \quad \mathbf{H}' = \mathbf{A}^{-1}\mathbf{G}',$$

so (26) implies that

$$V(\mathbf{L}(\mathbf{G}(C))) \leq \left\{ \max_{\mathbf{Y} \in C} |\mathbf{A}^{-1}\mathbf{G}'(\mathbf{Y})| \right\}^n V(C). \tag{27}$$

Since $\mathbf{L}$ is linear, Theorem 3.3 implies that

$$V(\mathbf{L}(\mathbf{G}(C))) = |\det \mathbf{A}^{-1}| V(\mathbf{G}(C)),$$

and this together with (27) implies (25).

Lemma 3.5 *If $\mathbf{G}: \mathcal{U}^n \to \mathcal{R}^n$ is regular on a cube C in $\mathcal{R}^n$, then*

$$V(\mathbf{G}(C)) \leq \int_C |J\mathbf{G}(\mathbf{U})| \, d\mathbf{U}. \tag{28}$$

Proof. Let $\mathbf{P}$ be a partition of C into subcubes $C_1, C_2, \ldots, C_k$ with centers $\mathbf{U}_1, \mathbf{U}_2, \ldots, \mathbf{U}_k$. Then

$$V(\mathbf{G}(C)) = \sum_{j=1}^{k} V(\mathbf{G}(C_j)). \tag{29}$$

Applying Lemma 3.4 to C_j with $\mathbf{A} = \mathbf{G}'(\mathbf{U}_j)$ yields

$$V(\mathbf{G}(C_j)) \leq |J\mathbf{G}(\mathbf{U}_j)| \left\{ \max_{\mathbf{Y} \in C_j} |(\mathbf{G}'(\mathbf{U}_j))^{-1}\mathbf{G}'(\mathbf{Y})| \right\}^n V(C_j). \tag{30}$$

Exercise 24, Section 6.1 implies that if $\varepsilon > 0$, there is a $\delta > 0$ such that

$$\max_{Y \in C_j} \left| (G'(U_j))^{-1} G'(Y) \right| < 1 + \varepsilon \qquad \text{if } \|P\| < \delta;$$

therefore, from (30),

$$V(G(C_j)) \leq (1 + \varepsilon)^n \left| JG(U_j) \right| V(C_j),$$

and so, from (29),

$$V(G(C)) \leq (1 + \varepsilon)^n \sum_{j=1}^{k} \left| JG(U_j) \right| V(C_j) \qquad \text{if } \|P\| < \delta. \tag{31}$$

Since the sum on the right is a Riemann sum for $\int_C |JG(U)| \, dU$ and ε can be taken arbitrarily small, this implies (28) (*Exercise 13*).

Proof of Theorem 3.4.   The result is trivial if $V(S) = 0$, since then both integrals in (24) vanish; hence, we assume that $V(S) > 0$.

From the continuity of JG and f on the compact sets S and $G(S)$, there are constants M_1 and M_2 such that

$$\left| JG(U) \right| \leq M_1 \quad \text{if} \quad U \in S \tag{32}$$

and

$$\left| f(X) \right| \leq M_2 \quad \text{if} \quad X \in G(S). \tag{33}$$

Moreover, $f \circ G$ is uniformly continuous on S, and so if $\varepsilon > 0$, there is a $\delta > 0$ such that

$$\left| f(G(U)) - f(G(U_0)) \right| < \varepsilon \quad \text{if} \quad |U - U_0| < \delta \text{ and } U \text{ and } U_0 \text{ are in } S.$$

Now let $C_1, C_2, \ldots, C_k$ be cubes contained in S, with edge lengths less than $\delta/\sqrt{n}$, such that $C_i^0 \cap C_j^0 = \varnothing$ if $i \neq j$ and

$$V(S) < \sum_{j=1}^{k} V(C_j) + \varepsilon$$

(*Exercise 5*). Then

$$S = \left(\bigcup_{j=1}^{k} C_j \right) \cup S_1,$$

where $V(S_1) < \varepsilon$, and $V(S_1 \cap C_j) = 0 \, (j = 1, 2, \ldots, k)$.
We first show that the quantity

$$Q(S) = \int_{G(S)} f(X) \, dX - \int_S f(G(U)) \left| JG(U) \right| \, dU \tag{34}$$

is nonpositive if

$$f(X) \geq 0 \qquad \text{for } X \in G(S). \tag{35}$$

Suppose $\mathbf{U}_1, \mathbf{U}_2, \ldots, \mathbf{U}_k$ are points in $C_1, C_2, \ldots, C_k$, and $\mathbf{X}_j = \mathbf{G}(\mathbf{U}_j)$; then

$$Q(S) = \int_{\mathbf{G}(S_1)} f(\mathbf{X}) \, d\mathbf{X} - \int_{S_1} f(\mathbf{G}(\mathbf{U})) |J\mathbf{G}(\mathbf{U})| \, d\mathbf{U}$$

$$+ \sum_{j=1}^{k} \int_{\mathbf{G}(C_j)} f(\mathbf{X}) \, d\mathbf{X} - \sum_{j=1}^{k} \int_{C_j} f(\mathbf{G}(\mathbf{U})) |J\mathbf{G}(\mathbf{U})| \, d\mathbf{U}$$

$$= \int_{\mathbf{G}(S_1)} f(\mathbf{X}) \, d\mathbf{X} - \int_{S_1} f(\mathbf{G}(\mathbf{U})) |J\mathbf{G}(\mathbf{U})| \, d\mathbf{U}$$

$$+ \sum_{j=1}^{k} \int_{\mathbf{G}(C_j)} [f(\mathbf{X}) - f(\mathbf{X}_j)] \, d\mathbf{X}$$

$$+ \sum_{j=1}^{k} \int_{C_j} [f(\mathbf{G}(\mathbf{U}_j)) - f(\mathbf{G}(\mathbf{U})] |J\mathbf{G}(\mathbf{U})| \, d\mathbf{U}$$

$$+ \sum_{j=1}^{k} f(\mathbf{X}_j) \left[V(\mathbf{G}(C_j)) - \int_{C_j} |J\mathbf{G}(\mathbf{U})| \, d\mathbf{U} \right].$$

From (35) and Lemma 3.5, the last sum is nonpositive; moreover, the following inequalities apply to the other terms:

$$\int_{\mathbf{G}(S_1)} f(\mathbf{X}) \, d\mathbf{X} \leq M_1 M_2 \varepsilon, \tag{36}$$

$$\int_{S_1} f(\mathbf{G}(\mathbf{U})) |J\mathbf{G}(\mathbf{U})| \, d\mathbf{U} \geq 0,$$

$$\sum_{j=1}^{k} \int_{\mathbf{G}(C_j)} [f(\mathbf{X}) - f(\mathbf{X}_j)] \, d\mathbf{X} \leq M_1 V(S) \varepsilon, \tag{37}$$

$$\sum_{j=1}^{k} \int_{C_j} [f(\mathbf{G}(\mathbf{U}_j)) - f(\mathbf{G}(\mathbf{U}))] |J\mathbf{G}(\mathbf{U})| \, d\mathbf{U} \leq M_1 V(S) \varepsilon \tag{38}$$

(*Exercise 14*). Since ε can be made arbitrarily small, this implies that $Q(S) \leq 0$; that is,

$$\int_{\mathbf{G}(S)} f(\mathbf{X}) \, d\mathbf{X} \leq \int_{S} f(\mathbf{G}(\mathbf{U})) |J\mathbf{G}(\mathbf{U})| \, d\mathbf{U} \tag{39}$$

if $f(\mathbf{X}) \geq 0$ for all $\mathbf{X}$ in $\mathbf{G}(S)$.

Since $\mathbf{G}^{-1}$ is regular on $\mathbf{G}(S)$ and $(|J\mathbf{G}|)f \circ \mathbf{G}$ is continuous on S, we can apply (39) with f, $\mathbf{G}$, and S replaced by $(|J\mathbf{G}|)f \circ \mathbf{G}$, $\mathbf{G}^{-1}$, and $\mathbf{G}(S)$, respectively, and then interchange $\mathbf{X}$ and $\mathbf{U}$. This yields the inequality

$$\int_{S} f(\mathbf{G}(\mathbf{U})) |J\mathbf{G}(\mathbf{U})| \, d\mathbf{U} \leq \int_{\mathbf{G}(S)} f(\mathbf{X}) \, d\mathbf{X}, \tag{40}$$

again if $f(\mathbf{X}) \geq 0$ for all $\mathbf{X}$ in $\mathbf{G}(S)$. With (39), this implies (24) for such functions. For a function f which may assume both positive and negative values, the conclusion follows on applying this result to $g = f - m$, where

$$m = \min_{\mathbf{X} \in \mathbf{G}(S)} f(\mathbf{X})$$

(*Exercise 16*).

The assumptions of Theorem 3.4 are too stringent for some very simple applications. For example, to find the area of the disc

$$\{(x, y) \,|\, x^2 + y^2 \leq 1\},$$

it is convenient to use polar coordinates and regard the circle as $\mathbf{G}(S)$, where

$$\mathbf{G}(r, \theta) = \begin{bmatrix} r \cos \theta \\ r \sin \theta \end{bmatrix} \tag{41}$$

and

$$S = \{(r, \theta) \,|\, 0 \leq r \leq 1, 0 \leq \theta \leq 2\pi\} \tag{42}$$

(Figure 3.4). Taking $f \equiv 1$ and $|J\mathbf{G}(r, \theta)| = r$ (*Exercise 26(a), Section 6.2*), we then have

$$A = \int_{\mathbf{G}(S)} d\mathbf{X} = \int_S r \, d(r, \theta) = \int_0^1 r \, dr \int_0^{2\pi} d\theta = \pi.$$

Although this is a familiar result, Theorem 3.4 does not apply here, since $\mathbf{G}$ is not regular on S. (Why not?) The next theorem shows that the assumptions of Theorem 3.4 can be relaxed so as to include this example, and also to permit f to have discontinuities.

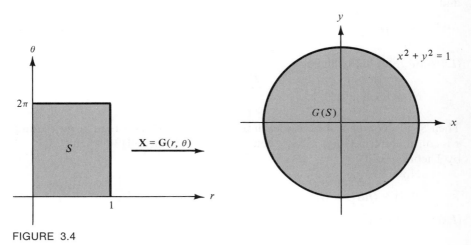

FIGURE 3.4

Theorem 3.5   *Suppose* $\mathbf{G}: \mathcal{U}^n \to \mathcal{R}^n$ *is continuously differentiable on an open set N containing the closed Jordan-measurable set S, and regular on S^0. Suppose also that $\mathbf{G}(S)$ is Jordan-measurable and f is continuous on $\mathbf{G}(S)$ except on a subset T of zero content. Then*

$$\int_{\mathbf{G}(S)} f(\mathbf{X}) \, d\mathbf{X} = \int_S f(\mathbf{G}(\mathbf{U})) |J\mathbf{G}(\mathbf{U})| \, d\mathbf{U}. \tag{43}$$

Proof.   The integral on the left of (43) exists by Corollary 3.1; since the set

$$D = \{\mathbf{U} \,|\, \mathbf{G}(\mathbf{U}) \in T\}$$

has zero content (*Exercise 17*) and contains the discontinuities of $(|J\mathbf{G}|) f \circ \mathbf{G}$, the integral on the right also exists by Corollary 3.1. Now let

$$\rho = \text{dist}(\partial S \cup \bar{D}, N^c)$$

(*Exercise 27, Section 5.1*), and

$$P = \{\mathbf{U} \,|\, \text{dist}(\mathbf{U}, \partial S \cup \bar{D})\} \le \frac{\rho}{2}$$

(*Exercise 28, Section 5.1*). Then P is a compact subset of N and

$$\partial S \cup \bar{D} \subset P^0$$

(Figure 3.5). If $\varepsilon > 0$, let $C_1, C_2, \ldots, C_k$ be cubes in P^0 such that

$$\partial S \cup \bar{D} \subset \bigcup_{j=1}^{k} C_j^0 \tag{44}$$

and

$$\sum_{j=1}^{k} V(C_j) < \varepsilon \tag{45}$$

(*Exercise 18*). Finally, define

$$S_1 = S \cap \left(\bigcup_{j=1}^{k} C_j \right)^c.$$

Then S_1 is a compact Jordan-measurable subset of S^0, $\mathbf{G}$ is regular on S_1, and f is continuous on $\mathbf{G}(S_1)$. Consequently, if Q is as defined in (34), then $Q(S_1) = 0$ by Theorem 3.4.

Now

$$Q(S) = Q(S_1) + Q(S - S_1) \tag{46}$$

(*Exercise 19*), and

$$|Q(S - S_1)| \le \left| \int_{\mathbf{G}(S - S_1)} f(\mathbf{X}) \, d\mathbf{X} \right| + \left| \int_{S - S_1} f(\mathbf{G}(\mathbf{U})) |J\mathbf{G}(\mathbf{U})| \, d\mathbf{U} \right|.$$

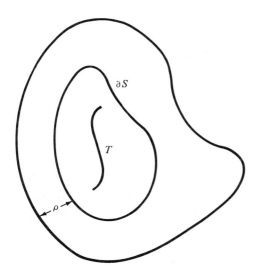

N = open set bounded by outer curve
S = closed set bounded by inner curve
FIGURE 3.5

But

$$\left| \int_{S-S_1} f(\mathbf{G}(\mathbf{U}))|J\mathbf{G}(\mathbf{U})|\, d\mathbf{U} \right| \le M_1 M_2 V(S - S_1) \le M_1 M_2 \varepsilon, \tag{47}$$

where M_1 and M_2 are as defined in (32) and (33), and

$$\left| \int_{\mathbf{G}(S-S_1)} f(\mathbf{X})\, d\mathbf{X} \right| \le M_2 V(\mathbf{G}(S - S_1)) \le M_2 \sum_{j=1}^{k} V(\mathbf{G}(C_j)). \tag{48}$$

From Lemma 3.4,

$$V(\mathbf{G}(C_j)) \le \left\{ \max_{\mathbf{Y} \in C_j} |\mathbf{G}'(\mathbf{Y})| \right\}^n V(C_j),$$

so (48) can be rewritten as

$$\left| \int_{\mathbf{G}(S-S_1)} f(\mathbf{X})\, d\mathbf{X} \right| \le M_2 \left\{ \max_{\mathbf{Y} \in P} |\mathbf{G}'(Y)| \right\}^n \varepsilon,$$

because of (45). Since ε can be made arbitrarily small, this and (47) imply that $Q(S - S_1) = 0$; hence, $Q(S) = 0$ and the proof is complete.

The transformation to polar coordinates to compute the area of the disc is now justified, since $\mathbf{G}$ and S as defined by (41) and (42) satisfy the assumptions of Theorem 3.5.

Polar coordinates

If **G** is the transformation from polar to rectangular coordinates

$$\begin{bmatrix} x \\ y \end{bmatrix} = \mathbf{G}(r, \theta) = \begin{bmatrix} r \cos \theta \\ r \sin \theta \end{bmatrix}, \tag{49}$$

then $J\mathbf{G}(r, \theta) = r$, and (43) becomes

$$\int_{\mathbf{G}(S)} f(x, y) \, d(x, y) = \int_S f(r \cos \theta, r \sin \theta) r \, d(r, \theta)$$

if we assume, as is conventional, that S is in the closed right half of the (r, θ) plane. This transformation is especially useful when the boundaries of S can be expressed conveniently in terms of polar coordinates, as in the example preceding Theorem 3.5. Two more examples follow.

Example 3.2 To evaluate

$$I = \int_T (x^2 + y) \, d(x, y)$$

where T is the annulus

$$T = \{(x, y) \,|\, 1 \le x^2 + y^2 \le 4\}$$

[Figure 3.6(b)], we write $T = \mathbf{G}(S)$, with $\mathbf{G}$ as in (49) and

$$S = \{(r, \theta) \,|\, 1 \le r \le 2, 0 \le \theta \le 2\pi\}$$

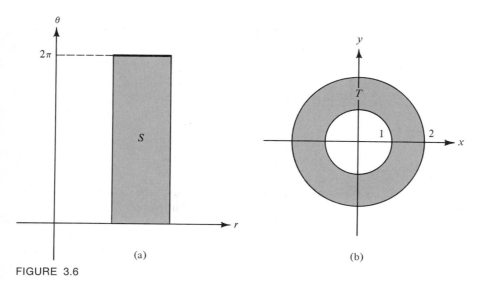

(a)   (b)

FIGURE 3.6

[Figure 3.6(a)]. Theorem 3.5 implies that

$$I = \int_S (r^2 \cos^2 \theta + r \sin \theta) r \, d(r, \theta),$$

which we evaluate as an iterated integral:

$$I = \int_1^2 r^2 \, dr \int_0^{2\pi} (r \cos^2 \theta + \sin \theta) \, d\theta$$

$$= \int_1^2 r^2 \, dr \int_0^{2\pi} \left(\frac{r}{2} + \frac{r}{2} \cos 2\theta + \sin \theta \right) d\theta \quad \left[\text{since } \cos^2 \theta = \frac{1}{2}(1 + \cos 2\theta) \right]$$

$$= \int_1^2 r^2 \left[\frac{r\theta}{2} + \frac{r}{4} \sin 2\theta - \cos \theta \right]\Big|_{\theta=0}^{2\pi} dr = \pi \int_1^2 r^3 \, dr$$

$$= \pi \frac{r^4}{4}\Big|_1^2 = \frac{15\pi}{4}.$$

Example 3.3   To evaluate

$$I = \int_T y \, d(x, y)$$

where T is the region in the xy plane bounded by the curve whose points have polar coordinates satisfying

$$r = 1 - \cos \theta, \qquad 0 \le \theta \le \pi$$

[Figure 3.7(b)], we write $T = \mathbf{G}(S)$, with $\mathbf{G}$ as in (49) and S the shaded region in Figure 3.7(a). From (43),

$$I = \int_S (r \sin \theta) r \, d(r, \theta),$$

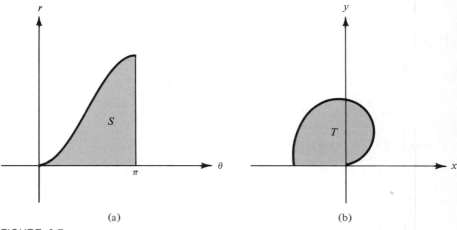

(a)   (b)

FIGURE 3.7

which we evaluate as an iterated integral:

$$I = \int_0^\pi \sin \theta \, d\theta \int_0^{1-\cos \theta} r^2 \, dr = \frac{1}{3} \int_0^\pi (1 - \cos \theta)^3 \sin \theta \, d\theta$$

$$= \frac{1}{12} (1 - \cos \theta)^4 \bigg|_0^\pi = \frac{4}{3}.$$

Spherical coordinates

If **G** is the transformation from spherical to rectangular coordinates,

$$\begin{bmatrix} x \\ y \\ z \end{bmatrix} = \mathbf{G}(r, \theta, \phi) = \begin{bmatrix} r \cos \theta \cos \phi \\ r \sin \theta \cos \phi \\ r \sin \phi \end{bmatrix}, \tag{50}$$

then $J\mathbf{G}(r, \theta, \phi) = r^2 \cos \phi$ (*Exercise 26(b), Section 6.2*), so (43) becomes

$$\int_{\mathbf{G}(S)} f(x, y, z) \, d(x, y, z)$$

$$= \int_S f(r \cos \theta \cos \phi, r \sin \theta \cos \phi, r \sin \phi) r^2 \cos \phi \, d(r, \theta, \phi), \tag{51}$$

if we assume, as is conventional, that $|\phi| \le \pi/2$ and $r \ge 0$.

Example 3.4   To find the volume of

$$T = \{(x, y, z) \mid x^2 + y^2 + z^2 \le a^2, x \ge 0, y \ge 0, z \ge 0\},$$

which is one eighth of a sphere [Figure 3.8(b)], we write $T = \mathbf{G}(S)$ with **G** as in (50) and

$$S = \{(r, \theta, \phi) \mid 0 \le r \le a, 0 \le \theta \le \pi/2, 0 \le \phi \le \pi/2\}$$

[Figure 3.8(a)], and let $f \equiv 1$ in (51). Theorem 3.5 implies that

$$V(T) = \int_{\mathbf{G}(S)} d\mathbf{X} = \int_S r^2 \cos \phi \, d(r, \theta, \phi)$$

$$= \int_0^a r^2 \, dr \int_0^{\pi/2} d\theta \int_0^{\pi/2} \cos \phi \, d\phi = \left(\frac{a^3}{3}\right)\left(\frac{\pi}{2}\right)(1) = \frac{\pi a^3}{6}.$$

Example 3.5   To evaluate the iterated integral

$$I = \int_0^a x \, dx \int_0^{\sqrt{a^2 - x^2}} dy \int_0^{\sqrt{a^2 - x^2 - y^2}} z \, dz \qquad (a > 0)$$

we first rewrite it as a multiple integral

$$I = \int_{\mathbf{G}(S)} xz \, d(x, y, z),$$

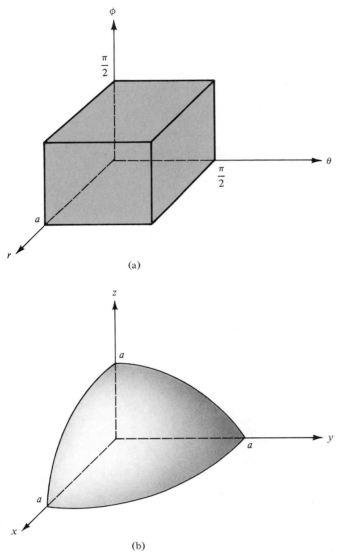

(a)

(b)

FIGURE 3.8

where **G** and S are as in Example 3.4. Theorem 3.5 implies that

$$I = \int_S (r \cos \theta \cos \phi)(r \sin \phi)(r^2 \cos \phi) \, d(r, \theta, \phi)$$

$$= \int_0^a r^4 \, dr \int_0^{\pi/2} \cos \theta \, d\theta \int_0^{\pi/2} \cos^2 \phi \sin \phi \, d\phi = \left(\frac{a^5}{5}\right)(1)\left(\frac{1}{3}\right) = \frac{a^5}{15}.$$

Other examples

We now consider other applications of Theorem 3.4.

Example 3.6   To evaluate

$$I = \int_T (x + 4y)\, d(x, y),$$

where T is the parallelogram bounded by the lines

$$x + y = 1, \quad x + y = 2, \quad x - 2y = 0, \quad \text{and} \quad x - 2y = 3$$

[Figure 3.9(b)], we define new variables u and v by

$$\begin{bmatrix} u \\ v \end{bmatrix} = \mathbf{F}(x, y) = \begin{bmatrix} x + y \\ x - 2y \end{bmatrix}.$$

Then

$$\begin{bmatrix} x \\ y \end{bmatrix} = \mathbf{F}^{-1}(u, v) = \begin{bmatrix} \dfrac{2u + v}{3} \\ \dfrac{u - v}{3} \end{bmatrix},$$

$$J\mathbf{F}^{-1}(u, v) = \begin{vmatrix} \frac{2}{3} & \frac{1}{3} \\ \frac{1}{3} & -\frac{1}{3} \end{vmatrix} = -\frac{1}{3},$$

and $T = \mathbf{F}^{-1}(S)$, where

$$S = \{(u, v)\,|\,1 \le u \le 2,\, 0 \le v \le 3\}$$

[Figure 3.9(a)]. Applying Theorem 3.4 with $\mathbf{G} = \mathbf{F}^{-1}$ yields

$$I = \int_S \left(\frac{2u + v}{3} + \frac{4u - 4v}{3} \right)\!\left(\frac{1}{3} \right) d(u, v) = \frac{1}{3} \int_S (2u - v)\, d(u, v)$$

$$= \frac{1}{3} \int_0^3 dv \int_1^2 (2u - v)\, du = \frac{1}{3} \int_0^3 (u^2 - uv)\Big|_{u=1}^{2} dv$$

$$= \frac{1}{3} \int_0^3 (3 - v)\, dv = \frac{1}{3}\left(3v - \frac{v^2}{2} \right)\Big|_0^3 = \frac{3}{2}.$$

Example 3.7   Let

$$I = \int_T e^{(x^2 - y^2)^2} e^{4x^2 y^2}(x^2 + y^2)\, d(x, y)$$

where T is the annulus

$$T = \{(x, y)\,|\,a^2 \le x^2 + y^2 \le b^2\}$$

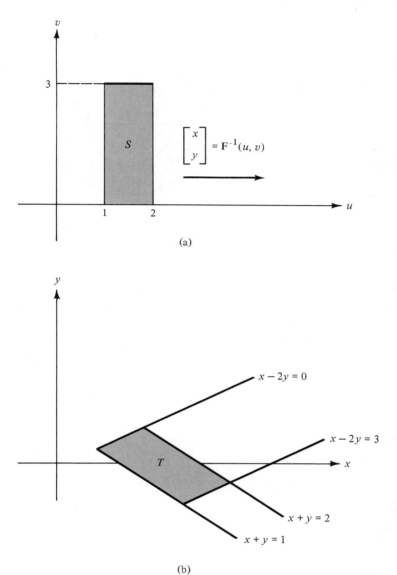

(a)

(b)

FIGURE 3.9

[Figure 3.10(a)]. The forms of the arguments of the exponential functions suggest that we introduce new variables u and v, defined by

$$\begin{bmatrix} u \\ v \end{bmatrix} = \mathbf{F}(x, y) = \begin{bmatrix} x^2 - y^2 \\ 2xy \end{bmatrix},$$

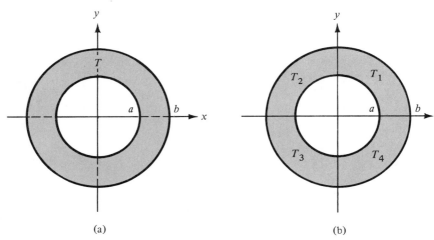

(a)   (b)

FIGURE 3.10

and apply Theorem 3.5 to $G = F^{-1}$; however, F is not one-to-one, and so does not have an inverse, on T^0 (Example 3.4, Section 6.3). To remove this difficulty we divide T into closed sections T_1, T_2, T_3, and T_4, as shown in Figure 3.10(b). Since the pairwise intersections of these sections have zero content,

$$I = I_1 + I_2 + I_3 + I_4, \tag{52}$$

where

$$I_j = \int_{T_j} e^{(x^2 - y^2)^2} e^{4x^2 y^2} (x^2 + y^2) \, d(x, y).$$

From *Exercise 20(c)*, $I_1 = I_2 = I_3 = I_4$, so (52) becomes

$$I = 4I_1.$$

To evaluate I_1 we observe that F is one-to-one on T_1 and

$$F(T_1) = S_1 = \{(u, v) \mid a^4 \le u^2 + v^2 \le b^4, v \ge 0\}$$

[Figure 3.11(b)], and a branch G of F^{-1} can be defined on S_1 (Example 3.8, Section 6.3). Now Theorem 3.4 implies that

$$I_1 = \int_{S_1} e^{(x^2 - y^2)^2} e^{4x^2 y^2} (x^2 + y^2) |J\mathbf{G}(u, v)| \, d(u, v),$$

where x and y must still be written in terms of u and v. Since it is easy to verify that

$$J\mathbf{F}(x, y) = 4(x^2 + y^2),$$

and so that

$$J\mathbf{G}(u, v) = \frac{1}{4(x^2 + y^2)},$$

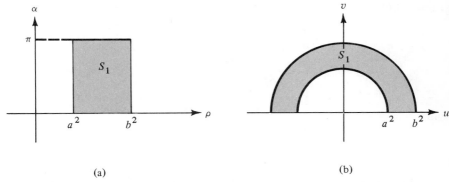

(a)   (b)

FIGURE 3.11

doing this yields

$$I_1 = \frac{1}{4} \int_{S_1} e^{u^2 + v^2} \, d(u, v). \tag{53}$$

To evaluate this we let ρ and α be polar coordinates in the uv plane (Figure 3.12) and define $\mathbf{H}$ by

$$\begin{bmatrix} u \\ v \end{bmatrix} = \mathbf{H}(\rho, \alpha) = \begin{bmatrix} \rho \cos \alpha \\ \rho \sin \alpha \end{bmatrix};$$

then $S_1 = H(\tilde{S}_1)$, where

$$\tilde{S}_1 = \{(\rho, \alpha) \,|\, a^2 \le \rho \le b^2, 0 \le \alpha \le \pi\}$$

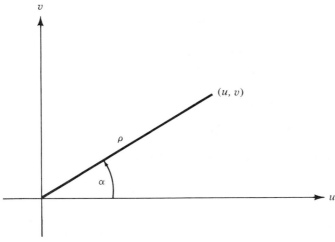

FIGURE 3.12

[Figure 3.11(a)]; hence, applying Theorem 3.4 to (53) yields

$$I_1 = \frac{1}{4} \int_{\tilde{S}_1} e^{\rho^2} |J\mathbf{H}(\rho, \alpha)| \, d(\rho, \alpha) = \frac{1}{4} \int_{\tilde{S}_1} \rho e^{\rho^2} \, d(\rho, \alpha)$$

$$= \frac{1}{4} \int_0^\pi d\alpha \int_{a^2}^{b^2} \rho e^{\rho^2} \, d\rho = \frac{\pi}{4} \frac{e^{b^4} - e^{a^4}}{2};$$

hence,

$$I = \frac{\pi}{2}(e^{b^4} - e^{a^4}).$$

It is legitimate to let a approach zero here and conclude that if C is the disc

$$C = \{(x, y) \,|\, 0 \le x^2 + y^2 \le b^2\},$$

then

$$\int_C e^{(x^2 - y^2)^2} e^{4x^2 y^2} (x^2 + y^2) \, d(x, y) = \frac{\pi}{2}(e^{b^4} - 1);$$

however, to start with this region of integration would lead to complications. (Why?) In this connection, see *Exercise 35*.

Example 3.8   To evaluate

$$I = \int_T e^{x_1 + x_2 + \cdots + x_n} \, d(x_1, x_2, \ldots, x_n)$$

where T is the region defined by

$$a_i \le x_1 + x_2 + \cdots + x_i \le b_i, \qquad i = 1, 2, \ldots, n,$$

we introduce the new variables $u_1, u_2, \ldots, u_n$ by $\mathbf{U} = \mathbf{F}(\mathbf{X})$, where

$$f_i(\mathbf{X}) = x_1 + x_2 + \cdots + x_i.$$

If $\mathbf{G} = \mathbf{F}^{-1}$, then $T = \mathbf{G}(S)$, where

$$S = [a_1, b_1] \times [a_2, b_2] \times \cdots \times [a_n, b_n],$$

and $J\mathbf{G}(\mathbf{U}) = 1$, since $J\mathbf{F}(\mathbf{X}) = 1$ (verify); hence, Theorem 3.4 implies that

$$I = \int_S e^{u_n} \, d(u_1, u_2, \ldots, u_n)$$

$$= \int_{a_1}^{b_1} du_1 \int_{a_2}^{b_2} du_2 \cdots \int_{a_{n-1}}^{b_{n-1}} du_{n-1} \int_{a_n}^{b_n} e^{u_n} \, du_n$$

$$= (b_1 - a_1)(b_2 - a_2) \cdots (b_{n-1} - a_{n-1})(e^{b_n} - e^{a_n}).$$

7.3   EXERCISES

1. Give a counterexample to the following statement: If S_1 and S_2 are disjoint subsets of a rectangle R, then either

$$\overline{\int_R} \psi_{S_1}(\mathbf{X}) \, d\mathbf{X} + \overline{\int_R} \psi_{S_2}(\mathbf{X}) \, d\mathbf{X} = \overline{\int_R} \psi_{S_1 \cup S_2}(\mathbf{X}) \, d\mathbf{X}$$

or

$$\underline{\int_R} \psi_{S_1}(\mathbf{X}) \, d\mathbf{X} + \underline{\int_R} \psi_{S_2}(\mathbf{X}) \, d\mathbf{X} = \underline{\int_R} \psi_{S_1 \cup S_2}(\mathbf{X}) \, d\mathbf{X}.$$

2. Show that a set has content zero according to Definition 1.4 if and only if it has Jordan content zero.

3. Show that if S_1 and S_2 are Jordan-measurable, then so are $S_1 \cup S_2$ and $S_1 \cap S_2$. (*Hint:* Use Theorem 3.1 and Exercise 18, Section 1.3.)

4. Prove: (a) If S is Jordan-measurable, then so is $\overline{S}$, and $V(\overline{S}) = V(S)$. (b) If T is a Jordan-measurable subset of a Jordan-measurable set S, then $S - T$ is Jordan-measurable. (*Hint:* Use Theorem 3.1 and Exercise 23, Section 1.3.) Must S be Jordan-measurable if $\overline{S}$ is?

5. Suppose S is Jordan-measurable, $\varepsilon > 0$, and $\rho > 0$. Show that there are cubes $C_1, C_2, \ldots, C_r$ in S with edges of length $\le \rho$ such that $C_i^0 \cap C_j^0 = \emptyset$ if $i \ne j$ and

$$V(S) \le \sum_{j=1}^{r} V(C_j) + \varepsilon.$$

(*Hint:* See the proof of Lemma 3.1.)

6. Suppose H is a subset of a closed Jordan-measurable set S such that the intersection of H with any closed subset of S^0 has zero content. Show that $V(H) = 0$. (*Hint:* Use Theorem 3.1.)

7. Suppose $\mathbf{E}$ is an $n \times n$ elementary matrix and $\mathbf{A}$ is an arbitrary $n \times p$ matrix. Show that $\mathbf{EA}$ is the matrix obtained by applying to $\mathbf{A}$ the operation by which $\mathbf{E}$ is obtained from the $n \times n$ identity matrix.

8. (a) Calculate the determinants of elementary matrices of types (a), (b), and (c) of Lemma 3.3.
   (b) Show that the inverse of an elementary matrix of type (a), (b), or (c) is an elementary matrix of the same type.
   (c) Verify the inverses given for $\hat{\mathbf{E}}_1, \ldots, \hat{\mathbf{E}}_6$ in Example 3.1.

9. Write as a product of elementary matrices:

(a) $\begin{bmatrix} 1 & 0 & 1 \\ 1 & 1 & 0 \\ 0 & 1 & 1 \end{bmatrix}$ (b) $\begin{bmatrix} 2 & 3 & -2 \\ 0 & -1 & 5 \\ 0 & -2 & 4 \end{bmatrix}$

10. Suppose $ad - bc \neq 0$. Find the area of the parallelogram bounded by the lines

$$ax + by = u_1, \qquad ax + by = u_2,$$
$$cx + dy = v_1, \qquad cx + dy = v_2.$$

11. Find the volume of the parallelepiped defined by

$$1 \leq 2x + 3y - 2z \leq 2$$
$$5 \leq \quad -x + 5y \quad \leq 7$$
$$1 \leq \quad -2x + 4y \quad \leq 6.$$

12. Given that Eq. (15) holds for all Jordan-measurable sets if $\mathbf{A}$ is an $n \times n$ elementary matrix, show that it holds if $\mathbf{A}$ is any nonsingular $n \times n$ matrix.

13. Show that Eq. (31) implies Eq. (28).

14. Verify the inequalities (36), (37), and (38).

15. Verify that if $f(\mathbf{X}) \geq 0$ for all $\mathbf{X}$ in $\mathbf{G}(S)$, then Eq. (40) can be obtained from Eq. (39) by the procedure suggested in the text following Eq. (39).

16. Given that Eq. (24) holds if $f(\mathbf{X}) \geq 0$ for all $\mathbf{X}$ in $\mathbf{G}(S)$, use the method suggested in the text immediately after Eq. (40) to show that it holds for a function which may assume negative values.

17. Prove: Under the assumptions of Theorem 3.5, if D is a subset of S such that $V(\mathbf{G}(D)) = 0$, then $V(D) = 0$. (*Hint:* Apply Lemma 3.2 to $\mathbf{G}^{-1}$.)

18. How do we know that there are cubes $C_1, C_2, \ldots, C_k$ in P^0 that satisfy Eqs. (44) and (45)?

19. In writing Eq. (46) we have assumed that

$$\int_{\mathbf{G}(S)} f(\mathbf{X}) \, d\mathbf{X} = \int_{\mathbf{G}(S_1)} f(\mathbf{X}) \, d\mathbf{X} + \int_{\mathbf{G}(S - S_1)} f(\mathbf{X}) \, d\mathbf{X}.$$

Justify this. [*Hint:* Show that $\mathbf{G}(S_1) \cap \mathbf{G}(S - S_1)$ has zero content.]

20. Let $|e_i| = 1$, $i = 0, 1, \ldots, n$. Let T be a bounded subset of $\mathscr{R}^n$ and

$$\tilde{T} = \{(e_1 x_1, e_2 x_2, \ldots, e_n x_n) \,|\, (x_1, x_2, \ldots, x_n) \in T\}.$$

Suppose f is defined on T and define g on $\tilde{T}$ by

$$g(e_1 x_1, e_2 x_2, \ldots, e_n x_n) = e_0 f(x_1, x_2, \ldots, x_n).$$

(a) Prove: f is integrable over T if and only if g is integrable over $\hat{T}$, and in this case

$$\int_{\tilde{T}} g(\mathbf{Y}) \, d\mathbf{Y} = e_0 \int_T f(\mathbf{X}) \, d\mathbf{X}.$$

(*Hint:* If f is continuous, this follows easily from Theorem 3.4; however, in the form stated, it must be proved directly from Definitions 1.2 and 1.5.)

(b) Show that if $\tilde{T} = T$ in part (a) and

$$f(e_1 x_1, e_2 x_2, \ldots, e_n x_n) = -f(x_1, x_2, \ldots, x_n),$$

then

$$\int_T f(\mathbf{X}) \, d\mathbf{X} = 0.$$

(c) Show that part (a) implies that $I_1 = I_2 = I_3 = I_4$ in Example 3.7.

21. In elementary calculus the "quantity" $r \, dr \, d\theta$ is called *the differential element of area in polar coordinates.* Explain why this is consistent with Theorem 3.5.

22. In elementary calculus the "quantity" $r^2 \cos \phi \, dr \, d\theta \, d\phi$ is called the *differential element of volume in spherical coordinates.* Explain why this is consistent with Theorem 3.5.

23. Find the area of

(a) $\{(x, y) | y \le x \le 4y, \, 1 \le x + 2y \le 3\}$
(b) $\{(x, y) | 2 \le xy \le 4, \, 2x \le y \le 5x\}$

24. Evaluate

$$\int_T (3x^2 + 2y + z) \, d(x, y, z),$$

where T is the cube defined by

$$|x - y| \le 1, \qquad |y - z| \le 1, \qquad |z + x| \le 1.$$

25. Evaluate

$$\int_T (y^2 + x^2 y - 2x^4) \, d(x, y),$$

where T is the region bounded by the curves

$$xy = 1, \quad xy = 2, \quad y = x^2, \quad y = x^2 + 1.$$

26. Evaluate

$$\int_T (x^4 - y^4) e^{xy} \, d(x, y),$$

where T is the region in the first quadrant bounded by the hyperbolas

$$xy = 1, \quad xy = 2, \quad x^2 - y^2 = 2, \quad x^2 - y^2 = 3.$$

27. Find the volume of the ellipsoid

$$\frac{x^2}{a^2} + \frac{y^2}{b^2} + \frac{z^2}{c^2} = 1.$$

28. Evaluate

$$\int_T \frac{e^{x^2 + y^2 + z^2}}{\sqrt{x^2 + y^2 + z^2}} \, d(x, y, z),$$

where

$$T = \{(x, y, z)|9 \leq x^2 + y^2 + z^2 \leq 25\}.$$

29. Find the volume of the set T bounded by the surfaces $z = 0$, $z = \sqrt{x^2 + y^2}$, and $x^2 + y^2 = 4$.

30. Evaluate

$$\int_T xyz(x^4 - y^4) \, d(x, y, z),$$

where

$$T = \{(x, y, z)|1 \leq x^2 - y^2 \leq 2, 3 \leq x^2 + y^2 \leq 4, 0 \leq z \leq 1\}.$$

31. Evaluate:

(a) $\int_0^{\sqrt{2}} dy \int_y^{\sqrt{4-y^2}} \frac{dx}{1 + x^2 + y^2}$ (b) $\int_0^2 dx \int_0^{\sqrt{4-x^2}} e^{(x^2 + y^2)} \, dy$

(c) $\int_{-1}^1 dx \int_{-\sqrt{1-x^2}}^{\sqrt{1-x^2}} dy \int_0^{\sqrt{1-x^2-y^2}} z^2 \, dz$

32. Use the change of variables

$$\begin{bmatrix} x_1 \\ x_2 \\ x_3 \\ x_4 \end{bmatrix} = \mathbf{G}(r, \theta_1, \theta_2, \theta_3) = \begin{bmatrix} r \sin \theta_1 \sin \theta_2 \sin \theta_3 \\ r \sin \theta_1 \sin \theta_2 \cos \theta_3 \\ r \sin \theta_1 \cos \theta_2 \\ r \cos \theta_1 \end{bmatrix}$$

to compute the content of the 4-sphere

$$T = \{(x_1, x_2, x_3, x_4)|x_1^2 + x_2^2 + x_3^2 + x_4^2 \leq a^2\}.$$

33. Suppose $\mathbf{A} = [a_{ij}]$ is a nonsingular $n \times n$ matrix and T is the region in $\mathcal{R}^n$ defined by

$$\alpha_i \leq a_{i1}x_1 + a_{i2}x_2 + \cdots + a_{in}x_n \leq \beta_i, \qquad 1 \leq i \leq n.$$

(a) Find $V(T)$.
(b) Show that if

$$\begin{bmatrix} c_1 \\ c_2 \\ \vdots \\ c_n \end{bmatrix} = \mathbf{A}^t \begin{bmatrix} d_1 \\ d_2 \\ \vdots \\ d_n \end{bmatrix},$$

then

$$\int_T \left(\sum_{j=1}^n c_j x_j \right) = \frac{V(T)}{2} \sum_{i=1}^n d_i(\alpha_i + \beta_i).$$

34. If V_n is the content of the n-sphere, $T = \{\mathbf{X} \, |\mathbf{X}| \le 1\}$, find the content of the n-dimensional ellipsoid defined by

$$\sum_{j=1}^n \frac{x_j^2}{a_j^2} = 1.$$

Leave the answer in terms of V_n.

35. Suppose $\mathbf{F} \colon \mathscr{R}^n \to \mathscr{U}^n$ is continuously differentiable on an open set M containing the closed Jordan-measurable set T, and one-to-one on T^0, with Jacobian $J\mathbf{F}$ which does not vanish on T except at points of a subset T_0 of T with content zero. Let $\mathbf{F}(T)$ be Jordan-measurable. Suppose f is continuous on T and the function

$$h(\mathbf{X}) = \frac{f(\mathbf{X})}{J\mathbf{F}(\mathbf{X})}$$

is bounded on $T - T_0$. Define h arbitrarily (but bounded) on T_0.
(a) Prove that

$$\int_T f(\mathbf{X}) \, d\mathbf{X} = \int_{\mathbf{F}(T)} h(F^{-1}(\mathbf{U})) \, d\mathbf{U}.$$

(b) Use part (a) to evaluate

$$\int_{T_1} e^{(x^2 - y^2)^2} e^{4x^2 y^2} (x^2 + y^2) \, d(x, y),$$

where $T_1 = \{(x, y) \,|\, x^2 + y^2 \le b^2, \, x, y \ge 0\}$. (See Example 3.7.)

36. Evaluate

$$I = \int_T 4xyz(x^4 - y^4) \, d(x, y, z),$$

where

$$T = \{(x, y, z) \,|\, 0 \le x^2 - y^2 \le 1, \, 0 \le xy \le 1, \, 0 \le z^2 + y^2 \le 1, \, x, y, z \ge 0\}.$$

(*Hint:* See Exercise 35.)

7.4 IMPROPER MULTIPLE INTEGRALS

Our study of the multiple integral $\int_S f(\mathbf{X}) \, d\mathbf{X}$ has so far been restricted to the case where S is a bounded set and f is bounded on S. To extend the definition to the case where one or both of these is false, we now develop the idea of the improper multiple integral. This development is only partially analogous to the development in Section 3.4 of the improper ordinary integral. If $n > 1$, the structure of $\mathscr{R}^n$ introduces complications not present when $n = 1$, where the improper integral of a function f locally integrable on $[a, b)$ is defined in the natural way as

$$\int_a^b f(x)\,dx = \lim_{c \to b-} \int_a^c f(x)\,dx$$

if the limit exists. If $n > 1$, there are many equally natural ways to define an analogous limiting procedure which—for some functions—produce different results.

Example 4.1   Suppose we wish to define the improper integral

$$\int_S (x - y)\,d(x, y)$$

where S is the first quadrant. By analogy with our definition of ordinary improper integrals, it would seem reasonable to integrate over a bounded subset of S and study the result as the subset expands in some way to cover S. The difficulty here is that there are many equally natural ways to choose the bounded subsets over which we integrate, and different choices lead to different results. For example, suppose we take the bounded sets to be the rectangles

$$R_a(\rho) = [0, \rho] \times [0, a\rho]$$

(Figure 4.1), where a and ρ are positive, and define

$$\int_S (x - y)\,d(x, y) = \lim_{\rho \to \infty} \int_{R_a(\rho)} (x - y)\,d(x, y)$$

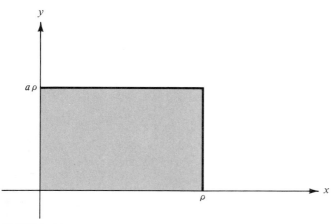

FIGURE 4.1

if the limit exists. Now,

$$\int_{R_a(\rho)} (x - y) \, d(x, y) = \int_{R_a(\rho)} x \, d(x, y) - \int_{R_a(\rho)} y \, d(x, y)$$

$$= \int_0^\rho x \, dx \int_0^{a\rho} dy - \int_0^\rho dx \int_0^{a\rho} y \, dy$$

$$= \tfrac{1}{2} a \rho^3 (1 - a);$$

hence,

$$\lim_{\rho \to \infty} \int_{R_a(\rho)} (x - y) \, d(x, y) = \begin{cases} \infty & \text{if } 0 < a < 1, \\ 0 & \text{if } a = 1, \\ -\infty & \text{if } a > 1. \end{cases} \tag{1}$$

There is no reason in this example to consider any particular shape (choice of a) to be more natural than any other, or even to prefer rectangular sets in the first place. A sensible definition of the improper multiple integral $\int_S f(\mathbf{X}) \, d\mathbf{X}$ must be independent of the particular kinds of sets chosen to "fill up" S. The following definition has this property. For convenience we let

$$f^+(\mathbf{X}) = \begin{cases} f(\mathbf{X}) & \text{if } f(\mathbf{X}) \geq 0, \\ 0 & \text{if } f(\mathbf{X}) < 0, \end{cases}$$

and

$$f^-(\mathbf{X}) = \begin{cases} -f(\mathbf{X}) & \text{if } f(\mathbf{X}) \leq 0, \\ 0 & \text{if } f(\mathbf{X}) > 0, \end{cases}$$

so that

$$f(\mathbf{X}) = f^+(\mathbf{X}) - f^-(\mathbf{X})$$

and

$$|f(\mathbf{X})| = f^+(\mathbf{X}) + f^-(\mathbf{X}).$$

Definition 4.1 *Suppose f is defined on a set S such that $S^0 \neq \varnothing$ and every bounded subset of ∂S has zero content. Let $J(S)$ be the collection of closed Jordan-measurable subsets of S^0 and suppose $\int_T f(\mathbf{X}) \, d\mathbf{X}$ exists for every T in $J(S)$. Then if*

$$A = \text{l.u.b.} \left\{ \int_T f^+(\mathbf{X}) \, d\mathbf{X} \,\middle|\, T \in J(S) \right\} \tag{2}$$

and

$$B = \text{l.u.b.} \left\{ \int_T f^-(\mathbf{X}) \, d\mathbf{X} \,\middle|\, T \in J(S) \right\}$$

are not both infinite, we define

$$\int_S f(\mathbf{X}) \, d\mathbf{X} = A - B. \tag{3}$$

If S is bounded and f is integrable—and therefore bounded—on S, then A and B are both finite and (3) holds (*Exercise 1*). Thus, Definition 4.1 is consistent with Definition 1.5 where the latter applies; however, (3) defines $\int_S f(\mathbf{X})\,d\mathbf{X}$ in some cases where S is unbounded or f is unbounded on S, or both. In these cases we say that $\int_S f(\mathbf{X})\,d\mathbf{X}$ is an *improper integral*, which *converges* to $A - B$ if A and B are finite, and *diverges* otherwise. When we wish to distinguish between improper integrals and integrals in the sense of Section 7.1, we will call the latter *proper* multiple integrals.

If $g(\mathbf{X}) \geq 0$ for all $\mathbf{X}$ in S, we will write

$$\int_S g(\mathbf{X})\,d\mathbf{X} < \infty$$

to indicate that $\int_S g(\mathbf{X})\,d\mathbf{X}$ converges.

The next theorem shows that Definition 4.1 does not admit an analog of conditional convergence.

Theorem 4.1   *The improper integral $\int_S f(\mathbf{X})\,d\mathbf{X}$ converges if and only if*

$$\int_S |f(\mathbf{X})|\,d\mathbf{X} < \infty.$$

Proof.   *Exercise 2.*

Definition 4.1 offers little practical help in evaluating a specific improper multiple integral. The next theorem is useful for this purpose.

Theorem 4.2   *If $\int_S f(\mathbf{X})\,d\mathbf{X}$ converges to a finite value or diverges to $\pm\infty$, and $\{T_j\}$ is a sequence of sets in $J(S)$ such that*

$$T_j \subset T_{j+1}\ (j = 1, 2, \ldots) \quad and \quad S^0 = \bigcup_{j=1}^{\infty} T_j^0, \tag{4}$$

then

$$\int_S f(\mathbf{X})\,d\mathbf{X} = \lim_{j\to\infty} \int_{T_j} f(\mathbf{X})\,d\mathbf{X}. \tag{5}$$

Proof.   It suffices to prove the theorem for the case where f is nonnegative on S, since the general case follows from this (*Exercise 3*). Then $B = 0$ and (2) implies that there is a sequence $\{U_m\}$ of sets in $J(S)$ such that

$$\lim_{m\to\infty} \int_{U_m} f(\mathbf{X})\,d\mathbf{X} = A. \tag{6}$$

From (4) and the Heine–Borel theorem there is for each m an integer M such that

$$U_m \subset T_j \quad \text{if} \quad j \geq M.$$

Therefore,

$$\int_{U_m} f(\mathbf{X}) \, d\mathbf{X} \leq \int_{T_j} f(\mathbf{X}) \, d\mathbf{X} \leq A \quad \text{if} \quad j \geq M.$$

This and (6) imply (5). ∎

The following corollary of Theorem 4.2 is usually more convenient for applications.

Corollary 4.1 *If $\int_S f(\mathbf{X}) \, d\mathbf{X}$ converges to a finite value or diverges to $\pm\infty$ and $\{T_\rho | a < \rho < b\}$ is a family of sets in $J(S)$ such that*

$$T_{\rho_1} \subset T_{\rho_2} \quad \text{if} \quad \rho_1 < \rho_2 \tag{7}$$

and

$$S^0 = \cup \{T_\rho^0 | a < \rho < b\},$$

then

$$\int_S f(\mathbf{X}) \, d\mathbf{X} = \lim_{\rho \to b-} \int_{T_\rho} f(\mathbf{X}) \, d\mathbf{X}. \tag{8}$$

If (7) is replaced by

$$T_{\rho_1} \supset T_{\rho_2} \quad \text{if} \quad \rho_1 < \rho_2$$

then (8) is replaced by

$$\int_S f(\mathbf{X}) \, d\mathbf{X} = \lim_{\rho \to a+} \int_{T_\rho} f(\mathbf{X}) \, d\mathbf{X}.$$

Proof. *Exercise 4.*

According to Definition 4.1, $\int_S f(\mathbf{X}) \, d\mathbf{X}$ either converges to a finite value or diverges to ∞ if $f(\mathbf{X}) \geq 0$ for all $\mathbf{X}$ in S; hence, Theorem 4.2 and Corollary 4.1 can always be applied in this case.

Example 4.2 To investigate the convergence of

$$I = \int_{\mathscr{R}^2} \frac{d(x, y)}{(1 + x^2 + y^2)^p}$$

in which the integrand is positive for all (x, y), we take

$$T_\rho = \{(x, y) \,|\, x^2 + y^2 \le \rho^2\}, \qquad 0 < \rho < \infty.$$

Changing to polar coordinates yields

$$\int_{T_\rho} \frac{d(x, y)}{(1 + x^2 + y^2)^p} = \int_0^{2\pi} d\theta \int_0^\rho \frac{r \, dr}{(1 + r^2)^p}$$

$$= \begin{cases} \dfrac{\pi}{p - 1}\left[1 - \dfrac{1}{(1 + \rho^2)^{p - 1}}\right] & \text{if } p \ne 1, \\ \Pi \log(1 + \rho^2) & \text{if } p = 1. \end{cases}$$

Therefore,

$$I = \lim_{\rho \to \infty} \int_{T_\rho} \frac{d(x, y)}{(1 + x^2 + y^2)^p} = \begin{cases} \dfrac{\pi}{p - 1} & \text{if } p > 1, \\ \infty & \text{if } p \le 1. \end{cases}$$

Example 4.3   The function

$$f(x, y) = \frac{1}{(1 - x^2 - y^2)^p} \qquad (p > 0)$$

is positive and unbounded on the open unit disc

$$S = \{(x, y) \,|\, x^2 + y^2 < 1\}.$$

To investigate the convergence of

$$I = \int_S f(x, y) \, d(x, y),$$

we take

$$T_\rho = \{(x, y) \,|\, x^2 + y^2 \le \rho^2\}, \tag{9}$$

where $0 < \rho < 1$. Introducing polar coordinates yields

$$\int_{T_\rho} f(x, y) \, d(x, y) = \int_0^{2\pi} d\theta \int_0^\rho \frac{r \, dr}{(1 - r^2)^p}$$

$$= \begin{cases} \dfrac{\pi}{1 - p}\left[1 - \dfrac{1}{(1 - \rho^2)^{p - 1}}\right] & \text{if } p \ne 1, \\ -\pi \log(1 - \rho^2) & \text{if } p = 1. \end{cases}$$

Therefore,

$$I = \lim_{\rho \to 1-} \int_{T_\rho} f(x, y) \, d(x, y) = \begin{cases} \dfrac{\pi}{1 - p} & \text{if } 0 < p < 1, \\ \infty & \text{if } p = 1. \end{cases}$$

If $p \le 0$, then I exists as a proper integral, and

$$I = \frac{\pi}{1 - p}.$$

Example 4.4   The convergent ordinary improper integral

$$I = \int_{-\infty}^{\infty} e^{-x^2}\, dx$$

is important in probability theory. To evaluate it, we define

$$I_R = \int_{-R}^{R} e^{-x^2}\, dx$$

and write

$$I_R^2 = \int_{-R}^{R} e^{-x^2}\, dx \int_{-R}^{R} e^{-y^2}\, dy.$$

This can be rewritten as a double integral:

$$I_R^2 = \int_{S_R} e^{-x^2 - y^2}\, d(x, y),$$

where S_R is the square with vertices (R, R), $(-R, R)$, $(R, -R)$, and $(-R, -R)$ (Figure 4.2). Corollary 4.1 implies that

$$I^2 = \lim_{R \to \infty} I_R^2 = \int_{\mathcal{R}^2} e^{-x^2 - y^2}\, d(x, y)$$

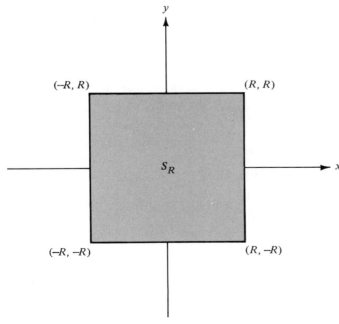

FIGURE 4.2

and that the improper integral on the right can be evaluated as

$$I^2 = \lim_{\rho \to \infty} \int_{T_\rho} e^{-x^2 - y^2} \, d(x, y), \tag{10}$$

where T_ρ is as defined in (9). Transforming to polar coordinates yields

$$\int_{T_\rho} e^{-x^2 - y^2} \, d(x, y) = \int_0^{2\pi} d\theta \int_0^\rho r e^{-r^2} \, dr = \pi(1 - e^{-\rho^2}).$$

Hence, from (10),

$$I^2 = \lim_{\rho \to \infty} \pi(1 - e^{-\rho^2}) = \pi,$$

and so

$$\int_{-\infty}^\infty e^{-x^2} \, dx = \sqrt{\pi}.$$

In applying Theorem 4.2 when f is nonnegative on S, it suffices to choose any convenient sequence $\{T_j\}$ from $J(S)$ which satisfies (4), and apply (5). However, if f changes sign on S, then a particular choice of $\{T_j\}$ may produce a finite limit in (5), from which it might be erroneously inferred that $\int_S f(\mathbf{X}) \, d\mathbf{X}$ converges even though it does not. [Thus, taking $a = 1$ in Example 4.1 could lead to the incorrect conclusion that $\int_S (x - y) \, d(x, y) = 0$; see (1).] The following procedure should be used to prevent this kind of mistake:

1. Use Theorem 4.2 with any convenient choice of suitable $\{T_j\}$ to determine whether

$$\int_S |f(\mathbf{X})| \, d\mathbf{X} < \infty. \tag{11}$$

2. If (11) holds, then use Theorem 4.2, again with any convenient choice of suitable $\{T_j\}$, to evaluate $\int_S f(\mathbf{X}) \, d\mathbf{X}$.

Similar comments apply to Corollary 4.1.

Example 4.5   Consider the improper integral

$$I = \int_S f(x, y) \, d(x, y)$$

where

$$f(x, y) = \frac{\sin \sqrt{x^2 + y^2}}{(x^2 + y^2)^p}$$

and

$$S = \{(x, y) \,|\, x^2 + y^2 \geq 1\}.$$

Let

$$T_\rho = \{(x, y) \mid 1 \le x^2 + y^2 \le \rho^2\}$$

(Figure 4.3). Introducing polar coordinates yields

$$\int_{T_\rho} f(x, y) \, d(x, y) = \int_0^{2\pi} d\theta \int_1^\rho \frac{\sin r}{r^{2p-1}} \, dr,$$

and so

$$\lim_{\rho \to \infty} \int_{T_\rho} f(x, y) \, d(x, y) = 2\pi \int_1^\infty \frac{\sin r}{r^{2p-1}} \, dr,$$

which exists if and only if $p > \frac{1}{2}$ (Example 4.14, Section 3.4). However, we cannot conclude that I converges if $p > \frac{1}{2}$, since applying the same argument to the improper integral

$$I_1 = \int_S |f(x, y)| \, d(x, y)$$

yields

$$\lim_{\rho \to \infty} \int_{T_\rho} |f(x, y)| \, d(x, y) = 2\pi \int_1^\infty \frac{|\sin r|}{r^{2p-1}} \, dr,$$

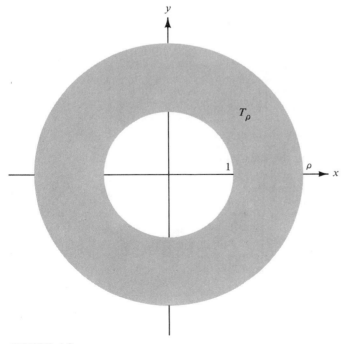

FIGURE 4.3

which converges if and only if $p > 1$ (Example 4.13, Section 3.4). Thus, I converges if and only if $p > 1$, in which case

$$I = 2\pi \int_1^\infty \frac{\sin r}{r^{2p-1}} \, dr.$$

Theorem 4.3 (Comparison Test)   *Suppose S satisfies the assumptions of Definition 4.1, f and g are properly integrable on members of $J(S)$, and $|f(\mathbf{X})| \leq |g(\mathbf{X})|$ if $\mathbf{X} \in S$. Then*
 (a) $\int_S f(\mathbf{X}) \, d\mathbf{X}$ converges if $\int_S g(\mathbf{X}) \, d\mathbf{X}$ converges;
 (b) $\int_S g(\mathbf{X}) \, d\mathbf{X}$ diverges if $\int_S f(\mathbf{X}) \, d\mathbf{X}$ diverges.

Proof.   *Exercise 10.*

We emphasize that this is a theorem about improper multiple integrals; if $n = 1$ and the integrals in (a) and (b) are interpreted as ordinary improper integrals according to the definitions of Section 3.4, then (a) and (b) are both false.

Example 4.6   Let

$$I = \int_S e^{-x} \sin y \, d(x, y)$$

where S is the semiinfinite strip

$$S = \{(x, y) \mid 0 \leq x + y \leq 2\pi, x \geq y\}$$

[Figure 4.4(a)]. Here it is convenient to take

$$T_\rho = \{(x, y) \mid 0 \leq x + y \leq 2\pi, 0 \leq x - y \leq \rho\}$$

[Figure 4.4(b)]. Introducing the variables

$$u = \frac{x + y}{2}, \qquad v = \frac{x - y}{2},$$

yields

$$\int_{T_\rho} e^{-x} \sin y \, d(x, y) = 2 \int_{\tilde{T}_\rho} e^{-u-v} \sin(u - v) \, d(u, v), \tag{12}$$

where

$$\tilde{T}_\rho = \{(u, v) \mid 0 \leq u \leq \pi, 0 \leq v \leq \rho/2\}.$$

Since

$$\left| e^{-u-v} \sin(u - v) \right| \leq e^{-u-v}$$

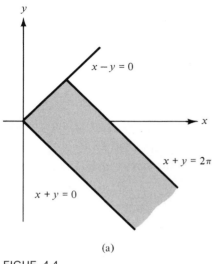

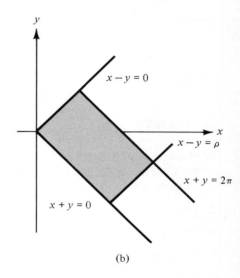

(a)   (b)

FIGUE 4.4

and

$$\lim_{\rho \to \infty} \int_{\tilde{T}_\rho} e^{-u-v}\, d(u, v) = \left(\int_0^\pi e^{-u}\, du \right) \lim_{\rho \to \infty} \int_0^{\rho/2} e^{-v}\, dv = 1 - e^{-\pi} < \infty,$$

Theorem 4.3 implies that I converges. To evaluate I, we write

$$\int_{\tilde{T}_\rho} e^{-u-v} \sin(u - v)\, d(u, v) = \int_{\tilde{T}_\rho} e^{-u-v}(\sin u \cos v - \cos u \sin v)\, d(u, v)$$

$$= \int_0^\pi e^{-u} \sin u\, du \int_0^{\rho/2} e^{-v} \cos v\, dv \qquad (13)$$

$$- \int_0^\pi e^{-u} \cos u\, du \int_0^{\rho/2} e^{-v} \sin v\, dv.$$

Using the elementary integration formulas

$$\int_0^\tau e^{-t} \cos t\, dt = \tfrac{1}{2}[1 - e^{-\tau}(\cos \tau - \sin \tau)]$$

and

$$\int_0^\tau e^{-t} \sin t\, dt = \tfrac{1}{2}[1 - e^{-\tau}(\cos \tau + \sin \tau)],$$

we infer from (13) that

$$\int_{\tilde{T}_\rho} e^{-u-v} \sin(u - v)\, d(u, v) = \left(\frac{1 + e^{-\pi}}{2} \right) e^{-\rho/2} \sin \rho/2;$$

therefore,

$$\lim_{\rho \to \infty} \int_{\tilde{T}_\rho} e^{-u-v} \sin(u - v)\, d(u, v) = 0$$

and so, from (12), $I = 0$.

7.4   EXERCISES

1. Show that if S is Jordan-measurable and $\int_S f(\mathbf{X})\, d\mathbf{X}$ exists as a proper multiple integral, then it also exists and has the same value in the sense of Definition 4.1.

2. Prove Theorem 4.1.

3. Complete the proof of Theorem 4.2.

4. Show that Theorem 4.2 implies Corollary 4.1.

5. Evaluate:

   (a) $\int_S \log(x^2 + y^2)\, d(x, y);\ S = \{(x, y) \,|\, 0 < x^2 + y^2 \le 1\}$

   (b) $\int_S \log xy\, d(x, y);\ S = (0, 1] \times (0, 1]$

   (c) $\int_S (x + y + z) \log(x^2 + y^2 + z^2)\, d(x, y, z);$

   $S = \{(x, y, z) \,|\, 0 < x^2 + y^2 + z^2 \le 1,\, z \ge 0\}$

   (d) $\int_S \dfrac{x[\log(x - y) + \log(x + y)]}{(x^2 - y^2)^2}\, d(x, y);$

   $S = \{(x, y) \,|\, 1 \le x^2 - y^2 \le 2,\, x \ge 0\}$

   (e) $\int_S x^{p-q} y^{p+q}\, d(x, y);\ S = \{(x, y) \,|\, x \ge 0,\, y \ge 0,\, 1 \le xy \le 2\}$

6. Evaluate:

   (a) $\int_S \dfrac{x^2 y^2}{(x^2 + y^2)^{11/4}}\, d(x, y);\ S = \{(x, y) \,|\, 0 < x^2 + y^2 \le 1\}$

   (b) $\int_S \dfrac{ze^{-y}\, d(x, y, z)}{(1 + x^2 + z^2)^2 (x^2 + z^2)^{1/2}};\ S = \{(x, y, z) \,|\, x^2 + z^2 > 0,\, y \ge 0,\, z \ge 0\}$

   (c) $\int_S (x + 2y)e^{-2x - 3y}\, d(x, y);\ S = \{(x, y) \,|\, x + y \ge 0,\, x + 2y \ge 1\}$

   (d) $\int_S \dfrac{xy(x^4 - y^4)}{(1 + x^4 - x^2 y^2 + y^4)^2}\, d(x, y);$

   $S = \{(x, y) \,|\, x \ge 0,\, y \ge 0,\, 1 \le x^2 - y^2 \le 2\}$

7. Find all values of p and q for which the integrals converge:

(a) $\displaystyle\int_S \frac{d(x, y)}{(x^2 + y^2)^p(1 + x^2 + y^2)^q}$; $S = \mathscr{R}^2 - \{(0, 0)\}$

(b) $\displaystyle\int_S \frac{d(x, y, z)}{(x^2 + y^2 + z^2)^p(1 + x^2 + y^2 + z^2)^q}$; $S = \mathscr{R}^3 - \{(0, 0, 0)\}$

(c) $\displaystyle\int_S \frac{d(x, y)}{(2x^2 + 6xy + 5y^2)^p}$; $S = \{(x, y)\,|\,x + y \geq 0\}$

[Hint: $2x^2 + 6xy + 5y^2 = (x + y)^2 + (x + 2y)^2$.]

(d) $\displaystyle\int_S \frac{d(x, y, z)}{(1 - x^2 - y^2)^p z^q}$; $S = \{(x, y, z)\,|\,x^2 + y^2 < 1, z \geq 1\}$

(e) $\displaystyle\int_S (1 + |x + y|)^p e^{q(x + 2y)}\, d(x, y)$; $S = \{(x, y)\,|\,x + 2y \geq 0\}$

8. If S is an unbounded set such that every bounded subset of ∂S has zero content, we define $\int_S d\mathbf{X}$ to be the content (area if $n = 2$, volume if $n = 3$) of S. Find the content of the following sets:

(a) $S = \{(x, y)\,|\,0 \leq y \leq e^{-x}, x \geq 0\}$

(b) $S = \{(x, y, z)\,|\,0 \leq z \leq e^{-x^2 - y^2}\}$

(c) $S = \{(x, y)\,|\,0 < x + y \leq 1, 0 \leq x - y \leq (x + y)^{-1/2}\}$

(d) $S = \left\{(x, y)\,\Big|\,0 \leq x \leq \dfrac{|\sin y|}{y}\right\}$

9. Suppose S satisfies the assumptions of Definition 4.1 and f and g are properly integrable on sets in $J(S)$. Show that if f and g are integrable over S, then so is $f + g$, and

$$\int_S (f + g)(\mathbf{X})\, d\mathbf{X} = \int_S f(\mathbf{X})\, d\mathbf{X} + \int_S g(\mathbf{X})\, d\mathbf{X}.$$

10. Prove Theorem 4.3.

11. Determine whether the following integrals converge:

(a) $\displaystyle\int_S \frac{1 + e^{-x} + e^{-y}}{1 - x^2 - y^2 - z^2}\, d(x, y, z)$; $S = \{(x, y, z)\,|\,x^2 + y^2 + z^2 < 1\}$

(b) $\displaystyle\int_{\mathscr{R}^2} \frac{\sin(x + y)}{(1 + x^2 + y^2)^{3/2}}\, d(x, y)$

(c) $\displaystyle\int_{\mathscr{R}^2} \frac{2 + \sin(x + y)}{(1 + x^2 + y^2)^{1/2}}\, d(x, y)$

(d) $\displaystyle\int_{\mathscr{R}^2} e^{-x^2 - y^2} e^{2x} \sin y\, d(x, y)$

(e) $\int_{\mathscr{R}^2} \dfrac{\sin x^2 + \sin y^2}{(1 - x^2 - y^2)^{1/2}}\, d(x, y)$

12. Suppose f is continuous on a closed Jordan-measurable subset S of $\mathscr{R}^2$ and $(x_0, y_0) \in S^0$. Show that

$$\int_S \frac{f(x, y)}{[(x - x_0)^2 + (y - y_0)^2]^p}\, d(x, y)$$

converges if $p < 1$.

13. Suppose f is continuous on a closed Jordan-measurable subset S of $\mathscr{R}^3$ and $(x_0, y_0, z_0) \in S^0$. Show that

$$\int_S \frac{f(x, y, z)}{[(x - x_0)^2 + (y - y_0)^2 + (z - z_0)^2]^p}\, d(x, y, z)$$

converges if $p < \frac{3}{2}$.

14. Let f be continuous on the disc

$$S = \{(x, y) \mid x^2 + y^2 \leq 1\}.$$

Show that

$$\int_S \frac{f(x, y)}{(1 - x^2 - y^2)^p}\, d(x, y)$$

converges if $p < 1$.

15. Suppose f is continuous on

$$S = \{(x, y, z) \mid x^2 + y^2 + z^2 \geq 1\}$$

and, for some M and α,

$$|f(x, y, z)| \leq M(x^2 + y^2 + z^2)^{-\alpha}$$

if $x^2 + y^2 + z^2$ is sufficiently large. Show that

$$\int_S \frac{f(x, y, z)}{(x^2 + y^2 + z^2 - 1)^p}\, d(x, y, z)$$

converges if $p < 1$ and $\alpha > -p + \frac{3}{2}$.

16. Show that the integral

$$\int_S \frac{\sin t}{t}\, dt, \qquad S = (0, \infty)$$

is not convergent according to Definition 4.1 (with $n = 1$).

17. Suppose S satisfies the assumptions of Definition 4.1 and f is properly integrable over sets in $J(S)$. Let $J_1(S)$ be a subfamily of $J(S)$ containing a

sequence $\{T_j\}$ such that

$$T_j \subset T_{j+1} \quad (j = 1, 2, \ldots) \quad \text{and} \quad S^0 = \bigcup_{j=1}^{\infty} T_j^0,$$

and suppose

$$\lim_{j \to \infty} \int_{T_j} f(\mathbf{X}) \, d\mathbf{X} = L \text{ (finite)}$$

whenever $\{T_j\}$ is such a sequence in $J_1(S)$. Then we say that $\int f(\mathbf{X}) \, d\mathbf{X}$ *converges to L relative to* $J_1(S)$. If, in addition, $\int_S f(\mathbf{X}) \, d\mathbf{X}$ diverges, then we say that the convergence relative to $J_1(S)$ is *conditional*.

Show that if $n = 1$, $S = [a, b)$, f is locally integrable on S, and

$$J_1(S) = \{[a, c) \,|\, a \le c < b\},$$

then these definitions reduce to the definitions of convergence and conditional convergence given in Section 3.4.

18. Prove that if S is the first quadrant, then the integral $\int_S \cos(x^2 + y^2) \, d(x, y)$ converges conditionally relative to the family of rectangles

$$J_1(S) = \{[0, a] \times [0, b] \,|\, a, b > 0\}.$$

See Exercise 17 for the definition. (*Hint:* To see that the integral diverges, integrate over the quarter circle $\{(x, y) \,|\, x^2 + y^2 \le R^2, x, y \ge 0\}$.)

19. Let S be the first quadrant. Show that $\int_S \cos xy \, d(x, y)$ converges conditionally to $\pi/2$ relative to the family of rectangles

$$J_1(S) = \{[0, a] \times [0, b] \,|\, a, b > 0\}.$$

(*Hint:* See Exercise 32, Section 7.2.)

7.5 ORDINARY INTEGRALS INVOLVING A PARAMETER

In Section 7.2 we considered functions of the form

$$F(y) = \int_a^b f(x, y) \, dx, \qquad c \le y \le d, \tag{1}$$

in which the variable y appears as a parameter in the integral. We saw that if f is continuous on $[a, b] \times [c, d]$ then F is continuous on $[c, d]$ (*Exercise 3, Section 7.2*), and the iterated integral

$$\int_c^d F(y) \, dy = \int_c^d dy \int_a^b f(x, y) \, dx$$

can also be evaluated as

$$\int_c^d F(y) \, dy = \int_a^b dx \int_c^d f(x, y) \, dy.$$

The following theorem presents another important property of F.

Theorem 5.1   *If f and f_y are continuous on $R = [a, b] \times [c, d]$, then it is permissible to differentiate (1) under the integral sign with respect to y; that is,*

$$F'(y) = \int_a^b f_y(x, y)\, dx, \qquad c \le y \le d. \tag{2}$$

Here $F'(a)$ and $f_y(x, a)$ are to be interpreted as derivatives from the right, $F'(b)$ and $f_y(x, b)$ as derivatives from the left.

Proof.   If y and $y + \Delta y$ are in $[c, d]$ and $\Delta y \ne 0$, then

$$\frac{F(y + \Delta y) - F(y)}{\Delta y} = \int_a^b \frac{f(x, y + \Delta y) - f(x, y)}{\Delta y}\, dx. \tag{3}$$

From the mean value theorem,

$$f(x, y + \Delta y) - f(x, y) = f_y(x, y(x))\, \Delta y, \tag{4}$$

where $y(x)$ is between y and $y + \Delta y$, so that

$$|y - y(x)| < |\Delta y|. \tag{5}$$

Rewriting (4) as

$$f(x, y + \Delta y) - f(x, y) = f_y(x, y)\, \Delta y + (f_y(x, y(x)) - f_y(x, y))\, \Delta y,$$

we see from (3) that

$$\left| \frac{F(y + \Delta y) - F(y)}{\Delta y} - \int_a^b f_y(x, y)\, dx \right| < \int_a^b |f_y(x, y(x)) - f_y(x, y)|\, dx. \tag{6}$$

Now suppose $\varepsilon > 0$. Since f_y is uniformly continuous on R, there is a $\delta > 0$ such that

$$|f_y(x, y) - f_y(x', y')| < \varepsilon \tag{7}$$

if (x, y) and (x', y') are in R and

$$\sqrt{(x - x')^2 + (y - y')^2} < \varepsilon. \tag{8}$$

If $(x', y') = (x, y(x))$ and $|\Delta y| < \varepsilon$, then (8) holds [see (5)] and therefore so does (7); that is,

$$|f_y(x, y) - f_y(x, y(x))| < \varepsilon \quad \text{if} \quad (x, y) \in R.$$

This and (6) imply that

$$\left| \frac{F(y + \Delta y) - F(y)}{\Delta y} - \int_a^b f_y(x, y)\, dx \right| < \varepsilon(b - a)$$

if y and $y + \Delta y$ are in $[c, d]$ and $|\Delta y| < \delta$. This implies (2).

Example 5.1   Since the functions

$$f(x, y) = \cos xy \text{ and } fy(x, y) = -x \sin xy$$

are continuous for all (x, y), Theorem 5.1 implies that if

$$F(y) = \int_0^\pi \cos xy \, dx \tag{9}$$

then

$$F'(y) = -\int_0^\pi x \sin xy \, dx, \qquad -\infty < y < \infty. \tag{10}$$

[In applying Theorem 5.1 to obtain (10) for a specific value of y, we take $R = [0, \pi] \times [-\rho, \rho]$, where $\rho > |y|$.] This provides a convenient way to evaluate the integral on the right of (10); thus, integrating (9) yields

$$F(y) = \frac{\sin xy}{y}\bigg|_{x=0}^\pi = \frac{\sin \pi y}{y}, \qquad y \neq 0.$$

Differentiating this and using (10) yields

$$\int_0^\pi x \sin xy \, dx = \frac{\sin \pi y}{y^2} - \frac{\pi \cos \pi y}{y}, \qquad y \neq 0.$$

This can be verified by integrating by parts.

Our main objective in this section is to study the continuity, differentiability, and integrability of functions F as defined by (1), where the integral is improper for some or all values of y in $[c, d]$. In this case mere convergence of $F(y)$ for each y in $[c, d]$ (*pointwise* convergence) does not imply that f is continuous or integrable on $[c, d]$, and the added assumption that f_y is continuous and $\int_a^b f_y(x, y) \, dx$ converges pointwise on $[c, d]$ does not imply (2).

Example 5.2   The function

$$f(x, y) = ye^{-|y|x}$$

is continuous on $[0, \infty) \times [c, d]$ for every c and d, and the integral

$$F(y) = \int_0^\infty f(x, y) \, dx = \int_0^\infty ye^{-|y|x} \, dx$$

converges for all y, with

$$F(y) = \begin{cases} 1 & \text{if } y > 0, \\ 0 & \text{if } y = 0, \\ -1 & \text{if } y < 0; \end{cases}$$

thus, F is discontinuous at $y = 0$.

Example 5.3   The function

$$f(x, y) = y^3 e^{-y^2 x}$$

is continuous on $[0, \infty) \times [c, d]$ for every c and d, and the integral

$$F(y) = \int_0^\infty f(x, y) \, dx = \int_0^\infty y^3 e^{-y^2 x} \, dx$$

converges for all y to

$$F(y) = y$$

(verify); therefore,

$$F'(y) = 1, \qquad -\infty < y < \infty.$$

However,

$$\int_0^\infty \frac{\partial}{\partial y}(y^3 e^{-y^2 x}) \, dx = \int_0^\infty (3y^2 - 2y^4 x) e^{-y^2 x} \, dx,$$

which converges for all y, with

$$\int_0^\infty (3y^2 - 2y^4 x) e^{-y^2 x} \, dx = \begin{cases} 1, & y \neq 0, \\ 0, & y = 0; \end{cases}$$

thus,

$$F'(y) \neq \int_0^\infty \frac{\partial f(x, y)}{\partial y} \, dx \quad \text{if} \quad y = 0.$$

Uniform convergence of improper integrals

To formulate assumptions that ensure continuity, differentiability, and integrability of the function

$$F(y) = \int_a^b f(x, y) \, dx$$

when the integral may be improper, we proceed by analogy with the corresponding development for infinite series (Section 4.4). Thus, since pointwise convergence is not strong enough, we introduce the idea of uniform convergence, as follows.

Definition 5.1   *Suppose that for each y in a set S the function f is locally integrable on $[a, b) \, (-\infty < a < b \leq \infty)$ with respect to x, and the improper integral*

$$\int_a^b f(x, y) \, dx = \lim_{r \to b-} \int_a^r f(x, y) \, dx$$

converges. Then $\int_a^b f(x, y)\,dx$ *is said to converge uniformly on S if for each* $\varepsilon > 0$ *there is an R in* (a, b) *such that*

$$\left| \int_a^b f(x, y)\,dx - \int_a^r f(x, y)\,dx \right| = \left| \int_r^b f(x, y)\,dx \right| < \varepsilon$$

for all y in S, provided only that $R \le r \le b$.

We leave it to the reader to formulate the corresponding definition of uniform convergence of the improper integral

$$\int_a^b f(x, y)\,dx = \lim_{r \to a+} \int_r^b f(x, y)\,dx,$$

applicable to the case where f is locally integrable on $(a, b]\,(-\infty \le a < b < \infty)$ for each y in S (*Exercise 4*). If f is locally integrable on $(a, b)\,(-\infty \le a < b \le \infty)$ for each y in S, then $\int_a^b f(x, y)\,dx$ is said to converge uniformly on S if $\int_a^\alpha f(x, y)\,dx$ and $\int_\alpha^b f(x, y)\,dx$ both converge uniformly on S, where α is any point in (a, b). This definition is independent of the particular choice of α (*Exercise 5*).

We leave it to the reader (*Exercise 6*) to show that if f is bounded on $[a, b] \times [c, d]$ and $\int_a^b f(x, y)\,dx$ exists as a proper integral for each y in $[c, d]$, then it converges uniformly on $[c, d]$ according to all three versions of Definition 5.1.

Example 5.4   For the improper integral of Example 5.2,

$$\left| \int_r^\infty f(x, y)\,dx \right| = \int_r^\infty |y| e^{-|y|x}\,dx = e^{-r|y|}, \quad y \ne 0.$$

If $|y| \ge \rho$, then

$$\left| \int_r^\infty f(x, y)\,dx \right| \le e^{-r\rho},$$

and so $\int_0^\infty f(x, y)\,dx$ converges uniformly on $S = (-\infty, -\rho] \cup [\rho, \infty)$; however, it does not converge uniformly on any neighborhood of $y = 0$, because for any $r > 0$, $e^{-r|y|} > \frac{1}{2}$ if $|y|$ is sufficiently small.

Example 5.5   To determine whether the integral

$$F(y) = \int_0^\infty x^{-1/2} e^{-yx}\,dx$$

converges uniformly, we consider separately the integrals

$$F_1(y) = \int_0^1 x^{-1/2} e^{-yx}\,dx \quad \text{and} \quad F_2(y) = \int_1^\infty x^{-1/2} e^{-yx}\,dx.$$

(The choice of 1 as the upper limit in F_1 and the lower limit in F_2 is arbitrary; any positive number would do as well.) Since

$$\left| \int_0^r x^{-1/2} e^{-yx} \, dx \right| < \int_0^r x^{-1/2} \, dx = 2r^{1/2}, \qquad y \geq 0,$$

$F_1(y)$ converges uniformly for $y \geq 0$. Since

$$\left| \int_r^\infty x^{-1/2} e^{-yx} \, dx \right| \leq r^{-1/2} \int_r^\infty e^{-yx} \, dx \leq \frac{e^{-ry}}{yr^{1/2}},$$

$F_2(y)$ converges uniformly for y in $[\rho, \infty)$, for any $\rho > 0$. Consequently, $F(y)$ converges uniformly for y in $[\rho, \infty)$ for any $\rho > 0$. It does not converge uniformly for y in $(0, \rho)$, since the change of variable $u = yx$ yields

$$\int_r^\infty x^{-1/2} e^{-yx} \, dx = y^{-1/2} \int_{ry}^\infty u^{-1/2} e^{-u} \, du,$$

which, for any fixed $r > 0$, can be made arbitrarily large by taking y sufficiently small.

Example 5.6   If $y > 0$, then

$$\int_0^\infty \frac{\sin yx}{x} \, dx = \frac{\pi}{2} \tag{11}$$

(*Exercise 32, Section 7.2; see also Exercise 30, this section*). To investigate uniform convergence of (11), we first observe that if $0 < r < r_1$ and $y > 0$, then

$$\int_r^{r_1} \frac{\sin yx}{x} \, dx = \int_{ry}^{r_1 y} \frac{\sin u}{u} \, du,$$

and letting $r_1 \to \infty$ yields

$$\int_r^\infty \frac{\sin yx}{x} \, dx = \int_{ry}^\infty \frac{\sin u}{u} \, du. \tag{12}$$

If $\varepsilon > 0$, there is an R_0 such that

$$\left| \int_R^\infty \frac{\sin u}{u} \, du \right| < \varepsilon \quad \text{if} \quad R \geq R_0;$$

hence (12) implies that if $y \geq \rho > 0$ and $r \geq R_0/\rho$, then

$$\left| \int_r^\infty \frac{\sin yx}{x} \, dx \right| < \varepsilon.$$

Therefore, (11) converges uniformly on $[\rho, \infty)$ for any $\rho > 0$. On the other hand, since

$$\int_0^\infty \frac{\sin u}{u} \, du = \frac{\pi}{2},$$

there is a $\delta > 0$ such that

$$\int_{u_0}^{\infty} \frac{\sin u}{u}\, du > \frac{\pi}{4}, \qquad 0 \leq u_0 < \delta.$$

This and (12) imply that

$$\int_{r}^{\infty} \frac{\sin yx}{x}\, dx > \frac{\pi}{4}$$

for any $r > 0$, provided $0 < y < \delta/r$. Hence, (11) does not converge uniformly on $(0, \infty)$.

Tests for uniform convergence

We will state and prove theorems in this section for the case where f is locally integrable with respect to x on $[a, b)$ for each y in S; however, analogous results hold for the other cases mentioned in Definition 4.1, and some of our examples will apply to the other cases.

Theorem 5.2 (Weierstrass's Test for Uniform Convergence of Improper Integrals)
Suppose f is locally integrable with respect to x on $[a, b)$ and

$$|f(x, y)| \leq M(x) \quad \text{if} \quad x_0 \leq x < b \tag{13}$$

for each y in S, where

$$\int_{a}^{b} M(x)\, dx < \infty. \tag{14}$$

Then

$$\int_{a}^{b} f(x, y)\, dy$$

converges uniformly on S.

Proof. If $\varepsilon > 0$, then (14) implies that there is an $R \geq a$ such that

$$\int_{r}^{b} M(x)\, dx < \varepsilon \quad \text{if} \quad R \leq r < b$$

(*Exercise 7, Section 3.4*). This and (13) imply that $\int_{a}^{b} |f(x, y)|\, dx$ converges for each y in S (Theorem 4.2, Section 3.4); therefore, so does $\int_{a}^{b} f(x, y)\, dx$ (Theorem 4.4, Section 3.4), and

$$\left| \int_{r}^{b} f(x, y)\, dx \right| \leq \int_{r}^{b} |f(x, y)|\, dx < \varepsilon \quad \text{if} \quad R \leq r \leq b \quad \text{and} \quad y \in S.$$

Therefore, $\int_{a}^{b} f(x, y)\, dx$ converges uniformly on S.

Example 5.7 Suppose g is locally integrable with respect to x on $[0, \infty)$ and, for some constants K, p_0, and x_0,

$$|g(x, y)| \leq K e^{p_0 x}, \qquad x \geq x_0, \qquad y \in S.$$

Then if $p > p_0$, the integral

$$\int_0^\infty e^{-px} g(x, y) \, dx$$

converges uniformly on S. To see this, we take

$$f(x, y) = e^{-px} g(x, y) \quad \text{and} \quad M(x) = K e^{-(p - p_0)x}$$

in Theorem 5.2. Since

$$|x^\alpha \sin yx| < e^{p_0 x} \quad \text{and} \quad |x^\alpha \cos yx| < e^{p_0 x}$$

for x sufficiently large if $p_0 > 0$, this result implies that

$$\int_0^\infty e^{-px} x^\alpha \sin yx \, dx \quad \text{and} \quad \int_0^\infty e^{-px} x^\alpha \cos yx \, dx$$

converge uniformly for y in $(-\infty, \infty)$, provided $p > 0$ and $\alpha \geq 0$.

Example 5.8 If g is locally integrable with respect to x on $(0, 1]$ and

$$|g(x, y)| \leq A x^{-\beta}, \qquad 0 < x \leq x_0$$

for each y in S, then

$$\int_0^1 x^\alpha g(x, y) \, dx$$

converges uniformly on S if $\alpha > \beta - 1$. To see this we take

$$f(x, y) = x^\alpha g(x, y) \quad \text{and} \quad M(x) = A x^{\alpha - \beta}$$

in Theorem 5.2. For example, this implies that

$$\int_0^1 x^\alpha \cos xy \, dx$$

converges uniformly for all y if $\alpha > -1$.

The conclusion of Theorem 5.2 is stronger than stated, since its proof actually shows that $\int_a^b |f(x, y)| \, dx$ converges uniformly for y in S. The next theorem applies in some cases where this is not so.

Theorem 5.3 (Dirichlet's Test for Uniform Convergence of Improper Integrals)
If g, g_x, and h are continuous on $[a, b) \times S$, then

$$\int_a^b g(x, y) h(x, y) \, dx$$

converges uniformly for y in S if:

(a) $\lim\limits_{x \to b-} \left\{ \text{l.u.b.}\limits_{y \in S} |g(x, y)| \right\} = 0.$

(b) *There is a constant M such that*

$\text{l.u.b.}\limits_{y \in S} \left| \int_a^x h(u, y) \, du \right| \le M, \qquad a \le x < b.$

(c) $\int_a^b |g_x(x, y)| \, dx$ *converges uniformly for y in S.*

Proof.   The proof is similar to that of Theorem 4.10, Section 4.4. If

$$H(x, y) = \int_a^x h(u, y) \, du,$$

then integration by parts yields

$$\int_a^r g(x, y)h(x, y) \, dx = \int_a^r g(x, y)H_x(x, y) \, dx$$

$$= g(r, y)H(r, y) - \int_a^r g_x(x, y)H(x, y) \, dx.$$

We leave the rest of the proof to the reader (*Exercise 12*).

Example 5.9   Theorem 5.3 implies that

$$\int_0^\infty \frac{\cos xy}{x + y} \, dx$$

converges uniformly for $y \ge \rho > 0$. To see this, let

$$g(x, y) = \frac{1}{x + y} \quad \text{and} \quad h(x, y) = \cos xy.$$

Then (a) of Theorem 5.3 holds because

$$g(x, y) \le \frac{1}{x + \rho}, \qquad y \ge \rho,$$

and (c) holds—by Theorem 5.2—since

$$|g_x(x, y)| = \frac{1}{(x + y)^2} \le \frac{1}{(x + \rho)^2}, \qquad y \ge \rho$$

and

$$\int_0^\infty \frac{dx}{(x + \rho)^2} < \infty, \qquad \rho > 0.$$

To see that (b) holds, we write

$$\left| \int_0^x h(u, y)\, du \right| = \left| \int_0^x \cos uy\, du \right| = \left| \frac{\sin xy}{y} \right| \leq \frac{1}{\rho}, \qquad y \geq \rho.$$

Consequences of uniform convergence

The next theorem is analogous to Theorem 4.11, Section 4.4.

Theorem 5.4   *Suppose f is continuous on $[a, b) \times [c, d]$ and the integral*

$$F(y) = \int_a^b f(x, y)\, dx$$

converges uniformly for y in $[c, d]$. Then F is continuous on $[c, d]$.

Proof.   Definition 5.1 implies that if $\varepsilon > 0$, there is an r in $[a, b)$ such that

$$\left| \int_r^b f(x, y)\, dx \right| < \varepsilon, \qquad c \leq y \leq d.$$

Therefore, if y and y_0 are in $[c, d]$, then

$$\begin{aligned}
|F(y) - F(y_0)| &= \left| \int_a^b f(x, y)\, dx - \int_a^b f(x, y_0)\, dx \right| \\
&\leq \left| \int_a^r [f(x, y) - f(x, y_0)]\, dx \right| + \left| \int_r^b f(x, y)\, dx \right| \\
&\quad + \left| \int_r^b f(x, y_0)\, dx \right| \\
&< \int_a^r |f(x, y) - f(x, y_0)|\, dx + 2\varepsilon.
\end{aligned} \tag{15}$$

Since f is uniformly continuous on $[a, r] \times [c, d]$, there is a $\delta > 0$ such that

$$|f(x, y) - f(x, y_0)| < \varepsilon$$

if (x, y) and (x, y_0) are in $[a, r] \times [c, d]$ and $|y - y_0| < \delta$. This and (15) imply that

$$|F(y) - F(y_0)| < (r - a)\varepsilon + 2\varepsilon < (b - a + 2)\varepsilon$$

if y and y_0 are in $[c, d]$ and $|y - y_0| < \delta$. This completes the proof.

If f is continuous on $[a, b) \times [c, d]$ and the improper integral

$$F(y) = \int_a^b f(x, y)\, dx \tag{16}$$

converges uniformly for y in $[c, d]$, then F is continuous and the integral

$$I_1 = \int_c^d dy \int_a^b f(x, y)\, dx$$

exists. On the other hand, we can also consider the *proper* integral

$$G(x) = \int_c^d f(x, y)\, dy,$$

which is continuous for x in $[a, b)$, and ask whether the *improper* integral

$$I_2 = \int_a^b dx \int_c^d f(x, y)\, dy$$

exists and equals I_1. Thus, we are asking whether we can reverse the order of integration and write

$$\int_c^d dy \int_a^b f(x, y)\, dx = \int_a^b dx \int_c^d f(x, y)\, dy. \tag{17}$$

The next theorem states that this is legitimate.

Theorem 5.5   *If f is continuous on $[a, b) \times [c, d]$ and (16) converges uniformly for y in $[c, d]$, then (17) holds.*

Proof.   *Exercise 19.*

Example 5.10   It is straightforward to verify that

$$\int_0^\infty e^{-xy}\, dx = \frac{1}{y}, \qquad y > 0,$$

and that the convergence is uniform on $[\rho, \infty)$ for any $\rho > 0$. Therefore, Theorem 5.5 implies that

$$\int_{y_1}^{y_2} \frac{dy}{y} = \int_{y_1}^{y_2} dy \int_0^\infty e^{-xy}\, dx = \int_0^\infty dx \int_{y_1}^{y_2} e^{-xy}\, dx$$

$$= \int_0^\infty \frac{e^{-xy_1} - e^{-xy_2}}{x}\, dx, \qquad y_2 \geq y_1 > 0.$$

Since

$$\int_{y_1}^{y_2} \frac{dy}{y} = \log \frac{y_2}{y_1}, \qquad y_2 \geq y_1 > 0$$

this implies that

$$\int_0^\infty \frac{e^{-xy_1} - e^{-xy_2}}{x}\, dx = \log \frac{y_1}{y_2}, \qquad y_2 \geq y_1 > 0.$$

Example 5.11   From Example 5.6,

$$\int_0^\infty \frac{\sin xy}{x}\, dx = \frac{\pi}{2}, \qquad y > 0,$$

and the convergence is uniform if $y \geq \rho > 0$. Therefore, Theorem 5.5 implies that

$$\frac{\pi}{2}(y_2 - y_1) = \int_{y_1}^{y_2} dy \int_0^\infty \frac{\sin xy}{x}\, dx = \int_0^\infty \frac{dx}{x} \int_{y_1}^{y_2} \sin xy\, dy$$

$$= \int_0^\infty \frac{\cos xy_1 - \cos xy_2}{x^2}\, dx, \qquad y_2 \geq y_1 > 0. \tag{18}$$

The last integral converges uniformly for all y_1 [*Exercise 10(h)*], and is therefore continuous with respect to y_1 on $(-\infty, \infty)$, by Theorem 5.4; in particular, we can let $y_1 \to 0^+$ in (18) and replace y_2 by y, to obtain

$$\int_0^\infty \frac{1 - \cos xy}{x^2}\, dx = \frac{\pi y}{2} \qquad \text{if } y \geq 0.$$

The next theorem is analogous to Theorem 4.13, Section 4.4.

Theorem 5.6   *Suppose f and f_y are continuous on $[a, b) \times [c, d]$, that the integral*

$$F(y) = \int_a^b f(x, y)\, dx$$

converges for some y in $[c, d]$, and that the integral

$$G(y) = \int_a^b f_y(x, y)\, dx$$

converges uniformly for y in $[c, d]$. Then F converges uniformly for y in $[c, d]$, and

$$F'(y) = G(y), \qquad c \leq y \leq d. \tag{19}$$

Here $F'(c)$ and $f_y(x, c)$ are to be interpreted as derivatives from the right, $F'(d)$ and $f_y(x, d)$ as derivatives from the left.

Proof.   Suppose $a < r < b$ and define

$$F_r(y) = \int_a^r f(x, y)\, dx, \qquad c \leq y \leq d.$$

From Theorem 5.1 and *Exercise 3, Section 7.2*, F_r is continuously differentiable on $[c, d]$, and

$$F_r'(y) = \int_a^r f_y(x, y)\, dx, \qquad c \leq y \leq d.$$

Integrating this yields

$$F_r(y) = F_r(y_0) + \int_{y_0}^y dt \int_a^r f_y(x, t)\, dx, \qquad c \le y \le d.$$

Therefore,

$$\left| F_r(y) - F(y_0) - \int_{y_0}^y G(t)\, dt \right| \le |F_r(y_0) - F(y_0)|$$

(20)

$$+ \left| \int_{y_0}^y \left| \int_r^b f_y(x, t)\, dx \right| dt \right|.$$

Our assumptions imply that if $\varepsilon > 0$, there is an R in (a, b) such that

$$|F_r(y_0) - F(y_0)| < \varepsilon$$

and

$$\left| \int_r^b f_y(x, t)\, dx \right| < \varepsilon \qquad \text{for all } t \text{ in } [c, d]$$

if $R \le r < b$; therefore, (20) yields

$$\left| F_r(y) - F(y_0) - \int_{y_0}^y G(t)\, dt \right| < \varepsilon(1 + |y - y_0|) \le \varepsilon(1 + d - c)$$

if $R \le r < b$ and $y \in [c, d]$. This implies that $F(y)$ converges uniformly for y in $[c, d]$ and

$$F(y) = F(y_0) + \int_{y_0}^y G(t)\, dt, \qquad c \le y \le d.$$

Since G is continuous on $[c, d]$ by Theorem 5.4, (19) follows from differentiating this (Theorem 3.10, Section 3.3).

Example 5.12   Consider the integral

$$\int_0^\infty e^{-yx^2}\, dx, \qquad y > 0.$$

Since

$$\int_0^r e^{-yx^2}\, dx = \frac{1}{\sqrt{y}} \int_0^{r\sqrt{y}} e^{-u^2}\, du$$

and

$$\int_0^\infty e^{-u^2}\, du = \frac{\sqrt{\pi}}{2}$$

(Example 4.4), it follows that

$$\int_0^\infty e^{-yx^2}\, dx = \frac{\sqrt{\pi}}{2\sqrt{y}} \qquad \text{if } y > 0,$$

and the convergence is uniform on $[\rho, \infty)$ for any $\rho > 0$ [*Exercise 10(i)*]. Differentiating n times with respect to y yields

$$\int_0^\infty x^{2n} e^{-yx^2} \, dx = \frac{1 \cdot 3 \cdots (2n-1)\sqrt{\pi}}{2^{n+1} y^{n+1/2}} \quad \text{if} \quad y > 0 \text{ and } n \geq 1,$$

and the differentiation is justified—by Theorem 5.6—because the integral on the left converges uniformly on $[\rho, \infty)$ for any $\rho > 0$ [*Exercise 10(i)*].

An application to Laplace transforms

The *Laplace transform* of a function f which is locally integrable on $[0, \infty)$ is defined to be

$$F(s) = \int_0^\infty e^{-sx} f(x) \, dx \tag{21}$$

for those values of s for which the integral converges. Laplace transforms are widely applied in mathematics, particularly in solving differential equations.

Theorem 5.7   *If f is continuous and the integral*

$$H(x) = \int_0^x e^{-s_0 t} f(t) \, dt$$

is bounded on $(0, \infty)$, then F as defined in (21) is infinitely differentiable on (s_0, ∞), with

$$F^{(n)}(s) = (-1)^n \int_0^\infty e^{-sx} x^n f(x) \, dx, \qquad n = 0, 1, 2, \ldots. \tag{22}$$

Proof.   The function

$$g(x, s) = e^{-(s-s_0)x} x^n$$

satisfies (a) and (c) of Theorem 5.3 with $y = s$, $[a, b) = [0, \infty)$, and $S = [s_1, \infty)$, if $s_1 > s_0$ (*Exercise 16*). By assumption the function

$$h(x, s) = e^{-s_0 x} f(x)$$

satisfies (b). Since

$$g(x, s)h(x, s) = e^{-sx} x^n f(x),$$

Theorem 5.3 implies that the integrals in (22) converge uniformly on $[s_1, \infty)$ if $s_1 > s_0$. Since every s in (s_0, ∞) is in such an interval, Theorem 5.6 and induction imply (22) for $s > s_0$ (*Exercise 27*). This completes the proof.

Example 5.13   If $a \neq 0$, then the integral

$$\int_0^x \cos at \, dt = \frac{\sin ax}{a}$$

is bounded on $(0, \infty)$; therefore, the Laplace transform

$$F(s) = \int_0^\infty e^{-sx} \cos ax \, dx$$

converges and

$$F^{(n)}(s) = (-1)^n \int_0^\infty e^{-sx} x^n \cos ax \, dx \quad \text{if } s > 0. \tag{23}$$

(Note that this is also true if $a = 0$.) Elementary integration shows that

$$F(s) = \frac{s}{s^2 + a^2},$$

and so, from (23),

$$\int_0^\infty e^{-sx} x^n \cos ax \, dx = (-1)^n \frac{d^n}{ds^n} \frac{s}{s^2 + a^2}, \qquad n = 0, 1, \ldots.$$

7.5   EXERCISES

1. Evaluate

$$F(y) = \int_0^1 \frac{dx}{1 + yx}, \qquad y > -1,$$

and use Theorem 5.1 to evaluate

$$I_1 = \int_0^1 \frac{x \, dx}{(1 + x)^2} \quad \text{and} \quad I_2 = \int_0^1 \frac{x^2 \, dx}{(1 + x)^3}.$$

2. Evaluate

$$F(y) = \int_0^1 y^x \, dx, \qquad y > 0,$$

and use Theorem 5.1 to evaluate

$$I_1(y) = \int_0^1 xy^x \, dx \quad \text{and} \quad I_2(y) = \int_0^1 x^2 y^x \, dx, \quad y > 0.$$

3. Suppose g and h are differentiable on $[a, b]$, with

$$a \leq g(y) \leq b \quad \text{and} \quad a \leq h(y) \leq b \qquad \text{if } c \leq y \leq d.$$

Let f and f_y be continuous on $[a, b] \times [c, d]$. Derive *Liebniz's rule*:

$$\frac{d}{dy} \int_{g(y)}^{h(y)} f(x, y) \, dy = f(h(y), y)h'(y) - f(g(y), y)g'(y) + \int_{g(y)}^{h(y)} f_y(x, y) \, dy$$

[*Hint*: Define $H(y, u, v) = \int_u^v f(x, y) \, dx$ and use the chain rule.]

4. Define uniform convergence of $\int_a^b f(x, y) \, dx$ for y in S, for the case where f is locally integrable on $(a, b]$ $(-\infty \leq a < b < \infty)$ for y in S.

5. If f is locally integrable on (a, b) for each y in S, then $\int_a^b f(x, y) \, dx$ is said to converge uniformly on S if $\int_a^\alpha f(x, y) \, dx$ and $\int_\alpha^b f(x, y) \, dx$ converge uniformly on S, where α is any point in (a, b). Show that this definition does not depend on the particular choice of α.

6. (a) Show that if f is bounded on $[a, b] \times [c, d]$ and $\int_a^b f(x, y) \, dx$ exists as a proper integral for each y in $[c, d]$, then it converges uniformly on $[c, d]$ according to all three versions of Definition 5.1.
   (b) Give an example showing that the boundedness of f is essential in part (a).

7. Working directly from Definition 5.1, discuss uniform convergence for the following integrals:

(a) $\int_0^\infty \dfrac{dx}{1 + y^2 x^2}$

(b) $\int_0^\infty e^{-xy} x^2 \, dx$

(c) $\int_0^\infty x^{2n} e^{-yx^2} \, dx$

(d) $\int_0^\infty \sin yx^2 \, dx$

(e) $\int_0^\infty (3y^2 - 2yx)e^{-y^2 x} \, dx$ (Example 5.3)

(f) $\int_0^\infty (2yx - y^2 x^2)e^{-xy} \, dx$

8. Suppose f is locally integrable on $[a, b)$ with respect to x for each y in S. Show that $\int_a^b f(x, y) \, dx$ converges uniformly on S if and only if for each $\varepsilon > 0$ there is an R in (a, b) such that

$$\left| \int_{r_1}^{r_2} f(x, y) \, dx \right| < \varepsilon$$

for all y in S if $R < r_1 < r_2 < b$. Whose name should be associated with this result?

9. State and prove a form of Theorem 5.2 for the case where f is locally integrable on $(a, b]$ for each y in S.

10. Use Weierstrass's test to show that the integral converges uniformly for y in S:

(a) $\int_0^\infty e^{-xy} \sin x \, dx$; $S = [\rho, \infty)$, $\rho > 0$

(b) $\int_0^\infty \frac{\sin x}{x^y} dx$; $S = [c, d]$, $1 < c < d < 2$

(c) $\int_0^\infty e^{-px} \frac{\sin yx}{x} dx$ $(p > 0)$; $S = (-\infty, \infty)$

(d) $\int_0^1 \frac{e^{xy}}{(1 - x)^y} dx$; $S = (-\infty, b)$, $b < 1$

(e) $\int_{-\infty}^\infty \frac{\cos xy}{1 + x^2 y^2} dx$; $S = (-\infty, -\rho] \cup [\rho, \infty)$, $\rho > 0$

(f) $\int_1^\infty e^{-x/y} dx$; $S = [\rho, \infty)$, $\rho > 0$

(g) $\int_{-\infty}^\infty e^{xy} e^{-x^2} dx$; $S = [-\rho, \rho]$, $\rho > 0$

(h) $\int_0^\infty \frac{\cos yx - \cos ax}{x^2} dx$; $S = (-\infty, \infty)$

(i) $\int_0^\infty x^{2n} e^{-yx^2} dx$; $S = [\rho, \infty)$, $\rho > 0$ (if $n \geq 0$)

11. (a) Show that the integral

$$\Gamma(y) = \int_0^\infty x^{y-1} e^{-x} dx$$

converges if $y > 0$, and uniformly on $[c, d]$ if $0 < c < d < \infty$.
(b) Use integration by parts to show that

$$\Gamma(y) = \frac{\Gamma(y + 1)}{y}, \qquad y > 0,$$

and then show by induction that for each integer $n \geq 0$,

$$\Gamma(y) = \frac{\Gamma(y + n)}{y(y + 1) \cdots (y + n - 1)}, \qquad y > 0.$$

How can this be used to define $\Gamma(y)$ in a natural way for all $y \neq 0, -1, -2, \ldots$? (This function is called the *gamma function*.)
(c) Show that $\Gamma(n + 1) = n!$ if $n = 1, 2, \ldots$.
(d) Show that $\Gamma(\frac{1}{2}) = \sqrt{\pi}$. (*Hint:* See Example 4.4.)
(e) Show that

$$\int_0^\infty e^{-st} t^\alpha dt = s^{-\alpha-1} \Gamma(\alpha + 1) \qquad \text{if } \alpha > -1 \quad \text{and} \quad s > 0.$$

12. Complete the proof of Theorem 5.3. (*Hint:* See the proof of Theorem 4.10, Section 4.4.)

13. State the analog of Theorem 5.3 for the case where f is continuous on $(a, b] \times S$.

14. Show that Theorem 5.3 remains valid with assumption (c) replaced by the assumption that $g_x(x, y) \le 0$ for y in S.

15. Use Dirichlet's test or Exercise 14 to show that the following integrals converge uniformly on $[\rho, \infty)$ if $\rho > 0$:

(a) $\int_1^\infty \frac{\sin xy}{x^y} \, dx$ (b) $\int_1^\infty \frac{\sin xy}{\log x} \, dx$

(c) $\int_0^\infty \frac{\cos xy}{x + y^2} \, dx$ (d) $\int_0^\infty \frac{\sin xy}{1 + xy} \, dx$

16. Show that the function

$$g(x, s) = e^{-(s - s_0)x} x^\alpha \qquad (\alpha > 0)$$

satisfies the assumptions (a) and (c) of Theorem 5.3 with $[a, b] = [0, \infty)$, $y = s$, and $S = [s_1, \infty)$ $(s_1 > s_0)$.

17. Show that the functions

$$C(y) = \int_{-\infty}^\infty f(x) \cos xy \, dx \quad \text{and} \quad S(y) = \int_{-\infty}^\infty f(x) \sin xy \, dx$$

are continuous on $(-\infty, \infty)$ if

$$\int_{-\infty}^\infty |f(x)| \, dx < \infty.$$

18. Suppose f is continuously differentiable on $[a, \infty)$,

$$\int_a^\infty |f'(x)| \, dx < \infty,$$

and $\lim_{x \to \infty} f(x) = 0$. Show that the functions

$$C(y) = \int_a^\infty f(x) \cos xy \, dx \quad \text{and} \quad S(y) = \int_a^\infty f(x) \sin xy \, dx$$

are continuous for all $y \ne 0$. Give examples showing that they need not be continuous at $y = 0$.

19. Prove Theorem 5.5.

20. Evaluate $F(y)$ and use Theorem 5.5 to evaluate I:

(a) $F(y) = \int_0^\infty \frac{dx}{1 + y^2x^2} (y \ne 0); I = \int_0^\infty \frac{\tan^{-1} ax - \tan^{-1} bx}{x} \, dx \, (a, b > 0)$

(b) $F(y) = \int_0^1 x^y \, dx \, (y > -1); I = \int_0^1 \frac{x^a - x^b}{\log x} \, dx \, (a, b > -1)$

(c) $F(y) = \int_0^\infty e^{-yx} \cos x \, dx \, (y > 0);$

$$I = \int_0^\infty \frac{e^{-ax} - e^{-bx}}{x} \cos x \, dx \, (a, b > 0)$$

(d) $F(y) = \int_0^\infty e^{-yx} \sin x \, dx \; (y > 0);$

$\qquad I = \int_0^\infty \dfrac{e^{-ax} - e^{-bx}}{x} \sin x \, dx \; (a, b > 0)$

(e) $F(y) = \int_0^\infty e^{-x} \sin xy \, dx; \; I = \int_0^\infty e^{-x} \dfrac{1 - \cos ax}{x} \, dx$

(f) $F(y) = \int_0^\infty e^{-x} \cos xy \, dx; \; I = \int_0^\infty e^{-x} \dfrac{\sin ax}{x} \, dx$

21. Use Theorem 5.6 to evaluate

   (a) $\int_0^1 (\log x)^n x^y \, dx \; (y > -1, n = 0, 1, 2, \dots)$

   (b) $\int_0^\infty \dfrac{dx}{(x^2 + y)^{n+1}} \; (y > 0, n = 0, 1, 2, \dots)$

   (c) $\int_0^\infty x^{2n} e^{-yx^2} \, dx \; (y > 0, n = 0, 1, 2, \dots)$

   (d) $\int_0^\infty x^{2n+1} e^{-yx^2} \, dx \; (y > 0, n = 0, 1, 2, \dots)$

   (e) $\int_0^\infty xy^x \, dx \; (0 < y < 1)$

22. (a) Use Theorem 5.6 and integration by parts to show that the function

   $$F(y) = \int_0^\infty e^{-x^2} \cos 2xy \, dx$$

   satisfies the differential equation

   $$F' + 2yF = 0.$$

   (b) Use part (a) to show that

   $$F(y) = \dfrac{\sqrt{\pi}}{2} e^{-y^2}$$

23. Show that

   $$\int_0^\infty e^{-x^2} \sin 2xy \, dx = e^{-y^2} \int_0^y e^{u^2} \, du.$$

   (*Hint:* See Exercise 22.)

24. State a condition guaranteeing that C and S of Exercise 17 are n times differentiable on $(-\infty, \infty)$. (Your condition should imply the hypotheses of Exercise 17.)

25. Suppose f is continuously differentiable on $[a, \infty)$,

   $$\int_a^\infty |(x^n f(x))'| \, dx < \infty,$$

and $\lim_{x \to \infty} x^n f(x) = 0$. Show that if C and S are as defined in Exercise 18, then $C, C', \dots, C^{(n)}$ and $S, S', \dots, S^{(n)}$ are continuous for all $y \neq 0$.

26. Differentiating the right side of

$$F(y) = \int_1^\infty \cos \frac{y}{x} \, dx$$

under the integral sign yields

$$\int_1^\infty -\frac{1}{x} \sin \frac{y}{x} \, dx,$$

which converges uniformly for y in any finite interval. (Why?) Does this mean that F is differentiable for all y?

27. Show that Theorem 5.6 and induction imply Eq. (22).

28. Prove: If f is continuous on $[0, \infty)$ and $\int_0^\infty e^{-s_0 x} f(x) \, dx$ converges, then

$$\lim_{s \to s_0 +} \int_0^\infty e^{-sx} f(x) \, dx = \int_0^\infty e^{-s_0 x} f(x) \, dx.$$

(*Hint:* See the proof of Theorem 5.9, Section 4.5.)

29. Suppose f is continuous on $[0, \infty)$ and the integral

$$F(s) = \int_0^\infty e^{-sx} f(x) \, dx$$

converges for $s \geq s_0$. Show that $\lim_{s \to \infty} F(s) = 0$. (*Hint:* Integrate by parts.)

30. (a) Starting from the result of Exercise 20(d), let $b \to \infty$ and invoke Exercise 29 to evaluate

$$\int_0^\infty e^{-ax} \frac{\sin x}{x} \, dx \qquad (a > 0).$$

(b) Use part (a) and Exercise 28 to show that

$$\int_0^\infty \frac{\sin x}{x} \, dx = \frac{\pi}{2}.$$

31. (a) Suppose f is continuously differentiable on $[0, \infty)$ and

$$|f(x)| \leq M e^{s_0 x}, \qquad x \geq x_1.$$

Show that the integral

$$G(s) = \int_0^\infty e^{-sx} f'(x) \, dx$$

converges uniformly on $[s_1, \infty)$ if $s_1 > s_0$. (*Hint:* Integrate by parts.)
(b) Show from part (a) that the integral

$$G(s) = \int_0^\infty e^{-sx} x e^{x^2} \sin e^{x^2} \, dx$$

converges uniformly on $[\rho, \infty)$ if $\rho > 0$. (Notice that this does not follow from Theorem 5.2 or 5.3.)

32. Suppose f is continuous on $[0, \infty)$,

$$\lim_{x \to 0+} \frac{f(x)}{x}$$

exists, and the integral

$$F(s) = \int_0^\infty e^{-sx} f(x) \, dx$$

exists for $s = s_0$. Show that

$$\int_{s_0}^\infty F(u) \, du = \int_0^\infty e^{-sx} \frac{f(x)}{x} \, dx.$$

[*Hint:* First show that $F(s)$ converges uniformly for s in $[s_0, \infty)$.]

line and surface integrals

8.1 VECTORS AND CURVES

The purpose of this section is to prepare for the rest of the chapter, which deals with line integrals in $\mathcal{R}^n$—mainly $\mathcal{R}^2$ and $\mathcal{R}^3$—and surface integrals in $\mathcal{R}^3$. To do this we will review some elementary vector analysis in $\mathcal{R}^3$, and consider vector and scalar fields and parametrically defined curves. We discuss parametrically defined surfaces in Section 8.3.

Brief review of vector analysis in $\mathcal{R}^3$

Until now we have identified the vector

$$\mathbf{X} = \begin{bmatrix} x_1 \\ x_2 \\ \vdots \\ x_n \end{bmatrix}$$

with the directed line segment from the origin in $\mathcal{R}^n$ to the point

$$\mathbf{X} = (x_1, x_2, \ldots ,x_n).$$

In the elementary vector analysis of $\mathcal{R}^3$ it is convenient to associate the vector

$$\mathbf{U} = \begin{bmatrix} u_1 \\ u_2 \\ u_3 \end{bmatrix}$$

with any of the infinitely many directed line segments, all with the same length and direction, connecting pairs of points $\mathbf{X}_0 = (x_0, y_0, z_0)$ and $\mathbf{X}_1 = (x_1, y_1, z_1)$, where

$$x_1 - x_0 = u_1, \qquad y_1 - y_0 = u_2, \qquad z_1 - z_0 = u_3$$

(Figure 1.1). We say in this case that $\mathbf{U}$ is the vector *from* $\mathbf{X}_0$ *to* $\mathbf{X}_1$. For example,

$$\mathbf{U} = \begin{bmatrix} 2 \\ -1 \\ -2 \end{bmatrix}$$

is the vector from $\mathbf{X}_0 = (1, -1, 2)$ to $\mathbf{X}_1 = (3, -2, 0)$; it is also the vector from $\mathbf{X}_2 = (5, 1, 0)$ to $\mathbf{X}_3 = (7, 0, -2)$.

As before, the angle θ between nonzero vectors

$$\mathbf{U} = \begin{bmatrix} u_1 \\ u_2 \\ u_3 \end{bmatrix} \quad \text{and} \quad \mathbf{V} = \begin{bmatrix} v_1 \\ v_2 \\ v_3 \end{bmatrix}$$

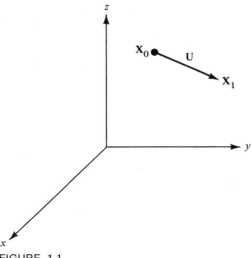

FIGURE 1.1

is defined by

$$\cos \theta = \frac{\mathbf{U} \cdot \mathbf{V}}{|\mathbf{U}| |\mathbf{V}|}, \qquad \text{where } 0 \le \theta \le \pi.$$

To visualize this angle we think of $\mathbf{U}$ and $\mathbf{V}$ as beginning from the same point (Figure 1.2).

We say that $\mathbf{U}$ and $\mathbf{V}$ *have the same direction* if one is a nonnegative multiple of the other, or *opposite directions* if one is a nonpositive multiple of the other; in either case $\mathbf{U}$ and $\mathbf{V}$ are *collinear*. If $\mathbf{U} \cdot \mathbf{V} = 0$, then $\mathbf{U}$ and $\mathbf{V}$ are *perpendicular;* if they are both nonzero, this means that the angle between them is $\theta = \pi/2$.

Notice that the *zero vector,*

$$\mathbf{0} = \begin{bmatrix} 0 \\ 0 \\ 0 \end{bmatrix},$$

is simultaneously collinear with and perpendicular to every vector in $\mathscr{R}^3$; it is the only vector in $\mathscr{R}^3$ with this property.

Example 1.4 of Section 5.1 illustrates the interpretation of vector addition in terms of the parallelogram law. The *vector plane* of two noncollinear vectors $\mathbf{U}$ and $\mathbf{V}$ is the set of all vectors of the form

$$a\mathbf{U} + b\mathbf{V}, \qquad -\infty < a, \quad b < \infty. \tag{1}$$

Just as a vector can be identified with infinitely many line segments with the same length and direction, the vector plane (1) can be identified with each of

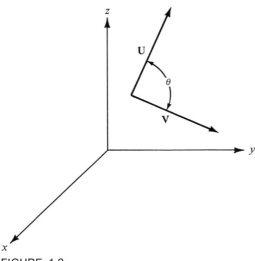

FIGURE 1.2

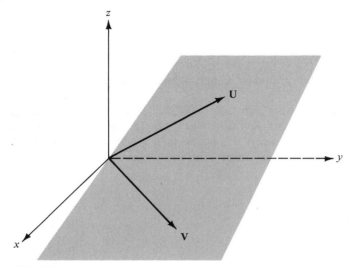

FIGURE 1.3

the infinitely many planes in $\mathscr{R}^3$ parallel to the plane determined by **U** and **V** considered as line segments initiating at the origin (Figure 1.3). Henceforth, we will follow standard practice and call (1) the *plane* (rather than the *vector plane*) *of* **U** *and* **V**.

If **U** and **V** are noncollinear and $\mathbf{W} \cdot \mathbf{U} = \mathbf{W} \cdot \mathbf{V} = 0$, then **W** is perpendicular to every vector in the plane of **U** and **V**, since

$$\mathbf{W} \cdot (a\mathbf{U} + b\mathbf{V}) = a(\mathbf{W} \cdot \mathbf{U}) + b(\mathbf{W} \cdot \mathbf{V}) = a0 + b0 = 0.$$

In this case we say that **W** is *perpendicular to the plane of* **U** *and* **V**.

An arbitrary vector

$$\mathbf{U} = \begin{bmatrix} u_1 \\ u_2 \\ u_3 \end{bmatrix}$$

can be written as

$$\mathbf{U} = u_1 \mathbf{i} + u_2 \mathbf{j} + u_3 \mathbf{k},$$

where

$$\mathbf{i} = \begin{bmatrix} 1 \\ 0 \\ 0 \end{bmatrix}, \qquad \mathbf{j} = \begin{bmatrix} 0 \\ 1 \\ 0 \end{bmatrix}, \qquad \mathbf{k} = \begin{bmatrix} 0 \\ 0 \\ 1 \end{bmatrix}.$$

The vectors **i**, **j**, and **k** can be represented as line segments in the positive directions of the x, y, and z axes, respectively (Figure 1.4). They are unit vectors, each

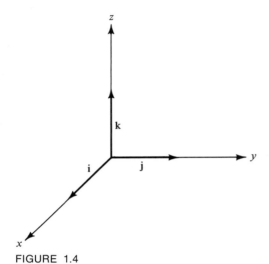

FIGURE 1.4

perpendicular to the other two; thus,

$$|\mathbf{i}| = |\mathbf{j}| = |\mathbf{k}| = 1$$

and

$$\mathbf{i} \cdot \mathbf{j} = \mathbf{j} \cdot \mathbf{k} = \mathbf{k} \cdot \mathbf{i} = 0. \tag{2}$$

The ideas we have discussed so far have analogs in $\mathscr{R}^n$ for any n; however, the next definition applies only to $\mathscr{R}^3$.

Definition 1.1   *If* $\mathbf{U}$ *and* $\mathbf{V}$ *are noncollinear vectors and* θ *is the angle between them, then the cross product* $\mathbf{U} \times \mathbf{V}$ *is the vector perpendicular to the plane of* $\mathbf{U}$ *and* $\mathbf{V}$, *in the direction in which a right-hand screw would move if rotated about its axis from* $\mathbf{U}$ *to* $\mathbf{V}$ *through the angle* θ *(Figure 1.5) and with length*

$$|\mathbf{U} \times \mathbf{V}| = |\mathbf{U}|\,|\mathbf{V}| \sin \theta.$$

If $\mathbf{U}$ *and* $\mathbf{V}$ *are collinear, then* $\mathbf{U} \times \mathbf{V} = \mathbf{0}$.

Example 1.1   From Figure 1.4,

$$\mathbf{i} \times \mathbf{j} = \mathbf{k}, \qquad \mathbf{j} \times \mathbf{k} = \mathbf{i}, \qquad \mathbf{k} \times \mathbf{i} = \mathbf{j},$$
$$\mathbf{j} \times \mathbf{i} = -\mathbf{k}, \qquad \mathbf{k} \times \mathbf{j} = -\mathbf{i}, \qquad \mathbf{i} \times \mathbf{k} = -\mathbf{j};$$

also,

$$\mathbf{i} \times \mathbf{i} = \mathbf{j} \times \mathbf{j} = \mathbf{k} \times \mathbf{k} = \mathbf{0}.$$

U X V

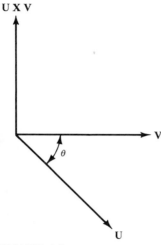

FIGURE 1.5

Figure 1.6 makes it easy to remember the first six cross products: the cross product of a pair of distinct vectors from $\{i, j, k\}$ equals the third vector if the second factor immediately follows the first in the clockwise sense shown in Figure 1.6; it is the negative of the third factor if the second factor precedes the first in this sense.

Theorem 1.1   *If* **U, V,** *and* **W** *are vectors in $\mathscr{R}^3$ and c is a real number, then*
    (a) $\mathbf{V} \times \mathbf{U} = -(\mathbf{U} \times \mathbf{V})$;
    (b) $(c\mathbf{U}) \times \mathbf{V} = c(\mathbf{U} \times \mathbf{V})$;
    (c) $\mathbf{U} \times (\mathbf{V} + \mathbf{W}) = (\mathbf{U} \times \mathbf{V}) + (\mathbf{U} \times \mathbf{W})$.

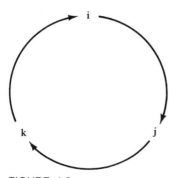

FIGURE 1.6

Proof.   Definition 1.1 immediately implies (a) and (b) (*Exercise 4*). We will prove (c) in the case where $\mathbf{U} \neq \mathbf{0}$ and $\mathbf{V}$ and $\mathbf{W}$ are noncollinear and perpendicular to $\mathbf{U}$; the rest of the proof is straightforward, as sketched in *Exercise 5*. For the case stated, $\mathbf{U} \times \mathbf{V}$, $\mathbf{U} \times \mathbf{W}$, and $\mathbf{U} \times (\mathbf{V} + \mathbf{W})$ are obtained by rotating $\mathbf{V}$, $\mathbf{W}$, and $\mathbf{V} + \mathbf{W}$ 90 degrees (all in the same sense) in the plane perpendicular to $\mathbf{U}$ and multiplying their lengths by the common factor $|\mathbf{U}|$. The situation is depicted in Figure 1.7, where $\mathbf{U}$ is perpendicular to the plane of the page, directed toward the reader, and the other vectors are in the plane of the page. By elementary geometric arguments, the triangles OAB and $OA'B'$ are similar, as are OCB and $OC'B'$. Since OAB and OCB are similar (in fact, congruent), $OA'B'$ and $OC'B'$ are similar. Moreover, since the similarity between $OA'B'$ and $OC'B'$ obviously makes the common side OB' correspond to itself, these two triangles are in fact congruent. Therefore, $OA'B'C'$ is a parallelogram with adjacent sides represented by $\mathbf{U} \times \mathbf{V}$ and $\mathbf{U} \times \mathbf{W}$, and diagonal represented by $\mathbf{U} \times (\mathbf{V} + \mathbf{W})$. Therefore, (c) holds in this case, from the geometric interpretation of vector addition.

Combining (a) and (b) yields

$$\mathbf{U} \times (c\mathbf{V}) = c(\mathbf{U} \times \mathbf{V})$$

(*Exercise 4*) This and (a), (b), and (c) lead to a convenient formula for $\mathbf{U} \times \mathbf{V}$. If

$$\mathbf{U} = u_1\mathbf{i} + u_2\mathbf{j} + u_3\mathbf{k} \quad \text{and} \quad \mathbf{V} = v_1\mathbf{i} + v_2\mathbf{j} + v_3\mathbf{k},$$

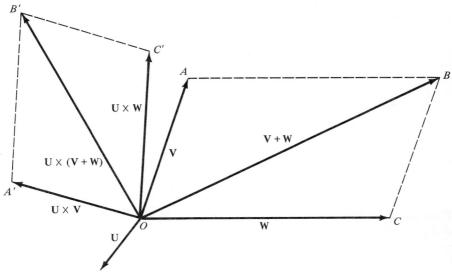

FIGURE 1.7

then

$$\begin{aligned}
\mathbf{U} \times \mathbf{V} &= (u_1\mathbf{i} + u_2\mathbf{j} + u_3\mathbf{k}) \times (v_1\mathbf{i} + v_2\mathbf{j} + v_3\mathbf{k}) \\
&= u_1[\mathbf{i} \times (v_1\mathbf{i} + v_2\mathbf{j} + v_3\mathbf{k})] \\
&\quad + u_2[\mathbf{j} \times (v_1\mathbf{i} + v_2\mathbf{j} + v_3\mathbf{k})] \\
&\quad + u_3[\mathbf{k} \times (v_1\mathbf{i} + v_2\mathbf{j} + v_3\mathbf{k})] \\
&= u_1[v_1(\mathbf{i} \times \mathbf{i}) + v_2(\mathbf{i} \times \mathbf{j}) + v_3(\mathbf{i} \times \mathbf{k})] \\
&\quad + u_2[v_1(\mathbf{j} \times \mathbf{i}) + v_2(\mathbf{j} \times \mathbf{j}) + v_3(\mathbf{j} \times \mathbf{k})] \\
&\quad + u_3[v_1(\mathbf{k} \times \mathbf{i}) + v_2(\mathbf{k} \times \mathbf{j}) + v_3(\mathbf{k} \times \mathbf{k})].
\end{aligned}$$

From (2) and Example 1.1 this can be written as

$$\begin{aligned}
\mathbf{U} \times \mathbf{V} &= (u_1v_2\mathbf{k} - u_1v_3\mathbf{j}) + (-u_2v_1\mathbf{k} + u_2v_3\mathbf{i}) + (u_3v_1\mathbf{j} - u_3v_2\mathbf{i}) \\
&= (u_2v_3 - u_3v_2)\mathbf{i} + (u_3v_1 - u_1v_3)\mathbf{j} + (u_1v_2 - u_2v_1)\mathbf{k}.
\end{aligned}$$

To remember this, we write it as

$$\mathbf{U} \times \mathbf{V} = \begin{vmatrix} \mathbf{i} & \mathbf{j} & \mathbf{k} \\ u_1 & u_2 & u_3 \\ v_1 & v_2 & v_3 \end{vmatrix},$$

where the "determinant" is to be expanded formally with respect to the cofactors of the first row, treating $\mathbf{i}, \mathbf{j},$ and $\mathbf{k}$ as if they were real numbers; thus,

$$\mathbf{U} \times \mathbf{V} = \begin{vmatrix} u_2 & u_3 \\ v_2 & v_3 \end{vmatrix}\mathbf{i} - \begin{vmatrix} u_1 & u_3 \\ v_1 & v_3 \end{vmatrix}\mathbf{j} + \begin{vmatrix} u_1 & u_2 \\ v_1 & v_2 \end{vmatrix}\mathbf{k}.$$

Example 1.2   If

$$\mathbf{U} = 2\mathbf{i} + 3\mathbf{j} - \mathbf{k} \quad \text{and} \quad \mathbf{V} = -\mathbf{i} + 2\mathbf{j} + 3\mathbf{k},$$

then

$$\begin{aligned}
\mathbf{U} \times \mathbf{V} &= \begin{vmatrix} \mathbf{i} & \mathbf{j} & \mathbf{k} \\ 2 & 3 & -1 \\ -1 & 2 & 3 \end{vmatrix} \\
&= \begin{vmatrix} 3 & -1 \\ 2 & 3 \end{vmatrix}\mathbf{i} - \begin{vmatrix} 2 & -1 \\ -1 & 3 \end{vmatrix}\mathbf{j} + \begin{vmatrix} 2 & 3 \\ -1 & 2 \end{vmatrix}\mathbf{k} \\
&= 11\mathbf{i} - 5\mathbf{j} + 7\mathbf{k}.
\end{aligned}$$

Example 1.3   The area of a parallelogram is bh, where b is the common length of a pair of parallel sides and h is the distance between them. If a parallelogram in $\mathcal{R}^3$ has vertices $\mathbf{X}_0, \mathbf{X}_0 + \mathbf{U}, \mathbf{X}_0 + \mathbf{V},$ and $\mathbf{X}_0 + \mathbf{U} + \mathbf{V}$ (Figure 1.8) and θ is the angle between $\mathbf{U}$ and $\mathbf{V}$, then we can take

$$b = |\mathbf{V}| \quad \text{and} \quad h = |\mathbf{U}| \sin \theta,$$

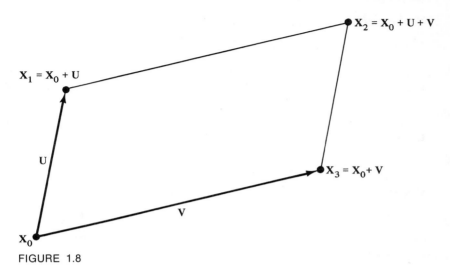

FIGURE 1.8

so that its area is

$$A = bh = |\mathbf{U}|\,|\mathbf{V}|\sin\theta = |\mathbf{U} \times \mathbf{V}|.$$

This can be put another way: if a parallelogram in $\mathcal{R}^3$ has vertices $\mathbf{X}_0$, $\mathbf{X}_1$, $\mathbf{X}_3$, and $\mathbf{X}_4$ as shown in Figure 1.8, then

$$A = |(\mathbf{X}_1 - \mathbf{X}_0) \times (\mathbf{X}_3 - \mathbf{X}_0)|.$$

For more on this, see *Exercise 7*.

The vector triple products

The cross product is not an associative operation; that is,

$$(\mathbf{U} \times \mathbf{V}) \times \mathbf{W} \neq \mathbf{U} \times (\mathbf{V} \times \mathbf{W}),$$

in general. For example,

$$(\mathbf{i} \times \mathbf{j}) \times \mathbf{j} = \mathbf{k} \times \mathbf{j} = -\mathbf{i}$$

while

$$\mathbf{i} \times (\mathbf{j} \times \mathbf{j}) = \mathbf{i} \times \mathbf{0} = \mathbf{0}.$$

Therefore, it is not sensible to write

$$\mathbf{U} \times \mathbf{V} \times \mathbf{W},$$

which is undefined.

We now derive formulas for the *vector triple products*

$$\mathbf{U} \times (\mathbf{V} \times \mathbf{W}) \quad \text{and} \quad (\mathbf{U} \times \mathbf{V}) \times \mathbf{W}.$$

Let

$$\mathbf{U} = u_1\mathbf{i} + u_2\mathbf{j} + u_3\mathbf{k}, \quad \mathbf{V} = v_1\mathbf{i} + v_2\mathbf{j} + v_3\mathbf{k}, \quad \mathbf{W} = w_1\mathbf{i} + w_2\mathbf{j} + w_3\mathbf{k}.$$

Then

$$\mathbf{U} \times (\mathbf{V} \times \mathbf{W}) = u_1(\mathbf{i} \times (\mathbf{V} \times \mathbf{W})) + u_2(\mathbf{j} \times (\mathbf{V} \times \mathbf{W})) + u_3(\mathbf{k} \times (\mathbf{V} \times \mathbf{W})) \quad (3)$$

and

$$\mathbf{V} \times \mathbf{W} = \begin{vmatrix} \mathbf{i} & \mathbf{j} & \mathbf{k} \\ v_1 & v_2 & v_3 \\ w_1 & w_2 & w_3 \end{vmatrix}$$

$$= (v_2w_3 - v_3w_2)\mathbf{i} + (v_3w_1 - v_1w_3)\mathbf{j} + (v_1w_2 - v_2w_1)\mathbf{k}.$$

This and Example 1.1 imply that

$$\begin{aligned} \mathbf{i} \times (\mathbf{V} \times \mathbf{W}) &= (v_2w_1 - v_1w_2)\mathbf{j} + (v_3w_1 - v_1w_3)\mathbf{k} \\ &= w_1(v_1\mathbf{i} + v_2\mathbf{j} + v_3\mathbf{k}) - v_1(w_1\mathbf{i} + w_2\mathbf{j} + w_3\mathbf{k}) \quad (4) \\ &= w_1\mathbf{V} - v_1\mathbf{W}. \end{aligned}$$

Similarly,

$$\mathbf{j} \times (\mathbf{V} \times \mathbf{W}) = w_2\mathbf{V} - v_2\mathbf{W} \quad (5)$$

and

$$\mathbf{k} \times (\mathbf{V} \times \mathbf{W}) = w_3\mathbf{V} - v_3\mathbf{W}. \quad (6)$$

Substituting (4), (5) and (6) into (3) yields

$$\mathbf{U} \times (\mathbf{V} \times \mathbf{W}) = (u_1w_1 + u_2w_2 + u_3w_3)\mathbf{V} - (u_1v_1 + u_2v_2 + u_3v_3)\mathbf{W},$$

or

$$\mathbf{U} \times (\mathbf{V} \times \mathbf{W}) = (\mathbf{U} \cdot \mathbf{W})\mathbf{V} - (\mathbf{U} \cdot \mathbf{V})\mathbf{W}. \quad (7)$$

To memorize this it helps to observe that $\mathbf{U} \times (\mathbf{V} \times \mathbf{W})$ is perpendicular to $\mathbf{V} \times \mathbf{W}$, and so in the plane of $\mathbf{V}$ and $\mathbf{W}$.

To obtain a formula for $(\mathbf{U} \times \mathbf{V}) \times \mathbf{W}$ we write

$$(\mathbf{U} \times \mathbf{V}) \times \mathbf{W} = -\mathbf{W} \times (\mathbf{U} \times \mathbf{V})$$

and apply (7) with $\mathbf{U}$, $\mathbf{V}$, and $\mathbf{W}$ replaced by $-\mathbf{W}$, $\mathbf{U}$, and $\mathbf{V}$, respectively; the result is

$$(\mathbf{U} \times \mathbf{V}) \times \mathbf{W} = (\mathbf{W} \cdot \mathbf{U})\mathbf{V} - (\mathbf{W} \cdot \mathbf{V})\mathbf{U}. \quad (8)$$

Example 1.4   Let

$$\mathbf{U} = 2\mathbf{i} + 3\mathbf{j} - \mathbf{k}, \quad \mathbf{V} = -\mathbf{i} + 2\mathbf{j} + 3\mathbf{k}, \quad \mathbf{W} = \mathbf{i} + \mathbf{j} + \mathbf{k}.$$

To obtain $U \times (V \times W)$ from (7), we calculate

$$U \cdot W = 4, \qquad U \cdot V = 1,$$

and

$$U \times (V \times W) = 4V - W = -5i + 7j + 11k.$$

To obtain $(U \times V) \times W$ from (8) we calculate

$$W \cdot U = 4, \qquad W \cdot V = 4,$$

and

$$(U \times V) \times W = 4V - 4U = -12i - 4j + 16k.$$

Vector and scalar fields

Until now we have regarded functions of the form

$$F(X) = \begin{bmatrix} f_1(x_1, x_2, \ldots, x_n) \\ f_2(x_1, x_2, \ldots, x_n) \\ \vdots \\ f_n(x_1, x_2, \ldots, x_n) \end{bmatrix} : \mathcal{R}^n \to \mathcal{R}^n$$

as transformations from $\mathcal{R}^n$ to $\mathcal{R}^n$. In connection with line and surface integrals it is useful to view such a function as associating with each point X in its domain the vector $F(X)$, with its initial point at X; we then say that F defines a *vector field* on its domain.

Example 1.5 Suppose a particle of mass m_0 is fixed at the origin of $\mathcal{R}^3$ and a particle of unit mass is placed at $X \neq 0$. According to Newton's law of gravitation, the force exerted by the former on the latter is inversely proportional to the distance between them and directed toward the origin; more specifically,

$$F(X) = -\frac{Gm_0 X}{|X|^3},$$

where G is a universal constant. This is the *inverse-square-law force field*.

Example 1.6 In *steady flow* of a fluid through a channel, the velocity $V(X)$ at each point X in the channel depends on X alone, and not on time. Associating $V(X)$ with X defines a *vector velocity field* in the channel (Figure 1.9).

It will also be convenient to think of a real-valued function $f: \mathcal{R}^n \to \mathcal{R}$ as associating with each point X in its domain the real number $f(X)$; we then say that f defines a *scalar field* on its domain. For example, the temperature at every point in a room defines a scalar field in the room.

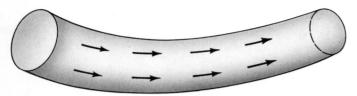

FIGURE 1.9

The gradient

If f is a differentiable scalar field defined on a domain D of $\mathscr{R}^n$, then the *gradient of f* at a point $\mathbf{X}_0$ in D is defined by

$$\nabla f(\mathbf{X}_0) = \begin{bmatrix} \dfrac{\partial f(\mathbf{X}_0)}{\partial x_1} \\[2mm] \dfrac{\partial f(\mathbf{X}_0)}{\partial x_2} \\[1mm] \vdots \\[1mm] \dfrac{\partial f(\mathbf{X}_0)}{\partial x_n} \end{bmatrix}; \tag{9}$$

in particular, if $n = 3$,

$$\nabla f(x_0, y_0, z_0) = \begin{bmatrix} f_x(x_0, y_0, z_0) \\ f_y(x_0, y_0, z_0) \\ f_z(x_0, y_0, z_0) \end{bmatrix}$$

$$= f_x(x_0, y_0, z_0)\mathbf{i} + f_y(x_0, y_0, z_0)\mathbf{j} + f_z(x_0, y_0, z_0)\mathbf{k}.$$

Thus, the gradient of a scalar field is a vector field, denoted by ∇f.

Not every vector field is the gradient of a scalar field, but those that are have important properties, as we will see later. If

$$\mathbf{F} = \nabla f$$

we say that $\mathbf{F}$ is *conservative* and f is a *potential function* for $\mathbf{F}$.

Example 1.7   The vector field of Example 1.5 is the gradient of

$$f(\mathbf{X}) = \frac{Gm_0}{|\mathbf{X}|}.$$

(Verify.)

The gradient is not an essentially new concept, since $\mathbf{V}f(\mathbf{X}_0)$ in (9) is simply the transpose of the differential matrix $f'(\mathbf{X}_0)$ (Section 6.1). Recognizing this, we can interpret $\mathbf{V}f(\mathbf{X}_0)$ as in the following theorem.

Theorem 1.2   *Suppose $f: \mathcal{R}^n \to \mathcal{R}$ is a differentiable scalar field and $\mathbf{U}$ is a unit vector. Then the directional derivative of f at $\mathbf{X}_0$ in the direction of $\mathbf{U}$ is given by*

$$\frac{\partial f(\mathbf{X}_0)}{\partial \mathbf{U}} = \mathbf{U} \cdot \mathbf{V}f(\mathbf{X}_0).$$

Moreover, if $\mathbf{V}f(\mathbf{X}_0) \neq \mathbf{0}$, then the direction in which the directional derivative attains its maximum value is that of the unit vector

$$\mathbf{U}_0 = \frac{\mathbf{V}f(\mathbf{X}_0)}{|\mathbf{V}f(\mathbf{X}_0)|},$$

and the maximum value is $|\mathbf{V}f(\mathbf{X}_0)|$.

Proof.   *Exercise 19.*

Example 1.8   Let $\mathbf{X}_0 = (1, 2, -1)$ and

$$f(\mathbf{X}) = x^2 y + 2xy + z;$$

then

$$\mathbf{V}f(\mathbf{X}) = (2xy + 2y)\mathbf{i} + (x^2 + 2x)\mathbf{j} + \mathbf{k}$$

and

$$\mathbf{V}f(\mathbf{X}_0) = 8\mathbf{i} + 3\mathbf{j} + \mathbf{k}.$$

If

$$\mathbf{U} = \begin{bmatrix} u_1 \\ u_2 \\ u_3 \end{bmatrix}$$

is a unit vector, then

$$\frac{\partial f(\mathbf{X}_0)}{\partial \mathbf{U}} = \mathbf{U} \cdot \mathbf{V}f(\mathbf{X}_0) = 8u_1 + 3u_2 + u_3.$$

The direction of the maximum rate of change of f is that of the unit vector

$$\mathbf{U}_0 = \frac{\mathbf{V}f(\mathbf{X}_0)}{|\mathbf{V}f(\mathbf{X}_0)|} = \frac{8\mathbf{i} + 3\mathbf{j} + \mathbf{k}}{\sqrt{74}},$$

and its value is

$$\frac{\partial f(\mathbf{X}_0)}{\partial \mathbf{U}_0} = |\nabla f(\mathbf{X}_0)| = \sqrt{74}.$$

Curves in $\mathscr{R}^n$

In elementary calculus a curve in $\mathscr{R}^2$ is defined as the locus of points that satisfy an equation relating x and y. For example, the curve defined by

$$x^2 + y^2 = 1$$

is the circle of unit radius centered at the origin [Figure 1.10(a)], while the curve defined by

$$y = x^2, \qquad -1 \le x \le 1,$$

is part of a parabola [Figure 1.10(b)]. This definition is not adequate for the theory of line integrals, in which a curve must be regarded not merely as a set of points, but as a set of points traversed in a specific order. For this reason we consider *parametric curves*.

A parametric curve C in $\mathscr{R}^n$ is a directed set of points of the form

$$\mathbf{X} = \mathbf{\Phi}(t) = \begin{bmatrix} \phi_1(t) \\ \phi_2(t) \\ \vdots \\ \phi_n(t) \end{bmatrix}, \qquad a \le t \le b, \tag{10}$$

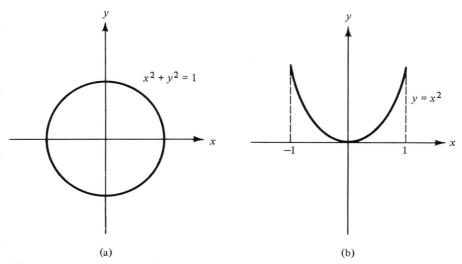

$x^2 + y^2 = 1$

$y = x^2$

(a)   (b)

FIGURE 1.10

where $\phi_1, \phi_2, \ldots, \phi_n$ are continuous on $[a, b]$; we say that the set is directed because $\mathbf{\Phi}$ dictates an order in which the points of C are traversed as t varies from a to b. The function $\mathbf{\Phi}: [a, b] \to \mathscr{R}^n$ is a *parametric representation* of C. If the function

$$\mathbf{X} = \mathbf{\Psi}(\tau) = \begin{bmatrix} \psi_1(\tau) \\ \psi_2(\tau) \\ \vdots \\ \psi_n(\tau) \end{bmatrix}, \qquad c \le \tau \le d, \tag{11}$$

is continuous on $[c, d]$ and there is an increasing function σ which maps $[c, d]$ onto $[a, b]$ such that

$$\mathbf{\Psi}(\tau) = \mathbf{\Phi}(\sigma(\tau)), \qquad c \le \tau \le d, \tag{12}$$

then (10) and (11) represent the same parametric curve C; that is, $\mathbf{\Phi}(t)$ and $\mathbf{\Psi}(\tau)$ traverse the same set of points, in the same order, as t varies from a to b and τ varies from c to d. We say in this case that $\mathbf{\Phi}$ and $\mathbf{\Psi}$ are *equivalent parametric representations* of C (*Exercise 23*), and that $\mathbf{\Psi}$ is obtained from $\mathbf{\Phi}$ by the *change of parameter $t = \sigma(\tau)$.*

Example 1.9   The parametric functions

$$\mathbf{\Phi}(t) = \begin{bmatrix} t^2 \\ t^4 \end{bmatrix}, \qquad 0 \le t \le \sqrt{2},$$

and

$$\mathbf{\Psi}(\tau) = \begin{bmatrix} 1 + \tau \\ (1 + \tau)^2 \end{bmatrix}, \qquad -1 \le \tau \le 1,$$

both represent the parabolic arc

$$y = x^2, \qquad 0 \le x \le 2,$$

traversed from $(0, 0)$ to $(2, 4)$ (Figure 1.11); $\mathbf{\Psi}$ is obtained from $\mathbf{\Phi}$ by the change of parameter

$$t = \sigma(\tau) = \sqrt{1 + \tau},$$

and $\mathbf{\Phi}$ from $\mathbf{\Psi}$ by the inverse change of parameter

$$\tau = \sigma^{-1}(t) = t^2 - 1.$$

Henceforth, we will refer to parametric curves simply as *curves*.

Example 1.10   The parametric functions

$$\mathbf{\Phi}(t) = \begin{bmatrix} t \\ |t| \end{bmatrix}, \qquad -1 \le t \le 1,$$

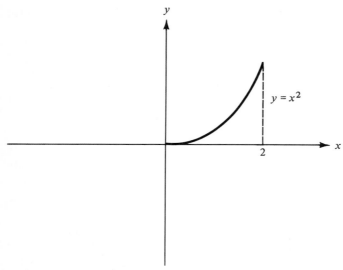

$y = x^2$

2

FIGURE 1.11

and

$$\Psi(\tau) = \begin{bmatrix} \tau^3 \\ \tau^2|\tau| \end{bmatrix}, \qquad -1 \leq \tau \leq 1, \tag{13}$$

both represent the curve

$$y = |x|, \qquad -1 \leq x \leq 1,$$

traversed from $(-1, 1)$ to $(1, 1)$ (Figure 1.12); Ψ is obtained from Φ by the change of parameter

$$t = \sigma(\tau) = \tau^3$$

and Φ from Ψ by

$$\tau = \sigma^{-1}(t) = t^{1/3}.$$

We say that a point X_0 is on the curve C defined by (10) if there is a t_0 in $[a, b]$ such that $X_0 = \Phi(t_0)$. The collection of all such points is the *trace* of C, denoted by $\text{tr}(C)$. The next example shows that distinct curves may have the same trace.

Example 1.11   The parametric functions

$$\Phi(t) = \begin{bmatrix} \cos t \\ \sin t \end{bmatrix}, \qquad 0 \leq t \leq 2\pi, \tag{14}$$

$$\Psi(\tau) = \begin{bmatrix} \cos(2\tau - \pi/2) \\ \sin(2\tau - \pi/2) \end{bmatrix}, \qquad \pi/4 \leq \tau \leq 5\pi/4, \tag{15}$$

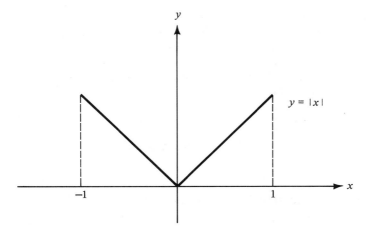

FIGURE 1.12

$$\mathbf{H}(r) = \begin{bmatrix} \sin 2r \\ \cos 2r \end{bmatrix}, \qquad 0 \le r \le 2\pi, \tag{16}$$

have the same trace: the unit circle

$$x^2 + y^2 = 1$$

[Figure 1.10(a)]. The first two are equivalent, since $\mathbf{\Phi}(t)$ and $\mathbf{\Psi}(\tau)$ both traverse the circle counterclockwise, beginning and ending at $(1, 0)$, as t traverses $[0, 2\pi]$ and τ traverses $[\pi/4, 5\pi/4]$. However, $\mathbf{H}$ is not equivalent to $\mathbf{\Phi}$ and $\mathbf{\Psi}$, since $\mathbf{H}(r)$ traverses the circle twice clockwise, starting and ending at $(0, 1)$, as r varies from 0 to 2π.

Definition 1.2 *Let C be the curve defined by (10). Then $\mathbf{\Phi}(a)$ and $\mathbf{\Phi}(b)$ are the initial and terminal end points of C; if they are equal, C is said to be closed. If $\mathbf{\Phi}(t_1) \ne \mathbf{\Phi}(t_2)$ when $a \le t_1 < t_2 < b$ or $a < t_1 < t_2 \le b$, then C is said to be simple. A simple closed curve is one that is both closed and simple.*

Since every curve has infinitely many parametric representations (why?), it is important to note that although Definition 1.1 is stated in terms of a specific representation $\mathbf{\Phi}$, each of its parts is actually independent of the particular representation chosen for C. For example, if $\mathbf{\Psi}: [c, d] \to \mathcal{R}^n$ is another representation of C, equivalent to $\mathbf{\Phi}$, then

$$\mathbf{\Psi}(c) = \mathbf{\Phi}(a) \quad \text{and} \quad \mathbf{\Psi}(d) = \mathbf{\Phi}(b);$$

that is, the end points of C are independent of the representation. (See *Exercise 25* for more on this point.)

Example 1.12 The curves in Examples 1.9 and 1.10 are simple; (14) and (15) define the same simple closed curve, while the curve defined by (16) is closed but not simple. If tr(C) is as in Figure 1.13, then C is not simple. (Can it be closed?)

If C and C_1 consist of the same points, but traversed in opposite orders, we say that C_1 is the *negative* of C, and write

$$C_1 = -C \quad \text{and} \quad C = -C_1.$$

If $\Phi: [a, b] \to \mathscr{R}^n$ represents C, then the parametric function

$$\Psi(\tau) = \Phi(\mu(\tau)), \qquad c \le \tau \le d,$$

represents $-C$ if μ is a decreasing function mapping $[c, d]$ onto $[a, b]$. For example, we could take $c = -b, d = -a,$ and

$$\mu(\tau) = -\tau, \qquad -b \le \tau \le -a,$$

so that

$$\Psi(\tau) = \Phi(-\tau), \qquad -b \le \tau \le -a.$$

Example 1.13 The parametric function

$$\Phi(t) = \begin{bmatrix} 2 \cos t \\ \sin t \end{bmatrix}, \qquad 0 \le t \le \pi,$$

represents the top half of the ellipse

$$x^2 + 2y^2 = 4,$$

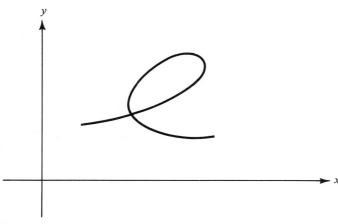

FIGURE 1.13

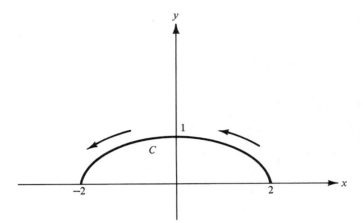

FIGURE 1.14

traversed from $(2, 0)$ to $(-2, 0)$ (Figure 1.14). If this curve is C, then the parametric function

$$\mathbf{\Psi}(\tau) = \mathbf{\Phi}(-\tau) = \begin{bmatrix} 2\cos(-\tau) \\ \sin(-\tau) \end{bmatrix} = \begin{bmatrix} 2\cos\tau \\ -\sin\tau \end{bmatrix}, \qquad -\pi \le \tau \le 0,$$

represents $-C$.

If the terminal point of C_i is the initial point of C_{i+1} for $i = 1, 2, \ldots,$ $k - 1$, we write

$$C = C_1 + C_2 + \cdots + C_k \tag{17}$$

to stand for the curve obtained by traversing $C_1, C_2, \ldots, C_k$ in that order. Any curve can be decomposed in this way; for example, if $\mathbf{\Phi} \colon [a, b] \to \mathscr{R}^n$ represents C and

$$P \colon a = t_0 < t_1 < \cdots < t_k = b$$

is a partition of $[a, b]$, then (17) holds if $C_1, C_2, \ldots, C_k$ are the curves defined by

$$\mathbf{\Phi} \colon [t_{i-1}, t_i] \to \mathscr{R}^n, \qquad i = 1, 2, \ldots, k.$$

However, we usually write (17) when it is natural to think of C as consisting of more than one piece to begin with; for example, we would think of the curve from $(1, 0)$ to $(-1, 0)$ along the upper half of the circle

$$x^2 + y^2 = 1,$$

followed by the line segment from $(-1, 0)$ to $(-1, 1)$ as

$$C = C_1 + C_2,$$

where C_1 is the semicircle and C_2 the line segment, traversed as in Figure 1.15.

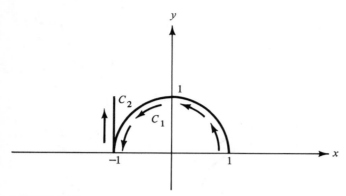

FIGURE 1.15

Smooth curves

The derivative of a parametric function $\mathbf{\Phi}\colon [a, b] \to \mathscr{R}^n$ is

$$\mathbf{\Phi}'(t) = \begin{bmatrix} \phi_1'(t) \\ \phi_2'(t) \\ \vdots \\ \phi_n'(t) \end{bmatrix}$$

for each t in $[a, b]$ such that the vector on the right exists. Here $\phi_i'(a)$ and $\phi_i'(b)$ are to be interpreted as derivatives from the right and left, respectively.

Definition 1.3   *A curve C is said to be smooth if it can be represented by*

$$\mathbf{X} = \mathbf{\Phi}(t), \qquad a \le t \le b,$$

where $\mathbf{\Phi}'$ is continuous and never $\mathbf{0}$ on $[a, b]$. A curve C is piecewise smooth if

$$C = C_1 + C_2 + \cdots + C_k,$$

where $C_1, C_2, \ldots, C_k$ are smooth.

To interpret smoothness geometrically, suppose $\mathbf{\Phi}$ satisfies the assumptions of Definition 1.3 and $a < t_0 < b$. (A similar argument applies if $t_0 = a$ or $t_0 = b$.) Then the vector

$$\mathbf{V}(h) = \frac{1}{|\mathbf{\Phi}'(t_0)|} \frac{\mathbf{\Phi}(t_0 + h) - \mathbf{\Phi}(t_0)}{h} \qquad (h \ne 0),$$

which is collinear with the secant vector $\mathbf{\Phi}(t_0 + h) - \mathbf{\Phi}(t_0)$ (Figure 1.16), approaches the unit vector

$$\mathbf{T}(\mathbf{\Phi}(t_0)) = \frac{\mathbf{\Phi}'(t_0)}{|\mathbf{\Phi}'(t_0)|}$$

as $h \to 0$; thus, the direction of motion along C approaches the direction of $\mathbf{T}(\mathbf{X}_0)$, where $\mathbf{X}_0 = \mathbf{\Phi}(t_0)$, as $h \to 0$. We call $\mathbf{T}(\mathbf{X}_0)$ the *unit tangent vector to C at* $\mathbf{X}_0$; any positive multiple of $\mathbf{T}(\mathbf{X}_0)$ is also said to be tangent to C at $\mathbf{X}_0$. We can think of $\mathbf{T}(\mathbf{X}_0)$ as the direction of C at $\mathbf{X}_0$; thus, the direction of a smooth curve varies continuously along the curve.

A function $\mathbf{\Phi}$ with the properties required in Definition 1.3 is a *smooth parametric function*. If $\mathbf{\Psi}: [c, d] \to \mathscr{R}^n$ is equivalent to $\mathbf{\Phi}$ and σ in (12) is continuously differentiable and σ' has no zeros on $[c, d]$, then $\mathbf{\Psi}$ is also smooth, since

$$\mathbf{\Psi}'(\tau) = \mathbf{\Phi}'(\sigma(\tau))\sigma'(\tau).$$

We say that $\mathbf{\Psi}$ is obtained from $\mathbf{\Phi}$ by the *smooth change of parameter $t = \sigma(\tau)$*. On the other hand, if $\mathbf{\Phi}$ and $\mathbf{\Psi}$ are smooth and related by (12), where σ is assumed only to be an increasing transformation from $[c, d]$ onto $[a, b]$, then σ must be continuously differentiable, with

$$\sigma'(\tau) = \frac{\mathbf{\Psi}'(\tau)}{|\mathbf{\Phi}'(\sigma(\tau))|} > 0$$

(*Exercise 30*).

Example 1.14 Let C be the unit circle

$$x^2 + y^2 = 1$$

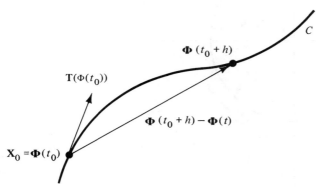

FIGURE 1.16

traversed counterclockwise from $(1, 0)$ to $(1, 0)$ (Figure 1.17). The representation

$$\mathbf{\Phi}(t) = \begin{bmatrix} \cos t \\ \sin t \end{bmatrix}, \qquad 0 \le t \le 2\pi,$$

(Example 1.11) is smooth, with

$$\mathbf{\Phi}'(t) = \begin{bmatrix} -\sin t \\ \cos t \end{bmatrix}. \tag{18}$$

Since

$$\mathbf{\Phi}'(t) \cdot \mathbf{\Phi}(t) = 0,$$

the tangent vector at every point is perpendicular to the radius vector $\mathbf{\Phi}(t)$ (surely a familiar property of the circle). It points in the direction of motion (Figure 1.17).

Example 1.15   The curve of Example 1.10 has the continuously differentiable representation (13), for which

$$\mathbf{\Psi}'(\tau) = \begin{bmatrix} 3\tau^2 \\ 3\tau|\tau| \end{bmatrix},$$

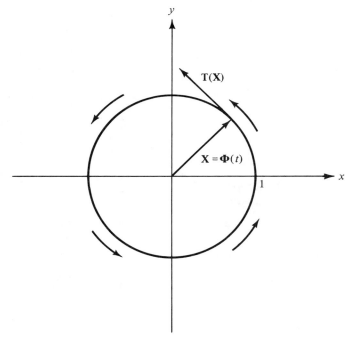

FIGURE 1.17

but this representation is not smooth, since $\mathbf{\Psi}'(0) = \mathbf{0}$. Obviously, C has no tangent at $\mathbf{\Psi}(0) = (0, 0)$ (Figure 1.12).

Arc length

By definition, the length of the line segment from $\mathbf{X}_1$ to $\mathbf{X}_2$ in $\mathscr{R}^n$ is $|\mathbf{X}_1 - \mathbf{X}_2|$. The next definition extends the notion of length to more complicated curves.

Definition 1.4   *Suppose C is a curve represented by $\mathbf{\Phi}: [a, b] \to \mathscr{R}^n$. Let*

$$P: a = t_0 < t_1 < \cdots < t_k = b \tag{19}$$

be a partition of $[a, b]$, and define

$$L_P(C) = \sum_{j=1}^{k} |\mathbf{\Phi}(t_j) - \mathbf{\Phi}(t_{j-1})|.$$

Then C is said to be rectifiable if the set

$$S(C) = \{L_P(C) | P \text{ is a partition of } [a, b]\}$$

is bounded above, in which case the arc length of C is defined to be

$$L(C) = \text{l.u.b. } S(C).$$

Figure 1.18 illustrates the geometric meaning of this definition: $L_P(C)$ is the sum of the lengths of the line segments in the inscribed polygonal path obtained by connecting the successive points

$$\mathbf{X}_0 = \mathbf{\Phi}(t_0), \quad \mathbf{X}_1 = \mathbf{\Phi}(t_1), \quad \ldots, \quad \mathbf{X}_k = \mathbf{\Phi}(t_k);$$

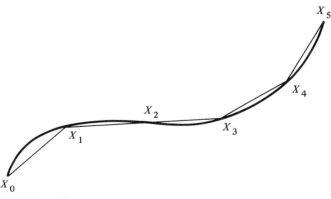

FIGURE 1.18

thus, $L(C)$ is the least upper bound of the total lengths of all such paths, provided the least upper bound is finite.

The existence and value of $L(C)$ is independent of the particular parametric representation chosen for C, because if $\boldsymbol{\Psi}$ is equivalent to $\boldsymbol{\Phi}$ as in (12), then every partition (19) of $[a, b]$ is associated with a partition

$$P: c = \tau_0 < \tau_1 < \cdots < \tau_k = d,$$

of $[c, d]$ and conversely, where

$$t_i = \sigma(\tau_i), \qquad 1 \leq i \leq k,$$

or, equivalently,

$$\tau_i = \sigma^{-1}(t_i), \qquad 1 \leq i \leq k.$$

Moreover, from (12),

$$\sum_{j=1}^{k} |\boldsymbol{\Phi}(t_j) - \boldsymbol{\Phi}(t_{j-1})| = \sum_{j=1}^{k} |\boldsymbol{\Psi}(\tau_j) - \boldsymbol{\Psi}(\tau_{j-1})|,$$

so that defining $L(C)$ to be the least upper bound of either the set of sums on the right or the set of sums on the left of this equation yields the same result.

We leave the proof of the next theorem to the reader (*Exercise 34*).

Theorem 1.3   *If $C = C_1 + C_2 + \cdots + C_k$, then C is rectifiable if and only if $C_1, C_2, \ldots, C_k$ are rectifiable, in which case*

$$L(C) = L(C_1) + L(C_2) + \cdots + L(C_k).$$

A general criterion for rectifiability of a curve is given in *Exercise 33*. The next theorem is sufficient for our purposes.

Theorem 1.4   *If C is a smooth curve with the smooth parametric representation $\mathbf{X} = \boldsymbol{\Phi}(t)$, $a \leq t \leq b$, then C is rectifiable and*

$$L(C) = \int_a^b |\boldsymbol{\Phi}'(t)| \, dt. \tag{20}$$

Proof.   Applying the mean value theorem to the components of $\boldsymbol{\Phi}$ yields

$$\phi_i(t_j) - \phi_i(t_{j-1}) = \phi_i'(\tau_{ij})(t_j - t_{j-1})$$
$$= [\phi_i'(t_j) + (\phi_i'(\tau_{ij}) - \phi_i'(t_j))](t_j - t_{j-1}), \tag{21}$$

where

$$t_{j-1} < \tau_{ij} < t_j. \tag{22}$$

Since (21) holds for $i = 1, 2, \ldots, n$,

$$\boldsymbol{\Phi}(t_j) - \boldsymbol{\Phi}(t_{j-1}) - (t_j - t_{j-1})\boldsymbol{\Phi}'(t_j) = (t_j - t_{j-1})\mathbf{E}_j, \tag{23}$$

where

$$\mathbf{E}_j = \begin{bmatrix} \phi_1'(\tau_{1j}) - \phi_1'(t_j) \\ \phi_2'(\tau_{2j}) - \phi_2'(t_j) \\ \vdots \\ \phi_n'(\tau_{nj}) - \phi_n'(t_j) \end{bmatrix}. \tag{24}$$

Applying the inequality

$$\big\| \mathbf{X} \big| - \big| \mathbf{Y} \big\| \leq |\mathbf{X} - \mathbf{Y}|$$

(Corollary 1.2, Section 5.1) to (23) with

$$\mathbf{X} = \boldsymbol{\Phi}(t_j) - \boldsymbol{\Phi}(t_{j-1}) \quad \text{and} \quad \mathbf{Y} = (t_j - t_{j-1})\boldsymbol{\Phi}'(t_j)$$

yields

$$\big\| \boldsymbol{\Phi}(t_j) - \boldsymbol{\Phi}(t_{j-1})\big| - \big|\boldsymbol{\Phi}'(t_j)\big|(t_j - t_{j-1})\big\| \leq |\mathbf{E}_j|(t_j - t_{j-1}),$$

and so

$$\left| \sum_{j=1}^{k} |\boldsymbol{\Phi}(t_j) - \boldsymbol{\Phi}(t_{j-1})| - \sum_{j=1}^{k} |\boldsymbol{\Phi}'(t_j)|(t_j - t_{j-1}) \right|$$

$$\leq \sum_{j=1}^{k} \big\| \boldsymbol{\Phi}(t_j) - \boldsymbol{\Phi}(t_{j-1})| - |\boldsymbol{\Phi}'(t_j)|(t_j - t_{j-1})\big\| \tag{25}$$

$$\leq \sum_{j=1}^{k} |\mathbf{E}_j|(t_j - t_{j-1}).$$

Since ϕ_1', ϕ_2', $\ldots$, ϕ_n' are uniformly continuous on $[a, b]$, there is for each $\varepsilon > 0$ a $\delta > 0$ such that

$$|\phi_j'(t') - \phi_j'(t)| < \frac{\varepsilon}{\sqrt{n}} \qquad \begin{array}{l} \text{for } 1 \leq j \leq n \text{ if } t \text{ and } t' \text{ are in } [a, b] \\ \text{and } |t - t'| < \delta. \end{array}$$

This, (22), and (24) imply that

$$|\mathbf{E}_j| < \varepsilon \qquad \text{for } 1 \leq j \leq n \text{ if } \|P\| < \delta.$$

Now (25) yields

$$\left| \sum_{j=1}^{k} |\boldsymbol{\Phi}(t_j) - \boldsymbol{\Phi}(t_{j-1})| - \sum_{j=1}^{k} |\boldsymbol{\Phi}'(t_j)|(t_j - t_{j-1}) \right| < \varepsilon(b - a) \quad \text{if} \quad \|P\| < \delta.$$

Since the second sum here is a Riemann sum for $\int_a^b |\boldsymbol{\Phi}'(t)| \, dt$, (20) follows from this.

Example 1.16   From (18) and (20) the arc length (circumference) of the unit circle is

$$L(C) = \int_0^{2\pi} \sqrt{(-\sin t)^2 + (\cos t)^2} \, dt$$

$$= \int_0^{2\pi} 1 \, dt = 2\pi,$$

a familiar result.

8.1   EXERCISES

Assume that all vectors in Exercises 1–17 are in $\mathcal{R}^3$.

1. Find the lengths of **U** and **V** and the cosine of the angle between them:
   (a) $\mathbf{U} = 2\mathbf{i} + 3\mathbf{j} - 4\mathbf{k}$, $\mathbf{V} = \mathbf{i} - \mathbf{j} + \mathbf{k}$
   (b) $\mathbf{U} = -\mathbf{i} + 3\mathbf{j} + 2\mathbf{k}$, $\mathbf{V} = -2\mathbf{j} + 3\mathbf{k}$
   (c) $\mathbf{U} = -2\mathbf{i} + \mathbf{j} + 4\mathbf{k}$, $\mathbf{V} = 4\mathbf{i} - 2\mathbf{j} - 8\mathbf{k}$
   (d) $\mathbf{U} = \mathbf{i} + \mathbf{j} + \mathbf{k}$, $\mathbf{V} = \mathbf{i} + 5\mathbf{j} - 3\mathbf{k}$

2. Express $|\mathbf{U} - \mathbf{V}|$ in terms of the length of **U** and **V** and the cosine of the angle between them.

3. If $\mathbf{U} \neq \mathbf{0}$ and **V** is an arbitrary vector, then the vector

$$\text{proj}_\mathbf{U} \mathbf{V} = \frac{(\mathbf{U} \cdot \mathbf{V})\mathbf{U}}{|\mathbf{U}|^2}$$

   is called the *projection of* **V** *on* **U**; if $\mathbf{U} = \mathbf{0}$, then $\text{proj}_\mathbf{U} \mathbf{V} = \mathbf{0}$.
   (a) Show that $\mathbf{V} - \text{proj}_\mathbf{U} \mathbf{V}$ is perpendicular to **U**.
   (b) Use (a) to show that if **U** and **V** are arbitrary vectors, then it is possible to write **V** in exactly one way as

   $$\mathbf{V} = \mathbf{V}_1 + a\mathbf{U},$$

   where a is a real number and $\mathbf{V}_1$ is perpendicular to **U**.

4. Show that Definition 1.1 implies that (a) $\mathbf{V} \times \mathbf{U} = -(\mathbf{U} \times \mathbf{V})$ and (b) $c\mathbf{U} \times \mathbf{V} = c(\mathbf{U} \times \mathbf{V})$. (c) Show that $\mathbf{U} \times (c\mathbf{V}) = c(\mathbf{U} \times \mathbf{V})$.

5. (a) Suppose $\mathbf{V} = \mathbf{V}_1 + a\mathbf{U}$, where $\mathbf{V}_1$ is perpendicular to **U** [Exercise 3(b)]. Show that $\mathbf{U} \times \mathbf{V} = \mathbf{U} \times \mathbf{V}_1$. (*Hint:* Trivial if $\mathbf{V}_1 = \mathbf{0}$; if $\mathbf{V}_1 \neq \mathbf{0}$, show that the plane of **U** and **V** is the same as the plane of **U** and $\mathbf{V}_1$, and— from Exercise 3—that $|\mathbf{V}_1| = |\mathbf{V}| \sin \theta$, where θ is the angle between **U** and **V**.)
   (b) Complete the proof of Theorem 1.1 for the case where **V** and **W** are not necessarily perpendicular to **U**.

6. Find $\mathbf{U} \times \mathbf{V}$ for the vectors of Exercise 1.

7. (a) Show that the area of the parallelogram in Example 1.3 can also be written as

$$A = |(\mathbf{X}_2 - \mathbf{X}_0) \times (\mathbf{X}_3 - \mathbf{X}_0)| \quad \text{or} \quad A = |(\mathbf{X}_1 - \mathbf{X}_0) \times (\mathbf{X}_2 - \mathbf{X}_0)|.$$

   (b) Show that, in fact, there are 12 formulas of the form

$$A = |(\mathbf{X}_i - \mathbf{X}_r) \times (\mathbf{X}_j - \mathbf{X}_r)|$$

   for the area of the parallelogram.

8. Verify that the given points form the vertices of a parallelogram and use the result of Example 1.3 or Exercise 7 to find its area:
   (a) $(1, 2, 1), (2, 3, 0), (4, 0, 1), (3, -1, 2)$
   (b) $(2, 1, 0), (5, -1, 1), (6, -2, 0), (3, 0, -1)$
   (c) $(7, -1, 2), (9, -4, 3), (10, -5, 4), (8, -2, 3)$

9. Find the area of the triangle with the given vertices:
   (a) $(1, 1, 0), (2, 3, 1), (-1, 3, 0)$
   (b) $(2, -1, 0), (7, 2, -3), (6, -1, 1)$
   (c) $(-3, 0, -1), (0, 1, 2), (1, 1, 1)$

10. (a) Prove: If Γ is a plane through $\mathbf{X}_0$, $\mathbf{N}$ is a nonzero vector perpendicular to Γ, and $\mathbf{X}_1$ is an arbitrary point in $\mathscr{R}^3$, then the distance from $\mathbf{X}_1$ to Γ is

$$d = \frac{|(\mathbf{X}_1 - \mathbf{X}_0) \cdot \mathbf{N}|}{|\mathbf{N}|}.$$

   (b) Find the distance from $\mathbf{X}_1 = (1, -1, 1)$ to the plane

$$3x + 2y + z = 5.$$

11. (a) Show that the distance from the point $\mathbf{X}_1$ to the line

$$\mathbf{X} = \mathbf{X}_0 + t\mathbf{U}, \qquad -\infty < t < \infty,$$

   is

$$d = \frac{|\mathbf{U} \times (\mathbf{X}_1 - \mathbf{X}_0)|}{|\mathbf{U}|}.$$

   (b) Find the distance from $\mathbf{X}_1 = (1, -1, 1)$ to the line

$$\frac{x - 1}{2} = \frac{y + 2}{3}, \qquad z = -2.$$

12. The distance between the lines L_1 and L_2 given by

$$\mathbf{X} = \mathbf{X}_1 + t\mathbf{U}_1, \qquad -\infty < t < \infty,$$

and

$$\mathbf{X} = \mathbf{X}_2 + s\mathbf{U}_2, \qquad -\infty < s < \infty,$$

is defined to be the minimum value of $|\mathbf{Y}_1 - \mathbf{Y}_2|$ as $\mathbf{Y}_1$ and $\mathbf{Y}_2$ vary over L_1 and L_2, respectively.

(a) Show that if L_1 and L_2 are not parallel, then

$$d = \frac{|(\mathbf{X}_2 - \mathbf{X}_1) \cdot (\mathbf{U}_1 \times \mathbf{U}_2)|}{|\mathbf{U}_1 \times \mathbf{U}_2|}.$$

(*Hint:* If $|\hat{\mathbf{Y}}_1 - \hat{\mathbf{Y}}_2| = d$ with $\hat{\mathbf{Y}}_1$ on L_1 and $\hat{\mathbf{Y}}_2$ on L_2, then $\hat{\mathbf{Y}}_1 - \hat{\mathbf{Y}}_2$ is perpendicular to L_1 and L_2.)

(b) Find the distance between the lines

$$\frac{x - 1}{2} = \frac{y + 1}{3} = z \quad \text{and} \quad \frac{x + 1}{3} = \frac{y - 1}{2}, \qquad z = 3.$$

13. (a) Suppose N_1 and N_2 are perpendicular to planes Γ_1 and Γ_2, respectively. Show that if Γ_1 and Γ_2 are not parallel, then their line of intersection L is given by

$$\mathbf{X} = \mathbf{X}_0 + t(\mathbf{N}_1 \times \mathbf{N}_2), \qquad -\infty < t < \infty,$$

where $\mathbf{X}_0$ is any point on L.

(b) Find a parametric equation for the line of intersection of the planes

$$3x + 2y + z = 4 \quad \text{and} \quad 2x + y + z = 1.$$

14. (a) Show that if $\mathbf{X}_1$, $\mathbf{X}_2$, and $\mathbf{X}_3$ are not on a line, then $(\mathbf{X}_2 - \mathbf{X}_1) \times (\mathbf{X}_3 - \mathbf{X}_1)$ is perpendicular to the plane of $\mathbf{X}_1, \mathbf{X}_2$, and $\mathbf{X}_3$.

(b) Find the equation of the plane of $(1, 2, 1)$, $(2, -1, 3)$, and $(0, 4, 3)$.

15. Calculate $\mathbf{U} \times (\mathbf{V} \times \mathbf{W})$ and $(\mathbf{U} \times \mathbf{V}) \times \mathbf{W}$ from Eqs. (7) and (8). Check your results by calculating the triple cross products without (7) and (8):

(a) $\mathbf{U} = \mathbf{i} - \mathbf{j} + \mathbf{k}, \mathbf{V} = 2\mathbf{i} - 3\mathbf{j} + \mathbf{k}, \mathbf{W} = \mathbf{i} + \mathbf{k}$

(b) $\mathbf{U} = 2\mathbf{i} + 3\mathbf{j} - \mathbf{k}, \mathbf{V} = \mathbf{i} + \mathbf{j} + \mathbf{k}, \mathbf{W} = \mathbf{j}$

(c) $\mathbf{U} = \mathbf{i} + \mathbf{j} + 3\mathbf{k}, \mathbf{V} = 2\mathbf{i} + \mathbf{j} - \mathbf{k}, \mathbf{W} = \mathbf{i} - \mathbf{j} - \mathbf{k}$

16. Prove the *Jacobi identity*:

$$(\mathbf{U} \times \mathbf{V}) \times \mathbf{W} + (\mathbf{V} \times \mathbf{W}) \times \mathbf{U} + (\mathbf{W} \times \mathbf{U}) \times \mathbf{V} = \mathbf{0}.$$

17. The quantity $\mathbf{U} \cdot \mathbf{V} \times \mathbf{W}$ is called the *scalar triple product* of $\mathbf{U}, \mathbf{V}$, and $\mathbf{W}$.

(a) Explain why no parentheses are necessary in $\mathbf{U} \cdot \mathbf{V} \times \mathbf{W}$.

(b) Express $\mathbf{U} \cdot \mathbf{V} \times \mathbf{W}$ as a 3×3 determinant.

(c) Explain why $\mathbf{U} \cdot \mathbf{V} \times \mathbf{W}$ could just as well be written as $\mathbf{U} \times \mathbf{V} \cdot \mathbf{W}$.

(d) Show that

$$\mathbf{U} \cdot \mathbf{V} \times \mathbf{W} = \mathbf{V} \cdot \mathbf{W} \times \mathbf{U} = \mathbf{W} \cdot \mathbf{U} \times \mathbf{V} = -\mathbf{V} \cdot \mathbf{U} \times \mathbf{W}$$
$$= -\mathbf{W} \cdot \mathbf{V} \times \mathbf{U} = -\mathbf{U} \cdot \mathbf{W} \times \mathbf{V}.$$

18. Let f and g be differentiable in a domain D in $\mathscr{R}^n$ and let a be a real number. Show that
  (a) $\mathbf{V}(af) = a\mathbf{V}f$
  (b) $\mathbf{V}(f + g) = \mathbf{V}f + \mathbf{V}g$
  (c) $\mathbf{V}(fg) = f\mathbf{V}g + g\mathbf{V}f$
  (d) $\mathbf{V}\left(\dfrac{f}{g}\right) = \dfrac{g\mathbf{V}f - f\mathbf{V}g}{g^2}$ for $\mathbf{X}$ such that $g(\mathbf{X}) \neq 0$.

19. Prove Theorem 1.2.

20. Find $\mathbf{V}f$, $f(\mathbf{X}_0)$, and $\partial f(\mathbf{X}_0)/\partial \mathbf{U}$:

  (a) $f(\mathbf{X}) = x^2 + xy + y^2$, $\mathbf{X}_0 = (1, 2)$, $\mathbf{U} = \begin{bmatrix} \dfrac{1}{\sqrt{2}} \\[2ex] -\dfrac{1}{\sqrt{2}} \end{bmatrix}$

  (b) $f(\mathbf{X}) = 2x^2 yz$, $\mathbf{X}_0 = (-1, 2, 1)$, $\mathbf{U} = \dfrac{2}{\sqrt{6}}\mathbf{i} + \dfrac{1}{\sqrt{6}}\mathbf{j} - \dfrac{1}{\sqrt{6}}\mathbf{k}$

  (c) $f(\mathbf{X}) = x^2 + y^2 + z^2$, $\mathbf{X}_0 = (-1, 1, 1)$, $\mathbf{U} = -\dfrac{1}{\sqrt{3}}\mathbf{i} + \dfrac{1}{\sqrt{3}}\mathbf{j} + \dfrac{1}{\sqrt{3}}\mathbf{k}$

  (d) $f(\mathbf{X}) = e^{-x_1 - x_2 - \cdots - x_n}$, $\mathbf{X}_0 = (0, 0, \ldots, 0)$, $\mathbf{U} = \dfrac{1}{\sqrt{n}}\begin{bmatrix} 1 \\ 1 \\ \vdots \\ 1 \end{bmatrix}$

21. By exhibiting a function σ with the necessary properties, show that $\mathbf{\Phi}$ and $\mathbf{\Psi}$ represent the same curve:

  (a) $\mathbf{\Phi}(t) = \begin{bmatrix} e^{2t} \\ e^{t} \\ e^{-t} \end{bmatrix}$, $0 \le t \le \log 2$; $\mathbf{\Psi}(\tau) = \begin{bmatrix} \tau^2 \\ \tau \\ 1/\tau \end{bmatrix}$, $1 \le \tau \le 2$

  (b) $\mathbf{\Phi}(t) = \begin{bmatrix} t \\ \sqrt{1 - t^2} \end{bmatrix}$, $-1 \le t \le 1$; $\mathbf{\Psi}(\tau) = \begin{bmatrix} \cos \tau \\ -\sin \tau \end{bmatrix}$, $\pi \le \tau \le 2\pi$

*22. Let C be represented by $\mathbf{\Phi}: [a, b] \to \mathscr{R}^n$, and suppose $-\infty < c < d < \infty$. Show that C can also be represented by a parametric function $\mathbf{\Psi}: [c, d] \to \mathscr{R}^n$ equivalent to $\mathbf{\Phi}$. (*Hint:* What is the simplest change of parameter which maps $[c, d]$ onto $[a, b]$?)

23. Suppose we say that a continuous parametric function $\mathbf{\Phi}: [a, b] \to \mathscr{R}^n$ is equivalent to $\mathbf{\Psi}: [c, d] \to \mathscr{R}^n$ if there is an increasing function σ which maps

$[c, d]$ onto $[a, b]$ such that

$$\Psi(\tau) = \Phi(\sigma(\tau)), \qquad c \leq \tau \leq d.$$

(a) Prove: Φ is equivalent to itself.
(b) Prove: If Φ is equivalent to Ψ, then Ψ is equivalent to Φ.
(c) Use part (b) to justify the terminology used in the text; that is, "Φ and Ψ are equivalent . . . ," rather than "Φ is equivalent to Ψ . . . ."
(d) Prove: If $\Phi: [a, b] \to \mathcal{R}^n$ is equivalent to $\Psi: [c, d] \to \mathcal{R}^n$ and Ψ is equivalent to $\Gamma: [e, f] \to \mathcal{R}^n$, then Φ is equivalent to Γ.

24. Describe the curves defined by the following parametric functions as closed, simple, both, or neither:

(a) $\Phi(t) = \begin{bmatrix} t \\ t^2 \end{bmatrix}, \ -1 \leq t \leq 1$

(b) $\Phi(t) = \begin{bmatrix} \cos t \\ \sin t \\ t \end{bmatrix}, 0 \leq t \leq \pi$

(c) $\Phi(t) = \begin{bmatrix} \phi_1(t) \\ \phi_2(t) \end{bmatrix}$, with

$$\phi_1(t) = \begin{cases} t, & 0 \leq t \leq 1, \\ 2 - t, & 1 < t \leq 2, \\ 0, & 2 < t \leq 3, \end{cases} \quad \text{and} \quad \phi_2(t) = \begin{cases} 0, & 0 \leq t \leq 1, \\ t - 1, & 1 < t \leq 2, \\ 3 - t, & 2 < t \leq 3. \end{cases}$$

(d) $\Phi(t) = \begin{bmatrix} \cos t \\ \sin t \\ 1 + \cos t \end{bmatrix}, 0 \leq t \leq 3\pi$

25. Suppose $\Phi: [a, b] \to \mathcal{R}^n$ and $\Psi: [c, d] \to \mathcal{R}^n$ are equivalent parametric representations of a curve C. State what needs to be proved about Φ and Ψ to justify the definition of a simple curve. Then prove it.

26. For each curve C in Exercise 24, find a parametric representation for $-C$.

27. Find the unit tangent vector to the curve C represented by $\Phi: [a, b] \to \mathcal{R}^n$, at the point $X_0 = \Phi(t_0)$:

(a) $\Phi(t) = \begin{bmatrix} -\sin t \\ \cos t \\ e^t \end{bmatrix}, \ -\pi \leq t \leq \pi, t_0 = 0$

(b) $\Phi(t) = \begin{bmatrix} t^2 + 1 \\ t^2 + t \\ t^4 \end{bmatrix}, 0 \leq t \leq 2, t_0 = 1$

(c) $\boldsymbol{\Phi}(t) = \begin{bmatrix} \sqrt{t} \\ t-1 \\ 1 \end{bmatrix}, 1 \le t \le 4, t_0 = 2$

28. Suppose $\boldsymbol{\Phi} : [a, b] \to \mathscr{R}^n$ is a smooth representation of a curve C and $\mathbf{X}_0 \in \mathrm{tr}(C)$. Show that there cannot be infinitely many distinct points $\{t_k\}$ in $[a, b]$ such that $\boldsymbol{\Phi}(t_k) = \mathbf{X}_0$. (*Hint:* Use Theorem 2.4, Section 4.2, to prove that if there are infinitely many such points, then $\boldsymbol{\Phi}'(\bar{t}) = \mathbf{0}$ for some $\bar{t}$ in $[a, b]$, a contradiction.)

29. Show that a piecewise smooth curve can be represented parametrically as

$$\mathbf{X} = \boldsymbol{\Phi}(t), \qquad 0 \le t \le 1,$$

where $\boldsymbol{\Phi}$ is continuous on $[0, 1]$ and there is a vector-valued function $\boldsymbol{\Phi}_1$, with piecewise continuous components, such that $\boldsymbol{\Phi}'(t) = \boldsymbol{\Phi}_1(t)$ except for finitely many values of t. (*Hint:* Use Exercise 22.)

30. Prove: If $\boldsymbol{\Phi} : [a, b] \to \mathscr{R}^n$ and $\boldsymbol{\Psi} : [c, d] \to \mathscr{R}^n$ are smooth parametric functions related by $\boldsymbol{\Psi}(\tau) = \boldsymbol{\Phi}(\sigma(\tau))$, $c \le \tau \le d$, where σ is an increasing transformation of $[c, d]$ onto $[a, b]$, then σ is continuously differentiable on $[c, d]$ and

$$\sigma'(\tau) = \frac{|\boldsymbol{\Psi}'(\tau)|}{|\boldsymbol{\Phi}'(\sigma(\tau))|}.$$

(*Hint:* You will need Theorem 2.8, Section 2.2.)

31. Let $\boldsymbol{\Phi} : [a, b] \to \mathscr{R}^n$ be differentiable. Show that

$$\frac{d}{dt}|\boldsymbol{\Phi}(t)|^2 = 2\boldsymbol{\Phi}'(t) \cdot \boldsymbol{\Phi}(t)$$

and conclude from this that if $|\boldsymbol{\Phi}(t)|$ is constant on $[a, b]$, then $\boldsymbol{\Phi}'(t)$ is perpendicular to $\boldsymbol{\Phi}(t)$ for $a \le t \le b$.

32. Suppose $\boldsymbol{\Phi} : [a, b] \to \mathscr{R}^3$ is differentiable and never zero on $[a, b]$. Show that

$$\frac{d}{dt}\frac{\boldsymbol{\Phi}(t)}{|\boldsymbol{\Phi}(t)|} = \frac{\boldsymbol{\Phi}(t) \times (\boldsymbol{\Phi}'(t) \times \boldsymbol{\Phi}(t))}{|\boldsymbol{\Phi}(t)|^3}.$$

[*Hint:* Use Exercise 31 and Eq. (7).]

33. Prove: If C is the curve represented by

$$\boldsymbol{\Phi} = \begin{bmatrix} \phi_1 \\ \phi_2 \\ \vdots \\ \phi_n \end{bmatrix} : [a, b] \to \mathscr{R}^n,$$

then C is rectifiable if and only if $\phi_1, \phi_2, \ldots, \phi_n$ are of bounded variation on $[a, b]$. (See Exercise 6, Section 3.2.)

34. Prove Theorem 1.3.

35. Deduce from Theorem 1.4 that if C is the curve in $\mathscr{R}^3$ defined by

$$y = f(x), \qquad a \le x \le b,$$

where f is continuously differentiable on $[a, b]$, then

$$L(C) = \int_a^b \sqrt{1 + [f'(x)]^2}\, dx.$$

8.2   LINE INTEGRALS

In this section we study the line integral. To motivate its definition we first define the work done on a particle as it moves along a curve in a vector field in $\mathscr{R}^2$.

The work done by a constant force

$$\mathbf{F} = \begin{bmatrix} u \\ v \end{bmatrix}$$

on an object that moves along the line segment from (x_0, y_0) to (x_1, y_1) is, by definition,

$$W = |\mathbf{F}|\sqrt{(x_1 - x_0)^2 + (y_1 - y_0)^2}\, \cos \theta,$$

where θ is the angle between $\mathbf{F}$ and the direction of motion (Figure 2.1); this can also be written as

$$W = \mathbf{F} \cdot (\mathbf{X}_1 - \mathbf{X}_0) = u(x_1 - x_0) + v(y_1 - y_0). \tag{1}$$

Now suppose the object moves along a curve C from (x_0, y_0) to (α, β), through the domain of a continuous vector field

$$\mathbf{F}(x, y) = \begin{bmatrix} u(x, y) \\ v(x, y) \end{bmatrix}$$

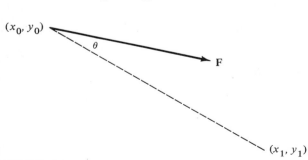

FIGURE 2.1

and let C be represented by the continuous parametric function

$$\mathbf{X} = \mathbf{X}(t) = \begin{bmatrix} x(t) \\ y(t) \end{bmatrix}, \qquad a \le t \le b. \tag{2}$$

Let

$$P: a = t_0 < t_1 < \cdots < t_k = b$$

be a partition of $[a, b]$, define

$$\mathbf{X}_j = (x(t_j), y(t_j)),$$

let L_j be the line segment from $\mathbf{X}_j$ to $\mathbf{X}_{j+1}$ ($j = 0, \ldots, k - 1$), and let C' be the polygonal path from (x_0, y_0) to (α, β) consisting of the line segments $L_0, \ldots, L_{k-1}$ (Figure 2.2). Since $\mathbf{F}$ and the parametric representation (2) are continuous, choosing $\|P\|$ sufficiently small ensures that C' approximates C as closely as we wish, and that $\mathbf{F}$ is practically constant on each segment of C'. (We could make this statement more precise, but it is adequate for our present purpose, which is to motivate a definition.)

If we assume that $\mathbf{F}$ is constant on L_j, and equal to $\mathbf{F}(x(\tau_j), y(\tau_j))$, where τ_j is a point in $[t_{j-1}, t_j]$, then the work done by $\mathbf{F}$ in moving the object from $\mathbf{X}_{j-1}$ to $\mathbf{X}_j$ along L_{j-1} is, from definition (1),

$$u(x(\tau_j), y(\tau_j))[x(t_j) - x(t_{j-1})] + v(x(\tau_j), y(\tau_j))[y(t_j) - y(t_{j-1})].$$

Therefore, the total work done by $\mathbf{F}$ as the object traverses C' is

$$\sum_{j=1}^{k} u(x(\tau_j), y(\tau_j))[x(t_j) - x(t_{j-1})] + \sum_{j=1}^{k} v(x(\tau_j), y(\tau_j))[y(t_j) - y(t_{j-1})]. \tag{3}$$

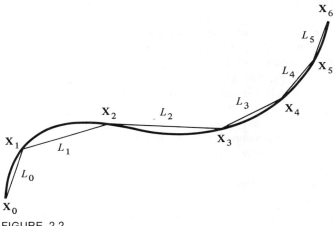

FIGURE 2.2

The reader who studied the Riemann–Stieltjes integral defined in Section 3.1 will recognize the sums in (3) as Riemann–Stieltjes sums for the integrals in the sum

$$\int_a^b u(x(t),\, y(t))\, dx(t) + \int_a^b v(x(t),\, y(t))\, dy(t);$$

therefore, we could define this sum to be the work done by $\mathbf{F}$ as the object traverses C from $(x_0,\, y_0)$ to $(\alpha,\, \beta)$, provided the integrals exist. However, since we have not emphasized the Riemann–Stieltjes integral, we give instead a definition in terms of Riemann integrals. If the parametric representation (2) is continuously differentiable on $[a,\, b]$, we can apply the mean value theorem to its components, and write

$$\begin{aligned} x(t_j) - x(t_{j-1}) &= [x'(\tau_j) + E_{1j}](t_j - t_{j-1}),\\ y(t_j) - y(t_{j-1}) &= [y'(\tau_j) + E_{2j}](t_j - t_{j-1}), \end{aligned} \tag{4}$$

where, if $\varepsilon > 0$, there is a $\delta > 0$ such that

$$|E_{1j}| < \varepsilon \quad \text{and} \quad |E_{2j}| < \varepsilon \qquad \text{for } j = 1,\, 2,\, \ldots,\, k \text{ if } \|P\| < \delta \tag{5}$$

(*Exercise 1*). Using (4), we can rewrite (3) as

$$\sum_{j=1}^{k} \left[u(x(\tau_j),\, y(\tau_j))x'(\tau_j) + v(x(\tau_j),\, y(\tau_j))y'(\tau_j) \right](t_j - t_{j-1})$$

$$+ \sum_{j=1}^{k} \left[u(x(\tau_j),\, y(\tau_j))E_{1j} + v(x(\tau_j),\, y(\tau_j))E_{2j} \right](t_j - t_{j-1}).$$

Because of (5) and the boundedness of $\mathbf{F}$ on C, the second sum can be made arbitrarily small by choosing $\|P\|$ sufficiently small. The first sum is a Riemann sum for

$$\int_a^b \left[u(x(t),\, y(t))x'(t) + v(x(t),\, y(t))y'(t) \right] dt.$$

Since we have assumed that u and v are continuous on $\operatorname{tr}(C)$ and that the parametric representation (2) is continuously differentiable, this integral exists; we define it to be the work done by $\mathbf{F}$ as the object traverses C from $(x_0,\, y_0)$ to $(\alpha,\, \beta)$. We will also call it the *line integral of* $\mathbf{F}$ *along* C.

The general definition of the line integral is as follows.

Definition 2.1   *Suppose C is a smooth curve in $\mathscr{R}^n$ and $\mathbf{F} = (f_1, f_2, \ldots, f_n)$ is continuous on $\operatorname{tr}(C)$. Then the line integral of $\mathbf{F}$ along C is defined by*

$$\int_C \mathbf{F} \cdot d\mathbf{X} = \int_a^b \mathbf{F}(\mathbf{\Phi}(t)) \cdot \mathbf{\Phi}'(t)\, dt = \int_a^b \left[\sum_{j=1}^{n} f_j(\mathbf{\Phi}(t))\phi_j'(t) \right] dt, \tag{6}$$

where $\mathbf{\Phi}: [a,\, b] \to \mathscr{R}^n$ is a smooth representation of C. (Here "·" stands for inner product.)

We observe in passing that this definition still makes sense with "smooth" replaced by "continuously differentiable."

To avoid cumbersome notation in this chapter, we will dispense with writing the arguments of functions appearing in integrands, except where clarity requires it; thus, we have written $\int_C \mathbf{F} \cdot d\mathbf{X}$ rather than $\int_C \mathbf{F}(\mathbf{X}) \cdot d\mathbf{X}$ on the left side of (6).

For Definition 2.1 to make sense, it must be shown that $\int_C \mathbf{F} \cdot d\mathbf{X}$ is independent of the particular smooth representation chosen for C. To this end, suppose $\boldsymbol{\Phi} : [a, b] \to \mathscr{R}^n$ and $\boldsymbol{\Psi} : [c, d] \to \mathscr{R}^n$ are representations of C, with

$$\boldsymbol{\Psi}(\tau) = \boldsymbol{\Phi}(\sigma(\tau)), \qquad c \le \tau \le d, \tag{7}$$

where σ is continuously differentiable and maps $[c, d]$ onto $[a, b]$. Applying the change of variable $t = \sigma(\tau)$ to the last integral in (6) yields

$$\int_C \mathbf{F} \cdot d\mathbf{X} = \int_c^d \left[\sum_{j=1}^n f_j(\boldsymbol{\Phi}(\sigma(\tau))) \phi_j'(\sigma(\tau)) \right] \sigma'(\tau) \, d\tau \tag{8}$$

(Theorem 3.13, Section 3.3). Since $\phi_j(\sigma(\tau)) = \psi_j(\tau)$ [see (7)], the chain rule implies that

$$\phi_j'(\sigma(\tau)) \sigma'(\tau) = \psi_j'(\tau),$$

so (8) can be rewritten as

$$\int_C \mathbf{F} \cdot d\mathbf{X} = \int_c^d \left[\sum_{j=1}^n f_j(\psi(\tau)) \psi_j'(\tau) \right] d\tau,$$

which is precisely the definition of $\int_C \mathbf{F} \cdot d\mathbf{X}$ in terms of $\boldsymbol{\Psi}$. Therefore, $\int_C \mathbf{F} \cdot d\mathbf{X}$ does not depend on the particular parametric representation chosen for C.

Definition 2.2   If $C = C_1 + C_2 + \cdots + C_k$ is piecewise smooth and $\mathbf{F}$ is continuous on $\operatorname{tr}(C)$, then

$$\int_C \mathbf{F} \cdot d\mathbf{X} = \sum_{i=1}^k \int_{C_i} \mathbf{F} \cdot d\mathbf{X}.$$

Theorem 2.1   Under the assumptions of Definition 2.2,

$$\int_{-C} \mathbf{F} \cdot d\mathbf{X} = -\int_C \mathbf{F} \cdot d\mathbf{X}.$$

Proof.   Exercise 2.

Example 2.1   To evaluate $\int_C \mathbf{F} \cdot d\mathbf{X}$, where $n = 3$,

$$\mathbf{F}(x, y, z) = \begin{bmatrix} xy \\ yz \\ xz \end{bmatrix},$$

and C is the line segment from $(0, 0, 0)$ to $(1, 2, -1)$ (Figure 2.3), we parametrize C with

$$\mathbf{X} = \mathbf{\Phi}(t) = \begin{bmatrix} t \\ 2t \\ -t \end{bmatrix}, \qquad 0 \le t \le 1;$$

then

$$\mathbf{\Phi}'(t) = \begin{bmatrix} 1 \\ 2 \\ -1 \end{bmatrix}, \qquad \mathbf{F}(\mathbf{\Phi}(t)) = \begin{bmatrix} t(2t) \\ 2t(-t) \\ t(-t) \end{bmatrix} = \begin{bmatrix} 2t^2 \\ -2t^2 \\ -t^2 \end{bmatrix},$$

and

$$\int_C \mathbf{F} \cdot d\mathbf{X} = \int_0^1 \mathbf{F}(\mathbf{\Phi}(t)) \cdot \mathbf{\Phi}'(t)\, dt$$
$$= \int_0^1 \left[(2t^2)(1) + (-2t^2)(2) + (-t^2)(-1) \right] dt$$
$$= -\int_0^1 t^2\, dt = -\tfrac{1}{3}.$$

Example 2.2   To evaluate $\int_C \mathbf{F} \cdot d\mathbf{X}$, where $n = 2$,

$$\mathbf{F}(x, y) = \begin{bmatrix} x^2 + y^2 \\ 2xy \end{bmatrix},$$

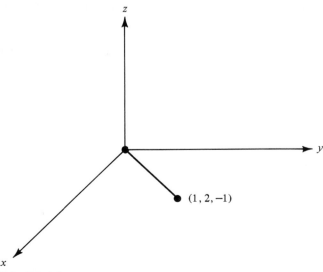

FIGURE 2.3

and C is the arc of the circle $x^2 + y^2 = 1$ from $(1, 0)$ to $(0, 1)$, followed by the line segment from $(0, 1)$ to $(1, 1)$ (Figure 2.4), we write

$$\int_C \mathbf{F} \cdot d\mathbf{X} = \int_{C_1} \mathbf{F} \cdot d\mathbf{X} + \int_{C_2} \mathbf{F} \cdot d\mathbf{X}, \tag{9}$$

where C_1 is the circular arc and C_2 is the line segment. Parametrizing C_1 as

$$\mathbf{X} = \mathbf{\Phi}_1(t) = \begin{bmatrix} \cos t \\ \sin t \end{bmatrix}, \qquad 0 \le t \le \pi/2,$$

yields

$$\mathbf{\Phi}_1'(t) = \begin{bmatrix} -\sin t \\ \cos t \end{bmatrix}$$

and

$$\mathbf{F}(\mathbf{\Phi}_1(t)) = \begin{bmatrix} \cos^2 t + \sin^2 t \\ 2 \cos t \sin t \end{bmatrix} = \begin{bmatrix} 1 \\ 2 \cos t \sin t \end{bmatrix},$$

so

$$\begin{aligned}
\int_{C_1} \mathbf{F} \cdot d\mathbf{X} &= \int_0^{\pi/2} \mathbf{F}(\mathbf{\Phi}_1(t)) \cdot \mathbf{\Phi}_1'(t) \, dt \\
&= \int_0^{\pi/2} \left[(1)(-\sin t) + (2 \cos t \sin t)(\cos t) \right] dt \\
&= \int_0^{\pi/2} (-\sin t + 2 \sin t \cos^2 t) \, dt \\
&= (\cos t - \tfrac{2}{3} \cos^3 t) \Big|_0^{\pi/2} = -\tfrac{1}{3}.
\end{aligned}$$

Parametrizing C_2 as

$$\mathbf{X} = \mathbf{\Phi}_2(t) = \begin{bmatrix} t \\ 1 \end{bmatrix}, \qquad 0 \le t \le 1,$$

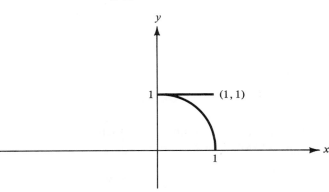

FIGURE 2.4

yields

$$\Phi'_2(t) = \begin{bmatrix} 1 \\ 0 \end{bmatrix} \quad \text{and} \quad F(\Phi_2(t)) = \begin{bmatrix} t^2 + 1 \\ 2t \end{bmatrix};$$

hence,

$$\int_{C_2} F \cdot dX = \int_0^1 [(t^2 + 1)(1) + 2t(0)] \, dt$$

$$= \int_0^1 (t^2 + 1) \, dt = \tfrac{4}{3}.$$

From (9),

$$\int_C F \cdot dX = -\tfrac{1}{3} + \tfrac{4}{3} = 1.$$

Henceforth, it should be assumed that curves mentioned in connection with line integrals are smooth or piecewise smooth; this assumption will not always be explicitly stated. If $\Phi: [a, b] \to \mathscr{R}^n$ represents such a curve, it is to be understood that Φ is smooth or piecewise smooth, as the case may be, and we will also write

$$\int_C F \cdot dX = \int_a^b F(\Phi(t)) \cdot \Phi'(t) \, dt \tag{10}$$

in the latter case, even though Φ' may fail to be defined at finitely many points $t_1 < t_2 < \cdots < t_{k-1}$, to stand for the more precise statement that

$$\int_C F \cdot dX = \sum_{j=1}^k \int_{t_{j-1}}^{t_j} F(\Phi(t)) \cdot \Phi'(t) \, dt,$$

where $t_0 = a$ and $t_k = b$. We will say in this case that we are interpreting the integral on the right of (10) in the *extended sense*.

It is convenient to write line integrals less formally. If $n = 2$, we will usually write

$$\int_C P \, dx + Q \, dy \tag{11}$$

to stand for $\int_C F \cdot dX$ with

$$F = \begin{bmatrix} P \\ Q \end{bmatrix},$$

and if P and Q are specific functions, we simply write them in; for example, the line integral in Example 2.2 would be written as

$$\int_C (x^2 + y^2) \, dx + 2xy \, dy.$$

Also, instead of introducing the symbol Φ to represent a parametrization of a curve, we will often write, for example,

$$\begin{bmatrix} x \\ y \end{bmatrix} = \begin{bmatrix} x(t) \\ y(t) \end{bmatrix}, \qquad a \le t \le b, \tag{12}$$

to represent C. With this notation the definition of (11) for a smooth curve C given by (12) is

$$\int_C P\,dx + Q\,dy = \int_a^b \left[P(x(t), y(t))x'(t) + Q(x(t), y(t))y'(t) \right] dt.$$

A similar convention applies for arbitrary n. Thus, the line integral of Example 2.1 can be written as

$$\int_C xy\,dx + yz\,dy + xz\,dz.$$

Example 2.3 To evaluate the integral

$$I = \int_C \frac{-y\,dx + x\,dy}{(x^2 + y^2)^\alpha},$$

where C is the circle

$$x^2 + y^2 = r^2 \quad (r > 0)$$

traversed counterclockwise from $(r, 0)$ to $(r, 0)$, we take

$$\begin{bmatrix} x \\ y \end{bmatrix} = \begin{bmatrix} r\cos t \\ r\sin t \end{bmatrix}, \qquad 0 \le t \le 2\pi.$$

This yields

$$I = \int_0^{2\pi} \frac{(-r\sin t)(-r\sin t) + (r\cos t)(r\cos t)}{r^{2\alpha}}\,dt$$

$$= r^{-2\alpha+2} \int_0^{2\pi} dt = (2\pi)r^{-2\alpha+2}.$$

To evaluate

$$I = \int_C P\,dx + Q\,dy$$

along the curve C defined by

$$y = y(x), \qquad a \le x \le b,$$

in the direction of increasing x, we could parametrize C as

$$\begin{bmatrix} x \\ y \end{bmatrix} = \begin{bmatrix} t \\ y(t) \end{bmatrix}, \qquad a \le t \le b,$$

and write

$$I = \int_a^b [P(t, y(t)) + Q(t, y(t))y'(t)] \, dt;$$

however, nothing is gained by introducing a new parameter t here, so we write

$$I = \int_a^b [P(x, y(x)) + Q(x, y(x))y'(x)] \, dx$$

instead. A similar comment applies to line integrals in $\mathscr{R}^n$; if C is defined by expressing $n - 1$ of the variables $x_1, x_2, \ldots, x_n$ in terms of the remaining one, then the latter can be taken as the parameter. For example, if C is defined by

$$x = x(y), \qquad z = z(y), \qquad c \le y \le d,$$

then the integral

$$I = \int_C P \, dx + Q \, dy + R \, dz$$

can be evaluated as

$$I = \int_c^d [P(x(y), y, z(y))x'(y) + Q(x(y), y, z(y)) + R(x(y), y, z(y))z'(y)] \, dy.$$

The next theorem follows directly from the definition of the line integral (*Exercise 6*).

Theorem 2.2   *If the vector fields* **F** *and* **G** *are continuous on* C *and* a *and* b *are constants, then*

$$\int_C (a\mathbf{F} + b\mathbf{G}) \cdot d\mathbf{X} = a \int_C \mathbf{F} \cdot d\mathbf{X} + b \int_C \mathbf{G} \cdot d\mathbf{X}.$$

Line integrals of real-valued functions

On occasion it is necessary to deal with line integrals of real-valued functions. If f is continuous on a curve C given by $\mathbf{\Phi}: [a, b] \to \mathscr{R}^n$, we define

$$\int_C f \, dx_i = \int_a^b f(\mathbf{\Phi}(t))\phi_i'(t) \, dt.$$

This is equivalent to regarding f as the ith component of a vector field $\mathbf{F}$ whose remaining $n - 1$ components vanish identically, and defining

$$\int_C f \, dx_i = \int_C \mathbf{F} \cdot d\mathbf{X}.$$

With this definition and Theorem 2.2, if $\mathbf{F} = (f_1, f_2, \ldots, f_n)$, then

$$\int_C \mathbf{F} \cdot d\mathbf{X} = \sum_{i=1}^n \int_C f_i \, dx_i.$$

Example 2.4   Writing

$$\mathbf{F} = \begin{bmatrix} P \\ Q \end{bmatrix}$$

as

$$\mathbf{F} = \begin{bmatrix} P \\ 0 \end{bmatrix} + \begin{bmatrix} 0 \\ Q \end{bmatrix} = \mathbf{F}_1 + \mathbf{F}_2$$

and applying Theorem 2.2 yields

$$\int_C \mathbf{F} \cdot d\mathbf{X} = \int_C \mathbf{F}_1 \cdot d\mathbf{X} + \int_C \mathbf{F}_2 \cdot d\mathbf{X}$$
$$= \int_C P\, dx + 0\, dy + \int_C 0\, dx + Q\, dy;$$

that is, we have shown that

$$\int_C P\, dx + Q\, dy = \int_C P\, dx + \int_C Q\, dy.$$

This may seem trivial, but it is not; the advantage here is that the integrals on the right may be evaluated by using different parametrizations of C, if convenient. This is crucial in the proof of Green's theorem (Theorem 2.6), presented below.

Integrals around closed curves

The common initial and terminal end point $\boldsymbol{\alpha}$ of a closed curve C seems special simply because it is the point at which the traversal of C begins and ends. However, shifting the parameter so that another point plays this role does not change line integrals along C. To make this precise, let $\boldsymbol{\Phi} \colon [a, b] \to \mathscr{R}^n$ represent C, suppose $a < c < b$, and let C_1 be the curve represented by $\boldsymbol{\Psi} \colon [c, c + b - a] \to \mathscr{R}^n$, where

$$\boldsymbol{\Psi}(\tau) = \begin{cases} \boldsymbol{\Phi}(\tau), & c \le \tau \le b, \\ \boldsymbol{\Phi}(\tau - b + a), & b \le \tau \le c + b - a \end{cases}$$

Thus, $\boldsymbol{\beta} = \boldsymbol{\Phi}(c)$ is the common initial and terminal end point of C_1 (Figure 2.5). However, C and C_1 are indistinguishable as far as line integrals are concerned, since

$$\int_{C_1} \mathbf{F} \cdot d\mathbf{X} = \int_c^{c+b-a} \mathbf{F}(\boldsymbol{\Psi}(\tau)) \cdot \boldsymbol{\Psi}'(\tau)\, d\tau$$
$$= \int_c^b \mathbf{F}(\boldsymbol{\Phi}(\tau)) \cdot \boldsymbol{\Phi}'(\tau)\, d\tau$$
$$+ \int_b^{c+b-a} \mathbf{F}(\boldsymbol{\Phi}(\tau - b + a)) \cdot \boldsymbol{\Phi}'(\tau - b + a)\, d\tau.$$

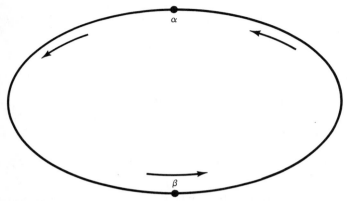

FIGURE 2.5

(These integrals are to be interpreted in the extended sense if C is piecewise smooth.) Replacing τ by t in the first integral on the right and $\tau - b + a$ by t in the second yields

$$\int_{C_1} \mathbf{F} \cdot d\mathbf{X} = \int_c^b \mathbf{F}(\Phi(t)) \cdot \Phi'(t) \, dt + \int_a^c \mathbf{F}(\Phi(t)) \cdot \Phi'(t) \, dt$$

$$= \int_a^b \mathbf{F}(\Phi(t)) \cdot \Phi'(t) \, dt = \int_C \mathbf{F} \cdot d\mathbf{X}.$$

Because of this we will no longer specify an initial point when speaking of a line integral around a closed curve. For example, we may speak of $\int_C \mathbf{F} \cdot d\mathbf{X}$, where C is the unit circle traversed once counterclockwise; then any of the infinitely many parametrizations

$$\mathbf{X} = \begin{bmatrix} \cos t \\ \sin t \end{bmatrix}, \qquad \lambda \leq t \leq \lambda + 2\pi,$$

(and others equivalent to them) may be used to evaluate the integral.

Line integrals independent of path

If $\mathbf{F}$ is continuous in a domain D in $\mathscr{R}^n$ and

$$\int_{C_1} \mathbf{F} \cdot d\mathbf{X} = \int_{C_2} \mathbf{F} \cdot d\mathbf{X}$$

whenever C_1 and C_2 are piecewise smooth curves in D with the same end points, we say that line integrals of $\mathbf{F}$ are *independent of path in D*. In this case we write

$$\int_{\mathbf{X}_0}^{\mathbf{X}_1} \mathbf{F} \cdot d\mathbf{X}$$

to denote the line integral of $\mathbf{F}$ along any piecewise smooth curve in D from $\mathbf{X}_0$ to $\mathbf{X}_1$.

We leave it to the reader (*Exercise 10*) to show that line integrals of **F** are independent of path in D if and only if

$$\int_C \mathbf{F} \cdot d\mathbf{X} = 0$$

whenever C is a closed curve in D.

Example 2.5   Line integrals of

$$\mathbf{F}(x, y) = \begin{bmatrix} x - y \\ x + y \end{bmatrix}$$

are not independent of path in $\mathscr{R}^2$. For example, suppose C_1 and C_2 are, respectively, the arc of the unit circle and the line segment from $(1, 0)$ to $(0, 1)$ (Figure 2.6). Parametrizing C_1 as

$$\mathbf{X} = \begin{bmatrix} \cos t \\ \sin t \end{bmatrix}, \qquad 0 \le t \le \pi/2,$$

yields

$$\int_{C_1} (x - y)\, dx + (x + y)\, dy =$$

$$\int_0^{\pi/2} \left[(\cos t - \sin t)(-\sin t) + (\cos t + \sin t)(\cos t) \right] dt$$

$$= \int_0^{\pi/2} dt = \pi/2,$$

while parametrizing C_2 as

$$\mathbf{X} = \begin{bmatrix} 1 - t \\ t \end{bmatrix}, \qquad 0 \le t \le 1,$$

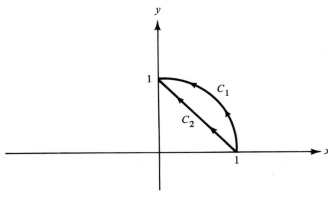

FIGURE 2.6

yields

$$\int_{C_2} (x - y)\, dx + (x + y)\, dy = \int_0^1 \left[(1 - t - t)(-1) + (1 - t + t)(1) \right] dt$$
$$= \int_0^1 2t\, dt = 1.$$

Example 2.6 Line integrals of

$$\mathbf{F}(x, y, z) = yz\mathbf{i} + xz\mathbf{j} + xy\mathbf{k}$$

are independent of path in $\mathcal{R}^3$. This becomes clear if we recognize that

$$yz = \frac{\partial}{\partial x}(xyz), \quad xz = \frac{\partial}{\partial y}(xyz), \quad xy = \frac{\partial}{\partial z}(xyz),$$

because if C is a smooth curve with end points (x_0, y_0, z_0) and (x_1, y_1, z_1), represented by

$$\mathbf{X} = x(t)\mathbf{i} + y(t)\mathbf{j} + z(t)\mathbf{k}, \quad a \le t \le b,$$

then

$$\int_C yz\, dx + xz\, dy + xy\, dz = \int_a^b \left[y(t)z(t)x'(t) + x(t)z(t)y'(t) + x(t)y(t)z'(t) \right] dt$$
$$= \int_a^b \frac{d}{dt}\left[x(t)y(t)z(t) \right] dt$$
$$= x(b)y(b)z(b) - x(a)y(a)z(a)$$
$$= x_1 y_1 z_1 - x_0 y_0 z_0.$$

Therefore, $\int_C \mathbf{F} \cdot d\mathbf{X}$ depends only on the end points of C if C is smooth. As we will see in the proof of the next theorem, this is also the case if C is piecewise smooth.

Theorem 2.3 *Suppose* $\mathbf{F} = (f_1, f_2, \ldots, f_n)$ *is continuous in an open region D. Then line integrals of $\mathbf{F}$ are independent of path in D if and only if there is a function f defined on D such that*

$$\mathbf{F} = \nabla f, \tag{13}$$

that is, if and only if $\mathbf{F}$ is conservative in D.

Proof. Suppose (13) holds throughout D and C is a smooth curve in D, represented by the smooth parametric function

$$\mathbf{X} = \mathbf{X}(t), \quad a \le t \le b,$$

with initial and terminal points $\boldsymbol{\alpha}$ and $\boldsymbol{\beta}$. Then

$$\int_C \mathbf{F} \cdot d\mathbf{X} = \int_a^b \mathbf{F}(\mathbf{X}(t)) \cdot \mathbf{X}'(t) \, dt$$

$$= \int_a^b \left[\sum_{j=1}^n f_j(\mathbf{X}(t)) x_j'(t) \right] dt$$

$$= \int_a^b \left[\sum_{j=1}^n \frac{\partial f(\mathbf{X}(t))}{\partial x_j} x_j'(t) \right] dt \qquad [\text{by (13)}]$$

$$= \int_a^b \frac{d}{dt} f(\mathbf{X}(t)) \, dt \qquad\qquad (\text{by the chain rule})$$

$$= f(\mathbf{X}(b)) - f(\mathbf{X}(a)) = f(\boldsymbol{\beta}) - f(\boldsymbol{\alpha}).$$

If $C = C_1 + C_2 + \cdots + C_k$, where C_i is a smooth curve with end points $\boldsymbol{\alpha}_{i-1}$ and $\boldsymbol{\alpha}_i$ $(i = 1, 2, \ldots, k)$ (Figure 2.7), then applying this result to C_i yields

$$\int_{C_i} \mathbf{F} \cdot d\mathbf{X} = f(\boldsymbol{\alpha}_i) - f(\boldsymbol{\alpha}_{i-1});$$

hence

$$\int_C \mathbf{F} \cdot d\mathbf{X} = \sum_{i=1}^k \left[f(\boldsymbol{\alpha}_i) - f(\boldsymbol{\alpha}_{i-1}) \right]$$

$$= f(\boldsymbol{\alpha}_k) - f(\boldsymbol{\alpha}_0) = f(\boldsymbol{\beta}) - f(\boldsymbol{\alpha}),$$

since $\boldsymbol{\alpha} = \boldsymbol{\alpha}_0$ and $\boldsymbol{\beta} = \boldsymbol{\alpha}_k$ in this case. This proves that (13) implies independence of path in D.

For the converse, suppose line integrals of $\mathbf{F}$ are independent of path in D, let $\mathbf{X}_0$ be an arbitrary point in D, and define

$$f(\mathbf{X}) = \int_{\mathbf{X}_0}^{\mathbf{X}} \mathbf{F} \cdot d\mathbf{Y},$$

where $\mathbf{X}$ is in D and the integration is along any piecewise smooth curve in D. (Here we use $\mathbf{Y}$ as the variable of integration to avoid confusion with the "upper limit" $\mathbf{X}$.) We will show that

$$\frac{\partial f(\mathbf{X})}{\partial x_1} = f_1(\mathbf{X}); \tag{14}$$

the proof that $f_{x_i}(\mathbf{X}) = f_i(\mathbf{X})$ for $i = 2, \ldots, n$ is similar.

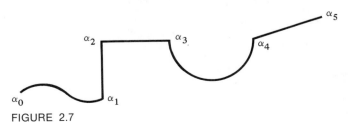

FIGURE 2.7

If

$$\mathbf{X} = (x_1, x_2, \ldots, x_n) \quad \text{and} \quad \mathbf{X}' = (x_1 + \Delta x_1, x_2, \ldots, x_n), \tag{15}$$

we can find $f(\mathbf{X}')$ by integrating along any piecewise smooth curve in D from $\mathbf{X}_0$ to $\mathbf{X}$, and then along the line segment L from $\mathbf{X}$ to $\mathbf{X}'$ (Figure 2.8 depicts the situation for $n = 2$), assuming that Δx_1 is so small that $L \subset D$; thus,

$$\begin{aligned} f(\mathbf{X}') &= \int_{\mathbf{X}_0}^{\mathbf{X}} \mathbf{F} \cdot d\mathbf{Y} + \int_{\mathbf{X}}^{\mathbf{X}'} \mathbf{F} \cdot d\mathbf{Y} \\ &= f(\mathbf{X}) + \int_{\mathbf{X}}^{\mathbf{X}'} \mathbf{F} \cdot d\mathbf{Y}. \end{aligned} \tag{16}$$

The line segment L can be represented by

$$\mathbf{Y} = \mathbf{Y}(t) = \begin{bmatrix} x_1 + t\,\Delta x_1 \\ x_2 \\ \vdots \\ x_n \end{bmatrix}, \qquad 0 \le t \le 1,$$

so (16) implies that

$$f(\mathbf{X}') - f(\mathbf{X}) = \Delta x_1 \int_0^1 f_1(x_1 + t\,\Delta x_1, x_2, \ldots, x_n)\,dt.$$

Therefore,

$$\begin{aligned} &\frac{f_1(\mathbf{X}') - f_1(\mathbf{X})}{\Delta x_1} - f_1(\mathbf{X}) \\ &= \int_0^1 \left[f_1(x_1 + t\,\Delta x_1, x_2, \ldots, x_n) - f_1(x_1, x_2, \ldots, x_n) \right] dt \end{aligned} \tag{17}$$

if $\Delta x_1 \ne 0$. Now suppose $\varepsilon > 0$. Since f is continuous at $\mathbf{X}$, there is a $\delta > 0$ such that

$$\left| f_1(x_1 + t\,\Delta x_1, x_2, \ldots, x_n) - f_1(x_1, x_2, \ldots, x_n) \right| < \varepsilon$$

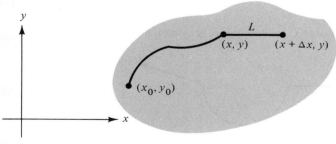

FIGURE 2.8

if $0 \leq t \leq 1$ and $|\Delta x_1| < \delta$; therefore, (15) and (17) imply that

$$\left| \frac{f(\mathbf{X}') - f(\mathbf{X})}{\Delta x_1} - f_1(\mathbf{X}) \right| < \varepsilon \quad \text{if} \quad 0 < |\Delta x_1| < \delta.$$

This implies (14) and completes the proof.

Without a method for determining whether $\mathbf{F}$ is conservative, Theorem 2.3 is of little practical value for computing specific line integrals. The next theorem gives easily applicable necessary conditions for $\mathbf{F}$ to be conservative.

Theorem 2.4 *If* $\mathbf{F} = (f_1, f_2, \ldots, f_n)$ *is a continuously differentiable conservative vector field in an open region D of $\mathscr{R}^n$, then*

$$\frac{\partial f_i(\mathbf{X})}{\partial x_j} = \frac{\partial f_j(\mathbf{X})}{\partial x_i} \quad \text{if } i, j = 1, \ldots, n \quad \text{and} \quad \mathbf{X} \in D. \tag{18}$$

Proof. If $\mathbf{F} = \nabla f$ in D, then

$$f_i(\mathbf{X}) = \frac{\partial f(\mathbf{X})}{\partial x_i} \quad \text{and} \quad f_j(\mathbf{X}) = \frac{\partial f(\mathbf{X})}{\partial x_j} \quad \text{if} \quad \mathbf{X} \in D.$$

Therefore,

$$\frac{\partial f_i(\mathbf{X})}{\partial x_j} = \frac{\partial^2 f(\mathbf{X})}{\partial x_j \, \partial x_i} \quad \text{and} \quad \frac{\partial f_j(\mathbf{X})}{\partial x_i} = \frac{\partial^2 f(\mathbf{X})}{\partial x_i \, \partial x_j} \quad \text{if} \quad \mathbf{X} \in D, \tag{19}$$

and, since $\mathbf{F}$ is continuously differentiable, the second partial derivatives in (19) are equal (Theorem 3.3, Section 5.3). Therefore, (19) implies (18).

If $n = 2$ and

$$\mathbf{F} = \begin{bmatrix} P \\ Q \end{bmatrix},$$

then (18) reduces to

$$\frac{\partial P}{\partial y} = \frac{\partial Q}{\partial x}. \tag{20}$$

If $n = 3$ and

$$\mathbf{F} = P\mathbf{i} + Q\mathbf{j} + R\mathbf{k}, \tag{21}$$

then (18) reduces to

$$\frac{\partial P}{\partial y} = \frac{\partial Q}{\partial x}, \qquad \frac{\partial Q}{\partial z} = \frac{\partial R}{\partial y}, \qquad \frac{\partial R}{\partial x} = \frac{\partial P}{\partial z}.$$

For this case there is a formulation in terms of a vector field derived from $\mathbf{F}$. If $\mathbf{F}$ in (21) is differentiable, we define its *curl* by

$$\operatorname{curl} \mathbf{F} = \left(\frac{\partial R}{\partial y} - \frac{\partial Q}{\partial z}\right)\mathbf{i} + \left(\frac{\partial P}{\partial z} - \frac{\partial R}{\partial x}\right)\mathbf{j} + \left(\frac{\partial Q}{\partial x} - \frac{\partial P}{\partial y}\right)\mathbf{k}.$$

This definition is easy to remember if we write it formally as

$$\operatorname{curl} \mathbf{F} = \begin{vmatrix} \mathbf{i} & \mathbf{j} & \mathbf{k} \\ \dfrac{\partial}{\partial x} & \dfrac{\partial}{\partial y} & \dfrac{\partial}{\partial z} \\ P & Q & R \end{vmatrix}.$$

Theorem 2.4 implies that if $\mathbf{F}$ is continuously differentiable and conservative in D, then $\operatorname{curl} \mathbf{F} = \mathbf{0}$ in D.

Example 2.7 If

$$\mathbf{F}(\mathbf{X}) = x^3 y\mathbf{i} + (x + z)\mathbf{j} + xyz\mathbf{k},$$

then

$$\operatorname{curl} \mathbf{F}(\mathbf{X}) = \begin{vmatrix} \mathbf{i} & \mathbf{j} & \mathbf{k} \\ \dfrac{\partial}{\partial x} & \dfrac{\partial}{\partial y} & \dfrac{\partial}{\partial z} \\ x^3 y & x + z & xyz \end{vmatrix}$$

$$= (xz - 1)\mathbf{i} - yz\mathbf{j} + (1 - x^3)\mathbf{k}.$$

Example 2.8 We saw in Example 2.6 that line integrals of

$$\mathbf{F}(\mathbf{X}) = yz\mathbf{i} + xz\mathbf{j} + xy\mathbf{k}$$

are independent of path in $\mathscr{R}^3$. By Theorem 2.3, $\mathbf{F}$ is therefore conservative in $\mathscr{R}^3$, and so Theorem 2.4 implies that $\operatorname{curl} \mathbf{F} = \mathbf{0}$ in $\mathscr{R}^3$. This is easily verified:

$$\operatorname{curl} \mathbf{F}(\mathbf{X}) = \begin{vmatrix} \mathbf{i} & \mathbf{j} & \mathbf{k} \\ \dfrac{\partial}{\partial x} & \dfrac{\partial}{\partial y} & \dfrac{\partial}{\partial z} \\ yz & xz & xy \end{vmatrix}$$

$$= (x - x)\mathbf{i} + (y - y)\mathbf{j} + (z - z)\mathbf{k} = \mathbf{0}.$$

Example 2.9 The conditions of Theorem 2.4 are *necessary*, but *not sufficient*, for $\mathbf{F}$ to be conservative in a region D. From Example 2.3,

$$\int_{C_r} \frac{-y \, dx + x \, dy}{x^2 + y^2} = 2\pi \neq 0 \tag{22}$$

if C_r is any circle

$$x^2 + y^2 = r^2 \qquad (r > 0)$$

around the origin, traversed counterclockwise. Therefore, the vector field

$$\mathbf{F}(x, y) = \begin{bmatrix} P(x, y) \\ Q(x, y) \end{bmatrix} = \begin{bmatrix} \dfrac{-y}{x^2 + y^2} \\[2ex] \dfrac{x}{x^2 + y^2} \end{bmatrix} \tag{23}$$

is not conservative in any neighborhood of $(0, 0)$, even though it is continuously differentiable and satisfies (20), with

$$\frac{\partial P(x, y)}{\partial y} = \frac{y^2 - x^2}{(x^2 + y^2)^2} = \frac{\partial Q(x, y)}{\partial x}, \qquad (x, y) \neq (0, 0).$$

Pursuing this further, we note by comparing (23) and Eq. (31), Section 6.3, that $\mathbf{F}$ in (23) is conservative—and therefore $\int_C \mathbf{F} \cdot d\mathbf{X}$ is independent of path, by Theorem 2.3—in any region D in which a branch of the argument function can be defined, because

$$\mathbf{F}(x, y) = \nabla \arg(x, y)$$

in such a region. Therefore, (22) implies that no branch of $\arg(x, y)$ can be defined in a neighborhood of $(0, 0)$. This is consistent with *Exercise 17, Section 6.3.*

Simply connected regions

To assert that $\int_C \mathbf{F} \cdot d\mathbf{X}$ is independent of path in D whenever $\mathbf{F}$ is continuously differentiable and satisfies (18) in D, we need an additional assumption on D: that it be *simply connected*. Intuitively, this means that if C_0 and C_1 are any piecewise smooth curves in D with common end points, then $\mathrm{tr}(C_0)$ can be deformed, without breaking it and without leaving D, so as to coincide with $\mathrm{tr}(C_1)$.

A region in $\mathscr{R}^2$ is simply connected if it has no holes; thus, D_1 in Figure 2.9 is simply connected, but D_2 is not, since $\mathrm{tr}(C_0)$ cannot be deformed into $\mathrm{tr}(C_1)$ without breaking it or leaving D_2. In $\mathscr{R}^3$ the region between concentric spheres is simply connected, but the region

$$D = \{(x, y, z) \mid r_1^2 < x^2 + y^2 < r_2^2, z_1 < z < z_2\}$$

is not, because the curves C_0 and C_1 shown in Figure 2.10 have common end points, but it is impossible to deform $\mathrm{tr}(C_0)$ into $\mathrm{tr}(C_1)$ without breaking it or leaving D.

More precisely, D is simply connected if whenever C_0 and C_1 are curves in D there is a function $\mathbf{H}$ continuous on $S = [0, 1] \times [0, 1]$ such that

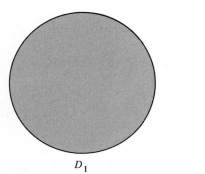

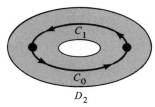

D_1

FIGURE 2.9

$\mathbf{H}(t, s) \in D$ if $(t, s) \in S$ and the functions

$$\boldsymbol{\Phi}(t) = \mathbf{H}(t, 0) \quad \text{and} \quad \boldsymbol{\Psi}(t) = \mathbf{H}(t, 1), \qquad 0 \le t \le 1, \tag{24}$$

are parametric representations of C_0 and C_1. For example, a *convex* set D—that is, a set such that if $\mathbf{X}_0$ and $\mathbf{X}_1$ are in D, then so is the line segment connecting them—is simply connected, as can be seen by taking

$$\mathbf{H}(t, s) = (1 - s)\boldsymbol{\Phi}(t) + s\boldsymbol{\Psi}(t),$$

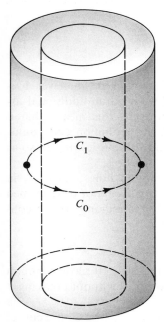

FIGURE 2.10

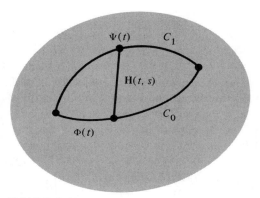

FIGURE 2.11

where $\Phi: [0, 1] \to \mathscr{R}^n$ and $\Psi: [0, 1] \to \mathscr{R}^n$ represent C_0 and C_1; see Figure 2.11. (From *Exercise 22, Section 8.1*, there is no loss of generality in assuming that the parametrizations of C_0 and C_1 are defined on $[0, 1]$.)

Theorem 2.5 *If a continuously differentiable vector field* **F** *satisfies* (18) *in a simply connected domain D of* $\mathscr{R}^n$, *then* **F** *is conservative in D; that is, line integrals of* **F** *are independent of path in D.*

Partial Proof. We consider the case where C_0 and C_1 are smooth curves with common initial point $\boldsymbol{\alpha}$ and terminal point $\boldsymbol{\beta}$, represented by the smooth parametric functions $\Phi: [0, 1] \to \mathscr{R}^n$ and $\Psi: [0, 1] \to \mathscr{R}^n$; moreover, we assume that the components of the function $\mathbf{H} = (h_1, h_2, \ldots, h_n)$ that satisfies (24) have continuous second partial derivatives on $[0, 1] \times [0, 1]$. Then, for each s in $[0, 1]$, $\mathbf{H}$ as a function of t defines a continuously differentiable curve C_s in D with end points $\boldsymbol{\alpha}$ and $\boldsymbol{\beta}$, so we can consider the line integral

$$I(s) = \int_{C_s} \mathbf{F} \cdot d\mathbf{X} = \int_0^1 \left[\sum_{j=1}^n f_j(\mathbf{H}(t, s)) \frac{\partial h_j(t, s)}{\partial t} \right] dt, \qquad 0 \le s \le 1. \quad (25)$$

We will show that (18) implies that $I'(s) = 0$ for $0 \le s \le 1$. This implies that

$$\int_{C_0} \mathbf{F} \cdot d\mathbf{X} = \int_{C_1} \mathbf{F} \cdot d\mathbf{X},$$

the desired conclusion.

Differentiating (25) yields

$$I'(s) = \int_0^1 \left[\sum_{i,j=1}^n \frac{\partial f_j(\mathbf{H}(t, s))}{\partial x_i} \frac{\partial h_i(t, s)}{\partial s} \frac{\partial h_j(t, s)}{\partial t} \right] dt$$

$$+ \int_0^1 \left[\sum_{j=1}^n f_j(\mathbf{H}(t, s)) \frac{\partial^2 h_j(t, s)}{\partial s \, \partial t} \right] dt. \qquad (26)$$

Since

$$\frac{\partial^2 h_j}{\partial s\, \partial t} = \frac{\partial^2 h_j}{\partial t\, \partial s}$$

because of the assumed continuity of the second partial derivatives of h_j (Theorem 3.3, Section 5.3), we can rewrite the second integral in (26) and integrate by parts, to obtain

$$\int_0^1 \left[\sum_{j=1}^n f_j(\mathbf{H}(t,s)) \frac{\partial^2 h_j(t,s)}{\partial t\, \partial s} \right] dt$$

$$= \sum_{j=1}^n f_j(\mathbf{H}(t,s)) \frac{\partial h_j(t,s)}{\partial s} \Big|_0^1 - \int_0^1 \left[\sum_{i,j=1}^n \frac{\partial f_j(\mathbf{H}(t,s))}{\partial x_i} \frac{\partial h_i(t,s)}{\partial t} \frac{\partial h_j(t,s)}{\partial s} \right] dt. \qquad (27)$$

Since

$$\mathbf{H}(0,s) = \boldsymbol\alpha \quad \text{and} \quad \mathbf{H}(1,s) = \boldsymbol\beta, \qquad 0 \le s \le 1,$$

it follows that

$$\frac{\partial h_j(1,s)}{\partial s} = \frac{\partial h_j(0,s)}{\partial s} = 0, \qquad 0 \le s \le 1,$$

so the first sum on the right of (27) vanishes. For convenience we interchange i and j in the second sum, and rewrite (27) as

$$\int_0^1 \left[\sum_{j=1}^n f_j(\mathbf{H}(t,s)) \frac{\partial^2 h_j(t,s)}{\partial t\, \partial s} \right] dt = -\int_0^1 \left[\sum_{i,j=1}^n \frac{\partial f_i(\mathbf{H}(t,s))}{\partial x_j} \frac{\partial h_j(t,s)}{\partial t} \frac{\partial h_i(t,s)}{\partial s} \right] dt.$$

Substituting this in (26) and collecting terms into one summation yields

$$I'(s) = \int_0^1 \left[\sum_{i,j=1}^n \left(\frac{\partial f_j(\mathbf{H}(t,s))}{\partial x_i} - \frac{\partial f_i(\mathbf{H}(t,s))}{\partial x_j} \right) \frac{\partial h_i(t,s)}{\partial s} \frac{\partial h_j(t,s)}{\partial t} \right] dt,$$

which vanishes because of (18). This completes the partial proof.

Finding potential functions

If $\mathbf{F} = \nabla f$ in a region D, then the symbol $\mathbf{F} \cdot d\mathbf{X}$ can be interpreted as df (the differential of f as defined in Section 5.3), since in this case

$$\mathbf{F} \cdot d\mathbf{X} = f_1\, dx_1 + f_2\, dx_2 + \cdots + f_n\, dx_n$$
$$= f_{x_1}\, dx_1 + f_{x_2}\, dx_2 + \cdots + f_{x_n}\, dx_n$$
$$= df.$$

We say then that $\mathbf{F} \cdot d\mathbf{X}$ is an *exact differential*, and write

$$\int_C \mathbf{F} \cdot d\mathbf{X} = \int_C df.$$

It may be possible to find f by inspection; for example, it is not hard to see that

$$(x + y)\, dx + (x - y)\, dy = x\, dx - y\, dy + (y\, dx + x\, dy)$$
$$= d(x^2/2) - d(y^2/2) + d(xy)$$
$$= d\left(\frac{x^2}{2} - \frac{y^2}{2} + xy\right),$$

and so we can write

$$\int_C (x + y)\, dx + (x - y)\, dy = \int_C d\left(\frac{x^2}{2} - \frac{y^2}{2} + xy\right).$$

Even if $\mathbf{F}$ is not conservative, it may be possible to simplify the evaluation of $\int_C \mathbf{F} \cdot d\mathbf{X}$ by writing $\mathbf{F}$ as the sum of conservative and nonconservative fields. For example, if C is an arbitrary closed curve in $\mathscr{R}^3$,

$$\int_C (x + \sin y)\, dx + (y + e^x)\, dy = \int_C \sin y\, dx + e^x\, dy + \int_C x\, dx + y\, dy,$$

and the second integral vanishes, since

$$x\, dx + y\, dy = d\left(\frac{x^2 + y^2}{2}\right).$$

The proof of Theorem 2.3 suggests a way to find a potential function for a conservative vector field for which inspection is not feasible: Pick a convenient point $\mathbf{X}_0$ in D and evaluate

$$f(\mathbf{X}) = \int_{\mathbf{X}_0}^{\mathbf{X}} \mathbf{F} \cdot d\mathbf{Y} \tag{28}$$

as a function of $\mathbf{X}$, along any convenient curve in D from $\mathbf{X}_0$ to $\mathbf{X}$. If D is convex, the integral can be taken along the line segment from $\mathbf{X}_0$ to $\mathbf{X}$, given parametrically by

$$\mathbf{Y} = t\mathbf{X} + (1 - t)\mathbf{X}_0, \qquad 0 \le t \le 1,$$

so that (28) becomes

$$f(\mathbf{X}) = \int_0^1 \mathbf{F}[t\mathbf{X} + (1 - t)\mathbf{X}_0] \cdot (\mathbf{X} - \mathbf{X}_0)\, dt. \tag{29}$$

Another method is suggested by *Exercises 15 and 17.*

Example 2.10   If

$$\mathbf{F}(x, y) = \begin{bmatrix} P(x, y) \\ Q(x, y) \end{bmatrix} = \begin{bmatrix} x^3 y^4 + x \\ x^4 y^3 + y \end{bmatrix},$$

then

$$\frac{\partial P(x, y)}{\partial y} = \frac{\partial Q(x, y)}{\partial x} = 4x^3 y^3,$$

so $\mathbf{F}$ is conservative in $\mathcal{R}^2$, by Theorem 2.5. With $\mathbf{X}_0 = (0, 0)$, (29) becomes

$$f(x, y) = \int_0^1 [P(tx, ty)x + Q(tx, ty)y]\, dt$$

$$= \int_0^1 [(t^7x^3y^4 + tx)x + (t^7x^4y^3 + ty)y]\, dt$$

$$= 2x^4y^4 \int_0^1 t^7\, dt + (x^2 + y^2) \int_0^1 t\, dt$$

$$= \frac{x^4y^4}{4} + \frac{x^2}{2} + \frac{y^2}{2}.$$

We can now use f to evaluate line integrals of $\mathbf{F}$; for example,

$$\int_{(1,1)}^{(2,0)} (x^3y^4 + x)\, dx + (x^4y^3 + y)\, dy = f(2, 0) - f(1, 1) = 2 - \tfrac{5}{4} = \tfrac{3}{4}.$$

In this example f can also be found by inspection; thus,

$$(x^3y^4 + x)\, dx + (x^4y^3 + y)\, dy = (x^3y^4\, dx + x^4y^3\, dy) + x\, dx + y\, dy$$

$$= d\left(\frac{x^4y^4}{4}\right) + d\left(\frac{x^2}{2}\right) + d\left(\frac{y^2}{2}\right)$$

$$= d\left(\frac{x^4y^4}{4} + \frac{x^2}{2} + \frac{y^2}{2}\right).$$

Example 2.11   If

$$\mathbf{F}(x, y) = \begin{bmatrix} P(x, y) \\ Q(x, y) \end{bmatrix} = \begin{bmatrix} x^2 + y^2 \\ 2xy \end{bmatrix},$$

then

$$\frac{\partial P(x, y)}{\partial y} = \frac{\partial Q(x, y)}{\partial x} = 2y,$$

so $\mathbf{F}$ is conservative in $\mathcal{R}^3$, by Theorem 2.5. With $\mathbf{X}_0 = (0, 0)$, (29) becomes

$$f(x, y) = \int_0^1 [P(tx, ty)x + Q(tx, ty)y]\, dt$$

$$= \int_0^1 t^2[(x^2 + y^2)x + (2xy)y]\, dt$$

$$= (x^3 + 3xy^2) \int_0^1 t^2\, dt$$

$$= \frac{x^3 + 3xy^2}{3}.$$

Hence,

$$\int_{(1,0)}^{(1,1)} (x^2 + y^2)\, dx + 2xy\, dy = f(1, 1) - f(1, 0) = \tfrac{4}{3} - \tfrac{1}{3} = 1,$$

which agrees with the result of Example 2.2.

To find f by inspection, we write

$$(x^2 + y^2) \, dx + (2xy) \, dy = x^2 \, dx + (y^2 \, dx + 2xy \, dy)$$

$$= d\left(\frac{x^3}{3}\right) + d(xy^2)$$

$$= d\left(\frac{x^3}{3} + xy^2\right).$$

Example 2.12 If

$$\mathbf{F}(x, y, z) = \begin{bmatrix} P(x, y, z) \\ Q(x, y, z) \\ R(x, y, z) \end{bmatrix} = \begin{bmatrix} x + z \\ y + z \\ x + y \end{bmatrix},$$

then

$$\frac{\partial P}{\partial y} = \frac{\partial Q}{\partial x} = 0, \qquad \frac{\partial Q}{\partial z} = \frac{\partial R}{\partial y} = 1, \qquad \frac{\partial R}{\partial x} = \frac{\partial P}{\partial z} = 1,$$

so $\mathbf{F}$ is conservative in $\mathcal{R}^3$. With $\mathbf{X}_0 = (0, 0, 0)$, (29) becomes

$$f(x, y, z) = \int_0^1 \left[P(tx, ty, tz)x + Q(tx, ty, tz)y + R(tx, ty, tz)z \right] dt$$

$$= \int_0^1 \left[(tx + tz)x + (ty + tz)y + (tx + ty)z \right] dt$$

$$= (x^2 + y^2 + 2xz + 2yz) \int_0^1 t \, dt$$

$$= \frac{x^2 + y^2 + 2xz + 2yz}{2},$$

and so, for example,

$$\int_{(1,-1,0)}^{(2,1,-1)} (x + z) \, dx + (y + z) \, dy + (x + y) \, dz = f(2, 1, -1) - f(1, -1, 0)$$

$$= -\tfrac{1}{2} - 1 = -\tfrac{3}{2}.$$

To find f by inspection, we write

$$(x + z) \, dx + (y + z) \, dy + (x + y) \, dz = x \, dx + y \, dy + (z \, dx + x \, dz)$$

$$\qquad\qquad + (z \, dy + y \, dz)$$

$$= d\left(\frac{x^2}{2}\right) + d\left(\frac{y^2}{2}\right) + d(xz) + d(yz)$$

$$= d\left(\frac{x^2}{2} + \frac{y^2}{2} + xz + yz\right).$$

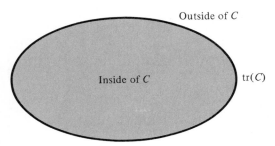

Outside of C

Inside of C

$\text{tr}(C)$

FIGURE 2.12

Green's theorem in $\mathscr{R}^2$

The *Jordan curve theorem* states that if C is a simple closed curve in $\mathscr{R}^2$, then the complement of $\text{tr}(C)$ consists of two open regions, one bounded and one unbounded. (Although intuitively obvious, this is quite difficult to prove, so we take it as given.) The former is the *inside of C*, the latter, the *outside of C* (Figure 2.12). The union of $\text{tr}(C)$ and the inside of C is called the *(closed) region bounded by C*, and C is *positively directed* if an observor standing on the xy plane and traversing C always sees the inside of C immediately to the left; C is *negatively directed* if $-C$ is positively directed. Thus, C_1 in Figure 2.13 is positively directed, while C_2 is negatively directed.

Theorem 2.6 (First Form of Green's Theorem)   *Suppose D is a region in $\mathscr{R}^2$ bounded by a simple closed curve C. Let P and Q be continuously differentiable on an open set containing D. Then*

$$\int_C P \, dx + Q \, dy = \int_D \left(\frac{\partial Q}{\partial x} - \frac{\partial P}{\partial y} \right) d(x, y) \qquad (30)$$

if C is positively directed.

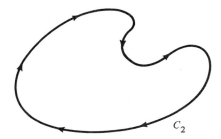

C_1

C_2

FIGURE 2.13

Partial Proof. We first consider the case where, in addition to satisfying the stated hypothesis, D can be written in both of the forms

$$D = \{(x, y) \mid u(x) \le y \le v(x), a \le x \le b\} \tag{31}$$

and

$$D = \{(x, y) \mid U(y) \le x \le V(y), c \le y \le d\} \tag{32}$$

(Figure 2.14).

From Example 2.4 it is permissible to write

$$\int_C P \, dx + Q \, dy = \int_C P \, dx + \int_C Q \, dy$$

and evaluate the integrals on the right with different parametrizations of C. From Figure 2.14(a), we can write

$$C = C_1 - C_2,$$

where C_1 and C_2 are given by

$$y = u(x) \quad \text{and} \quad y = v(x), \qquad a \le x \le b.$$

Taking x as the parameter yields

$$
\begin{aligned}
\int_C P \, dx &= \int_{C_1} P \, dx - \int_{C_2} P \, dx \\
&= \int_a^b P(x, u(x)) \, dx - \int_a^b P(x, v(x)) \, dx \\
&= -\int_a^b dx \int_{u(x)}^{v(x)} \frac{\partial P}{\partial y} \, dy,
\end{aligned}
\tag{33}
$$

where the last equality follows from Theorem 3.11, Section 3.3, applied to P as a function of y for each fixed x in $[a, b]$. From Theorem 2.6, Section 7.2,

$$\int_a^b dx \int_{u(x)}^{v(x)} \frac{\partial P}{\partial y} \, dx = \int_D \frac{\partial P}{\partial y} \, d(x, y);$$

therefore, (33) implies that

$$\int_C P \, dx = -\int_D \frac{\partial P}{\partial y} \, d(x, y). \tag{34}$$

From Figure 2.14(b) we can write

$$C = C_3 - C_4,$$

where C_3 and C_4 are given by

$$x = V(y) \quad \text{and} \quad x = U(y), \qquad c \le y \le d.$$

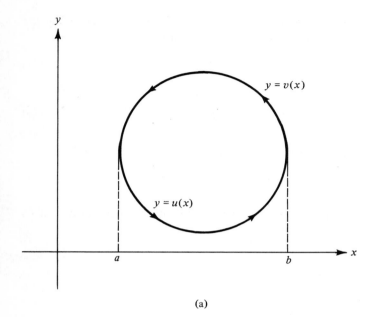

(a)

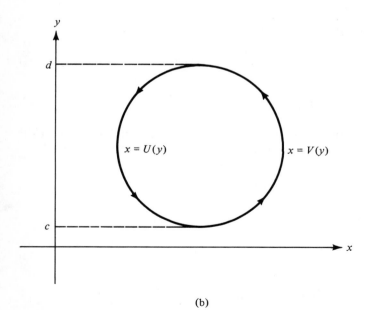

(b)

FIGURE 2.14

We leave it to the reader to verify from this that

$$\int_C Q \, dy = \int_D \frac{\partial Q}{\partial x} \, d(x, y)$$

(*Exercise 20*). This and (34) imply (30) for regions that satisfy (31) and (32).

Some regions that cannot be written in the forms (31) and (32) are decomposable into finitely many subregions that can be. For example, inserting line segments as shown in Figure 2.15(b) divides the region D of Figure 2.15(a) into seven subregions that satisfy (31) and (32). Applying our known result to each of these yields

$$\int_{C_i} P \, dx + Q \, dy = \int_{D_i} \left(\frac{\partial Q}{\partial x} - \frac{\partial P}{\partial y} \right) d(x, y), \qquad i = 1, 2, \dots, 7, \qquad (35)$$

where C_i is the positively directed curve bounding the ith region D_i. Among the line integrals on the left the contributions from the line segments introduced in the subdivision cancel, because each is traversed twice, in opposite directions. Therefore, summing (35) from $i = 1$ to $i = 7$ yields (30) in this case also. The same argument yields (30) for the case where D can be subdivided into any finite number of regions satisfying (31) and (32); however, this is as far as we will take the proof.

Example 2.13   If

$$\frac{\partial P}{\partial y} = \frac{\partial Q}{\partial x}$$

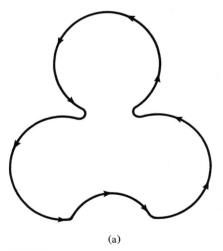

(a)

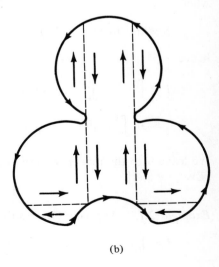

(b)

FIGURE 2.15

in addition to the assumptions of Theorem 2.6, then (30) implies that

$$\int_C P \, dx + Q \, dy = 0.$$

This is consistent with Theorem 2.5.

Example 2.14   Let

$$I = \int_C (x^2 - y^3) \, dx + (x^2 + y^2) \, dy,$$

where C is the unit circle,

$$x^2 + y^2 = 1,$$

traversed counterclockwise. Here

$$\frac{\partial Q(x, y)}{\partial x} - \frac{\partial P(x, y)}{\partial y} = 2x + 3y^2,$$

so (30) implies that

$$I = \int_D (2x + 3y^2) \, d(x, y),$$

where $D = \{(x, y) \, | \, x^2 + y^2 \le 1\}$. Introducing polar coordinates yields

$$I = \int_S (2r \cos \theta + 3r^2 \sin^2 \theta) r \, d(r, \theta),$$

where

$$S = \{(r, \theta) \, | \, 0 \le r \le 1, 0 \le \theta \le 2\pi\};$$

hence

$$I = \int_0^1 r^2 \, dr \int_0^{2\pi} (2 \cos \theta + 3r \sin^2 \theta) \, d\theta$$

$$= 3\pi \int_0^1 r^3 \, dr = \frac{3\pi}{4}.$$

We leave it to the reader to verify this by evaluating the line integral directly [*Exercise 3(e)*].

Example 2.15   Applying Green's theorem to

$$I = \int_C (x^2 + y^2) \, dx + (y^2 - x^2) \, dy, \tag{36}$$

where C is the perimeter of the square $D = [0, 1] \times [0, 1]$, traversed counter-clockwise (Figure 2.16), yields

$$I = \int_D (-2x - 2y) \, d(x, y) = -2 \int_0^1 dx \int_0^1 (x + y) \, dy$$

$$= -2 \int_0^1 \left(xy + \frac{y^2}{2} \right) \Big|_0^1 dx = -2 \int_0^1 \left(x + \frac{1}{2} \right) dx$$

$$= -2 \left(\frac{x^2}{2} + \frac{x}{2} \right) \Big|_0^1 = -2.$$

The advantage of applying Green's theorem here is that it avoids the necessity of parametrizing the four sides of D to evaluate I from (36) [*Exercise 4(g)*].

Corollary 2.1   *If C is a positively directed simple closed curve in $\mathscr{R}^2$, then the area of the region bounded by C is*

$$A = \int_C x \, dy = -\int_C y \, dx = \tfrac{1}{2} \int_C -y \, dx + x \, dy.$$

Proof.   *Exercise 23.*

Example 2.16   To find the area enclosed by the ellipse

$$\frac{x^2}{a^2} + \frac{y^2}{b^2} = 1 \qquad (a, b > 0)$$

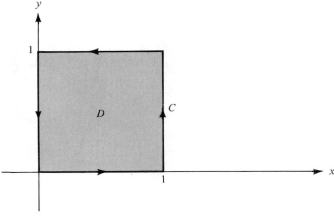

FIGURE 2.16

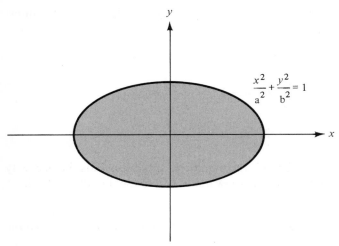

FIGURE 2.17

(Figure 2.17), we parametrize the ellipse as

$$\begin{bmatrix} x \\ y \end{bmatrix} = \begin{bmatrix} a \cos t \\ b \sin t \end{bmatrix}, \qquad 0 \le t \le 2\pi,$$

and calculate

$$A = \int_C x \, dy = \int_0^{2\pi} a \cos t (b \cos t) \, dt$$

$$= ab \int_0^{2\pi} \cos^2 t \, dt = \frac{ab}{2} \int_0^{2\pi} (1 + \cos 2t) \, dt = \pi ab.$$

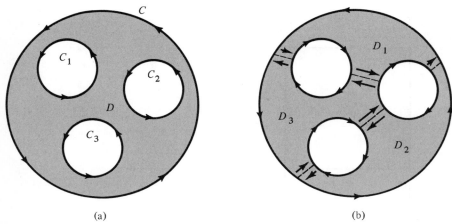

(a)   (b)

FIGURE 2.18

Also,

$$A = -\int_C y\, dx = -\int_0^{2\pi} b \sin t(-a \sin t)\, dt$$

$$= ab \int_0^{2\pi} \sin^2 t\, dt = \frac{ab}{2} \int_0^{2\pi} (1 - \cos 2t)\, dt = \pi ab.$$

Theorem 2.7 (Second Form of Green's Theorem)   *Let C be a piecewise smooth curve in $\mathcal{R}^2$ and suppose $C_1, C_2, \ldots, C_k$ are piecewise smooth curves inside C, no two of which intersect. Let P and Q be continuously differentiable on the closed region D bounded on the outside by C and on the inside by $C_1, C_2, \ldots, C_k$ [Figure 2.18(a)]. Then*

$$\int_D \left(\frac{\partial Q}{\partial x} - \frac{\partial P}{\partial y} \right) d(x, y) = \int_C P\, dx + Q\, dy - \sum_{j=1}^{k} \int_{C_j} P\, dx + Q\, dy, \qquad (37)$$

if C and $C_1, C_2, \ldots, C_k$ are positively directed.

Partial Proof.   We give an intuitive proof. By introducing finitely many auxiliary line segments as shown in Figure 2.18(b), we divide D into simply connected subregions $D_1, D_2, \ldots, D_r$ bounded by simple closed curves $\gamma_1, \gamma_2, \ldots, \gamma_r$. Applying Theorem 2.6 to each of these yields

$$\int_{D_i} \left(\frac{\partial Q}{\partial x} - \frac{\partial P}{\partial y} \right) d(x, y) = \int_{\gamma_i} P\, dx + Q\, dy, \qquad i = 1, 2, \ldots, r, \qquad (38)$$

where γ_i is positively directed. Each auxiliary line occurs twice, traversed in opposite directions, among $\gamma_1, \gamma_2, \ldots, \gamma_r$, while C occurs once, positively directed, and $C_1, C_2, \ldots, C_k$ each occur once, negatively directed. Therefore, summing (38) from $i = 1$ to $i = r$ yields (37).

Example 2.17   If the origin is inside the simple closed curve C, then

$$\int_C \frac{-y\, dx + x\, dy}{x^2 + y^2} = 2\pi$$

if C is traversed in the positive direction. To see this, let C_1 be a circle centered at $(0, 0)$ and not intersecting C [Figure 2.19(a) or (b)]. Since $P_y(x, y) = Q_x(x, y)$ in this example if (x, y) is between C and C_1, Theorem 2.7 implies that

$$\int_C \frac{-y\, dx + x\, dy}{x^2 + y^2} = \int_{C_1} \frac{-y\, dx + x\, dy}{x^2 + y^2} = 2\pi,$$

from Example 2.3.

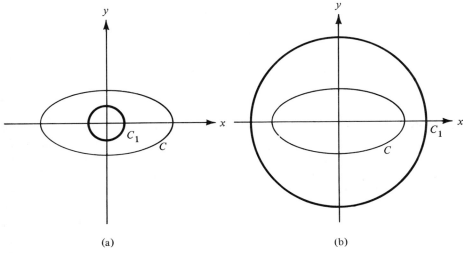

(a)   (b)

FIGURE 2.19

Line integrals with respect to arc length

Definition 2.3   *Let C be a smooth curve in $\mathscr{R}^n$, suppose f is continuous on $\mathrm{tr}(C)$, and let $\boldsymbol{\Phi}: [a, b] \to \mathscr{R}^n$ be a smooth representation of C. Then*

$$\int_a^b f(\boldsymbol{\Phi}(t))|\boldsymbol{\Phi}'(t)|\, dt$$

is called the line integral of f, with respect to arc length, along C. It is denoted by

$$\int_C f\, ds.$$

If $C = C_1 + C_2 + \cdots + C_k$, where $C_1, C_2, \ldots, C_k$ are smooth, then

$$\int_C f\, ds = \sum_{j=1}^{k} \int_{C_j} f\, ds.$$

In particular, $\int_C ds$, obtained by taking $f = 1$, is the arc length of C, as defined in Section 8.1.

We leave it to the reader to verify that $\int_C f\, ds$ is independent of the representation chosen for C, and that

$$\int_C f\, ds = \int_{-C} f\, ds \qquad\qquad\qquad (39)$$

(*Exercise 28*). Notice that the minus sign that appears in Theorem 2.1 does not appear in (39).

Example 2.18   Any line integral $\int_C \mathbf{F} \cdot d\mathbf{X}$ can be written as a line integral with respect to arc length. To see this we write

$$\int_C \mathbf{F} \cdot d\mathbf{X} = \int_a^b \mathbf{F}(\mathbf{X}(t)) \cdot \mathbf{X}'(t)\, dt = \int_a^b \mathbf{F}(\mathbf{X}(t)) \cdot \frac{\mathbf{X}'(t)}{|\mathbf{X}'(t)|} |\mathbf{X}'(t)|\, dt \qquad (40)$$

and recall from Section 8.1 that

$$\frac{\mathbf{X}'(t)}{|\mathbf{X}'(t)|} = \mathbf{T}(\mathbf{X}(t)),$$

where $\mathbf{T}(\mathbf{X})$ is the unit tangent vector to C at $\mathbf{X}$, in the direction of C. Therefore, (40) implies that

$$\int_C \mathbf{F} \cdot d\mathbf{X} = \int_a^b \mathbf{F}(\mathbf{X}(t)) \cdot \mathbf{T}(\mathbf{X}(t)) |\mathbf{X}'(t)|\, dt = \int_C \mathbf{F} \cdot \mathbf{T}\, ds.$$

An application to harmonic functions

Let C be a smooth simple closed curve, positively directed, in $\mathscr{R}^2$, and for each $\mathbf{X}$ in tr(C) let

$$\mathbf{T}(\mathbf{X}) = \begin{bmatrix} T_1(\mathbf{X}) \\ T_2(\mathbf{X}) \end{bmatrix}$$

be the unit tangent vector in the direction of C; that is, if C is represented by

$$\mathbf{X}(t) = \begin{bmatrix} x(t) \\ y(t) \end{bmatrix}, \qquad a \le t \le b, \qquad (41)$$

then

$$\mathbf{T}(\mathbf{X}(t)) = \frac{1}{|\mathbf{X}'(t)|} \begin{bmatrix} x'(t) \\ y'(t) \end{bmatrix}.$$

Then the unit vector

$$\mathbf{n}(\mathbf{X}) = \begin{bmatrix} T_2(\mathbf{X}) \\ -T_1(\mathbf{X}) \end{bmatrix},$$

obtained by rotating $\mathbf{T}(\mathbf{X})$ through 90 degrees clockwise (Figure 2.20) is called the *outward normal to C at $\mathbf{X}$*; in terms of the parametrization (41) it is given by

$$\mathbf{n}(\mathbf{X}(t)) = \frac{1}{|\mathbf{X}'(t)|} \begin{bmatrix} y'(t) \\ -x'(t) \end{bmatrix}.$$

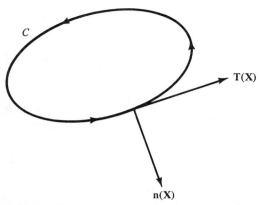

FIGURE 2.20

Now suppose u is continuously differentiable on $\operatorname{tr}(C)$ and let

$$\frac{\partial u(\mathbf{X})}{\partial \mathbf{n}}$$

denote its directional derivative in the direction of the outward normal to C at $\mathbf{X}$; thus,

$$\frac{\partial u(\mathbf{X})}{\partial \mathbf{n}} = \mathbf{n}(\mathbf{X}) \cdot \nabla u(\mathbf{X})$$

(Theorem 1.2), and

$$\int_C \frac{\partial u}{\partial \mathbf{n}}\, ds = \int_C \mathbf{n} \cdot \nabla u\, ds = \int_a^b \mathbf{n}(\mathbf{X}(t)) \cdot \nabla u(\mathbf{X}(t))\, |\mathbf{X}'(t)|\, dt, \qquad (42)$$

by Definition 2.3. Since

$$\mathbf{n}(\mathbf{X}(t))\,|\mathbf{X}'(t)| = \begin{bmatrix} y'(t) \\ -x'(t) \end{bmatrix} \quad \text{and} \quad \nabla u = \begin{bmatrix} u_x \\ u_y \end{bmatrix},$$

(42) can be rewritten as

$$\int_C \frac{\partial u}{\partial \mathbf{n}}\, ds = \int_a^b \left[-u_y(\mathbf{X}(t))x'(t) + u_x(\mathbf{X}(t))y'(t) \right] dt,$$

and so

$$\int_C \frac{\partial u}{\partial \mathbf{n}}\, ds = \int_C -u_y\, dx + u_x\, dy. \qquad (43)$$

If interpreted in the appropriate extended sense, this also holds if C is piecewise smooth rather than smooth.

If u_x and u_y are continuously differentiable in the closed region D bounded by C, then applying Theorem 2.6 to the right side of (43) yields

$$\int_C \frac{\partial u}{\partial \mathbf{n}}\, ds = \int_D (u_{xx} + u_{yy})\, d(x, y). \tag{44}$$

Now suppose u is *harmonic* in D, by which we mean that u_x and u_y are continuously differentiable and u satisfies *Laplace's equation*,

$$u_{xx} + u_{yy} = 0,$$

in an open set containing D. For example, it is easily verified that

$$u(x, y) = x^2 - y^2$$

is harmonic in $\mathcal{R}^2$ and

$$u(x, y) = \log(x^2 + y^2)$$

is harmonic in $\mathcal{R}^2 - \{(0, 0)\}$. Clearly, (44) implies the following theorem.

Theorem 2.8 *If C is a piecewise smooth simple closed curve and u is harmonic on the closed region D bounded by C, then*

$$\int_C \frac{\partial u}{\partial \mathbf{n}}\, ds = 0.$$

8.2 EXERCISES

1. Derive Eqs. (4) and (5), assuming that the parametric representation in Eq. (2) is continuously differentiable on $[a, b]$.

2. Prove Theorem 2.1.

3. Use Definition 2.1 to evaluate:
   (a) $\int_C y\, dx + x\, dy + (x^2 + y^2)\, dz$; from $(0, 2, 0)$ to $(2, 0, 2)$, with $x^2 + y^2 = 4$, $z = x$.
   (b) $\int_C (x + y)^2\, dx + (x - y)^2\, dy$; C is the line segment from $(0, 1)$ to $(2, 1)$.
   (c) $\int_C (x^2 + y)\, dx + (x - y^2)\, dy$; along $y = x^{3/2}$, $0 \le x \le 1$
   (d) $\int_C (x^2 - y^2)\, dx + xy\, dy$; along $x = y^2$, $-1 \le y \le 1$
   (e) $\int_C (x^2 - y^3)\, dx + (x^2 + y^2)\, dy$; from $(1, 0)$ to $(1, 0)$, counterclockwise around the circle $x^2 + y^2 = 1$
   (f) $\int_C (x + y)\, dx + xy\, dy + z^2\, dz$; C is the line segment from $(0, 0, 0)$ to $(1, 2, 1)$
   (g) $\int_C (x^2 + y)\, dx + (x^2 + y^2)\, dy$; from $(2, 0)$ to $(-2, 0)$, with $9x^2 + 4y^2 = 36$, $y \ge 0$

4. Evaluate:

   (a) $\int_C (4x + y + 3z) \, dx + (x + 3y + 2z) \, dy + (3x + 2y + 5z) \, dz$; along line segments from $(1, 4, 2)$ to $(4, 2, 3)$ to $(1, 1, 2)$

   (b) $\int_C (3x + y + z) \, dx + (x + 3y - z) \, dy + (x - y + 3z) \, dz$; along line segments from $(3, 1, 2)$ to $(5, 4, 3)$ to $(1, 1, 1)$

   (c) $\int_C (x^2 + y^2) \, dx + (y^2 - x^2) \, dy$; along the line segment from $(0, 0)$ to $(\sqrt{3}, 1)$, then to $(0, 2)$ with $x = \sqrt{4 - y^2}$

   (d) $\int_C (x^2 y + x) \, dx + (x^2 - y) \, dy$; from $(0, 0)$ to $(1, 1)$ with $y = x^2$, then to $(0, 0)$ with $x = y^2$

   (e) $\int_C (3x^2 + 2y) \, dx + (2y + 2x) \, dy$; from $(0, 0)$ to $(1, 1)$ with $y = x^2$, then along a line segment to $(0, 0)$

   (f) $\int_C 3y^2 \, dx + (2x^3 + z) \, dy + y \, dz$; along line segments from $(0, 0, 0)$ to $(1, 1, 2)$ to $(-3, 1, 0)$

   (g) $\int_C (x^2 + y^2) \, dx + (y^2 - x^2) \, dy$; from $(1, 1)$ to $(1, 1)$, counterclockwise on the boundary of $[0, 1] \times [0, 1]$.

5. Find the work done in moving a particle along the given path subject to the force $\mathbf{F}$:

   (a) $\mathbf{F}(x, y, z) = \begin{bmatrix} 2x^2 \\ y + z \\ z \end{bmatrix}$; $\mathbf{X}(t) = \begin{bmatrix} 1 \\ t \\ t^2 \end{bmatrix}$, $0 \le t \le 1$

   (b) $\mathbf{F}(x, y) = -(x^2 + y^2)^{-3/2} \begin{bmatrix} x \\ y \end{bmatrix}$; twice clockwise around $x^2 + y^2 = 1$, starting and ending at $(1, 0)$

   (c) $\mathbf{F}(x, y) = \begin{bmatrix} -2x \\ 4 \end{bmatrix}$; from $(1, 0)$ to $(0, -1)$ along $y = x^2 - 1$

6. Prove Theorem 2.2.

7. Let C be a piecewise smooth curve in $\mathscr{R}^n$ with arc length L and suppose $\mathbf{F}$ is continuous $|\mathbf{F}(\mathbf{X})| \le M$ on C. Show that

$$\left| \int_C \mathbf{F} \cdot d\mathbf{X} \right| \le ML.$$

8. Evaluate:

   (a) $\int_C (x^2 + y^2) \, dx$; clockwise along $x^2 + y^2 = 1$ from $(-1, 0)$ to $(-1, 0)$

   (b) $\int_C (x^2 + y^2 + z^2) \, dy$; along the line segment from $(-1, 0, 1)$ to $(0, 1, 2)$

   (c) $\int_C y \sin x \, dz$; along $y = x$, $z = x^2$, $0 \le x \le \pi$

   (d) $\int_C (x^2 - y^2) \, dy$; from $(0, 0)$ to $(1, 2)$ along $y = 2x^2$

   (e) $\int_C (x^2 e^{x^3} + z) \, dy$; along the line segment from $(0, 1, 2)$ to $(-1, 1, 3)$

9. Evaluate:

(a) $\int_C (x + 2y) \, dx + y \, dy$; counterclockwise around the ellipse $x^2 + 4y^2 = 1$

(b) $\int_C (x^2 + y) \, dx + 2xy \, dy$; counterclockwise around the boundary of $[-1, 1] \times [-1, 1]$

(c) $\int_C (x^2 + y^2) \, dx + 2xy \, dy + z \, dz$; along the line segment from $(0, 0, 1)$ to $(1, 0, 0)$, then to $(0, 1, 0)$ along the circular arc $x^2 + y^2 = 1$, $x \geq 0$, then along a line segment to $(0, 0, 1)$.

10. Suppose the vector field $\mathbf{F}$ is continuous on a domain D. Show that line integrals of $\mathbf{F}$ are independent of path in D if and only if $\int_C \mathbf{F} \cdot d\mathbf{X} = 0$ whenever C is a closed piecewise smooth curve in D.

11. Determine which of the line integrals in Exercise 3 are independent of path in $\mathcal{R}^2$ or $\mathcal{R}^3$, as the case may be. Find potential functions for those that are, and use them to evaluate the integrals specified in the exercise.

12. Follow the instructions of Exercise 11 for the line integrals of Exercise 4.

13. Follow the instructions of Exercise 11 for the line integrals of Exercise 9.

14. Show that the following integrals are independent of path, and evaluate them:

(a) $\int_{(0,1)}^{(1,0)} x \exp(x^2 + y^2) \, dx + y \exp(x^2 + y^2) \, dy$

(b) $\int_{(1,-1,1)}^{(2,0,1)} 3x^2 y^2 z \, dx + 2x^3 yz \, dy + x^3 y^2 \, dz$

(c) $\int_{(1,-1,2)}^{(1,1,-1)} (2x + 2yz) \, dx + (2y + 2xz) \, dy + 2xy \, dz$

(d) $\int_{(1,-1)}^{(2,1)} \frac{x \, dx + y \, dy}{(x^2 + y^2)^{3/2}}$; along any curve not through $(0, 0)$

(e) $\int_{(1,0)}^{(0,1)} y(e^{xy} + 1) \, dx + x(e^{xy} + 1) \, dy$

(f) $\int_{(1,1,2)}^{(3,1,2)} (2x^2 + y^2) \, dx + (2xy + z^2) \, dy + 2yz \, dz$

(g) $\int_{(1,-1,0)}^{(2,0,1)} (3x^2 \sin xyz + x^3 yz \cos xyz) \, dx + x^4 z \cos xyz \, dy + x^4 y \cos xyz \, dz$

15. Suppose P and Q are continuously differentiable and $P \, dx + Q \, dy$ is an exact differential in a suitable domain D in $\mathcal{R}^2$. Then a potential function f such that $P = f_x$ and $Q = f_y$ can be found as follows:
(i) Find a function f_1 such that

$$\frac{\partial f_1(x, y)}{\partial x} = P(x, y)$$

(that is, f_1 is an indefinite integral of P with respect to x), and let

$$f(x, y) = f_1(x, y) + \phi(y), \tag{A}$$

where ϕ is to be determined.

(ii) Differentiate (A) with respect to y and require that the result equal Q; thus,

$$\frac{\partial f_1(x, y)}{\partial y} + \phi'(y) = Q(x, y),$$

or

$$\phi'(y) = Q(x, y) - \frac{\partial f_1(x, y)}{\partial y}. \tag{B}$$

(iii) Integrate (B) with respect to y to obtain ϕ. Substituting the result in (A) yields a potential function.

Question: What does the assumption that $P \, dx + Q \, dy$ is an exact differential have to do with this procedure? What assumption must be made on D?

16. Use the method of Exercise 15 to find potential functions associated with the following exact differentials:
(a) $(x + y) \, dx + (x - y) \, dy$
(b) $(x^2 + y) \, dx + (x - y^2) \, dy$
(c) $(3x^2 + 2y) \, dx + (2y + 2x) \, dy$
(d) $\exp(x^2 + y^2)(x \, dx + y \, dy)$

(e) $\dfrac{x \, dx + y \, dy}{(x^2 + y^2)^{3/2}}$

17. Devise a method analogous to that of Exercise 15 for finding a potential function for $\mathbf{F} = P\mathbf{i} + Q\mathbf{j} + R\mathbf{k}$, given that P, Q, and R are continuously differentiable and $P \, dx + Q \, dy + R \, dz$ is exact in a region D in $\mathscr{R}^3$.

18. Use the method you devised in Exercise 17 to find the potential functions associated with the following exact differentials:
(a) $(4x + y + 3z) \, dx + (x + 3y + 2z) \, dy + (3x + 2y + 5z) \, dz$
(b) $(3x + y + z) \, dx + (x + 3y - z) \, dy + (x - y + 3z) \, dz$
(c) $(2x^3y^6 + 3x^2yz) \, dx + (3x^4y^5 + x^3z) \, dy + (x^3y + z) \, dz$
(d) $(\sin z + y \cos x) \, dx + (\sin x + z \cos y) \, dy + (x \cos z + \sin y) \, dz$
(e) $(2x + 2yz^2 \cos xz) \, dx + (2z \sin xz) \, dy + (2y \sin xz + 2xyz \cos xz) \, dz$

19. (a) The *kinetic energy* of a particle of mass in moving with velocity $\mathbf{v}$ is $m|\mathbf{v}|^2/2$. Use Newton's second law of motion,

$$\mathbf{F} = m\mathbf{a},$$

to show that the work done on the particle as it moves along a twice continuously differentiable curve C in the domain of a continuously differentiable force field $\mathbf{F}$ equals the change in its kinetic energy.

(b) If $\mathbf{F} = \nabla f$, where f is continuously differentiable in D, then $-f(\mathbf{X})$ is the *potential energy* of the particle at a point $\mathbf{X}$ in D. The sum of the potential and kinetic energies of the particle is its *total energy*. Show that if $\mathbf{F}$ is conservative in D, then the total energy of the particle remains constant as it traverses any twice continuously differentiable curve in D; that is, its total energy is *conserved*.

20. Assuming that D can be represented as in Eq. (32) and C is its positively directed piecewise smooth boundary, verify that

$$\int_D \frac{\partial Q}{\partial x}\, d(x, y) = \int_C Q\, dy.$$

[*Hint:* See the derivation of Eq. (34).]

21. Use Theorem 2.6 to evaluate the following line integrals, with C positively directed:

(a) $\int_C (x^2 y + y^3 - \log x)\, dx + (xy^2 + x^2 y + x^2 + \log y^2)\, dy$; C is the boundary of $[1, 2] \times [1, 2]$.

(b) $\int_C (\sin y - x^2 y)\, dx + (x \cos y + xy^2)\, dy$; C is the circle $x^2 + y^2 = 1$.

(c) $\int_C (e^x \sin y - xy^2)\, dx + (e^x \cos y + x^2 y)\, dy$; C is the ellipse $9x^2 + 4y^2 = 36$.

(d) $\int_C (x^2 y + y^3 x)\, dx + (y^4 x^2 - x^3)\, dy$; C is the boundary of $[1, 3] \times [0, 1]$

(e) $\int_C (3x^2 y^2 - 3xy + y^4/4)\, dx + (xy^3 + x^3/3)\, dy$; C is the boundary of the triangle with vertices $(0, 1)$, $(1, 0)$, and $(-1, 0)$.

22. Suppose u_x, u_y, v_x, and v_y are differentiable and satisfy the Cauchy–Riemann equations, $u_x = v_y$, $u_y = -v_x$, on an open set containing the region bounded by a simple closed curve C. Show that

$$\int_C u\, dx - v\, dy = \int_C v\, dx + u\, dy = 0.$$

23. Prove Corollary 2.1.

24. Use Corollary 2.1 to find the area of the region bounded by the given curves:

(a) $y = x^2$, $x + y = 2$, $x + 3y = 0$

(b) $y = -x$, $y = x(x - 1)$, $x^2 + y^2 = 1$

(c) $x = 1$, $x + y = 3$, $4x + y = 9$, $x + y = 0$

(d) $x = y^2$, $y = x^2$

25. Suppose a region D in $\mathscr{R}^2$ is defined in terms of polar coordinates as

$$D = \{(r \cos \theta, r \sin \theta) \mid 0 \le r \le f(\theta),\ \alpha \le \theta \le \beta\},$$

where $\beta \leq \alpha + 2\pi$ and f is continuously differentiable on $[\alpha, \beta]$. Use Corollary 2.1 to show that the area of D is

$$A = \tfrac{1}{2} \int_\alpha^\beta f^2(\theta) \, d\theta.$$

26. Suppose the transformation

$$\begin{bmatrix} x \\ y \end{bmatrix} = \begin{bmatrix} x(u, v) \\ y(u, v) \end{bmatrix}$$

maps the closed region S in the uv plane onto the closed region D in the xy plane. Let γ and C be the positively directed boundaries of S and D, respectively, and assume that they are both simple closed piecewise smooth curves. Use Theorem 2.6 and Corollary 2.1 to prove formally (that is, by purely manipulative methods, assumed to be valid) that the area of D is

$$A = \int_S \left| \frac{\partial(x, y)}{\partial(u, v)} \right| d(u, v).$$

[*Hint:* First show that

$$\int_C x \, dy = \pm \int_\gamma x(u, v) \frac{\partial y(u, v)}{\partial u} \, du + x(u, v) \frac{\partial y(u, v)}{\partial v} \, dv$$

and then apply Theorem 2.6.]

27. (a) Suppose P and Q are continuously differentiable and

$$Q_x(x, y) = P_y(x, y) \tag{A}$$

for all (x, y) in the open region D except at a single point (x_0, y_0), and let P and Q be bounded on a deleted neighborhood of (x_0, y_0). Show that

$$\int_C P \, dx + Q \, dy = 0 \tag{B}$$

if C is a simple closed curve in D which does not pass through (x_0, y_0). (*Hint:* Use Theorems 2.6 and 2.7 and Exercise 7.)

(b) Use part (a) to show that

$$\int_C \frac{x \, dx + y \, dy}{\sqrt{x^2 + y^2}} = 0$$

if C is any simple closed curve which does not pass through $(0, 0)$. How else can this be shown?

(c) Show that (B) remains valid if (A) holds for all (x, y) except for a finite or infinite sequence $\{(x_0, y_0), (x_1, y_1), \ldots\}$ of points in D, provided the sequence has no limit point in D, P and Q are bounded in a deleted neighborhood of each (x_j, y_j), and C does not pass through any (x_j, y_j).

28. (a) Show that $\int_C f\, ds$ is independent of the representation chosen for C.
   (b) Show that

$$\int_C f\, ds = \int_{-C} f\, ds.$$

29. Compute $\int_C f\, ds$:
   (a) $f(x, y, z) = x^2 + y^2 + z^2$; $\mathbf{X}(t) = a\cos t\mathbf{i} + a\sin t\mathbf{j} + bt\mathbf{k}$, $0 \le t \le 2\pi$
   (b) $f(x, y) = 3x^2 - 2xy + y^2$; C is the semicircle $x^2 + y^2 = 4$, $y \ge 0$.
   (c) $f(x, y, z) = z^2$; $\mathbf{X}(t) = \cos t\mathbf{i} + \sin t\mathbf{j} + e^t\mathbf{k}$, $0 \le t \le \pi/2$

   (d) $f(x, y) = xy$; $\mathbf{X}(t) = \begin{bmatrix} e^t \cos t \\ e^t \sin t \end{bmatrix}$, $0 \le t \le \pi/4$

   (e) $f(x, y, z) = xy$; $\mathbf{X}(t) = \dfrac{t^2}{2}\mathbf{i} + \dfrac{\sqrt{2}}{3}t^3\mathbf{j} + \dfrac{t^4}{4}\mathbf{k}$, $0 \le t \le 1$

In Exercises 30–33 let C be a positively directed simple closed curve bounding a region D in $\mathscr{R}^2$. Derive the indicated relationships under appropriate assumptions on the functions involved.

30. Prove: If $\mathbf{F} = (f_1, f_2)$, then

$$\int_D \left(\frac{\partial f_1}{\partial x} + \frac{\partial f_2}{\partial y} \right) d(x, y) = \int_C \mathbf{F} \cdot \mathbf{n}\, ds.$$

   (Hint: Use Theorem 2.6 with $Q = f_1$ and $P = -f_2$.)

31. Prove:

$$\int_D \left[p\left(\frac{\partial f_1}{\partial x} + \frac{\partial f_2}{\partial y} \right) + \mathbf{F} \cdot \nabla p \right] d(x, y) = \int_C p(\mathbf{F} \cdot \mathbf{n})\, ds.$$

   (Hint: Use Exercise 30 with $\mathbf{F}$ replaced by $p\mathbf{F}$.)

*32. If $p = p(x, y)$, the *Laplacian of p* is defined by

$$\nabla^2 p = p_{xx} + p_{yy}.$$

   (a) Use Exercise 31 to derive *Green's identities*:

$$\int_D [u\, \nabla^2 v + (\nabla u) \cdot (\nabla v)]\, d(x, y) = \int_C u\frac{\partial v}{\partial \mathbf{n}}\, ds$$

   and

$$\int_D (u\, \nabla^2 v - v\, \nabla^2 u)\, d(x, y) = \int_C \left(u\frac{\partial v}{\partial \mathbf{n}} - v\frac{\partial u}{\partial \mathbf{n}} \right) ds.$$

   (b) Show how Theorem 2.8 follows from either of these identities.

(c) Show that if u and v are harmonic in D, then

$$\int_C u \frac{\partial v}{\partial \mathbf{n}} \, ds = \int_C v \frac{\partial u}{\partial \mathbf{n}} \, ds.$$

33. Prove: If u is harmonic in D, then

$$\int_C u \frac{\partial u}{\partial \mathbf{n}} \, ds = \int_D \left[(u_x)^2 + (u_y)^2 \right] d(x, y).$$

(*Hint:* See Exercise 32.)

34. If $u = u(x, y)$ is differentiable, then its *conjugate differential* is defined by

$$d^*u = -u_y \, dx + u_x \, dy.$$

Prove: If u is harmonic in a simply connected region D in $\mathscr{R}^2$, then d^*u is an exact differential in D; that is, there is a differentiable function v defined in D such that $dv = d^*u$. Moreover, v is harmonic in D.

35. If u and v are harmonic and $dv = d^*u$ as in Exercise 34, then v is said to be *a harmonic conjugate of* v. Use methods developed in this section for finding potential functions of a conservative field to find harmonic conjugates of the following functions:
(a) $x^2 - y^2$ (b) $e^x \cos y$ (c) $x^3 - 3xy^2$
(d) $\cos x \cosh y$ (e) $\sin x \cosh y$

8.3   SURFACES IN $\mathscr{R}^3$

We will restrict our study of surface integrals to surfaces in $\mathscr{R}^3$. In elementary calculus a surface in $\mathscr{R}^3$ is the set of points (x, y, z) satisfying a single equation, which may be in one of the *explicit* forms

$$z = f(x, y), \qquad y = g(x, z), \qquad x = h(y, z), \tag{1}$$

or in *implicit* form:

$$F(x, y, z) = 0. \tag{2}$$

Clearly any of (1) can be rewritten as (2).
    Suppose S is defined by

$$z = f(x, y), \tag{3}$$

where f is differentiable at (x_0, y_0), and let $z_0 = f(x_0, y_0)$. In Section 5.3 we defined the tangent plane to S at (x_0, y_0, z_0) to be the plane given by

$$z = T(x, y) = f(x_0, y_0) + f_x(x_0, y_0)(x - x_0) + f_y(x_0, y_0)(y - y_0), \tag{4}$$

which intersects S at (x_0, y_0, z_0) and approximates S so well near there that

$$\lim_{(x,y)\to(x_0,y_0)} \frac{f(x, y) - T(x, y)}{\sqrt{(x - x_0)^2 + (y - y_0)^2}} = 0.$$

If F is continuously differentiable near (x_0, y_0, z_0) and $F_z(x_0, y_0, z_0) \neq 0$, then (2) defines z as a function of (x, y) near (x_0, y_0) as in (3), with

$$f_x = -\frac{F_x}{F_z} \quad \text{and} \quad f_y = -\frac{F_y}{F_z}$$

(Corollary 4.1, Section 6.4). In this case we can rewrite (4) as

$$z = z_0 - \frac{F_x(x_0, y_0, z_0)}{F_z(x_0, y_0, z_0)}(x - x_0) - \frac{F_y(x_0, y_0, z_0)}{F_z(x_0, y_0, z_0)}(y - y_0)$$

or

$$F_x(x_0, y_0, z_0)(x - x_0) + F_y(x_0, y_0, z_0)(y - y_0) + F_z(x_0, y_0, z_0)(z - z_0) = 0. \quad (5)$$

This equation for the tangent plane can also be obtained by assuming that $F_x(x_0, y_0, z_0) \neq 0$ or $F_y(x_0, y_0, z_0) \neq 0$, so that the equation of S can be written in one of the other forms in (1) (*Exercise 2*). This leads to the following definition.

Definition 3.1 *Suppose the surface S is defined by*

$$F(x, y, z) = 0,$$

where F is differentiable at a point (x_0, y_0, z_0) and at least one of $F_x(x_0, y_0, z_0)$, $F_y(x_0, y_0, z_0)$, and $F_z(x_0, y_0, z_0)$ is nonzero. Then the plane defined by (5) is called the tangent plane to S at (x_0, y_0, z_0).

We can write (5) in vector form as

$$(\mathbf{X} - \mathbf{X}_0) \cdot \nabla F(\mathbf{X}_0) = 0,$$

from which it is clear that $\mathbf{X}$ is in the tangent plane if and only if $\nabla F(\mathbf{X}_0)$ is perpendicular to the vector from $\mathbf{X}_0$ to $\mathbf{X}$ (Figure 3.1); therefore, $\nabla F(\mathbf{X}_0)$ is said to be *normal* to the tangent plane at $\mathbf{X}_0$, as is any nonzero multiple of $\nabla F(\mathbf{X}_0)$. We also say that these vectors are normal to S at $\mathbf{X}_0$, since if C is any smooth curve on the surface and through $\mathbf{X}_0$, then these vectors are perpendicular to the tangent vector to C at $\mathbf{X}_0$. (See Figure 3.1 and *Exercise 3*.)

Example 3.1 To find the equation of the tangent plane to the sphere

$$(x - 1)^2 + (y - 2)^2 + z^2 = 25$$

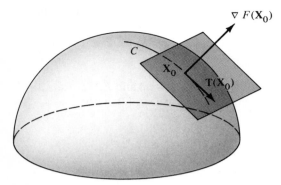

FIGURE 3.1

at $\mathbf{X}_0 = (5, 5, 0)$, we take

$$F(x, y, z) = (x - 1)^2 + (y - 2)^2 + z^2 - 25;$$

then

$$F_x(x, y, z) = 2(x - 1), \qquad F_y(x, y, z) = 2(y - 2), \qquad F_z(x, y, z) = 2z,$$

so that

$$F_x(\mathbf{X}_0) = 8, \qquad F_y(\mathbf{X}_0) = 6, \qquad F_z(\mathbf{X}_0) = 0.$$

From (5) the equation of the tangent plane is

$$8(x - 5) + 6(y - 5) + 0(z) = 0$$

or

$$4x + 3y = 35.$$

The gradient

$$\nabla F(\mathbf{X}_0) = 8\mathbf{i} + 6\mathbf{j}$$

is normal to the tangent plane.

Example 3.2 To find the equation of the tangent plane to the surface

$$z = x^2 + 2xy - y^2$$

at $\mathbf{X}_0 = (1, -1, -2)$, we take

$$F(x, y, z) = x^2 + 2xy - y^2 - z;$$

then

$$F_x(x, y, z) = 2(x + y), \qquad F_y(x, y, z) = 2(x - y), \qquad F_z(x, y, z) = -1,$$

so that

$$F_x(\mathbf{X}_0) = 0, \qquad F_y(\mathbf{X}_0) = 4, \qquad F_z(\mathbf{X}_0) = -1.$$

From (5) the equation of the tangent plane is

$$0(x - 1) + 4(y + 1) + (-1)(z + 2) = 0$$

or

$$4y - z = -2.$$

The gradient

$$\mathbf{\nabla} F(\mathbf{X}_0) = 4\mathbf{j} - \mathbf{k}$$

is normal to the tangent plane.

Smooth parametric functions

We are interested in surfaces which have a tangent plane at each point, with a normal whose direction varies continuously with the point. Because of this and because it is not always possible to write the equation of a surface in one of the explicit forms (1), we consider parametrically defined smooth surfaces.

In this and the next section it is convenient to identify the point $\mathbf{X} = (x, y, z)$ with the vector $\mathbf{X} = x\mathbf{i} + y\mathbf{j} + z\mathbf{k}$.

Definition 3.2 *A function*

$$\mathbf{\Phi}(u, v) = \phi_1(u, v)\mathbf{i} + \phi_2(u, v)\mathbf{j} + \phi_3(u, v)\mathbf{k}, \qquad (u, v) \in D,$$

is said to be a smooth parametric function from $\mathscr{R}^2$ to $\mathscr{R}^3$ if:

(a) Its domain D is a closed region in $\mathscr{R}^2$ with boundary consisting of finitely many piecewise smooth simple closed curves.

(b) $\mathbf{\Phi}(u, v) \neq \mathbf{\Phi}(u_1, v_1)$ *if* $(u, v) \in D^0$, $(u_1, v_1) \in D$, *and* $(u, v) \neq (u_1, v_1)$.

(c) $\mathbf{\Phi}$ *is continuously differentiable and the vector*

$$\mathbf{N}(u, v) = \frac{\partial \mathbf{\Phi}(u, v)}{\partial u} \times \frac{\partial \mathbf{\Phi}(u, v)}{\partial v}$$

is nonzero in D^0.

(d) $\displaystyle \lim_{(u,v) \to (u_0, v_0)} \frac{\mathbf{N}(u, v)}{|\mathbf{N}(u, v)|}$

exists for each (u_0, v_0) in ∂D.

The set of points

$$S = \{\mathbf{X} \mid \mathbf{X} = \mathbf{\Phi}(u, v), (u, v) \in D\}$$

is the *surface represented by* $\mathbf{\Phi}: D \to \mathscr{R}^2$; it is called a *smooth surface*. We will soon see that different parametric functions may represent the same surface.

Notice that requirement (b) implies that $\mathbf{\Phi}$ is one-to-one on D^0 and that points in D^0 and ∂D have distinct images under $\mathbf{\Phi}$; however, distinct points in ∂D may have the same image.

Points on S can be written as

$$\mathbf{X} = \mathbf{\Phi}(u, v) = \phi_1(u, v)\mathbf{i} + \phi_2(u, v)\mathbf{j} + \phi_3(u, v)\mathbf{k};$$

we will often write this less formally as

$$\mathbf{X} = x(u, v)\mathbf{i} + y(u, v)\mathbf{j} + z(u, v)\mathbf{k}. \tag{6}$$

We will then write

$$\mathbf{N}(u, v) = \frac{\partial \mathbf{X}(u, v)}{\partial u} \times \frac{\partial \mathbf{X}(u, v)}{\partial v}.$$

Since

$$\mathbf{N} = \begin{vmatrix} \mathbf{i} & \mathbf{j} & \mathbf{k} \\ \dfrac{\partial x}{\partial u} & \dfrac{\partial y}{\partial u} & \dfrac{\partial z}{\partial u} \\ \dfrac{\partial x}{\partial v} & \dfrac{\partial y}{\partial v} & \dfrac{\partial z}{\partial v} \end{vmatrix} \tag{7}$$

$$= \frac{\partial(y, z)}{\partial(u, v)}\mathbf{i} + \frac{\partial(z, x)}{\partial(u, v)}\mathbf{j} + \frac{\partial(x, y)}{\partial(u, v)}\mathbf{k},$$

the requirement that $\mathbf{N}$ be nonzero in D^0 means that at least one of the Jacobians on the right is nonzero at each point of D^0. If $(u_0, v_0) \in D^0$ and, for example,

$$\frac{\partial(x, y)}{\partial(u, v)}\bigg|_{(u_0, v_0)} \neq 0,$$

then Theorem 4.1 of Section 6.4 implies that the system

$$x = x(u, v), \qquad y = y(u, v)$$

can be solved near (x_0, y_0) for u and v as differentiable functions of (x, y),

$$u = u(x, y), \qquad v = v(x, y), \tag{8}$$

with

$$\begin{bmatrix} \dfrac{\partial u}{\partial x} & \dfrac{\partial u}{\partial y} \\ \dfrac{\partial v}{\partial x} & \dfrac{\partial v}{\partial y} \end{bmatrix} = \begin{bmatrix} \dfrac{\partial x}{\partial u} & \dfrac{\partial x}{\partial v} \\ \dfrac{\partial y}{\partial u} & \dfrac{\partial y}{\partial v} \end{bmatrix}^{-1} = \frac{1}{\dfrac{\partial(x, y)}{\partial(u, v)}} \begin{bmatrix} \dfrac{\partial y}{\partial v} & -\dfrac{\partial x}{\partial v} \\ -\dfrac{\partial y}{\partial u} & \dfrac{\partial x}{\partial u} \end{bmatrix}. \tag{9}$$

Substituting (8) in the equation

$$z = z(u, v)$$

[see the third component of (6)] yields

$$z = z(u(x, y), v(x, y))$$

near (x_0, y_0); that is, the surface can be represented near (x_0, y_0, z_0) in the form

$$z = f(x, y) = z(u(x, y), v(x, y)). \tag{10}$$

From (4) the equation of the tangent plane to the surface at (x_0, y_0, z_0) is

$$f_x(x_0, y_0)(x - x_0) + f_y(x_0, y_0)(y - y_0) - (z - z_0) = 0, \tag{11}$$

which we have written in this form to emphasize that the vector

$$\mathbf{n} = f_x(x_0, y_0)\mathbf{i} + f_y(x_0, y_0)\mathbf{j} - \mathbf{k}$$

is normal to the tangent plane.

From (10) and the chain rule,

$$\frac{\partial f}{\partial x} = \frac{\partial z}{\partial u}\frac{\partial u}{\partial x} + \frac{\partial z}{\partial v}\frac{\partial v}{\partial x},$$

$$\frac{\partial f}{\partial y} = \frac{\partial z}{\partial u}\frac{\partial u}{\partial y} + \frac{\partial z}{\partial v}\frac{\partial v}{\partial y}.$$

Substituting from (9) in these equations for u_x, v_x, u_y, and v_y and collecting terms yields

$$\frac{\partial f}{\partial x} = -\frac{\dfrac{\partial(y, z)}{\partial(u, v)}}{\dfrac{\partial(x, y)}{\partial(u, v)}}, \qquad \frac{\partial f}{\partial y} = -\frac{\dfrac{\partial(z, x)}{\partial(u, v)}}{\dfrac{\partial(x, y)}{\partial(u, v)}}.$$

Evaluating these at $\mathbf{X}_0 = (x_0, y_0, z_0)$ and substituting in (11) shows that

$$\left(\frac{\partial(y, z)}{\partial(u, v)}\bigg|_{\mathbf{X}_0}\right)(x - x_0) + \left(\frac{\partial(z, x)}{\partial(u, v)}\bigg|_{\mathbf{X}_0}\right)(y - y_0) + \left(\frac{\partial(x, y)}{\partial(u, v)}\bigg|_{\mathbf{X}_0}\right)(z - z_0) = 0.$$

Comparing this and (7) shows that $\mathbf{N}(u_0, v_0)$ is normal to the surface at $\mathbf{X}_0 = \mathbf{X}(u_0, v_0)$.

Because of this the tangent plane to S at $\mathbf{X}_0$ can be represented parametrically by

$$\mathbf{X} = \boldsymbol{\Psi}(u, v) = \mathbf{X}_0 + (u - u_0)\mathbf{A} + (v - v_0)\mathbf{B}, \tag{12}$$

where **A** and **B** are any noncollinear vectors perpendicular to $\mathbf{N}(u_0, v_0)$. However, there is exactly one choice of **A** and **B** such that

$$\lim_{(u,v)\to(u_0,v_0)} \frac{\mathbf{\Phi}(u, v) - \mathbf{\Psi}(u, v)}{\sqrt{(u - u_0)^2 + (v - v_0)^2}} = \mathbf{0}; \tag{13}$$

that is,

$$\mathbf{A} = \frac{\partial \mathbf{\Phi}(u_0, v_0)}{\partial u} \quad \text{and} \quad \mathbf{B} = \frac{\partial \mathbf{\Phi}(u_0, v_0)}{\partial v},$$

which are noncollinear because of requirement (c) of Definition 3.2. Therefore, (12) becomes

$$\mathbf{X} = \mathbf{\Psi}(u, v) = \mathbf{X}_0 + (u - u_0) \frac{\partial \mathbf{\Phi}(u_0, v_0)}{\partial u} + (v - v_0) \frac{\partial \mathbf{\Phi}(u_0, v_0)}{\partial v}$$

(*Exercise 5*).

Example 3.3   The surface S defined by the continuously differentiable function

$$\mathbf{X} = \mathbf{\Phi}(\theta, z) = a \cos \theta \mathbf{i} + a \sin \theta \mathbf{j} + z\mathbf{k} \qquad (a > 0) \tag{14}$$

for (θ, z) in

$$D = \{(\theta, z) \,|\, 0 \le \theta \le 2\pi, |z| \le 1\} \tag{15}$$

[Figure 3.2(b)] is the part of the cylinder $x^2 + y^2 = a^2$ bounded by the planes

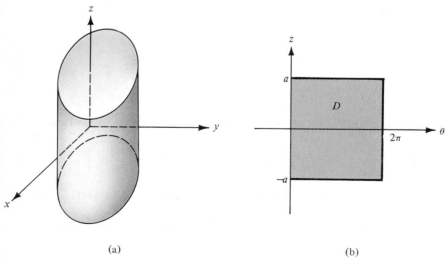

(a)   (b)

FIGURE 3.2

$z = 1$ and $z = -1$ [Figure 3.2(a)]. Here

$$\mathbf{N}(\theta, z) = \begin{vmatrix} \mathbf{i} & \mathbf{j} & \mathbf{k} \\ -a \sin \theta & a \cos \theta & 0 \\ 0 & 0 & 1 \end{vmatrix} \tag{16}$$

$$= a \cos \theta \mathbf{i} + a \sin \theta \mathbf{j}$$

is nonzero for all (θ, z) in D and

$$\frac{\mathbf{N}(\theta, z)}{|\mathbf{N}(\theta, z)|} = \cos \theta \mathbf{i} + \sin \theta \mathbf{j}$$

is continuous on D, and therefore satisfies requirement (d) of Definition 3.2. It is in the direction of the outward normal to the cylinder. If (θ_1, z_1) and (θ_2, z_2) are distinct points of D such that $\mathbf{\Phi}(\theta_1, z_1) = \mathbf{\Phi}(\theta_2, z_2)$, then $z_1 = z_2$ and $|\theta_1 - \theta_2| = 2\pi$, so that (θ_1, z_1) and (θ_2, z_2) are in ∂D; hence requirement (b) is also satisfied. Therefore, (14) is a smooth parametric function from $\mathcal{R}^2$ to $\mathcal{R}^3$, and S is a smooth surface.

Example 3.4 The sphere S,

$$x^2 + y^2 + z^2 = a^2$$

of radius a centered at the origin [Figure 3.3(a)] can be represented by the continuously differentiable function

$$\mathbf{X} = \mathbf{\Psi}(\theta, \phi) = a \cos \theta \cos \phi \mathbf{i} + a \sin \theta \cos \phi \mathbf{j} + a \sin \phi \mathbf{k}$$

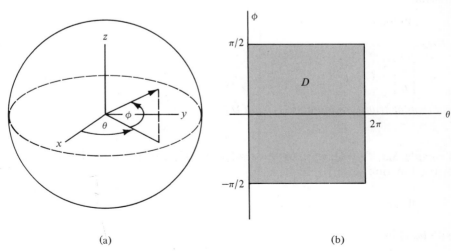

(a)   (b)

FIGURE 3.3

for (θ, ϕ) in

$$D = \{(\theta, \phi)|0 \leq \theta \leq 2\pi, |\phi| \leq \pi/2\} \tag{17}$$

[Figure 3.3(b)]. Here

$$\mathbf{N}(\theta, \phi) = \begin{vmatrix} \mathbf{i} & \mathbf{j} & \mathbf{k} \\ -a \sin \theta \cos \phi & a \cos \theta \cos \phi & 0 \\ -a \cos \theta \sin \phi & -a \sin \theta \sin \phi & a \cos \phi \end{vmatrix}$$

$$= a^2 \cos \theta \cos^2 \phi \mathbf{i} + a^2 \sin \theta \cos^2 \phi \mathbf{j} + a^2 \sin \phi \cos \phi \mathbf{k} \tag{18}$$

$$= (a \cos \phi) \mathbf{X}$$

is nonzero in D^0 and

$$\frac{\mathbf{N}(\theta, \phi)}{|\mathbf{N}(\theta, \phi)|} = \cos \theta \cos \phi \mathbf{i} + \sin \theta \cos \phi \mathbf{j} + \sin \phi \mathbf{k}$$

is continuous on D. If (θ_1, ϕ_1) and (θ_2, ϕ_2) are distinct points in D such that $\mathbf{\Phi}(\theta_1, \phi_1) = \mathbf{\Phi}(\theta_2, \phi_2)$, then $\phi_1 = \phi_2$ and $|\theta_1 - \theta_2| = 2\pi$, so that (θ_1, ϕ_1) and (θ_2, ϕ_2) are in ∂D; hence, requirement (b) of Definition 3.2 is also satisfied. Again we see that S is a smooth surface.

Different smooth parametric functions may represent the same smooth surface, as in the next definition.

Definition 3.3   *Two smooth parametric functions* $\mathbf{\Phi}: D_1 \to \mathscr{R}^3$ *and* $\mathbf{\Psi}: D_2 \to \mathscr{R}^3$ *are said to be equivalent if they are related by*

$$\mathbf{\Psi}(\xi(u, v), \eta(u, v)) = \mathbf{\Phi}(u, v), \tag{19}$$

where

$$\begin{bmatrix} \xi \\ \eta \end{bmatrix} = \mathbf{G}(u, v) = \begin{bmatrix} \xi(u, v) \\ \eta(u, v) \end{bmatrix}$$

is a regular transformation of D_1 *onto* D_2.

Example 3.5   We leave it to the reader (*Exercise 6*) to show that the parametric function

$$\mathbf{\Phi}_1(\alpha, \zeta) = a \cos 2\alpha \mathbf{i} + a \sin 2\alpha \mathbf{j} + \frac{\zeta}{2} \mathbf{k},$$

with (α, ζ) in

$$D_1 = \{(\alpha, \zeta)|0 \leq \alpha \leq \pi, |\zeta| \leq 2\}$$

is equivalent to $\boldsymbol{\Phi}$ of Example 3.3, and that the parametric function

$$\boldsymbol{\Psi}_1(\alpha, \beta) = a \cos 2\alpha \cos \frac{\beta}{2} \mathbf{i} + a \sin 2\alpha \cos \frac{\beta}{2} \mathbf{j} + a \sin \frac{\beta}{2} \mathbf{k}$$

with (α, β) in

$$D_1 = \{(\alpha, \beta) \,|\, 0 \le \alpha \le \pi, |\beta| \le \pi\}$$

is equivalent to $\boldsymbol{\Psi}$ of Example 3.4.

Surface area

Earlier we defined the arc length of a rectifiable curve C to be the least upper bound of the lengths of polygonal paths inscribed in C, and then derived the formula in Eq. (20) of Section 8.1 for the case where C is smooth. There is no analogous elementary approach to the definition of surface area; consequently, we simply define the area of a piecewise smooth surface by means of a certain integral. The definition is motivated by intuitive arguments and reduces to the usual definition of area for surfaces that lie in a plane.

To ease the "discovery" of our definition, we assume first that S is smooth and the region D of Definition 3.2 is a rectangle $D = [a, b] \times [c, d]$. Let $\mathbf{P} = P_1 \times P_2$ be the partition of D determined by

$$P_1: a = u_0 < u_1 < \cdots < u_m = b \text{ and } P_2: c = v_0 < v_1 < \cdots < v_n = d,$$

and define

$$D_{ij} = [u_{i-1}, u_i] \times [v_{j-1}, v_j], \qquad 1 \le i \le m, 1 \le j \le n.$$

Let S_{ij} be the part of S given by

$$\mathbf{X} = \boldsymbol{\Phi}(u, v), \qquad (u, v) \in D_{ij}.$$

Clearly, for any sensible definition of surface area $A(S)$,

$$A(S) = \sum_{i=1}^m \sum_{j=1}^n A(S_{ij}). \tag{20}$$

Now let $\mathbf{U}_{ij} = (u_{ij}, v_{ij})$ be an arbitrary point in D_{ij}, define $\mathbf{X}_{ij} = \boldsymbol{\Phi}(\mathbf{U}_{ij})$, and let T_{ij} be the part of the tangent plane to S at $\mathbf{X}_{ij}$ given parametrically by

$$\mathbf{X} = \boldsymbol{\Psi}_{ij}(u, v) = \mathbf{X}_{ij} + (u - u_{ij})\frac{\partial \boldsymbol{\Phi}(\mathbf{U}_{ij})}{\partial u} + (v - v_{ij})\frac{\partial \boldsymbol{\Phi}(\mathbf{U}_{ij})}{\partial v}, (u, v) \in D_{ij}$$

(Figure 3.4). Since $\boldsymbol{\Psi}_{ij}$ approximates $\boldsymbol{\Phi}$ so well near $\mathbf{U}_{ij}$ [see (13)], it is reasonable that

$$A(S_{ij}) \cong A(T_{ij}) \tag{21}$$

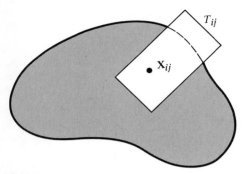

FIGURE 3.4

for a sensible definition of surface area, provided $\|\mathbf{P}\|$ is sufficiently small. To obtain the vertices of T_{ij}, we set (u, v) equal to (u_{i-1}, v_{j-1}), (u_i, v_{j-1}), (u_{i-1}, v_j), and (u_i, v_j); this yields

$$\mathbf{X}_{0,ij} = \mathbf{X}_{ij} + (u_{i-1} - u_{ij})\frac{\partial \mathbf{\Phi}(\mathbf{U}_{ij})}{\partial u} + (v_{j-1} - v_{ij})\frac{\partial \mathbf{\Phi}(\mathbf{U}_{ij})}{\partial v},$$

$$\mathbf{X}_{1,ij} = \mathbf{X}_{ij} + (u_i - u_{ij})\frac{\partial \mathbf{\Phi}(\mathbf{U}_{ij})}{\partial u} + (v_{j-1} - v_{ij})\frac{\partial \mathbf{\Phi}(\mathbf{U}_{ij})}{\partial v},$$

$$\mathbf{X}_{2,ij} = \mathbf{X}_{ij} + (u_{i-1} - u_{ij})\frac{\partial \mathbf{\Phi}(\mathbf{U}_{ij})}{\partial u} + (v_j - v_{ij})\frac{\partial \mathbf{\Phi}(\mathbf{U}_{ij})}{\partial v},$$

$$\mathbf{X}_{3,ij} = \mathbf{X}_{ij} + (u_i - u_{ij})\frac{\partial \mathbf{\Phi}(\mathbf{U}_{ij})}{\partial u} + (v_j - v_{ij})\frac{\partial \mathbf{\Phi}(\mathbf{U}_{ij})}{\partial v}.$$

From Example 1.3,

$$A(T_{ij}) = \left|(\mathbf{X}_{1,ij} - \mathbf{X}_{0,ij}) \times (\mathbf{X}_{2,ij} - \mathbf{X}_{0,ij})\right|,$$

which can be written as

$$A(T_{ij}) = \left|\frac{\partial \mathbf{\Phi}(\mathbf{U}_{ij})}{\partial u} \times \frac{\partial \mathbf{\Phi}(\mathbf{U}_{ij})}{\partial v}\right|(u_i - u_{i-1})(v_j - v_{j-1}).$$

(Verify.) Now (20) and (21) suggest that

$$A(S) \cong \sum_{i=1}^{m} \sum_{j=1}^{n} \left|\frac{\partial \mathbf{\Phi}(\mathbf{U}_{ij})}{\partial u} \times \frac{\partial \mathbf{\Phi}(\mathbf{U}_{ij})}{\partial v}\right|(u_i - u_{i-1})(v_j - v_{j-1})$$

if $\|\mathbf{P}\|$ is sufficiently small. Since the right side is a Riemann sum for an easily recognizable integral, this suggests the following definition.

Definition 3.4 *The surface area of the smooth surface of Definition 3.2 is*

$$A(S) = \int_D |\mathbf{N}(u, v)| \, d(u, v) = \int_D \left| \frac{\partial \mathbf{\Phi}(u, v)}{\partial u} \times \frac{\partial \mathbf{\Phi}(u, v)}{\partial v} \right| d(u, v).$$

Example 3.6 From (16) the surface area of the cylinder of Example 3.3 is

$$A = \int_D \sqrt{a^2 \cos^2 \theta + a^2 \sin^2 \theta} \, d(\theta, z) = a \int_D d(\theta, z),$$

with D as defined in (15); thus,

$$A = a \int_0^{2\pi} d\theta \int_{-1}^1 dz = a(2\pi)(2) = 4\pi a.$$

This is easily checked: cutting the cylinder along a line parallel to its axis and flattening it out yields a rectangle with sides of length 2 and $2\pi a$, and therefore with area $4\pi a$ (Figure 3.5).

Example 3.7 From (18), the surface area of the sphere of Example 3.4 is

$$A = \int_D [a^4 \cos^2 \theta \cos^4 \phi + a^4 \sin^2 \theta \cos^4 \phi + a^4 \sin^2 \phi \cos^2 \phi]^{1/2} \, d(\theta, \phi)$$

$$= \int_D a^2 \cos \phi \, d(\theta, \phi)$$

with D as in (17); thus,

$$A = a^2 \int_0^{2\pi} d\theta \int_{-\pi/2}^{\pi/2} \cos \phi \, d\phi = a^2(2\pi)(2) = 4\pi a^2.$$

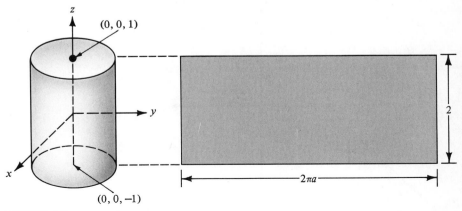

FIGURE 3.5

Example 3.8   If S is defined explicitly by

$$z = f(x, y), \qquad (x, y) \in D,$$

then it can be represented parametrically as

$$\mathbf{X} = \mathbf{X}(x, y) = x\mathbf{i} + y\mathbf{j} + f(x, y)\mathbf{k},$$

with

$$\frac{\partial \mathbf{X}}{\partial x} \times \frac{\partial \mathbf{X}}{\partial y} = \begin{vmatrix} \mathbf{i} & \mathbf{j} & \mathbf{k} \\ 1 & 0 & f_x \\ 0 & 1 & f_y \end{vmatrix}$$

$$= -f_x\mathbf{i} - f_y\mathbf{j} + \mathbf{k}.$$

The area of the surface is

$$A = \int_D \left| \frac{\partial \mathbf{X}}{\partial x} \times \frac{\partial \mathbf{X}}{\partial y} \right| d(x, y) = \int_D \sqrt{1 + f_x^2 + f_y^2}\, d(x, y). \tag{22}$$

Example 3.9   Let S be the part of the paraboloid

$$z = 1 - x^2 - y^2$$

on or above the xy plane (Figure 3.6). Taking

$$f(x, y) = 1 - x^2 - y^2$$

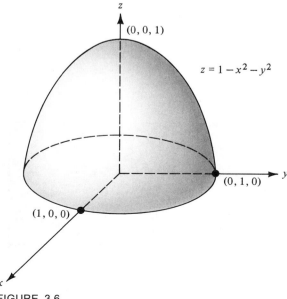

FIGURE 3.6

yields

$$f_x(x, y) = -2x, \qquad f_y(x, y) = -2y,$$

so (22) implies that

$$A = \int_D \sqrt{1 + 4x^2 + 4y^2} \, d(x, y),$$

where

$$D = \{(x, y) \mid x^2 + y^2 \le 1\}.$$

Changing to polar coordinates yields

$$A = \int_0^{2\pi} d\theta \int_0^1 \sqrt{1 + 4r^2} \, r \, dr$$

$$= 2\pi \left[\frac{1}{12} (1 + 4r^2)^{3/2} \Big|_0^1 \right] = \frac{\pi}{6} (5^{3/2} - 1).$$

For Definition 3.4 to make sense, the area of a smooth surface must be independent of which of the infinitely many equivalent smooth parametric functions is chosen to represent it. The next theorem shows that this is so.

Theorem 3.1 *Let $\Phi: D_1 \to \mathscr{R}^2$ and $\Psi: D_2 \to \mathscr{R}^2$ be equivalent smooth parametric functions. Then*

$$\int_{D_1} \left| \frac{\partial \Phi(u, v)}{\partial u} \times \frac{\partial \Phi(u, v)}{\partial v} \right| d(u, v) = \int_{D_2} \left| \frac{\partial \Psi(\xi, \eta)}{\partial \xi} \times \frac{\partial \Psi(\xi, \eta)}{\partial \eta} \right| d(\xi, \eta). \quad (23)$$

Proof. From (19) and the chain rule,

$$\frac{\partial \Phi}{\partial u} = \frac{\partial \xi}{\partial u} \frac{\partial \Psi}{\partial \xi} + \frac{\partial \eta}{\partial u} \frac{\partial \Psi}{\partial \eta} \quad \text{and} \quad \frac{\partial \Phi}{\partial v} = \frac{\partial \xi}{\partial v} \frac{\partial \Psi}{\partial \xi} + \frac{\partial \eta}{\partial v} \frac{\partial \Psi}{\partial \eta}, \quad (24)$$

where the partial derivatives of Φ, ξ, and η are evaluated at (u, v), and those of Ψ are evaluated at $(\xi(u, v), \eta(u, v))$. Using (24) and the relations

$$\frac{\partial \Psi}{\partial \xi} \times \frac{\partial \Psi}{\partial \xi} = \frac{\partial \Psi}{\partial \eta} \times \frac{\partial \Psi}{\partial \eta} = 0$$

and

$$\frac{\partial \Psi}{\partial \eta} \times \frac{\partial \Psi}{\partial \xi} = -\frac{\partial \Psi}{\partial \xi} \times \frac{\partial \Psi}{\partial \eta}$$

yields

$$\frac{\partial \Phi}{\partial u} \times \frac{\partial \Phi}{\partial v} = \frac{\partial(\xi, \eta)}{\partial(u, v)} \left(\frac{\partial \Psi}{\partial \xi} \times \frac{\partial \Psi}{\partial \eta} \right).$$

(Verify.) Therefore,

$$\int_{D_1} \left| \frac{\partial \Phi}{\partial u} \times \frac{\partial \Phi}{\partial v} \right| d(u, v) = \int_{D_1} \left| \frac{\partial \Psi}{\partial \xi} \times \frac{\partial \Psi}{\partial \eta} \right| \left| \frac{\partial(\xi, \eta)}{\partial(u, v)} \right| d(u, v)$$

$$= \int_{D_2} \left| \frac{\partial \Psi}{\partial \xi} \times \frac{\partial \Psi}{\partial \eta} \right| d(\xi, \eta),$$

where the last equality follows from Theorem 3.4, Section 7.3. This proves (23).

Boundary of a surface

Even without a formal definition the idea of the boundary of a smooth surface is intuitively clear. Thus, the boundary of the hemisphere

$$\{(x, y, z) \,|\, x^2 + y^2 + z^2 = a^2, z \geq 0\} \tag{25}$$

is the circle

$$\{(x, y, z) \,|\, x^2 + y^2 = a, z = 0\}, \tag{26}$$

while the boundary of the cylinder

$$\{(x, y, z) \,|\, x^2 + y^2 = a^2, |z| \leq 1\} \tag{27}$$

(Figure 3.2) consists of the circles

$$\{(x, y, z) \,|\, x^2 + y^2 = a^2, z = 1\}$$

and

$$\{(x, y, z) \,|\, x^2 + y^2 = a^2, z = -1\}. \tag{28}$$

The sphere

$$\{(x, y, z) \,|\, x^2 + y^2 + z^2 = a^2\} \tag{29}$$

[Figure 3.3(a)] has no boundary.

Since a precise definition of the boundary of a surface is surprisingly difficult to formulate and deal with, our approach to problems related to this question will be considerably less rigorous than heretofore.

In the examples we have just considered it is clear that the meaning of "boundary" in "boundary of a surface" differs from its meaning in "boundary of a set" (Definition 3.3, Section 1.3), according to which every point on each of the surfaces (25), (27), and (29) is a boundary point. (Why?) Although it would seem appropriate to avoid this inconsistency by using another term for the boundary of a surface, this is not the common practice.

The following definition describes intuitively what we mean by the boundary of a smooth surface.

Definition 3.5 *The boundary of a smooth surface S, denoted by ∂S, is the set of points $\mathbf{X}_0$ on S for which all sets of the form*

$$U_\varepsilon(\mathbf{X}_0) = \{\mathbf{X} \,\big|\, |\mathbf{X} - \mathbf{X}_0| < \varepsilon \text{ and } \mathbf{X} \notin S\} \qquad (\varepsilon > 0)$$

are connected.

It is easy to see that the boundary points (26) of the hemisphere (25) and those of the cylinder (27) satisfy this definition, and that if $\mathbf{X}_0$ is any other point on either of these surfaces, then $U_\varepsilon(\mathbf{X}_0)$ is disconnected for sufficiently small ε. Since we have shown that if $\mathbf{X}_0 = \mathbf{\Phi}(u_0, v_0)$ with $(u_0, v_0) \in D^0$, then S can be represented in one of the forms (1) near $\mathbf{X}_0$, it follows that every point on ∂S must be of the form $\mathbf{X}_0 = \mathbf{\Phi}(u_0, v_0)$ with $(u_0, v_0) \in \partial D$ (*Exercise 15*); however, a point on ∂D need not map onto ∂S. Thus, in Example 3.3, points of the form $(0, z)$ and $(2\pi, z)$ are in ∂D if $|z| \le 1$, but $\mathbf{\Phi}(0, z)$ and $\mathbf{\Phi}(2\pi, z)$ are not boundary points of the cylinder in Figure 3.2(a) if $|z| < 1$. The situation is more extreme in Example 3.4, where no point on ∂D corresponds to a boundary point of the sphere, which has no boundary.

We will say that a surface without boundary, such as the sphere, is *closed*. (This is not standard terminology, but it is adequate for the level of rigor that we are maintaining.) It is intuitively clear—but not easy to prove—that the complement of a closed surface consists of two open sets, one bounded and one unbounded. The former is the *inside* and the latter the *outside* of the surface. A normal vector to a closed surface S is an *outward* normal to S if it points toward the outside of S, or an *inward* normal if it points toward the inside. Thus,

$$\mathbf{n}(\theta, \phi) = \cos\theta \cos\phi\mathbf{i} + \sin\theta \cos\phi\mathbf{j} + \sin\phi\mathbf{k}$$

is a unit outward normal to the sphere of Example 3.4.

We will consider only closed surfaces or surfaces with boundaries consisting of finitely many piecewise smooth simple closed curves. From what we have seen, if S is represented by $\mathbf{\Phi}: D \to \mathscr{R}^3$, then these curves are the images under $\mathbf{\Phi}$ of some—but not necessarily all—of the piecewise smooth curves that make up ∂D.

Piecewise smooth and orientable surfaces

If $S_1, S_2, \ldots, S_k$ are smooth surfaces and each of the boundaries $\partial S_1, \partial S_2, \ldots, \partial S_k$ intersects at least one of the others in a piecewise smooth

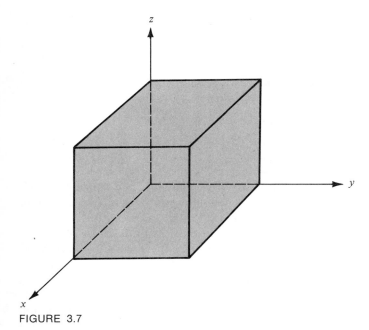

FIGURE 3.7

curve, while no two of the surfaces have nonboundary points in common, we say that $S_1, S_2, \ldots, S_k$ form a *piecewise smooth surface* with *smooth sections* $S_1, S_2, \ldots, S_k$, and we write

$$S = S_1 + S_2 + \cdots + S_k.$$

Example 3.10   The surface of the cube $D = [0, 1] \times [0, 1] \times [0, 1]$ (Figure 3.7) is piecewise smooth, with six smooth sections joined along the edges of the cube. This surface is closed and has a unique outward unit normal at each of its points except those on its edges.

Example 3.11   The surface comprised of the hemisphere

$$x^2 + y^2 + z^2 = 1, \qquad z \le 0,$$

and the cylinder

$$x^2 + y^2 = 1, \qquad 0 \le z \le 1$$

(Figure 3.8), is piecewise smooth and not closed.

We define the surface area of a piecewise smooth surface

$$S = S_1 + S_2 + \cdots + S_k$$

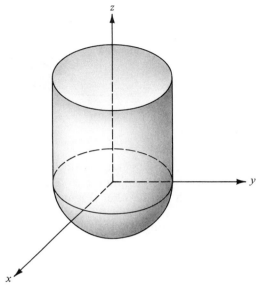

FIGURE 3.8

as

$$A(S) = A(S_1) + A(S_2) + \cdots + A(S_k).$$

Definition 3.6   *A smooth surface S is said to be orientable if it has a representation* $\Phi \colon D \to \mathscr{R}^3$ *such that if*

$$\mathbf{N}(u, v) = \frac{\partial \mathbf{\Phi}(u, v)}{\partial u} \times \frac{\partial \mathbf{\Phi}(u, v)}{\partial v},$$

then

$$\lim_{(u,v) \to (u_0,v_0)} \frac{\mathbf{N}(u, v)}{\left| \mathbf{N}(u, v) \right|} = \lim_{(u,v) \to (u_1,v_1)} \frac{\mathbf{N}(u, v)}{\left| \mathbf{N}(u, v) \right|} \tag{30}$$

whenever (u_0, v_0) *and* (u_1, v_1) *are points on* ∂D *such that*

$$\mathbf{\Phi}(u_0, v_0) = \mathbf{\Phi}(u_1, v_1). \tag{31}$$

Example 3.12   If $\mathbf{\Phi}$ is one-to-one on D, then S is orientable, since then (31) implies that $(u_0, v_0) = (u_1, v_1)$, and (30) surely holds. The cylinder and sphere of Examples 3.3 and 3.4 are orientable.

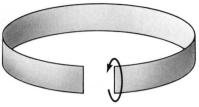
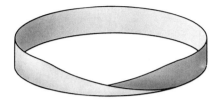

FIGURE 3.9

Example 3.13 Perhaps the simplest and most famous nonorientable surface is the *Möbius strip*, which can be constructed by subjecting one end of a narrow strip of paper to a half-twist and pasting it to the other (Figure 3.9). It is intuitively clear that a unit normal vector moved continuously around the strip from a given point X_0 back to X_0 will end up pointing in the direction opposite to that with which it started (*Exercise 17*).

If S is orientable and $\boldsymbol{\Phi} \colon D \to \mathscr{R}^3$ represents S and satisfies Definition 3.6, then we can define a vector field $\mathbf{n} = \mathbf{n}(X)$ on S as follows: If $X = \boldsymbol{\Phi}(u, v)$, where $(u, v) \in D^0$, and

$$\mathbf{N}(u, v) = \frac{\partial \boldsymbol{\Phi}(u, v)}{\partial u} \times \frac{\partial \boldsymbol{\Phi}(u, v)}{\partial v},$$

then

$$\mathbf{n}(X) = \frac{\mathbf{N}(u, v)}{|\mathbf{N}(u, v)|},$$

while if $X = \boldsymbol{\Phi}(u_0, v_0)$ for some (u_0, v_0) in ∂D, then

$$\mathbf{n}(X) = \lim_{(u,v) \to (u_0, v_0)} \frac{\mathbf{N}(u, v)}{|\mathbf{N}(u, v)|}.$$

From (30), $\mathbf{n}(X)$ is uniquely defined even if X is the image of more than one point in ∂D.

The vector field $\mathbf{n}$ is called the *orientation of S induced by $\boldsymbol{\Phi}$*. Clearly, $\mathbf{n}(X)$ is a unit vector for each X on S, and we have shown that it is normal to S at X if $X \in \boldsymbol{\Phi}(D^0)$; we *define* the normal to S to be in the direction of $\mathbf{n}(X)$ if $X \in \boldsymbol{\Phi}(\partial D)$. The orientation can be visualized by drawing an arrow of unit length in the direction of $\mathbf{n}(X)$ at each point X on S.

Although an orientable surface has infinitely many parametric representations satisfying Definition 3.6, it can have only two orientations. We leave the proof of the following theorem to the reader (*Exercise 18*).

Theorem 3.2   *Let* $\Phi: D_1 \to \mathscr{R}^3$ *and* $\Psi: D_2 \to \mathscr{R}^3$ *be equivalent parametric representations of S which satisfy Definition 3.6, and suppose they are related by* (19). *Let* $\mathbf{n}_1$ *and* $\mathbf{n}_2$ *be the orientations induced on S by* Φ *and* Ψ, *respectively. Then* $\mathbf{n}_1 = \mathbf{n}_2$ *if*

$$\frac{\partial(\xi, \eta)}{\partial(u, v)} > 0$$

and $\mathbf{n}_1 = -\mathbf{n}_2$ *if*

$$\frac{\partial(\xi, \eta)}{\partial(u, v)} < 0.$$

To discuss orientability of a piecewise smooth surface, we must introduce the notion of a *positively directed* boundary of a smooth surface. If $\mathbf{n}$ is an orientation of a smooth surface S and C is a smooth curve in ∂S, we say that C is *positively directed with respect to* $\mathbf{n}$ if a person walking in the direction of C with his head in the direction of $\mathbf{n}(\mathbf{X})$ always has S to his left (Figure 3.10). We say that ∂S is positively directed with respect to $\mathbf{n}$ if every smooth curve in ∂S is.

Example 3.14   If S is the section of the paraboloid

$$z = 4 - x^2 - y^2$$

bounded by the planes $z = 0$ and $z = 1$, then ∂S traversed in the direction shown in Figure 3.11(a) is positively directed with respect to the indicated orientation of S. The same is true in Figure 3.11(b). In Figure 3.11(c), C_1 is positively directed with respect to $\mathbf{n}$, while C_2 is not.

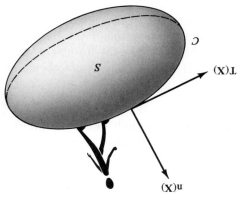

FIGURE 3.10

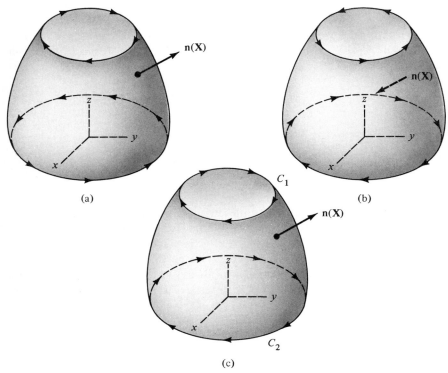

FIGURE 3.11

Definition 3.7 *A piecewise smooth surface*

$$S = S_1 + S_2 + \cdots + S_k$$

is said to be orientable if its smooth sections $S_1, S_2, \ldots, S_k$ are orientable and it is possible to choose orientations $\mathbf{n}_1, \mathbf{n}_2, \ldots, \mathbf{n}_k$ on them so that if C is any curve common to ∂S_i and ∂S_j $(i \neq j)$, then the positive direction of C with respect to $\mathbf{n}_i$ is its negative direction with respect to $\mathbf{n}_j$. If the orientations are chosen in this way, then S is said to be oriented.

An orientable piecewise smooth surface has exactly two orientations, since the orientation chosen for one smooth section determines the orientations of the rest. We take this without proof.

Example 3.15 The piecewise smooth surface in Figure 3.12 is oriented. The positive direction of the common boundary curve with respect to $\mathbf{n}_1$ is the negative direction with respect to $\mathbf{n}_2$, and vice versa.

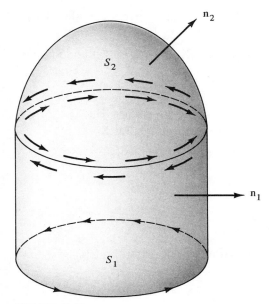

FIGURE 3.12

8.3 EXERCISES

1. Find the equation of the tangent plane to the surface at the given point:
   (a) $z = \sqrt{1 - x^2 - y^2}$; $\mathbf{X}_0 = (\frac{1}{2}, -\frac{1}{2}, 1/\sqrt{2})$
   (b) $z = 1 + 4x^2 - y^2$; $\mathbf{X}_0 = (1, 1, 4)$
   (c) $x = 9 - y^2$; $\mathbf{X}_0 = (5, 2, 4)$
   (d) $y = z^2 + 2x^2$; $\mathbf{X}_0 = (-2, 9, 1)$
   (e) $x^2 + y^2 + z^2 = 1$; $\mathbf{X}_0 = (1/\sqrt{3}, -1/\sqrt{6}, 1/\sqrt{2})$
   (f) $x^2 + y^2 + e^{z-2} = 3$; $\mathbf{X}_0 = (1, -1, 2)$
   (g) $y + z + \log(1 + y^2 + z^2) + x^2 + 2x = 0$; $\mathbf{X}_0 = (0, 0, 0)$

2. (a) Suppose F is continuously differentiable near $\mathbf{X}_0 = (x_0, y_0, z_0)$ and
   $F_y(\mathbf{X}_0) \neq 0$, so that the equation $F(x, y, z) = 0$ defines $y = y(x, z)$ near
   (x_0, y_0) (Corollary 4.1, Section 6.4). Show that if

   $$T_1(x, z) = y_0 - \frac{F_x(\mathbf{X}_0)}{F_y(\mathbf{X}_0)}(x - x_0) - \frac{F_z(\mathbf{X}_0)}{F_y(\mathbf{X}_0)}(y - y_0),$$

   then

   $$\lim_{(x,z) \to (x_0, z_0)} \frac{T_1(x, z) - y(x, z)}{\sqrt{(x - x_0)^2 + (z - z_0)^2}} = 0.$$

(b) Formulate a conclusion analogous to that of part (a) under the assumption that $F_x(\mathbf{X}_0) \neq 0$.

3. Let F be continuously differentiable near $\mathbf{X}_0 = (x_0, y_0, z_0)$ and suppose the smooth curve C given by

$$\mathbf{X} = x(t)\mathbf{i} + y(t)\mathbf{j} + z(t)\mathbf{k}, \qquad a \leq t \leq b,$$

is on the surface defined by $F(x, y, z) = 0$. Let $\mathbf{X}_0 = \mathbf{X}(t_0)$, with $a < t_0 < b$. Show that $\nabla F(\mathbf{X}_0)$ is perpendicular to the tangent to C at $\mathbf{X}_0$.

4. Find smooth parametric representations for the following surfaces.
   (a) $4x^2 + 9(y - 1)^2 + 36z^2 = 36$
   (b) $x + y + z = 1, x \geq 0, y \geq 0, z \geq 0$
   (c) $x^2 + y^2 + z^2 = 1, -1/\sqrt{2} \leq z \leq 1/\sqrt{2}$
   (d) $x + y + z = 2, x^2 + y^2 \leq 4$
   (e) $x^2 + y^2 = z^2, 1 \leq z \leq 2$

5. Suppose $\boldsymbol{\Phi}: D \to \mathscr{R}^3$ is a smooth parametric function and $(u_0, v_0) \in D^0$. Show that

$$\lim_{(u,v) \to (u_0,v_0)} \frac{\boldsymbol{\Phi}(u, v) - \boldsymbol{\Phi}(u_0, v_0) - (u - u_0)\mathbf{A} - (v - v_0)\mathbf{B}}{\sqrt{(u - u_0)^2 + (v - v_0)^2}} = 0$$

if and only if

$$\mathbf{A} = \frac{\partial \boldsymbol{\Phi}(u_0, v_0)}{\partial u} \quad \text{and} \quad \mathbf{B} = \frac{\partial \boldsymbol{\Phi}(u_0, v_0)}{\partial v}.$$

6. (a) Show that $\boldsymbol{\Phi}_1$ of Example 3.5 is equivalent to $\boldsymbol{\Phi}$ of Example 3.3.
   (b) Show that $\boldsymbol{\Psi}_1$ of Example 3.5 is equivalent to $\boldsymbol{\Psi}$ of Example 3.4.

7. Find the area of the given surface:
   (a) $z^2 = x^2 + y^2, 0 < R_1 \leq z \leq R_2$
   (b) $\mathbf{X} = r \cos \theta \mathbf{i} + r \sin \theta \mathbf{j} + \theta \mathbf{k}, 0 \leq r \leq 1, 0 \leq \theta \leq 2\pi$
   (c) $x + y - 2z = 0, 0 \leq x + y \leq 4, 2 \leq x - y \leq 6$
   (d) $x^2 + y^2 + z^2 = a^2, |z| \leq \rho \ (0 < \rho < a)$
   (e) $ax + by + z = d, x^2 + y^2 \leq \rho^2$
   (f) $x^2 + y^2 = 2x, 0 \leq z \leq \sqrt{x^2 + y^2}$
   (g) $x^2 + y^2 + z^2 = 1, z \geq \sqrt{x^2 + y^2}$
   (h) $x^2 + y^2 = 9, 2x \leq z \leq 3x + 7$
   (i) The part of the cylinder $x^2 + y^2 = a^2$ inside the cylinder $x^2 + z^2 = a^2$
   (j) The part of the cylinder $x^2 + y^2 = ax$ inside the sphere $x^2 + y^2 + z^2 = a^2$

8. Suppose $ad - bc \neq 0$. Find the area of the part of the plane

$$\alpha x + \beta y + z = \rho$$

for which

$$u_1 \leq ax + by \leq u_2, \qquad v_1 \leq cx + dy \leq v_2.$$

9. Find the surface area of the part of the cylinder $x^2 + y^2 = \rho^2$ bounded above by the plane

$$z = A_2 x + B_2 y + C_2$$

and below by the plane

$$z = A_1 x + B_1 y + C_1.$$

Assume that

$$C_2 - C_1 \geq \rho[(A_2 - A_1)^2 + (B_2 - B_1)^2]^{1/2}.$$

10. If $0 < b < a$, then the surface defined by

$$\mathbf{X} = (a + b \cos \phi) \cos \theta \mathbf{i} + (a + b \cos \phi) \sin \theta \mathbf{j} + b \sin \phi \mathbf{k},$$

where $0 \leq \theta \leq 2\pi$, $0 \leq \phi \leq 2\pi$, is a *torus*. Sketch it and find its surface area.

11. Let the smooth surface S be defined by $z = f(x, y)$, $(x, y) \in D$. Show that if $\gamma(x, y)$ is the acute angle between the z axis and the normal to S at $(x, y, f(x, y))$, then

$$A(S) = \int_D \sec \gamma(x, y)\, d(x, y).$$

(*Hint:* See Example 3.8.)

12. Suppose the smooth surface S is given by $\mathbf{X} = \mathbf{X}(u, v)$, $(u, v) \in D$, and define

$$E = \left| \frac{\partial \mathbf{X}}{\partial u} \right|^2, \qquad F = \frac{\partial \mathbf{X}}{\partial u} \cdot \frac{\partial \mathbf{X}}{\partial v}, \qquad G = \left| \frac{\partial \mathbf{X}}{\partial v} \right|^2.$$

Show that

$$A(S) = \int_D \sqrt{EG - F^2}\, d(x, y).$$

13. Prove: If the smooth surface S defined by $\mathbf{X} = \mathbf{X}(u, v)$, $(u, v) \in D$, lies in the xy plane, then Definition 3.4 yields

$$A(S) = \int_S d(x, y),$$

consistent with our previous definition of the area of a region in the xy plane.

14. Let $f = f(x)$ be positive and continuously differentiable on $[a, b]$. Show that the area of the surface S obtained by rotating the curve

$$y = f(x), \qquad a \leq x \leq b,$$

through a complete revolution about the x axis is

$$A(S) = 2\pi \int_a^b f(x)\sqrt{1 + [f'(x)]^2}\, dx.$$

(This is the *theorem of Pappus.*)

15. Suppose $\mathbf{X} = \mathbf{X}(u, v)$, $(u, v) \in D$, defines a smooth surface, and let $\mathbf{X}_0 = \mathbf{X}(u_0, v_0)$, where $(u_0, v_0) \in D^0$. Use the fact (established earlier) that S can be represented in one of the explicit forms of Eq. (1) near $\mathbf{X}_0$ to show that the set $U_\varepsilon(\mathbf{X}_0)$ of Definition 3.5 is disconnected for sufficiently small ε, and so $\mathbf{X}_0$ cannot be a boundary point of S.

16. Find the area of
    (a) the surface of the region

    $$x^2 + y^2 \le 1, \qquad 0 \le z \le 2 - x - y;$$

    (b) the surface of the region obtained by removing from the parallelopiped $[-2, 2] \times [-2, 2] \times [1, 2]$ all (x, y, z) such that

    $$x^2 + y^2 < z \le 2.$$

17. A Möbius strip can be represented by

    $$\mathbf{X} = \mathbf{X}(\theta, v) = \left(1 + v \sin \frac{\theta}{2}\right) \cos \theta \mathbf{i} + \left(1 + v \sin \frac{\theta}{2}\right) \sin \theta \mathbf{j} + v \cos \frac{\theta}{2} \mathbf{k},$$

    $0 \le \theta \le 2\pi$, $|v| \le 1$. Verify that $\mathbf{X}(0, 0) = \mathbf{X}(2\pi, 0)$, but

    $$\lim_{(\theta, v) \to (0, 0)} \frac{\mathbf{N}(\theta, v)}{\mathbf{N}(\theta, v)} = -\lim_{(\theta, v) \to (2\pi, 0)} \frac{\mathbf{N}(\theta, v)}{\mathbf{N}(\theta, v)} = \mathbf{i},$$

    which is consistent with the assertion that the strip is not orientable (Example 3.13).

18. Prove Theorem 3.2.

8.4 SURFACE INTEGRALS

In Section 8.2 we first defined the line integral of a vector field, then defined the line integral of a scalar field, and finally showed how they are related. In this section we consider surface integrals of scalar and vector fields, but here we consider scalar fields first.

Definition 4.1 *Let S be a smooth surface represented by $\boldsymbol{\Phi}: D \to \mathscr{R}^3$ and suppose f is a real-valued function continuous on S. Then the surface integral of f over S is defined by*

$$\int_S f \, dA = \int_D f(\boldsymbol{\Phi}(u, v)) \left| \frac{\partial \boldsymbol{\Phi}(u, v)}{\partial u} \times \frac{\partial \boldsymbol{\Phi}(u, v)}{\partial v} \right| d(u, v).$$

If $S = S_1 + S_2 + \cdots + S_k$ is a piecewise smooth surface, then

$$\int_S f \, dA = \sum_{i=1}^{k} \int_{S_i} f \, dA.$$

A proof like that of Theorem 3.1 shows that $\int_S f\, dA$ is independent of the particular representation chosen for S (*Exercise 1*). Notice that if $f = 1$, $\int_S f\, dA$ reduces to the area of S, according to Definition 3.4.

Example 4.1   Let S be the part of the paraboloid

$$z = 1 - x^2 - y^2$$

above the xy plane (Figure 4.1), and

$$f(x, y, z) = x^2.$$

Parametrizing S as

$$\mathbf{X} = \mathbf{X}(x, y) = x\mathbf{i} + y\mathbf{j} + (1 - x^2 - y^2)\mathbf{k},$$

where (x, y) is in

$$D = \{(x, y)\,|\,x^2 + y^2 \le 1\},$$

yields

$$\left|\frac{\partial \mathbf{X}(x, y)}{\partial x} \times \frac{\partial \mathbf{X}(x, y)}{\partial y}\right| = \sqrt{1 + 4x^2 + 4y^2}$$

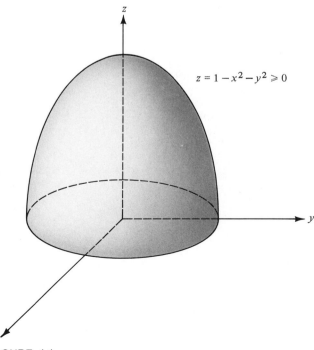

$z = 1 - x^2 - y^2 \ge 0$

FIGURE 4.1

(Examples 3.8 and 3.9), so

$$\int_S x^2 \, dA = \int_D x^2 \sqrt{1 + 4x^2 + 4y^2} \, d(x, y).$$

Changing to polar coordinates yields

$$\int_S x^2 \, dA = \int_0^{2\pi} \cos^2 \theta \, d\theta \int_0^1 r^3 \sqrt{1 + 4r^2} \, dr. \tag{1}$$

The integration with respect to r requires integration by parts:

$$\int_0^1 r^3 \sqrt{1 + 4r^2} \, dr = \int_0^1 r^2 [r(1 + 4r^2)^{1/2}] \, dr$$

$$= \frac{r^2}{12} (1 + 4r^2)^{3/2} \Big|_0^1 - \frac{1}{6} \int_0^1 r(1 + 4r^2)^{3/2} \, dr$$

$$= \left[\frac{r^2}{12} (1 + 4r^2)^{3/2} - \frac{1}{120} (1 + 4r^2)^{5/2} \right] \Big|_0^1$$

$$= \frac{25\sqrt{5} + 1}{120}.$$

Since

$$\int_0^{2\pi} \cos^2 \theta \, d\theta = \frac{1}{2} \int_0^{2\pi} (1 + \cos 2\theta) \, d\theta = \pi,$$

(1) yields

$$\int_S x^2 \, dA = \frac{\pi(25\sqrt{5} + 1)}{120}.$$

Example 4.2   If S consists of the hemisphere

$$S_1 = \{(x, y, z) \,|\, x^2 + y^2 + z^2 = a^2, z \geq 0\}$$

and the disc

$$S_2 = \{(x, y, z) \,|\, x^2 + y^2 = a^2, z = 0\}$$

(Figure 4.2), and

$$f(x, y, z) = x^2 + y^2,$$

then

$$\int_S f \, dA = \int_{S_1} (x^2 + y^2) \, dA + \int_{S_2} (x^2 + y^2) \, dA.$$

Representing S_1 as

$$\mathbf{X} = \mathbf{X}(\theta, \phi) = a \cos \theta \cos \phi \mathbf{i} + a \sin \theta \cos \phi \mathbf{j} + a \sin \phi \mathbf{k}$$

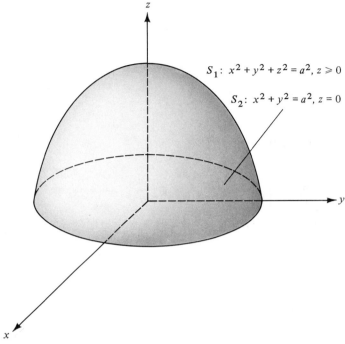

S_1: $x^2 + y^2 + z^2 = a^2$, $z \geqslant 0$

S_2: $x^2 + y^2 = a^2$, $z = 0$

x

FIGURE 4.2

with (θ, ϕ) in

$$D_1 = \{(\theta, \phi)\,|\,0 \leq \theta \leq 2\pi, 0 \leq \phi \leq \pi/2\}$$

yields

$$\left|\frac{\partial \mathbf{X}(\theta, \phi)}{\partial x} \times \frac{\partial \mathbf{X}(\theta, \phi)}{\partial y}\right| = a^2 \cos \phi$$

(see Example 3.4), so

$$\int_{S_1} (x^2 + y^2)\, dA = \int_{D_1} (a^2 \cos^2 \phi)(a^2 \cos \phi)\, d(\theta, \phi)$$

$$= a^4 \int_0^{2\pi} d\theta \int_0^{\pi/2} \cos^3 \phi\, d\phi.$$

Since

$$\int_0^{\pi/2} \cos^3 \phi\, d\phi = \int_0^{\pi/2} (1 - \sin^2 \phi) \cos \phi\, d\phi$$

$$= \left(\sin \phi - \frac{\sin^3 \phi}{3}\right)\Bigg|_0^{\pi/2} = \frac{2}{3},$$

this yields

$$\int_{S_1} (x^2 + y^2) \, dA = a^4 (2\pi) \left(\frac{2}{3} \right) = \frac{4}{3} \pi a^4.$$ (2)

Representing S_2 as

$$\mathbf{X} = \mathbf{X}(r, \alpha) = r \cos \alpha \mathbf{i} + r \sin \alpha \mathbf{j}$$

with (r, α) in

$$D_2 = \{(r, \alpha) | 0 \le r \le a, 0 \le \alpha \le 2\pi\}$$

yields

$$\frac{\partial \mathbf{X}(r, \alpha)}{\partial r} \times \frac{\partial \mathbf{X}(r, \alpha)}{\partial \alpha} = \begin{vmatrix} \mathbf{i} & \mathbf{j} & \mathbf{k} \\ \cos \alpha & \sin \alpha & 0 \\ -r \sin \alpha & r \cos \alpha & 0 \end{vmatrix}$$

$$= r\mathbf{k},$$

so

$$\int_{S_2} (x^2 + y^2) \, dA = \int_{D_2} (r^2) r \, d(r, \alpha) = \int_0^{2\pi} d\alpha \int_0^a r^3 \, dr$$

$$= (2\pi) \left(\frac{a^4}{4} \right) = \frac{\pi a^4}{2}.$$

This and (2) imply that

$$\int_S (x^2 + y^2) \, dA = \frac{4}{3} \pi a^4 + \frac{1}{2} \pi a^4 = \frac{11}{6} \pi a^4.$$

We now use Definition 4.1 to define the surface integral of a vector field.

Definition 4.2 *Let*

$$\mathbf{F} = P\mathbf{i} + Q\mathbf{j} + R\mathbf{k}$$

be continuous on the smooth orientable surface S, and suppose $\mathbf{n}$ *is an orientation of S. Then the surface integral of* $\mathbf{F}$ *over the oriented surface S is defined to be*

$$\int_S \mathbf{F} \cdot \mathbf{n} \, dA.$$

If $\mathbf{n}$ is induced on S by the parametric representation

$$\mathbf{X} = \mathbf{X}(u, v), \qquad (u, v) \in D,$$

then

$$\mathbf{n}(\mathbf{X}(u, v)) = \frac{\dfrac{\partial \mathbf{X}(u, v)}{\partial u} \times \dfrac{\partial \mathbf{X}(u, v)}{\partial v}}{\left| \dfrac{\partial \mathbf{X}(u, v)}{\partial u} \times \dfrac{\partial \mathbf{X}(u, v)}{\partial v} \right|},$$

and Definition 4.1 implies that

$$\int_S \mathbf{F} \cdot \mathbf{n} \, dA = \int_D \mathbf{F}(\mathbf{X}(u, v)) \cdot \left(\frac{\partial \mathbf{X}(u, v)}{\partial u} \times \frac{\partial \mathbf{X}(u, v)}{\partial v} \right) d(u, v). \tag{3}$$

Example 4.3 To integrate

$$\mathbf{F}(x, y, z) = y\mathbf{j} + z\mathbf{k}$$

over the sphere

$$x^2 + y^2 + z^2 = a^2$$

oriented with the outward normal, we parametrize S as

$$\mathbf{X} = \mathbf{X}(\theta, \phi) = a \cos \theta \cos \phi \mathbf{i} + a \sin \theta \cos \phi \mathbf{j} + a \sin \phi \mathbf{k},$$

with (θ, ϕ) in

$$D = \{(\theta, \phi) \,|\, 0 \le \theta \le 2\pi, \, |\phi| \le \pi/2\}.$$

This induces the outward normal, since we saw in Example 3.4 that

$$\frac{\partial \mathbf{X}(\theta, \phi)}{\partial \theta} \times \frac{\partial \mathbf{X}(\theta, \phi)}{\partial \phi} = (a \cos \phi)\mathbf{X}(\theta, \phi);$$

therefore,

$$
\begin{aligned}
\int_S (y\mathbf{j} + z\mathbf{k}) \cdot \mathbf{n}(\mathbf{X}) \, dA &= a \int_D (y\mathbf{j} + z\mathbf{k}) \cdot (x\mathbf{i} + y\mathbf{j} + z\mathbf{k}) \cos \phi \, d(\theta, \phi) \\
&= a \int_D (y^2 + z^2) \cos \phi \, d(\theta, \phi) \\
&= a \int_D (a^2 \sin^2 \theta \cos^2 \phi + a^2 \sin^2 \phi) \cos \phi \, d(\theta, \phi) \\
&= a^3 \int_0^{2\pi} \sin^2 \theta \, d\theta \int_{-\pi/2}^{\pi/2} \cos^3 \phi \, d\phi \\
&\quad + a^3 \int_0^{2\pi} d\theta \int_{-\pi/2}^{\pi/2} \sin^2 \phi \cos \phi \, d\phi \\
&= a^3(\pi)\left(\frac{4}{3}\right) + a^3(2\pi)\left(\frac{2}{3}\right) = \frac{8}{3}\pi a^3.
\end{aligned}
$$

Example 4.4 Let

$$I = \int_S (x\mathbf{i} + y^2\mathbf{j} + z\mathbf{k}) \cdot \mathbf{n}(\mathbf{X}) \, dA,$$

where S is the triangle determined by the plane

$$x + y + z = 1$$

and the coordinate planes, and $\mathbf{n}(\mathbf{X})$ has a positive z component [Figure 4.3(a)]. We parametrize S as

$$\mathbf{X} = x\mathbf{i} + y\mathbf{j} + (1 - x - y)\mathbf{k} \tag{4}$$

with (x, y) in

$$D = \{(x, y) \,|\, 0 \le x + y \le 1, x \ge 0, y \ge 0\}$$

[Figure 4.3(b)]. Then

$$\frac{\partial \mathbf{X}}{\partial x} \times \frac{\partial \mathbf{X}}{\partial y} = \begin{vmatrix} \mathbf{i} & \mathbf{j} & \mathbf{k} \\ 1 & 0 & -1 \\ 0 & 1 & -1 \end{vmatrix}$$

$$= \mathbf{i} + \mathbf{j} + \mathbf{k},$$

which has a positive z component; therefore, (4) induces the required orientation, and

$$I = \int_D [x\mathbf{i} + y^2\mathbf{j} + (1 - x - y)\mathbf{k}] \cdot (\mathbf{i} + \mathbf{j} + \mathbf{k}) \, d(x, y)$$

$$= \int_D (x + y^2 + 1 - x - y) \, d(x, y) = \int_D (y^2 - y + 1) \, d(x, y)$$

$$= \int_0^1 dx \int_0^{1-x} (y^2 - y + 1) \, dy = \int_0^1 \left[\frac{(1 - x)^3}{3} - \frac{(1 - x)^2}{2} + (1 - x) \right] dx$$

$$= \left[-\frac{(1 - x)^4}{12} + \frac{(1 - x)^3}{6} - \frac{(1 - x)^2}{2} \right] \Big|_0^1$$

$$= \frac{1}{12} - \frac{1}{6} + \frac{1}{2} = \frac{5}{12}.$$

If $S = S_1 + S_2 + \cdots + S_k$ is piecewise smooth, where $S_1, S_2, \ldots, S_k$ are smooth orientable surfaces with orientations $\mathbf{n}_1, \mathbf{n}_2, \ldots, \mathbf{n}_k$, then it may seem natural to define

$$\int_S \mathbf{F} \cdot \mathbf{n} \, dA = \sum_{i=1}^{k} \int_{S_i} \mathbf{F} \cdot \mathbf{n}_i \, dA; \tag{5}$$

however, a moment's reflection shows that this is not sensible. Since each of $\mathbf{n}_1, \mathbf{n}_2, \ldots, \mathbf{n}_k$ can be chosen in two ways, there are 2^k possible interpretations

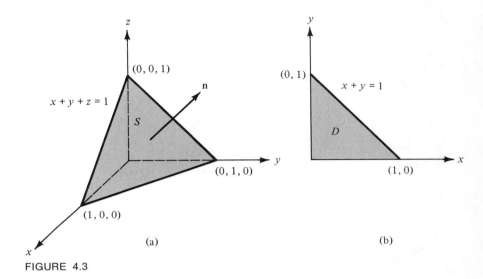

FIGURE 4.3

of the right side of (5), in the absence of any restriction on the relationships among $n_1, n_2, \ldots, n_k$. This is too unwieldy to be useful. As we will see when we consider Stokes' theorem, it is natural to require that S be orientable, with $n_1, n_2, \ldots, n_k$ related as in Definition 3.7. Subject to this, $\int_S \mathbf{F} \cdot \mathbf{n}\, dA$ is defined as in (5). There are then only two possible values for the integral—one the negative of the other—corresponding to the two possible orientations of S.

Example 4.5   Let S be the piecewise smooth closed surface consisting of

$$S_1 = \{(x, y, z)\,|\,z = 1 - x^2 - y^2 \geq 0\},$$
$$S_2 = \{(x, y, z)\,|\,x^2 + y^2 = 1,\, -1 \leq z \leq 0\},$$

and

$$S_3 = \{(x, y, z)\,|\,x^2 + y^2 \leq 1,\, z = -1\}$$

(Figure 4.4), with the outward orientation, and suppose we wish to evaluate

$$I = \int_S (x\mathbf{i} + y\mathbf{j} + z\mathbf{k}) \cdot \mathbf{n}(\mathbf{X})\, dA.$$

Parametrizing S_1 as

$$\mathbf{X} = x\mathbf{i} + y\mathbf{j} + (1 - x^2 - y^2)\mathbf{k} \tag{6}$$

with (x, y) in

$$D_1 = \{(x, y)\,|\,0 \leq x^2 + y^2 \leq 1\}$$

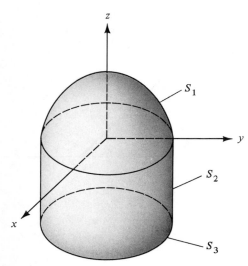

FIGURE 4.4

yields

$$\frac{\partial \mathbf{X}}{\partial x} \times \frac{\partial \mathbf{X}}{\partial y} = \begin{vmatrix} \mathbf{i} & \mathbf{j} & \mathbf{k} \\ 1 & 0 & -2x \\ 0 & 1 & -2y \end{vmatrix}$$

$$= 2x\mathbf{i} + 2y\mathbf{j} + \mathbf{k};$$

therefore, (6) induces the required orientation on S_1 and

$$\int_{S_1} (x\mathbf{i} + y\mathbf{j} + z\mathbf{k}) \cdot \mathbf{n}(\mathbf{X}) \, dA$$

$$= \int_{D_1} [x\mathbf{i} + y\mathbf{j} + (1 - x^2 - y^2)\mathbf{k}] \cdot (2x\mathbf{i} + 2y\mathbf{j} + \mathbf{k}) \, d(x, y)$$

$$= \int_{D_1} (2x^2 + 2y^2 + 1 - x^2 - y^2) \, d(x, y)$$

$$= \int_{D_1} (1 + x^2 + y^2) \, d(x, y).$$

Changing to polar coordinates yields

$$\int_{S_1} (x\mathbf{i} + y\mathbf{j} + z\mathbf{k}) \cdot \mathbf{n}(\mathbf{X}) \, dA = \int_0^{2\pi} d\theta \int_0^1 (1 + r^2) r \, dr$$

$$= 2\pi \left(\frac{r^2}{2} + \frac{r^4}{4} \right) \Big|_0^1 = \frac{3\pi}{2}. \tag{7}$$

Parametrizing S_2 as

$$\mathbf{X} = \cos \theta \mathbf{i} + \sin \theta \mathbf{j} + z\mathbf{k}$$

with (θ, z) in

$$D_2 = \{(\theta, z) \,|\, 0 \le \theta \le 2\pi,\ -1 \le z \le 0\}$$

yields

$$\frac{\partial \mathbf{X}}{\partial \theta} \times \frac{\partial \mathbf{X}}{\partial z} = \begin{vmatrix} \mathbf{i} & \mathbf{j} & \mathbf{k} \\ -\sin\theta & \cos\theta & 0 \\ 0 & 0 & 1 \end{vmatrix}$$

$$= \cos\theta\mathbf{i} + \sin\theta\mathbf{j},$$

which also points toward the outside of S, so

$$\int_{S_2} (x\mathbf{i} + y\mathbf{j} + z\mathbf{k}) \cdot \mathbf{n}(\mathbf{X})\, dA$$

$$= \int_{D_2} (\cos\theta\mathbf{i} + \sin\theta\mathbf{j} + z\mathbf{k}) \cdot (\cos\theta\mathbf{i} + \sin\theta\mathbf{j})\, d(\theta, z)$$

$$= \int_{D_2} d(\theta, z) = \int_0^{2\pi} d\theta \int_{-1}^0 dz = 2\pi. \tag{8}$$

Parametrizing S_3 as

$$\mathbf{X} = r\cos\theta\mathbf{i} + r\sin\theta\mathbf{j} - \mathbf{k}$$

with (r, θ) in

$$D_3 = \{(r, \theta) \,|\, 0 \le r \le 1,\ 0 \le \theta \le 2\pi\}$$

yields

$$\frac{\partial \mathbf{X}}{\partial r} \times \frac{\partial \mathbf{X}}{\partial \theta} = \begin{vmatrix} \mathbf{i} & \mathbf{j} & \mathbf{k} \\ \cos\theta & \sin\theta & 0 \\ -r\sin\theta & r\cos\theta & 0 \end{vmatrix}$$

$$= r\mathbf{k}.$$

Since this vector points inside rather than outside of S,

$$\int_{S_3} (x\mathbf{i} + y\mathbf{j} + z\mathbf{k}) \cdot \mathbf{n}(\mathbf{X})\, dA = -\int_{D_3} (r\cos\theta\mathbf{i} + r\sin\theta\mathbf{j} - \mathbf{k}) \cdot r\mathbf{k}\, d(r, \theta)$$

$$= \int_{D_3} r\, d(r, \theta) = \int_0^{2\pi} d\theta \int_0^1 r\, dr$$

$$= (2\pi)\left(\frac{1}{2}\right) = \pi.$$

From (7), (8), and this,

$$I = \frac{3\pi}{2} + 2\pi + \pi = \frac{9\pi}{2}.$$

We will soon discuss a simpler way to obtain this result.

Example 4.6   Suppose the vector field

$$\mathbf{F} = P\mathbf{i} + Q\mathbf{j} + R\mathbf{k}$$

is continuously differentiable on the rectangular parallelepiped

$$T = [a_1, a_2] \times [b_1, b_2] \times [c_1, c_2],$$

and let $\mathbf{n}$ be the positive orientation of S, the surface of T. We will show that

$$\int_S \mathbf{F} \cdot \mathbf{n} \, dA = \int_T \left(\frac{\partial P}{\partial x} + \frac{\partial Q}{\partial y} + \frac{\partial R}{\partial z} \right) d\mathbf{X}. \tag{9}$$

First we write, by definition,

$$\int_S \mathbf{F} \cdot \mathbf{n} \, dA = \sum_{i=1}^{6} \int_{S_i} \mathbf{F} \cdot \mathbf{n}_i \, dA,$$

where $S_1, S_2, \ldots, S_6$ are the faces of T and $\mathbf{n}_i$ is the unit normal to S_i, pointed toward the outside of T.

Let S_1 be the face of T on which $x = a_2$ (Figure 4.5); it can be represented by

$$\mathbf{X} = a_2\mathbf{i} + y\mathbf{j} + z\mathbf{k}$$

with (y, z) in

$$D = [b_1, b_2] \times [c_1, c_2].$$

Since $\mathbf{n}_1 = \mathbf{i}$,

$$\int_{S_1} \mathbf{F} \cdot \mathbf{n} \, dA = \int_D P(a_2, y, z) \, d(y, z). \tag{10}$$

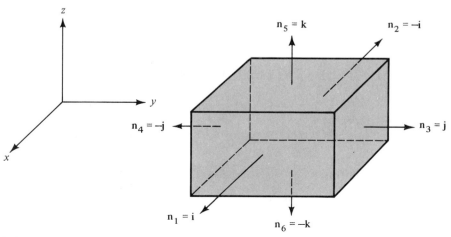

FIGURE 4.5

The face S_2 of T on which $x = a_1$ can be represented by

$$\mathbf{X} = a_1\mathbf{i} + y\mathbf{j} + z\mathbf{k}$$

with (y, z) in D, and $\mathbf{n}_2 = -\mathbf{i}$ (Figure 4.5); hence,

$$\int_{S_2} \mathbf{F} \cdot \mathbf{n}\, dA = -\int_D P(a_1, y, z)\, d(y, z). \tag{11}$$

Adding (10) and (11) yields

$$\int_{S_1} \mathbf{F} \cdot \mathbf{n}\, dA + \int_{S_2} \mathbf{F} \cdot \mathbf{n}\, dA = \int_D [P(a_2, y, z) - P(a_1, y, z)]\, d(y, z)$$

$$= \int_D d(y, z) \int_{a_1}^{a_2} \frac{\partial P(x, y, z)}{\partial x}\, dx,$$

so that

$$\int_{S_1} \mathbf{F} \cdot \mathbf{n}\, dA + \int_{S_2} \mathbf{F} \cdot \mathbf{n}\, dA = \int_T \frac{\partial P}{\partial x}\, d\mathbf{X}. \tag{12}$$

If S_3 and S_4 are the faces of T on which $y = b_2$ and $y = b_1$, a similar argument yields

$$\int_{S_3} \mathbf{F} \cdot \mathbf{n}\, dA + \int_{S_4} \mathbf{F} \cdot \mathbf{n}\, dA = \int_T \frac{\partial Q}{\partial y}\, d\mathbf{X}. \tag{13}$$

Also,

$$\int_{S_4} \mathbf{F} \cdot \mathbf{n}\, dA + \int_{S_5} \mathbf{F} \cdot \mathbf{n}\, dA = \int_T \frac{\partial R}{\partial z}\, d\mathbf{X}. \tag{14}$$

where S_4 and S_5 are the faces of T on which $z = c_2$ and $z = c_1$.
Adding (12), (13), and (14) yields (9).

Example 4.6 is a special case of an important result called the *divergence theorem*. The *divergence* of a differentiable vector field

$$\mathbf{F} = P\mathbf{i} + Q\mathbf{j} + R\mathbf{k}$$

is defined by

$$\operatorname{div} \mathbf{F} = \frac{\partial P}{\partial x} + \frac{\partial Q}{\partial y} + \frac{\partial R}{\partial z}.$$

Theorem 4.1 (Divergence Theorem)   *Let D be a closed region in $\mathscr{R}^3$ bounded by a piecewise smooth orientable closed surface S, and suppose the vector field $\mathbf{F}$ is continuously differentiable on D and $\mathbf{n}$ is the positive orientation for S. Then*

$$\int_S \mathbf{F} \cdot \mathbf{n}\, dA = \int_D \operatorname{div} \mathbf{F}\, d\mathbf{X}. \tag{15}$$

We give only a partial proof, under the additional assumption that D can be *projected* into the three coordinate planes. We say that D can be *projected into the xy plane* if every line parallel to the z axis that intersects D intersects it in exactly one point or in a line segment; then the *projection of D into the xy plane* is

$$D_{xy} = \{(x, y) | (x, y, z) \in D \text{ for some } z\}.$$

The corresponding definitions of projections into the yz and zx planes are similar.

Example 4.7 The hemispherical shell

$$D = \{\mathbf{X} | 0 < r_1 < |\mathbf{X}| < r_2, z \geq 0\}$$

can be projected into the xy plane, but not into the yz or xz planes (Figure 4.6). However, D can be decomposed into four pieces—the parts of D in each of the four octants of $\mathscr{R}^3$ in which $z \geq 0$—which can be projected into all three coordinate planes.

Now for the partial proof of Theorem 4.1. First we write

$$\int_S \mathbf{F} \cdot \mathbf{n} \, dA = \int_S (P\mathbf{i}) \cdot \mathbf{n} \, dA + \int_S (Q\mathbf{j}) \cdot \mathbf{n} \, dA + \int_S (R\mathbf{k}) \cdot \mathbf{n} \, dA. \tag{16}$$

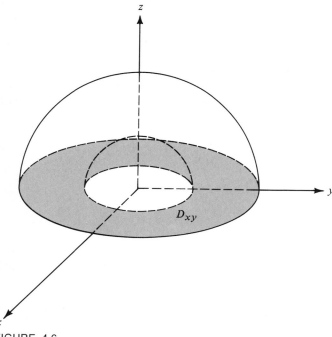

FIGURE 4.6

We will evaluate the last integral and leave the other two to the reader (*Exercise* 7).

For each (x, y) in D_{xy} there are unique numbers $z_1(x, y)$ and $z_2(x, y)$ such that $(x, y, z) \in D$ if and only if $(x, y) \in D_{xy}$ and $z_1(x, y) \le z \le z_2(x, y)$ (Figure 4.7). We assume that z_1 and z_2 are differentiable. Now S can be divided into three parts:

1. A lower surface S_1, defined by

$$\mathbf{X} = \mathbf{\Phi}_1(x, y) = x\mathbf{i} + y\mathbf{j} + z_1(x, y)\mathbf{k},$$

for which

$$\mathbf{N}_1 = \frac{\partial \mathbf{\Phi}_1}{\partial y} \times \frac{\partial \mathbf{\Phi}_1}{\partial x} = \begin{vmatrix} \mathbf{i} & \mathbf{j} & \mathbf{k} \\ 0 & 1 & \dfrac{\partial z_1}{\partial y} \\ 1 & 0 & \dfrac{\partial z_1}{\partial x} \end{vmatrix} = \frac{\partial z_1}{\partial x}\mathbf{i} + \frac{\partial z_1}{\partial y}\mathbf{j} - \mathbf{k}$$

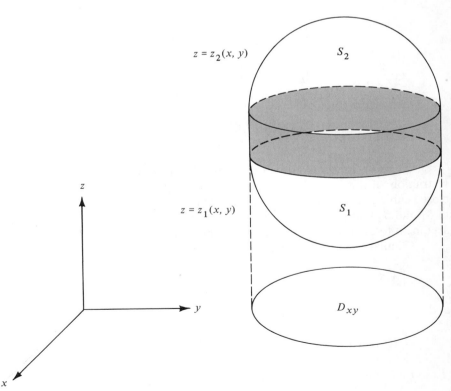

FIGURE 4.7

is in the direction of the outward normal.

2. An upper surface S_2, represented by

$$\mathbf{X} = \mathbf{\Phi}_2(x, y) = x\mathbf{i} + y\mathbf{j} + z_2(x, y)\mathbf{k},$$

for which

$$\mathbf{N}_2 = \frac{\partial \mathbf{\Phi}_2}{\partial x} \times \frac{\partial \mathbf{\Phi}_2}{\partial y} = -\frac{\partial z_2}{\partial x}\mathbf{i} - \frac{\partial z_2}{\partial y}\mathbf{j} + \mathbf{k}$$

is in the direction of the outward normal.

3. Possibly one or more vertical sections (shaded in Figure 4.7) on which the normal is parallel to the xy plane.

Since $\mathbf{k} \cdot \mathbf{n} = 0$ on any surface of the kind described in (3), these surfaces contribute nothing to $\int_S (R\mathbf{k}) \cdot \mathbf{n} \, dA$, which is given by

$$
\begin{aligned}
\int_S (R\mathbf{k}) \cdot \mathbf{n} \, dA &= \int_{S_2} (R\mathbf{k}) \cdot \mathbf{n} \, dA + \int_{S_1} (R\mathbf{k}) \cdot \mathbf{n} \, dA \\
&= \int_{D_{xy}} R(x, y, z_2(x, y))\mathbf{k} \cdot \mathbf{N}_2(x, y) \, d(x, y) \\
&\quad + \int_{D_{xy}} R(x, y, z_1(x, y))\mathbf{k} \cdot \mathbf{N}_1(x, y) \, d(x, y) \\
&= \int_{D_{xy}} [R(x, y, z_2(x, y)) - R(x, y, z_1(x, y))] \, d(x, y) \\
&= \int_{D_{xy}} d(x, y) \int_{z_1(x,y)}^{z_2(x,y)} \frac{\partial R}{\partial z} \, dz,
\end{aligned}
$$

which can be rewritten as

$$\int_S (R\mathbf{k}) \cdot \mathbf{n} \, dA = \int_D \frac{\partial R}{\partial z} \, d\mathbf{X}. \tag{17}$$

Here we have assumed that S_1 and S_2 are smooth, but this is no real restriction; if they are only piecewise smooth, then D_{xy} can be divided into closed subregions, intersecting only in boundary curves, such that the portions of S_1 and S_2 that project into each subregion are smooth. Applying the above argument to each such subregion of D_{xy} and adding the results then yields (17).

Similar arguments show that

$$\int_S (P\mathbf{i}) \cdot \mathbf{n} \, dA = \int_D \frac{\partial P}{\partial x} d\mathbf{X} \tag{18}$$

and

$$\int_S (Q\mathbf{j}) \cdot \mathbf{n} \, dA = \int_D \frac{\partial Q}{\partial y} d\mathbf{X} \tag{19}$$

(*Exercise 7*), and, from (16), adding (17), (18), and (19) yields (15).

Example 4.8   In Example 4.3 we evaluated

$$I = \int_S (y\mathbf{j} + z\mathbf{k}) \cdot \mathbf{n}(X) \, dA,$$

where S is the sphere defined by

$$x^2 + y^2 + z^2 = a^2.$$

Since

$$\operatorname{div}(y\mathbf{j} + z\mathbf{k}) = 2,$$

applying Theorem 4.1 with

$$D = \{(x, y, z) \, | \, x^2 + y^2 + z^2 \le a^2\}$$

yields

$$I = 2 \int_D d\mathbf{X} = 2V(D) = 2\left(\frac{4}{3}\pi a^3\right) = \frac{8}{3}\pi a^3,$$

which agrees with the result of Example 4.3.

Example 4.9   In Example 4.5 we evaluated

$$I = \int_S (x\mathbf{i} + y\mathbf{j} + z\mathbf{k}) \cdot \mathbf{n}(X) \, dA,$$

where S is the surface in Figure 4.4. Since

$$\operatorname{div}(x\mathbf{i} + y\mathbf{j} + z\mathbf{k}) = 3,$$

Theorem 4.1 implies that

$$I = 3 \int_D d\mathbf{X} = 3V(D), \tag{20}$$

where D is the region bounded by S, consisting of the two parts

$$D_1 = \{(x, y, z) \, | \, 0 \le z \le 1 - x^2 - y^2\}$$

and

$$D_2 = \{(x, y, z) \, | \, x^2 + y^2 \le 1, \, -1 \le z \le 0\}.$$

Straightforward calculation shows that

$$V(D) = V(D_1) + V(D_2) = \frac{\pi}{2} + \pi = \frac{3\pi}{2};$$

hence (20) implies that $I = 9\pi/2$, consistent with the result of Example 4.5.

The partial proof given for Theorem 4.1 can be extended to any region decomposable into subregions which can be projected into the three coordinate planes; in fact, the region may be bounded by more than one surface.

Example △ 10   If D is the punctured ball

$$D = \{\mathbf{X} \,|\, 0 < r_1 \leq |\mathbf{X}| \leq r_2\},$$

we can write

$$\int_D \operatorname{div} \mathbf{F} \, d\mathbf{X} = \sum_{i=1}^{8} \int_{D_i} \operatorname{div} \mathbf{F} \, d\mathbf{X}, \tag{21}$$

where $D_1, \ldots, D_8$ are the parts of D in each of the octants of $\mathscr{R}^3$. Since D_i can be projected into each of the coordinate planes, our proof of Theorem 4.1 implies that

$$\int_{D_i} \operatorname{div} \mathbf{F} \, d\mathbf{X} = \int_{S_i} \mathbf{F} \cdot \mathbf{n}_i \, dA, \tag{22}$$

where $\mathbf{n}_i$ is the outward orientation on S_i, the surface bounding D_i. Now S_i consists of an eighth of the sphere

$$\sigma_1 = \{\mathbf{X} \,|\, |\mathbf{X}| = r_1\},$$

an eighth of the sphere

$$\sigma_2 = \{\mathbf{X} \,|\, |\mathbf{X}| = r_2\},$$

and three planar surfaces, one in each of the coordinate planes (Figure 4.8).

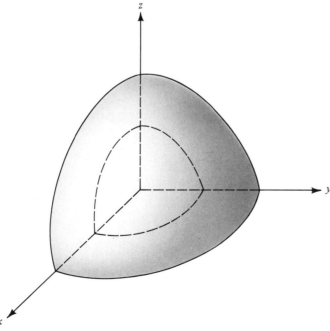

FIGURE 4.8

Each of the latter is common to exactly two of $S_1, \ldots, S_8$, with opposite orientations; hence, their contributions to the surface integrals on the right of (22) cancel on summing (22) from $i = 1$ to $i = 8$. From this and (21),

$$\int_D \operatorname{div} \mathbf{F} \, d\mathbf{X} = \int_{\sigma_1} \mathbf{F} \cdot \mathbf{n} \, dA + \int_{\sigma_2} \mathbf{F} \cdot \mathbf{n} \, dA,$$

where

$$\mathbf{n}(\mathbf{X}) = \begin{cases} \dfrac{\mathbf{X}}{|\mathbf{X}|} & \text{if } |\mathbf{X}| = r_2, \\[2mm] -\dfrac{\mathbf{X}}{|\mathbf{X}|} & \text{if } |\mathbf{X}| = r_1. \end{cases}$$

Stokes' theorem

If a vector field $\mathbf{F} = P\mathbf{i} + Q\mathbf{j} + R\mathbf{k}$ has continuous second partial derivatives on an open set T in $\mathscr{R}^3$, then

$$\operatorname{div}(\operatorname{\mathbf{curl}} \mathbf{F}) = 0$$

on T (*Exercise 10*). Therefore, Theorem 4.1 implies that if a closed orientable surface S and the region D enclosed by S are contained in T, then

$$\int_S (\operatorname{\mathbf{curl}} \mathbf{F}) \cdot \mathbf{n} \, dA = \pm \int_D \operatorname{div}(\operatorname{\mathbf{curl}} \mathbf{F}) \, d\mathbf{X} = 0, \tag{23}$$

where $\mathbf{n}$ is either orientation of S.

Now suppose C is a simple closed curve in T, and S_1 and S_2 are surfaces which have C as their common boundary. For simplicity assume that S_1 and S_2 do not intersect except along C, and that they are oriented so that C is positively directed with respect to both (Figure 4.9).

Now let S be the closed surface consisting of S_1 and S_2, with the orientation

$$\mathbf{n}(\mathbf{X}) = \begin{cases} \mathbf{n}_1(\mathbf{X}) & \text{if } \mathbf{X} \in S_1, \\ -\mathbf{n}_2(\mathbf{X}) & \text{if } \mathbf{X} \in S_2 - C, \end{cases}$$

and assume that S and the region D that it bounds are in T. Since

$$\int_S (\operatorname{\mathbf{curl}} \mathbf{F}) \cdot \mathbf{n} \, dA = \int_{S_1} (\operatorname{\mathbf{curl}} \mathbf{F}) \cdot \mathbf{n}_1 \, dA - \int_{S_2} (\operatorname{\mathbf{curl}} \mathbf{F}) \cdot \mathbf{n}_2 \, dA,$$

(23) implies that

$$\int_{S_1} (\operatorname{\mathbf{curl}} \mathbf{F}) \cdot \mathbf{n}_1 \, dA = \int_{S_2} (\operatorname{\mathbf{curl}} \mathbf{F}) \cdot \mathbf{n}_2 \, dA.$$

This suggests that integrals of F over all orientable surfaces in T with a given boundary C are equal (up to a factor of ± 1). If this is so, then it should be

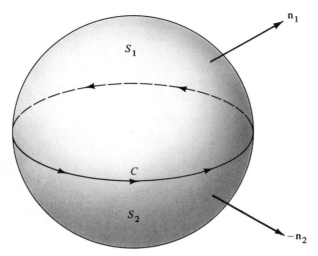

FIGURE 4.9

possible to calculate their common value from the values of $\mathbf{F}$ on C alone. Stokes' theorem shows how to do this. We first consider the following restricted form of this theorem.

Theorem 4.2 (Stokes' Theorem)   *Suppose S is a smooth orientable surface defined by*

$$\mathbf{X} = \mathbf{X}(u, v), \qquad (u, v) \in D, \tag{24}$$

where D is a closed region in the (u, v) plane, bounded by a simple closed curve γ. Suppose also that all second partial derivatives of x, y, and z with respect to u and v in (24) are continuous in D and the vector field $\mathbf{F} = P\mathbf{i} + Q\mathbf{j} + R\mathbf{k}$ is continuously differentiable on S. Let $\mathbf{n}$ be the orientation of S defined by

$$\mathbf{n}(\mathbf{X}(u, v)) = \frac{\dfrac{\partial \mathbf{X}(u, v)}{\partial u} \times \dfrac{\partial \mathbf{X}(u, v)}{\partial v}}{\left| \dfrac{\partial \mathbf{X}(u, v)}{\partial u} \times \dfrac{\partial \mathbf{X}(u, v)}{\partial v} \right|} \tag{25}$$

for $(u, v) \in D^0$, and by continuity on γ. Finally, suppose $\mathbf{X}(u, v)$ traverses ∂S exactly once as (u, v) traverses γ. Then

$$\int_S (\mathbf{curl}\ \mathbf{F}) \cdot \mathbf{n}\ dA = \int_{\partial S} \mathbf{F} \cdot d\mathbf{X}, \tag{26}$$

where the direction of ∂S is the one in which $\mathbf{X}(u, v)$ traverses it as (u, v) traverses γ in the positive sense.

Proof.   We first write

$$\int_{\partial S} \mathbf{F} \cdot d\mathbf{X} = \int_{\partial S} P\, dx + \int_{\partial S} Q\, dy + \int_{\partial S} R\, dz,$$

and consider the integrals on the right separately. Parametrizing γ in the positive direction with

$$u = u(t), \qquad v = v(t), \qquad a \le t \le b, \tag{27}$$

induces a parametrization

$$\mathbf{X} = x(u(t), v(t))\mathbf{i} + y(u(t), v(t))\mathbf{j} + z(u(t), v(t))\mathbf{k}, \qquad a \le t \le b,$$

of ∂S, and we can write

$$\int_{\partial S} P\, dx$$

$$= \int_a^b P(\mathbf{X}(u(t), v(t))) \frac{d}{dt} x(u(t), v(t))\, dt \tag{28}$$

$$= \int_a^b P(\mathbf{X}(u(t), v(t))) \left[\frac{\partial x(u(t), v(t))}{\partial u} \frac{du(t)}{dt} + \frac{\partial x(u(t), v(t))}{\partial v} \frac{dv(t)}{dt} \right] dt,$$

which can be rewritten as

$$\int_{\partial S} P\, dx = \int_\gamma P(\mathbf{X}(u, v)) \frac{\partial x}{\partial u}\, du + P(\mathbf{X}(u, v)) \frac{\partial x}{\partial v}\, dv, \tag{29}$$

since the last integral in (28) is precisely the one obtained from the integral on the right of (29) on parametrizing γ as in (27). Applying Theorem 2.6 to the integral on the right of (29) yields

$$\int_{\partial S} P\, dx = \int_D \left[\frac{\partial}{\partial u}\left(P \frac{\partial x}{\partial v} \right) - \frac{\partial}{\partial v}\left(P \frac{\partial x}{\partial u} \right) \right] d(u, v), \tag{30}$$

where $P = P(\mathbf{X})$ on the left and $P = P(\mathbf{X}(u, v))$ on the right. Our assumptions on the continuity of the second partial derivatives of x imply that $x_{uv} = x_{vu}$ (Theorem 3.2, Section 5.3); therefore, the integrand on the right of (30) can be rewritten as

$$\frac{\partial}{\partial u}\left(P \frac{\partial x}{\partial v} \right) - \frac{\partial}{\partial v}\left(P \frac{\partial x}{\partial u} \right) = \left(\frac{\partial P}{\partial x} \frac{\partial x}{\partial u} + \frac{\partial P}{\partial y} \frac{\partial y}{\partial u} + \frac{\partial P}{\partial z} \frac{\partial z}{\partial u} \right) \frac{\partial x}{\partial v}$$

$$- \left(\frac{\partial P}{\partial x} \frac{\partial x}{\partial v} + \frac{\partial P}{\partial y} \frac{\partial y}{\partial v} + \frac{\partial P}{\partial z} \frac{\partial z}{\partial v} \right) \frac{\partial x}{\partial u}$$

$$= \frac{\partial P}{\partial y}\left(\frac{\partial y}{\partial u} \frac{\partial x}{\partial v} - \frac{\partial y}{\partial v} \frac{\partial x}{\partial u} \right) + \frac{\partial P}{\partial z}\left(\frac{\partial z}{\partial u} \frac{\partial x}{\partial v} - \frac{\partial z}{\partial v} \frac{\partial x}{\partial u} \right)$$

$$= -\frac{\partial P}{\partial y} \frac{\partial(x, y)}{\partial(u, v)} + \frac{\partial P}{\partial z} \frac{\partial(z, x)}{\partial(u, v)}.$$

Substituting this in (30) yields

$$\int_{\partial S} P \, dx = \int_D \left[-\frac{\partial P}{\partial y} \frac{\partial(x, y)}{\partial(u, v)} + \frac{\partial P}{\partial z} \frac{\partial(z, x)}{\partial(u, v)} \right] d(u, v).$$

Similar arguments yield

$$\int_{\partial S} Q \, dy = \int_D \left[-\frac{\partial Q}{\partial z} \frac{\partial(y, z)}{\partial(u, v)} + \frac{\partial Q}{\partial x} \frac{\partial(x, y)}{\partial(u, v)} \right] d(u, v)$$

and

$$\int_{\partial S} R \, dz = \int_D \left[-\frac{\partial R}{\partial x} \frac{\partial(z, x)}{\partial(u, v)} + \frac{\partial R}{\partial y} \frac{\partial(y, z)}{\partial(u, v)} \right] d(u, v).$$

Adding the last three equations yields

$$\int_{\partial S} \mathbf{F} \cdot d\mathbf{X} = \int_D \left[\left(\frac{\partial R}{\partial y} - \frac{\partial Q}{\partial z} \right) \frac{\partial(y, z)}{\partial(u, v)} + \left(\frac{\partial P}{\partial z} - \frac{\partial R}{\partial x} \right) \frac{\partial(z, x)}{\partial(u, v)} \right.$$
$$\left. + \left(\frac{\partial Q}{\partial x} - \frac{\partial P}{\partial y} \right) \frac{\partial(x, y)}{\partial(u, v)} \right] d(u, v). \tag{31}$$

Since

$$\mathbf{curl\ F} = \left(\frac{\partial R}{\partial y} - \frac{\partial Q}{\partial z} \right) \mathbf{i} + \left(\frac{\partial P}{\partial z} - \frac{\partial R}{\partial x} \right) \mathbf{j} + \left(\frac{\partial Q}{\partial x} - \frac{\partial P}{\partial y} \right) \mathbf{k}$$

and

$$\frac{\partial \mathbf{X}}{\partial u} \times \frac{\partial \mathbf{X}}{\partial v} = \frac{\partial(y, z)}{\partial(u, v)} \mathbf{i} + \frac{\partial(z, x)}{\partial(u, v)} \mathbf{j} + \frac{\partial(x, y)}{\partial(u, v)} \mathbf{k},$$

(31) implies that

$$\int_{\partial S} \mathbf{F} \cdot d\mathbf{X} = \int_D (\mathbf{curl\ F}) \cdot \left(\frac{\partial \mathbf{X}}{\partial u} \times \frac{\partial \mathbf{X}}{\partial v} \right) d(u, v)$$
$$= \int_S (\mathbf{curl\ F}) \cdot \mathbf{n} \, dA,$$

from (3) and (25). This completes the proof.

Example 4.11   The surface

$$S = \{(x, y, z) | x^2 + y^2 = 1, \, y \geq 0, \, 0 \leq z \leq 1\}$$

[Figure 4.10(a)] can be parametrized as

$$\mathbf{X} = \cos \theta \mathbf{i} + \sin \theta \mathbf{j} + z \mathbf{k}$$

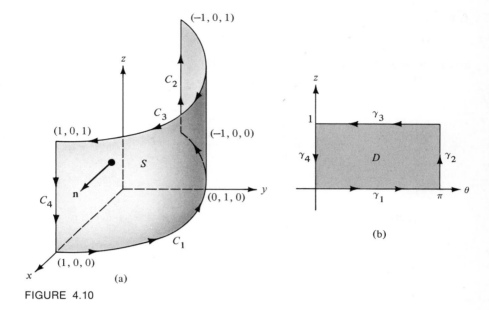

FIGURE 4.10

with (θ, z) in

$$D = \{(\theta, z) \,|\, 0 \leq \theta \leq \pi, 0 \leq z \leq 1\}$$

[Figure 4.10(b)]. As (θ, z) traverses ∂D in the positive sense [as in Figure 4.10(b)], $\mathbf{X}(\theta, z)$ traverses ∂S as in Figure 4.10(a). The orientation (25) of Theorem 4.2 is given by

$$\mathbf{n}(\mathbf{X}(\theta, z)) = \begin{vmatrix} \mathbf{i} & \mathbf{j} & \mathbf{k} \\ -\sin\theta & \cos\theta & 0 \\ 0 & 0 & 1 \end{vmatrix}$$

$$= \cos\theta\,\mathbf{i} + \sin\theta\,\mathbf{j}$$

$$= x(\theta, z)\mathbf{i} + y(\theta, z)\mathbf{j},$$

so

$$\mathbf{n}(\mathbf{X}) = x\mathbf{i} + y\mathbf{j},$$

as shown in Figure 4.10(a).

Now Theorem 4.2 implies that

$$\int_S (\textbf{curl } \mathbf{F}) \cdot (x\mathbf{i} + y\mathbf{j})\, dA = \int_{\partial S} \mathbf{F} \cdot d\mathbf{X} \tag{32}$$

if $\mathbf{F}$ is any continuously differentiable vector field on S and ∂S is directed as in Figure 4.10(a). Let us verify this for the vector field

$$\mathbf{F}(\mathbf{X}) = z^2\mathbf{i} + x\mathbf{j} + y\mathbf{k},$$

by calculating and comparing the integrals in (32).

Since

$$\mathbf{curl}\ \mathbf{F}(\mathbf{X}) = \begin{vmatrix} \mathbf{i} & \mathbf{j} & \mathbf{k} \\ \dfrac{\partial}{\partial x} & \dfrac{\partial}{\partial y} & \dfrac{\partial}{\partial z} \\ z^2 & x & y \end{vmatrix}$$

$$= \mathbf{i} + 2z\mathbf{j} + \mathbf{k},$$

$$\int_S [\mathbf{curl}\ \mathbf{F}(\mathbf{X})] \cdot (x\mathbf{i} + y\mathbf{j})\ dA = \int_S (\mathbf{i} + 2z\mathbf{j} + \mathbf{k}) \cdot (x\mathbf{i} + y\mathbf{j})\ dA$$

$$= \int_S (x + 2yz)\ dA$$

$$= \int_D (\cos \theta + 2z \sin \theta)\ d(\theta, z) \qquad (33)$$

$$= \int_0^\pi \cos \theta\ d\theta \int_0^1 dz + 2 \int_0^\pi \sin \theta\ d\theta \int_0^1 z\ dz$$

$$= (0)(1) + (2)(2)(\tfrac{1}{2}) = 2.$$

To calculate $\int_{\partial S} \mathbf{F} \cdot d\mathbf{X}$, we write

$$\int_{\partial S} \mathbf{F} \cdot d\mathbf{X} = \sum_{i=1}^4 \int_{C_i} \mathbf{F} \cdot d\mathbf{X},$$

where C_1, C_2, C_3, and C_4 are as in Figure 4.10(a). We leave it to the reader to fill in the details of the following calculations:

$$\int_{C_1} \mathbf{F} \cdot d\mathbf{X} = \int_{C_1} 0\ dx + x\ dy + y\ dz = \int_0^\pi \cos^2 \theta\ d\theta = \pi/2;$$

$$\int_{C_2} \mathbf{F} \cdot d\mathbf{X} = \int_{C_2} z^2\ dx + (-1)\ dy + 0\ dz = 0;$$

$$\int_{C_3} \mathbf{F} \cdot d\mathbf{X} = \int_{C_3} dx + x\ dy + y\ dz$$

$$= -\int_0^\pi (-\sin \theta + \cos^2 \theta)\ d\theta = 2 - \pi/2;$$

$$\int_{C_4} \mathbf{F} \cdot d\mathbf{X} = \int_{C_4} z^2\ dx + dy + 0\ dz = 0.$$

Adding these equations yields

$$\int_C \mathbf{F} \cdot d\mathbf{X} = 2,$$

consistent with (33).

If (26) is true for a given orientation **n** and direction on ∂S, then it remains true if both are reversed; what is important is the relationship between **n** and the direction of ∂S. This relationship is clear in Example 4.11 [see Figure 4.10(a)]: ∂S is positively directed with respect to **n**, as defined earlier. This is also true in general, but we will not prove it except for the case where the equation of S can be written explicitly as

$$z = f(x, y) \geq 0, \qquad (x, y) \in D,$$

where we still assume that D is bounded by a simple closed curve γ. In this case we can represent S parametrically as

$$\mathbf{X} = x\mathbf{i} + y\mathbf{j} + f(x, y)\mathbf{k}, \qquad (x, y) \in D,$$

and **n** as defined in (25) becomes

$$\mathbf{n}(\mathbf{X}) = \frac{-f_x(\mathbf{X})\mathbf{i} - f_y(\mathbf{X})\mathbf{j} + \mathbf{k}}{\sqrt{1 + [f_x(\mathbf{X})]^2 + [f_y(\mathbf{X})]^2}}.$$

Now the situation is as shown in Figure 4.11: as (x, y) traverses γ in the positive direction, **X** traverses ∂S in the positive direction with respect to **n**.

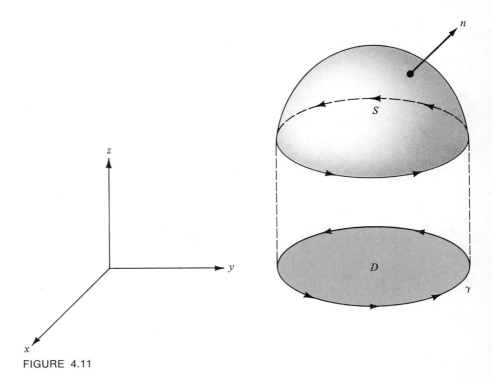

FIGURE 4.11

Example 4.12   Theorem 4.2 does not apply directly to the surface

$$S = \{(x, y, z) \,|\, x^2 + y^2 + z^2 = a^2, 0 \le z \le b\}$$

where $0 < b < a$ [Figure 4.12(a)], since ∂S consists of two circles

$$C_1 = \{(x, y, z) \,|\, x^2 + y^2 = 1, z = 0\}$$

and

$$C_2 = \{(x, y, z) \,|\, x^2 + y^2 = a^2 - b^2, z = b\}.$$

Nevertheless, the theorem implies a useful result for this surface: If $\mathbf{F}$ is continuously differentiable on S, and $\mathbf{n}$ and the directions of C_1 and C_2 are as shown in Figure 4.12(a), then

$$\int_S (\mathbf{curl\ F}) \cdot \mathbf{n}\, dA = \int_{C_1} \mathbf{F} \cdot d\mathbf{X} + \int_{C_2} \mathbf{F} \cdot d\mathbf{X}. \tag{34}$$

To see this we partition S into surfaces S_1 and S_2, where, for example, S_1 is the part of S on which $y \ge 0$ and S_2 is the part on which $y \le 0$ [Figure 4.12(b)]. Then S_1 and S_2 satisfy the hypotheses of Theorem 4.2, so

$$\int_{S_1} (\mathbf{curl\ F}) \cdot \mathbf{n}\, dA = \int_{\partial S_1} \mathbf{F} \cdot d\mathbf{X} \tag{35}$$

and

$$\int_{S_2} (\mathbf{curl\ F}) \cdot \mathbf{n}\, dA = \int_{\partial S_2} \mathbf{F} \cdot d\mathbf{X}, \tag{36}$$

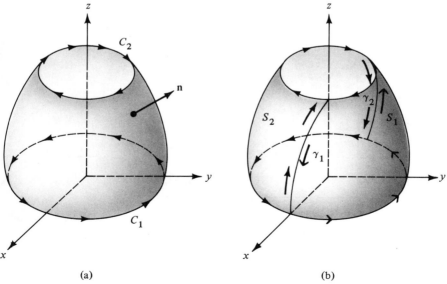

(a)   (b)

FIGURE 4.12

with ∂S_1 and ∂S_2 oriented as in Figure 4.12(b). Adding (35) and (36) yields

$$\int_S (\text{curl } \mathbf{F}) \cdot \mathbf{n} \, dA = \int_{\partial S_1} \mathbf{F} \cdot d\mathbf{X} + \int_{\partial S_2} \mathbf{F} \cdot d\mathbf{X}. \tag{37}$$

The right side of (37) equals the sum of the line integrals of $\mathbf{F}$ over the curves in Figure 4.12(b), where the arrows $\rightarrow$ and $\rightarrow$ refer to ∂S_1 and ∂S_2, respectively. Since the two integrals along

$$\gamma_1 = \{(x, y, z) \,|\, x^2 + z^2 = a^2, x \geq 0, 0 \leq z \leq b, y = 0\}$$

are taken in opposite directions, they cancel. The same is true of the two integrals along

$$\gamma_2 = \{(x, y, z) \,|\, x^2 + z^2 = a^2, x \leq 0, 0 \leq z \leq b, y = 0\}.$$

The remaining integrals add up to

$$\int_{C_1} \mathbf{F} \cdot d\mathbf{X} + \int_{C_2} \mathbf{F} \cdot d\mathbf{X},$$

with C_1 and C_2 directed as in Figure 4.12(a). This proves (34).

Example 4.13   Theorem 4.2 does not apply directly to the hemisphere

$$x^2 + y^2 + z^2 = a^2, z \geq 0,$$

represented in the usual way by

$$\mathbf{X}(\theta, \phi) = a \cos \theta \cos \phi \mathbf{i} + a \sin \theta \cos \phi \mathbf{j} + a \sin \phi \mathbf{k},$$

with (θ, ϕ) in

$$D = \{(\theta, \phi) \,|\, 0 \leq \theta \leq 2\pi, 0 \leq \phi \leq \pi/2\},$$

since $\mathbf{X}(\theta, \phi)$ does not merely traverse ∂S as (θ, ϕ) traverses ∂D; it also traverses the quarter circle

$$x^2 + z^2 = a^2, x \geq 0, z \geq 0, \quad y = 0,$$

twice in opposite directions, and the point $(0, 0, a)$ is the image of infinitely many points on ∂D. (Verify.) However, a limiting argument shows that if F is continuously differentiable on S, then (26) holds. To see this, we apply the result of Example 4.12 to

$$S_b = \{(x, y, z) \,|\, x^2 + y^2 + z^2 = a^2, 0 \leq z \leq b\},$$

where $b < a$ [Figure 4.12(a)], to obtain

$$\int_{S_b} (\text{curl } \mathbf{F}) \cdot \mathbf{n} \, dA = \int_{\partial S} \mathbf{F} \cdot d\mathbf{X} + \int_{C_b} \mathbf{F} \cdot d\mathbf{X}, \tag{38}$$

where

$$C_b = \{(x, y, z) \,|\, x^2 + y^2 = a^2 - b^2, z = b\}.$$

Since

$$\lim_{b \to a-} \int_{S_b} (\mathbf{curl\ F}) \cdot \mathbf{n} \, dA = \int_S (\mathbf{curl\ F}) \cdot \mathbf{n} \, dA \qquad (39)$$

(*Exercise 17*) and

$$\lim_{b \to a-} \int_{C_b} \mathbf{F} \cdot d\mathbf{X} = 0, \qquad (40)$$

(*Exercise 18*), (26) follows from (38) on letting $b \to a-$.

We now state without proof the general form of Stokes' theorem for surfaces in $\mathscr{R}^3$.

Theorem 4.3 *Let S be a piecewise smooth orientable surface with boundary consisting of piecewise smooth curves $C_1, C_2, \ldots, C_k$, each positively directed with respect to the orientation* $\mathbf{n}$ *of S, and suppose the vector field* $\mathbf{F}$ *is continuously differentiable on S. Then*

$$\int_{\partial S} (\mathbf{curl\ F}) \cdot \mathbf{n} \, dA = \sum_{j=1}^{k} \int_{C_j} \mathbf{F} \cdot d\mathbf{X}.$$

In this theorem the importance of orientability of a piecewise smooth surface is particularly apparent. For example, suppose S consists of the cylindrical shell

$$S_1 = \{(x, y, z) \,|\, x^2 + y^2 = 1, \, -1 \le z \le 0\}$$

surmounted by the hemisphere

$$S_2 = \{(x, y, z) \,|\, x^2 + y^2 + z^2 = 1, z \ge 0\},$$

and oriented as in Figure 4.13. Let $\mathbf{F}$ be continuously differentiable on S. From an argument like that of Example 4.12,

$$\int_{S_1} (\mathbf{curl\ F}) \cdot \mathbf{n} \, dA = \int_{C_1} \mathbf{F} \cdot d\mathbf{X} + \int_{C_2} \mathbf{F} \cdot d\mathbf{X}, \qquad (41)$$

with C_1 and C_2 directed as in Figure 4.13. Since $-C_1$ is positively directed when regarded as ∂S_2, Example 4.13 implies that

$$\int_{S_2} (\mathbf{curl\ F}) \cdot \mathbf{n} \, dA = -\int_{C_1} \mathbf{F} \cdot d\mathbf{X}. \qquad (42)$$

Adding (41) and (42) yields

$$\int_S (\mathbf{curl\ F}) \cdot \mathbf{n} \, dA = \int_{C_2} \mathbf{F} \cdot d\mathbf{X},$$

as implied by Theorem 4.3.

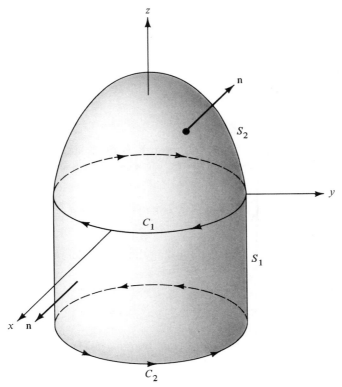

FIGURE 4.13

8.4 EXERCISES

1. Show that $\int_S f \, dA$ is independent of the particular representation chosen for S. (*Hint*: See the proof of Theorem 3.1.)

2. Evaluate $\int_S f \, dA$:
   (a) $f(\mathbf{X}) = (x^2 + y^2)z$; $S = \{\mathbf{X} \mid \|\mathbf{X}\| = a, z \geq 0\}$
   (b) $f(\mathbf{X}) = y^2$; S is the cylinder $x^2 + y^2 = 1$, $0 \leq z \leq 1$, and its top and bottom
   (c) $f(\mathbf{X}) = x - y$; $S = \{\mathbf{X} \mid x + y + z = 1, x^2 + y^2 \leq 1\}$
   (d) $f(\mathbf{X}) = x^2 + 2y^2$; $S = \{\mathbf{X} \mid 0 \leq z \leq 1 - x^2 - y^2\}$
   (e) $f(\mathbf{X}) = x^2 - 2y^2$; $S = \{\mathbf{X} \mid z^2 = x^2 + y^2, 0 \leq z \leq 2\}$
   (f) $f(\mathbf{X}) = y$; S is the part of the cylinder $x^2 + y^2 = 1$ for which $0 \leq z \leq x + 2$, and its top and bottom.

3. Evaluate $\int_S (\alpha x + \beta y + \gamma z) \, dA$, where
$$S = \{\mathbf{X} \mid ax + by + z = 0, x^2 + y^2 \leq r^2, y \geq 0\}.$$

4. The surface integral

$$\int_S (P\mathbf{i} + Q\mathbf{j} + R\mathbf{k}) \cdot \mathbf{n} \, dA$$

is often denoted by

$$\int_S P \, d(y, z) + Q \, d(z, x) + R \, d(x, y).$$

Why?

5. Let $\mathbf{n}$ be an orientation of a surface S, suppose $\mathbf{X}$ is on S, and let $\alpha(\mathbf{X})$, $\beta(\mathbf{X})$, and $\gamma(\mathbf{X})$ be the angles between $\mathbf{n}(\mathbf{X})$ and the x, y, and z axes, respectively. Show that

$$\int_S (P\mathbf{i} + Q\mathbf{j} + R\mathbf{j}) \cdot \mathbf{n} \, dA = \int_S (P \cos \alpha + Q \cos \beta + R \cos \gamma) \, dA.$$

6. Evaluate $\int_S \mathbf{F} \cdot \mathbf{n} \, dA$:
   (a) $\mathbf{F}(\mathbf{X}) = y\mathbf{i} - x\mathbf{j} + \mathbf{k}$: S is the surface bounding the region defined by

   $$-\sqrt{1 - x^2 - y^2} \le z \le 1 - x^2 - y^2;$$

   $\mathbf{n}$ is the outward normal.
   (b) $\mathbf{F}(\mathbf{X}) = x\mathbf{i} + y\mathbf{j} - z\mathbf{k}$; $S = \{\mathbf{X} \,|\, x^2 + y^2 = z^2, 1 \le z \le 3\}$;
   $\mathbf{n}$ has a negative $\mathbf{k}$ component.
   (c) $\mathbf{F}(\mathbf{X}) = \mathbf{X}$; S is the surface bounding the region $x^2 + y^2 + z^2 \le 1$, $|z| \le r \, (<1)$; $\mathbf{n}$ is the outward normal.
   (d) $\mathbf{F}(\mathbf{X}) = x^2\mathbf{i} + y^2\mathbf{j} + z^2\mathbf{k}$;

   $$S = \{\mathbf{X} \,|\, x + y - 2z = 0, 0 \le x + y \le 4, 2 \le x - y \le 6\};$$

   $\mathbf{n}$ has a positive $\mathbf{i}$ component.
   (e) $\mathbf{F}(\mathbf{X}) = 3(x^2 - y^2)\mathbf{i} + 2(y^2 - z^2)\mathbf{j} + 6(z^2 - x^2)\mathbf{k}$;

   $$S = \{\mathbf{X} \,|\, 2x + 3y + z = 2, x^2 + y^2 \le 9\};$$

   $\mathbf{n}$ has a negative $\mathbf{i}$ component.
   (f) $\mathbf{F}(\mathbf{X}) = (x + 1)\mathbf{i} + (y - 2)\mathbf{j} + z\mathbf{k}$;

   $$S = \{\mathbf{X} \,|\, x^2 + y^2 = 2x, 0 \le z \le \sqrt{x^2 + y^2}\};$$

   $\mathbf{n}$ points away from the axis of the cylinder.
   (g) $\mathbf{F}(\mathbf{X}) = x\mathbf{i} + y\mathbf{j}$;

   $$S = \{\mathbf{X} \,|\, x^2 + y^2 + z^2 = 1, z \ge \sqrt{x^2 + y^2}\};$$

   $\mathbf{n}$ points away from the origin.
   (h) $\mathbf{F}(\mathbf{X}) = 2x\mathbf{i} + 3y\mathbf{j} + z^2\mathbf{k}$; S is the surface of the cylindrical region defined by $x^2 + y^2 \le a^2$, $|z| \le 1$; $\mathbf{n}$ is the outward orientation.
   (i) $\mathbf{F}(\mathbf{X}) = y\mathbf{i} - x\mathbf{j} + xyz\mathbf{k}$; S is the surface bounding the region defined by $x^2 + y^2 \le 1$, $0 \le z \le x + 2$; $\mathbf{n}$ is the outward orientation.

7. Under the assumptions of Theorem 4.1:
   (a) Show that Eq. (18) holds if D can be projected into the yz plane.
   (b) Show that Eq. (19) holds if D can be projected into the xz plane.

8. Use the divergence theorem to evaluate the integrals in:
   (a) Exercise 6(a)   (b) Exercise 6(c)   (c) Exercise 6(h)
   (d) Exercise 6(i)

9. Use the divergence theorem to calculate $\int_S \mathbf{F} \cdot \mathbf{n}\, dA$, where S is the surface bounding D and $\mathbf{n}$ is the outward orientation of S:
   (a) $\mathbf{F}(\mathbf{X}) = 3x^2\mathbf{i} + xy\mathbf{j} + z\mathbf{k}$; $D = \{\mathbf{X}\,|\,x + y + z \le 1, x \ge 0, y \ge 0, z \ge 0\}$
   (b) $\mathbf{F}(\mathbf{X}) = 2x\mathbf{i} + 3y\mathbf{j} + z\mathbf{k}$; $D = \{\mathbf{X}\,|\,0 \le z \le 9 - x^2 - y^2\}$
   (c) $\mathbf{F}(\mathbf{X}) = x^2\mathbf{i} + y^2\mathbf{j} + z^2\mathbf{k}$; $D = \{\mathbf{X}\,|\,x^2 + y^2 \le 9, 0 \le z \le x + 1\}$
   (d) $\mathbf{F}(\mathbf{X}) = 2yz\mathbf{i} + 2xz\mathbf{j} + 2xy\mathbf{k}$; $D = \{\mathbf{X}\,|\,\sqrt{x^2 + y^2} \le z \le 2\}$
   (e) $\mathbf{F}(\mathbf{X}) = (2x + 3y)\mathbf{i} + (3y - 2z)\mathbf{j} + (x + 2y + z)\mathbf{k}$;
   $\quad D = \{\mathbf{X}\,|\,2x + y + z \le 2, x \ge 0, y \ge 0, z \ge 0\}$
   (f) $\mathbf{F}(\mathbf{X}) = x\mathbf{i} + y^2\mathbf{j} + z\mathbf{k}$; $D = \{\mathbf{X}\,|\,2 + x^2 + y^2 \le z \le 6\}$
   (g) $\mathbf{F}(\mathbf{X}) = x^2\mathbf{i} + y^2\mathbf{j} + z^2\mathbf{k}$; $D = [a_1, a_2] \times [b_1, b_2] \times [c_1, c_2]$

10. Prove: If the second partial derivatives of P, Q, and R are continuous in a region T and $\mathbf{F} = P\mathbf{i} + Q\mathbf{j} + R\mathbf{k}$, then $\operatorname{div}(\mathbf{curl}\,\mathbf{F}) = 0$ in T.

11. Suppose S is an orientable surface bounding D and $\mathbf{n}$ is the outward orientation on S. Show that the volume of D is

$$V = \tfrac{1}{3} \int_S \mathbf{X} \cdot \mathbf{n}\, dA.$$

12. (a) Suppose $S_r(\mathbf{X}_0) = \{\mathbf{X}\,|\,|\mathbf{X} - \mathbf{X}_0| = r\}$ and $\mathbf{n}$ is the outward orientation of $S_r(\mathbf{X}_0)$. Evaluate

$$\int_{S_r(\mathbf{X}_0)} \frac{(\mathbf{X} - \mathbf{X}_0) \cdot \mathbf{n}}{|\mathbf{X} - \mathbf{X}_0|^\alpha}\, dA.$$

   (b) Prove: If S is a piecewise smooth orientable surface bounding a region D and $\mathbf{X}_0 \in D^0$, then

$$\int_S \frac{(\mathbf{X} - \mathbf{X}_0) \cdot \mathbf{n}}{|\mathbf{X} - \mathbf{X}_0|^3}\, dA = 4\pi.$$

13. (a) Suppose D is a region in $\mathscr{R}^3$ to which the divergence theorem applies, let S be its surface, and assume that u, v, v_x, v_y and v_z are continuously differentiable on S and D. By applying the divergence theorem to $\mathbf{F} = u\,\nabla v$, show that

$$\int_S u\,\frac{\partial v}{\partial \mathbf{n}}\, dA = \int_D (\nabla u \cdot \nabla v + u\,\nabla^2 v)\, d\mathbf{X},$$

   where $\partial v/\partial \mathbf{n}$ is the directional derivative of v in the direction of the outward normal to S and $\nabla^2 v = v_{xx} + v_{yy} + v_{zz}$, the Laplacian of v. (This is *Green's first theorem*; see Exercise 32, Section 8.2.)

(b) In addition to the assumptions of part (a), suppose u_{xx}, u_{yy}, and u_{zz} are continuous on S and D. Show that

$$\int_S \left(u \frac{\partial v}{\partial \mathbf{n}} - v \frac{\partial u}{\partial \mathbf{n}} \right) dA = \int_D \left(u \nabla^2 v - v \nabla^2 u \right) d\mathbf{X}.$$

(This is *Green's second theorem*; see Exercise 32, Section 8.2.)

14. A function u is said to be harmonic in a region D in $\mathcal{R}^3$ if u_x, u_y, and u_z are continuously differentiable on an open set containing D, and

$$\nabla^2 u = u_{xx} + u_{yy} + u_{zz} = 0$$

in D.
   (a) Show that the function $v(\mathbf{X}) = |\mathbf{X} - \mathbf{X}_0|^{-1}$ is harmonic in every region in $\mathcal{R}^3$ which does not contain $\mathbf{X}_0$.
   (b) Let u be harmonic in a neighborhood N of $\mathbf{X}_0$ and let S_r be a sphere in N of radius r, with center at $\mathbf{X}_0$. Show that

$$u(\mathbf{X}_0) = \frac{1}{4\pi r^2} \int_{S_r} u \, dA;$$

that is, the value of u at the center of S_r is the average of its values on S_r. [*Hint*: Apply Exercise 13(b) to the region $D = \{\mathbf{X} | 0 < r_1 \le |\mathbf{X} - \mathbf{X}_0| \le r\}$ and let $r_1 \to 0+$.]

15. Suppose that $\mathbf{F}$ is continuously differentiable in a neighborhood of $\mathbf{X}_0$ in $\mathcal{R}^3$, let S_r be the sphere of radius r about $\mathbf{X}_0$, and let $\mathbf{n}$ be the outward orientation of S_r. Show that

$$\text{div } \mathbf{F}(\mathbf{X}_0) = \lim_{r \to 0+} \frac{1}{4\pi r^2} \int_{S_r} \mathbf{F} \cdot \mathbf{n} \, dA.$$

16. Verify Stokes' theorem for the given $\mathbf{F}$ and S:
   (a) $\mathbf{F}(\mathbf{X}) = 2z\mathbf{i} + 3x\mathbf{j} + 4y\mathbf{k}$; $S = \{\mathbf{X} | 0 \le z = 9 - x^2 - y^2\}$
   (b) $\mathbf{F}(\mathbf{X}) = (4x^2 + y)\mathbf{i} + 2y\mathbf{j} + (x - 3z^2)\mathbf{k}$;
       $S = \{\mathbf{X} | 3x + y + 3z = 4, x \ge 0, y \ge 0, z \ge 0\}$
   (c) $\mathbf{F}(\mathbf{X}) = 2x\mathbf{i} + 3z\mathbf{j} - 2y\mathbf{k}$; $S = \{\mathbf{X} | |\mathbf{X}| = 3, y \ge 0\}$
   (d) $\mathbf{F}(\mathbf{X}) = 4x^3\mathbf{i} + 2y^2\mathbf{j}$; $S = \{\mathbf{X} | z = x^2 + y^2 \le 4\}$
   (e) $\mathbf{F}(\mathbf{X}) = 2y\mathbf{i} + 3x\mathbf{j} + z\mathbf{k}$; $S = \{\mathbf{X} | x^2 + y^2 = 4, 1 \le z \le x + 4\}$
   (f) $\mathbf{F}(\mathbf{X}) = (y + x)\mathbf{i} + (x + z)\mathbf{j} + z^2\mathbf{k}$; $S = \{\mathbf{X} | z = \sqrt{x^2 + y^2} \le 1\}$

17. Verify Eq. (39) under the assumptions of Example 4.13.

18. Verify Eq. (40) under the assumptions of Example 4.13.

answers to selected exercises

CHAPTER 1

Section 1.1, pages 11–12

5. (a) ∞ (no); -1 (yes)   (b) 3 (no); -3 (no)   (c) $\sqrt{7}$ (no); $-\sqrt{7}$ (yes)
   (d) 2 (no); -3 (no)   (e) 1 (no); -1 (no)   (f) $\sqrt{7}$ (no); $-\sqrt{7}$ (no)

Section 1.2, pages 18–20

7. (a) $2^n/(2n)!$ (b) $2 \cdot 3^n/(2n + 1)!$ (c) $2^{-n}(2n)!/(n!)^2$ (d) $n^n/n!$
8. (b) no   9. (b) no

Section 1.3, pages 29–31

1. (a) $[\frac{1}{2}, 1)$; $(-\infty, \frac{1}{2}) \cup [1, \infty)$; $(-\infty, 0] \cup (\frac{3}{2}, \infty)$; $(0, \frac{3}{2}]$; $(-\infty, 0] \cup (\frac{3}{2}, \infty)$;
   $(-\infty, \frac{1}{2}] \cup [1, \infty)$
   (b) $(-3, -2) \cup (2, 3)$; $(-\infty, -3] \cup [-2, 2] \cup [3, \infty)$; ϕ; $(-\infty, \infty)$; ϕ;
   $(-\infty, -3] \cup [-2, 2] \cup [3, \infty)$

(c) $\phi; (-\infty, \infty); \phi; (-\infty, \infty); \phi; (-\infty, \infty)$

(d) $\phi; (-\infty, \infty); [-1, 1]; (-\infty, -1) \cup (1, \infty); [-1, 1]; (-\infty, \infty)$

2. (a) $(0, 3]$ (b) $[1, 2]$ (c) $(-\infty, 1) \cup (2, \infty)$ (d) $(-\infty, 0] \cup (3, \infty)$

3. (a) $\frac{1}{4}$ (b) $\frac{1}{6}$ (c) 6   (d) 1

4. (a) neither; $(-1, 2) \cup (3, \infty); (-\infty, -1) \cup (2, 3); (-\infty, -1] \cup [2, 3]$

   (b) open; $S; (1, 2); [1, 2]$

   (c) closed; $(-3, -2) \cup (7, 8); (-\infty, -3) \cup (-2, 7) \cup (8, \infty);$
$(-\infty, -3] \cup [-2, 7] \cup [8, \infty)$

   (d) closed; $\phi; \bigcup \{(n, n + 1) | n = \text{integer}\}; (-\infty, \infty)$

18. (a) $\{x | x = 1/n, n = 1, 2, \ldots\}$ (b) ϕ

   (c), (d) $S_1 = \text{rationals}, S_2 = \text{irrationals}$

   (e) any set whose least upper bound is an isolated point of the set

   (f), (g) the rationals   (h) $S_1 = \text{rationals}, S_2 = \text{irrationals}$

CHAPTER 2

Section 2.1, pages 50–56

2. (a), (b) $\{x | x \neq (2k + 1)\pi/2, \text{ where } k = \text{integer}\}$ (c) $\{x | x \neq 0, 1\}$

   (d) $\{x | x \neq 0\}$ (e) $(1, \infty)$

3. $D_f = [-2, 1) \cup [3, \infty), D_g = (-\infty, -3] \cup [3, 7) \cup (7, \infty),$
$D_{f \pm g} = D_{fg} = [3, 7) \cup (7, \infty), D_{f/g} = (3, 4) \cup (4, 7) \cup (7, \infty)$

4. (a) 4   (b) 12   (c) -1 (d) no limit

6. (a) $\frac{11}{17}$ (b) $-\frac{2}{3}$ (c) $\frac{1}{3}$ (d) 2

7. (a) 0, 2   (b) 0, none   (c) $-\frac{1}{3}, \frac{1}{3}$

   (d) none, 0   (e) $-\log 2$, none   (f) $-1, 1$

15. (a) 0   (b) 0   (c) none   (d) 0   (e) none   (f) 0

18. (a) 0   (b) 0   (c), (d), (e) none   (f) 0

20. (a) ∞ (b) $-\infty$ (c) ∞ (d) ∞ (e) ∞ (f) $-\infty$

22. (a) none   (b) ∞ (c) ∞ (d) none

24. (a) ∞ (b) ∞ (c) ∞ (d) $-\infty$ (e) none   (f) ∞

30. (a) $\frac{3}{2}$ (b) $\frac{3}{2}$ (c) ∞ (d) $-\infty$ (e) ∞ (f) $\frac{1}{2}$

31. $\lim_{x \to \infty} r(x)$ is ∞ if $n > m$ and $a_n/b_m > 0$, $-\infty$ if $n > m$ and $a_n/b_m < 0$, a_n/b_m if $n = m$, 0 if $n < m$. $\lim_{x \to -\infty} r(x) = (-1)^{n-m} \lim_{x \to \infty} r(x)$

32. $\lim_{x \to x_0} f(x) = \lim_{x \to x_0} g(x)$

35. (c) $\overline{\lim}_{x \to x_0-} (f - g)(x) \leq \overline{\lim}_{x \to x_0-} f(x) - \underline{\lim}_{x \to x_0-} g(x),$
$\underline{\lim}_{x \to x_0-} (f - g)(x) \geq \underline{\lim}_{x \to x_0-} f(x) - \overline{\lim}_{x \to x_0-} g(x)$

Section 2.2, pages 72–76

3. (a) from the right   (b) continuous   (c) none   (d) continuous

   (e) none   (f) continuous   (g) from the left

4. $[0, 1), (0, 1), [1, 2), (1, 2), (1, 2]$ 5. $[0, 1), (0, 1), (1, \infty)$
13. (b) $\tanh x$ is continuous for all x, $\coth x$ for all $x \neq 0$
16. (b) no
21. (a) $[-1, 1], [0, \infty)$ (b) $\bigcup_{n=-\infty}^{\infty} (2n\pi, (2n + 1)\pi), (0, \infty)$
     (c) $\bigcup_{n=-\infty}^{\infty} (k\pi, (k + 1)\pi), (-\infty, -1) \cup (-1, 1) \cup (1, \infty)$
     (d) $\bigcup_{n=-\infty}^{\infty} [n\pi, (n + \frac{1}{2})\pi], [0, \infty)$
23. (a) $(-1, 1)$ (b) $(-\infty, \infty)$
     (c) $x_0 \neq (2k + \frac{3}{2})\pi, k = $ integer   (d) $x \neq \frac{1}{2}$
     (e) $x \neq 1$ (f) $x \neq (k + \frac{1}{2})\pi, k = $ integer
     (g) $x \neq (k + \frac{1}{2})\pi, k = $ integer   (h) $x \neq 0$
     (i) $x \neq 0$

Section 2.3, pages 90–93

4. $f^{(k)}(x) = n(n - 1) \cdots (n - k - 1)x^{n-k-1}|x|$ if $1 \leq k \leq n - 1$; $f^{(n)}(x) = n!$ if $x > 0$, $-n!$ if $x < 0$; $f^{(k)}(x) = 0$ if $k > n$ and $x \neq 0$; $f^{(k)}(0)$ does not exist if $k \geq n$.
5. (b) $p'_-(c)$ and $\lim_{x \to c+} (q(x) - p(c))/(x - c)$ must exist and be equal.
7. (a) $c' = ac - bs$, $s' = bc + as$ (b) $c(x) = e^{ax} \cos bx$, $s(x) = e^{ax} \sin bx$
15. (b) $f(x) = -1$ if $x \leq 0$, $f(x) = 1$ if $x > 0$; then $f'(0+) = 0$, but $f'_+(0)$ does not exist.
     (c) continuous from the right
19. $f(x) = x^{2n} \sin(1/x)$ if $x \neq 0$, $f(0) = 0$.
23. There is no such function (Theorem 3.6).
25. Counterexample: Let $x_0 = 0$, $f(x) = |x|^{3/2} \sin(1/x)$ if $x \neq 0$, $f(0) = 0$.
33. Counterexample: Let $x_0 = 0$, $f(x) = x/|x|$ if $x \neq 0$, $f(0) = 0$.

Section 2.4, pages 101–104

2. 1	3. $\frac{1}{2}$	4. ∞	5. $(-1)^{n-1}n$
6. 1	7. 0	8. 1	9. 0
10. 0	11. 0	12. ∞	13. 0
14. $-\frac{1}{2}$	15. 0	16. 0	17. 1
18. 1	19. 1	20. e	21. 1
22. 0	23. $-\infty$ if $\alpha \leq 0$, 0 if $\alpha > 0$		24. $1/e$
25. e^2	26. 1	27. 0	28. 0
29. ∞ if $\alpha > 0$, $-\infty$ if $\alpha \leq 0$		30. ∞	31. 1
32. $1/120$	33. ∞		34. $-\infty$
35. $-\infty$ if $\alpha \leq 0$, 0 if $\alpha > 0$			36. $e/2$
37. ∞	38. 1		39. e
40. 0 if $m - n$ is even; no limit if $m - n$ is odd			
41. ∞	42. 0	43. ∞	44. ∞
45. 2	46. 1	47. 1	48. 1

49. 0   50. $e^{-1/4}$ 51. ∞
52. 0 if $k \le n$; $1/(n + 1)!$ if $k = n + 1$; ∞ if $k - n - 1$ is even and positive; no limit if $k - n - 1$ is odd and positive
53. 0 if $k \le n$; $(-1)^{n+1}/(2n + 3)!$ if $k = n + 1$; $(-1)^{n+1}\infty$ if $k > n + 1$
55. (b) Suppose g' is continuous at x_0 and $f(x) = g(x)$ if $x \le x_0$, $f(x) = 1 + g(x)$ if $x > x_0$.
58. (a) 1   (b) e (c) 1   59. e^L

Section 2.5, pages 113–118

2. (b) Counterexample: Let $x_0 = 0$ and $f(x) = x|x|$.
3. (b) Let $x_0 = 0$, $f(x) = x^k(1 + |x|)$. (See Exercise 4, Sec. 2.3.)
6. (b) (i) 1, 2, 2, 0   (ii) 0, $-\pi$, $3\pi/2$, $-4\pi + \pi^3/2$
    (iii) $-\pi^2/4$, -2π, $-6 + \pi^2/4$, 4π (iv) $-2, 5, -16, 65$
7. (b), (d) 0, -1, 0, 5
8. (b) (i) 0, 1, 0, 5   (ii) $-1, 0, 6, -24$ (iii) $\sqrt{2}, 3\sqrt{2}, 11\sqrt{2}, 57\sqrt{2}$
    (iv) $-1, 3, -14, 88$
10. (a) 1, 0, 1, 0   (b) $-\sqrt{2}, -2, -8\sqrt{2}, -96$ (c) $1, -2, 11, -100$
     (d) $1, -2, 9, -40$ (e) $\frac{1}{2}, -\frac{1}{2}, 1, -3$ (f) $1, -2, 13, -128$
11. (a) min   (b) neither   (c) min   (d) max
     (e) min   (f) neither   (g) neither   (h) min
12. $f(x) = e^{-1/x^2}$ if $x \ne 0$, $f(0) = 0$ (see Exercise 1)
13. None if $b^2 - 4c < 0$; local min at $x_1 = (-b + \sqrt{b^2 - 4c})/2$ and local max at $x_2 = (-b - \sqrt{b^2 - 4c})/2$ if $b^2 - 4c > 0$; if $b^2 = 4c$, then $x = -b/2$ is a critical point, but not a local extreme point.
14. $aA\alpha$, $aB\beta$, $aC\gamma$, $bA\beta$, $bB\beta$, $bC\gamma$, $cA\gamma$, $cB\gamma$, $cC\gamma$
15. (a) .000102   (b) .002   (c) .014   (d) $.16 \times 10^{-8}$
19. (a) $\le M_3 h/3$, where $M_3 = $ l.u.b.$\{|f^{(3)}(c)| \, | \, x - h < c < x + h\}$
     (b) $\le M_4 h^2/12$, where $M_4 = $ l.u.b.$\{|f^{(4)}(c)| \, | \, x - h < c < x + h\}$
20. $k = -h/2$

CHAPTER 3

Section 3.1, pages 133–134

6. (b) $s(P)$ is the g.l.b. of the set of all Riemann sums of f over P.
7. (b) monotonic functions
    (c) Let $[a, b] = [0, 1]$ and P be the two points 0 and 1. Let $f(0) = f(1) = \frac{1}{2}$ and $f(x) = x$ if $0 < x < 1$. Then $s(P) = 0$ and $S(P) = 1$, but neither is a Riemann sum of f over P.
8. (a) $\frac{1}{2}, -\frac{1}{2}$ (b) $\frac{1}{2}, 1$ 9. $e^b - e^a$

10. $1 - \cos b$ 11. $\sin b$

14. $f(a)(g_1 - g(a)) + f(b)(g(b) - g_2) + \sum_{j=1}^{n-1} f(a_j)(g_{j+1} - g_j)$

15. (a) If $g \equiv 1$ and f is arbitrary on $[a, b]$, then $\int_a^b f(x)\,dg(x) = 0$.

Section 3.3, pages 154–157

5. (b) Yes; let $f(x) = 1$ if x is rational, $f(x) = -1$ if x is irrational

7. (a) $\bar{u} = c = \frac{2}{3}$ (b) $\bar{u} = c = 0$ (c) $\bar{u} = (e - 2)/(e - 1)$, $c = (\bar{u})^{1/2}$

Section 3.4, pages 170–176

4. (a) (i) $p \geq 2$ (ii) $p > 0$ (b) (i) $p \geq 2$ (ii) $p > 0$
   (c) (i) none (ii) $p > 0$ (d) (i) $p \leq 0$ (ii) $0 < p < 1$
   (e) (i) none (ii) none

5. (a) $n!$ (b) $\frac{1}{2}$ (c) div   (d) 1   (e) -1 (f) 0

6. For the second part: The first equation is valid if its right side is not indeterminate. The second is valid if $k \neq 0$.

11. (a) div.   (b) conv.   (c) div.   (d) conv.   (e) conv.   (f) div.

12. Let $f(x) = x^{-1}$ and $g(x) = x^{-2}$; then $\lim_{x \to \infty} f(x)/g(x) = \infty$ and $\lim_{x \to \infty} g(x)/f(x) = 0$, while $\int_1^\infty f(x)\,dx = \infty$ and $\int_1^\infty g(x)\,dx < \infty$.

13. (a) $p < 2$ (b) $p < 1$ (c) $p > -1$
   (d) $-1 < p < 2$ (e) none   (f) none
   (g) $p < 1$

15. (a) $p - q < 1$ (b) $p, q < 1$ (c) $-1 < p < 2q - 1$
   (d) $q > -1$, $p + q > 1$ (e) $p + q > 1$ (f) $q + 1 < p < 3q + 1$

16. $\deg g - \deg f \geq 2$

22. (a) (i) $p > 1$ (ii) $0 < p \leq 1$ (b) (i) $p > 1$ (ii) $p \leq 1$
   (c) (i) $p > 1$ (ii) $0 \leq p \leq 1$ (d) (i) $p > 0$ (ii) none
   (e) (i) $1 < p < 4$ (ii) $0 < p \leq 1$ (f) (i) $p > \frac{1}{2}$ (ii) $0 < p \leq \frac{1}{2}$
   (g) (i) none (ii) $0 < p < 1$

30. (a) (i) $p > -1$ (ii) $-2 < p \leq -1$ (b) (i) $p > -1$ (ii) none
   (c) (i) $p < -1$ (ii) none   (d) (i) none (ii) none
   (e) (i) $p < 0$ (ii) $p > 1$ (f) (i) $p < -1$ (ii) $p > 1$

CHAPTER 4

Section 4.1, pages 199–202

2. (a) 2   (b) 1   (c) 0

11. (a) 1   (b) 1   (c) 1
   (d) $-\infty$ (e) ∞ if $a > 1$; $-\infty$ if $a \leq 1$ (f) 0

17. (a) 1   (b) 0 if $0 < r < 1$; $\frac{1}{2}$ if $r = 1$; 1 if $r > 1$
    (c) ∞ (d) ∞
    (e) $-\infty$ (f) ∞ if $r > s$; a_s/b_s if $r = s$; 0 if $r < s$
19. If $s_n = 1$, $t_n = -1/n$, then $(\lim_{n \to \infty} s_n)/(\lim_{n \to \infty} t_n) = 1/0 = \infty$, but $\lim_{n \to \infty} s_n/t_n = -\infty$.
21. (a) $\infty, 0$
    (b) $\infty, -\infty$ if $|r| > 1$; $2, -2$ if $r = -1$; $0, 0$ if $r = 1$; $1, -1$ if $|r| < 1$
    (c) $\infty, -\infty$ if $r < -1$; $0, 0$ if $|r| < 1$; $\frac{1}{2}, \frac{1}{2}$ if $r = 1$; ∞, ∞ if $r > 1$
    (d) $\infty, -\infty$ (e) $|t|, -|t|$
29. If $\{s_n\} = \{1, 0, 1, 0, \ldots\}$, then $\lim_{n \to \infty} t_n = \frac{1}{2}$.

Section 4.2, pages 207–208

2. (a) $\lim_{m \to \infty} s_{2m} = \infty$, $\lim_{m \to \infty} s_{2m+1} = -\infty$
   (b) $\lim_{m \to \infty} s_{4m} = 1$, $\lim_{m \to \infty} s_{4m+2} = -1$, $\lim_{m \to \infty} s_{2m+1} = 0$
   (c) $\lim_{m \to \infty} s_{2m} = 0$, $\lim_{m \to \infty} s_{4m+1} = 1$, $\lim_{m \to \infty} s_{4m+3} = -1$
   (d) $\lim_{n \to \infty} s_n = 0$ (e) $\lim_{m \to \infty} s_{2m} = \infty$, $\lim_{m \to \infty} s_{2m+1} = 0$
   (f) $\lim_{m \to \infty} s_{8m} = \lim_{m \to \infty} s_{8m+2} = 1$, $\lim_{m \to \infty} s_{8m+1} = \sqrt{2}$,
      $\lim_{m \to \infty} s_{8m+3} = \lim_{m \to \infty} s_{8m+7} = 0$, $\lim_{m \to \infty} s_{8m+5} = -\sqrt{2}$,
      $\lim_{m \to \infty} s_{8m+4} = \lim_{m \to \infty} s_{8m+6} = -1$
3. $\{1, 2, 1, 2, 3, 1, 2, 3, 4, 1, 2, 3, 4, 5, \ldots\}$
8. Let $\{t_n\}$ be any convergent sequence and $\{s_n\} = \{t_1, 1, t_2, 2, \ldots, t_n, n, \ldots\}$.

Section 4.3, pages 239–244

5. (b) no; consider $\sum 1/n$
10. (a) conv.   (b) conv.   (c) div.   (d) div.
    (e) conv.   (f) conv.   (g) div.   (h) conv.
12. (a) $p > 1$ (b) $p > 1$ (c) $p > 1$
17. (a) conv.   (b) conv. if $0 < r < 1$, div. if $r \geq 1$ (c) div.
    (d) conv.   (e) div.   (f) conv.
21. (a) conv.   (b) conv.   (c) conv.   (d) conv.
23. (a) div.   (b) conv. if $0 < r < 1$, div. if $r \geq 1$ (c) conv.
    (d) conv.   (e) conv.
24. (a) div.   (b) conv.   (c) conv.
    (d) conv. if $\alpha < \beta - 1$, div. if $\alpha \geq \beta - 1$
25. (a) div.   (b) conv.   (c) div.   (d) conv.

26. (a) $\sum (-1)^n$ (b) $\sum (-1)^n/n$, $\sum \left[\dfrac{(-1)^n}{n} + \dfrac{1}{n \log n} \right]$

    (c) $\sum (-1)^n 2^n$ (d) $\sum (-1)^n$
31. (a) cond. conv.   (b) cond. conv.   (c) abs. conv.   (d) abs. conv.

32. Let k and s be the degrees of the numerator and denominator, respectively. If $|r| = 1$, the series converges absolutely if and only if $s \geq k + 2$. It diverges if $r = 1$; if $r = -1$, it converges conditionally if and only if $s = k + 1$.

34. (b) $\sum (-1)^n/\sqrt{n}$

Section 4.4, pages 265–268

1. (a) $F(x) = 0,\ |x| \leq 1$ (b) $F(x) = 0,\ |x| \leq 1$
   (c) $F(x) = 0,\ -1 < x \leq 1$ (d) $F(x) = \sin x,\ -\infty < x < \infty$
   (e) $F(x) = 1,\ -1 < x \leq 1;\ F(x) = 0,\ |x| > 1$
   (f) $F(x) = x,\ -\infty < x < \infty$ (g) $F(x) = x^2/2,\ -\infty < x < \infty$
   (h) $F(x) = 0,\ -\infty < x < \infty$ (i) $F(x) = 1,\ -\infty < x < \infty$

5. (a) $F(x) = 0$ (b) $F(x) = 1,\ |x| < 1;\ F(x) = 0,\ |x| > 1$
   (c) $F(x) = (\sin x)/x$

6. (c) $F_n(x) = x^n;\ S_k = [1/k,\ 1 - 1/k],\ k = 1, 2, \ldots$

7. (a) $S = [-1, 1]$ (b) $[-r, r] \cup \{1\} \cup \{-1\},\ 0 < r < 1$
   (c) $[-r, r] \cup \{1\},\ 0 < r < 1$ (d) $[-r, r],\ r > 0$
   (e) $(-\infty, -1/r] \cup [-r, r] \cup [1/r, \infty) \cup \{1\},\ 0 < r < 1$
   (f) $[-r, r],\ r > 0$ (g) $[-r, r],\ r > 0$
   (h) $(-\infty, -r] \cup [r, \infty) \cup \{0\},\ r > 0$ (i) $[-r, r],\ r > 0$

13. (b) Let $S = (0, 1]$, $F_n(x) = \sin(x/n)$, $G_n(x) = x^{-2}$; then $F = 0$, $G = x^{-2}$, and the convergence is uniform, but $\|F_n G_n\|_S = \infty$.

16. (a) 3   (b) 1   (c) $\frac{1}{2}$ (d) $e - 1$

20. (a) compact subsets of $(-\frac{1}{2}, \infty)$ (b) $(-\frac{1}{2}, \infty)$

   (c) closed subsets of $\left(\dfrac{1 - \sqrt{5}}{2},\ \dfrac{1 + \sqrt{5}}{2} \right)$ (d) $(-\infty, \infty)$

   (e) $[r, \infty),\ r > 1$
   (f) compact subsets of $(-\infty, 0) \cup (0, \infty)$

22. (a) Let $S = (-\infty, \infty)$, $f_n = a_n$ (constant), where $\sum a_n$ converges conditionally, and $g_n = |a_n|$.
   (b) "absolutely"

23. (a) (i) means that $\sum |f_n(x)|$ converges pointwise and $\sum f_n(x)$ converges uniformly on S;   (ii) means that $\sum |f_n(x)|$ converges uniformly on S

31. (a) $\displaystyle\sum_{n=0}^{\infty} (-1)^n \frac{x^{2n+1}}{n!(2n+1)!}$ (b) $\displaystyle\sum_{n=0}^{\infty} (-1)^n \frac{x^{2n+1}}{(2n+1)(2n+1)!}$

Section 4.5, pages 288–293

2. (a) $1/3e$ (b) 1   (c) $\frac{1}{3}$ (d) 1   (e) ∞
8. (a) 1   (b) $\frac{1}{2}$ (c) $\frac{1}{4}$ (d) 4   (e) $1/e$ (f) 1

10. $x(1 + x)(1 - x)^{-3}$ 12. e^{-x^2}

16. $\displaystyle\sum_{n=1}^{\infty} \frac{(-1)^{n-1}}{n^2}(x - 1)^n; \ R = 1$

17. $\displaystyle\mathrm{Tan}^{-1} x = \sum_{n=0}^{\infty} (-1)^n \frac{x^{2n+1}}{(2n + 1)}; \ f^{(2n)}(0) = 0; \ f^{(2n+1)}(0) = (-1)^n(2n)!;$

$\displaystyle\frac{\pi}{6} = \mathrm{Tan}^{-1} \frac{1}{\sqrt{3}} = \sum_{n=0}^{\infty} \frac{(-1)^n}{(2n + 1)3^{n+1/2}}$

22. $\displaystyle\cosh x = \sum_{n=0}^{\infty} \frac{x^{2n}}{(2n)!}, \ \sinh x = \sum_{n=0}^{\infty} \frac{x^{2n+1}}{(2n + 1)!}$

23. $(1 - x) \sum_{n=0}^{\infty} x^n = 1$ converges for all x

24. (a) $x + x^2 - \dfrac{x^3}{3} - \dfrac{x^5}{30} + \cdots$ (b) $1 - x - \dfrac{x^2}{2} + \dfrac{5x^3}{6} + \cdots$

   (c) $1 - \dfrac{x^2}{2} + \dfrac{x^4}{24} - \dfrac{721}{720} x^6 + \cdots$ (d) $x^2 - \dfrac{x^3}{2} + \dfrac{x^4}{6} - \dfrac{x^5}{6} + \cdots$

27. (a) $1 + x + \dfrac{2x^2}{3} + \dfrac{x^3}{3} + \cdots$ (b) $1 - x - \dfrac{x^2}{2} + \dfrac{3x^3}{2} + \cdots$

   (c) $1 + \dfrac{x^2}{2} + \dfrac{5x^4}{24} + \dfrac{61x^5}{720} + \cdots$ (d) $1 + \dfrac{x^2}{6} + \dfrac{7x^4}{360} + \dfrac{31x^6}{15120} + \cdots$

   (e) $2 - x^2 + \dfrac{x^4}{12} - \dfrac{x^6}{360} + \cdots$

28. $F(x) = \dfrac{5}{(1 - 3x)(1 + 2x)} = \dfrac{3}{1 - 3x} + \dfrac{2}{1 + 2x} = \sum_{n=0}^{\infty} [3^{n+1} - (-2)^{n+1}]x^n$

29. 1

CHAPTER 5

Section 5.1, pages 314–317

1. (a) $(3, 0, 3, 3)$ (b) $(-1, -1, 4)$ (c) $(\frac{1}{6}, \frac{11}{12}, \frac{23}{24}, \frac{5}{36})$
3. (a) $\sqrt{15}$ (b) $\sqrt{65}/12$ (c) $\sqrt{31}$ (d) $\sqrt{3}$
4. (a) $\sqrt{89}$ (b) $\sqrt{166}/12$ (c) 3   (d) $4\sqrt{2}$
5. (a) 12   (b) $\frac{1}{32}$ (c) 27
9. $\mathbf{X} = \mathbf{X}_0 + t\mathbf{U}(-\infty < t < \infty)$ in all cases
10. $\ldots \mathbf{U}$ and $\mathbf{X} - \mathbf{X}_0$ are scalar multiples of $\mathbf{V}$.

11. (a) $\mathbf{X} = (1, -3, 4, 2) + t(1, 3, -5, 3)$
    (b) $\mathbf{X} = (3, 1, -2, 1, 4) + t(-1, -1, 1, 3, -7)$
    (c) $\mathbf{X} = (1, 2, -1) + t(-1, -3, 0)$
12. (a) 5   (b) 2   (c) $1/2\sqrt{2}$
13. (a) (i) $\{(x_1, x_2, x_3, x_4) | |x_i| \le 3 \ (i = 1, 2, 3) \text{ with at least one equality}\}$
    (ii) $\{(x_1, x_2, x_3, x_4) | |x_i| \le 3 \ (i = 1, 2, 3)\}$ (iii) S
    (iv) $\{(x_1, x_2, x_3, x_4) | |x_i| > 3 \text{ for at least one of } i = 1, 2, 3\}$
    (b) (i) S (ii) S (iii) $\varnothing$ (iv) $\{(x, y, z) | z \ne 1 \text{ or } x^2 + y^2 > 1\}$
14. (a) open   (b) neither   (c) closed
20. (a) $(\pi, 1, 0)$ (b) $(1, 0, e)$
21. (a) 6   (b) 6   (c) $2\sqrt{5}$ (d) $2L\sqrt{n}$ (e) ∞
31. $\{(x, y) | x^2 + y^2 = 1\}$
35. ... if for each A there is a K such that $|\mathbf{X}_k| > A$ if $k \ge K$.

Section 5.2, pages 331–333

1. (a) 10   (b) 3   (c) 1   (d) 0   (e) 0   (f) 0
3. (b) $a/(1 + a^2)$ (c) $\lim_{x \to 0+} g(x, \sqrt{x}) = \infty$
5. (a) ∞ (b) ∞ (c) no   (d) $-\infty$ (e) no
6. (a) 0   (b) 0   (c) none   (d) 0   (e) none
7. (a) ... if D_f is unbounded and for each M there is an R such that $f(\mathbf{X}) > M$
    if $\mathbf{X} \in D_f$ and $|\mathbf{X}| > R$.
    (b) replace "$>M$" by "$<M$" in (a)
8. $\lim_{\mathbf{X} \to \mathbf{0}} f(\mathbf{X}) = 0$ if $a_1 + a_2 + \cdots + a_n > b$; no limit if $a_1 + a_2 + \cdots + a_n \le b$
    and $a_1^2 + a_2^2 + \cdots + a_n^2 \ne 0$; $\lim_{\mathbf{X} \to \mathbf{0}} f(\mathbf{X}) = \infty$ if $a_1 = a_2 = \cdots = a_n = 0$
    and $b > 0$.
9. No; for example, $\lim_{x \to \infty} g(x, \sqrt{x}) = 0$
14. (a) $\mathscr{R}^3$ (b) $\mathscr{R}^2$ (c) $\mathscr{R}^3$
    (d) $\mathscr{R}^2 - \{(0, 0)\}$ (e) $\{(x, y) | x \ge y\}$ (f) $\mathscr{R}^n - \{\mathbf{0}\}$
15. (a) $\mathscr{R}^3 - \{(0, 0, 0)\}$ (b) $\mathscr{R}^2$ (c) $\mathscr{R}^2$ (d) $\mathscr{R}^2$ (e) $\mathscr{R}^2$
16. $f(x, y) = xy/(x^2 + y^2)$ if $(x, y) \ne (0, 0)$; $f(0, 0) = 0$

Section 5.3, pages 352–356

1. (a) $\dfrac{2}{\sqrt{3}}(x + y \cos x - xy \sin x) - 2\sqrt{\dfrac{2}{3}}(x \cos x)$

    (b) $\dfrac{(1 - 2y)}{\sqrt{3}} e^{-x+y^2+2z}$ (c) $\dfrac{2}{\sqrt{n}}(x_1 + x_2 + \cdots + x_n)$

    (d) $(1 + x + y + z)^{-1}$

2. $\phi_1^2\phi_2$

3. (a) $-5\pi/\sqrt{6}$ (b) $-2e$ (c) 0   (d) 0

5. (a) $f_x = f_y = (x + y + 2z)^{-1}$, $f_z = 2(x + y + 2z)^{-1}$

   (b) $f_x = 2x + 3yz + 2y$, $f_y = 3xz + 2x$, $f_z = 3xy$

   (c) $f_x = e^{yz}$, $f_y = xze^{yz}$, $f_z = xye^{yz}$

   (d) $f_x = 2xy \cos x^2 y$, $f_y = x^2 \cos x^2 y$, $f_z = 1$

7. (a) $f_{xx} = f_{yy} = f_{xy} = f_{yx} = -(x + y + 2z)^{-2}$, $f_{xz} = f_{zx} = f_{yz} = f_{zy} = -2(x + y + 2z)^{-2}$, $f_{zz} = -4(x + y + 2z)^{-2}$

   (b) $f_{xx} = 2$, $f_{yy} = f_{zz} = 0$, $f_{xy} = f_{yx} = 3z + 2$, $f_{xz} = f_{zx} = 3y$, $f_{yz} = f_{zy} = 3x$

   (c) $f_{xx} = 0$, $f_{yy} = xz^2 e^{yz}$, $f_{zz} = xy^2 e^{yz}$, $f_{xy} = f_{yx} = ze^{yz}$, $f_{xz} = f_{zx} = ye^{yz}$, $f_{yz} = f_{zy} = xe^{yz}$

   (d) $f_{xx} = 2y \cos x^2 y - 4x^2 y^2 \sin x^2 y$, $f_{yy} = -x^4 \sin x^2 y$, $f_{zz} = 0$, $f_{xy} = f_{yx} = 2x \cos x^2 y - 2x^3 y \sin x^2 y$, $f_{xz} = f_{zx} = 0$, $f_{yz} = f_{zy} = 0$

8. (a) $f_{xx}(0, 0) = f_{yy}(0, 0) = 0$, $f_{xy}(0, 0) = -1$, $f_{yx}(0, 0) = 1$

   (b) $f_{xx}(0, 0) = f_{yy}(0, 0) = 0$, $f_{xy}(0, 0) = -1$, $f_{yx}(0, 0) = 1$

9. $f(x, y) = g(x, y) + h(y)$, where g_{xy} exists everywhere, but h is nowhere differentiable

18. (b) If all but one of $f_{x_1}, f_{x_2}, \ldots, f_{x_n}$ are continuous at $\mathbf{X}_0$, while the remaining one exists at $\mathbf{X}_0$, then f is differentiable at $\mathbf{X}_0$.

19. (a) $df = (3x^2 + 4y^2 + 2y \sin x + 2xy \cos x) dx + (8xy + 2x \sin x) dy$, $d_{\mathbf{X}_0}f = 16 dx$, $(d_{\mathbf{X}_0}f)(\mathbf{X} - \mathbf{X}_0) = 16x$

   (b) $df = -e^{-x-y-z}(dx + dy + dz)$, $d_{\mathbf{X}_0}f = -dx - dy - dz$, $(d_{\mathbf{X}_0}f)(\mathbf{X} - \mathbf{X}_0) = -x - y - z$

   (c) $df = (1 + x_1 + 2x_2 + \cdots + nx_n)^{-1} \sum_{j=1}^{n} j\, dx_j$, $d_{\mathbf{X}_0}f = \sum_{j=1}^{n} j\, dx_j$, $(d_{\mathbf{X}_0}f)(\mathbf{X} - \mathbf{X}_0) = \sum_{j=1}^{n} jx_j$

   (d) $df = 2r|\mathbf{X}|^{r-1} \sum_{j=1}^{n} x_j\, dx_j$, $d_{\mathbf{X}_0}f = 2rn^{r-1} \sum_{j=1}^{n} dx_j$, $(d_{\mathbf{X}_0}f)(\mathbf{X} - \mathbf{X}_0) = 2rn^{r-1} \sum_{j=1}^{n} (x_j - 1)$

20. (b) The unit vector in the direction of $(f_{x_1}(\mathbf{X}_0), f_{x_2}(\mathbf{X}_0), \ldots, f_{x_n}(\mathbf{X}_0))$, provided this is not $\mathbf{0}$; if it is $\mathbf{0}$, then $\partial f(\mathbf{X}_0)/\partial\mathbf{\Phi} = 0$ for every $\mathbf{\Phi}$.

21. 3.74

26. (a) $z = 2x + 4y - 6$ (b) $z = 2x + 3y + 1$

    (c) $z = (\pi/2)x + y - \pi$ (d) $z = x + 10y + 8$

Section 5.4, pages 373–376

2. (a) $5\, du + 34\, dv$ (b) 0   (c) $6\, du - 18\, dv$ (d) $8\, du$

3. $h_r = f_x \cos\theta + f_y \sin\theta$, $h_\theta = r(-f_x \sin\theta + f_y \cos\theta)$, $h_z = f_z$

4. $h_r = f_x \sin\phi \cos\theta + f_y \sin\phi \sin\theta + f_z \cos\phi$, $h_\theta = r \sin\phi(-f_x \sin\theta + f_y \cos\theta)$, $h_\phi = r(f_x \cos\phi \cos\theta + f_y \cos\phi \sin\theta - f_z \sin\phi)$

6. $h_y = g_x x_y + g_y + g_w w_y$, $h_z = g_x x_z + g_z + g_w w_z$

7. $\frac{19}{6}$

15. $h_{rr} = f_{xx} \sin^2 \phi \cos^2 \theta + f_{yy} \sin^2 \phi \sin^2 \theta + f_{zz} \cos^2 \phi + f_{xy} \sin^2 \phi \sin 2\theta + f_{yz} \sin 2\phi \sin \theta + f_{xz} \sin 2\phi \cos \theta,$

$h_{r\theta} = \sin \phi(-f_x \sin \theta + f_y \cos \theta) + \dfrac{r}{2}(-f_{xx} + f_{yy}) \sin^2 \phi \sin 2\theta +$

$rf_{xy} \sin^2 \phi \cos 2\theta - rf_{xz} \cos \phi \sin^2 \theta - \dfrac{r}{2} f_{xz} \sin 2\phi \sin \theta + \dfrac{r}{2} f_{yz} \sin 2\phi \cos \theta$

18. (a) $1 + x + \dfrac{x^2}{2} - \dfrac{y^2}{2} + \dfrac{x^3}{6} - \dfrac{xy^2}{2}$

    (b) $1 - x - y + \dfrac{x^2}{2} + xy + \dfrac{y^2}{2} - \dfrac{x^3}{6} - \dfrac{x^2 y}{2} - \dfrac{xy^2}{2} - \dfrac{y^3}{6}$

    (c) 0   (d) xyz

23. (a) $(d^2_{(0,0)}p)(x, y) = 2(x - y)^2 = (d^2_{(0,0)}q)(x, y)$

24. (a) any (x, y) such that $x^2 + y^2 = 2k\pi$ ($k = $ integer ≥ 0) is a local maximum point; any (x, y) such that $x^2 + y^2 = (2k + 1)\pi$ is a local minimum point

    (b) $(0, 0)$ is the minimum point

    (c) $(0, 0)$ is the minimum point

    (d) $(0, 0)$ and (x, y) such that $x^2 + y^2 = (2k + \frac{3}{2})\pi$ are local minimum points; (x, y) such that $x^2 + y^2 = (2k + \frac{1}{2})\pi$ are local maximum points

CHAPTER 6

Section 6.1, pages 393–395

3. (a) $\begin{bmatrix} 3 & 4 & 6 \\ 2 & -4 & 2 \\ 7 & 2 & 3 \end{bmatrix}$ (b) $\begin{bmatrix} 2 & 4 \\ 3 & -2 \\ 7 & -4 \\ 6 & 1 \end{bmatrix}$

4. (a) $\begin{bmatrix} 8 & 8 & 16 & 24 \\ 0 & 0 & 4 & 12 \\ 12 & 16 & 28 & 44 \end{bmatrix}$ (b) $\begin{bmatrix} -2 & -6 & 0 \\ 0 & -2 & -4 \\ -2 & 2 & -6 \end{bmatrix}$

5. (a) $\begin{bmatrix} -2 & 2 & 6 \\ 5 & 7 & -3 \\ 0 & -2 & 6 \end{bmatrix}$ (b) $\begin{bmatrix} -1 & 7 \\ 3 & 5 \\ 5 & 14 \end{bmatrix}$

6. (a) $\begin{bmatrix} 13 & 25 \\ 16 & 31 \\ 16 & 25 \end{bmatrix}$ (b) $\begin{bmatrix} 29 \\ 50 \end{bmatrix}$

10. **A** and **B** square, of the same order

12. (a) $\begin{bmatrix} 7 & 3 & 3 \\ 4 & 7 & 7 \\ 6 & -9 & 1 \end{bmatrix}$ (b) $\begin{bmatrix} 14 & 10 \\ 6 & -2 \\ 14 & 2 \end{bmatrix}$

13. $\begin{bmatrix} -7 & 6 & 4 \\ -9 & 7 & 13 \\ 5 & 0 & -14 \end{bmatrix}, \begin{bmatrix} -5 & 6 & 0 \\ 4 & -12 & 3 \\ 4 & 0 & 3 \end{bmatrix}$

15. (a) $[6xyz \quad 3xz^2 \quad 3x^2y]; [-6 \quad 3 \quad -3]$
    (b) $\cos(x + y)[1 \quad 1]; [0 \quad 0]$
    (c) $[(1 - xz)ye^{-xz} \quad xe^{-xz} \quad -x^2ye^{-xz}]; [2 \quad 1 \quad -2]$
    (d) $\sec^2(x + 2y + z)[1 \quad 2 \quad 1]; [2 \quad 4 \quad 2]$

    (e) $|\mathbf{X}|^{-1}[x_1 \quad x_2 \quad \cdots \quad x_n]; \left[\dfrac{1}{\sqrt{n}} \quad \dfrac{1}{\sqrt{n}} \quad \cdots \quad \dfrac{1}{\sqrt{n}} \right]$

21. (a) $(2, 3, -2)$ (b) $(2, 3, 0)$ (c) $(-2, 0, -1)$ (d) $(3, 1, 3, 2)$

22. (a) $\dfrac{1}{10}\begin{bmatrix} 4 & 2 \\ -3 & 1 \end{bmatrix}$ (b) $\dfrac{1}{2}\begin{bmatrix} -1 & 1 & 2 \\ 3 & 1 & -4 \\ -1 & -1 & 2 \end{bmatrix}$

    (c) $\dfrac{1}{25}\begin{bmatrix} 4 & 3 & -5 \\ 6 & -8 & 5 \\ -3 & 4 & 10 \end{bmatrix}$ (d) $\dfrac{1}{2}\begin{bmatrix} 1 & -1 & 1 \\ -1 & 1 & 1 \\ 1 & 1 & -1 \end{bmatrix}$

    (e) $\dfrac{1}{7}\begin{bmatrix} 3 & -2 & 0 & 0 \\ 2 & 1 & 0 & 0 \\ 0 & 0 & 2 & -3 \\ 0 & 0 & 1 & 2 \end{bmatrix}$ (f) $\dfrac{1}{10}\begin{bmatrix} -1 & -2 & 0 & 5 \\ -14 & -18 & 10 & 20 \\ 21 & 22 & -10 & -25 \\ 17 & 24 & -10 & -25 \end{bmatrix}$

Section 6.2, pages 408–412

12. (a) $\mathbf{G}(\mathbf{X}) = \begin{bmatrix} 0 \\ 1 \\ 1 \end{bmatrix} + \begin{bmatrix} 1 & 1 & 2 \\ 0 & 0 & 0 \\ 0 & 0 & -1 \end{bmatrix}\begin{bmatrix} x - 1 \\ y + 1 \\ z \end{bmatrix}$

    (b) $\mathbf{G}(\mathbf{X}) = \begin{bmatrix} 0 \\ 1 \end{bmatrix} + \begin{bmatrix} 0 & -1 \\ 1 & 0 \end{bmatrix}\begin{bmatrix} x \\ y - \pi/2 \end{bmatrix}$

    (c) $\mathbf{G}(\mathbf{X}) = \begin{bmatrix} 2 & -2 & 0 \\ 0 & 2 & -2 \\ -2 & 0 & 2 \end{bmatrix}\begin{bmatrix} x - 1 \\ y - 1 \\ z - 1 \end{bmatrix}$

13. (a) $F'(X) = \begin{bmatrix} 2x & 1 & 2 \\ -\sin(x+y+z) & -\sin(x+y+z) & -\sin(x+y+z) \\ yze^{xyz} & xze^{xyz} & xye^{xyz} \end{bmatrix}$,

$$JF(X) = e^{xyz}\sin(x+y+z)[x(1-2x)(y-z) - z(x-y)]$$

(b) $F'(X) = \begin{bmatrix} e^x \cos y & -e^x \sin y \\ e^x \sin y & e^x \cos y \end{bmatrix}$, $JF(X) = e^{2x}$

(c) $F'(X) = \begin{bmatrix} 2x & -2y & 0 \\ 0 & 2y & -2z \\ -2x & 0 & 2z \end{bmatrix}$, $JF = 0$

14. (a) $F'(X) = \begin{bmatrix} (x+y+z+1)e^x & e^x & e^x \\ (2x - x^2 - y^2)e^{-x} & 2ye^{-x} & 0 \end{bmatrix}$

(b) $F'(X) = \begin{bmatrix} g_1'(x) \\ g_2'(x) \\ \vdots \\ g_n'(x) \end{bmatrix}$ (c) $F'(X) = \begin{bmatrix} e^x \sin yz & ze^x \cos yz & ye^x \cos yz \\ ze^y \cos xz & e^y \sin xz & xe^y \cos xz \\ ye^z \cos xy & xe^z \cos xy & e^z \sin xy \end{bmatrix}$

22. (a) $\begin{bmatrix} 0 & 0 & 4 \\ 0 & -\frac{1}{2} & 0 \end{bmatrix}$ (b) $\begin{bmatrix} -18 & 0 \\ 2 & 0 \end{bmatrix}$ (c) $\begin{bmatrix} 9 & -3 \\ 3 & -8 \\ 1 & 0 \end{bmatrix}$

(d) $\begin{bmatrix} 4 & -3 & 1 \\ 0 & 1 & 1 \end{bmatrix}$ (e) $\begin{bmatrix} 2 & 0 \\ 2 & 0 \end{bmatrix}$ (f) $\begin{bmatrix} 5 & 10 \\ 9 & 18 \\ -4 & -8 \end{bmatrix}$

26. (a) $F'(r, \theta) = \begin{bmatrix} \cos\theta & -r\sin\theta \\ \sin\theta & r\cos\theta \end{bmatrix}$, $JF(r, \theta) = r$

(b) $F'(r, \theta, \phi) = \begin{bmatrix} \cos\theta\cos\phi & -r\sin\theta\cos\phi & -r\cos\theta\sin\phi \\ \sin\theta\cos\phi & r\cos\theta\cos\phi & -r\sin\theta\sin\phi \\ \sin\phi & 0 & r\cos\phi \end{bmatrix}$,

$$JF(r, \theta, \phi) = r^2 \cos\phi$$

(c) $F'(r, \theta, z) = \begin{bmatrix} \cos\theta & -r\sin\theta & 0 \\ \sin\theta & r\cos\theta & 0 \\ 0 & 0 & 1 \end{bmatrix}$, $JF(r, \theta, z) = r$

Section 6.3, pages 432–436

1. (a) false; let $F(x, y) = (x, x + y)$
5. (a) $[1, \pi/2]$ (b) $[1, 2\pi]$ (c) $[1, \pi]$
   (d) $[2\sqrt{2}, 9\pi/4]$ (e) $[\sqrt{2}, 3\pi/4]$

6. (a) $[1, -3\pi/2]$ (b) $[1, -2\pi]$ (c) $[1, -\pi]$
    (d) $[2\sqrt{2}, -7\pi/4]$ (e) $[\sqrt{2}, -5\pi/4]$

7. (b) Let $f(x) = x$ $(0 \le x \le \frac{1}{2})$, $f(x) = x - \frac{1}{2}$ $(\frac{1}{2} < x \le 1)$; then f is locally invertible, but not invertible, on $[0, 1]$.

8. $F(S) = \{(u, v) \mid -\pi + 2\phi < \arg(u, v) < \pi + 2\phi\}$, where ϕ is an argument of (a, b);

$$F_S^{-1}(u, v) = \begin{bmatrix} (u^2 + v^2)^{1/4} \cos[\frac{1}{2} \arg(u, v)] \\ (u^2 + v^2)^{1/4} \sin[\frac{1}{2} \arg(u, v)] \end{bmatrix}, \quad 2\phi - \pi < \arg(u, v) < 2\phi + \pi$$

11. (a) $\begin{bmatrix} x \\ y \end{bmatrix} = \dfrac{1}{10} \begin{bmatrix} x - 2y \\ 3x + 4y \end{bmatrix}$; $(F^{-1})' = \dfrac{1}{10} \begin{bmatrix} 1 & -2 \\ 3 & 4 \end{bmatrix}$

    (b) $\begin{bmatrix} x \\ y \\ z \end{bmatrix} = \dfrac{1}{2} \begin{bmatrix} x + 2y + 3z \\ x - z \\ x + y + 2z \end{bmatrix}$; $(F^{-1})' = \dfrac{1}{2} \begin{bmatrix} 1 & 2 & 3 \\ 1 & 0 & -1 \\ 1 & 1 & 2 \end{bmatrix}$

14. $\begin{bmatrix} x \\ y \end{bmatrix} = G(u, v) = \begin{bmatrix} \pm[(u + v)/2]^{1/2} \\ \pm[(u - v)/2]^{1/2} \end{bmatrix}$, with four possible choices for $(\pm, \pm)$;

$$G'(u, v) = -\frac{1}{8xy} \begin{bmatrix} -2y & -2y \\ -2x & 2x \end{bmatrix}$$

15. Let $e_i = \pm 1 (i = 1, 2, \ldots, n)$ and $S = \{X \mid e_i x_i > 0 \ (i = 1, 2, \ldots, n)\}$. There are 2^n such sets and F is regular on each. If $A^{-1} = [b_{ij}]$, then $R(F) = \{U \mid \sum_{j=1}^{n} b_{ij} u_j > 0, i = 1, 2, \ldots, n\}$ and $X = F_S^{-1}(U)$ is defined by $(A)x_i = e_i(\sum_{j=1}^{n} b_{ij} u_j)^{1/2}$ $(i = 1, 2, \ldots, n)$;

$$(F_S^{-1})'(U) = 2 \begin{bmatrix} a_{11}x_1 & a_{12}x_2 & \cdots & a_{1n}x_n \\ a_{21}x_1 & a_{22}x_2 & \cdots & a_{2n}x_n \\ \vdots & \vdots & & \vdots \\ a_{n1}x_1 & a_{n2}x_2 & \cdots & a_{nn}x_n \end{bmatrix}$$

with $x_1, x_2, \ldots, x_n$ as in (A).

19. From solving $x = r \cos\theta$, $y = r \sin\theta$ for $\theta = \arg(x, y)$. Each equation is satisfied by angles that are not arguments of (x, y), since none of the formulas identifies the quadrant of (x, y) uniquely. Moreover, (c) does not hold if $x = 0$.

20. $\begin{bmatrix} x \\ y \end{bmatrix} = G(u, v) = \begin{bmatrix} (u^2 + v^2)^{1/4} \cos[\frac{1}{2} \arg(u, v)] \\ (u^2 + v^2)^{1/4} \sin[\frac{1}{2} \arg(u, v)] \end{bmatrix}$,

where $\beta - \pi/2 < \arg(u, v) < \beta + \pi/2$ and β is an argument of (a, b):

$$G'(u, v) = \frac{1}{2(x^2 + y^2)} \begin{bmatrix} x & y \\ -y & x \end{bmatrix}$$

21. $\begin{bmatrix} x \\ y \end{bmatrix} = G(u, v) = \begin{bmatrix} \log(u^2 + v^2)^{1/2} \\ \arg(u, v) \end{bmatrix}$,

where $\beta - \pi/2 < \arg(u, v) < \beta + \pi/2$ and β is an argument of (a, b):

$$\mathbf{G}'(u, v) = \begin{bmatrix} e^{-x} \cos y & e^{-x} \sin y \\ -e^{-x} \sin y & e^{-x} \cos y \end{bmatrix}$$

24. If $\mathbf{F}(x_1, x_2, \ldots, x_n) = (x_1^3, x_2^3, \ldots, x_n^3)$, then $\mathbf{F}$ is invertible, but $J\mathbf{F}(\mathbf{0}) = 0$.

26. (a) $\mathbf{A}(\mathbf{U}) = \begin{bmatrix} 1 \\ -1 \end{bmatrix} - \dfrac{1}{25} \begin{bmatrix} 5 & 5 \\ 3 & 8 \end{bmatrix} \begin{bmatrix} u + 5 \\ v - 4 \end{bmatrix}$

(b) $\mathbf{A}(\mathbf{U}) = \begin{bmatrix} 1 \\ 1 \end{bmatrix} + \dfrac{1}{6} \begin{bmatrix} 4 & -2 \\ -3 & 3 \end{bmatrix} \begin{bmatrix} u - 2 \\ v - 3 \end{bmatrix}$

(c) $\mathbf{A}(\mathbf{U}) = \begin{bmatrix} 0 \\ 1 \\ 1 \end{bmatrix} + \begin{bmatrix} 0 & -1 & 1 \\ -1 & 1 & 0 \\ 1 & 0 & 0 \end{bmatrix} \begin{bmatrix} u - 1 \\ v - 1 \\ w - 2 \end{bmatrix}$

(d) $\mathbf{A}(\mathbf{U}) = \begin{bmatrix} 1 \\ \pi/2 \\ \pi \end{bmatrix} + \begin{bmatrix} 0 & -1 & 0 \\ 1 & 0 & 0 \\ 0 & 0 & -1 \end{bmatrix} \begin{bmatrix} u \\ v + 1 \\ w \end{bmatrix}$

27. $\mathbf{G}'(x, y, z) = \begin{bmatrix} \cos\theta \cos\phi & \sin\theta \cos\phi & \sin\phi \\[2mm] -\dfrac{\sin\theta}{r \cos\phi} & \dfrac{\cos\theta}{r \cos\phi} & 0 \\[2mm] -\dfrac{1}{r}\cos\theta \sin\phi & -\dfrac{1}{r}\sin\theta \sin\phi & \dfrac{1}{r}\cos\phi \end{bmatrix}$

28. $\mathbf{G}'(x, y, z) = \begin{bmatrix} \cos\theta & \sin\theta & 0 \\[2mm] -\dfrac{1}{r}\sin\theta & \dfrac{1}{r}\cos\theta & 0 \\[2mm] 0 & 0 & 1 \end{bmatrix}$

Section 6.4, pages 452–455

1. (a) $\begin{bmatrix} u \\ v \end{bmatrix} = \dfrac{1}{2} \begin{bmatrix} -3 & 4 \\ 1 & -2 \end{bmatrix} \begin{bmatrix} x \\ y \end{bmatrix}$

(b) $u = -\frac{3}{2}x - \frac{3}{2}y$, $v = \frac{1}{2}x - y$, $w = -x - \frac{3}{2}y$

(c) $\begin{bmatrix} u \\ v \end{bmatrix} = \dfrac{1}{5} \begin{bmatrix} 2 & -1 \\ -1 & 3 \end{bmatrix} \begin{bmatrix} -y + \sin x \\ -x + \sin y \end{bmatrix}$

(d) $u = -x$, $v = -y$, $w = -z$

6. Let $f_i(\mathbf{X}, \mathbf{U}) = [\sum_{j=1}^{n} a_{ij}x_j]^r - u_i^s$ $(1 \le i \le m)$, where not all $a_{ij} = 0$, $\mathbf{X}_0 = \mathbf{0}$, $\mathbf{U}_0 = \mathbf{0}$, and (a) $r = s = 3$ (b) $r = 1$, $s = 3$ (c) $r = s = 2$.

7. $u_x(1, 1) = -\frac{5}{8}, u_y(1, 1) = -\frac{1}{2}$

8. $u_x(1, 1, 1) = \frac{5}{8}, u_y(1, 1, 1) = -\frac{9}{8}, u_z(1, 1, 1) = \frac{1}{2}$

9. (a) $u(1, 2) = 0, u_x(1, 2) = u_y(1, 2) = -4$

   (b) $u(-1, -2) = 2, u_x(-1, -2) = 1, u_y(-1, -2) = -\frac{1}{2}$

   (c) $u(\pi/2, \pi/2) = u_x(\pi/2, \pi/2) = u_y(\pi/2, \pi/2) = 0$

   (d) $u(1, 1) = 1, u_x(1, 1) = u_y(1, 1) = -1$

10. (a) $u(1, 1) = 2, u_x(1, 1) = -14, u_y(1, 1) = -2$; or $u(1, 1) = 1, u_x(1, 1) = 5$,
    $u_y(1, 1) = 2$

    (b) $u(0, \pi) = (k + \frac{1}{2})\pi \ (k = \text{integer}), u_x(0, \pi) = 0, u_y(0, \pi) = -1$

11. $\dfrac{1}{5}\begin{bmatrix} -1 & -2 & 1 \\ -1 & -2 & 1 \end{bmatrix}$ 12. $u'(0) = 3, v'(0) = -1$ 13. $\begin{bmatrix} \frac{5}{6} & \frac{5}{6} \\ -\frac{5}{6} & -\frac{5}{6} \\ 1 & 1 \end{bmatrix}$

14. $\mathbf{U}(1, 1) = \begin{bmatrix} 3 \\ 1 \end{bmatrix}, \mathbf{U}'(1, 1) = \begin{bmatrix} 7 & -9 \\ 1 & -2 \end{bmatrix}$ or $\mathbf{U}(1, 1) = \begin{bmatrix} -3 \\ -1 \end{bmatrix}, \mathbf{U}'(1, 1) = \begin{bmatrix} -7 & 9 \\ -1 & 2 \end{bmatrix}$

15. $u_x(0, 0, 0) = 2, v_x(0, 0, 0) = w_x(0, 0, 0) = -2$

$$y_x = -\dfrac{\dfrac{\partial(f, g, h)}{\partial(x, z, u)}}{\dfrac{\partial(f, g, h)}{\partial(y, z, u)}}, \quad y_v = -\dfrac{\dfrac{\partial(f, g, h)}{\partial(v, z, u)}}{\dfrac{\partial(f, g, h)}{\partial(y, z, u)}}, \quad z_x = -\dfrac{\dfrac{\partial(f, g, h)}{\partial(y, x, u)}}{\dfrac{\partial(f, g, h)}{\partial(y, z, u)}}$$

$$z_v = -\dfrac{\dfrac{\partial(f, g, h)}{\partial(y, v, u)}}{\dfrac{\partial(f, g, h)}{\partial(y, z, u)}}, \quad u_x = -\dfrac{\dfrac{\partial(f, g, h)}{\partial(y, z, x)}}{\dfrac{\partial(f, g, h)}{\partial(y, z, u)}}, \quad u_v = -\dfrac{\dfrac{\partial(f, g, h)}{\partial(y, z, v)}}{\dfrac{\partial(f, g, h)}{\partial(y, z, u)}}$$

19. $x = -2y - u, z = -2v; x = -2y - u, v = -\dfrac{z}{2}; y = -\dfrac{x}{2} - \dfrac{u}{2}, z = -2v;$

$y = -\dfrac{x}{2} - \dfrac{u}{2}, v = -\dfrac{z}{2}; z = -2v, u = -x - 2y; u = -x - 2y, v = -\dfrac{z}{2}$

20. $y_x(1, -1, -2) = -\frac{1}{2}, v_u(1, -1, -2) = 1$

21. $x_y(0, -1) = -1, u_y(0, -1) = 0, v_y(0, -1) = 0, x_w(0, -1) = 1, u_w(0, -1) = \frac{5}{6}$,
    $v_w(0, -1) = -\frac{5}{6}$

23. $u_x(1, 1) = u_y(1, 1) = 0, v_x(1, 1) = v_y(1, 1) = v_{xy}(1, 1) = -1$,
    $u_{xx}(1, 1) = u_{yy}(1, 1) = 2, v_{xx}(1, 1) = v_{yy}(1, 1) = -2, u_{xy}(1, 1) = 1$

24. $u_x(1, -1) = v_y(1, -1) = 0, u_y(1, -1) = v_x(1, -1) = -\frac{1}{2}, u_{xx}(1, -1) =$
    $v_{xx}(1, -1) = u_{xy}(1, -1) = -\frac{1}{8}, u_{yy}(1, -1) = v_{yy}(1, -1) = v_{xy}(1, -1) = \frac{1}{8}$

Section 6.5, pages 472–476

1. $(\frac{15}{7}, -\frac{2}{7}, \frac{25}{7})$

2. $\frac{14}{17}$, attained at $(\frac{10}{34}, \frac{5}{34}, \frac{28}{34}, -\frac{3}{34})$

4. $\pm\sqrt{53/6}$

5. $A^{3/2}/6\sqrt{3}$

6. $A^{3/2}/6\sqrt{6}$

7. length $=$ width $= (2V)^{1/3}$; height $= (2V)^{1/3}/2$

8. $x = a/3$, $y = b/3$, $z = c/3$ 9. ab

11. $d = |\sigma - ax_0 - by_0 - cz_0|(a^2 + b^2 + c^2)^{-1/2}$

12. $x = \dfrac{1}{n}\left[\lambda a + \displaystyle\sum_{i=1}^{n} x_i\right]$, $y = \dfrac{1}{n}\left[\lambda b + \displaystyle\sum_{i=1}^{n} y_i\right]$, $z = \dfrac{1}{n}\left[\lambda c + \displaystyle\sum_{i=1}^{n} z_i\right]$,

where $\lambda = (a^2 + b^2 + c^2)^{-1}\displaystyle\sum_{i=1}^{n} (\sigma - ax_i - by_i - cz_i)$

13. $7/4\sqrt{2}$; attained with $(-\tfrac{1}{2}, \tfrac{5}{4})$ on the parabola and $(-\tfrac{11}{8}, \tfrac{3}{8})$ on the line

14. $10\sqrt{6}/3$; attained with $(1, \tfrac{1}{3}, 1)$ on the ellipsoid and $(\tfrac{13}{3}, \tfrac{11}{3}, \tfrac{23}{3})$ on the plane

16. $18/\sqrt{62}$; the point on the ellipse nearest the plane is $(-2, 0, 8)$

17. $2, -1$ 18. $4, 2$ 19. $2, -2$

20. (a) $\|\mathbf{A}\| = 3$, $\mathbf{X}_1 = a\begin{bmatrix} 1 \\ 1 \end{bmatrix}$ $(a \neq 0)$; $m(\mathbf{A}) = 1$, $\mathbf{X}_2 = b\begin{bmatrix} 1 \\ -1 \end{bmatrix}$ $(b \neq 0)$

(b) $\|\mathbf{A}\| = 2$, $\mathbf{X}_1 = \begin{bmatrix} a \\ b \\ b \\ a \end{bmatrix}$ $(a^2 + b^2 \neq 0)$; $m(\mathbf{A}) = 0$, $\mathbf{X}_2 = \begin{bmatrix} a \\ b \\ -b \\ -a \end{bmatrix}$ $(a^2 + b^2 \neq 0)$

23. $x_r = \dfrac{4n + 2 - 6r}{n(n - 1)}$ 24. $6, -6$

25. $(bc)^{-1/2}$; attained at $\pm(0, 1/\sqrt{b}, -1/\sqrt{c}, 0)$ if $ad > bc$, and at any point $(x, y, -y\sqrt{b/c}, x\sqrt{a/d})$ such that $ax^2 + by^2 = 1$ if $ad = bc$.

CHAPTER 7

Section 7.1, pages 503–507

1. (a) 28 (b) $\tfrac{1}{4}$ 5. $3(b - a)(d - c), 0$

16. $\{(m, n)\,|\,m, n = \text{integers}\}$

Section 7.2, pages 523–528

1. (a) 12 (b) $\tfrac{79}{20}$ (c) -1 (d) $(1 - \log 2)/2$

5. (a) $\tfrac{7}{4}$ (b) 17 (c) $\tfrac{2}{3}(\sqrt{2} - 1)$ (d) $1/4\pi$

7. (a) $\tfrac{5}{8}, \tfrac{3}{8}$ (b) $\tfrac{5}{8}, \tfrac{3}{8}$

8. (a) $\tfrac{5}{4}, \tfrac{3}{4}$ (b) $z + \tfrac{5}{8}, z + \tfrac{3}{8}$ (c) $z + \tfrac{1}{2}, 1$

14. (a) -285 (b) 0 (c) 0 (d) $\tfrac{1}{4}(e - \tfrac{5}{2})$

17. (a) 324 (b) 0 (c) $\tfrac{1}{6}$ (d) 1

18. $\tfrac{52}{15}$

19. (a) 36   (b) 1   (c) $\frac{64}{3}$ (d) $(e^6 + 17)/2$
25. (a) $\frac{2}{27}$ (b) $\frac{1}{2}(e - \frac{5}{2})$ (c) $\frac{1}{24}$ (d) $\frac{1}{36}$
26. (a) 16π (b) $\frac{1}{6}$ (c) $\frac{128}{21}$ (d) π
27. (a) $\frac{1}{2}(b_1 - a_1) \cdots (b_n - a_n) \sum_{j=1}^n (a_j + b_j)$
    (b) $\frac{1}{3}(b_1 - a_1) \cdots (b_n - a_n) \sum_{j=1}^n (b_j^2 + a_j b_j + a_j)$
    (c) $2^{-n}(b_1^2 - a_1^2) \cdots (b_n^2 - a_n^2)$
28. $\int_{-\sqrt{3}/2}^{\sqrt{3}/2} dx \int_{1/2}^{\sqrt{1-x^2}} f(x, y)\, dy$ 30. (a) $\frac{1}{2}$

Section 7.3, pages 557–561

1. Let S_1 and S_2 be dense subsets of $\mathcal{R}$ such that $S_1 \cup S_2 = \mathcal{R}$.
8. (a) -1; c (where c is the constant); 1
10. $|u_2 - u_1|\,|v_2 - v_1|/|ad - bc|$ 11. $\frac{5}{6}$
23. (a) $\frac{4}{9}$ (b) $\log \frac{5}{2}$ 24. 3
25. $\frac{1}{2}$ 26. $\frac{5}{4}e(e - 1)$
27. $\frac{4}{3}\pi abc$ 28. $2\pi(e^{25} - e^9)$
29. $16\pi/3$ 30. 21
31. (a) $(\pi/8) \log 5$ (b) $(\pi/4)(e^4 - 1)$ (c) $2\pi/15$
32. $\pi^2 a^4/2$ 33. (a) $(\beta_1 - \alpha_1) \cdots (\beta_n - \alpha_n)/|\det \mathbf{A}|$
34. $|a_1 a_2 \cdots a_n| V_n$ 36. $\frac{1}{4}$

Section 7.4, pages 572–575

5. (a) $-\pi$ (b) -2 (c) $-\pi/8$ (d) ∞ (e) ∞
6. (a) $\pi/2$ (b) 1   (c) $2/e$ (d) $\frac{1}{4} \log \frac{5}{2}$
7. (a) $p < 1, q > \frac{1}{2}$ (b) $p < \frac{3}{2}, q > \frac{1}{2}$ (c) none
    (d) $p < 1, q > 1$ (e) $p < -1, q > 0$
8. (a) 1   (b) π (c) 1   (d) ∞
11. (b), (d), and (e) converge

Section 7.5, pages 589–595

1. $F(y) = \dfrac{1}{y} \log(1 + y); I_1 = \log 2 - \frac{1}{2}; I_2 = \log 2 - \frac{5}{8}$

2. $F(y) = (y - 1)/(\log y); I_1(y) = (y \log y - y + 1)/(\log y)^2;$
    $I_2(y) = (\log y)^{-3}[y(\log y)^2 - 2y \log y + 2y - 2]$
6. (b) If $f(x, y) = 1/y$ for $y \neq 0$ and $f(x, 0) = 1$, then $\int_a^b f(x, y)\, dx$ does not
    converge uniformly on $[0, d]$ for any $d > 0$.

7. (a), (d), and (e) converge uniformly on $(-\infty, -\rho] \cup [\rho, \infty)$ if $\rho > 0$;
   (b), (c), and (f) converge uniformly on $[\rho, \infty)$ if $\rho > 0$.
8. Cauchy's

18. Let $C(y) = \int_1^\infty \dfrac{\cos xy}{x}\,dx$, $S(y) = \int_1^\infty \dfrac{\sin xy}{x}\,dx$. Then $C(0) = \infty$ and $S(0) =$

   0, while $S(y) = \pi/2$ if $y \neq 0$.

20. (a) $F(y) = \dfrac{\pi}{2|y|}$, $I = \dfrac{\pi}{2}\log\dfrac{a}{b}$ (b) $F(y) = \dfrac{1}{y+1}$, $I = \log\dfrac{a+1}{b+1}$

   (c) $F(y) = \dfrac{y}{y^2+1}$, $I = \dfrac{1}{2}\log\dfrac{b^2+1}{a^2+1}$

   (d) $F(y) = \dfrac{1}{y^2+1}$, $I = \tan^{-1} b - \tan^{-1} a$

   (e) $F(y) = \dfrac{y}{y^2+1}$, $I = \tfrac{1}{2}\log(1+a^2)$ (f) $F(y) = \dfrac{1}{y^2+1}$, $I = \tan^{-1} a$

21. (a) $(-1)^n n!(y+1)^{-n-1}$ (b) $\pi 2^{-2n-1}\dbinom{2n}{n} y^{-n-1/2}$

   (c) $\sqrt{\pi}\,\dfrac{(2n)!}{2^{2n+1}n!}\,y^{-n-1/2}$ (d) $\dfrac{n!}{2y^{n+1}}$ (e) $(\log y)^{-2}$

24. $\int_{-\infty}^\infty |x^n f(x)|\,dx < \infty$
26. No; the integral defining F diverges for all y.

30. (a) $\dfrac{\pi}{2} - \tan^{-1} a$

CHAPTER 8

Section 8.1, pages 622–628

1. (a) $\sqrt{29}, \sqrt{3}, -5/\sqrt{87}$ (b) $\sqrt{14}, \sqrt{13}, 0$
   (c) $\sqrt{21}, \sqrt{84}, -1$ (d) $\sqrt{3}, \sqrt{35}, \sqrt{3/35}$
2. $|\mathbf{U}|^2 + |\mathbf{V}|^2 - 2|\mathbf{U}|\,|\mathbf{V}|\cos\theta$
6. (a) $-\mathbf{i} - 6\mathbf{j} - 5\mathbf{k}$ (b) $13\mathbf{i} + 3\mathbf{j} + 2\mathbf{k}$ (c) $\mathbf{0}$ (d) $-8\mathbf{i} + 4\mathbf{j} + 4\mathbf{k}$
8. (a) $\sqrt{38}$ (b) $\sqrt{26}$ (c) $\sqrt{6}$
9. (a) $\sqrt{11}$ (b) $\sqrt{442}/2$ (c) $\sqrt{38/2}$
10. (b) $3/\sqrt{14}$ 11. (b) $\sqrt{121/13}$ 12. (b) $5/\sqrt{38}$
13. (b) $x = -2 + t$, $y = 5 - t$, $z = -t$
14. (b) $10x + 4y + z = 19$

15. (a) $-2\mathbf{i} - 6\mathbf{j} - 4\mathbf{k}, \mathbf{i} - 3\mathbf{j} - \mathbf{k}$ (b) $3\mathbf{i} - \mathbf{j} + 3\mathbf{k}, \mathbf{i} + 4\mathbf{k}$
    (c) $-6\mathbf{i} - 3\mathbf{j} + 3\mathbf{k}, -8\mathbf{i} - 5\mathbf{j} - 3\mathbf{k}$

16. (b) $\begin{vmatrix} u_1 & u_2 & u_3 \\ v_1 & v_2 & v_3 \\ w_1 & w_2 & w_3 \end{vmatrix}$

20. (a) $\begin{bmatrix} 2x + y \\ x + 2y \end{bmatrix}, \begin{bmatrix} 4 \\ 5 \end{bmatrix}, -1/\sqrt{2}$

    (b) $4xyz\mathbf{i} + 2x^2z\mathbf{j} + 2x^2y\mathbf{k}, -8\mathbf{i} + 2\mathbf{j} + 4\mathbf{k}, -18/\sqrt{6}$
    (c) $2x\mathbf{i} + 2y\mathbf{j} + 2z\mathbf{k}, -2\mathbf{i} + 2\mathbf{j} + 2\mathbf{k}, 2\sqrt{3}$

    (e) $e^{-x_1 - x_2 - \cdots - x_n} \begin{bmatrix} 1 \\ 1 \\ \vdots \\ 1 \end{bmatrix}, \begin{bmatrix} 1 \\ 1 \\ \vdots \\ 1 \end{bmatrix}, \sqrt{n}$

21. (a) $\sigma(\tau) = \log \tau$ (b) $\sigma(\tau) = \cos \tau$

22. Let $\sigma(\tau) = \left(\dfrac{\tau - c}{d - c}\right) b + \left(\dfrac{\tau - d}{c - d}\right) a$ and $\mathbf{\Psi}(\tau) = \mathbf{\Phi}(\sigma(\tau)), c \le \tau \le d$

24. (a) simple   (b) simple   (c) both   (d) neither
25. In each case replace t by $-t$ and inequalities like $a \le t \le b$ by $-b \le t \le -a$.

27. (a) $-\dfrac{1}{\sqrt{2}}\mathbf{i} + \dfrac{1}{\sqrt{2}}\mathbf{k}$ (b) $\dfrac{1}{\sqrt{29}}(2\mathbf{i} + 3\mathbf{j} + 4\mathbf{k})$ (c) $\dfrac{1}{3}\mathbf{i} + \dfrac{2\sqrt{2}}{3}\mathbf{j}$

Section 8.2, pages 663–670

3. (a) 8   (b) $\frac{26}{3}$ (c) 1   (d) 0
   (e) $3\pi/4$ (f) $\frac{19}{6}$ (g) $-3\pi - \frac{16}{3}$
4. (a) $-\frac{75}{2}$ (b) $-\frac{45}{2}$ (c) $-8/\sqrt{3}$ (d) $\frac{3}{14}$
   (e) 0   (f) $-\frac{21}{2}$ (g) -2
5. (a) $\frac{4}{3}$ (b) 0   (c) -3
8. (a) 0   (b) 3   (c) $2\pi^2 - 8$ (d) $-\frac{5}{3}$ (e) 0

9. (a) $-\pi$ (b) -4 (c) 0   11. (c) $\dfrac{x^3}{3} + xy - \dfrac{y^3}{3}$

12. (a) $\frac{1}{2}(4x^2 + 3y^2 + 5z^2 + 2xy + 4yz + 6xz)$
    (b) $\frac{1}{2}(3x^2 + 3y^2 + 3z^2 + 2xy - 2yz + 2xz)$
    (e) $x^3 + 2xy + y^2$

13. (c) $\dfrac{x^3}{3} + xy^2 + \dfrac{z^2}{2}$

14. (a) 0   (b) -1 (c) 2   (d) $\dfrac{1}{\sqrt{2}} - \dfrac{1}{\sqrt{5}}$ (e) 0   (f) $\frac{58}{3}$ (g) 0

16. (a) $\dfrac{x^2}{2} - \dfrac{y^2}{2} + xy$ (b) $\dfrac{x^3}{3} - \dfrac{y^3}{3} + xy$ (c) $x^3 + y^2 + 2xy$

(d) $\frac{1}{2}\exp(x^2 + y^2)$ (e) $-(x^2 + y^2)^{-1/2}$

18. (a) $2x^2 + \frac{3}{2}y^2 + \frac{5}{2}z^2 + xy + 3xz + 2yz$

(b) $\frac{3}{2}x^2 + \frac{3}{2}y^2 + \frac{3}{2}z^2 + xy + xz - yz$ (c) $\frac{1}{2}(x^4y^6 + 2x^3yz + z^2)$

(d) $x \sin z + y \sin x + z \sin y$ (e) $x^2 + 2yz \sin xz$

21. (a) $\frac{1}{2}$ (b) $\pi/2$ (c) 0   (d) $-\frac{556}{15}$ (e) $-\frac{1}{30}$

24. (a) $\frac{11}{6}$ (b) $\pi/8 - \frac{1}{6}$ (c) $\frac{9}{2}$ (d) $\frac{1}{3}$

29. (a) $\left(2\pi a^2 + \dfrac{8\pi^3 b^2}{3}\right)(a^2 + b^2)^{1/2}$ (b) 16π

(c) $\frac{1}{3}[(1 + e^{\pi})^{3/2} - 2^{3/2}]$ (d) $\dfrac{\sqrt{2}}{26}(3e^{3\pi/4} + 2)$ (e) $8\sqrt{2}/189$

35. (a) $2xy + c$ (b) $e^x \sin y + c$ (c) $3x^2 y - y^3 + c$

(d) $-\sin x \sinh y + c$ (e) $\cos x \sinh y + c$

Section 8.3, pages 691–694

1. (a) $z = \dfrac{1}{\sqrt{2}}(y - x + 2)$ (b) $z = 8x - 2y - 2$ (c) $x + 4y = 13$

(d) $8x + y - 2z = -9$ (e) $\dfrac{x}{\sqrt{3}} - \dfrac{y}{\sqrt{6}} + \dfrac{z}{\sqrt{2}} = 1$

(f) $2x - 2y + z = 6$ (g) $2x + y + z = 0$

4. (a) $\mathbf{X} = 3\cos\theta\cos\phi\mathbf{i} + (1 + 4\sin\theta\cos\phi)\mathbf{j} + \sin\phi\mathbf{k}, 0 \le \theta \le 2\pi,$
$|\phi| \le \pi/2$

(b) $\mathbf{X} = x\mathbf{i} + y\mathbf{j} + (1 - x - y)\mathbf{k}, 0 \le x + y \le 1, x \ge 0, y \ge 0$

(c) $\mathbf{X} = \cos\theta\cos\phi\mathbf{i} + \sin\theta\cos\phi\mathbf{j} + \sin\phi\mathbf{k}, 0 \le \theta \le 2\pi, |\phi| \le \pi/4$

(d) $\mathbf{X} = r\cos\theta\mathbf{i} + r\sin\theta\mathbf{j} + (1 - r\cos\theta - r\sin\theta)\mathbf{k}, 0 \le r \le 2, 0 \le \theta \le 2\pi$

(e) $\mathbf{X} = z(\cos\theta\mathbf{i} + \sin\theta\mathbf{j} + \mathbf{k}), 0 \le \theta \le 2\pi, 1 \le z \le 2$

7. (a) $\pi\sqrt{2}(R_2^2 - R_1^2)$ (b) $\pi[\sqrt{2} + \log(1 + \sqrt{2})]$ (c) $4\sqrt{6}$

(d) $4\pi a\rho$ (e) $\pi\rho^2(1 + a^2 + b^2)^{1/2}$ (f) 8

(g) $\pi(2 - \sqrt{2})$ (h) 42π (i) $8a^2$

(j) $4a^2$

8. $(u_2 - u_1)(v_2 - v_1)|ad - bc|^{-1}(1 + \alpha^2 + \beta^2)^{1/2}$

9. $2\pi\rho(C_2 - C_1)$ 10. $4\pi^2 ab$

16. (a) $(5 + \sqrt{3})\pi$ (b) $48 - 3\pi + \dfrac{\pi}{6}[(17)^{3/2} - 5^{3/2}]$

Section 8.4, pages 721–724

2. (a) $\dfrac{\pi a^5}{2}$ (b) $3\pi/2$ (c) 0

     (d) $(5^{5/2} + 1)\pi/40$ (e) $-4\pi\sqrt{2}$ (f) π

3. $\frac{2}{3}r^3(\beta - \gamma b)(1 + a^2 + b^2)^{1/2}$

6. (a) 0   (b) $104\pi/3$ (c) $4\pi r$ (d) $104/3$ (e) 0

     (f) $\frac{40}{3}$ (g) $\dfrac{\pi}{6}(8 - 5\sqrt{2})$ (h) $10\pi a^2$ (i) 0

9. (a) $\frac{11}{24}$ (b) 243π (c) $279\pi/4$ (d) 0   (e) 4   (f) 16π

     (g) $(a_2 - a_1)(b_2 - b_1)(c_2 - c_1)(a_1 + a_2 + b_1 + b_2 + c_1 + c_2)$

12. (a) $4\pi r^{3-\alpha}$

16. (a) $\pm 27\pi$ (b) $\pm\frac{32}{9}$ (c) 0   (d) 0   (e) 0   (f) 0

index

78 79 80 9 8 7 6 5 4 3 2 1